Handbuch der experimentellen Pharmakologie

Vol. 45/II Heffter-Heubner New Series

Handbook of Experimental Pharmacology

Drug Addiction II

Amphetamine, Psychotogen, and Marihuana Dependence

Contributors

E. Änggård · W. L. Dewey · J. D. Griffith · L.-M. Gunnc · L. S. Harris · T. Lewander · W. R. Martin · R. K. Razdan · J. W. Sloan

Editor

William R. Martin

Springer-Verlag Berlin Heidelberg New York 1977

William R. Martin, M.D., U.S. Department of Health, Education, and Welfare, Public Health Service, National Institute on Drug Abuse, Addiction Research Center, P.O. Box 12390, Lexington, KY 40511/USA

With 15 Figures

ISBN-13: 978-3-642-66711-4 e-ISBN-13: 978-3-642-66709-1
DOI: 10.1007/978-3-642-66709-1

Library of Congress Cataloging in Publication Data. Main entry under title: Drug addiction. (Handbook of experimental pharmacology: New series; v. 45, pts. 1—2). Contributors, v. 2: E. Änggård and others. Includes bibliographies and indexes. CONTENTS: 1. Morphine, sedative-hypnotic, and alcohol dependence. — 2. Amphetamine, psychotogen, and marihuana dependence. 1. Narcotic habit. 2. Drug abuse. 3. Mental illness—Physiological aspects. I. Conrad, Harold, T. II. Martin, William R., 1921- III. Änggård, E. IV. Series: Handbuch der experimentellen Pharmakologie: New series; v. 45, pts. 1—2. QP905.H3 vol. 45, pts. 1—2 [RC566] 615'.1'08s
[615'.782] 77-24381

Softcover reprint of the hardcover 1st edition 1977

2122/3130-543210

Contents

SECTION I

Amphetamine and Psychotogen Dependence

CHAPTER 1

General Pharmacology of Amphetamine-Like Drugs.

A. Pharmacokinetics and Metabolism. E. ÄNGGÅRD. With 7 Figures

A. Structure-Activity Relationships Among the Amphetamines 3
I. Effect of Aromatic Substitution 3
II. Effect of Length of the Side Chain 4
III. Effect of Substitution of the Primary Nitrogen 4
IV. Effect of Substitution of the Side Chain 4
V. Stereoisomers 5
B. Metabolism 5
I. Pathways of Metabolism 5
1. Hydroxylation of Amphetamine 8
2. Aliphatic Hydroxylation 9
3. Oxidative Deamination 9
4. *N*-Dealkylation 12
5. *N*-Oxidation 13
II. Stereoselectivity in Amphetamine Metabolism 13
III. Factors Influencing the Metabolism of Amphetamine 14
1. Species 14
2. Urinary pH 15
3. Other Drugs 15
4. Other Factors 17
IV. Metabolism of Amphetamine in Dependent Subjects 18
C. Pharmacokinetics of Amphetamine 20
D. Relationship Between Blood Levels and Effect in Man 23
References 25

B. Effects of Amphetamine in Animals. T. LEWANDER

A. Introduction . . . 33
B. Effects of Amphetamine on Neuronal Transmission . . . 33
I. Biochemical Neuropharmacology . . . 33
1. Adrenergic Transmission . . . 33
a) Storage of CA . . . 34
b) Uptake of CA . . . 38
c) Release of CA . . . 40
d) Enzymatic Inactivation of CA . . . 45
e) Synthesis of CA . . . 47
f) Turnover of CA . . . 48
g) Uptake into and Release from Adrenergically Innervated Tissues of Amphetamine Itself . . . 49
h) Amphetamine Metabolites and Tissue CA . . . 50
i) Summary and Conclusions . . . 52
2. Serotoninergic (5HT) Transmission . . . 53
a) Storage of 5HT . . . 53
b) Uptake of 5HT into Neurons and Platelets . . . 54
c) Release of 5HT . . . 54
d) Enzymatic Inactivation of 5HT . . . 55
e) Synthesis and Turnover of Brain 5HT . . . 55
f) Summary and Conclusions . . . 56
3. Cholinergic Transmission . . . 56
4. Other Putative Transmittors . . . 57
5. Secondary Transmission; Cyclic Adenosine-3′,5′-monophosphate (cAMP) . . . 57
6. Summary and Conclusions . . . 58
II. Functional Neuropharmacology . . . 58
1. Effects of Amphetamine on Peripheral Noradrenergically Innervated and Other Tissues in Vivo and in Vitro . . . 58
a) The Cat Nictitating Membrane and the Rat Inferior Eyelid . . . 58
b) The Cat Iris . . . 60
c) The Guinea Pig Vas Deferens . . . 60
d) The Rabbit and Guinea Pig Intestines . . . 60
e) The Guinea Pig Bladder . . . 61
f) Other Peripheral Effects of Amphetamine Possibly Mediated via Serotoninergic Mechanisms . . . 61
g) On Tachyphylaxis to the Sympathomimetic Effects of Amphetamine and its Metabolites . . . 62
h) Summary of Conclusions . . . 64
2. Effect of Amphetamine on Sympathetic Ganglia . . . 64
3. Effects of Amphetamine on the Nerve Impulse Activity of Single Neurons in the Central Nervous System . . . 65
a) Brain Stem Reticular Formation . . . 65
b) Hypothalamus . . . 65
c) Locus Ceruleus . . . 65

d) Substantia Nigra 66
e) Caudate Nucleus 66
f) Raphé Nuclei 67
g) Summary and Conclusions 67
4. Functional Effects of Amphetamine Directly Related to Release of DA, NA, and 5HT in the CNS 67
a) Amphetamine Induced Turning and Rotation in Rats and Mice after Unilateral Lesions in the Dopamine Neuronal Systems . 68
b) The Flexor Reflex 69
c) The Extensor Reflex 69
d) Summary and Conclusions 69
5. General Summary and Conclusions 69

C. Pharmacologic Effects of Amphetamine 70

I. Behavioral Effects of Amphetamine 70
1. Gross Behavior 70
a) Stereotyped Behavior 71
b) Motor Activity 81
c) Behavior Observed in the Open Field 93
d) Exploratory Behavior 94
e) Gross Behavior Effects in Birds 95
f) General Summary and Conclusions 95
2. Social Behaviors 95
a) Social Interaction 96
b) Aggressive Behavior 96
c) Sexual Behavior 97
3. Conditioned and Operant Behaviors 98
a) The Conditioned Avoidance Response 98
b) Positively Reinforced Operant Behavior 108
c) Intracranial Self-Stimulation 121
d) Self-Administration of Amphetamine in Animals 124
e) Amphetamine as an Aversive Stimulus 126
f) Maze Performance 127
g) Conditioned Reflexes 128
h) Miscellaneous Behavioral Studies 128
4. Effects of Amphetamine on Learning and Memory 129
a) Experiments on Learning and Memory 129
b) State-Dependent Learning 130
c) Summary and Conclusions 131

II. Effects of Amphetamine on Food and Water Intake 131
1. Food Intake 131
a) Amphetamine-Induced Anorexia 131
b) Effects of Other Drugs on Amphetamine-Induced Anorexia . 132
c) Effects of Brain Lesions on Amphetamine-Induced Anorexia . 135
d) Effects of Brain Stimulations on Amphetamine-Induced Anorexia 137
e) Effects of Amphetamine on the Electrical Activity of Neurons Within the Hypothalamus 137

f) Intracerebral Injections of Amphetamine and Food Intake . . 137
g) Other Factors Affecting Amphetamine-Induced Anorexia . . . 137
h) Effects of Chronic Administration of Amphetamine on Food Intake and Body Weight 138
i) Summary of Proposed Mechanisms for the Anorexigenic Effect of Amphetamine 139
2. Hoarding Behavior 140
3. Water Intake 140

III. Autonomic, Metabolic, and Endocrine Effects of Amphetamine; Amphetamine Toxicity 141
1. Cardiovascular Effects of Amphetamine 141
a) Arterial Blood Pressure 141
b) Effect of Amphetamine on Pulmonary Circulation 143
c) Effect of Amphetamine on Cerebral Circulation 143
d) Effects of Surgical Manipulations on the Pressor Response to Amphetamine 144
e) Effects of Other Drugs on the Pressor Effect of Amphetamine . 144
f) Effects of Repeated Doses of Amphetamine on Blood Pressure . 148
g) Summary and Conclusions 148
2. Respiratory Effects of Amphetamine 148
3. Amphetamine Induced Salivation 149
4. Body Temperature 149
a) Amphetamine-Induced Hyperthermia 149
b) Effects of Other Drugs on Amphetamine-Induced Hyperthermia 150
c) Effects of Brain Lesions on Amphetamine-Induced Hyperthermia 155
d) Other Factors Influencing Amphetamine-Induced Hyperthermia 155
e) Physiologic Correlates to Amphetamine-Induced Hyperthermia 156
f) Biochemical Correlates to Amphetamine-Induced Hyperthermia 156
g) Effects of Chronic Amphetamine Administration on the Hyperthermic Response 157
h) Amphetamine Induced Hypothermia 158
i) Behavior Thermoregulation 158
j) Summary of Mechanisms Involved in Amphetamine-Induced Changes in Body Temperature 159
5. Effects of Amphetamine on General Metabolic Processes in Brain and Peripheral Tissues 160
a) Oxygen Consumption 160
b) Brain Adenosine Nucleotide and Phosphate Metabolism . . . 161
c) Carbohydrate Metabolism 162
d) Lipid Metabolism 163
e) Protein Metabolism 164
f) Nucleic Acid Metabolism 165
g) Miscellaneous Studies on General Metabolism 165
6. Endocrinologic Effects of Amphetamine 165
a) Insulin 165
b) ACTH-Corticosteroids 165

c) Growth Hormone . . . 166
d) Prolactin . . . 166
e) Pineal Gland Hormones . . . 166
f) Thyroid Hormones . . . 166
7. Toxicity of Amphetamine . . . 166
a) Lethality After Amphetamine Administration . . . 166
b) Factors of Importance for Amphetamine Induced Toxicity in Aggregated and Isolated Mice and Rats . . . 167
c) Effects of Other Drugs on Amphetamine-Induced Toxicity in Mice and Rats . . . 169
d) Chronic Toxicity of Amphetamine . . . 173
e) Pathologic-Anatomical Findings in Amphetamine-Intoxicated Animals . . . 173
f) On the Cause of Death in Amphetamine Intoxication . . . 174
g) Summary and Conclusions . . . 175
IV. Miscellaneous Pharmacologic Effects on Amphetamine . . . 176
1. Electroencephalographic Changes Caused by Amphetamine . . . 176
2. Effects of Amphetamine on Seizure Susceptibility . . . 177
3. Sleep and Wakefulness . . . 178
4. Amphetamine-Induced Analgesia . . . 179
5. Effects of Amphetamine on Physical Performance in Animals . . . 179
D. Conclusion . . . 180
Abbreviations Used . . . 180
References . . . 181

C. Effects of Amphetamines in Humans. L.-M. GUNNE. With 4 Figures

A. General Considerations . . . 247
B. Effects of Single Doses . . . 248
I. Mood . . . 248
II. Performance . . . 249
III. Wakefulness . . . 250
IV. Anorexia . . . 250
V. Analgesia . . . 251
VI. Libido . . . 251
VII. EEG . . . 251
VIII. Evoked Responses . . . 252
IX. REM Sleep . . . 252
X. Cerebral Blood Flow . . . 252
XI. Auditory Thresholds . . . 252
XII. Critical Flicker Frequency . . . 253
XIII. Pupil Size . . . 253
XIV. Cardiovascular Effects . . . 253
XV. Respiration . . . 254

XVI. Metabolic Changes 254
XVII. Blood Cells 255
XVIII. Various Smooth-Muscle Effects 255
XIX. Hormones and Neurotransmitters 255
C. Effects of Chronic Administration 255
I. Mood 255
II. Abuse Potential 257
III. Performance 258
IV. Wakefulness 258
V. Anorexia 259
VI. Analgesia 260
VII. Libido 260
VIII. EEG 260
IX. REM Sleep 260
X. Cerebral Blood Flow, Auditory Thresholds, Critical Flicker Frequency, and Pupil Size 261
XI. Cardiovascular Effects 261
XII. Metabolic Effects 261
XIII. Respiration, Blood Cells, and Smooth-Muscle Effects 261
XIV. Hormones and Neurotransmitters 261
D. Acute and Subacute Intoxication 262
I. Paranoid Psychosis 262
II. Stereotyped Behavior 263
III. The Choreic Syndrome 263
IV. Excitation Syndrome 264
V. Dysautonomic Syndrome 264
VI. Cerebrovascular Accidents 265
E. Blockade of Amphetamine Effects 265
References 266

CHAPTER 2

Amphetamine Dependence; Clinical Features. J. D. GRIFFITH

A. Naturally-Occurring Psychostimulants 277
I. Khat 277
II. Xanthines 277
III. Ephedrine 278
IV. Absinthe 278
V. Cocaine 278
B. Exogenous Phenethylamine Congeners 279
C. Synthetic Psychostimulants 279
I. Amphetamine and Methamphetamine 279
II. Other Amphetamine Congeners 281
III. Psychodelic Amphetamines 281

D. Changing Medical Views on Amphetamine Abuse and Psychotoxicity . . 281
E. Contemporary Medical Indications 294
I. Obesity 284
II. Narcolepsy 285
III. Hyperkinetic Syndromes of Children 286
IV. Emotional Depression 287
V. Analgesia 288
VI. Miscellaneous Applications 288
F. Psychotoxicity 288
I. Acute Intoxication 288
II. Dependency Syndromes 289
III. Amphetamine Psychosis 291
1. Clinical Syndrome 291
2. Clinical Features 292
3. Withdrawal, Hospital Course and Treatment 294
G. Experimentally Induced Amphetamine Psychoses 295
References 296

CHAPTER 3

Pharmacology and Classification of LSD-like Hallucinogens. W.R. Martin and J.W. Sloan. With 1 Figure

A. Introduction 305
B. Classification of LSD-like Hallucinogens 306
I. Problem 306
II. Effects of LSD-like Hallucinogens, Mescaline, and Other Hallucinogens 306
1. Man 306
a) Behavioral and Subjective Changes 306
b) Physiologic Changes 310
c) Effects of Related Drugs 310
α) Indoleamines and LSD Derivatives 310
β) β-Phenethylamines and Amphetamines 311
d) Interactions with LSD 312
2. Other Species 312
a) Mouse 312
b) Rat 313
c) Rabbit 315
d) Cat 315
e) Dog 316
f) Monkey 317
g) Others 317
III. Tolerance and Cross-Tolerance 318
1. Man 318
2. Rat 320

3. Rabbit . . . 320
4. Cat . . . 320
5. Dog . . . 321
IV. Criteria for LSD-like Hallucinogens . . . 321
Characteristics . . . 321
V. Chemical Structures of LSD-like Hallucinogens . . . 323
1. β-Phenethylamines . . . 323
2. Amphetamines . . . 328
3. Indoleamines . . . 328
4. Lysergic Acid Amides . . . 328
C. Mode of Action . . . 328
I. Serotonin Antagonist Theory . . . 328
II. Serotonin Agonist Theory . . . 329
III. Tryptamine Agonist Theory . . . 331
IV. Other Brain Amines and Polypeptides . . . 332
V. Distribution . . . 333
D. Sites of Action . . . 334
I. Spinal Cord . . . 334
II. Pons Medulla . . . 335
1. Cardiovascular Changes . . . 335
2. Respiration . . . 336
3. Chemoreceptive Trigger Zone . . . 336
4. Tolerance . . . 336
III. Mesencephalon . . . 336
1. Effects on the Eye, Optic Tract, and Lateral Geniculate . . . 336
2. Pupils . . . 337
IV. Diencephalon-thermal Regulations . . . 337
IV. Cortex . . . 338
1. Effects of LSD-like Hallucinogens on Cortical Neurons . . . 338
2. Seizure Activity . . . 339
3. Transcallosal Potentials . . . 339
4. EEG, Cortical Evoked Potentials, and Sleep . . . 339
a) Rat . . . 339
b) Rabbit . . . 339
c) Cat . . . 341
d) Monkey . . . 342
e) Man . . . 343
E. Conclusions and Discussion . . . 343
I. Pharmacologic Characteristics of LSD-like Hallucinogens . . . 343
II. LSD Hallucinogens — Agonists . . . 344
III. Tolerance . . . 344
IV. Modes of Action . . . 345
V. Mechanisms of Action . . . 346
VI. Abuse Potential and Identification of LSD-like Drugs . . . 348
Most Usual Abbreviations . . . 349
References . . . 349

SECTION II

Marihuana Dependence

CHAPTER 1

Cannabis. Its Chemistry, Pharmacology, and Toxicology. L. S. HARRIS, W. L. DEWEY, and R. K. RAZDAN. With 3 Figures

A. Introduction . . . 371
B. Chemistry . . . 372
Isolation . . . 374
C. Distribution and Fate . . . 380
D. Pharmacology . . . 384
I. Central Nervous System . . . 384
1. General Effects in Man . . . 384
2. Overt Behavior . . . 386
3. Aggression . . . 387
4. Operant or Conditioned Behavior . . . 388
5. Analgesia . . . 389
II. Biogenic Amines . . . 391
1. NE, E, and DA . . . 391
2. 5-HT . . . 392
3. ACh . . . 392
III. Cardiovascular and Autonomic Effects . . . 397
IV. Endocrine Effects . . . 400
V. Cellular Effects . . . 401
E. Toxicology . . . 403
F. Tolerance . . . 409
Physical Dependence . . . 412
G. Therapeutic Potential . . . 413
References . . . 415

Autor Index . . . 431

Subject Index . . . 491

Contents
Part I: Morphine, Sedative/Hypnotic and Alcohol Dependence

SECTION I

Problems of Drug Dependence

CHAPTER 1

General Problems of Drug Abuse and Drug Dependence. W. R. MARTIN. With 1 Figure

SECTION II

Morphine Dependence

CHAPTER 1

Neuropharmacology and Neurochemistry of Subjective Effects, Analgesia, Tolerance and Dependence Produced by Narcotic Analgesics. W. R. MARTIN and J. W. SLOAN. With 9 Figures

CHAPTER 2

Assessment of the Abuse Potential of Narcotic Analgesics in Animals. W. R. MARTIN and D. R. JASINSKI. With 2 Figures

CHAPTER 3

Assessment of the Abuse Potentiality of Morphine Like Drugs (Methods Used in Man). D. R. JASINSKI. With 1 Figure

CHAPTER 4

Psychiatric Treatment of Narcotic Addiction. H. T. CONRAD

CHAPTER 5

Chemotherapy of Narcotic Addiction. W. R. MARTIN. With 4 Figures

CHAPTER 6

Detection of Drugs of Abuse in Biological Fluids. C. W. GORODETZKY

SECTION III

Sedative-Hypnotics and Alcohol Dependence

CHAPTER 1

The Pharmacology of Sedative Hypnotics, Alcohol, and Anesthetics: Sites and Mechanisms of Action. C. M. SMITH

CHAPTER 2

The Assessment of the Abuse Potentiality of Sedative-Hypnotics (Depressants) (Methods Used in Animals and Man). H. F. FRASER and D. R. JASINSKI. With 3 Figures

CHAPTER 3

Clinical Aspects of Alcohol Dependence. N. K. MELLO and J. H. MENDELSON. With 8 Figures

CHAPTER 4

Abuse of Non-Narcotic Analgesics. P. KIELHOLZ and D. LADEWIG. With 1 Figure

Author Index

Subject Index

List of Contributors

E. ÄNGGÅRD, M. D., Department of Pharmacology, Karolinska Institutet, S-104 01 Stockholm 60, Sweden and Psychiatric Research Centre, Ulleråker Hospital, S-750 17 Uppsala, Sweden

W. L. DEWEY, M. D., Department of Pharmacology, Medical College of Virginia, Richmond, VA 23298 USA

J. D. GRIFFITH, M. D., NIDA Addiction Research Center, Lexington, KY 40511, USA

L.-M. GUNNE, Prof., University of Uppsala, Psychiatric Research Center, Ulleråker Hospital, S-750 17 Uppsala, Sweden

L. S. HARRIS, Prof. Dr., Department of Pharmacology, Medical College of Virginia, Richmond, VA 23298 USA

T. LEWANDER, M.D., Psychiatric Research Center, University of Uppsala, Ulleråker Hospital, S-750 17 Uppsala, Sweden

W. R. MARTIN, M. D., NIDA Addiction Research Center, Lexington, KY 40511, USA

R. K. RAZDAN, M. D., SISA, Inc. Cambridge, MA 02138 USA

J. W. SLOAN, M. D., NIDA Addiction Research Center, Lexington, KY 40511, USA

SECTION I

Amphetamine and Psychotogen Dependence

CHAPTER 1

General Pharmacology of Amphetamine-Like Drugs
A. Pharmacokinetics and Metabolism

E. ÄNGGÅRD

A. Structure-Activity Relationships Among the Amphetamines

Amphetamine and related central stimulant drugs are derivatives of β-phenylethylamine (Table 1) and are thus relatively simple organic bases. The general β-phenylethylamine skeleton is one which amphetamine shares with the neurotransmitters noradrenaline, adrenaline, and dopamine. Phenylethylamine itself has central stimulant properties, but has an extremely short half-life in the body due to rapid metabolism by monoamine oxidase (MAO). Amphetamine has, due to steric hindrance by the α-methyl group, much less affinity for MAO and therefore has a longer half-life.

Most amphetamines have cardiovascular, psychomotor stimulant, hyperthermic, and anorexigenic actions. All of the structural features of amphetamine are important for its spectrum of pharmacologic activity. Any alteration may enhance, diminish, or attenuate one or several components in the actions of the parent drug. (1) Substitution of the phenyl ring, (2) alteration of the length of the side chain, (3) substitution on the primary nitrogen group, (4) substitution of the α- and β-carbon atoms, and (5) their absolute configuration have been studied and will be briefly discussed here. For a more thorough review on the subject the reader is referred to the papers by BIEL (1970), VAN ROSSUM and SIMONS (1969) and VREE (1973). The effect of ring and side-chain substitution on distribution and elimination of the amphetamines has been discussed by BECKETT and BROOKES (1970). CALDWELL and SEVER (1974) recently reviewed the biochemical pharmacology of the amphetamines. The metabolism and disposition of methylphenidate were reported by FARAJ et al. (1974). A selected bibliography on detection of dependence-producing drugs (CHRUSCIEL and CHRUSCIEL, 1975) has 93 references relevant to the analysis of amphetamines and amphetamine-like substances.

I. Effect of Aromatic Substitution

Introduction of electron withdrawing groups in the ring, such as in chlorphentermine and fenfluramine, results in reduction or abolition of the psychomotor stimulant effects whereas the anorexigenic properties remain (LE DOUAREC et al., 1966; HOLLAND et al., 1963). These compounds have also been shown to cause depletion of brain serotonin levels without affecting the catecholamine levels (PLETSCHER et al., 1964; FULLER et al., 1965; VAN PRAAG et al., 1968; DUHAULT and VERDAVAIENNE, 1967). *p*-Chloroamphetamine has been particularly well studied in this respect (SULSER and SANDERS-BUSH, 1973). Fenfluramine and related CF_3-substituted amines

(BEREGI et al., 1970) are of particular interest since they demonstrate an apparent complete separation of the central stimulant from the anorexigenic actions in the amphetamine series (LE DOUAREC and NEVEU, 1970). The mechanism of the weight-reducing effects of the amphetamines is not clear, although certain authors (see OPITZ, 1970; PAWAN, 1970) favor a peripheral mechanism.

p-Hydroxyamphetamine is a major metabolite of amphetamine in the rat (AXELROD, 1954a) and may thus be responsible for part of the amphetamine action in this species. Ring hydroxylation decreases the CNS stimulant properties of amphetamine, presumably due to poor passage through the blood-brain barrier of the polar hydroxyamphetamine (Table 1). The phenyl ring may be replaced by isosteric planar aromatic systems (tiophen, furan) with retention of its pharmacological activity (ALLES and FEIGEN, 1941).

Polyalkoxylation of the phenyl ring, e.g., 3,4,5-trimethoxy-amphetamine yields compounds with psychotomimetic properties (SHULGIN and SARGENT, 1967). The most potent compound in this series appears to be 2,5-dimethoxy, 4-methylamphetamine (STP, DOM), causing hallucinations in doses from 3–10 mg in man (SNYDER et al., 1967). For a review on psychotomimetic phenylethylamines see NIEFORTH (1971) and Section I, Chapter 3 of this volume.

II. Effect of Length of the Side Chain

The phenylethylamine structure is essential. Shortening or lengthening the side chain greatly diminishes both locomotor stimulation (VAN ROSSUM and SIMONS, 1969) and, on inhibition, neuronal noradrenaline uptake (DALY et al., 1966).

III. Effect of Substitution of the Primary Nitrogen

With the exception of N-methylamphetamine, progressive alkylation and substitution with bulky groups (benzyl, furfuryl) leads to diminished effects of the central nervous and cardiovascular systems although some anorexigenic efficacy is retained.

IV. Effect of Substitution of the Side Chain

The α-amino group of amphetamine is essential since it confers resistance to oxidative deamination and therefore an increased metabolic half-life. Introduction of a further methyl group on the α-carbon, as in the phentermines (Table 1), provides an added protection against deamination. Shifting the α-methyl group to the β-carbon abolishes effects of amphetamine (DALY et al., 1966).

Amphetamine has moderate to weak MAO-inhibitory properties depending on the substrate and tissue (GLOWINSKI et al., 1966). It is possible to achieve very potent MAO-inhibitors by modification of the amphetamine structure. In tranylcypromine, the α-methyl group is bent toward the ring by making it part of a cyclopropane system. This compound is about 5000 times more potent as an MAO-inhibitor than amphetamine, possibly because the exposed nitrogen can interact more efficiently at the active center of the MAO (ZIRKLE and KAISER, 1964).

Substitution of the two β-hydrogens in amphetamine and *p*-chloroamphetamine with fluorine increases lipid solubility and accumulation in adipose tissue (FULLER et al., 1972, 1973).

V. Stereoisomers

D-Amphetamine has been found to be 2–4 times more potent on locomotor activity, stereotype behavior, gnawing (TAYLOR and SNYDER, 1971; SVENSSON, 1971), anorexia (LAWLOR et al., 1969), and self-administration (YOKEL and PICKENS, 1973), in rats. Intracranial self-stimulation was facilitated 7–10 times more by the D- than the L-isomer. Snyder and his group have presented evidence that D-amphetamine is about 10 times more active than L-amphetamine in eliciting release and inhibiting reuptake of noradrenaline from brain slices, whereas the isomers were about equal in affecting dopamine neurons (for ref. see SNYDER 1972).

In humans, D-amphetamine is generally thought to be more active than the L-isomer (ALLES and FEIGEN, 1941). However, about equal doses of D- and L-amphetamine were needed to elicit amphetamine psychosis (ANGRIST et al., 1971).

B. Metabolism

I. Pathways of Metabolism

The central stimulant drugs carrying the basic β-phenylethylamine structure can be metabolized by:

1. *Aromatic hydroxylation* to the corresponding *p*-hydroxy derivative, which may be excreted unchanged, be further hydroxylated, or conjugated
2. *Aliphatic hydroxylation* of the α- and β-carbon to the nitrogen
3. *Oxidative deamination*, which by way of numerous intermediates yields benzoic acid and benzoic acid conjugates as the end products
4. *N-dealkylation* of N-substituted amphetamines to the primary amine
5. *N-oxidation* to form hydroxylamines or N-oxides.

For a general survey of the comparative metabolism of amphetamine in various species see WILLIAMS et al. (1973) and for the fate of different amphetamines in man see GORROD (1973).

A large amount of information is available on amphetamine and methamphetamine as they are the most extensively studied drugs (AXELROD, 1954a, b, 1955; ALLEVA, 1963; ELLISON et al., 1966; DRING et al., 1966, 1970). The major metabolic pathways for these drugs are illustrated in Figure 1. N-oxidation is not included since the occurrence of this reaction has not been demonstrated in vivo.

Methamphetamine is excreted partly unchanged and partly demethylated to amphetamine. Amphetamine is excreted either unchanged or is metabolized along two pathways, namely hydroxylation and deamination. The relative importance of these pathways varies considerably among species. This is summarized in Table 2. In most animals and in man, deamination predominates over hydroxylation. This is especially noticeable in the rabbit and guinea pig, where deamination accounts for over 55% of the excreted radioactivity. The rat, on the other hand, metabolizes amphetamine mainly by hydroxylation.

Table 1. Structures of some central stimulants

Name	Structure	pK_a[1]	Partition coefficient heptane/water
Amphetamine (+)2S	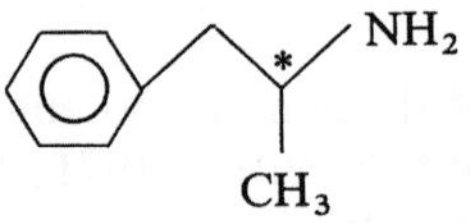	9.90	1.88
Methamphetamine (+)2S	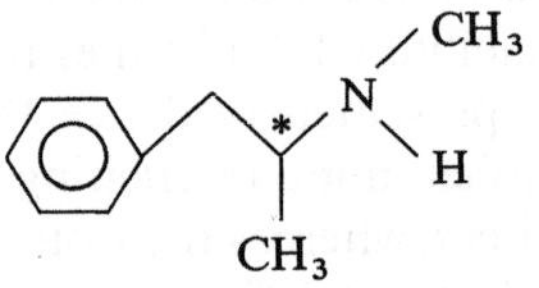	10.11	5.14
Phenmetrazine (threo(+)1S, 2S)	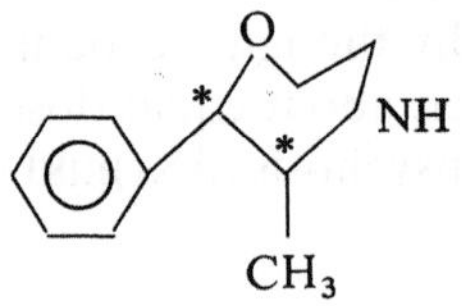	8.45	1.35
Methylphenidate (+)threo 1R; 2R	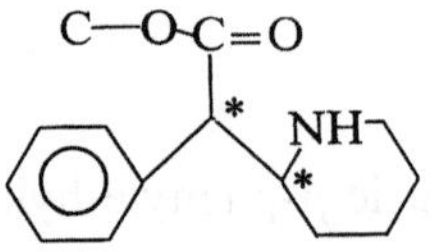	—	—
Hydroxy-amphetamine,	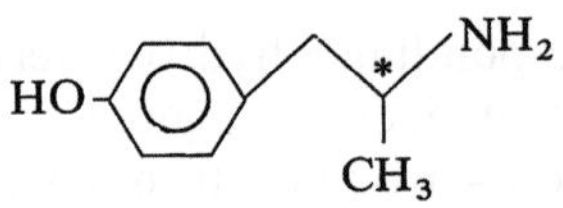	—	—
Ephedrine (−)1R; 2S	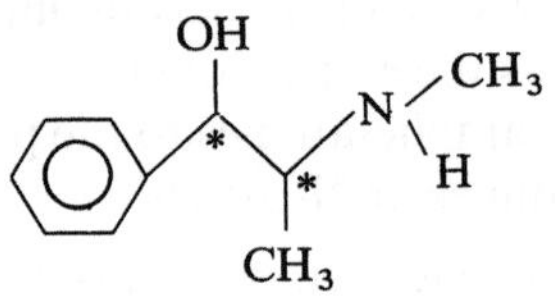	9.60	0.001
Norpseudo-ephedrine (+)1S; 2S, cathine	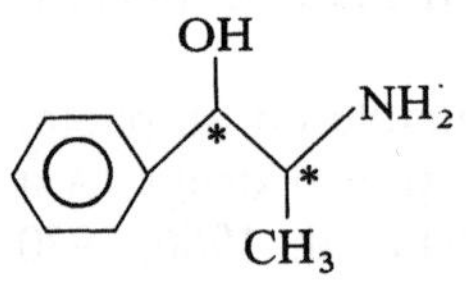	9.40	0.001
Benzphetamine (+)2S	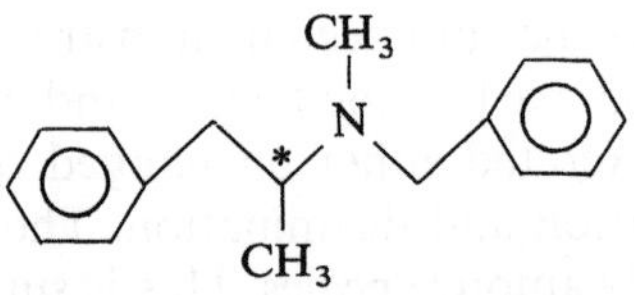	6.55	74.8
Phentermine	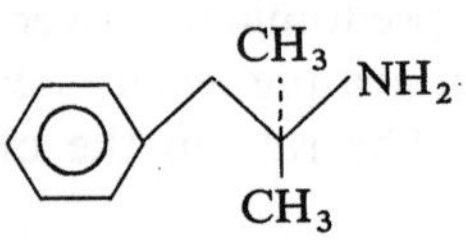	10.11	63.2

Name	Structure	pK_a	Partition coefficient heptane/water
Chlorphen-termine	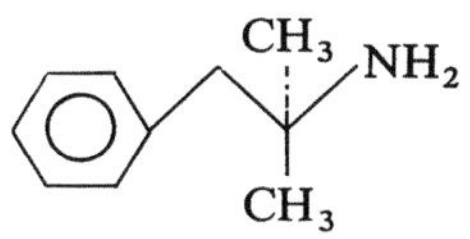	9.60	17.5
β, β-Difluoro-amphetamine	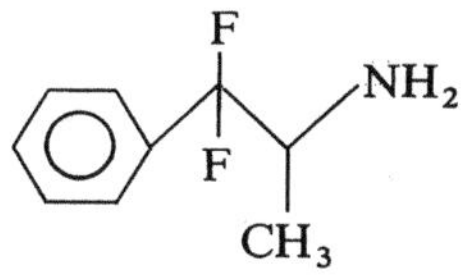	6.97	
Diethyl-propion	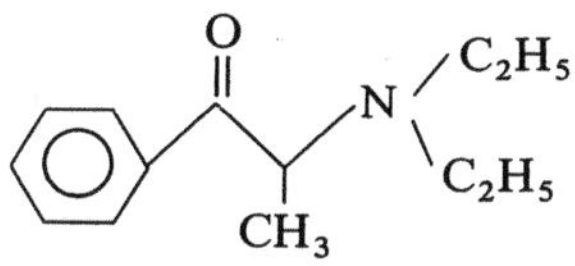	—	—
Aminorex	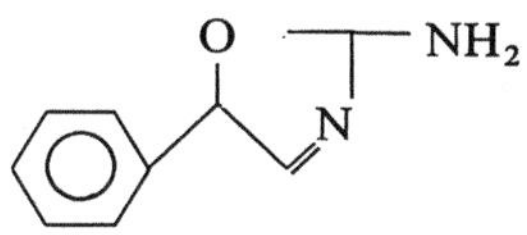	—	—
Fenfluramine	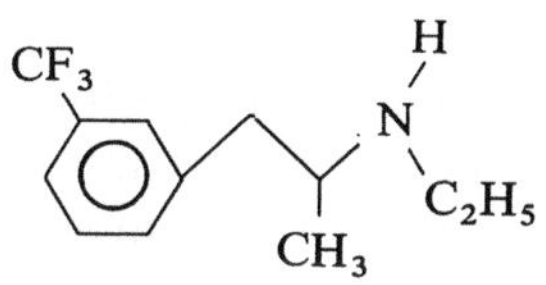	9.10	678

[1] After VREE and VAN ROSSUM, 1969.

The amount of amphetamine excreted unchanged is also subject to a considerable variation between species. Since it has been shown in man (ASATOOR et al., 1965; BECKETT and ROWLAND, 1965a–d, Table 2) that amphetamine, like many other bases (MILNE et al., 1958; WEINER and MUDGE, 1964), is subject to the mechanism of nonionic diffusion, it is conceivable that the differences in urinary excretion could be due to species variations in the urinary pH. Rabbits and guinea pigs are herbivorous and produce an alkaline urine, whereas the omnivorous and carnivorous species produce a slightly more acidic urine. The proposal by DRING et al. (1970), that there is a relationship between dietary habit and metabolic pattern, is in line with this suggestion. Herbivorous species metabolize the amphetamine-like compounds largely by oxidative deamination. Carnivorous and omnivorous species utilize hydroxylation as well as oxidative deamination and the degree to which either of these two reactions are utilized varies with the structure of the compound and the species.

AXELROD (1955) suggested that the species variation could be due to the presence of inhibitors. Thus, when dog and rat liver microsomes were added to rabbit liver microsomes, amphetamine deamination was markedly suppressed.

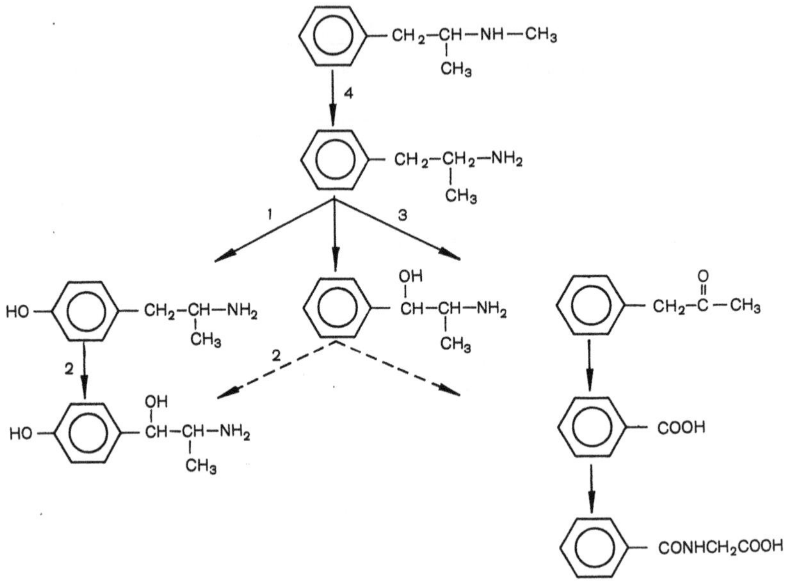

Fig. 1. Major metabolic pathways for methamphetamine and amphetamine

1. Hydroxylation of Amphetamine

Aromatic hydroxylation of amphetamine in the *para* position is the major pathway of amphetamine metabolism in the rat, but is relatively minor in man (Table 2). The occurrence of this reaction has been extensively documented in vivo (AXELROD, 1954a; DRING et al., 1966; 1970; SMITH and DRING, 1970; ALLEVA, 1963; ELLISON et al., 1966; ÄNGGÅRD et al., 1973; SEVER et al., 1973). Curiously, it has proved difficult to study this reaction in the usual microsomal mixed function oxidase system until quite recently (JONSSON, 1974; ROMMELSPACHER et al., 1974). The interaction between amphetamines and rat liver cytochrome P-450 has recently been demonstrated (FRANKLIN, 1974). It is not known, however, whether this binding is associated with aromatic or aliphatic hydroxylation.

p-Hydroxyamphetamine is pharmacologically active and has been used as a sympathomimetic drug. Its main acute effects are an elevation of blood pressure and tachycardia due to release of catecholamines (ABBOTT and HENRY, 1937; CHIDSEY et al., 1962). After prolonged administration of *p*-hydroxyamphetamine a *depression* of the blood pressure is observed, purportedly due to the accumulation of *p*-hydroxy-norephedrine (PHNE) as false transmitters in the adrenergic neurons (GILL et al., 1967; RANGNO et al., 1973). Due to the reduced lipid solubility conferred by the phenolic group, *p*-hydroxyamphetamine does not readily pass into the brain (JORI and CACCIA, 1974) and thus has only weak central stimulant effects (TAYLOR and SULSER, 1973). *p*-Hydroxy-amphetamine and amphetamine have similar potency in inhibiting uptake and deamination of labelled norepinephrine in chopped cerebral cortex (WENGER and RUTLEDGE, 1974).

2. Aliphatic Hydroxylation

The dopamine β-hydroxylase is localized in adrenergic neurons. This enzyme catalyzes the β-hydroxylation of several amphetamines in vitro (GOLDSTEIN and CONTRERA, 1962; CREVELING et al., 1962). *p*-Hydroxyamphetamine is a better substrate for the β-hydroxylase than amphetamine. Amphetamines contain a side chain of three carbon atoms each of which could be hydroxylated. Hydroxylation of the α-carbon to the nitrogen gives an unstable carbinolamine (BRODIE et al., 1958) which has been suggested as an intermediate in oxidative deamination (HUCKER et al., 1971; PARLI et al., 1971 a, b, see below).

The hydroxylation of the β-carbon has been definitely proved, inasmuch as β-hydroxylated metabolites of amphetamine and *p*-hydroxyamphetamine have been identified. It has been shown that hydroxyamphetamine is converted to *p*-hydroxynorephedrine (PHNE) in vivo in man (SJOERDSMA and VON STUDNITZ, 1963). It was also shown that H^3-amphetamine was partly converted to H^3-PHNE in animals (GOLDSTEIN and ANAGNOSTE, 1965) and also in man (CAVANAUGH et al., 1970; CALDWELL et al., 1972a; ÄNGGÅRD et al., 1973). The PHNE is stored in adrenergic neurons and can be liberated by nerve stimulation (THOENEN et al., 1966). Amphetamine is presumably converted to *p*-hydroxyamphetamine in the liver and then to PHNE by β-hydroxylase, which is stereospecific for D-amphetamine (GOLDSTEIN et al., 1964; TAYLOR, 1974).

In the last few years several groups have independently shown that the persistent decrease of the levels of noradrenaline in heart and brain caused by D-amphetamine is due to accumulation of roughly stochiometric amounts of PHNE as a false transmitter (BRODIE et al., 1970; COSTA and GROPPETTI, 1970; LEWANDER, 1970, 1971 a, b; TAYLOR and SULSER, 1973). The persistence of large concentrations of *p*-hydroxynorephedrine was particularly noticeable in noradrenaline-containing areas of the rat brain (EL GUEDRI et al., 1973; JORI and CACCIA, 1974). It has been suggested that this accumulation plays a role in the development of tolerance shown by rats to amphetamine (HARRISON et al., 1952) after long-term administration. This is discussed in detail by LEWANDER (Sect. I, Chap. 1. B).

β-Hydroxylation of amphetamine to norephedrine occurs less readily, presumably due to lower affinity of this substrate for the dopamine β-hydroxylase (CREVELING et al., 1962). It has, however, been possible to document the formation of this metabolite in animals (CALDWELL et al., 1972a) and in dependent humans taking large doses of the drug (ÄNGGÅRD et al., 1973; SEVER et al., 1973).

3. Oxidative Deamination

The enzymatic oxidation of primary amines was first reported by AXELROD (1954a, 1955) who found that amphetamine was converted to phenylacetone and ammonia by rabbit liver microsomes. The reaction required molecular oxygen and NADPH. Subsequent work (Table 2) has shown that amphetamine is deaminated in vivo. The major metabolic products are hippuric acid and phenylacetone. The rate of deamination varies between species, being predominant in rabbits and nearly absent in the rat.

It is necessary in this context to point out that drugs having the basic phenylethylamine structure can be deaminated by two different enzyme systems, the monoamine oxidase (MAO) and the mixed function oxidase in the liver. MAO is widely spread throughout the tissues and has a mitochondrial location. It catalyzes the oxidative deamination of dietary phenylethylamines and indolethyl amines such as catecholamines, serotonin, tyramine, and tryptamine. β-Phenylethylamine is very rapidly metabolized in vivo by MAO and has very transient and weak central stimulant effects. However, following MAO inhibition, phenylethylamine behaves like amphetamine (STEIN, 1964). The presence of an α-methyl group such as in the amphetamines confers a resistance to the MAO. These drugs are deaminated by the hepatic microsomal mixed function oxidase system first described by AXELROD (1954a, 1955).

Early observations showed that amphetamine is not only resistant to MAO but also inhibits the enzyme in vitro (BLASCHKO et al., 1937; MANN and QUASTEL, 1940; PRATESI and BLASCHKO, 1959). Pretreatment of animals with amphetamine has also been observed to inhibit oxidative deamination of intraventricularly injected ^{3}H-norepinephrine (GLOWINSKI and AXELROD, 1965; GLOWINSKI et al., 1966). However, as suggested by RUTLEDGE (1970), this effect of amphetamine may be due to an inhibition of the adrenergic neuron membrane pump, thus limiting access of the substrate to intraneuronal MAO.

The mechanism of amphetamine deamination has recently been the subject of several studies. The alternative pathways are indicated in Figure 2. The mechanism postulated by BRODIE et al. (1958) involved an initial hydroxylation (1) of the α-carbon of an unstable intermediary carbinolamine, which then loses ammonia (2) to form phenyl acetone. An alternative route was suggested by DRING et al. (1966) who proposed that the carbinolamine might be directly dehydrogenated (3) to an imine,

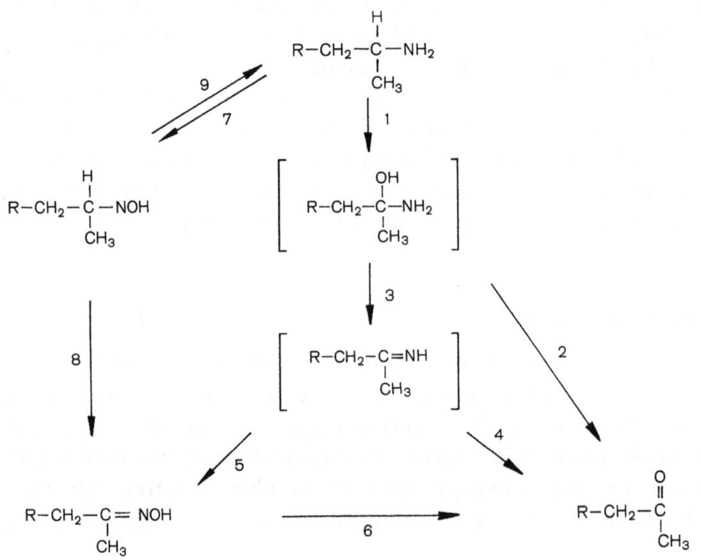

Fig. 2. Mechanisms of oxidative deamination of amphetamine R-phenyl

which would readily hydrolyze (4) to the ketone. In both mechanisms it is likely that the removal of the hydrogen on the α-carbon represents an initial rate-limiting step. Accordingly, deuterium isotope effects on the rate of metabolism have been observed when amphetamines are deaminated both in vitro (FOREMAN et al., 1969) and in vivo (VREE et al., 1971 a, b). On the basis of in vitro experiments with rat liver microsomes, HENDERSON et al. (1974) suggested that the removal of the hydrogen is rate-limiting in D-isomers of N-alkylated amphetamines, whereas in the L-isomers N-oxidation is more important. More recent studies have provided additional information on the initial steps in the deamination of amphetamine.

HUCKER et al. (1970, 1971) found phenylacetone oxime to be a major in-vitro metabolite of amphetamine in a rabbit liver microsomal system. These investigators raised the possibility that this oxime (5) arises from an intermediary imine. Phenylacetone is formed by the hydrolysis (6) of the oxime. Further evidence for this pathway was obtained by PARLI et al. (1971 a) using $^{18}O_2$. The major portion (about 75%) of phenylacetone contained ^{16}O rather than ^{18}O. This supports the notion that most phenylacetone is formed from phenylacetone oxime since this would result in the incorporation of oxygen from water. The minor portion (about 25%) of phenylacetone containing ^{18}O was most likely formed by the mechanism proposed by BRODIE et al. (1958), namely hydroxylation to the carbinolamine intermediate followed by loss of a molecule of ammonia to yield phenylacetone (retaining the oxygen).

In the experiments by PARLI et al. (1971 a) incorporation of ^{18}O into the phenylacetone oxime was almost quantitative. This indicated that the oxime was formed by the direct hydroxylation of an intermediate unstable imine or by N-hydroxylation (7, 8). That an imine might be an intermediate was supported by a further study by the same authors who showed that a chemically stable imine, 2,4,6-trimethylacetophenone imine, was N-hydroxylated by the rabbit liver microsomal system (PARLI et al., 1971 b).

The N-hydroxylation pathway derives its support from the isolation of N-hydroxyamphetamine from incubation with liver microsomes (BECKETT et al., 1973; LINDEKE et al., 1973; PARLI and MCMAHON, 1973; BECKETT and BELANGER, 1974; CHO et al., 1974 b). N-hydroxyamphetamine could be formed by direct N-oxidation of amphetamine or by hydration of the postulated intermediate imine. N-hydroxyamphetamine could then be converted into the oxime by metabolic dehydrogenation (8). A major difficulty in proving whether N-hydroxyamphetamine is an intermediate in the formation of the oxime is the ready nonenzymatic dehydrogenation (8) of the hydroxylamine to the oxime, especially under alkaline condition (LINDEKE et al., 1973). The mechanisms involved in the formation of the oxime are thus not clear at present. For a thorough discussion of this point the reader is referred to the communication by PARLI and MCMAHON (1973) and the discussion following that paper.

The pharmacology of N-hydroxyamphetamine is similar to that of the parent compound (BENINGTON et al., 1965). This is probably due to the in vivo reduction (9) of the hydroxylamine to the parent amine. In line with this notion is the finding that similar plasma levels of phentermine were found in the rat following administration of N-hydroxyphentermine and phentermine itself (CHO et al., 1974 a), and the same brain levels of *p*-chloroamphetamine and 5-hydroxyindoles were found following N-hydroxy-*p*-chloroamphetamine and *p*-chloroamphetamine (FULLER et al., 1974).

Phenylacetone may be reduced to the corresponding alcohol which is then conjugated with glucuronic acid (SMITH and DRING, 1970). It can also be oxidized via unknown intermediates to benzoic acid, which conjugated with glycine, as hippuric acid, constitutes the major end-product of amphetamine metabolism in most species (Fig. 1, Table 2).

4. N-Dealkylation

Several of the amphetamine-like drugs carry substituents on the nitrogen (Table 1). Methamphetamine, ethylamphetamine, ephedrine, diethylpropion, and fenfluramine are alkyl-substituted. In others, e.g., phenmetrazine, methylphenidate, and aminorex (Table 1), the nitrogen is joined either to the α- or β-carbon forming a six-membered ring (piperidine or oxazolidine).

AXELROD (1954b) found that N-demethylation was an important pathway for the metabolism of methamphetamine. This was one of the first papers to describe this general pathway of drug metabolism, which later has been shown to be common for many psychotropic drugs like the neuroleptics and tricyclic antidepressants. It was also one of the first demonstrations that drug-metabolizing enzymes are present in the liver microsomes and require oxygen and NADPH. This mixed-function oxidase system has since been the subject of intensive study in many laboratories (for review see LADU et al., 1971).

CALDWELL et al. (1972b, c) in a study on the metabolism of methamphetamine in man, the guinea pig, and the rat found marked species variation. In the rat, the main pathway was aromatic hydroxylation, in the guinea pig demethylation and deamination, whereas in man about half of the drug was excreted unchanged. In man, 15% of the dose administered was found as *p*-hydroxymethamphetamine. Minor metabolites were hippuric acid, norephedrine, *p*-hydroxyamphetamine, *p*-hydroxynorephedrine, and an acid labile precursor of phenylacetone.

The metabolism of N-alkyl substituted D-amphetamines was studied in man by VREE and VAN ROSSUM (1970) and BECKETT and BROOKES (1970). It was found that the rate of dealkylation increased with the size of the substituent. Thus isopropyl-amphetamine was metabolized faster than ethylamphetamine, which in turn was dealkylated more rapidly than methamphetamine. Since the lipid solubility of the amine increases with the degree of substitution on the nitrogen (VREE and VAN ROSSUM, 1969) the increased rate of metabolism might be a reflection of increased rate of entry into the lipid membranes of the liver microsomes. This would be in agreement with recent data demonstrating that cytochrome P-450 exhibits an increased affinity for the D-alkyl amphetamines as the size of the alkyl group increases (HOFFSTRÖM and ORRENIUS, 1973). These authors concluded, however, that the stereoselectivity in the cytochrome P-450–substrate interaction seemed more significant than lipid solubility.

Tertiary amines such as amitriptyline and propoxyphene are in general rapidly N-dealkylated compared to the corresponding secondary amines (MCMAHON, 1963; GRAM, 1971). Similarly, dialkyl-substituted amphetamines, e.g., dimethylamphetamine, are rapidly dealkylated to the monoalkyl derivative, which is converted to amphetamine at a slower rate (VREE and VAN ROSSUM, 1970). The ethyl group of ethylmethylamphetamine is removed faster than the methyl group.

Amphetamines with 23 different aryl substituents on the nitrogen have been examined using a rat liver microsomal preparation (BOISSIER et al., 1970). All of these substituents were removed to a varying degree resulting in the generation of free amphetamine and *p*-hydroxylated and deaminated amphetamine metabolites. From these data it seems likely that most drugs with the basic amphetamine structure (Table 1) but carrying substituents on the nitrogen will be degraded to the parent drug.

5. N-Oxidation

The oxidation of the nitrogen in aliphatic and aromatic amines is well known. For a review the reader is referred to a recent volume edited by BRIDGES et al. (1972). Formation of N-oxygenated amphetamines have been demonstrated in liver microsomes, although considerable differences exist between different species. These metabolites, which have never been isolated in vivo, could be intermediates in the deamination of amphetamine as discussed above.

II. Stereoselectivity in Amphetamine Metabolism

Early studies indicated that a certain stereospecificity existed in the fate of amphetamines in man. ALLES and WISEGARVER (1961) analyzed the optical activity of urinary extracts from an addict receiving large doses of racemic amphetamine and found it to be levorotatory. BECKETT and ROWLAND (1965c) reported that after separate administration of D- and L-amphetamine, less D-amphetamine was excreted in a comparable time period. Direct evidence for a difference in the fate of the optical antipodes of amphetamine and methylamphetamine was provided by simultaneous gas chromatographic analysis of the diastereoisomers in the urine following adminis-

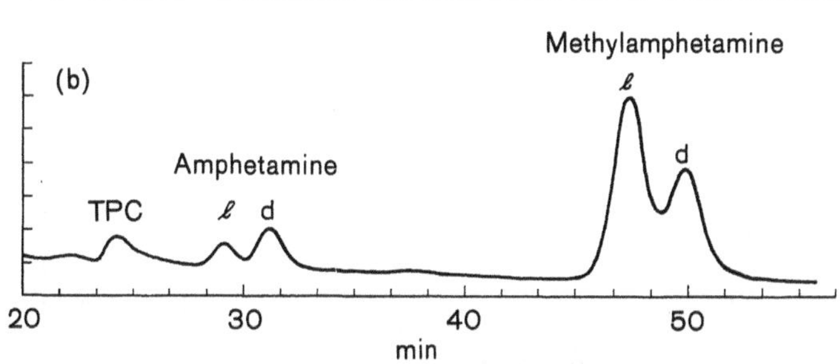

Fig. 3a and b. Gas chromatographic separation of the diastereoisomers of methylamphetamine and amphetamine present in urine 24–36 h after ingestion of 10 mg of D,L-methylamphetamine by a human subject. *Upper curve:* control. *Lower curve:* urinary extract

tration of the racemate (GUNNE, 1967). It was found that the ratio between D- and L-amphetamine decreased from about 1 to less than 0.5 over a period of 48 h. Since deamination is the major route of metabolism of amphetamine in man, the stereoselectivity is most likely to be in the mechanism of deamination. Interestingly, HENDERSON et al. (1974) described in vitro experiments indicating that D-amphetamine is deaminated by means of the carbinolamine pathway (Fig. 2) whereas L-amphetamine is deaminated by the N-oxidation pathway. However, differences in the distribution and excretion could provide alternative explanations for the different rates of urinary excretion of D- and L-amphetamine in man.

After the administration of D,L-methamphetamine the urine was found to contain decreasing amounts of D-methamphetamine but increasing amounts of D-amphetamine (Fig. 3). Thus a stereo specificity of the demethylation exists favoring the D-isomer. The stereospecificity of N-dealkylation of the D-isomer was further documented by BECKETT and BROOKES (1970) and VREE et al. (1971 a, b), who also showed that this difference is more pronounced for ethylamphetamine than for methylamphetamine. Curiously, the introduction of a *p*-trifluoromethyl group in the phenyl ring abolishes the stereoselectivity in N-dealkylation (BECKETT and BROOKES, 1970).

III. Factors Influencing the Metabolism of Amphetamine

1. Species

A considerable species variation has been shown to exist in the metabolism of amphetamine. This has been studied by R. T. WILLIAMS and his group (WILLIAMS et al., 1973) and is illustrated in Table 2. The rat is unique in that it preferentially *p*-hydroxylates amphetamine. Most other species, including man, predominantly deaminate the drug, as discussed above.

Table 2. Comparative metabolism of amphetamine

Species	Percentage of drug excreted as			Reference
	Unchanged %	Hydroxylated %	Deaminated %	
Man	30	3	21	DRING et al., 1970; WILLIAMS et al., 1973
Man acidic urine	65	5	15	DAVIS et al., 1971
Man alkaline urine	22	9	27	DAVIS et al., 1971
Man acidic urine	68	5	17	ÄNGGÅRD et al., 1970a, 1973
Man alkaline urine	27	9	37	ÄNGGÅRD et al., 1970a, 1973
Monkey, squirrel	29.1–41.7	1.4–2.0	7.0–8.9	ELLISON et al., 1966
Monkey	34	6	27	DRING et al., 1970
Dog, beagle	42.0–49.3	10.5–17.1	17.8–19.8	ELLISON et al., 1966
Dog, greyhound	30	5.9	30	DRING et al., 1966, 1970
Rabbit	3.5	6.2	55	DRING et al., 1966, 1970
Guinea pig	22	0	62	DRING et al., 1966, 1970
Rat	12.3–13.3	30.9–33.7	2.9–3.0	ELLISON et al., 1966
Rat	14	48	3.3	DRING et al., 1966, 1970
Mouse	33	14	31	DRING et al., 1966, 1970

2. Urinary pH

It has long been known that the rate of excretion of amphetamine in man is dependent on urinary pH (ASATOOR et al., 1965; BECKETT et al., 1965d, Table 2). About 65% is excreted unchanged at an acid pH (pH 5–6) and 25% at a neutral pH (pH 7–8). When the urinary pH is higher, a compensatory increase in the rate of metabolism is observed (DAVIS et al., 1971; ÄNGGÅRD et al., 1970a, 1973). The rate of excretion of deaminated and hydroxylated metabolites increased about twofold (Table 2). The increased excretion of *p*-hydroxyamphetamine (OHA) at the higher urinary pH is of interest since this metabolite is a key intermediate in the formation of *p*-hydroxynorephedrine (GOLDSTEIN and ANAGNOSTE, 1965; BRODIE et al., 1970; COSTA and GROPPETTI, 1970; LEWANDER, 1970) and *p*-methoxyamphetamine (SMYTHIES et al., 1967). The amphetamine psychosis is also more intense and long-lasting in dependent subjects with higher urinary pH (ÄNGGÅRD et al., 1970a). It was found that an increasing proportion of the drug was excreted as hydroxyamphetamine in these patients. This is curious since hydroxyamphetamine is more polar than amphetamine (Table 1) and has a shorter plasma half-life in nondependent subjects. One possible explanation may be that in these concentrations of hydroxyamphetamine in dependent subjects, hydroxyamphetamine is more tightly bound to tissues. Evidence exists for an increased tissue affinity of amphetamine in dependent subjects (ÄNGGÅRD et al., 1970a). It is of interest to note similar findings in amphetamine-tolerant cats (ELLISON et al., 1971).

3. Other Drugs

The observation that tricyclic antidepressant drugs potentiate and prolong the effects of amphetamine (CARLTON, 1961; STEIN, 1964) led to the suggestion that amphetamine-induced animal behavior might be a suitable model for the screening of antidepressant drugs. However, it was later found that this potentiation was not due to enhancement of the central effects of amphetamine, but to higher levels of the drug in the brain after pretreatment with desipramine (SULSER et al., 1966; CONSOLO et al., 1967; LEWANDER, 1968). This could be shown to be due to an inhibition of aromatic hydroxylation of amphetamine (DINGELL and BASS, 1969; LEWANDER, 1968, 1969). Subsequently a great number of psychotropic and other drugs have been examined with respect to their influence on amphetamine metabolism. These studies are summarized in Table 3. It is seen that a wide range of thymoleptic and neuroleptic drugs potentiate amphetamine by inhibition of its metabolism and thereby increasing amphetamine brain levels and prolonging its effects. Amphetamine excitation in rats is therefore unsuitable as a model for screening new antidepressant drugs.

The blockade of amphetamine-induced stereotype behavior has frequently been used to evaluate new neuroleptic drugs (JANSSEN, 1970). The finding that some of these inhibit the metabolic elimination of amphetamine (Table 3) as well as blocking its pharmacological effects at the receptor makes it difficult to evaluate the results of such screenings. From a clinical point of view it would have been valuable to have information on the interaction of psychotropic drugs with amphetamine metabolism in man or in a species with a similar mode of metabolism. However, since most studies have been performed in rats, which hydroxylate the drug, the results obtained

Table 3. Effects of drugs on metabolism of amphetamine

Drug	Metabolic pathway	Species	Effect	Reference
Tricyclic anti-depressants				
Desipramine	*p*-Hydroxylation	Rat	Decrease increased brain levels	VALZELLI et al., 1967 CONSOLO et al., 1967
Imipramine Desipramine	*p*-hydroxylation	Rat	Increased brain level	SULSER et al., 1966
Amitriptyline, nortriptyline			Increase brain level, prolonged action	SULSER and DINGELL 1968 VALZELLI, et al. 1968
Desipramine	*p*-Hydroxylation, deamination	Rat	Decreased metabolism; increased effect	LEWANDER, 1968, 1969
Nortriptyline, protriptyline, imipramine	*p*-Hydroxylation	Rat	Decreased metabolism	LEWANDER, 1969
Neuroleptics				
Chlorpromazine	*p*-Hydroxylation	Rat	Increased brain level	VALZELLI et al., 1968
Chlorpromazine	*p*-Hydroxylation	Rat	Decrease of PHA, prolongation of effects	BORELLA, 1969 BORELLA et al., 1969
Chlorpromazine	*p*-Hydroxylation	Rat	Increase brain $T_{\frac{1}{2}}$, decreased metabolism	LEMBERGER et al., 1970
Haloperidol	*p*-Hydroxylation	Rat	Increase brain $T_{\frac{1}{2}}$, decreased metabolism	LEMBERGER et al., 1970
Haloperidol	*p*-Hydroxylation	Rat	No effect on *p*-hydroxylation in vivo	LEWANDER, 1969
Chlorpromazine, promethazine, chlorprothixine, prochlorperazine	*p*-Hydroxylation	Rat	Inhibition	LEWANDER, 1969
Inducers and inhibitors of drug metabolism				
Phenobarbital	Deamination and *p*-Hydroxylation	Mouse	Small increase	BENAKIS and THOMASSET, 1970
Phenobarbital	*p*-Hydroxylation	Rat	No change in rate of elimination	GROPPETTI and COSTA, 1969
Phenobarbital	Deamination	Rabbit	Increased activity in vitro	PARLI et al., 1971a, b
Phenobarbital	*p*-Hydroxylation		No change after acute increase after repeated dosage	LEWANDER, 1969

Table 3 (continued)

Drug	Metabolic pathway	Species	Effect	Reference
Diphenyl-hydantoin	*p*-Hydroxylation	Rat	Decrease in rate of elimination	GROPPETTI and COSTA, 1969
SKF-525A	DEAMINATION	Rabbit	Decrease in liver slices	COOPER et al., 1954
SKF-525A	*p*-Hydroxylation, deamination	Mouse	Increase in amphetamine excretion	BENAKIS and THOMASSET, 1970
Drugs affecting adrenergic function				
Reserpine			No change	LEWANDER, 1969
α-Methyl-*p*-tyrosine				
Phenelzine (MAOI)				
Nialamide	*p*-Hydroxylation	Rat	Inhibition	LEWANDER, 1969
Cocaine	*p*-Hydroxylation	Rat	Inhibition	LEWANDER, 1969
Other psychotropic drugs				
Secobarbital	*p*-Hydroxylation	Rat	No change	LEWANDER, 1969
Meprobamate	*p*-Hydroxylation	Rat	No change	LEWANDER, 1969
Chlordiazepoxide	*p*-Hydroxylation	Rat	Inhibition	LEWANDER, 1969
Diazepam	*p*-Hydroxylation	Rat	Inhibition	LEWANDER, 1969
Atropine	*p*-Hydroxylation	Rat	No change	LEWANDER, 1969
Ethanol	*p*-Hydroxylation	Rat	Inhibition	CREAVEN and BARBER, 1969
Disulfiram, diethyldithio-carbamate	*p*-Hydroxylation	Rat	Inhibition	CREAVEN et al., 1970 JONSSON and LEWANDER, 1973
Tetrahydrocan-nobinol	*p*-Hydroxylation	Rat	No change	RATING et al., 1971

cannot be extrapolated to humans, who predominantly use the deamination pathway (Table 2). Amphetamine has been found to have no effect on the various components of the microsomal drug-metabolizing enzyme system (PLATT and COCKRILL, 1969).

4. Other Factors

Aromatic hydroxylation of drugs by liver microsomes has been shown to vary with age, sex, and hormone administration (CONNEY, 1967). In rats, the rate of amphetamine metabolism is faster in males than females and is decreased by treatment with estradiol (GROPPETTI and COSTA, 1969). The metabolism was also found to be slower in young animals as compared with adults. Tumors and aggregation of animals have been shown to increase and prolong tissue levels of amphetamine as well as increasing amphetamine toxicity (CONSOLO et al., 1965; ROSSO et al., 1968).

IV. Metabolism of Amphetamine in Dependent Subjects

Although the fate of amphetamine has been well investigated in normal subjects, relatively little is known of how chronic administration of high doses affects its metabolism quantitatively and qualitatively. A recent investigation on psychotic amphetamine addicts indicated that the rate of metabolism is not changed in psychotic addicts and that the proportion of amphetamine excreted as metabolite was similar to that of drug-naive subjects (ÄNGGÅRD et al., 1973, Table 2). SEVER et al. (1973) in a study of two dependent subjects also noticed that the composition of amphetamine metabolites in urine was essentially normal. In cats, ELLISON et al. (1971) found that tolerant animals excreted significantly more amphetamine than their controls. It is therefore unlikely that amphetamine tolerance is due to an enhanced metabolism.

Some changes in the metabolic pattern seem, however, to occur as a consequence of the high tissue concentrations present in amphetamine-tolerant subjects. Thus, norephedrine (phenylpropanolamine) is not normally a metabolite of amphetamine in man, since *p*-hydroxylation is a relative requirement for the β-hydroxylase (CREVELING et al., 1962) and man is a poor *p*-hydroxylator (Table 2). However, after high doses it is possible to demonstrate the presence of norephedrine in urine (ÄNGGÅRD et al., 1973; SEVER et al., 1973). It should also be mentioned that recent studies indicate that norephedrine is further metabolized to benzoic acid possibly by way of a 1-hydroxy-1-phenyl propan-2-one intermediate (FELLER and MALSPEIS, 1971).

It has been suggested that certain metabolites of amphetamine might be psychotoxic and therefore responsible for the development of the amphetamine psychosis (COSTA and GROPPETTI, 1970; SMYTHIES et al., 1967). Arguments for the "metabolic hypothesis" of amphetamine psychosis are:

1. *The amphetamine psychosis does not occur immediately.* In the studies on induced amphetamine psychosis there was a time period before the amphetamine psychosis developed. During the course of our studies on amphetamine-blocking drugs, 100–200 mg of D,L-amphetamine was injected i.v. to detoxified amphetamine-dependent subjects (JÖNSSON et al., 1971). In no case was an amphetamine psychosis precipitated. In the few reports of a sudden onset of amphetamine psychosis (CONNELL, 1958) it is possible that the patients had been using amphetamine before. An alternative explanation could be the activation of a latent schizophrenia as discussed by DAVIS and JANOWSKY (1973).

2. *Amphetamine psychosis is not related to the amphetamine plasma level.* In a study on hospitalized subjects having an amphetamine psychosis, no correlation was obtained between the plasma level of amphetamine and the intensity of the psychosis (ÄNGGÅRD et al., 1970a).

3. *The amphetamine psychosis is more intense under conditions which lead to increased metabolism.* Under conditions of alkaline urine there is an increased metabolism of amphetamine and also prolonged and intensified psychotic behavior (ÄNGGÅRD et al., 1970a, 1973). The ratio between the total metabolites and unchanged amphetamine increases for every day following withdrawal from amphetamine. It would therefore seem possible, as suggested by VREE and VAN ROSSUM (1970), that minor metabolites with slow rates of elimination could accumulate to cause psychotoxic effects.

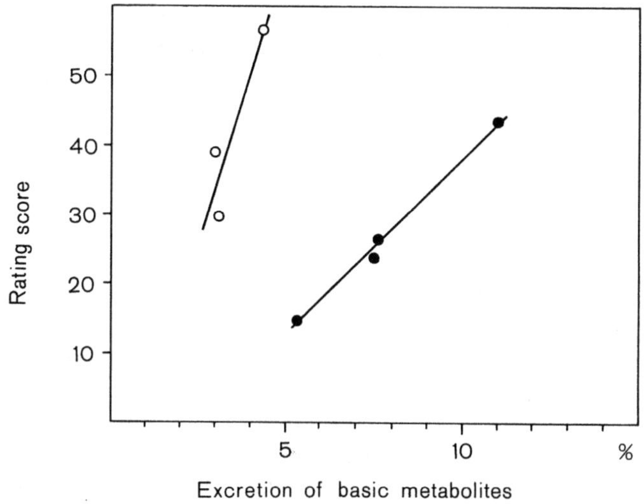

Fig. 4. Relationship between the excretion of basic hydroxylated metabolites of ^{3}H-*d,l*-amphetamine on the first day following admission and the total psychosis rating score. Subjects treated with ammonium chloride (○—○—○), subjects given sodium bicarbonate (●—●—●)

4. Positive correlation between excretion of hydroxylated amphetamine metabolites and the intensity of the psychosis. In a study on seven psychotic amphetamine-dependent subjects, the urinary metabolites were separated into unchanged amphetamine, hydroxylated basic metabolites, acidic metabolites, and aqueous residue. A positive correlation was found between the proportion of drug excreted as hydroxylated metabolites and the intensity of the psychosis (Fig. 4).

In the discussion of possible psychotoxic amphetamine metabolites two compounds *p*-hydroxynorephedrine (PHNE) and *p*-methoxyamphetamine (PMA) have received considerable attention. PHNE has been shown to accumulate in the rat brain as a false transmitter following administration of large doses of amphetamine (BRODIE et al., 1970; COSTA and GROPPETTI, 1970; LEWANDER 1970, 1971 b; TAYLOR and SULSER 1973, EL GUEDRI et al., 1973; JORI and CACCIA, 1974). PHNE has also been identified as a minor metabolite in urine from subjects with amphetamine psychosis (CAVANAUGH et al., 1970; ÄNGGÅRD et al., 1973). However, the amounts isolated have been so small that it is unlikely that it plays a role as a false transmitter in man. Hydroxylation is also a relatively minor metabolic pathway in man as compared to the rat (Table 2). Psychotic symptoms are not observed even after chronic administration of hydroxy-amphetamine to man (GILL et al., 1967). About 3% of the administered dose is excreted as PHNE (SJOERDSMA and VON STUDNITZ, 1963). *p*-Methoxyamphetamine (PMA) has been stated to be a relatively potent "psychotomimetic" compound in the rat (SMYTHIES et al., 1967). However, PMA administered in doses of 10–65 mg to normal humans did not cause hallucinations (SCHWEITZER et al., 1971) and PMA has not been found as a metabolite of amphetamine in amphetamine psychosis (ANGRIST et al., 1971; ÄNGGÅRD et al. 1973). The available information does not therefore support a role of PHNE or PMA in the genesis of the amphetamine psychosis. Whether or not amphetamine metabolites are partly or wholly responsible for the amphetamine psychosis is still an open question. Before metabolites of amphetamine can be definitely linked with the development of the psychosis, it would be necessary to identify a psychotomimetic amphetamine metabolite in the body fluids of a psychotic subject and to establish a quantitative correlation between the levels of the metabolite and the intensity of psychotic behavior.

C. Pharmacokinetics of Amphetamine

A recent review on the pharmacokinetics of amphetamine has been presented by VREE (1973). The absorption, distribution, and elimination of the amphetamines are, like many other drugs, governed by their physicochemical properties (SCHANKER, 1962), although evidence has been presented for a saturable transport system across the blood-brain barrier of the rat (PARDRIDGE and CONNOR, 1973). Briefly, the rate of penetration of an ionizable drug through a biomembrane is dependent on the amount of the drug present as the nonionized moiety and the concentration gradient. The biomembranes are of a lipid nature and readily permit the diffusion of the lipophilic nonionized molecules. The amount of the drug present as the nonionized moiety, is, in turn, dependent on the pH of the biophase in which it is dissolved and the pK_a of the drug. The lipid solubility of the nonionized drug can be determined by measuring the partition coefficients between an organic phase such as chloroform, octanol, or heptane and an aqueous phase. A comprehensive study on the acid dissociation constants (pK_a) and the partition coefficients of the amphetamines was reported by VREE and VAN ROSSUM (1969). Some of these values are also found in Table 1.

The amphetamines are weak to moderately strong bases with pK_a values between 8 and 11. Amphetamine has a pK_a of 9.9. At physiological pH less than 1% is nonionized and capable of diffusion. Phenmetrazine, on the other hand, has a pK_a of 7.45 and 50% is nonionized at physiological pH. There is therefore reason to expect that the phenmetrazine would have a more rapid absorption, distribution, and onset of action as compared to amphetamine. This has also been experimentally demonstrated (VAN ROSSUM and SIMONS, 1969). It is probable that the rapid onset of action of the more lipid soluble central stimulants, such as cocaine and phenmetrazine, contributes to their high addiction liability.

The most extensively investigated drug, both in animals and man is amphetamine. Following i.p. administration to rats, peak tissue levels are achieved after 30 min (MAICKEL et al., 1969). The drug is extensively bound to tissues, as evidenced by a brain to plasma ratio of 7–8:1 (MAICKEL et al., 1969; LEWANDER, 1971b). Similar tissue to plasma distribution ratios have been reported in the dog (AXELROD, 1954a) and cat (ELLISON et al., 1971). The plasma levels in the rat have been found to decline in a biphasic fashion with the first phase ($T_{1/2}=0.5$ h) lasting for about 2 h and the second phase ($T_{1/2}=3$ h) for up to 24 h (MAICKEL et al., 1969; LEWANDER, 1971b; JONSSON and LEWANDER, 1973).

A comparative study of the pharmacokinetics of amphetamine in domesticated animals was performed by BAGGOT and DAVIS (1973). These authors concluded that a relationship existed between dietary habit of different animals and the elimination rate and metabolic pattern of amphetamine. The half-life was significantly longer in carnivorous (dog and cat) than in herbivorous (pony, rabbit, and goat) species.

Absorption of amphetamine, metamphetamine, phenmetrazine, and ephedrines occurs readily in man following oral administration of the hydrochlorides or other salts (ASATOOR et al., 1965; BECKETT and ROWLAND, 1965a, b; BECKETT et al., 1965; QUINN et al., 1967; WILKINSON and BECKETT, 1968; ÄNGGÅRD et al., 1970b). The maximum blood level following an oral dose of 10 mg amphetamine sulfate was 30–40 ng/ml (CAMPBELL, 1969; ROWLAND, 1969). Following i.v. injection of 20–160 mg, peak blood levels of 50–590 ng/ml of D,L-amphetamine sulfate were recorded in tolerant amphetamine-dependent subjects (ÄNGGÅRD et al., 1970a, b; JÖNSSON et al., 1971). The levels in human cerebrospinal fluid were about 80% of the corresponding blood level in subjects with amphetamine psychosis. This indicates that cerebrospinal fluid is in equilibrium with plasma water rather than brain. The difference in concentration between cerebrospinal fluid and plasma may well be explained by the plasma protein binding of amphetamine, which is about 20% in man (FRANKSSON and ÄNGGÅRD, 1970). A similar degree of plasma protein binding ($<45\%$) was found in various animal species (BAGGOT et al., 1972).

Amphetamine has an extensive extravascular distribution. The apparent volume of distribution (V_D) is 3.5–4.6 l/kg in nontolerant subjects (ROWLAND, 1969). In tolerant addicts the V_D is somewhat larger, 6.1 ± 0.2 (ÄNGGÅRD et al., 1970a). It appears therefore that amphetamine might be sequestered in tissue compartments as a consequence of tolerance development. Further support for this contention was obtained in a direct comparison of the rates of amphetamine elimination between groups of drug-dependent and drug-naive subjects of comparable age groups and at the same dose level. The results showed that the mean plasma $T_{1/2}$ was significantly ($P<0.05$) slower in the group of chronic abusers. This is shown in Figure 5. Since the metabolism does not differ between tolerant and nontolerant individuals (DAVIS et al., 1971; ÄNGGÅRD et al., 1973; ELLISON et al., 1971) the results are indicative of a higher V_D and thus higher amphetamine tissue binding in tolerant amphetamine abusers. However this difference was not demonstrable in plasma protein binding studies (FRANKSSON and ÄNGGÅRD, 1970).

Amphetamine is eliminated partly by metabolism and partly by renal excretion of the unchanged drug. The degree by which the drug is cleared by metabolism or by renal excretion depends on the urinary pH. A low pH favors renal clearance of the unchanged drug and a high urinary pH favors metabolic clearance. The dependence of amphetamine excretion on urinary pH was first demonstrated by ASATOOR et al. (1965) and BECKETT et al. (1965). BECKETT and his associates characterized a number of amphetamines and related compounds in this respect (BECKETT and ROWLAND, 1965a–c; BECKETT et al., 1968; WILKINSON and BECKETT, 1968; BECKETT and BROOKES, 1967, 1970). The kinetics of metabolism and excretion of amphetamines have been discussed by VREE and VAN ROSSUM (1970) and by VREE (1973).

A more complete picture of the pharmacokinetics of amphetamines has subsequently emerged with the use of radioactively labelled amphetamine (BECKETT et al., 1969; DAVIS et al., 1971; ÄNGGÅRD et al., 1970a, 1973) and the development of new methods for the analysis of amphetamine in plasma (ROWLAND, 1969; BRUCE and MAYNARD, 1969; CAMPBELL, 1969; ÄNGGÅRD et al., 1970b). During conditions of acidic urine, 68–69% of the administered amphetamine is excreted in urine over a

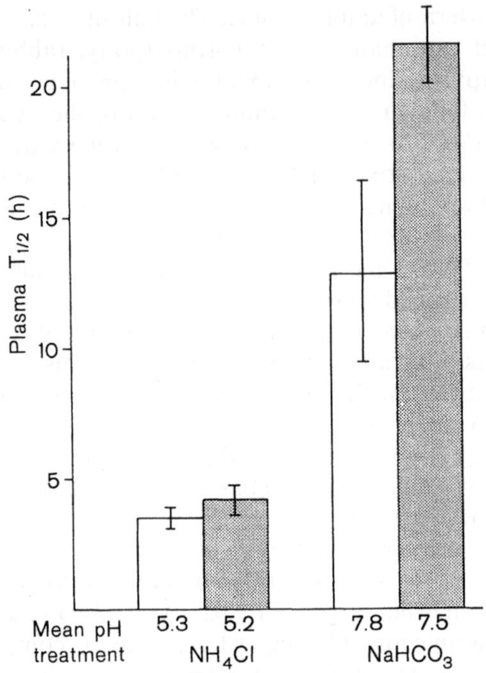

Fig. 5. Plasma half lives ($T_{1/2}$) of amphetamine in a group of drug-naive subjects *(open bars)* and amphetamine-dependent subjects *(shaded bars)* after treatment with ammonium chloride and sodium bicarbonate

period of 4 days and the rest is metabolized (Table 2). The plasma $T_{1/2}$ was 5–6 h (DAVIS et al., 1971; ÄNGGÅRD et al., 1970a, b). When the urine is rendered less acidic by oral $NaHCO_3$, the amphetamine excretion is reduced to about 30% with a compensatory increase in the rate of metabolism. The plasma $T_{1/2}$ increases to 20–30 h (ÄNGGÅRD et al., 1970a). The relationship between plasma $T_{1/2}$ and urinary pH is shown in Figure 6. Each unit of increase in urinary pH increases the plasma $T_{1/2}$ by about 7 h.

From the information on amphetamine blood and urine levels in man under conditions of varying urinary pH (ÄNGGÅRD et al., 1970a) the relative contributions of renal excretion and metabolism to the total drug clearance can be calculated. The total clearance is maximal at acidic pH, being about 500 ml/min ($T_{1/2}$ = 5–6 h), of which renal clearance is 370 ml/min and metabolic clearance 130 ml/min. At an alkaline pH (>7.5) the renal clearance of the drug becomes very small and the metabolic clearance of 130 ml/min is the major factor responsible for the elimination of the drug. The plasma $T_{1/2}$ increases to 25 h (Fig. 6).

At an acidic urinary pH the renal clearance of amphetamine is also dependent on the urinary output. Thus, in the pH range 5.2–6.0 the clearance of amphetamine at a urine production of <30 ml/h was about half that observed during a urinary output of 30–125 ml/h. During water diuresis (>125 ml/h) the renal clearance was increased (ÄNGGÅRD et al., 1970a). In the treatment of amphetamine-intoxicated subjects it is therefore important to keep a high flow of acidic urine to achieve a maximum renal

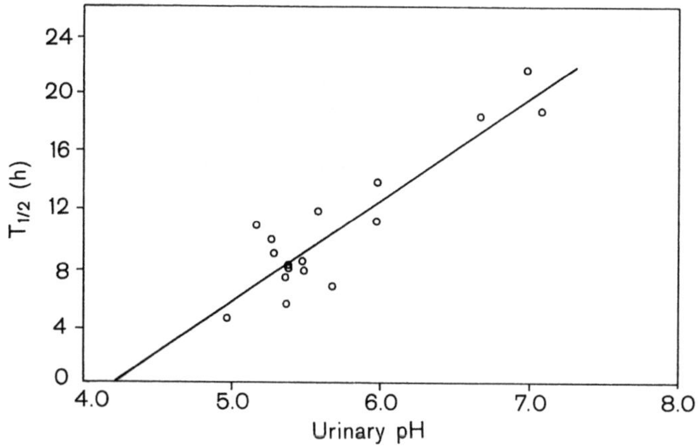

Fig. 6. The relationship between plasma elimination rate of amphetamine ($T_{t/2}$) and urinary pH in 17 cases of amphetamine psychosis

clearance of the drug from the body. Administration of fluid is especially important after prolonged use of amphetamine associated with low fluid intake and enhanced insensible loss of water from the body surface.

The renal clearance of amphetamine with acidic urine (300–400 ml/min) is higher than the glomerular filtration rate (125 ml/min; BECKETT et al., 1969; ROWLAND, 1969; ÄNGGÅRD et al., 1970a). Therefore the major part of the drug in urine must have originated from the tubules. Probably as the primary urine flows down the tubule the amphetamine passes from the plasma to the urine because of the greater concentration gradient of nonionized drug across the cell membrane.

D. Relationship Between Blood Levels and Effect in Man

There is a considerable difference between the ways in which amphetamine is used therapeutically and nonmedically. The usual therapeutic dose of amphetamine, 5–15 mg in a drug-naive subject, produces a moderate psychomotor stimulation (see Sect. I, Chap. 1.C and Chap. 2). The peak plasma level after this dose is 25–50 ng/ml. Doses in excess of 20 mg may produce toxic symptoms. Subjects who habitually use amphetamine develop a tolerance to most of these effects. In chronic abusers in Sweden, the usual i.v. self-administered dose is about 200 mg of D,L-amphetamine sulfate, which produces peak plasma levels of about 500 ng/ml. This dose produces a mood elevation, which is most intense immediately after the injection and then gradually disappears over the next 8 h (JÖNSSON et al., 1969, 1971). A dose of 20 mg D,L-amphetamine sulfate hardly produces any effect. In chronic amphetamine abusers there is thus a marked tolerance to the mood elevating as well as the toxic effects of amphetamine.

The time course of the effect of amphetamine on mood indicates that the effect disappears more quickly than could be expected from the plasma elimination curves. About 10 h after the dose, when the euphoria has disappeared, the plasma level is still about 150 ng/ml. At this time there are more frequent ratings for negative mood

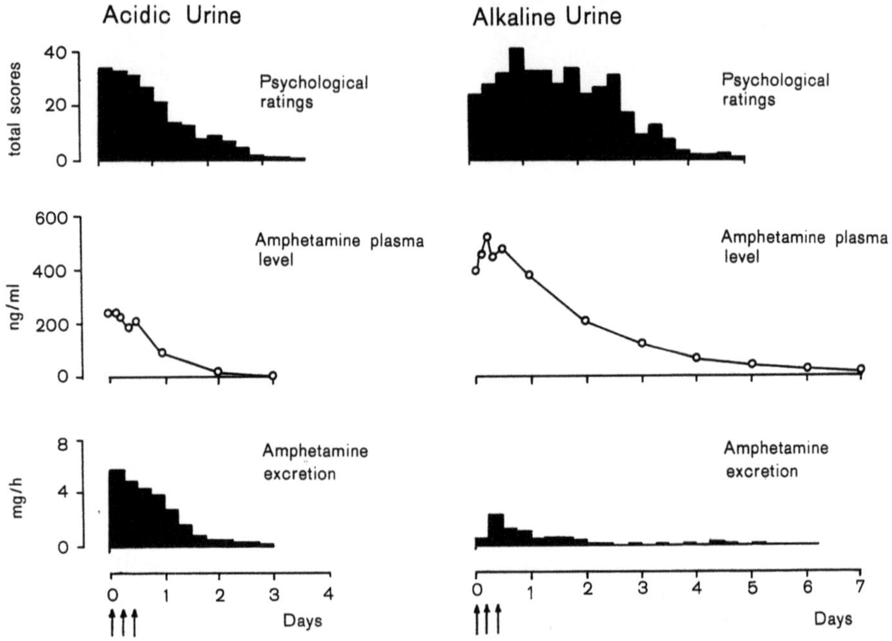

Fig. 7. Psychological ratings of amphetamine psychosis and mean blood and urine levels of amphetamine in conditions of acidic *(left)* and alkaline urine *(right)*. *Arrows* denote administration of 50 mg of D,L-amphetamine sulfate

parameters such as restlessness, anxiety, and depression (GUNNE and ÄNGGÅRD, 1973). Normally the addict can relieve these symptoms by another amphetamine injection.

In a study of 18 cases of amphetamine psychosis we found that three patients with a less acidic urine had a more intense and long-lasting psychosis. This was correlated with higher plasma levels of amphetamine and delayed elimination rates from plasma. This is shown in Figure 7. However, there was a lack of correlation between the intensity of the psychosis and the plasma level of amphetamine. The probable explanation for this surprising finding is that the degree of the amphetamine psychosis is more related to the total previous amphetamine exposure during the "run" rather than the actual body content of the drug at the time of the psychological rating.

The amphetamine psychosis is clearly not the consequence of a withdrawal reaction as is for instance alcohol delirium, since administration of amphetamine fails to improve the condition (ÄNGGÅRD et al., 1970a, Fig. 7). It seems more likely that the psychotic symptoms seen are related to a changing pharmacological spectrum of amphetamine as tolerance develops to some actions of the drug, but not to others, during repeated administration of high doses.

Acknowledgments: This work was supported by the Swedish Medical Research Council and the Bank of Sweden Bicentennial Foundation. I am grateful to Mrs. S. BYQVIST for her excellent secretarial assistance.

References

Abbott, W. O., Henry, C. M.: Paredrine (β-4-hydroxyphenylisopropylamine); a clinical investigation of a sympathomimetic drug. Amer. J. med. Sci. **193**, 661—673 (1937)

Alles, G. A., Feigen, G. A.: Comparative actions of phenylthienyl- and furylisopropylamines. J. Pharmacol. exp. Ther. **72**, 265—275 (1941)

Alles, G. A., Wisegarver, B. B.: Amphetamine excretion studies in man. J. Toxicol. appl. Pharmacol. **3**, 678—688 (1961)

Alleva, J. J.: Metabolism of tranylcypromine-^{14}C and D,L-amphetamine-^{14}C in the rat. J. med. Chem. **6**, 621—624 (1963)

Änggård, E., Gunne, L. M., Jönsson, L. E., Niklasson, F.: Pharmacokinetic and clinical studies on amphetamine dependent subjects. Europ. J. clin. Pharmacol. **3**, 3—11 (1970a)

Änggård, E., Gunne, L. M., Niklasson, F.: Gas chromatographic determination of amphetamine in blood, tissue, and urine. Scand. J. clin. Lab. Invest. **26**, 137—143 (1970b)

Änggård, E., Jönsson, L. E., Hogmark, A. L., Gunne, L. M.: Amphetamine metabolism in amphetamine psychosis. J. clin. Pharm. Ther. **14** (5), 870—880 (1973)

Angrist, B. M., Shopsin, B., Gershon, S.: Comparative psychotomimetic effects of stereoisomers of amphetamine. Nature (Lond.) **234**, 152—153 (1971)

Asatoor, A. M., Galman, B. R., Johnson, J. R., Milne, M. D.: The excretion of dexamphetamine and its derivatives. Brit. J. Pharmacol. **24**, 293—300 (1965)

Axelrod, J.: Studies on sympathomimetic amines. II. The biotransformation and physiological disposition of D-amphetamine, D-*p*-hydroxyamphetamine and D-amphetamine. J. Pharmacol. exp. Ther. **110**, 315—326 (1954a)

Axelrod, J.: Enzymatic demethylation of sympathomimetic amines. Fed. Proc. **13**, 332 (1954b)

Axelrod, J.: The enzymatic deamination of amphetamine (Benzedrine). J. biol. Chem. **214**, 753—763 (1955)

Baggot, J. D., Davis, L. E.: A comparative study of the pharmacokinetics of amphetamine. Res. Vet. Sci. **14**, 207—215 (1973)

Baggot, J. D., Davis, L. E., Neff, C. A.: Extent of plasma protein binding of amphetamine in different species. Biochem. Pharmacol. **21**, 1813—1816 (1972)

Beckett, A. H., Bélanger, P. M.: The identification of three new metabolic products of phentermine after liver microsomal incubation. Xenobiotica **4** (8), 509—519 (1974)

Beckett, A. H., Boyes, R. N., Tucker, Q. T.: Use of the analogue computer to examine the quantitative relation between urinary pH and kidney reabsorption of drugs partially ionized at physiological pH. J. Pharm. (Lond.) **90**, 277—282 (1968)

Beckett, A. H., Brookes, L. G.: The absorption and urinary excretion in man of fenfluramine and its main metabolite. J. Pharm. (Lond.) **19**, 42S—49S (1967)

Beckett, A. H., Brookes, L. G.: The effect of chain and ring substitution on the metabolism, distribution and biological action of amphetamines. In: Costa, F., Garattini, S. (Eds.): Amphetamines and Related Compounds, pp. 109—120. New York: Raven Press 1970

Beckett, A. H., Coutts, R. T., Ogunbona, F. A.: Metabolism of amphetamines. Identification of *N*-oxygenated products by gas chromatography and mass spectrometry. J. Pharm. (Lond.) **25**, 708—717 (1973)

Beckett, A. H., Rowland, M.: Urinary excretion of methylamphetamine in man. Nature (Lond.) **206**, 1260—1261 (1965a)

Beckett, A. H., Rowland, M.: Urinary excretion kinetics of methylamphetamine in man. J. Pharm. (Lond.) **17**, Suppl., pp. 109S—114S (1965b)

Beckett, A. H., Rowland, M.: Urinary excretion kinetics of amphetamine in man. J. Pharm. (Lond.) **17**, 628—639 (1965c)

Beckett, A. H., Rowland, M., Turner, P.: Influence of urinary pH on excretion of amphetamine. Lancet **1**, 303 (1965)

Beckett, A. H., Salmon, J. A., Mitchard, M.: The relation between blood levels and urinary excretion of amphetamine under controlled acidic and under fluctuating urinary pH values using [^{14}C] amphetamine. J. Pharm. (Lond.) **21**, 251—258 (1969)

Benakis, A., Thomasset, M.: Metabolism of amphetamines and their interaction with barbiturates and SKF-525A. In: Costa, E., Garattini, S. (Eds.): Amphetamines and Related Compounds, pp. 153—164. New York: Raven Press 1970

Benington, F., Morin, R. D., Clark, L. C.: Behavioral and neuropharmacological actions of N-alkylhydroxylamines and their O-methyl ethers. J. med. Chem. **8**, 100—104 (1965)
Beregi, L. G., Hugon, P., LeDourarec, J. C., Laubie, M., Duhault, J.: Structure activity relationships in CF_3 substituted phenethylamines. In: Costa, E., Garattini, S. (Eds.): Amphetamines and Related Compounds, pp. 21—61. New York: Raven Press 1970
Biel, J. H.: Structure-activity relationships of amphetamine and derivatives. In: Costa, E., Garattini, S. (Eds.): Amphetamines and Related Compounds, pp. 3—19. New York: Raven Press 1970
Blaschko, H., Richter, D., Schlossman, H.: The oxidation of adrenaline and other amines. Biochem. J. **31**, 2187—2196 (1937)
Boissier, J. R., Hirtz, J., Dumont, C., Gérardin, A.: Some aspects of the metabolism of anorexic phenyl isopropylamines in the rat. In: Costa, E., Garattini, S. (Eds.): Amphetamines and Related Compounds, pp. 141—152. New York: Raven Press 1970
Borella, L. F.: Effect of chlorpromazine on the urinary excretion of amphetamine in rats. Pharmacologist **11**, 292 (1969)
Borella, L., Herr, F., Wejdan, A.: Prolongation of certain effects of amphetamine by chlorpromazine. Canad. J. Physiol. Pharmacol. **47**, 7—13 (1969)
Bridges, J. W., Gorrod, J. W., Parke, D. V.: Biological oxidation of nitrogen in organic molecules. London: Taylor and Francis 1972
Brodie, B. B., Cho, A. K., Gessa, G. L.: Possible role of *p*-hydroxynorephedrine in the depletion of norepinephrine induced by D-amphetamine and in tolerance to this drug. In: Costa, E., Garattini, S. (Eds.): Amphetamines and Related Compounds, pp. 217—230. New York: Raven Press 1970
Brodie, B. B., Gillette, J. R., LaDu, B. N.: Enzymatic metabolism of drugs and other foreign compounds. Ann. Rev. Biochem. **27**, 427—454 (1958)
Bruce, R. B., Maynard, W. R. Jr.: Determination of amphetamine and related amines in blood by gas chromatography. Analyt. Chem. **41**, 977—979 (1969)
Caldwell, J., Dring, L. G., Williams, R. T.: Norephedrines as metabolites of ^{14}C amphetamine in urine in man. Biochem. J. **129**, 23—24 (1972a)
Caldwell, J., Dring, L. G., Williams, R. T.: Biliary excretion of amphetamine and methamphetamine in the rat. Biochem. J. **129**, 25—29 (1972b)
Caldwell, J., Dring, L. G., Williams, R. T.: Metabolism of (^{14}C)methamphetamine in man, the guinea pig and the rat. Biochem. J. **129**, 11—22 (1972c)
Caldwell, J., Sever, P. S.: The biochemical pharmacology of abused drugs. Clin. Pharmacol. Ther. **16**, 625—638 (1974)
Campbell, D. B.: A method for the measurement of therapeutic levels of D-amphetamine in human plasma. J. Pharm. (Lond.) **21**, 129—130 (1969)
Carlton, P. L.: Potentiation of the behavioural effects of amphetamine by imipramine. Psychopharmacologia (Berl.) **2**, 364 (1961)
Cavanaugh, J. H., Griffith, J. D., Oates, J. A.: Effect of amphetamine on the pressor response to tyramine. Formation of *p*-hydroxynorephedrine from amphetamine in man. Clin. Pharmacol. Ther. **11**, 656—664 (1970)
Chidsey, C. A., Harrison, D. C., Braunwald, E.: Release of norepinephrine from the heart by vasoactive amines. Proc. Soc. exp. Biol. (N.Y.) **109**, 488—490 (1962)
Cho, A. K., Lindeke, B., Jenden, D. J.: Gas chromatography—Mass spectrometry in the distribution and metabolism of phentermin. In: Frigerio, A., Caraguoli, N. (Eds.): Mass Spectrometry in Biochemistry and Medicine. New York: Raven Press 1974a
Cho, A. K., Lindeke, B., Sum, C. Y.: The *N*-hydroxylation of phentermine (2-methyl-2-amino-l-phenylpropane). Properties of the enzyme system. Drug Metab. Dispos. **2**, 1—8 (1974b)
Chrusciel, T. L., Chrusciel, M.: Selected bibliography on detection of dependence—producing drugs in body fluids. WHO offset publication No. 17, Geneva 1975
Connell, P. H.: Amphetamine psychosis. Maudsley Monographs No. 5. London: Chapman & Hall 1958
Conney, A. H.: Pharmacological implications of microsomal enzyme induction. Pharmacol. Rev. **19**, 317—366 (1967)
Consolo, S., Dolfini, F., Garattini, S., Valzelli, L.: Desipramine and amphetamine metabolism. J. Pharm. (Lond.) **19**, 253—256 (1967)

Consolo, S., Garattini, S., Ghielmetti, R., Valzelli, L.: Concentrations of amphetamine in the brain of normal or aggressive mice. J. Pharm. (Lond.) **17**, 666 (1965)

Cooper, J. R., Axelrod, J., Brodie, B. B.: Inhibitory effects of β-diethylaminoethyl diphenylpropylacetate on a variety of drug metabolic pathways in vitro. J. Pharmacol. exp. Ther. **112**, 55—63 (1954)

Costa, E., Groppetti, A.: Biosynthesis and storage of catecholamines in tissues of rats injected with various doses of D-amphetamine. In: Costa, E., Garattini, S. (Eds.): Amphetamines and Related Compounds, pp. 231—255. New York: Raven Press 1970

Creaven, P. J., Barber, T.: The effect of ethanol on the metabolism of amphetamine by the rat. J. Pharm. (Lond.) **21**, 859—860 (1969)

Creaven, P. J., Barber, T., Roach, M. K.: The interaction of ethanol and amphetamine metabolism. J. Pharm. (Lond.) **22**, 828—831 (1970)

Creveling, C. R., Daly, J. W., Witkop, B., Udenfriend, S.: Substrates and inhibitors of dopamine-β-oxidase. Biochim. biophys. Acta (Amst.) **64**, 125—134 (1962)
olism. Ann. N.Y. Acad. Sci. **179**, 493—501 (1971)

Daly, J. W., Creveling, C. R., Witkop, B.: The chemorelease of norepinephrine from mouse hearts. Structure-activity relationships. I. Sympathomimetic and related amines. J. med. Chem. **9**, 280—284 (1966)

Davis, J. M., Janowsky, D. S.: Amphetamine and methylphenidate psychosis. In: Usdin, E., Snyder, S. (Eds.): Frontiers in Catecholamine Research, pp. 977—981. Oxford: Pergamon Press 1973

Davis, J. M., Kopin, I. K., Lemberger, L., Axelrod, J.: Effect of urinary pH on amphetamine metabolism. Ann. N.Y. Acad. Sci. **179**, 493–501 (1971)

Dingell, J. V., Bass, A. D.: Inhibition of the hepatic metabolism of amphetamine by the rat. Biochem. Pharmacol. **18**, 1535—1538 (1969)

Dring, L. G., Smith, R. L., Williams, R. T.: The fate of amphetamine in man and other animals. J. Pharm. (Lond.) **18**, 402—404 (1966)

Dring, L. G., Smith, R. L., Williams, R. T.: The metabolic fate of amphetamine in man and other species. Biochem. J. **116**, 425—435 (1970)

Duhault, J., Verdavaienne, C.: Modification du taux de serotonine cérébrale chez le rat par les trifluoromethyl-phenyl-2-ethylamino-propane (fenfluramine 768S). Arch. int. Pharmacodyn. **170**, 276—286 (1967)

El Guedri, H., Jacquot, C., Rapin, J., Cohen, Y.: Métabolisme, distribution régionale et pharmacocinétique de la parahydroxy-noréphédrine dans le cerveau du rat. J. Pharmacol. (Paris) 4, 453 (1973)

Ellison, T., Gutzait, L., van Loon, E. J.: The comparative metabolism of D-amphetamine-^{14}C in the rat, dog and monkey. J. Pharmacol. exp. Ther. **152**, 383—387 (1966)

Ellison, T., Okun, R., Silverman, A., Siegel, M.: Metabolic fate of amphetamine in the cat during development of tolerance. Arch. int. Pharmacodyn. **190**, 135—149 (1971)

Faraj, B. A., Israili, Z. H., Perel, J. M., Jenkins, M. L., Holtzman, S. G., Cucinell, S. A., Dayton, P. G.: Metabolism and disposition of methylphenidate-^{14}C: studies in man and animals. J. Pharmacol. exp. Ther. **191**, 535—547 (1974)

Feller, D. R., Malspeis, L.: Metabolism of D(−)-ephedrine and L(+)-ephedrine in the microsomal and 9000 × g supernatant of the rabbit liver. Fed. Proc. **30**, 225abs. (1971)

Foreman, R. L., Siegel, F. P., Mrtek, R. G.: Synthesis of deutero-l-amphetamine sulfate. J. pharm. Sci. **58**, 189—192 (1969)

Franklin, M. R.: Complexes of metabolites of amphetamines with hepatic cytochrome, Xenobiotica **4**, 133—142 (1974)

Franksson, G., Änggård, E.: The plasma protein binding of amphetamine, catecholamines and related compounds. Acta pharmacol. (Kbh.) **28**, 209—214 (1970)

Fuller, R. W., Hines, C. W., Mills, J.: Lowering of brain serotonin levels by chloroamphetamines. Biochem. Pharmacol. **14**, 483—488 (1965)

Fuller, R. W., Molloy, B. B., Rousch, B. W., Hauser, K. M.: Disposition and behavioral effects of amphetamine and β, β-difluoroamphetamine in mice. Biochem. Pharmacol. **21**, 1299—1307 (1972)

Fuller, R. W., Perry, K. W., Baker, J. C., Parli, J., Lee, N., Day, W. A., Molloy, B. B.: Comparison of the active and the hydroxylamine derivatives of 4-chloroamphetamine as depletors of brain 5-hydroxyindoles. Biochem. Pharmacol. **23**, 3267—3272 (1974)

Fuller, R. W., Snoddy, H. D., Molloy, B. B.: Effect of β, β-difluoro substitution on the disposition and pharmacological effects of 4-chloroamphetamine in rats. J. Pharmacol. exp. Ther. **184**, 278—284 (1973)

Gill, J. R., Mason, D. T., Bartter, F. C.: Effects of hydroxy-amphetamine (Paredrine) on the function of the sympathetic nervous system in normotensive subjects. J. Pharmacol. exp. Ther. **155**, 288—295 (1967)

Glowinski, J., Axelrod, J.: Effect of drugs on the uptake, release and metabolism of ^{3}H-norepinephrine in the rat brain. J. Pharmacol. exp. Ther. **149**, 43—49 (1965)

Glowinski, J., Axelrod, J., Iversen, L. L.: Regional studies of catecholamines in the rat brain. IV. Effects of drugs on the disposition and metabolism. of H^{3}-norepinephrine and H^{3}-dopamine. J. Pharmacol. exp. Ther. **153**, 30—41 (1966)

Goldstein, M., Anagnoste, B.: The conversion in vivo of D-amphetamine to (+)-*p*-hydroxynorephedrine. Biochim. biophys. Acta (Amst.) **107**, 166 (1965)

Goldstein, M., Contrera, J. F.: The substrate specificity of phenylamine β-hydroxylase. J. biol. Chem. **237**, 1898—1902 (1962)

Goldstein, M., McKereghan, M. R., Lauber, E.: The stereospecificity of the enzymatic amphetamine β-hydroxylation. Biochim. biophys. Acta (Amst.) **89**, 191—193 (1964)

Gorrod, J. W.: The metabolism and excretion of "amphetamines" in man. In: Usdin, E., Snyder, S. (Eds.): Frontiers in Catecholamine Research, pp. 945—950. Oxford: Pergamon Press 1973

Gram, T. E.: Concepts in biochemical pharmacology. In: Brodie, B. B., Gillette, J. R. (Eds.): Handbook of Experimental Pharmacology, Vol. XXVIII, p. 334. New York: Springer 1971

Groppetti, A., Costa, E.: Factors affecting the rate of disappearance of amphetamine in rats. Int. J. Neuropharmacol. **8**, 209—215 (1969)

Gunne, L.-M.: The urinary output of D- and L-amphetamine in man. Biochem. Pharmacol. **16**, 863—869 (1967)

Gunne, L.-M., Änggård, E.: Pharmacokinetic studies with amphetamines—Relationship to neuropsychiatric disorders. J. Pharmacokinet. Biopharm. **1**, 481—495 (1973)

Harrison, J. W. E., Ambrus, C. M., Ambrus, J. L.: Tolerance of rats toward amphetamine and methamphetamine. J. Amer. Pharm. **41**, 539—541 (1952)

Henderson, P. T., Vree, T. B., van Ginneken, C. A., van Rossum, J. M.: Activation energies of α-C-oxidation and *N*-oxidation of *N*-alkyl-substituted amphetamines by rat liver microsomes. Stereochemistry and deuterium isotope effects. Xenobiotica **4** (2), 121—130 (1974)

Hoffström, I., Orrenius, S.: The interaction of various *N*-substituted amphetamines with cytochrome P-450 of rabbit liver microsomes. FEBS Letters **31** (2), 205—208 (1973)

Holland, G. F., Buck, C. J., Weissman, A.: Anorexigenic agents. Aromatic substituted 1-phenyl-2-propylamines. J. med. Chem. **6**, 519—524 (1963)

Hucker, H. B., Michniewicz, B. M., Rhodes, R. E.: Phenyl-2-propanone oxime—an intermediate in the oxidative deamination of D-(+)-amphetamine. Pharmacologist **12**, 255 (1970)

Hucker, H. B., Michniewicz, B. M., Rhodes, R. E.: Phenylacetone oxime—an intermediate in the oxidative deamination of amphetamine. Biochem. Pharmacol. **20**, 2123—2128 (1971)

Janssen, P. A. J.: Chemical and pharmacological classification of neuroleptics. In: Bobon, D. P., Janssen, P. A. J. (Eds.): Neuroleptics, p. 33. Basel-Munich-Paris-New York: S. Karger 1970

Jönsson, L. E., Gunne, L.-M., Änggård, E.: Effects of α-methyl-tyrosine in amphetamine dependent subjects. Europ. J. clin. Pharmacol. **2**, 27—29 (1969)

Jönsson, L. E., Änggard, E., Gunne, L.-M.: Blockade of intravenous amphetamine euphoria in man. Clin. Pharmacol. Ther. **12**, 889—896 (1971)

Jonsson, J.: Hydroxylation of amphetamine to parahydroxyamphetamine by rat liver microsomes. Biochem. Pharmacol. **23**, 3191—3197 (1974)

Jonsson, J., Lewander, T.: Effects of diethyldithiocarbamate and ethanol on the in vivo metabolism and pharmacokinetics of amphetamine in the rat. J. Pharm. (Lond.) **25**, 589—591 (1973)

Jori, A., Caccia, S.: Distribution of amphetamine and its hydroxylated metabolites in various areas of the rat brain. J. Pharm. (Lond.) **26**, 746—748 (1974)

La Du, B. N., Mandel, H. G., Way, E. L. (Eds.): Fundamentals of drug metabolism and drug disposition. Baltimore, Md.: Williams and Wilkins 1971

Lawlor, R. B., Trivedi, M. C., Yelnosky, J.: A determination of the anorexigenic potential of D,L-amphetamine, D-amphetamine, L-amphetamine and phentermine. Arch. int. Pharmacodyn. **179**, 401—407 (1969)

LeDouarec, J.C., Neveu, C.: Pharmacology and biochemistry of fenfluramine. In: Costa, E., Garattini, S. (Eds.): Amphetamines and Related Compounds, pp. 75—105. New York: Raven Press 1970

LeDouarec, J.C., Schmitt, H., Laubie, M.: Etude pharmacologique de la fenfluramine et de ses isomers optiques. Arch. int. Pharmacodyn. **161**, 206—232 (1966)

Lemberger, L., Witt, E.D., Davis, J.M., Kopin, I.J.: The effects of haloperidol and chlorpromazine on amphetamine metabolism and amphetamine stereotype behaviour in the rat. J. Pharmacol. exp. Ther. **174**, 428—433 (1970)

Lewander, T.: Effects of amphetamine on urinary and tissue levels of catecholamines in rats after inhibition of its metabolism by desmethylimipramine. Europ. J. Pharmacol. **5**, 1—9 (1968)

Lewander, T.: Influence of various psychoactive drugs on the in vivo metabolism of D-amphetamine in the rat. Europ. J. Pharmacol. **6**, 38—44 (1969)

Lewander, T.: Catecholamine turnover studies in chronic amphetamine intoxication. In: Costa, E., Garattini, S. (Eds.): Amphetamines and Related Compounds, pp. 317—329. New York: Raven Press 1970

Lewander, T.: Displacement of brain and heart noradrenaline by *p*-hydroxynorephedrine after administration of *p*-hydroxyamphetamine. Acta pharmacol. (Kbh.) **29**, 20—32 (1971 a)

Lewander, T.: On the presence of *p*-hydroxynorephedrine in the rat brain and heart in relation to changes in catecholamine levels after administration of amphetamine. Acta pharmacol. (Kbh.) **29**, 33—48 (1971 b)

Lindeke, B., Cho, A.K., Thomas, T.L., Michelsen, L.: Microsomal *N*-hydroxylation of phenylalkylamines. Acta pharm. suec. **10**, 493—506 (1973)

Maickel, R.P., Cox, R.H., Jr., Miller, F.P., Segal, D.S., Russell, R.W.: Correlation of brain levels of drugs with behavioral effects. J. Pharmacol. exp. Ther. **165**, 216—224 (1969)

Mann, P.J.G., Quastel, J.H.: Benzedrine (β-phenylisopropylamine) and brain metabolism. Biochem. J. **34**, 414—431 (1940)

McMahon, R.E.: The demethylation in vitro of *N*-methyl barbiturates and related compounds by mammalian liver microsomes. Biochem. Pharmacol. **12**, 1225 (1963)

Milne, M.D., Scribner, B.H., Crawford, M.A.: Non-ionic diffusion and the excretion of weak acids and bases. Amer. J. Med. **24**, 709—729 (1958)

Nieforth, K.A.: Psychotomimetic phenetylamines. J. pharm. Sci. **60**, 655—665 (1971)

Opitz, K.: Adipokinetic action of amphetamine—a study in the beagle dog. In: Costa, E., Garattini, S. (Eds.): Amphetamines and Related Compounds, pp. 627—639. New York: Raven Press 1970

Pardridge, W.M., Connor, J.D.: Saturable transport of amphetamine across the blood–brain barrier. Experientia (Basel) **29** (3), 302—304 (1973)

Parli, C.J., McMahon, R.E.: The mechanism of microsomal deamination: heavy isotope studies. Drug. Metab. Dispos. **1**, 337—341 (1973)

Parli, C.J., Wang, N; McMahon, R.E.: The mechanism of the oxidation of D-amphetamine by rabbit liver oxygenase. Oxygen-18 studies. Biochem. biophys. Res. Commun. **43**, 1204—1209 (1971 a)

Parli, C.J., Wang, N., McMahon, R.E.: The enzymatic *N*-hydroxylation of an imine. A new cytochrome P-450-dependent reaction catalyzed by hepatic microsomal monooxygenases. J. biol. Chem. **246**, 6953—6955 (1971 b)

Pawan, G.L.S.: Metabolic studies on the effect of fenfluramine in man and the mouse. In: Costa, E., Garattini, S. (Eds.): Amphetamines and Related Compounds, pp. 641—651. New York: Raven Press 1970

Platt, D.S., Cockrill, B.L.: Biochemical changes in rat liver in response to treatment with drugs and other agents. III. Effects of centrally acting drugs. Biochem. Pharmacol. **18**, 459—473 (1969)

Pletscher, A., Bartholini, G., Bruderer, H., Burkard, W.P., Gey, K.F.: Chlorinated arylalkylamines affecting the cerebral metabolism of 5-hydroxytryptamine. J. Pharmacol. exp. Ther. **145**, 344—350 (1964)

Pratesi, P., Blaschko, H.: Specificity of amine oxidase for optically active substrates and inhibitors. Brit. J. Pharmacol. **14**, 256—260 (1959)

Quinn, G. P., Cohn, M. M., Reid, M. B., Greenhard, P., Weiner, M.: The effect of formulation on phenmetrazine levels in man studied by a sensitive analytical method. Clin. Pharmacol. Ther. **8**, 369—373 (1967)
Rangno, R. E., Kaufman, J. S., Cavanaugh, J. H., Island, D., Watson, J. T., Oates, J.: Effects of a false neurotransmittor, *p*-hydroxynorephedrine, on the function of adrenergic neurons in hypertensive patients. J. clin. Invest. **52**, 952—960 (1973)
Rating, D., Honecker, H., Broermann, J., Strauss, S.: Hexobarbital-sleeping time and amphetamine motility after subchronic tetrahydrocannabinol treatment. Acta pharmacol. (Kbh.) **29**, suppl. 4, p. 92 (1971)
Rommelspacher, H., Honecker, H., Schulze, G., Strauss, S. M.: The hydroxylation of D-amphetamine by liver microsomes of the small rat. Biochem. Pharmacol. **23**, 1065—1071 (1974)
Rosso, R., Dolfini, L., Franchi, G.: Metabolism of amphetamine in tumour bearing rats. Biochem. Pharmacol. **17**, 633—634 (1968)
Rossum, J. M. van, Simons, F.: Locomotor activity and anorexogenic action. Psychopharmacologia (Berl.) **14**, 248—254 (1969)
Rowland, M.: Amphetamine blood and urine levels in man. J. pharm. Sci. **58**, 508—509 (1969)
Rutledge, C. O.: The mechanisms by which amphetamine inhibits oxidative deamination of norepinephrine in brain. J. Pharmacol. exp. Ther. **171**, 188—195 (1970)
Schanker, L. S.: The passage of drugs across body membranes. Pharmacol. Rev. **14**, 501—530 (1962)
Schweitzer, J. W., Friedhoff, A. J.: In: Zarafonetis, C. J. D. (Ed.): Proc. Int. Conf. on Drug Abuse. Philadelphia, Pa., 1972
Schweitzer, J. W., Friedhoff, A. J., Angrist, B. M., Gershon, S.: Excretion of *p*-methoxyamphetamine administered to humans. Nature (Lond.) **229**, 133—134 (1971)
Sever, P. S., Caldwell, J., Dring, L. G., Williams, R. T.: The metabolism of amphetamine in dependent subjects. Europ. J. clin. Pharmacol. **6**, 177—180 (1973)
Shulgin, A. T., Sargent, F.: Structure–activity relationships in the amphetamine series. In: Ethnopharmacological Search for Psychoactive Drugs, U.S. Public Health Service Publication No. 1645, p. 228. Washington: U.S. Government Printing Office 1967
Sjoerdsma, A., Studnitz, W. von: Dopamine-β-oxidase activity in man, using hydroxyamphetamine as substrate. Brit. J. Pharmacol. **20**, 278—284 (1963)
Smith, R. L., Dring, L. G.: Patterns of metabolism of the β-phenylisopropylamines in man and other species. In: Costa, E., Garattini, S. (Eds.): Amphetamines and Related Compounds, pp. 121—139. New York: Raven Press 1970
Smythies, J. R., Johnsson, V. S., Bradley, R. J., Benington, F., Morin, R. D., Clark, L. C.: Some new behaviour disrupting amphetamines and their significance. Nature (Lond.) **216**, 128—129 (1967)
Snyder, S. H.: Catecholamines in the brain as mediators of amphetamine psychosis. Arch. gen. Psychiat. **27**, 169—179 (1972)
Snyder, S. H., Faillace, L., Hollister, L.: 2,5-Dimethoxy-4-methylamphetamine (STP): A new hallucinogenic drug. Science **158**, 669—670 (1967)
Stein, L.: Self-stimulation of the brain and the central stimulant action of amphetamine. Fed. Proc. **23**, 836—850 (1964)
Sulser, F., Dingell, J. V.: The potentiation and blockade of the central action of amphetamine by chlorpromazine. Biochem. Pharmacol. **17**, 634—636 (1968)
Sulser, F., Owens, M. C., Dingell, J. V.: On the mechanism of amphetamine potentiation by desipramine (DMI), Life Sci. **5**, 2005—2010 (1966)
Sulser, F., Sanders-Bush, E.: Halogen substitution of amphetamine. Biochemical and pharmacological consequences, Frontiers in Catecholamine Research, pp. 995—1001. Pergamon Press 1973
Svensson, T. H.: Functional and biochemical effects of D- and L-amphetamine on central catecholamine neurons. Naunyn-Schmiedebergs Arch. Pharmacol. **271**, 170—180 (1971)
Taylor, K. B.: Dopamine-β-hydroxylase. Stereochemical course of the reaction. J. biol. Chem. **249**, 454—458 (1974)
Taylor, K. M., Snyder, S. H.: Differential effects of D- and L-amphetamine on behavior and on catecholamine disposition in dopamine and norepinephrine containing neurons of the rat brain. Brain. Res. **28**, 295—309 (1971)

Taylor, W. A., Sulser, F.: Effects of amphetamine and its hydroxylated metabolites on central noradrenergic mechanisms. J. Pharmacol. exp. Ther. **185**, 620—632 (1973)

Thoenen, H., Huerlimann, A., Gey, K. F., Haefely, W.: Liberation of *p*-hydroxynorephedrine from cat spleen by sympathetic nerve stimulation after pretreatment with amphetamine. Life Sci. **5**, 1715—1722 (1966)

Valzelli, L., Consolo, S., Morpurgo, C.: Influence of imipramine-like drugs on the metabolism of amphetamine. In: Proc. of the First Int. Symp. on Antidepressant Drugs. Excerpta Medica, Amsterdam, p. 61, 1967

Valzelli, L., Dolfini, E., Tansella, M., Garattini, S.: Activity of centrally acting drugs on amphetamine metabolism. J. Pharm. (Lond.) **20**, 595—599 (1968)

van Praag, H. M., Korf, J., van Wandenberg, F., Kits, F. P.: Influencing the human indolamine metabolism by means of chlorinated amphetamine derivatives with antidepressive action. Psychopharmacologica (Berl.) **13**, 145—160 (1968)

Vree, T. B.: Pharmacokinetics and metabolism of amphetamines. Ph. D. Thesis, Univ. of Nijwegen, Netherlands (1973)

Vree, T. B., Gorgels, J. P. M. C., Muskens, M., van Rossum, J. M.: Deuterium isotope effects in the metabolism of *N*-alkyl substituted amphetamines in man. Clin. chim. Acta **34**, 333—344 (1971 b)

Vree, T. B., Muskens, M. J., van Rossum, J. M.: Deuterium isotope effects and stereochemistry in the dealkylation and deamination of amphetamines and ephedrines in man. Xenobiotica **1**, 74 (1971 a)

Vree, T. B., van Rossum, J. M.: Some physicochemical properties of amphetamines and related drugs. J. Pharm. (Lond.) **21**, 774 (1969)

Vree, T. B., van Rossum, J. M.: Kinetics of metabolism and excretion of amphetamines in man. In: Costa, E., Garattini, S. (Eds.): Amphetamines and Related Compounds, pp. 165—190. New York: Raven Press 1970

Weiner, J. M., Mudge, G. H.: Renal tubular mechanisms for excretion of organic acids and bases. Amer. J. Med. **36**, 743—762 (1964)

Wenger, G. R., Rutledge, C. O.: A comparison of the effects of amphetamine and its metabolites, *p*-hydroxy-amphetamine and *p*-hydroxynorepinephrine on uptake release and catabolism of ^{3}H-norepinephrine in cerebral cortex of rat brain. J. Pharmacol. exp. Ther. **189**, 725—732 (1974)

Wilkinson, G. R., Beckett, A. H.: Absorption, metabolism and excretion of the ephedrines in man. I. The influence of urinary pH and urine volume output. J. Pharmacol. exp. Ther. **162**, 139—147 (1968)

Williams, R. T., Caldwell, J., Dring, L. G.: Comparative metabolism of some amphetamines in various species. In: Usdin, E., Snyder, S. (Eds.): Frontiers in Catecholamine Research, Pergamon Press, Oxford, pp. 927—932 (1973)

Yokel, R. A., Pickens, R.: Self-administration of optical isomers of amphetamine and methylamphetamine by rats. J. Pharmacol. exp. Ther. **187**, 27—33 (1973)

Zirkle, C. L., Kaiser, C.: Monoamine oxidase inhibitors (nonhydrazines). In: Psychopharmacological Agents, I., p. 474. New York: Academic Press 1964

B. Effects of Amphetamine in Animals

T. LEWANDER

A. Introduction

The present section of this review is subdivided into two major parts; the first part being concerned with the biochemical neuropharmacology, and the second part dealing with the modes and sites of action for the various peripheral and central effects of amphetamine. In addition to the acute effects of amphetamine, chronic effects of the drug have been emphasized, particularly in relation to discussions on amphetamine tolerance and withdrawal.

Because of the abundant literature on the amphetamines, this section has been restricted to studies on the animal pharmacology of dextro-, laevo-, and racemic amphetamine, its *para*- and *β*-hydroxylated metabolites, and methamphetamine. Occasionally, key references for other amphetamine derivatives have been given.

Comprehensive reviews of the earlier literature on the pharmacology of amphetamine in animals have appeared (IVY and GOETZL, 1943; BONHOFF and LEWRENZ, 1954; LEAKE, 1958). It has been the ambition of the present reviewer to focus on and to cover the amphetamine literature from the late 1950s up to and including 1974. (Any omission of relevant articles from this period is quite unintentional and is due to the difficulties encountered in trying to obtain a complete bibliography in spite of the access to computerized searches of the literature.) A number of general reviews of amphetamine have appeared during this period (VAN ROSSUM, 1970; SULSER and SANDERS-BUSH, 1971; GROSSMAN and SCLAFANI, 1971; WEINER, 1972). Reviews restricted to particular effects of amphetamine will be referred to under the respective headings.

B. Effects of Amphetamine on Neuronal Transmission

I. Biochemical Neuropharmacology

1. Adrenergic Transmission

The report by BURN and RAND in 1958 that amphetamine and a series of other phenethylamine derivatives exerted their peripheral sympathomimetic actions indirectly via endogenous catecholamines (CA) seems to have prompted a series of studies on the biochemical effects of amphetamine on adrenergic neurons. It soon became obvious that amphetamine interacted with the storage, uptake, release, and metabolic degradation of the CA both in the peripheral and the central nervous system. In the following each of these monoaminergic mechanisms will be considered separately. Short reviews on these matters have appeared previously (GLOWINSKI and BALDESSARINI, 1966; BEAUVALLET, 1968; ANDÉN et al., 1969; CARLSSON, 1969, 1970; AXELROD, 1970; VAN ROSSUM, 1970; SULSER and SANDERS-BUSH, 1971; LEWANDER, 1972; WEINER, 1972).

a) Storage of CA

α) CA in Peripheral Tissues

Rat: In acute experiments, amphetamine, administered i.p., s.c., or i.v. to rats causes a decrease in heart NA (AXELROD and TOMCHICK, 1960; AXELROD et al., 1962; POTTER and AXELROD, 1963; LEWANDER, 1968a, b, 1971a, b; BRODIE et al., 1968, 1969, 1970; GROPPETTI and COSTA 1969a, b; COSTA and GROPPETTI, 1970). The same result was found in rat stomach (BRODIE et al., 1970). Adipose tissue was depleted to 25% of the control level by amphetamine (BRODIE et al., 1969; COSTA and Groppetti, 1970).

Chronic administration of *dl*-amphetamine sulphate (dose schedule, see next sect.) caused a maximal depletion of heart NA to 40–50% of the control level (LEWANDER, 1968b, 1971a). Similar results were reported by BRODIE et al. (1970) for the heart and the stomach. In the adrenal gland, BEAUVALLET et al. (1962a) reported a decrease in A to 70% and NA to 60% of the control levels after 3.5 mg/kg of *d*-amphetamine tartrate. After a single injection of 20 mg/kg of *dl*-amphetamine sulphate, the adrenal gland A was decreased to 80% of the control level (LEWANDER, 1968a, b), but after *chronic administration* of the drug, the levels were unchanged (LEWANDER, 1968b).

Mouse: Amphetamine caused a dose-dependent decrease in the heart NA concentration (MOORE, 1963, 1964, 1965; BEAUVALLET et al., 1963) or had no effect (BEAUVALLET et al., 1964). The spleen NA level also decreased after amphetamine (MOORE, 1965). Aggregation of mice potentiated the effect of amphetamine on heart NA (MOORE, 1963, 1964). In hyperthyroid mice, amphetamine reduced the spleen NA level more effectively than the heart NA level (MOORE, 1965).

D-amphetamine did not change the adrenal gland content of A and NA in intact (MOORE, 1963; BEAUVALLET et al., 1965) or hyperthyroid (MOORE, 1965) mice.

Rabbit: The concentration of NA in the superior cervical ganglion decreased to 70% of the control level in the experiments by SANAN and VOGT (1962).

Guinea Pig: Heart NA decreased to 65% of the control level 4 h after 20 mg/kg *dl*-amphetamine (LEWANDER, 1971c). Chronic treatment of guinea pigs with this dose twice daily for 18 days caused an additional decrease to 52% of the control level.

β) Brain NA

Rat: Amphetamine, administered i.p., s.c., or i.v. in acute experiments to rats, caused a decrease in brain NA (MCLEAN and MCCARTNEY, 1961; MOORE and LARIVIERE, 1963, 1964; BAIRD and LEWIS, 1963, 1964; CARLSSON et al., 1965; GUNNE and LEWANDER, 1967; CORRODI et al., 1967; LEWANDER, 1968a, b; BAIRD, 1968; BEAUVALLET, 1968; BRODIE et al., 1969; GROPPETTI and COSTA, 1969b; SALAMA and GOLDBERG, 1969; TAYLOR and SNYDER, 1970, 1971; COSTA and GROPPETTI, 1970; SIMON et al., 1970b; LEWANDER, 1971a, b; LEONARD and SHALLICE, 1971; LEONARD, 1972; COSTA et al., 1972; GLICK et al., 1973, 1974). The decrease in brain NA is dose dependent (MOORE and LARIVIERE, 1963; BAIRD and LEWIS, 1964; LEONARD and SHALLICE, 1971). After sublethal doses (10–20 mg/kg *d*-amphetamine sulphate) the maximal NA depletion is about 50% of the control level and doses of *d*- or *dl*-amphetamine below 2 mg/kg (free base) seem to give small, variable, or no effect (BEAUVALLET et al., 1962a; BAIRD and LEWIS, 1964; LITTLETON, 1967; CORRODI et al., 1967; BRODIE et al., 1970; LEONARD and SHALLICE, 1971). Intracisternal injection of 40 μg *d*-amphetamine does not change brain NA (or DA) levels (BREESE et al., 1970; TAYLOR and SULSER, 1973). Direct comparisons of the effect of different doses of *d*-, *l*-amphetamine (BAIRD and LEWIS, 1964; TAYLOR and SNYDER, 1970, 1971; BRODIE et al., 1970; LEWANDER, 1971a; COSTA et al., 1972) show that the *d*-isomer is 2–3 times more effective than the *l*-isomer with respect to brain NA-depletion. The racemic compound lies in between (LEWANDER, 1971a). The NA depletion seems to be unequal between different brain areas (JAVOY et al., 1968; TAYLOR and SNYDER, 1970). The time response curves show that the NA depletion persists for 1–4 days, depending on the dose, after single injections of *d*- or *dl*-amphetamine (MCLEAN and MCCARTNEY, 1961; BRODIE et al., 1969; GROPPETTI and COSTA, 1969b; BRODIE et al., 1970; COSTA and GROPPETTI, 1970; LEWANDER, 1970a, 1971a). The duration of the NA-depletion after *l*-amphetamine is shorter (BRODIE et al., 1970).

Amphetamine in combination with grid shock stress (MOORE and LARIVIERE, 1963, 1964), forced muscular work (BEAUVALLET et al., 1962b), cold stress (SALAMA and GOLDBERG, 1969), or increased room temperature 33° C (BEAUVALLET et al., 1967; BEAUVALLET, 1968) caused a more pronounced decrease in brain NA levels than amphetamine alone. Methamphetamine also depleted brain NA (COOK and SCHANBERG, 1970; LEONARD, 1972), and abolished the daily rhythm of hypothalamic NA levels (ASANO and MOROJI, 1974).

Repeated injections of amphetamine or *chronic administration* of as much as 20 mg/kg of the *dl*-isomer of the drug once or twice daily for as long as 1 month have been shown to induce an additional decrease in brain NA levels but never exeeding about 40–50% of the control level (MC LEAN and MC CARTNEY, 1961; GUNNE and LEWANDER, 1967; BEAUVALLET, 1968; JAVOY et al., 1968; LEWANDER, 1968b, 1971b; BRODIE et al., 1970). The NA levels in the brain stem and hippocampus were reduced, but unchanged in other areas of rat brain after 3 mg/kg of *dl*-amphetamine sulphate daily in the food for 6 months (HERMAN et al., 1971a). No difference between controls and chronically treated rats was found after 9 months with the same dose schedule (HERMAN et al., 1971b). Acute and chronic treatment of rats with methamphetamine caused reduced brain NA levels (COOK and SCHANBERG, 1970; SCHANBERG and COOK, 1972).

Mouse: The reported effect of amphetamine on brain NA levels in mice varies between different authors due to differences in dose, time of injection before sacrifice, and, more importantly, the number of animals in the same cage. (For the discussion of effect of aggregation on the pharmacology of amphetamine in mice see Subsect. E.7). MOORE (1963) showed that there was a dose-dependent decrease in brain NA after administration of 3–30 mg/kg of *d*-amphetamine sulphate to isolated mice 2 h before sacrifice. A 3–4 times higher dose of the *l*-isomer than the *d*-isomer had to be given in order to achieve an equal depression of brain NA. Lower doses were without effect. Similar findings were reported by MOORE (1964, 1965), BEAUVALLET and SOLIER (1964), BEAUVALLET et al. (1962b, 1963, 1964, 1965), LAL and CHESSICK (1964), SMITH (1964), DOLFINI et al. (1969a, b), EVERETT and YELLIN (1971), WELCH and WELCH (1970), ALHAVA and KLINGE (1972), ALHAVA (1973), HITZE-

MANN et al. (1973), and LITTLETON (1974). Differences between mouse strains with respect to NA depletion have been shown (DOLFINI et al., 1969b, 1970; JORI and GARATTINI, 1973; CACCIA et al., 1973). Aggregation of mice in groups of five or more animals per cage was followed by a potentiation of the NA-depleting action of amphetamine (MOORE, 1963, 1964; HALPERN et al., 1962b; LAL and CHESSICK, 1964). No difference between isolated and aggregated mice was found by CONSOLO et al. (1965a). LITTLETON (1974) showed a more pronounced depletion of NA in grouped mice at 25 mg/kg of amphetamine than at higher doses. Mice kept in light were more sensitive than those kept in darkness (LITTLETON, 1974). Hyperthyroidism (MOORE, 1964; DOLFINI et al., 1970) and fighting (WELCH and WELCH, 1970) in combination with amphetamine also potentiated the depletion of brain NA caused by the drug. Amphetamine in doses between 1 and 15 mg/kg given 15–60 min before killing of the animals by freezing in a CO_2-acetone mixture caused an increase by 20–25% of the NA concentration in several areas, while 45 mg/kg caused a decrease (WELCH and WELCH, 1970). An increase in brain NA after amphetamine was also reported by SIMON et al. (1970b), SVENSSON (1971), and WELCH and WELCH (1972). No effect on brain NA was found at 30 min after injection in three different strains of mice treated with similar doses of the drug at three different environmental temperatures (LAL and CHESSICK, 1964; DOLFINI et al., 1969a).

In conclusion, *d*-amphetamine according to most but not all studies, seem to cause a dose-dependent decrease in brain NA in doses greater than 3 mg/kg given more than 1 h before killing the mice.

Rabbit: d-Amphetamine sulphate, 10 mg/kg, followed by a second dose of 10 mg/kg 2 h later caused a decrease in NA in the hypothalamus to approximately 70% of the control level at 4 h after the first injections (SANAN and VOGT, 1962). HIGUCHI et al. (1962), LAVERTY and SHARMAN (1965), and LEWANDER (1974b) confirmed this effect of *d*-amphetamine on brain NA concentration in rabbits.

Guinea Pig: Brain NA was decreased to 63% of the control levels both in the brain stem and telencephalon at 4–8 h after a single injection of 20 mg/kg *dl*-amphetamine sulphate (LEWANDER, 1971c). Control levels were not reattained within 24 h (brain stem). *Chronic* amphetamine treatment of guinea pigs with 20 mg/kg twice daily for 18 days caused a decrease to 40% of the control levels. Similar findings have been reported by GROPPETTI and COSTA (1969a) and GROPPETTI et al. (1972b).

Cat: D-amphetamine, 10 mg/kg at 4 h, caused a decrease of NA in the hypothalamus to 80% of the control level (LAVERTY and SHARMAN, 1965). ELLINWOOD and ESCALANTE (1970a, b) have reported decreased levels of NA in brain from cats treated *acutely* and *chronically* with methamphetamine. The decrease in brain NA was observed by histochemical techniques to be present in both NA cell bodies and terminals.

Dog: Hypothalamic NA decreased to 55% of the control level after 10 mg/kg of *d*-amphetamine 4 h before sacrifice (LAVERTY and SHARMAN, 1965).

γ) Brain DA

Rat: The *acute* effects of amphetamine on whole brain, or striatal, DA in the rat are not consistent. Some authors have reported unchanged or slightly (<15%) decreased levels (MOORE and LARIVIERE, 1963, 1964; BAIRD and LEWIS, 1963, 1964; CARLSSON et al., 1965; GUNNE and LEWANDER, 1967; CORRODI et al., 1967; BAIRD,

1968; BREESE et al., 1970; TAYLOR and SNYDER, 1970, 1971; LEONARD and SHALLICE, 1971; LEONARD, 1972; COSTA et al., 1972; GLICK et al., 1974), while others have reported an increase by up to 24% (LITTLETON, 1967; LEWANDER, 1968a, b, 1971a, b; BEAUVALLET, 1968; BOULU et al., 1972) with doses between 10–20 mg/kg about 3 h after the administration of either of the isomers or the racemate of the drug. The increase in brain DA after amphetamine was potentiated by keeping the animals at 33° C, which induced hyperthermia (40° C colonic temperature; BEAUVALLET, 1968). Lesion of the nigrostriatal neuronal pathway abolished the amphetamine-induced increase in brain DA (BOULU et al., 1972). Various forms of stress (grid shock, restraint, swim, sound) in combination with amphetamine were without additional effect on brain DA concentrations (MOORE and LARIVIERE, 1963). Injection of 40 μg of amphetamine into the cerebral cisternae did not affect brain DA concentrations (BREESE et al., 1970).

Methamphetamine (15 mg/kg) caused an increase in striatal DA, which became less prominent on repeated administration of the drug (KODA and GIBB, 1973).

Chronic administration of *dl*-amphetamine sulphate, 20 mg/kg twice daily for 1 week or 16–32 mg/kg twice daily for 1 month, caused a decrease to 85% and 50%, respectively of the control whole brain DA level (GUNNE and LEWANDER, 1967; LEWANDER, 1968b, 1971e). Daily administration of 3 mg/kg *dl*-amphetamine in the food for 6 months decreased hippocampal DA to 60% of the control level but did not change the DA concentration in other areas including striatum (HERMAN et al., 1971a). According to histochemical studies, the DA depletion after chronic amphetamine treatment seems to be confined to the nigrostriatal DA system and not to involve the hypothalamic DA neurons (FUXE and UNGERSTEDT, 1970).

Mouse: Amphetamine in doses up to 30 mg/kg given 1–2 h before killing did not change brain DA levels (MOORE, 1963; BEAUVALLET et al., 1964, 1965; WELCH and WELCH, 1970; ALHAVA and KLINGE, 1972). In mice killed within 1 h after doses of 1–45 mg/kg amphetamine increased brain DA by a maximum of 50% in various brain areas and this effect was more marked in grouped than isolated mice (WELCH and WELCH, 1970, 1972). EVERETT and YELLIN (1971) and HITZEMANN et al. (1973) reported a decrease in brain DA after *d*-amphetamine. The decrease was more pronounced at 25 mg/kg than at higher doses (LITTLETON, 1974) in grouped mice.

Rabbit: DA in the striatum increased by 13% 4 h after 15 mg/kg of *d*-amphetamine sulphate (LAVERTY and SHARMAN, 1965).

Guinea Pig: Telencephalic DA decreased to 86% of the control level at 4 h after a single dose of 20 mg/kg of *dl*-amphetamine (LEWANDER, 1971c). A further decrease to 50% was found after *chronic treatment* for 18 days. A lower dose (12 mg/kg) as a single injection or in chronically treated animals was without effect on brain DA (GROPPETTI et al., 1972b).

Cat: DA in the caudate nucleus decreased to 56% after 10 mg/kg of *dl*-amphetamine given 4 h before sacrifice (LAVERTY and SHARMAN, 1965).

Dog: No change in the DA concentration of the caudate nucleus was found after 10 mg/kg of *d*-amphetamine (LAVERTY and SHARMAN, 1965).

In summary, amphetamine in *acute* experiments causes a depletion of NA in brain and peripheral tissues in most species studied. The duration of the NA depletion varies between species. Brain DA seem to increase, decrease, or remain unchanged depending on the dose, time after injection and species.

Chronic amphetamine administration depletes both NA and DA in peripheral and central stores, but not to less than approximately 40–50% of the control levels.

b) Uptake of CA

α) Peripheral NA Neurons

AXELROD and TOMCHICK (1960) first demonstrated *d*-amphetamine to "prevent the protective binding" of intravenously injected radiolabeled NA and A in mice. In cats amphetamine prevented uptake and retention of ^{3}H-NA in the heart, spleen, adrenal gland, and liver but not in skeletal muscle (AXELROD et al., 1961; HERTTING et al., 1961). Similar results were obtained with the rat heart (AXELROD et al., 1962). Amphetamine also inhibited uptake of ^{3}H-metaraminol, a CA analogue, into the mouse heart (CARLSSON and WALDECK, 1965). Similarly, the depletion of heart NA in rats by guanethidine and bretylium, which drugs are also taken up over the axonal membrane, was prevented by amphetamine pretreatment (CHANG, 1965). Later in vitro studies showed NA uptake over the neuronal membrane pump to be inhibited in the cat spleen by concentrations of amphetamine of approximately 10^{-7} M (ID 50) (DENGLER et al., 1961), in isolated rat heart (BURGEN and IVERSEN, 1965[1]), or heart slices from rats and rabbits (HENDLEY and SNYDER, 1972), slices of vas deferens from reserpine-nialamide pretreated rats (CARLSSON et al., 1965, 1966a, b; HÄGGENDAL and HAMBERGER, 1967), isolated iris from rats and rabbits (HAMBERGER and MALMFORS, 1967; FARNEBO, 1971; HENDLEY and SNYDER, 1972), and rabbit aortal strips (FERRIS et al., 1972; HENDLEY and SNYDER, 1972). The two isomers of amphetamine were equipotent with regard to inhibition of uptake of NA into minces of rat heart (HARRIS and BALDESSARINI, 1973a).

LEITZ (1970) showed *d*-amphetamine to inhibit metaraminol-induced release of heart NA by inhibiting the uptake of metaraminol. Uptake of NA into isolated bovine splenic nerve NA storing granules was inhibited by amphetamine and methamphetamine (EULER and LISHAJKO, 1968).

β) Brain NA Neurons

Inhibition of the uptake mechanism of brain NA neurons in vivo by amphetamine has been shown by different procedures. The uptake and retention of intraventricularly injected radioactively labeled NA into different brain areas was reduced in amphetamine pretreated rats (GLOWINSKI and AXELROD, 1964; GLOWINSKI et al., 1966a, b; FUXE and UNGERSTEDT, 1968, 1970; STRADA et al., 1970). Methamphetamine had the same property (COOK and SCHANBERG, 1970; SCHANBERG and COOK, 1972). In these studies an increased uptake of NA was observed 3–5 h after the methamphetamine injection. Dextroamphetamine was found to be a more potent inhibitor of NA uptake in various brain areas than the levoisomer (TAYLOR and SNYDER, 1970, 1971). Amphetamine also antagonized the brain NA depletion caused by 4, α-dimethyl-m-tyramine (H77/77), which indicates that the uptake of this compound into rat brain NA neurons was inhibited (CARLSSON et al., 1969). In contrast, a potentiation of the H77/77-induced NA-depletion by amphetamine was reported by LEONARD and SHALLICE (1971). The differences in results might, however, be ex-

[1] Uptake 2 according to IVERSEN (1967) was also inhibited by amphetamine at an ID 50 concentration of 10^{-4} M.

plained by different dosage schedules. The *l*-isomer did not alter the effect of H 77/77 in the mouse brain (SVENSSON, 1971). Cortex slices from mice treated in vivo with amphetamine, which were incubated in vitro, took up less NA than slices from control mice (ROSS and RENYI, 1966a, 1967b).

Various in vitro systems have been used for studies of the effects of amphetamine on NA uptake. Thus, *d*-, *l*-, and racemic amphetamine have been shown to inhibit uptake of NA into slices of cerebral cortex in mice (ROSS and RENYI, 1964, 1967b; REITZ et al., 1972) and rats (CARLSSON et al., 1965, 1966a, b; HÄGGENDAL and HAMBERGER, 1967; RUTLEDGE et al., 1972a; ZIANCE and RUTLEDGE, 1972: FERRIS et al., 1972), in chopped cortex (ZIANCE et al., 1972), in homogenates of cerebral cortex, hypothalamus, medulla oblongata, cerebellum (COYLE and SNYDER, 1969b), and in synaptosomal preparations from cortex and hypothalamus (HORN et al., 1971; HORN and SNYDER, 1972; FERRIS and MAXWELL, 1972; FERRIS et al., 1972; HARRIS and BALDESSARINI, 1973a; THORNBURG and MOORE, 1973a). Methamphetamine inhibited uptake of NA into slices from hypothalamus and brain stem but not from telencephalon of rats (MORGAN et al., 1972b).

γ) Brain DA Neurons

Amphetamine has been shown to inhibit the reserpine-resistant uptake of DA and NA into the tuberoinfundibular DA neurons in reserpine-nialamide treated rats in vivo (FUXE et al., 1967). The uptake of intraventricularly administered DA into nigrostriatal DA cell bodies and terminals was inhibited by amphetamine (FUXE and UNGERSTEDT, 1968, 1970). TAYLOR and SNYDER (1970, 1971) showed that *d*- but not *l*-amphetamine (10 mg/kg s.c.) inhibited the uptake of intraventricularly administered ^{3}H-NA into the striatum. However, in a previous study by GLOWINSKI et al. (1966a), *d*-amphetamine (15 mg/kg i.p.) did not affect the uptake of intraventricular ^{3}H-NA or ^{3}H-DA into the striatum. A series of in vitro studies shows unequivocally that *d*- and *l*-amphetamine inhibit uptake of DA into rat brain slices from striatum (CARLSSON et al., 1966a, b; HÄGGENDAL and HAMBERGER, 1967; FARNEBO, 1971), homogenates of striatum (COYLE and SNYDER, 1969a, b), and synaptosomal preparations of striatal tissue (HORN et al., 1971; HORN and SNYDER, 1972; FERRIS and MAXWELL, 1972; FERRIS et al., 1972; HARRIS and BALDESSARINI, 1973a; THORNBURG and MOORE, 1973a). Methamphetamine has similar actions (MORGAN, et al., 1972b). Amphetamine inhibits the uptake of DA (and NA) also into isolated vesicles from striatal tissue of pig brain (PHILIPPU and BEYER, 1973).

Kinetic analysis of the uptake inhibition exerted by amphetamine showed it to be competitive and reversible in both hypothalamus and striatum (HORN et al., 1971). K_i values for amphetamine are in the order of 10^{-7} M in in vitro studies.

Controversial findings with regard to the differential potencies of *d*- and *l*-amphetamine as uptake inhibitors on NA and DA neurons have been reported. COYLE and SNYDER (1969b) and TAYLOR and SNYDER (1970, 1971), in in vitro and in vivo experiments, reported *d*-amphetamine to be up to 10 times more potent than *l*-amphetamine as uptake inhibitors for brain NA neurons, while the two isomers were equipotent with regard to striatal DA neurons. In contrast, FERRIS et al. (1972), HARRIS and BALDESSARINI (1973a), and THORNBURG and MOORE (1973a) reported *d*- and *l*-amphetamine to be about equipotent in inhibiting NA uptake into brain NA neurons and *d*-amphetamine to be 4–5 times more effective than *l*-amphetamine on brain DA (striatal) neurons. A four- to sixfold difference between *d*- and *l*-amphetamine was found with regard to DA uptake into mesolimbic and striatal DA nerve terminals (HORN et al., 1974). Differences in experimental conditions, e.g., the use of preparations from reserpine-treated rats (COYLE and SNYDER, 1969b; HENDLEY and SNYDER, 1972) might explain the differences in results. However, HARRIS and BALDESSARINI (1973a) point out that reserpine does not affect the *d*/*l*-quotient but reduces the total amounts of CA taken up, which makes the results unreliable.

Other important factors affecting the outcome of stereoselective mechanisms are: thickness of slices, linearity of the kinetics and nonsaturating concentrations of amines (HENDLEY and SNYDER, 1972). Thus, these authors found *d*-amphetamine to be more potent than *l*-amphetamine as inhibitors of uptake over the neuronal cell membrane of NA neurons in the rat heart, rabbit vas deferens, and iris; *d*- to be equal to *l*-amphetamine in rabbit heart and retina (DA cells), and *l*-amphetamine to be more potent than *d*-amphetamine in rabbit thoracic aorta. FERRIS et al. (1972) reported that the isomers were equipotent in the thoracic aorta of rabbits not treated with reserpine; all animals were pretreated with reserpine in the study by HENDLEY and SNYDER (1972).

Theoretical considerations and experimental data suggest that the preferred conformation of amphetamine at the CA uptake site on NA neurons should be the anticlinical conformation (HORN and SNYDER, 1972; MILLER et al., 1973).

c) *Release of CA*

α) Peripheral CA

Amphetamine has been shown to release NA or ^{3}H-NA previously taken up into the mouse or rat heart in vivo (AXELROD et al., 1962; POTTER and AXELROD, 1963; BENINGTON and MORIN, 1968) and ^{3}H-NA, α-methyl-NA or metaraminol from the mouse heart (CARLSSON and WALDECK, 1966a, b). The releasing activity of amphetamine does not seem to be due to inhibition of the uptake mechanisms, since more potent uptake inhibitors, like desmethylimipramine (DMI) (CARLSSON and WALDECK, 1968; LUNDBORG, 1969) or protriptyline (CARLSSON and WALDECK, 1966a; LUNDBORG and WALDECK, 1972), do not release ^{3}H-NA from the mouse heart. Amphetamine has to be transported into the NA neuron in order to exert its action (LUNDBORG and WALDECK, 1972), and it has been claimed that it releases mainly, but not exclusively, extravesicular monoamines from the mouse heart (LUNDBORG, 1969).

Studies in vitro showed that amphetamine increases the release of unchanged NA from the isolated rat heart by reducing NA metabolism by deamination (LEITZ and STEFANO, 1971). Amphetamine facilitated release of NA by nerve stimulation from the isolated rabbit heart (LÖFFELHOLZ and MUSCHOLL, 1970) and the superfused rat iris (FARNEBO, 1971). Also in the iris extragranular NA, i.e., ^{3}H-NA taken up after pretreatment of the animal with reserpine plus nialamide, was readily released by the drug (FARNEBO, 1971). Similar results were obtained in heart slices of rats pretreated by reserpine plus pheniprazine (ENNA and SHORE, 1974). The effect of amphetamine was inhibited by 6-OH-DA pretreatment or guanethidine, α-methyl-NA, or amantadine added to the slices before amphetamine, indicating that amphetamine releases NA from an extragranular storage pool. Electric field stimulation was without releasing effect in the reserpine+MAOI preparations; thus amphetamine and nerve-impulse-induced release of NA seems to be different (ENNA and SHORE, 1974). In heart slices, amphetamine caused release of ^{3}H-NA newly synthesized from ^{3}H-DA (CHEVILLARD et al., 1971) and this effect was potentiated by angiotensin II (CHEVILLARD and ALEXANDRE, 1972). Amphetamine also releases newly synthesized NA from heart (RAPIN et al., 1972) and spleen slices (RAPIN et al., 1973b). Rather high (10^{-4} M) concentrations of amphetamine are needed for release of NA from isolated bovine splenic nerve vesicles (EULER and LISHAJKO, 1968).

Catecholamines in adrenal glands from cats (RUBIN and JAANUS, 1966) but not rats (CESSION-FOSSION, 1967) are released on perfusion with amphetamine. In this tissue amphetamine has an acetylcholinelike action, since the presence of hexamethonium or denervation does not inhibit this effect. Both in perfused cat and cow adrenals amphetamine also causes the release of nucleotides, protein, and dopamine-β-hydroxylase in addition to the catecholamines, indicating release by exocytosis at certain drug concentrations (RUBIN and JAANUS, 1967; SCHNEIDER, 1972). This release is dependent on the presence of calcium ions (cf., RUBIN, 1970). At high concentrations of amphetamine a direct effect on the catecholamine storage vesicles seems to be at work (SCHÜMANN and PHILIPPU, 1962; RUBIN and JAANUS, 1967; SCHNEIDER, 1972; WAGNER et al., 1973, 1974).

CA in Plasma. An amphetamine-induced increase of NA and A in plasma has been reported in dogs (CHIDSEY et al., 1962; HARVEY et al., 1968). In the study by CHIDSEY et al. (1962), NA was shown by coronary sinus cannulation to be released from the heart and in the study by HARVEY et al. (1968), increased concentrations of A in the cannulated adrenal veins were observed after amphetamine. Dose-response relationships were established for the increase in plasma CA and the effects of amphetamine on the blood pressure (HARVEY et al., 1968).

Urinary Excretion of CA. The urinary excretion of NA and A reflects primarily the release of these substances caused from peripheral noradrenergic neurons and the adrenal medulla caused by nerve impulses or drugs and also the release of CA from noninnervated chromaffin cells.

An increase in the urinary excretion of NA and adrenaline in the rat after *acute* amphetamine administration has been demonstrated (BISCARDI et al., 1964; BEAUVALLET et al., 1966b; GUNNE and LEWANDER, 1967; LEWANDER, 1968a, b; BEAUVALLET, 1968; BASSO et al., 1970). The ganglion blocking agent, mecamylamine, blocked the effect of amphetamine on the NA excretion but not on the A excretion; adrenal demedullation (by expression) did not completely block the increase in the A excretion after amphetamine (BISCARDI et al., 1964). The increase in the urinary A excretion was, however, absent after bilateral cutting of the splenchnic nerves (GUNNE and LEWANDER, 1967; LEWANDER, 1968b). Later BASSO et al. (1970) could show that the adrenal demedullation could be performed more efficiently if the whole gland was removed and pieces of its cortex were grafted under the capsule of the kidney. Amphetamine administered to rats demedullated in this manner did not cause an increase in the urinary excretion of A. It was also found (LEWANDER, 1968b; BASSO et al., 1970) that the amphetamine-induced increase in the urinary excretion of NA was more pronounced in rats with denervated or demedullated glands than in intact rats.

During *chronic administration* of *dl*-amphetamine, 16–32 mg/kg twice daily for approximately 14 days per dose level, there was an initial increase in the urinary excretion of both A and NA followed by a successive decrease toward control excretion values (LEWANDER, 1968b). This pattern was most evident at the two higher dose levels and was considered to be a sign of tolerance to these effects of amphetamine in the rats. The urinary excretion of A and NA was also increased by amphetamine in the guinea pig (LEWANDER, 1971c). Chronic experiments have not been performed in this species.

Amphetamine causes a decrease in the urinary DA excretion in rats between 4 and 8 h after administration (BEAUVALLET et al., 1966b).

β) Brain NA and DA

Various methods have been used for the demonstration of amphetamine-induced release of CA in the brain in vivo. Several of these approaches utilize the hitherto accepted view that extraneuronally liberated CA are metabolized by COMT to the 3-0-methylated amines (NM and MTA, respectively), which are subsequently deaminated by MAO to neutral and acidic metabolites, while CA liberated intraneuronally are directly deaminated by MAO to neutral or acidic catechol metabolites (see reviews by AXELROD, 1959, 1965; KOPIN, 1964; CARLSSON, 1966; IVERSEN, 1967; ANDÉN et al., 1969).

Measurements of Endogenous CA and 3-0-methylated CA Metabolites after MAO Inhibition. Amphetamine causes an increase of brain NM and MTA after a single dose or chronic administration in rats (GUNNE and LEWANDER, 1967, 1968). The levels of NM and MTA are more easily detected by conventional methods in control rat and guinea pig brains after pretreatment of the animals with an MAO inhibitor, i.e., nialamide. Administration of amphetamine after nialamide pretreatment causes a marked increase of the NM and MTA accumulation in brains of rats (RANDRUP and JONAS, 1967; GUNNE and LEWANDER, 1968; JONAS and SCHEEL-KRÜGER, 1969; SCHEEL-KRÜGER, 1971, 1972a), mice (SVENSSON, 1971), and guinea pigs (LEWANDER, 1971c) indicating extraneuronal release of both NA and DA in the brain.

In another series of reports, rats were pretreated with reserpine plus nialamide plus L-dopa in order to study selectively the extravesicularly but intraneuronally located CA (CARLSSON et al., 1965, 1966a, b). In these rats very low doses of amphetamine (0.1–1 mg/kg) caused a depletion of the CA, as shown with biochemical and histochemical methods, and an increase in brain NM and MTA indicated that amphetamine has a marked releasing effect on such an artificially enlarged pool of extravesicular CA. Similar results have been reported by ANDÉN and SVENSSON (1973) in several brain regions in reserpine-nialamide pretreated rats. Histochemical evidence for the amphetamine-induced depletion of α-methyl-NA and α-methyl-DA was reported in rats given tetrabenazine plus α-methyl-dopa (CARLSSON et al., 1968). Potent and selective uptake inhibitors like DMI do not cause depletion of the α-methylated amines in similar experiments (CARLSSON et al., 1966b).

Sectioning of brain DA neurons or treatment of rats with γ-hydroxybutyric acid causes an increase in brain DA. Amphetamine given before or after these treatments counteracts this increase, which indicates that amphetamine releases DA from the brain DA neurons (ANDÉN et al., 1973; ROTH et al., 1973). The effects of *d*- and *l*-amphetamine on the accumulation of NM and reduction of NA, i.e., release of NA, were equal, while *d*-amphetamine was 1.7 times more potent than *l*-amphetamine on the accumulation of MTA, i.e., release of DA (SCHEEL-KRÜGER, 1972a).

Measurement of HVA. At 1–2 h after amphetamine (10–20 mg/kg i.p.), HVA levels increase in cats (LAVERTY and SHARMAN, 1965), rats, and mice (JORI and BERNARDI, 1969, 1972; BIZZI et al., 1970; ROFFLER-TARLOV et al., 1971; FUENTES and DEL RIO, 1972; JORI and DOLFINI, 1974). Amphetamine was more potent than *l*-amphetamine in this respect (JORI et al., 1973; JORI and DOLFINI, 1974). In some mouse strains, brain HVA levels do not increase after amphetamine (JORI and GARATTINI, 1973; CACCIA et al., 1973). The increase in striatal HVA in mice was antagonized by amantadine (MENON et al., 1973) and γ-hydroxybutyric acid (MENON et al., 1974). However, reduced HVA levels after amphetamine have been found in dogs (LAVERTY and SHARMAN, 1965) and guinea pigs (LEWANDER, 1971c).

Chronically amphetamine-treated rats (but not mice) became tolerant to the effect of amphetamine on HVA (JORI and BERNARDI, 1972); chronically treated guinea pigs had lower HVA concentrations in brain as compared with a single dose (LEWANDER, 1971c).

Accumulation of ^{3}H-NM and ^{3}H-MTA Synthesized from ^{3}H-dopa. Whether injected before or after i.v. injection of ^{3}H-dopa, amphetamine (20 mg/kg i.p.) increases the accumulation in the rat brain of ^{3}H-NM and ^{3}H-MTA synthesized from the precursor (even without MAO inhibitors), indicating an amphetamine-induced release of CA (LEWANDER, 1970a, 1971e).

Fate of NA and DA Taken up Into Brain CA Neurons after Intraventricular Injection. ^{3}H-NA or ^{3}H-DA injected into the lateral ventricle or cisterna magna of rats is taken up by central NA and DA neurons. Subsequent administration of *d*-amphetamine (10–15 mg/kg) accelerates the disappearance of the labeled NA and causes an increase in ^{3}H-NM and a decrease in deaminated catechol metabolites (GLOWINSKI and AXELROD, 1964; GLOWINSKI et al., 1966a, b; STRADA et al., 1970; TILSON and SPARBER, 1972). Very similar findings were reported for methamphetamine (COOK and SHANBERG, 1970). Chronic methamphetamine treatment did not change this pattern of NA metabolites (SCHANBERG and COOK, 1972). Small doses (5 mg/kg) of amphetamine have no effect on ^{3}H-NM except after repeated injections (GLOWINSKI, 1970b). Labeled MOPEG, however, was increased after 3 mg/kg of amphetamine in such experiments (LADISCH et al., 1970b); progesterone mitigated this effect of amphetamine (LADISCH and BAUMANN, 1971).

Intracerebroventricular administration of amphetamine also caused a decrease in labeled NA and an increase in ^{3}H-NM (TAYLOR and SULSER, 1973). Intraventricularly administered unlabeled CA to reserpine-nialamide pretreated rats are also released from CA neurons by amphetamine as demonstrated by histochemical methods (FUXE and UNGERSTEDT, 1968, 1970).

Perfusion of the Cerebral Ventricular System. CARR and MOORE (1969, 1970a, b) perfused the cerebroventricular system in cats with ^{3}H-NA, which was taken up by the brain CA neurons. Administration of amphetamine (25–400 µg/ml) added to the perfusion fluid or injected i.v. (1 mg/kg) was followed by increased amounts of ^{3}H-NA and ^{3}H-NM in the perfusate indicating amphetamine-induced NA release. Levoamphetamine was less effective than *d*-amphetamine in this respect. Release of NA has been reported in rats equipped with a push-pull cannula in the ventricular system (SPARBER and TILSON, 1972b). Later, cerebroventricular perfusion experiments in cats have shown that ^{3}H-DA previously taken up or newly synthesized from ^{3}H-tyrosine or ^{3}H-dopa is released by amphetamine (CHIUEH and MOORE,

1973, 1974b). D-amphetamine was 3–4 times more potent than *l*-amphetamine (VOIGTLÄNDER and MOORE, 1973a, b; CHIUEH and MOORE, 1974a).

Amphetamine-induced release of DA was reduced by sectioning of the nigrostriatal DA neurons. It was also demonstrated that amphetamine potentiated the release of DA by electrical stimulation of the nigrostriatal bundle (VOIGTLÄNDER and MOORE, 1973b). Haloperidol did not antagonize amphetamine-induced DA-release (VOIGTLÄNDER and MOORE, 1973a). The amphetamine effect on CA release in these experiments was not changed by α-MT (CHIUEH and MOORE, 1974c) or amantadine (GUDETSKY et al., 1974). The lack of inhibition of amphetamine-induced release of DA is thus in disagreement with the study by ENNA et al. (1973, below).

Perfusion or Superfusion of Different Brain Areas in vivo with Push-Pull Cannulas or Cups. The output of endogenous DA from cat striatum was found to increase after addition of amphetamine (0.5 mg/ml) to the perfusate through a push-pull cannula (MCKENZIE and SZERB, 1968). Increased amounts of newly synthesized ^{14}C-DA from ^{14}C-dopa or ^{14}C-tyrosine was found in the perfusate after administration of amphetamine in similar experiments (RIDDELL and SZERB, 1971). In rats, STEIN and WISE (1967, 1969) showed that amphetamine (3 mg/kg i.p.) increased the output of ^{3}H-NA in push-pull cannula perfusates placed in the amygdaloid nucleus and in the hypothalamus. GLOWINSKI (1970b) and BESSON et al. (1971a) superfused the caudate nucleus of cats with ^{3}H-tyrosine by use of cup technique in order to label the DA stores with ^{3}H-DA. Addition of amphetamine (10^{-6} M) to the superfusate or i.v. injection of *l*-amphetamine (2 mg/kg) caused a release of ^{3}H-DA into the superfusate. This result was confirmed in a similar study in monkeys (GAUCHY et al., 1974).

Release of Newly Synthesized CA in in vitro Systems. Isolated corpus striatum (BESSON et al., 1969a) or slices of striatum or medulla oblongata (BESSON et al., 1969b; GLOWINSKI, 1970b) from rats were superfused or incubated in the presence of ^{3}H-tyrosine. ^{3}H-CA synthesized during the incubation were measured in the tissues and in the media. Amphetamine, either given to the animals before killing or added to the incubations, caused a release of newly synthesized ^{3}H-CA from the tissue into the superfusion fluid or media.

Release of CA Previously Taken up Into in vitro Tissue Preparations. NG et al. (1970) showed that *d*-amphetamine was more potent than *l*-amphetamine in releasing ^{3}H-NA taken up by rat striatal slices. FARNEBO (1971) demonstrated *d*-amphetamine (10^{-6} M) to release ^{3}H-NA taken up into cortex slices and ^{3}H-DA taken up into neostriatal slices. EBSTEIN et al. (1972) have reported that amphetamine caused a release of NA more efficiently in slices from the amygdaloid nucleus than from hypothalamus. The effect was more evident when slices from reserpine plus nialamide pretreated rats were used (FARNEBO, 1971). Electrical stimulation of brain slices caused an increased overflow of ^{3}H-CA in the presence of *d*-amphetamine than in its absence (FARNEBO, 1971).

^{3}H-CA taken up into synaptosomal preparations from rat cerebral cortex, hypothalamus, and striatum were released four times easier by *d*- than *l*-amphetamine (FERRIS and MAXWELL, 1972; FERRIS et al., 1972). D-Amphetamine (10^{-6} M) caused release of ^{3}H-NA taken up either by chopped tissue, homogenates, or synaptosomes from rat hypothalamus (PHILIPPU et al., 1974) and cerebral cortex (RUTLEDGE et al., 1972a, b, 1973; ZIANCE et al., 1972; ZIANCE and RUTLEDGE, 1972; AZZARO and RUTLEDGE, 1973; AZZARO et al., 1974; TSENG et al., 1974). L-amphetamine was less active than *d*- or *dl*-amphetamine. The release was temperature dependent and amphetamine did not cause unspecific release of, e.g., inulin or urea. The release process was calcium dependent only at amphetamine concentrations below 10^{-3} M; at or above that concentration release could be demonstrated in calcium-free media. Thus, amphetamine-induced release differs from electrically induced, potassium-induced, and tyramine-induced release in not being completely calcium dependent (cf., RUBIN, 1970). Only a slight releasing effect of amphetamine was noticed when synaptosomes from rat brain were superfused with a solution containing amphetamine (RAITERI et al., 1974). These authors therefore concluded that the main action of amphetamines in the brain is uptake inhibition. P-hydroxyamphetamine and p-hydroxynorephedrine also released NA from chopped cortex tissue in vitro: p-hydroxynorephedrine was weaker in effect than amphetamine and p-hydroxyamphetamine (WENGER and RUTLEDGE, 1974).

Alpha-methyl-p-tyrosine, an inhibitor of tyrosine hydroxylase, has been found to inhibit amphetamine-induced release of NA and some CA analogues from brain slices (ENNA et al., 1973). Cocaine and desipramine prevent amphetamine-induced release of NA from chopped brain tissue by inhibiting the uptake of amphetamine into the neurons (RUTLEDGE et al., 1973; AZZARO et al., 1974). Incorporation of α-methyl-m-tyramine into brain DA neurons occurs after administration of α-methyl-m-tyrosine (DORRIS and SHORE, 1974). Amphetamine caused a depletion of this false transmitter indicating release, but by a mechanism different from release by nerve impulses (DORRIS and SHORE, 1974). Amantadine prevented amphetamine-induced release of α-methyl-m-tyramine (DORRIS and SHORE, 1974).

d) Enzymatic Inactivation of CA

α) Monoamine Oxidase (MAO)

Since the early demonstration (BLASCHKO et al., 1937; PUGH and QUASTEL, 1937) that amphetamine, not being a substrate of MAO itself, inhibits this enzyme (MAO) competitively and reversibly (MANN and QUASTEL, 1940; cf. review by BLASCHKO, 1952), a number of studies have confirmed this finding for a number of species, tissues, and substrates. Thus, preparations from rat, mouse, and rabbit brain (GRANA and LILLA, 1959; GLOWINSKI et al., 1966a; NIELSEN et al., 1967; GREEN, 1971; MÖLLER-Nielsen and DUBNIK, 1970; MORGAN et al., 1972b), rat, rabbit, and guinea pig liver (ORZECHOWSKI, 1941; PRATESI and BLASCHKO, 1959; GRANA and LILLA, 1959; FULLER and WALTERS, 1964; PARMAR, 1966; NIELSEN et al., 1967; YASUNOBU and OI, 1972), rat kidney (GRANA and LILLA, 1959), rat heart (LEITZ and STEFANO, 1971), pig plasma (BUFFONI and DELLA CORTE, 1972), and human brain and salivary glands (BLASCHKO and STRÖMBLAD, 1960) have been used as enzyme sources. Concentrations of amphetamine ranging from 10^{-5}–10^{-2} M seem to be needed for inhibition of deamination of typical MAO substrates: adrenaline, noradrenaline, tyramine, oxedrine, kynuramine, benzylamine, serotonin, tryptamine, and phenylethylamine. Recent comparative studies have shown that IC50 concentrations are lowest with serotonin, tryptamine, and tyramine as substrates (MÖLLER et al., 1970; EVERETT and YELLIN, 1971; FULLER, 1972), while phenylethylamine requires 25–50 times higher concentrations. This is in line with recent findings of multiple forms of MAO with selective substrate and inhibitor preferences.

PARMAR (1966) made the observation that amphetamine protects MAO in rat liver mitochondria from other MAO inhibitors like iproniazid, tranylcypromine, and nialamide. Dextro-amphetamine is generally a more potent MAO inhibitor in vitro than *l*-amphetamine (PRATESI and BLASCHKO, 1959; GRANA and LILLA, 1959; BLASCHKO and STRÖMBLAD, 1960; FULLER and WALTERS, 1964; PARMAR, 1966; MÖLLER-NIELSEN and DUBNIK, 1970) except in the rabbit liver (PRATESI and BLASCHKO, 1959) where the isomers are equally active.

SALAMA and GOLDBERG (1969) found that systemically administered amphetamine in combination with cold stress, but not other forms of stress, caused a reduced MAO activity in the rat brain.

After 50 days of chronic administration of methamphetamine to guinea pigs, there was a decrease in brain MAO activity to 75% of the controls. After withdrawal there was a compensatory increase in MAO-activity (UTENA et al., 1959).

RUTLEDGE (1970b) and RUTLEDGE et al. (1972b), have shown that amphetamine, in addition to causing release of ^{3}H-NA and ^{3}H-DA previously taken up into brain slices or chopped tissue, reduces the oxidative deamination of the CA. A quite different pattern of release and metabolism was found for fenfluramine as compared with amphetamine (ZIANCE and RUTLEDGE, 1972). RUTLEDGE (1970b) also found that amphetamine does not inhibit rat brain MAO from lysed synaptosomes or mitochondria from the rat brain and concluded that amphetamine causes a functional MAO inhibition by inhibiting access of the CA to intraneuronal MAO by blockade of the uptake mechanisms for CA at membrane levels (cf. Uptake of CA, above). In addition, the metabolic pattern of CA metabolites is changed toward nondeaminated metabolites after amphetamine (GLOWINSKI et al., 1966a, b; COOK and SCHANBERG, 1970; TSENG et al., 1974), methamphetamine (STRADA et al., 1970), and p-hydroxyamphetamine (WENGER and RUTLEDGE, 1974). Similar findings were reported with the isolated perfused rat heart. It has been concluded that amphetamine protects intraneuronal CA from deamination.

Studies on the endogenous concentrations of brain HVA, the 3-0-methylated and deaminated product of DA, as an indicator of the in vivo importance of MAO inhibition in brain caused by amphetamine have given conflicting results. Thus, amphetamine causes an increase in brain HVA in cats (LAVERTY and SHARMAN, 1965), rats, and mice (JORI and BERNARDI, 1969, 1972; BIZZI et al., 1970; FUENTES and DEL RIO, 1972; MENON et al., 1973; JORI and DOLFINI, 1974) while amphetamine causes a reduction of HVA in the dog (LAVERTY and SHARMAN, 1965) and the guinea pig brain (LEWANDER, 1971c). *Dextro*-amphetamine was more potent than *l*-amphetamine in increasing rat brain HVA (JORI et al., 1973; JORI and DOLFINI, 1974). ROFFLER-TARLOV et al. (1971) confirmed that amphetamine increased HVA in mouse brain and showed that Dopac, the intraneuronally formed deaminated metabolite of DA, was decreased. The rat brain Dopac levels were also decreased after amphetamine (ROTH et al., 1973; BUNNEY et al., 1973b). Tolerance to the amphetamine-induced increase in HVA in rats, but not in mice, has been reported after repeated administration of amphetamine (JORI and BERNARDI, 1972). Chronic amphetamine administration to guinea pigs caused a more pronounced fall in brain HVA than a single injection of the drug (LEWANDER, 1971c).

It appears that the brain concentration of HVA might be influenced in two directions by amphetamine: (1) increased by release of DA from DA neuron terminals and (2) decreased as a result of functional MAO inhibition and/or a negative feed-back modulation of DA synthesis. The resultant effective HVA concentration may thus be dependent on the concentration of amphetamine within the brain and, possibly, the animal species.

The early suggestion by MANN and QUASTEL (1940) that the central stimulant effect of amphetamine should be due to MAO inhibition is not compatible with the fact, that other, more efficient MAO inhibitors do not share the central stimulant action of amphetamine (PRATESI and BLASCHKO, 1959; GRANA and LILLA, 1959). However, recent evidence (GLOWINSKI et al., 1966a; GREEN, 1971; RUTLEDGE, 1970b; LEITZ and STEFANO, 1971; and RUTLEDGE et al., 1972b) suggest that any interpretation of the mechanism of action of amphetamine in relation to central stimulation that ignores the effect of high tissue concentrations of the drug on deamination of central and peripheral catecholamine might be incomplete.

β) Catechol-0-methyltransferase (COMT)

Amphetamine has not been found to change liver COMT activity in vitro (AXELROD and TOMCHICK, 1960). The increase in 3-0-methylated metabolites of CA in various tissues (see Release of CA, above) is most probably due to increase of the COMT substrates caused by amphetamine-induced release of catecholamines.

e) Synthesis of CA

Tyrosine hydroxylase, partly purified from beef or cat brain (BAGCHI and MCGEER, 1964; BESSON et al., 1973), or in a crude homogenate of rat brain (MCGEER and MCGEER, 1967), was not inhibited by *d*-amphetamine in vitro in concentrations of 10^{-4} M. Tyrosine hydroxylase activity was decreased in striatal slices in vitro by 10^{-7}–10^{-4} M concentrations of amphetamine (BESSON et al., 1971b).

In other studies rats have been injected with amphetamine and the tyrosine hydroxylase activity has been determined in tissue homogenates or brain slices in vitro, either as formation of ^3H-H_2O from tyrosine-3,5-^3H or as accumulation of ^3H-CA from the precursor. Thus, BESSON et al. (1969b) found that 5 mg/kg *d*-amphetamine increased CA synthesis in brain slices from the striatum and the brain stem, while in later studies (BESSON et al., 1971b, 1973) a reduction of the tyrosine hydroxylase activity was reported. No effect was found on the ^3H-NA accumulation from ^3H-tyrosine after chronic administration of 1 mg/kg *d*-amphetamine during 8 days (BESSON et al., 1973). The amphetamine-induced inhibition of tyrosine hydroxylase was noncompetitive for tyrosine (BESSON et al., 1973). Amphetamine has been reported to change the physical state of striatal but not midbrain tyrosine hydroxylase, i.e., it becomes more membrane bound and is thereby activated (MANDELL et al., 1972).

HARRIS and BALDESSARINI (1973b) reported that in vivo administration of amphetamine (5 mg/kg) to rats caused a decrease in tyrosine hydroxylase activity in striatal homogenates for 0.5–2 h. This effect of amphetamine was blocked by chlorpromazine or bicucculline pretreatment, indicating that it might be secondary to indirect stimulation of DA and GABA receptors. Amphetamine added in vitro did not cause a decrease in the activity of the enzyme.

Administration of amphetamine in drinking water for 3–7 days did not, however, change tyrosine hydroxylase activity in the rat forebrain structures (HULME et al., 1974).

By another in vivo technique, i.e., interstriatal injection of labeled tyrosine and subsequent determination of tritiated water and DA in the striatal tissue, amphetamine was found to decrease the synthesis of DA at the tyrosine hydroxylase step (JAVOY et al., 1974).

FIBIGER and MCGEER (1971) found no effect of 10 mg/kg methamphetamine on tyrosine hydroxylase in brain and adrenal glands except after repeated injection, when the activity of the caudate nucleus enzyme decreased and that of the adrenal gland increased. Similar results were reported by KODA and GIBB (1973). The decrease in striatal and increase in adrenal gland tyrosine hydroxylase caused by amphetamine could be antagonized by haloperidol BUENING and GIBB, 1974). Chlorpromazine antagonized amphetamine in the striatum but not in the adrenal gland (BLUENING and GIBB, 1974). MANDELL and MORGAN (1970) and MANDELL and KNAPP (1972) also reported increased tyrosine hydroxylase activity in adrenal glands of chicken after repeated administration of methamphetamine.

The effect of amphetamine on dopa decarboxylase, the next step in the CA synthesis, does not seem to have been investigated. Dopamine-β-hydroxylase in the beef adrenal gland has been reported to be inhibited by amphetamine, *d*-amphetamine, and *dl*-amphetamine (GOLDSTEIN et al., 1964). Similarly, DBH in spleen homogenates was inhibited by amphetamine, methamphetamine, and p-hydroxyamphetamine (JACQUOT et al., 1974).

f) Turnover of CA

GLOWINSKI and AXELROD (1965) injected ^{3}H-NA intraventricularly into the rat brain and found that *d*-amphetamine (20 mg/kg) retarded its disappearance. As reported by JAVOY et al. (1968) and later reviewed by GLOWINSKI (1970a), smaller doses of amphetamine (5 mg/kg) accelerated brain NA turnover, while repeated or higher doses have no effect on NA turnover as measured by this method.

Determination of the rate of disappearance of CA after inhibition of CA synthesis at the tyrosine hydroxylase step, by giving α-methyltyrosine (CORRODI et al., 1967; ANDÉN, 1970) or 3-iodotyrosine, or at the DA-β-hydroxylase step, by diethyldithiocarbamate (LITTLETON, 1967) have also been used as methods for turnover measurement. In these studies 2 mg/kg *d*-amphetamine accelerated NA depletion after DDC and retarded DA depletion after 3-iodo-tyrosine (LITTLETON, 1967). Large doses of *d*-amphetamine (15–20 mg/kg) accelerated the depletion of NA in the whole rat or mouse brain (CORRODI et al., 1967; ALHAVA, 1973), but had no effect on NA in the rat spinal cord (ANDÉN, 1970). In the two rat studies, brain DA depletion was not affected by amphetamine. No change of the degree of depletion of NA or DA was found after amphetamine in α-methyltyrosine pretreated rats by ANDÉN et al. (1967). Methamphetamine, but not amphetamine, has been reported to antagonize α-methyltyrosine-induced depletion of NA (LEONARD, 1972).

A third method for determination of brain CA turnover is the accumulation of labeled CA after i.v. injection of ^{3}H- or ^{14}C-tyrosine or dopa, two CA precursors. By this technique COSTA and GROPPETTI (1970, 1972), PERSSON (1970), COSTA et al. (1972), and GROPPETTI et al. (1972a) found consistently that small doses of amphetamine caused an increase in DA turnover in the striatum but had no effect on NA turnover, FULGINITI and ORSINGHER (1971), however, reported an increased accumulation of labeled NA after 2 mg/kg *dl*-amphetamine. In mice implanted with amphetamine containing pellets there was an increased accumulation of both NA and DA from ^{3}H-tyrosine. A large dose (10 mg/kg) of amphetamine counteracted this effect (HITZEMAN et al., 1973).

Acute or chronic administration of large doses of the drug (16–32 mg/kg i.v.) reduced the accumulation of labeled NA in rat brain and heart synthesized from tyrosine or dopa, which was suggested to be due to release of the newly synthesized NA (LEWANDER, 1971e). No effect on brain DA was found (LEWANDER, 1970a, 1971e). When amphetamine was given 2 h after labeling the rat brain CA stores by ^{3}H-tyrosine administration, it was found that the disappearance of ^{3}H-NA was accelerated and that of ^{3}H-DA retarded (LEWANDER, 1970a). JAVOY et al. (1970) injected ^{3}H-tyrosine into various parts of the rat brain and measured the accumulation of ^{3}H-CA. D-Amphetamine (5 mg/kg) caused an increase in DA accumulation in cell bodies (substantia nigra) and a decrease in the terminals (caudate nucleus) of DA neurons, while no effect on NA accumulation in the ventromedial nucleus of the hypothalamus was found.

It is pointed out in this connection that in none of the turnover studies where labeled tyrosine was used, did amphetamine affect plasma or brain tyrosine-^{3}H and endogenous concentrations, as has also been reported by RAPIN et al. (1971). However, LEONARD and SHALLICE (1971) have reported decreased levels of endogenous tyrosine in rat brain and plasma in contrast to an increase in brain tyrosine reported by TAGLIAMONTE et al. (1971a, b).

It is evident from this overview that differences exist in the reported effects of amphetamine on CA turnover. These differences may be due to the amphetamine doses used. Thus, it seems that low doses of amphetamine in rats cause an increase while higher doses cause a decrease or no change of brain CA-turnover. This is probably due to the fact that one or another of the effects of amphetamine on either uptake, release, synthesis, or degradation of CA or feedback modulation of CA neuron impulse flow (Subsect. B.3) may predominate at certain dose levels. Regional variations in turnover within the brain might be added to the list of causes for the differences in the results.

g) Uptake into and Release from Adrenergically Innervated Tissues of Amphetamine Itself

Amphetamine accumulates in slices of the mouse heart (ROSS and RENYI, 1966b) and perfused rat heart (THOENEN et al., 1968). Subcellular fractionation of the mouse heart showed that only 12% of the total amphetamine content is particle-bound 1 h after the amphetamine injection (OBIANWU et al., 1968). Mouse cerebral cortex slices (ROSS and RENYI, 1966c; ROSS et al., 1968) and rat brain homogenate (BALDESSARINI and VOGT, 1971) also accumulate amphetamine in vitro. Subcelluar fractionation of rat brains by differential or gradient centrifugation after systemic administration of amphetamine showed that amphetamine is mainly found in the high speed supernatant (PFEIFFER et al., 1969; WONG et al., 1972; FULLER et al., 1972b); only 7% is located in the synaptosomal fraction and less than 30% all together is particle bound (WONG et al., 1972).

Several studies indicate that the accumulation of the drug in the tissues might be largely extraneuronal. Thus, denervation (THOENEN et al., 1968; OBIANWU et al., 1968), incubation temperature (BALDESSARINI and VOGT, 1971), and metabolic inhibitors like ouabain (ROSS and RENYI, 1966c; THOENEN et al., 1968) do not affect amphetamine accumulation. Efficient neuronal membrane pump blockers for NA uptake like cocaine, desmethylimipramine (DMI), and protriptyline have little or no effect on the uptake of amphetamine (ROSS and RENYI, 1966b, c; ROSS et al., 1968; THOENEN et al., 1968; OBIANWU et al., 1968). Amphetamine-induced NA depletion in the rat brain and heart in vivo is poorly inhibited by DMI (BRODIE et al., 1968; LEWANDER, 1968a). Protriptyline only partially prevents amphetamine-induced release of α-methyl-NA from mouse heart in vivo (LUNDBORG and WALDECK, 1972). Recently, however, it has been reported that cocaine and DMI inhibits uptake of amphetamine into central NA neurons, at low concentrations of amphetamine (RUTLEDGE et al., 1973; AZZARO et al., 1974). It appears that amphetamine, to a small extent at least, is dependent on the NA transport mechanisms for its uptake by the adrenergic cell membrane.

Adrenal gland chromaffin vesicles take up amphetamine but to a slighter extent than tyramine and by a mechanism which is different from adrenaline (WAGNER et al., 1974). Tritiated amphetamine taken up into peripheral tissues or cerebral cortex slices was only poorly released by electrical stimulation (KOPIN and GORDON, 1963; BALDESSARINI, 1971).

h) Amphetamine Metabolites and Tissue CA

The metabolism of amphetamine is subject to large species variations (see Sect. I, Chap. 1.A). In rats amphetamine is metabolized by p-hydroxylation mainly in the liver and by subsequent β-hydroxylation in noradrenergic neurons. The p-hydroxylation step does not seem to be stereoselective, while hydroxylation in the β-position occurs only for *d*-p-hydroxyamphetamine in vivo (GOLDSTEIN and ANAGNOSTE, 1965; LEWANDER, 1971a). Direct β-hydroxylation of amphetamine can be demonstrated in vitro (GOLDSTEIN et al., 1964; TAYLOR, 1974), but is only detected in very small amounts in tissues in vivo and does not hitherto seem to be of importance for the pharmacologic actions of amphetamine (GOLDSTEIN and ANAGNOSTE, 1965; THOENEN et al., 1966; GROPPETTI and COSTA, 1969b; BRODIE et al., 1970; LEWANDER, 1971a).

p-Hydroxynorephedrine was first demonstrated in the rat heart (GOLDSTEIN and ANAGNOSTE, 1965) and cat spleen (THOENEN et al., 1966). Later this metabolite was also detected in adipose tissue (GESSA et al., 1969) and in the rat brain after larger doses of amphetamine (> 1–2 mg/kg) or repeated administration of small doses of the drug (GROPPETTI and COSTA, 1969a, b; COSTA and GROPPETTI, 1970; BRODIE et al., 1970; LEWANDER, 1970a, b, 1971a; CLAY et al., 1971; FREEMAN and SULSER, 1972). p-Hydroxynorephedrine is believed to be present in NA neurons and stored in NA storage vesicles. This is based on the findings that this metabolite of amphetamine is not detected in the rat liver (GOLDSTEIN and ANAGNOSTE, 1965), that it can be liberated from the splenic nerve by sympathetic nerve stimulation in cats (THOENEN et al., 1966), and that reserpine inhibits its accumulation into rat brain and heart, whether it is formed in vivo from amphetamine (LEWANDER, 1971a) or from p-hydroxyamphetamine (LEWANDER, 1971b). Intracerebroventricular administration of 6-hydroxydopamine reduces the brain content of p-hydroxynorephedrine after amphetamine (CATTABENI et al., 1973). Methamphetamine is dealkylated to amphetamine which in turn is p-hydroxylated in the rat. No evidence of the presence of p-hydroxynorephedrine nor of p-hydroxyephedrine was found in the rat brain following administration of methamphetamine (MORGAN et al., 1972a).

Studies on the duration of the NA depletion after *d*- or *dl*-amphetamine in rats showed that control levels of NA were not reached until several hours or days after the disappearance of amphetamine from the tissues and it was demonstrated that a maintained NA depletion correlated in time with the presence of p-hydroxynorephedrine in the brain, heart, and fat tissues (BRODIE et al., 1969, 1970; GROPPETTI and COSTA, 1969b; COSTA and GROPPETTI, 1970; LEWANDER, 1970a, 1971a).

Investigation of the detailed stoichiometrical relationship between NA depletion and the amount of p-hydroxynorephedrine present in rat brain and heart showed that these parameters were related after, but not before, the first 24 h after the amphetamine injection (GROPPETTI and COSTA, 1969b; COSTA and GROPPETTI, 1970; LEWANDER, 1971a). Thus, the initial decrease in tissue NA levels seemed to be caused by other mechanisms than NA displacement due to p-hydroxynorephedrine accumulation, while the persistent NA depletion might be exclusively caused by the presence of this metabolite. Supporting evidence for this interpretation has been presented: (1) *l*-amphetamine, which is not converted to p-hydroxynorephedrine, causes only a transient (<24 h) and less pronounced NA depletion than *d*-amphetamine (BRODIE et al., 1970; LEWANDER, 1971a; CLAY et al., 1971). (2) After pretreatment of rats with drugs that inhibit p-hydroxylation of amphetamine in the rat liver like desipramine (LEWANDER, 1969, 1971a; GROPPETTI and COSTA, 1969a, b; COSTA and GROPPETTI, 1970), iprindole (FREEMAN and SULSER, 1972), or SKF-525A (CLAY et al., 1971), no p-hydroxynorephedrine was present in the tissues after *d*- or *dl*-amphetamine administration and the NA depletion was of shorter duration (<24 h), than after *d*- or *dl*-amphetamine alone. (3) In the guinea pig p-hydroxylation of amphetamine does not occur. Thus, p-hydroxynorephedrine is not formed in this species and the brain NA depletion after amphetamine corresponds in time with the presence of amphetamine itself (COSTA and GROPPETTI, 1970; LEWANDER, 1970b, 1971c). (4) In peripheral noradrenergically innervated tissues, p-hydroxynorephedrine follows the NA containing vesicular fraction in density gradients, while amphetamine is found in the supernatant (JACQUOT et al., 1971).

Administration of p-hydroxyamphetamine or p-hydroxynorephedrine (in sufficient doses) to rats and mice causes depletion of heart, spleen, adipose tissue, and brain NA (MAITRE and BRUNNER, 1967; LEWANDER, 1969, 1971b; GROPPETTI and COSTA, 1969b; GESSA et al., 1969; BRODIE et al., 1969, 1970; COSTA and GROPPETTI, 1970; MCCULLOUGH et al., 1970; CLAY et al., 1971; RAPIN et al., 1973b). In rats chemically sympathectomized by 6-hydroxydopamine, less p-hydroxynorephedrine accumulated in peripheral sympathetically innervated tissues (JACQUOT et al., 1973a). Inhibition of NA uptake (DMI, cocaine, etc.) inhibits the accumulation of pOH-amphetamine into peripheral NA neurons (JACQUOT et al., 1973b).

p-Hydroxynorepinephrine does not seem to accumulate in DA neurons, which lack DA-β-hydroxylase, the enzyme required for its formation. p-Hydroxyamphetamine, on the other hand, is present in the rat brain, although for a short (<24 h) period of time after administration of amphetamine (COSTA and GROPPETTI, 1970; LEWANDER, 1971a). After administration of amphetamine, the highest concentration of the p-hydroxylated metabolite is found in the caudate nucleus (CATTABENI et al., 1973). Brain DA is decreased after p-hydroxyamphetamine administration in rats (LEWANDER, 1969, 1971b) and mice (MCCULLOUGH et al., 1970) and therefore p-hydroxyamphetamine might possibly be of importance for the depletion of brain DA, particularly after chronic amphetamine treatment of rats (LEWANDER, 1968b, 1971a). The rat brain HVA concentration does not change after peripheral p-hydroxyamphetamine injections (JORI and BERNARDI, 1972; JORI and DOLFINI, 1974). After methamphetamine injections brain and heart NA is decreased, but the depletion of brain NA does not last for several days as after amphetamine. Further, there is no evidence for the presence of β-hydroxylated metabolites of methamphetamine in the rat brain (MORGAN et al., 1972a).

Intracisternally injected p-hydroxyamphetamine or p-hydroxynorephedrine causes a sustained decrease in rat brain NA (BREESE et al., 1970; MCCULLOUGH et al., 1970; HITZEMAN and LOH, 1972; TAYLOR and SULSER, 1973) and, in some experiments, in the brain DA concentration (HITZEMAN and LOH, 1972). Release of brain NA and DA by intracisternally administered p-hydroxynorephedrine has been shown by a decrease in labeled CA and an increase in methoxylated and deaminated CA metabolites (HITZEMAN and LOH, 1972). Similar results for brain NA were obtained after intraventricular administration of both p-hydroxyamphetamine and p-hydroxynorephedrine by TAYLOR and SULSER (1973). In their experiments, however, deaminated NA metabolites decreased. The tyrosine hydroxylase activity was reported to increase in the rat brain after p-hydroxynorephedrine (HITZEMAN and LOH, 1972). NA synthesis was, however, decreased in vivo after p-hydroxynorephedrine alone or in combination with amphetamine (HITZEMAN and LOH, 1974). In this study, marked effects of p-hydroxynorephedrine on both synthesis, release, and metabolism of brain NA were observed. The half-life of p-hydroxynorephedrine in the rat brain after intracerebral injections seems to be shorter (1.5–5 h; EL-GUEDRI et al., 1973; TAYLOR and SULSER, 1973) than after peripheral administration of the drug or its precursor (appr. 22 h; COSTA and GROPPETTI, 1970; LEWANDER, 1971a). It seems reasonable to assume that the free drug disappears more rapidly than the portion that is incorporated into the NA neurons.

It is not likely that formation of p-hydroxyamphetamine or p-hydroxynorephedrine are required for amphetamine to exert its acute CA mediated effects on, e.g., blood pressure, body temperature, motor activity. However, p-hydroxynorephedrine can be released from NA neurons by nerve impulses (THOENEN et al., 1966) and by amphetamine itself (BRODIE et al., 1970; LEWANDER, 1971b) as a false transmitter, and this metabolite has been implicated in the development of tolerance to amphetamine in rats (see Subsect. C.III.4 below).

i) Summary and Conclusions

There is ample biochemical evidence that amphetamine causes release of CA from peripheral and central NA and DA neuron terminals. The exact mechanism for the release is not known, but it seems to be different from the release induced by nerve impulses. Inhibition of (re-)uptake of CA by the axonal membranes of NA and DA neurons and by vesicular membranes within the CA neuron terminals might contribute to the increase in the amounts of transmitters in the synaptic clefts. Similarly, decreased inactivation of the CA by deamination, at least at high tissue concentrations of amphetamine, leads to the same results.

Increased stimulation of pre- and/or postsynaptic CA receptors caused by the increased amounts of extraneuronal CA might affect the rate of synthesis and turnover of NA and DA in the brain CA neurons by negative feedback regulatory mechanisms.

The transient depletion of the tissue levels of NA and DA might thus be due to both enhanced release and reduced resynthesis and refilling of the storage particles. The accumulation of p-hydroxynorephedrine, a metabolite of amphetamine, into NA neurons of rats and certain other species further delays the restoration of control tissue levels of NA. Para-hydroxynorephedrine, incorporated into the NA storage sites, is released as a false transmitter on electrical stimulation of sympathetic neurons or by subsequent doses of amphetamine. These facts provide the biochemical basis for the proposed false transmitter hypothesis as one mechanism of amphetamine tolerance (cf. Subsect. B.II.1.g and C.III.4).

Although the hydroxylated metabolites of amphetamine, when injected, release and displace tissue CA, there is no evidence in support of the conclusion that these metabolites are normally mediating the acute central and peripheral pharmacologic effects of amphetamine.

2. Serotonergic (5HT) Transmission

a) Storage of 5HT

Reports of amphetamine effects on brain 5HT levels are not all together consistent. Thus, *in rats*, *d-* or *dl*-amphetamine in doses between 0.6 and 30 mg/kg given 0.5–16 h before sacrifice of the animals did not change the whole brain 5HT concentrations in the majority of the studies (PLETSCHER et al., 1964; CARLSSON et al., 1965; BEAUVALLET et al., 1967; SALAMA and GOLDBERG, 1969; GANGULI, 1969; COSTA and GROPPETTI, 1970; MAICKEL et al., 1970; REID, 1970a, b; SIMON et al., 1970b; TAGLIAMONTE et al., 1971a, b; LEONARD and SHALLICE, 1971; GÖRLITZ and FREY, 1972; MORGAN et al., 1972a; FULLER et al., 1972b). SALAMA and GOLDBERG (1969) and LEONARD (1972) reported a decrease in 5HT after amphetamine. An increase in 5HT in the whole brain or in the striatum, pons, and mesencephalon but no change in cortex, medulla oblongata respectively was found after 15 mg/kg amphetamine by BEAUVALLET et al. (1969, 1970). MCLEAN and MCCARTNEY (1961), however, reported a 20–30% increase in 5HT levels after repeated injections of 16–32 mg/kg or *chronic* treatment with *d*-amphetamine. Similar results were later reported by DIAZ and HUTTUNEN (1972). Addition of 3 mg/kg of amphetamine daily in the food for 6 months did not change brain 5HT (HERMAN et al., 1971b).

Methamphetamine in doses of 7–14.5 mg/kg given 1–6 h before sacrifice reduced brain 5HT by 25–47%; *d-* and *l*-methamphetamine were equally effective in this respect (MORGAN et al., 1972a; LEONARD, 1972). Determination of 5HT in different brain areas (OTT et al., 1971) showed that methamphetamine (5 mg/kg, 40 min before death) caused a reduction in cortex, olfactory bulb and hippocampus, while no change occurred in six other regions. GANGULY (1969) reported an increase in brain

5HT after methamphetamine, which was abolished by chlorpromazine pretreatment.

In mice LAL and CHESSICK (1964) reported *d*-amphetamine (25 mg/kg) to cause a decrease in brain 5HT by 30–40%. Subsequent studies reported brain 5HT to be increased by 20–30% after 4–30 mg/kg of *d*- or *dl*-amphetamine (SMITH, 1965; WELCH and WELCH, 1970; EVERETT and YELLIN, 1971). SCHUBERTH et al. (1970) found no change in brain 5HT 30 min after 15 mg/kg *d*-amphetamine, while there was a small increase at 3 h. No effect was found after 10 mg/kg *d*- or 20 mg/kg *l*-amphetamine (DUBNICK et al., 1973), or after 2.5 mg/kg *d*-amphetamine (CARLSSON et al., 1969). There was no change in *chicken* brain 5HT after 6 mg/kg *d*-amphetamine (SCHROLD and SQUIRES, 1971). *In dogs* PAASONEN and VOGT (1956) found a decrease in 5HT in the hypothalamus after 15–30 mg/kg of *d*-amphetamine and similar results were reported by LAVERTY and SHARMAN (1965) for dogs and *cats*. Methamphetamine in acute and chronic experiments either had varying effects on cat brain 5HT in different brain regions (FUNATOGAWA, 1965; UTENA, 1966) or did not change this amine (ELLINWOOD and ESCALANTE, 1970a).

The contents of tryptophan and 5HT in oesophageal ganglia of snail *(Helix pomatia)* were not changed by amphetamine (OSBORNE et al., 1974).

b) Uptake of 5 HT into Neurons and Platelets

Total body uptake and retention of 5HT in the mouse was not affected by amphetamine (AXELROD and INSCOE, 1963). Uptake and accumulation of 5HT into mouse cortex slices in vitro were inhibited by amphetamine in a concentration of 10^{-5} M (ROSS and RENYI, 1967a). Similarly, WONG et al. (1973) found that amphetamine competitively inhibits the high affinity uptake of 5HT into isolated rat brain synaptosomes. In vivo models for uptake studies include histochemical evaluations of 5HT accumulation into tuberoinfundibular DA-neurons in reserpine treated rats, which was blocked by amphetamine (FUXE et al., 1967) and accumulation of 5HT into central neurons after administration of 5HT intraventricularly in the brain, which was not blocked by amphetamine (FUXE and UNGERSTEDT, 1968). Depletion of brain 5HT by 4-methyl-α-ethyl-m-tyramine (H75/12; CARLSSON et al., 1969; LEONARD, 1972) or *l*-p-acetyldeoxyephedrine (DUBNICK et al., 1973) was not blocked by *d*- or *l*-amphetamine. Methamphetamine, on the other hand, potentiated the H75/12 depletion of brain 5HT (LEONARD, 1972). Methamphetamine caused a decrease in the accumulation of 5 HT into striatal slices, but not in slices of telencephalon, hypothalamus, or brain stem (MORGAN et al., 1972b).

In conclusion, amphetamines seem to affect the 5HT uptake mechanism in vitro but a higher concentration is required ($>10^{-5}$ M) than is required for inhibition of CA uptake ($>10^{-6}$ M; ROSS and RENYI, 1967a).

Blood platelets have been utilized as models for 5HT uptake—storage and release mechanisms. Amphetamine has been reported not to inhibit 5HT accumulation into rabbit platelets (DAPRADA et al., 1965) but to competitively inhibit 5HT uptake into human platelets ($K_i = 10^{-5}$ M) (LEMMER, 1973).

c) Release of 5HT

Only a few techniques for the demonstration of amphetamine-induced release of brain 5HT in vivo have been applied. FUXE and UNGERSTEDT (1970) reported that amphetamine caused a change toward extraneuronal accumulation of 5HT in reserpine-treated rats as evaluated by a histochemical fluorescence microscopic technique for visualization of monoamines. The use of cerebrointraventricularly located push-pull cannula in rats showed that after a single dose of amphetamine there was no increase in the release of 5HT–^{14}C, previously taken up into brain 5HT neurons, into the perfusion fluid (SPARBER and TILSON, 1972a). However, in chronically amphetamine-treated rats, amphetamine caused a release of 5HT.

Superfusion of isolated corpora striata from rats in vivo has been used for the demonstration of release of CA after amphetamine (see Subsect. I.1). In similar experiments 10^{-4} M concentrations of amphetamine in the perfusate caused a small release of 5HT (GLOWINSKI, 1970a, b). *d*-Amphetamine but not *l*- was shown to release 5HT taken up into central cortex slices (NG et al., 1970). D-amphetamine, 10^{-4} M, also induces release of 5HT from chopped pieces of tissue from the caudate nucleus, medulla oblongata, and cortex of the rat brain (RUTLEDGE et al., 1972b; AZZARO and RUTLEDGE, 1973). Amphetamine-induced liberation of 5HT from rabbit blood platelets has been reported by DA PRADA et al. (1965) and from human blood platelets by LEMMER (1973).

d) Enzymatic Inactivation of 5 HT

TAGLIAMONTE et al. (1971a, b) showed that amphetamine induces an increase in the concentration of brain 5HIAA, which might be secondary to the endogenous tryptophan loading of the brain caused by the drug (see above). In chicken, SCHROLD and SQUIRES (1971) obtained similar results. LEONARD (1972), however, found no change in rat brain 5HIAA after a single injection or after chronic administration of amphetamine (DIAZ and HUTTUNEN, 1972). The effect of amphetamine on brain 5-HIAA levels might be dependent on several factors including dose, time, and environmental temperature. Thus, SCHEEL-KRÜGER and HASSELAGER (1974) reported an increase in 5-HIAA in rat brain at 3 h after 5–15 mg/kg of amphetamine at 21° C but not at 13° C ambient temperature. An initial (1 h) decrease and a late (4 h) increase in rat brain 5HIAA was observed after 10 mg/kg *d*-amphetamine; these fluctuations were absent in chronically amphetamine-treated rats (LEWANDER, 1974b). A time dependency of the effect of amphetamine on brain 5HIAA might explain the inconsistent results in the earlier literature. No change in the brain 5HIAA concentration was found in dogs and cats (LAVERTY and SHARMAN, 1965). In vitro-reduced deamination of 5HT on incubation of chopped brain tissue was found with amphetamine (10^{-3} M; RUTLEDGE et al., 1972b). Amphetamine also inhibited the deamination of 5HT in aggregated rabbit blood platelets (DA PRADA et al., 1965).

e) Synthesis and Turnover of Brain 5 HT

In rats, *d*-amphetamine (5 mg/kg × 2) has been shown to cause an increase in 5HT turnover as measured by the slope of the decrease in brain 5HIAA after the MAO-inhibitor pargyline (REID, 1970b). This effect of amphetamine was interpreted possibly to be an effect of amphetamine-induced hyperthermia. Amphetamine (1 mg/kg) decreased the rate of accumulation of 5HIAA after probenecid, indicating a decrease in brain 5HT turnover in rats (GÖRLITZ and FREY, 1972).

In mice, SCHUBERTH et al. (1970) found that amphetamine (15 mg/kg) reduced the accumulation of brain 5HT after i.v. administration of labeled tryptophan. Amphetamine also retarded the disappearance from the brain of ^{3}H-5HIAA after intracerebral injection of labeled tryptophan in amphetamine-treated mice (HITZEMANN et al., 1970) and rats (DIAZ and HUTTUNEN, 1972). In contrast, determination of the decrease in brain 5HT after inhibition of its synthesis by p-chlorophenylalanine showed that amphetamine and methamphetamine accelerated the turnover as measured by this method (LEONARD, 1972). SCHUBERTH et al. (1970) were first to report that amphetamine causes an increase in the brain tryptophan concentration at 3 h after its administration in mice. This finding was confirmed with dose- and time-response curves and extended to include rats as well (SCHUBERTH and SEDVALL, 1972). In addition there was an increase in plasma tryptophan concentration after amphetamine, while there was no change in the liver. Independently, TAGLIAMONTE et al. (1971 a, b) and LEONARD and SHALLICE (1971) have confirmed that amphetamine, most markedly in rats reacting with extreme hyperthermia, induced increased brain and plasma tryptophan concentrations. This effect on brain tryptophan concentration is not confined to amphetamine but is seen after a number of various drugs (TAGLIAMONTE et al., 1971 b). Methamphetamine is also effective in this respect (LEONARD and SHALLICE, 1971; MORGAN et al., 1972b). No definite explanation for the effect of amphetamine on brain tryptophan has been put forward. However, the amphetamine-induced changes in brain tryptophan certainly complicate measurements of 5HT turnover. Reports dealing with the in vitro determination of the tryptophan hydroxylase activity in the rat brain tissue showed that methamphetamine or amphetamine had no effect (MORGAN et al., 1972b; KNAPP et al., 1974). Chronic administration of amphetamine in the drinking water for 3–7 days did not change the tryptophan hydroxylase activity in the rat forebrain tissue (HULME et al., 1974). However, striatal synaptosomes from amphetamine-treated rats showed a reduced conversion of tryptophan into 5HT when assayed in vitro (KNAPP et al., 1974).

f) Summary and Conclusions

The effects of amphetamine on brain 5HT storage, uptake, release, and metabolism were not very consistent and, in any case, far less pronounced than its effects on brain CA. However, there is evidence that amphetamine, at least in high tissue concentrations, causes a release of 5HT, a change in its turnover (via an increase in brain tryptophan), and a change of the 5HT synthesizing capacity of striatal synaptosomes. The findings of changes in brain tryptophan concentrations after amphetamine have to be further investigated with respect to its functional consequences.

In this connection it might be pointed out that several halogenated derivatives of amphetamine have profound effects on brain 5HT neurons (PLETSCHER et al., 1963, 1970; FULLER et al., 1965; DUHAULT and VERDAVAINNE, 1967; OPITZ, 1967; BOISSIER et al., 1969; MAICKEL et al., 1970; SANDERS-BUSH and SULSER, 1970; COSTA et al., 1971; EVERETT and YELLIN, 1971; CATTABENI et al., 1972; FULLER et al., 1972b; MORGAN et al., 1972a, b; WONG et al., 1973).

3. Cholinergic Transmission

ACh release has been measured by superfusion of an area of cat cerebral cortex in vivo by a cup technique. By this method amphetamine has been shown to increase

ACh release from the cat cortex (PEPEU and BARTOLINI, 1967; BEANI et al., 1968; HEMSWORTH and NEAL, 1968a, b). This effect of amphetamine seems to be due to an indirect activation of ACh neurons rather than to a direct presynaptic effect, since no effect of amphetamine was obtained after undercutting of the cerebral cortex (DEFFENU et al., 1970; PEPEU et al., 1970) or septal lesions (NISTRI et al., 1972). Administration of chlorpromazine (PEPEU and BARTOLINI, 1968; NISTRI et al., 1972), haloperidol (DEFFENU et al., 1970) or α-methyltyrosine (PEPEU et al., 1970; NISTRI et al., 1972) blocks the amphetamine-induced ACh release, while α-receptor blocking agents like phenoxybenzamine or phentolamine are without effect (DEFFENU et al., 1970; PEPEU et al., 1970) and the β-receptor blocking agent propranolol has varying effects (BARTOLINI and PEPEU, 1970; DEFFENU et al., 1970; PEPEU et al., 1970). These findings strongly suggest that amphetamine-induced ACh release is a CA-mediated phenomenon. ACh release, as measured with a caudate nucleus cup in cats, was not altered by amphetamine given i.p. (5 mg/kg) or in the superfusing medium (JONES et al., 1973). Brain acetylcholine concentrations have been reported to be decreased in rats after 2 mg/kg amphetamine (DOMINO and OLDS, 1972; DOMINO and WILSON, 1972) but only in young or very old animals (VASKO et al., 1974). The acetylcholine levels increased after amphetamine (10–20 mg/kg) in striatal tissue from adult female rats (CONSOLO et al., 1974; GLICK et al., 1974). CONSOLO et al. (1972), however, found no effect of amphetamine in brain acetylcholine and choline in mice. Chronically methamphetamine treated mice had increased acetylcholine levels (UTENA, 1966).

Single doses of amphetamine (5 mg/kg) to rats do not change the activities of brain choline acetylase or acetylcholinesterase (HO and GERSHON, 1972; MANDELL and KNAPP, 1972). However, chronic treatment with amphetamine for 1–2 weeks caused an increased choline acetylase activity (HO and GERSHON, 1972; MANDELL and KNAPP, 1972). In cats an increase in acetylcholine esterase activity in the substantia nigra has been reported (ELLINWOOD and ESCALANTE, 1970a). In vitro, amphetamine in concentrations of 10^{-3} M or more activated both enzymes (ELLINWOOD and ESCALANTE, 1970a). The choline-acetylase activity in the adrenal glands of chicken increased after methamphetamine administration (MANDELL and MORGAN, 1970; MANDELL and KNAPP, 1972).

4. Other Putative Transmittors

Brain γ-aminobutyric acid (GABA)-concentrations have been reported to be unchanged in mice after amphetamine (SKOLNIK et al., 1962). However, LEONARD and SHALLICE (1971) found an increase after 5 mg/kg of amphetamine or methamphetamine, but no change or a decrease after 1 mg/kg. Substance P levels in the dog brain were not changed by 15–30 mg/kg of amphetamine (PAASONEN and VOGT, 1956).

5. Secondary Transmission; Cyclic Adenosine-3′,5′-monophosphate (cAMP)

Since brain cAMP levels are increased by NA in brain slices, it might be expected that amphetamine should induce an increase in brain cAMP via release of NA. However, several authors have been unable to show an amphetamine-induced increase in whole brain cAMP in mice or rats, in spite of sophisticated methods of rapid sacrifice using coagulation or freezing of the brain (PAUL et al., 1970; SCHMIDT et al., 1972; FERENDELLI et al., 1972). In fact, amphetamine and p-hydroxyamphetamine, its main metabolite in the rat, were found to inhibit NA-induced activation of adenylcyclase in brain slices in vitro (PALMER, 1973). COSTA et al. (1973), however, reported that amphetamine increased cAMP in the rat caudate nucleus simultaneously with an increase in the DA turnover in that brain area.

Cyclic guanosine-3',5'-monophosphate (cGMP) was increased in mouse cerebellum by 10 mg/kg amphetamine (FERENDELLI et al., 1972).

Activation of phosphorylase b kinase and the phosphorylase activity (enzymes which are involved in β-receptor mediated glycogenolysis) in brain have been reported to be increased by amphetamine (BRECKENRIDGE and NORMAN, 1965; DRUMMOND and BELLWARD, 1970). Dextro- and levo-amphetamine were equipotent in increasing phosphorylase activity in the rat heart (MCNEILL and MUSCHEK, 1972). High doses of the drug, however, blocked the stimulating effect of NA or isoprenaline on this cardiac enzyme (MCNEILL and MUSCHEK, 1972).

6. Summary and Conclusions

Biochemical evidence has accumulated during the last 10–15 years showing that amphetamine interferes with storage, uptake, release, and inactivation of NA and DA in peripheral and central neurons. The functional consequences of the amphetamine-induced release of neuronal and adrenal CA will be described in later sections of this review. There is also some evidence indicating that high doses of amphetamine cause similar changes in 5HT containing neurons in the CNS.

Amphetamine-induced release of ACh from brain cortex in vivo, on the contrary, seems to be a secondary phenomenon mediated by the release of CA.

II. Functional Neuropharmacology

1. Effects of Amphetamine on Peripheral Noradrenergically Innervated and Other Tissues in Vivo and in Vitro

The effects of indirectly acting sympathomimetic amines including amphetamine on peripheral noradrenergically innervated tissues have been reviewed previously by TRENDELENBURG (1963) and MUSCHOLL (1966). In this section only the amphetamines and their basic metabolites will be considered.

a) The Cat Nictitating Membrane and the Rat Inferior Eyelid

Amphetamine (or methamphetamine) caused a prolonged contraction of the cat nictitating membrane and the response was potentiated after preganglionic (superior cervical ganglion) denervation (LOCKETT, 1950a) or at short times (<24 h) after postganglionic denervation (FLECKENSTEIN and BURN, 1953; RYALL, 1961). Between 30 and 72 h after postganglionic denervation the amphetamine-induced contraction gradually decreased and the response was absent at four or more days after denervation when the noradrenergic neuron had degenerated (LOCKETT, 1950b; FLECKENSTEIN and BURN, 1953; FLECKENSTEIN and STÖCKLE, 1955). The dose-response curve for the effect of amphetamine on the nictitating membrane at 7 days after decentralization was shifted to the left and the maximal response was elevated, showing that decentralization caused supersensitivity to amphetamine. However, 1 week after denervation of the nictitating membrane, the dose response curve was flattened and the maximal response was reduced, showing the importance of intact postganglionic noradrenergic neurons for this effect of amphetamine (TRENDELENBURG et al., 1962b).

BURN and RAND (1958) showed that reserpine treatment of cats prevented the effect of amphetamine on the nictitating membrane and this finding was soon confirmed (SCHMITT and SCHMITT, 1960; TRENDELENBURG et al., 1962a). However, if reserpine was given less than 4 h before amphetamine, a potentiation of its effect was observed, as was also the case with the pressor response to amphetamine (SCHMITT and SCHMITT, 1960). After dopa administration, the sympathomimetic effect of amphetamine was restored in reserpinized cats (MCGRATH and KETTELER, 1963). Most of the above-mentioned studies were performed on spinal preparations, which rule out a central effect of amphetamine as an explanation for the induced contraction of the nictitating membrane. Removal of the superior cervical ganglion did not immediately abolish the effect of amphetamine, nor did ganglionic blocking agents (RYALL, 1961). Close arterial injection of amphetamine into the superior cervical ganglion caused a contraction of the nicitating membrane (Subsect. B.2; KEWITZ and REINERT, 1954). However, even if amphetamine might have some intraganglionic effect, most of the evidence supports the concept that amphetamine is an indirectly acting sympatomimetic agent, i.e., the drug exerts its effect on the nictitating membrane via release of NA from the nerve terminals.

In addition to having an effect of its own on the nictitating membrane, amphetamine (<0.5 mg/kg) also potentiates the contraction caused by electrical stimulation of the efferent nerve (NAKAMURA and SHIMAMOTO, 1960; RYALL, 1961; DAY and RAND, 1963a; DODA et al., 1966). Interestingly, 0.5 mg/kg of amphetamine (and p-hydroxyamphetamine) blocked the effect of nerve stimulation in one study; there is, however, no ready explanation for this sympatholytic effect (DAY and RAND, 1963a). The effects of injected NA, A, and tyramine on the nictitating membrane were also potentiated by amphetamine (NAKAMURA and SHIMAMOTO, 1960; DAY and RAND, 1963a). Amphetamine also prevented the adrenergic neuron blocking effects of some experimental drugs (DODA et al., 1966), including epsilon-amino-caproic acid (ANDÉN et al., 1968).

The rat inferior eyelid has also been used as a model for studies of the effect of amphetamine on noradrenergic transmission. Thus the amphetamine-induced potentiation of the response elicited by nerve stimulation has been confirmed in this preparation (OBIANWU, 1969a, b; FOLLENFANT and ROBSON, 1970; BABULOVA et al., 1972). Similarly amphetamine prevented the effect of the adrenergic neuron blocking agent guanethidine (idem).

A most interesting study (OBIANWU, 1969a) with regard to the mechanism of action of amphetamine showed that the effect of amphetamine on the eyelid was inhibited by α-methyl-p-tyrosine (a TH inhibitor) and potentiated by nialamide (an MAO inhibitor). The effect of nerve stimulation, however, was not affected by any of these drugs. The effects of tyramine and p-hydroxyamphetamine (α-methyltyramine) were potentiated by nialamide but remained unaffected by α-methyltyrosine. OBIANWU (1969a), therefore, suggested that nerve stimulation, amphetamine, and p-hydroxyamphetamine (and tyramine) release NA from the adrenergic nerve terminals by three partially different mechanisms. Previous studies have provided indirect evidence for this contention. In amphetamine tachyphylactic cats, nerve stimulation of the nictitating membrane still caused contraction of it, as least at higher stimulation frequencies (COWAN et al., 1961; DAY and RAND, 1963b). After tachyphylaxis to p-hydroxyamphetamine, however, the effect of nerve stimulation on the nictitating membrane was almost abolished (DAY and RAND, 1963b). In addition there is a lack of cross tachyphylaxis between amphetamine and tyraminelike sympathomimetics (see below).

b) The Cat Iris

A single study describes in detail the effects of amphetamine on the iris (MARLEY, 1962). It was concluded that amphetamine dilated the iris by releasing peripheral NA and by inhibition of the parasympathetic tone. After reserpine administration this effect was lost. In chronically denervated irises (sympathetic + parasympathetic ganglionectomy) amphetamine caused miosis, which effect was blocked by BOL, suggesting it to be due to stimulation of serotonergic receptors.

c) The Guinea Pig Vas Deferens

Amphetamine in low concentrations potentiated the contraction of the vas deferens on electrical stimulation of the hypogastric nerve (DAY and RAND, 1963a; MORAES et al., 1973b), while the response was blocked by higher concentrations of the drug (DAY and RAND, 1963b). Amphetamine and p-hydroxyamphetamine also caused contractions of the vas deferens without simultaneous nerve stimulation in vitro (PATIL et al., 1967a; LAPORTE et al., 1969; BABULOVA et al., 1972). The two isomers of amphetamine and p-hydroxyamphetamine, respectively, were equipotent on this preparation (idem). Reserpine treatment of the animals blocked the effect of p-hydroxyamphetamine (PATIL et al., 1967b). In amphetamine treated preparations, reserpine caused a contraction of the vas deferens; this effect was abolished by previous reserpine administration in vivo or by addition of α-adrenergic blockers to the bath suggesting the release of NA (LAPORTE et al., 1969). Amphetamine also potentiated the contraction of the vas deferens induced by NA in vitro, both in the intact (MORAES et al., 1970b, 1973b; MORAES and CARVALHO, 1971) and in the denervated preparation (MORAES et al., 1970b; MORAES and CARVALHO, 1971). This is not surprising, since the vas deferens contains so-called short adrenergic neurons and sectioning of the hypogastric nerve is in fact a preganglionic denervation. Reserpine further potentiated the amphetamine-induced NA potentiation (MORAES et al., 1973a).

d) The Rabbit and Guinea Pig Intestines

DETRICK et al. (1937) were first to show effects of amphetamine on various portions of the gut, which were blocked by atropine. ÅSTRÖM (1949) and DAY and RAND (1963a) have demonstrated a blockade by amphetamine of the effects of nerve stimulation of the rabbit and guinea pig ileum. The nature of this antisympatholytic action of amphetamine remains unexplained. A similar sympatholytic effect has been noticed both in the cardiovascular system, the nictitating membrane, and the vas deferens (Subsect. C.III.1 and above).

e) The Guinea Pig Bladder

Amphetamine seems to have a direct effect on the smooth muscle of the bladder, since the amphetamine-induced contraction of this in vitro preparation could not be blocked by a series of various receptor blocking agents or by denervation, and no tachyphylaxis developed to this effect of the drug (HAWKE and CHESHER, 1970).

f) Other Peripheral Effects of Amphetamine Possibly Mediated Via Serotonergic Mechanisms

In mammals, VANE (1960) found that amphetamine, like 5HT, but unlike the CA, contracted the isolated rat fundus (ventriculi) strip. The *d*- to *l*-amphetamine ratio for this effect was approximately $^1/_2$ on a molar basis. LSD and tryptamine desensitization antagonized the 5HT-like effect of amphetamine. INNES (1963) and INNES and KOHLI (1969) investigated the 5HT-like effect of amphetamine on several in vitro preparations (e.g., rat fundus, guinea pig ileum, dog m. retractor penis, rabbit uterus) and confirmed VANE'S (1960) observation that the 5HT receptor blocking agent, BOL, and 5HT receptor desensitization abolished the amphetamine-induced contraction of these organs. Similarly, methysergide, another 5HT receptor blocking agent, blocked (NIELSEN and FREY, 1969a, b), while phenoxybenzamine, reserpine, cocaine, hexamethonium, atropine, and morphine did not alter (INNES, 1963; INNES and KOHLI, 1969; NIELSEN and FREY, 1969a, b) the actions of amphetamine. These authors, therefore, concluded that amphetamine acts on 5HT receptors of the D-type in the gut. VACCARI and VERTUA (1970) reconfirmed the antiamphetamine effect of LSD on the fundus preparation and could also demonstrate that the amphetamine-induced contraction of the rat fundus preparation was abolished by destruction of 5HT receptors with neuraminidase plus EDTA. Tachyphylaxis to the contraction of the rat fundus was observed after repeated applications of amphetamine and the responsiveness was restored by 5HT administration to the tachyphylactic preparation (NIELSEN and FREY, 1969a). These findings add up to the view that amphetamine might act through release of 5HT in the gut. The dog's dorsal metatarsal vein was shown to respond to amphetamine with a contraction in two phases: a rapid initial component being antagonized by cinanserin (5HT receptor blocking agent) and a delayed component being reduced by α-receptorblocking agents (CHENG and LONG, 1973). These authors also concluded that amphetamine acts directly or indirectly on 5HT receptors.

A series of biochemical studies have been devoted to the question, whether amphetamine acts on the same receptor as 5HT or not; the results so far seem to be suggesting that the proposed 5HT receptors do not accept amphetamine and 5HT in a similar manner (VACCARI, 1968; VACCARI and CUGURRA, 1968; VACCARI and VERTUA, 1970; VACCARI et al., 1971; DEANGELIS et al., 1974).

The frequency and amplitude of the contractions of the liver fluke *(fasciola hepatica)* increased after amphetamine, as after 5HT, and the amphetamine-induced effects were antagonized by BOL or methergoline, indicating a serotonergic action of amphetamine (CHANCE and MANSOUR, 1949; MANSOUR, 1957; BERETTA and LOCATELLI, 1968).

In conclusion, there seems to be evidence for some peripheral effects of amphetamine to be mediated via 5HT, receptors. The question, whether amphetamine and

5 HT act on the same receptor or amphetamine causes a release of 5 HT, has not been studied in sufficient detail, however, some findings seem congruent with an indirect mechanism of action.

g) *On Tachyphylaxis to the Sympathomimetic Effects of Amphetamine and its Metabolites*

Tachyphylaxis to the pressor effect of amphetamine (Subsect. C.III.1) develops rapidly; already after a second injection of the drug the response was considerably reduced (ALLES, 1933, 1939; DETRICK et al., 1937; CRISMON and TAINTER, 1938; PINKSTON et al., 1939; GUNN and GURD, 1940; MALORNY and ORZECHOWSKI, 1940; ORZECHOWSKI, 1941; WINDER et al., 1948; REINERT, 1958; HANNA, 1960; COWAN et al., 1961; DAY and RAND, 1963b; CESSION-FOSSION, 1963a; BHAGAT, 1965a, b; EBLE and RUDZIK, 1965; DAY, 1967; TAKASAKI et al., 1973). After 8–10 doses of the drug administered during a couple of hours, complete tachyphylaxis was present (WINDER et al., 1948).

HARVEY et al. (1968) found tachyphylaxis both to the pressor response and to the increase in plasma CA levels in anesthethized dogs. No tachyphylaxis developed in the conscious dog, however (BUYNISKI et al., 1974). Tachyphylaxis was also observed to the pressor and heart rate effects of p-hydroxyamphetamine (ALLES, 1933; CRISMON and TAINTER, 1938; DAY and RAND, 1963b; VALETTE et al., 1971), but not to p-hydroxynorephedrine (VALETTE et al., 1971).

Tachyphylaxis also develops to the amphetamine-induced increase in heart rate (COWAN et al., 1961; DAVEY and FARMER, 1963), and to the amphetamine-induced contraction of the aortic strip (HANNA, 1960; MAENGWYN-DAVIES et al., 1966) and the contraction of the nictitating membrane (COWAN et al., 1961; DAY and RAND, 1963b). The duration of the fully developed tachyphylactic stage for the amphetamine-induced pressor response in dogs has been reported to be more than 8 h (HANNA, 1960). Tachyphylaxis developed to both *d*- and *l*-amphetamine on blood pressure in dogs and a complete cross tachyphylaxis between the two isomers could be demonstrated (HARVEY et al., 1968). L-amphetamine tachyphylaxis was also observed on the rabbit aortic strip preparation (MAENGWYN-DAVIES et al., 1966).

Pretreatment with MAO inhibitors decreased by a factor of 10 the tachyphylactic dose of amphetamine on the rabbit aortic strip (MAENGWYN-DAVIES et al., 1966). The levels of NA in amphetamine tachyphylactic isolated guinea pig hearts did not differ from control levels (DAVEY and FARMER, 1963). Similarly, the heart levels of NA were unchanged after development of tachyphylaxis to the pressor response to amphetamine in cats (BHAGAT, 1965b) and rats (TAKASAKI et al., 1973), whereas the NA stores were decreased in tyramine tachyphylactic animals. Thus, amphetamine tachyphylaxis is apparently not due to depletion of the NA stores. Intravenous infusion of NA to amphetamine tachyphylactic cats, rats, and dogs only partially (COWAN et al., 1961; MAENGWYN-DAVIES and KOPPANYI, 1966; HARVEY et al., 1968) or failed to (CESSION-FOSSION, 1963a) restore the amphetamine response. The responses to tyramine in tyramine tachyphylactic animals in the same studies, however, were fully restored after NA administration. Thus, it seems that amphetamine and tyramine tachyphylaxis differ with regard to the mechanisms by which this phenomenon is produced.

This contention is supported by a series of studies on *cross tachyphylaxis* between different sympathomimetic agents. Although the results are somewhat contradictory, it seems that monophenolic phenylethylamines, like tyramine and α-methyltyramine (p-hydroxyamphetamine), on the one hand, and nonhydroxylated phenylethylamines, like amphetamine, ephedrine, mephentermine, on the other hand, show cross tachyphylaxis only within the respective groups (DAY and RAND, 1963b; DAY, 1967; MORAES and CARVALHO, 1968; MORAES et al., 1970a). Tyramine tachyphylactic animals do not show cross tachyphylaxis toward amphetamine (BHAGAT, 1965b; DAY, 1967; HARVEY et al., 1968) and vice versa (BHAGAT, 1965a, b; MORAES and CARVALHO, 1968). In animals treated with an MAO inhibitor, however, cross tachyphylaxis between amphetamine and tyramine was observed (MORAES et al., 1970a).

An interesting question is of course whether or not there is cross tachyphylaxis between amphetamine and its metabolites p-hydroxyamphetamine and p-hydroxynorephedrine. In the two available studies on this subject (DAY, 1967; VALETTE et al., 1971) the results are not in agreement. DAY (1967) showed cross tachyphylaxis to amphetamine in p-hydroxyamphetamine tachyphylactic pithed rats (pressor response), but no cross tachyphylaxis was observed if the drugs were given in the reverse order. However, VALETTE et al. (1971), studying the pressor response in the intact rat, observed complete cross tachyphylaxis between amphetamine and p-hydroxyamphetamine in both directions. Rats pretreated with p-hydroxynorephedrine did not develop direct tachyphylaxis but the responses to amphetamine and p-hydroxyamphetamine were markedly reduced; amphetamine or p-hydroxyamphetamine tachyphylactic rats, however, showed a full pressor response after administration of p-hydroxynorephedrine (VALETTE et al., 1971). From these experiments there is reason to suspect that p-hydroxynorephedrine might be acting directly on NA receptors, rather than indirectly, and this is supported by the finding that reserpine treatment does not reduce the pressor response to p-hydroxynorephedrine (SCHMITT and SCHMITT, 1970). This matter should be further investigated.

The studies on tachyphylaxis and cross tachyphylaxis as well as the studies on the mechanism of release of NA discussed in a previous Subsection (B.I.1) strongly suggest that nerve impulses and groups of indirectly sympathomimetic agents release NA from partly different intraneuronal pools (DAY, 1967; OBIANWU, 1969b).

The number of possible *mechanisms of tachyphylaxis* is not exhausted by considering depletion of different pools of presynaptic stores of NA, however. An extended list of possible mechanisms, reviewed in part previously by PATIL et al. (1970) might include:

1. Depletion of stores of NA, as mentioned above;

2. the agent inhibits its own uptake into the nerve terminals (FAWAS and SIMAAN, 1965);

3. incorporation of the agent, or a metabolite of it, as a false transmitter, which is released by subsequent doses of the drug. This mechanism might be of importance in the case of p-hydroxyamphetamine or amphetamine in the in vivo situation, at least in some species such as rats, cats, dogs, man (KOPIN et al., 1965; THOENEN et al., 1966; GROPPETTI and COSTA, 1969b; BRODIE et al., 1969; LEWANDER, 1968b, 1971a, d);

4. saturation or desensitization of the postsynaptic receptors (depolarization block; ORZECHOWSKI, 1941; WINDER et al., 1948; POHLE and MATTHIES, 1967);

5. the administration intervals of the sympathomimetic agent are too short for a complete metabolic disposition or elimination of the drug to occur; the tissue responses add to each other up to a maximal response, which cannot be augmented further (WINDER et al., 1948);

6. in the case of the intact animal, an increased blood pressure for instance, might be antagonized by reflex compensatory mechanisms. In addition, amphetamine seems to have a central depressor effect, which might contribute to a reduction of its peripheral pressor effect.

From the preceding discussion of amphetamine tachyphylaxis it is obvious that several of the listed mechanisms might be at work simultaneously. Further studies are required for a more complete understanding of these problems.

h) Summary of Conclusions

It appears to be established beyong doubt that amphetamine belongs to one of two, or several, groups of indirectly acting sympathomimetic amines. The drug has effects on all adrenergically innervated tissues, that have been studied. The mechanism for the amphetamine-induced release of NA from the nerve terminals seems to differ from the mechanism by which nerve impulses and certain other sympathomimetic amines (tyramines) release NA.

The development of tachyphylaxis to the sympathomimetic effects of amphetamine is common on repeated administration of the drug. Several different mechanisms of tachyphylaxis, rather than one, seem to explain this phenomenon.

In addition, there is evidence, that amphetamine might be a direct or, more probably, an indirectly acting serotonergic agent in the gastrointestinal and vascular systems.

2. Effect of Amphetamine on Sympathetic Ganglia

Amphetamine seems to stimulate the perfused superior cervical ganglion in cats, i.e., injection of 50–100 μg of the drug into the perfusion fluid caused contraction of the ipsilateral nictitating membrane (KEWITZ and REINERT, 1954). Tachyphylaxis developed to this effect. The tachyphylactic preparation became insensitive to a subsequent injection of acetylcholine, which also caused nictitating membrane contraction, but not to electrical stimulation of the preganglionic nerve. Later GOLD and REINERT (1960) reported a nicotinelike effect of amphetamine, namely blockade of ganglionic transmission. REINERT (1958) summarized and elaborated on this subject. Amphetamine induced reversal of the depression of the postganglionic nerve impulses caused by high frequency preganglionic electrical stimulation (DOWNING, 1972). This effect of amphetamine seemed to be due to stimulation of α-receptors within the ganglion. The effect of low-frequency electrical stimulation of the preganglionic neurons was blocked by amphetamine (DOWNING, 1972). No changes of the demarcation potential of the postganglionic nerve cell or of the depolarization or hyperpolarization potentials induced by carbachol of the postganglionic cell were caused by the drug. Therefore, DOWNING (1972) concluded that amphetamine acts in the ganglion on presynaptic α-receptors, which regulate the release of acetylcholine.

Evidently amphetamine affects the ganglionic transmission as well as monoaminergic transmission elsewhere in the body, and this might have to be taken into consideration in the interpretation of the mechanisms of action of the peripheral pharmacologic effect of the drug, including tachyphylaxis and tolerance phenomena. Although significant, the ganglionic effects of amphetamine are not clearly understood with regard to their detailed mechanisms of events. A reevaluation of the above findings in the light of the recent concepts of ganglionic mechanisms, particularly the dopamine-containing interneurons modulating ganglionic transmission, reviewed by GREENGARD and KEBABIAN (1974), might turn out to be fruitful.

3. Effects of Amphetamine on the Nerve Impulse Activity of Single Neurons in the Central Nervous System

a) Brain Stem Reticular Formation

Iontophoretic application of amphetamine onto single neurons in the rat brain stem was found to cause inhibition of activity in about 50% of the neurons tested (BRADLEY and WOLSTENCROFT, 1965). In subsequent studies it was found that amphetamine excited 11%, while approx. 40% were unaffected (BRADLEY, 1968). In 90% of the cases amphetamine and NA elicited the same response when the two compounds were tested on the same neuron (BOAKES et al., 1971). In reserpine-treated rats (BOAKES et al., 1971) or rats treated with a dopamine-β-hydroxylase inhibitor, FLA-63, or a tyrosine-hydroxylase inhibitor, α-methyltyrosine, amphetamine lost its effect, while the effect of NA was still maintained (BOAKES et al., 1972). The inhibitory effect of α-methyltyrosine on amphetamine-elicited excitation of neurons could be reversed by pretreatment of the rats with *l*-dopa (BOAKES et al., 1973). Taken together, these findings are in accordance with the contention that amphetamine acts by releasing presynaptically stored NA.

Systemic administration of amphetamine in rats stimulated the activity of most reticular single cell units (GROVES et al., 1974). Single neurons in the tectum opticum, excited by visual stimuli, became even more activated after i.v. injection of amphetamine in cats (STRASCHILL and HOFFMANN, 1969).

b) Hypothalamus

In the ventromedial and lateral hypothalamus, amphetamine also inhibits some neurons and excites others. About 50% were unaffected (KREBS et al., 1969). Very often these neurons showed the same response to NA as to amphetamine (KREBS and BINDRA, 1971). These observations are discussed further in relation to amphetamine-induced anorexia (Subsect. C.II.1).

c) Locus Ceruleus

The cell bodies within the locus ceruleus are mainly NA-containing and these cell bodies send their axons to cortical, limbic, hypothalamic, and cerebellar areas (UNGERSTEDT, 1971 b). Recordings of the impulse activity of single locus ceruleus cells showed that all tested neurons were depressed after intravenous administration of amphetamine, while neurons outside this area were either excited or unaffected (GRAHAM and AGHAJANIAN, 1971). Chlorpromazine antagonized this effect of amphetamine (GRAHAM and AGHAJANIAN, 1971). The inhibition of locus ceruleus cells might be due to feedback regulation of the impulse flow, elicited by stimulation of postsynaptic cells by NA released from the locus ceruleus cell terminals, which is in accordance with other data (CORRODI et al., 1967). Also *l*-amphetamine (approximately in similar doses as *d*-amphetamine) decreased the impulse activity of locus ceruleus cells (ROTH et al., 1973). Naloxone, morphine inhibition antagonizer of locus ceruleus cell activity, was without effect on the amphetamine-induced inhibition of these cells (KORF et al., 1974). This finding, and the fact that chlorpromazine antagonized amphetamine but not morphine (KORF et al., 1974) in these experiments, suggests that amphetamine and morphine affects the locus cells by different mechanisms.

d) Substantia Nigra

The cell bodies of DA-containing neurons are mainly located in the substantia nigra and the ventral tegmentum (see UNGERSTEDT, 1971b). Intravenous injection of 1.60 mg/kg of *d*-amphetamine caused a decrease by 50% of the impulse activity of all recorded nigral neurons (AGHAJANIAN and BUNNEY, 1973; BUNNEY and AGHAJANIAN, 1973; BUNNEY et al., 1973a). A *d*-/*l*-amphetamine ratio of 6:10 was observed for this effect of the drug (BUNNEY and AGHAJANIAN, 1973). The inhibition was seen after iontophoretic application of amphetamine onto these cells (AGHAJANIAN and BUNNEY, 1973). Lesioning of the DA neurons cranial to the DA neuron cell bodies abolished the effect of intravenous amphetamine (BUNNEY and AGHAJANIAN, 1973). Pretreatment of the rats with α-methyltyrosine antagonized amphetamine on this measure and addition of *l*-dopa reversed the effect of the tyrosine hydroxylase inhibitor (BUNNEY and AGHAJANIAN, 1973). A series of neuroleptic drugs including chlorpromazine, thioridazine, and haloperidol, antagonized the inhibition of the nerve impulse activity caused by amphetamine (BUNNEY and AGHAJANIAN, 1973; BUNNEY et al., 1973a, b). As for the effect of amphetamine on the locus ceruleus cells, these findings regarding the nigral cells have been interpreted as to be consistent with release of DA from the DA nerve terminals, which in turn causes a feedback inhibition of the impulse flow of the DA neuron. The neurophysiologic findings are thus in agreement with earlier biochemical data on, e.g., NA and DA turnover in the whole brain (see Subsect. A.1), and might also explain the decrease in axonal transport of leucine-^{3}H-labeled proteins in nigrostriatal neurons (VOIGTLÄNDER, 1974).

e) Caudate Nucleus

Cells in the caudate nucleus receive afferent input from the nigral DA neurons. Amphetamine, like iontophoretically applied DA, causes hyperpolarization of caudate cells in cats (FELTZ, 1970; FELTZ and DECHAMPLAIN, 1973). This hyperpolarization seems to be due to amphetamine-induced release of DA stored presynaptically in the DA nerve terminals. However, amphetamine continued to induce hyperpolarization even after destruction of the nigroneostriatal pathway by intracerebral administration of 6-OHDA or after pretreatment with reserpine plus α-methyltyrosine, i.e., after depletion and inhibition of resynthesis of the DA stores (FELTZ and DE CHAMPLAIN, 1973). These observations might indicate a direct effect of amphetamine on the postsynaptic striatal cells. However, in spite of the pharmacologic deafferentation of the striatal cells, it is not possible to rule out that small amounts of DA might remain in the terminals and be left available for release by amphetamine. It was not reported if the stereotyped activity after amphetamine could be observed in these pharmacologic deafferented cats.

Systemic administration of amphetamine (0.5–4 mg/kg) in rats caused an initial increase and 30 min later a depression of the spontaneous activity of single cell units in the caudate nucleus (GROVES et al., 1974). Haloperidol antagonized the depressant effect of amphetamine (GROVES et al., 1974).

f) Raphé Nuclei

The raphé nuclei in the brain stem contain the cell bodies of serotonin-containing neurons. Amphetamine caused almost a twofold increase of the firing rate in about 50% of these neurons, while the remaining cell were insensitive or only slightly excited (FOOTE et al., 1969). The amphetamine-sensitive neurons were mainly located in the caudal portion of the dorsal raphé nucleus. In LSD-treated rats all raphé neurons were silent and insensitive to amphetamine (FOOTE et al., 1969). This shows that amphetamine and LSD have different effects on the raphé neurons and that LSD is antagonistic to amphetamine on a certain proportion of the rat brain serotonin neurons.

g) Summary and Conclusions

Amphetamine excites, inhibits, or has no effect on single neurons within the reticular formation of the brain stem and hypothalamus. In some instances the amphetamine-induced effects are similar to those of iontophoretically applied NA. Studies using pharmacologic interaction with NA storage and synthesis and NA receptors indicate that these amphetamine effects are due to release of endogenous NA rather than to a direct effect on NA receptors. Amphetamine induces hyperpolarization of caudate neurons either directly or indirectly. The inhibition of the nerve impulses in central DA and NA neurons seems to be mediated by release of the endogenous transmitters and negative feedback regulatory mechanisms.

Available data indicate that amphetamine excites half the 5HT neuron population in the brain stem, while having no effect on the other half. Whether this excitatory effect is secondary to interference with 5 HT releasing mechanisms or a direct action on 5HT receptors and positive feedback regulation of the impulse flow or to yet another mechanism is not known.

4. Functional Effects of Amphetamine Directly Related to Release of DA, NA, and 5HT in the CNS

Three models for the study of the effect of drugs on specific monoaminergic functions in the central nervous system have been developed. Turning and rotation are seen in animals with unilateral lesions of nigrostriatal DA projections, and is used as a functional model for change in central DA transmission. The spinal flexor reflex is used for studies of NA transmission, and the spinal extensor reflex for the study of 5 HT transmission (see reviews by ANDÉN, 1970; FUXE and UNGERSTEDT, 1970).

a) *Amphetamine Induced Turning and Rotation in Rats and Mice after Unilateral Lesions in the Dopamine Neuronal Systems*

The nigroneostriatal dopamine neurons are uncrossed (see FUXE et al., 1969) and unilateral lesions of the dopamine system in combination with drugs affecting brain dopamine transmission cause turning of the animals to the ipsilateral or the contralateral side (ANDÉN et al., 1966; ANDÉN, 1970). Selective destruction of dopaminergic pathways within the brain was later made by the intracerebral injection of 6-hydroxydopamine (UNGERSTEDT, 1971a). Turning or rotation are either counted under visual observation (turns per minute), or registered automatically in a special rotometer (UNGERSTEDT and ARBUTHNOTT, 1970). This animal model permits direct studies of drugs affecting dopaminergic transmission within the brain.

Administration of amphetamine to unilaterally striatectomized animals caused ipsilateral turning and rotation in rats (ANDÉN et al., 1967; CHRISTIE and CROW, 1971a, b; BOULU et al., 1972; NAYLOR and OLLEY, 1972; GUILLEUX and PETERFALVI, 1974) and mice (VOIGTLÄNDER and MOORE, 1973a). D- and *l*-amphetamine were equipotent on this behaviour in one study (CHRISTIE and CROW, 1971a, b), but in another *d*-amphetamine was more potent than *l*-amphetamine (VOIGTLÄNDER and MOORE, 1973a). The para- and β-hydroxylated metabolites of amphetamine were without effect (BOULU et al., 1972). Methamphetamine was more potent than amphetamine (CHRISTIE and CROW, 1971a, b). Intrastriatal injection of amphetamine in rats caused turning of the head to the contralateral side (MCKENZIE et al., 1972).

A detailed study showed that restricted electrolytic lesions within the caudate nucleus, the putamen, and globus pallidus, were all effective in producing turning (NAYLOR and OLLEY, 1972). Lesions in substantia nigra electrolytically or by 6-hydroxydopamine injections also caused ipsilateral turning and rotation after amphetamine injection (UNGERSTEDT and ARBUTHNOTT, 1970; UNGERSTEDT, 1971a; MARSDEN and GULDBERG, 1973). Similar results were observed in rats after unilateral lesions in the mesencephalic reticular area (MARSDEN and GULDBERG, 1973) or the frontal cortex (GLICK and GREENSTEIN, 1973). Lesions in areas containing 5HT neurons (raphé nuclei) do not cause turning or rotation (MARSDEN and GULDBERG, 1973). Recently, JERUSSI and GLICK (1974) reported that amphetamine caused spontaneous rotation in nonlesioned rats, indicating a left-right imbalance in dopaminergic activity. Unlesioned rats rotating after amphetamine had lower levels of DA in the ipsilateral striatum, indicating an unequal DA release between the two sides (GLICK et al., 1974).

Experiments with drugs interfering with MA mechanisms given in combination with amphetamine in unilaterally lesioned rats and mice showed that turning or rotation is inhibited by α-methyltyrosine (ANDÉN et al., 1967; UNGERSTEDT, 1971a; MARSDEN and GULDBERG, 1973; VOIGTLÄNDER and MOORE, 1973a). FLA63, a DBH inhibitor, either potentiated (UNGERSTEDT, 1971a) or was without effect (MARSDEN and GULDBERG, 1973). Caffeine, which is presumed to supersensitize DA receptors by inhibition of the degradation of cAMP, potentiates amphetamine (FUXE and UNGERSTEDT, 1974). Dopamine receptor blocking agents like chlorpromazine, haloperidol, or spiriperidol (UNGERSTEDT, 1971a) inhibited the response. Reserpine, 48 h before test, potentiated the amphetamine-induced rotation (UNGERSTEDT, 1971a). Protriptyline and desmethylimipramine as usual (see ÄNGGÅRD, Chap. 1.A) potentiated and prolonged the effect of amphetamine (CROW and GILLBE, 1970) as did procyclidine (an anticholinergic agent) (MARSDEN and GULDBERG, 1973). Parachlorphenylalanine, a 5HT synthesis inhibitor, was without effect (MARSDEN and GULDBERG, 1973). Chlorpromazine and α-methyltyrosine blocked methamphetamine rotation, while FLA63 was without effect (CHRISTIE and CROW, 1971b). Taken together these results seem to agree that DA released by amphetamine is functionally active within the neostriatum, causing rotation in this model.

b) The Flexor Reflex

The flexor reflex in spinalized rats has been developed as a model for investigation of functional NA receptor stimulation in the central nervous system (ANDÉN, 1970). Amphetamine causes a marked increase in the flexor reflex (ANDÉN et al., 1967; ANDÉN, 1970; FUXE and UNGERSTEDT, 1970). Pretreatment of animals with α-methyltyrosine, reserpine, or the combination of the two drugs reduced or inhibited completely the amphetamine effect, indicating that amphetamine acts on NA receptors via release of NA from spinal neurons (ANDÉN, 1970). In addition, amphetamine facilitates flexor reflex activity also in spinalized cats (VALDECASAS et al., 1958; VYKLICKY and TABIN, 1964) and rabbits (VIALA and BUSER, 1971).

c) The Extensor Reflex

Facilitation of the extensor reflex in spinal rats pretreated with an MAOI is an indicator of functional 5HT receptor stimulation in the central nervous system. Sometimes an α-receptor blocking agent is also administered in order to prevent stimulation of the flexor reflex (ANDÉN, 1970; FUXE and UNGERSTEDT, 1970). With this preparation, controversial results have been reported for amphetamine; the drug facilitated the reflex in one study (FUXE and UNGERSTEDT, 1970) and was completely without effect in another (ANDÉN, 1970). In another model for studying functional 5HT receptor stimulation, the inhibition of the lordosis response in female rats (MEYERSON, 1968; ELIASSON et al., 1972), amphetamine has been shown to be active, possibly indirectly by releasing brain 5HT (see Subsect. C.II.3). Thus, the evidence for 5HT receptor-mediated effects of amphetamine are so far equivocal.

d) Summary and Conclusions

The effects of amphetamine in the models used for studying brain DA and NA mechanisms support the contention that amphetamine acts indirectly via release of DA and NA within the CNS.

It does not seem established yet whether or not amphetamine has a similar effect on central 5HT-mediated functions.

5. General Summary and Conclusions

Available studies on the functional neuropharmacology of amphetamine show that amphetamine exerts pharmacologic effects both in the periphery and in the CNS indirectly by the release of NA and DA. These results are thus in agreement with biochemical studies reviewed above, showing that amphetamine induces release of these transmitter substances. There is very little evidence of the previously held view that amphetamine acts directly on central CA receptors.

There are indications both in the peripheral and central nervous system that amphetamine might cause some functional effects by the release of 5HT from serotonergic neurons. Such a mechanism of action of the drug is, however, not well established as yet, but warrants further investigation.

C. Pharmacologic Effects of Amphetamine

I. Behavioral Effects of Amphetamine

1. Gross Behavior

Central stimulation caused by amphetamine is a term partly used for designating its effects on the level of the general motor activity of animals and its antagonism of the reserpine syndrome (TRIPOD et al., 1954; EVERETT et al., 1957; VAN ROSSUM et al., 1962; FISCHER and HELLER, 1967), and partly with reference to its effect on natural or drug-induced sleep or EEG activity. Generally there is an initial alerting effect of the drug accompanied by increased exploratory behavior, grooming, rearing, and forward locomotion, later by a decrease in these activities concomitant with various stereotype behaviors, with marked species varations (see RANDRUP and MUNKVAD, 1967a, b, 1968, 1970). On disappearance of amphetamine, these symptoms reappear in reverse order. Sometimes, when all signs of stimulation are gone, a slight after-sedation might be observed (UTENA, 1966). The above description stems from studies in rats (EHRICH and KRUMBHAAR, 1937; EHRICH et al., 1939; HAUSCHILD, 1938; SEARLE and BROWN, 1938; HEIMSTRA, 1962a, b; SILVERMAN, 1966; RANDRUP and MUNKVAD, 1967a, b; 1970; FOG, 1969; SCHIØRRING, 1971; NORTON, 1973), but similar gross behavioral effects have been observed in mice (e.g., GUNN et al., 1939; UTENA, 1966; RANDRUP and MUNKVAD, 1967a; SATINDER et al., 1970; RICHARDSON et al., 1972), guinea pigs (UTENA et al., 1959; UTENA, 1966; RANDRUP and MUNKVAD, 1967a; LEWANDER, 1971c; KLAWANS et al., 1972), rabbits (EHRICH et al., 1939; BRODIE and SHORE, 1957; LAVERTY and SHARMAN, 1965), cats (EHRICH et al., 1939; WHITE et al., 1961; LAVERTY and SHARMAN, 1965; NORTON, 1967; RANDRUP and MUNKVAD, 1967a; ELLINWOOD and ESCALANTE, 1970a; ELLINWOOD, 1971b; WALLACH and GERSHON, 1971; BELOZERTSEV, 1974), dogs (EHRICH et al., 1939; CHISTONI and BECCARI, 1940; ACCORNERO, 1947; LAVERTY and SHARMAN, 1965; WILLNER et al., 1970; BUYNISKI et al., 1974), monkeys (EHRICH et al., 1939; FITZGERALD, 1967; RANDRUP and MUNKVAD, 1967a, b; MACHIYAMA et al., 1970; ELLINWOOD, 1971b; KJELLBERG and RANDRUP, 1969, 1972) and sheep (EHRICH et al., 1939). Other conspicuous symptoms are salivation, mydriasis, panting, pilo-erection (e.g., EHRICH et al., 1939; CHISTONI and BECCARI, 1940; ACCORNERO, 1947; BRODIE and SHORE, 1957).

In chicken a peculiar behavioral syndrome has been observed composed of initial sleeplike behavior interrupted by eating, wing droop, certain postural changes, twittering, foot-shuffle, and aggressiveness (CLYMER and SEIFTER, 1947; DEWHURST and MARLEY, 1964, 1965a, b; SCHROLD and SQUIRES, 1971; SPOONER and FLAMING, 1972).

A paradoxical finding is the "sedative" or calming effect of amphetamine in hyperactive (or hyperkinetic) dogs (CORSON et al., 1971; ARNOLD et al., 1973), but not genetically nervous dogs (MURPHREE et al., 1974), and rats in which hyperactivity has been induced through pretreatment with desmethylimipramine plus benzoquinolizine (Ro 4-1284) (MATUSSEK and LINSMAYER, 1968; LADISCH et al., 1970a, b), or by morphine (AYHAN and RANDRUP, 1973). These conditions might be suitable animal models for deepening our studies on the effects of amphetamine in hyperactive children. No definite explanation for these "calming" effects or amphetamine has so far been advanced. Young rats seem to react differently to amphetamine than older or adult rats (HEIMSTRA and MCDONALD, 1962; ZIEM, 1970).

The effects of amphetamine on operant behavior, conditioned avoidance response, as well as social behaviors (i.e., aggressive behavior and sexual behavior), will be treated in later sections. First, however, the sites and mode of action of amphetamine with regard to its effects on the various aspects of motor behaviors in individual rats will be discussed. A subdivision of the various aspects of the excitatory effects of amphetamine is undertaken for two related reasons. According to brain lesion studies and differential effects of other drugs (pharmacologic tools) on the amphetamine-induced motor symptoms, different sites and mode of action have been suggested for different behavioral effects. Subsequently, specific methods have been developed in order to resolve these problems. In the following, "motor activity" refers to studies, where motility has been electronically recorded. Where methods claimed specific for studying "stereotyped behavior," "locomotor activity," or "exploratory behavior" have been used, these reports have been referred to under the appropriate headings.

a) Stereotyped Behavior

Stereotyped behavior has been studied by observation of the animals after amphetamine has been administrated systematically, intraventricularly or intracerebrally. Either the percentage of animals showing stereotypies in a group has been recorded or the animals have been scored according to a rating scale (e.g., QUINTON and HALLIWELL, 1963; COOK and SHANBERG, 1970; NAYLOR and COSTALL, 1971; ELLINWOOD and BALSTER, 1974; KLAWANS et al., 1974). The rat has, however, become the preferred species for the detailed analysis of the biochemical, neuroanatomical, and neuropharmacologic studies of the site and mechanism of action of amphetamine in relation to the induced stereotypies. The motor behavior effects of amphetamine seem to be related to an amphetamine-induced stimulation of dopaminergic receptors in the brain as suggested by VAN ROSSUM (1963). The early conclusions (VAN ROSSUM et al., 1962, 1963; VAN DER SCHOOTT et al., 1962; SMITH, 1963, 1965) that amphetamine acted directly on central NA or DA receptors have largely been abandoned. Now, 10 years later, the stereotypies are believed to be due to an amphetamine-induced release of dopamine from presynaptic dopaminergic nerve terminals in the neostriatum, mesolimbic areas, and the paleostriatum. The evidence for this view will be presented below.

Previous reviews on stereotyped behavior have been compiled by MUNKVAD et al. (1968), RANDRUP and MUNKVAD (1968, 1969, 1970, 1972a, b), and FOG (1972). Reviews of stereotyped behavior induced by other central stimulants in addition to amphetamine have appeared (FOG, 1972; SCHEEL-KRÜGER, 1972b).

α) Behavior that Becomes Stereotyped or Compulsive after Amphetamine Administration

As mentioned above, increased exploration, locomotion, sniffing, repetitive head-movements (swaying), licking, and gnawing are the most prominent signs of amphetamine intoxication in rodents. At the height of the effects of amphetamine only sniffing, licking, and gnawing are observed, while exploration and locomotion disappear.

Whereas rats usually display compulsive stereotyped behavior in any environment without interruptions or disturbances, guinea pigs (Lewander, 1971c) and dogs (Wallach et al., 1971) cease with their stereotypies when distracted. Similar constrictions of the behavioral repertoires are also observed in other species. Stereotypies also appear in more highly integrated behaviors such as lever pressing in operant conditioning situations (Randrup and Munkvad, 1970; Lyon and Randrup, 1972). Thus, rats might continue to press the lever even if the reinforcing consequences are removed, and, occasionally, stereotyped gnawing of the lever occurred. Evidently, stereotypies might interfere with the interpretation of the effects of amphetamine in experiments with lever-operated reinforcements. Rats of various ages react differently to amphetamine. Below 12 days of age only sniffing and licking are present and the full adult pattern of stereotyped behavior is not observed until 35 days of age (McGeer et al., 1971; Lal and Sourkes, 1973). In addition, potentiation of amphetamine-induced stereotyped behavior by scopolamine (see below) does not occur until 30 days of age (McGeer et al., 1971). With medium doses of amphetamine, 2–5 mg/kg, even the locomotor activity becomes stereotyped (Lat, 1965).

β) Difference in Potency Between *d*- and *l*-Amphetamine

There is general agreement that *d*-amphetamine is more potent than the *l*-isomer in inducing stereotyped behavior, but the reported quantitative differences vary widely. Twice the amount of *l*-amphetamine was found equivalent to the *d*-form in MAO treated rats (Taylor and Snyder, 1970, 1971; Snyder et al., 1972), while in unpretreated rats there was a 4–10-fold difference (Scheel-Krüger, 1972a; Ellinwood and Balster, 1974; Costall and Naylor, 1974). Intracerebroventricularly administered *d*-amphetamine is also more potent than *l*-amphetamine (Segal et al., 1974). In dogs there was a *d*/*l*-ratio of about 1.4 (Wallach et al., 1971) and in cats about 1 (North et al., 1974).

γ) Effects of Other Drugs on Amphetamine Induced Stereotyped Behavior

The results pertinent to this topic are summarized in Table 1. The results with catecholaminergic receptor blocking agents show that neuroleptic drugs and particularly those with DA-receptor blocking actions are very effective antagonists of amphetamine-induced stereotyped behavior in rats, mice, guinea pigs, cats, and dogs. Adrenergic α- and β-receptor antagonists are mainly without effect. The few instances of disagreement are probably due to differences in dosage of the blockers in comparison with the dose of amphetamine given and further due to the problem of metabolic interactions between the drugs sometimes resulting in potentiation and prolongation of the amphetamine effect. The introduction of selective inhibitors of CA synthesis has shown that only drugs that inhibit tyrosine hydroxylase (α-methyltyrosine) are antagonistic, while dopamine-β-hydroxylase inhibitors are ineffective, or potentiate and prolong the action of amphetamine due to inhibition of its metabo-

Table 1. Effects of other drugs on intensity and duration of amphetamine-induced stereotyped behavior (SB). Symbols: No effect: 0, intensity increased: ↑, decreased: ↓, duration reduced: ←, prolonged: →

Drug	Intensity	Duration	References: Rat	Mouse	Guinea pig	Cat	Dog
Neuroleptic drugs							
Chlorpromazine[l]	↓	←	11, 14, 26, 28, 29, 30, 31, 37[m], 41[l], 51, 52, 55, 63, 75[p]	49 88[q]	79	23	92
			26[i]				
Chloraciazine	↑	→	40				
Fluphenazine	↓		8[e], 29, 30				
Trifluoperazine	↓		11[a], 30, 38, 39[l], 50[a]	3 89[q]	79		
Perphenazine	↓		1, 14, 29, 30, 36, 38, 39[e] 51, 59, 60	49			35
Thioridazine	↓	→	14, 45, 51	49			
Thioproperazine	↓		14, 38, 39[e], 43, 59				
Chlorprothixene[l]	↓	←	26, 28, 30, 36, 38	88[q]			
	↑	→	26, 36, 38[i, l], 51				
Haloperidol	↓		8[e], 9, 10, 11, 13, 14, 29, 30, 31, 38, 41[l], 51, 63	49 89[q]	79	4 5	92
Aceperone	0		29, 30				
	↑			3			
Pimozide	↓		15, 31				35
Spiramide	↓		29, 30	3			
Chlothiapine	↓		8[e], 30				
Clozapine	↓		8[f]				
	0		45				
Metoclopramide	↓		8[f], 11, 30				
Oxypertine	↓		8[e], 30, 58, 59				
Promazine[l]	(↓)	→	14, 29, 30				
Others	↓		14, 28, 29[k], 30[k], 36[l], 38[l], 39[l], 51, 63	49			
α-receptor blocking agents							
Phenoxybenzamine	0		14, 45, 53, 63				
	(↑)			3			
Phentolamine	0		11, 14, 28, 63				
Dihydro-ergotamine	0		63				
Yohimbine	↓		77				
	0						67
β-receptor blocking agents							
Pronethanol	0		53, 63				
Dichlorisopro-terenol	0		28, 63				
Propranolol	0	←	11, 14				
	↑	→	28, 76				
Prindolol	↑	→	76				
Practolol	0		76				

Table 1 (continued)

Drug	Intensity	Duration	References: Rat	Mouse	Guinea pig	Cat	Dog
5HT-receptor blocking agents							
Methysergide	0		58	78			
	↑				83		
Cyproheptadine	0		58				
BOL-148	↓		58				
Benanserine	0		58				
Others	0		58				
Drugs affecting storage of monoamines							
Reserpine	↑	←	6[b], 19, 20, 28, 57, 69, 72	49	25		
	0		29, 38, 50, 62, 63, 64	66 88[q]			
	↓					81	
Tetrabenazine	↓		29				
		←	28				
	↑[o]		74[o], 24[o]				
	0			66 89[q]			
Reserpine + FLA 63	↑		6				
Reserpine + α-MD	↓		57				
Reserpine + α-MT	↓		69, 72				
CA synthesis inhibitors							
Tyrosine-hydroxylase inhibitors							
α-methyltyrosine (αMT)	↓		17, 18, 22, 56, 61, 62, 68[n] 69[n], 70[n], 71[n], 72, 84, 85	3,91	79	5,81	
mI-tyrosine	↓		85				
mBr-tyrosine	↓		85				
Dopadecarboxylase inhibitor							
α-methyldopa	↓		17, 18, 57, 84, 85		79		
Dopamine-β-hydroxylase inhibitors							
Disulfiram[l]	↑	→	16, 38[l]			16	
Diethyl-dithiocarbamate	0	→	38[l], 39[l], 44, 47, 62, 65		65		
FLA 63[l]	0		6				
Drug combinations							
α-MT + *l*-dopa	0		84				
	↑		61				
α-MT + nialamide	0		22				

Table 1 (continued)

Drug	Intensity	Duration	References				
			Rat	Mouse	Guinea pig	Cat	Dog
MA precursors							
l-tyrosine	0		84				
l-dopa	↓		12				
5HTP	0		84				
	↓			83	25		
αMMT	0		28, 84, 85				
MAO inhibitors							
Pargyline	↑		11, 28				
Iproniazid	↑		11, 17, 73				
Nialamide[l]	↑		11, 28, 73				
Tranylcypromine	↑		11		79		
Others	↑		11				
		→	38[l]				
Tricyclic antidepressants							
Imipramine	↑	→	26, 40, 47, 51, 75[p]	49			
Desmethyl-imipramine[l]	↑	→	26, 51	49			
Chlorimipramine[l]	↑	→	26, 51				
Amitriptyline[l]	↑	→	51	49			
Nortriptyline[l]		→	51				
Iprindole[l]	↑	→	11[g]				
Others		→	51				
	0			49			
Cholinergics							
Physostigmine	↓		1		33		
Oxotremorine	↓[(h)]		1				
	0[(h)]		14				
Arecoline	↓		1, 9, 10				
Pilocarpine	0		14	89[q]			
Anticholinergics							
Atropine[l]	↑		7[d], 10, 11				
	0						86
Scopolamine	↑		1, 48		33		
Hyoscyamine	↑		1, 11				
	0						86
Methylatropine	0		1				
Benztropine	↑		11, 54				
Benactyzine	↑		1, 11				
Benzhexol	↑		1, 11, 54				
Caramiphen	↑		1				
Orphenadrine	↓–↑		7[d], 11, 54				
Phenglutarimide	↓–↑		7[d], 11				
Others	↑		11				
Antihistaminics							
Promethazine	↑	→	26, 38[l]				
Chlorpheniramine	↑		11, 54				
Diphenhydramine	↑		11, 54				
Others	↑		11				

Table 1 (continued)

Drug	Intensity	Duration	References: Rat	Mouse	Guinea pig	Cat	Dog
Sedatives and hypnotics							
Barbiturates	0		11, 28, 52, 53				
Meprobamate	0		11, 28, 52, 53				
Chloral hydrate	0 (↑)		11		79		
Chlordiazepoxide[l]	↑	→	2, 38[l], 39[l]				
	0		11, 28, 52, 53				
Diazepam[l]	↑	→	2, 38[l], 39[l]				
	0		11, 28, 52				
Nitrazepam	↑	→	2				
	0		11, 52				
Oxazepam	↑	→	2[c]				
Miscellaneous							
Li^+	0		38, 46				
Morphine	↓		11, 21[g], 93		79		
	↑	→	82				
Bulbocapnine	↓		80[p]				
	0			87, 89[q]	89[q]		
SKF 525 A[1]	(↑)	→	2				
Ethynyl estradiol		→	38				
Amantadine	↓:0		12, 54				
Fenfluramine	↓	→	3, 32[l]				
p-OH-amphetamine	0		42				
Caffeine	↑				34		

[a] Scopolamine or benzhexol restored stereotyped behavior.
[b] Reserpine + FLA-63: intense SB.
[c] And several other bensodiazepines.
[d] Intracerebral administration of drugs, initial inhibition, late potentiation.
[e] Reduced antiamphetamine effect by lesions in central amygdaloid nucleus.
[f] Not reduced by this lesion (cf. [e]).
[g] Rats became cataleptic.
[h] Reference 14 used less than 0.5 mg/kg, Reference 1 used 2 mg/kg.
[i] Depending on dose and time after injection.
[k] Tested 40 (Reference 29) and 63 (Reference 30) different potential neuroleptics.
[l] Shown to inhibit amphetamine metabolism in rats (Sect. I, Chap. 1.A.; LEWANDER and JONSSON, 1972).
[m] And several metabolites of chlorpromazine.
[n] α-MT + amphetamine cataleptic state; anticholinergics before amphetamine antagonizes catalepsia and causes hyperactivity, see text.
[o] Tetrabenazine given after disappearance of amphetamine stereotyped behavior causes reappearance of SB, which can be blocked by a series of neuroleptics, etc.
[p] Tested 26 different neuroleptic and antidepressant drugs (translation not available).
[q] Animals became cataleptic; anticataleptic properties of amphetamine against this drug were tested.

References

1. ARNFRED and RANDRUP, 1968
2. BABBINI et al., 1971
3. BERGER et al., 1973
4. COOLS and VAN ROSSUM, 1970
5. COOLS, 1971
6. CORRODI et al., 1970
7. COSTALL and NAYLOR, 1972
8. COSTALL and NAYLOR, 1973
9. COSTALL, et al. 1972a
10. COSTALL et al., 1972b
11. COSTALL et al., 1972c
12. COX and THA, 1973
13. DIVAC, 1972
14. DELRIO and FUENTES, 1969
15. ELLINWOOD and BALSTER, 1974
16. ELLINWOOD et al., 1972
17. ERNST, 1967a
18. ERNST, 1967b
19. EVERETT et al., 1957
20. FISCHER and HELLER, 1967
21. FOG, 1970
22. FOG, et al., 1967
23. FUNATOGAWA, 1964
24. FUXE and UNGERSTEDT, 1970
25. GOETZ and KLAWANS, 1974
26. HALLIWELL et al., 1964
27. HASSELAGER, et al., 1972
28. HERMAN, 1967
29. JANSSEN et al., 1965
30. JANSSEN et al., 1967
31. JANSSEN et al., 1968
32. JONSSON and GUNNE, 1972
33. KLAWANS et al., 1972
34. KLAWANS, et al., 1974
35. KLINGENSTEIN, et al., 1973
36. LAL, et al., 1971
37. LAL and SOURKES, 1972a
38. LAL and SOURKES, 1972b
39. LAL, et al., 1974
40. LAPIN and SCHELKUNOV, 1965
41. LEMBERGER et al., 1970
42. LEWANDER, 1971d
43. LESLIE and MAXWELL, 1964
44. MAJ and PRZEGALINSKI, 1967
45. MAJ et al., 1974
46. MATUSSEK and LINSMAYER, 1968
47. MAYER and EYBL, 1971
48. MCGEER, et al., 1971
49. MENGE and BRAND, 1971
50. MORPURGO and THEOBALD, 1964
51. MORPURGO and THEOBALD, 1967
52. MÜLLER and SCHOETENSACK, 1969
53. MUNKVAD and RANDRUP, 1966
54. NAYLOR and COSTALL, 1971
55. PHILLIPS and BRADLEY, 1969
56. PAPESCHI and RANDRUP, 1973
57. QUINTON and HALLIWELL, 1963
58. RANDRUP and MUNKVAD, 1964
59. RANDRUP and MUNKVAD, 1965a
60. RANDRUP and MUNKVAD, 1965b
61. RANDRUP and MUNKVAD, 1966
62. RANDRUP and MUNKVAD, 1967c
63. RANDRUP et al., 1963
64. RANDRUP and JONAS, 1967
65. RANDRUP and SCHEEL-KRÜGER, 1966
66. ROSS and RENYI, 1967b
67. SANGHVI and GERSHON, 1969
68. SAYERS, 1972
69. SAYERS and HANDLEY, 1973a
70. SAYERS and HANDLEY, 1973b
71. SAYERS and SPENCER, 1971
72. SCHEEL-KRÜGER, 1971
73. SCHEEL-KRÜGER, 1972a
74. SCHEEL-KRÜGER and JONAS, 1974
75. SCHELKUNOV, 1964
76. SIMON et al., 1972
77. SINGER et al., 1971
78. SOUTHGATE et al., 1971
79. SRIMAL and DHAWAN, 1970
80. TSENG and WALASZEK, 1972
81. WALLACH and GERSHON, 1972
82. VEDERNIKOV, 1970
83. WEINER et al., 1973
84. WEISSMAN and KOE, 1965
85. WEISSMAN et al., 1966
86. WHITE et al., 1961
87. ZETLER and MOOG, 1958
88. ZETLER et al., 1960
89. ZETLER, 1970
90. ZETLER and THÖRNER, 1973
91. ZIEGLER et al., 1972
92. WILLNER et al., 1970
93. SCHEEL-KRÜGER, 1973

lism. The amphetamine blocking effects of α-methyltyrosine may be overcome by administration of *l*-dopa, which by-passes the tyrosine hydroxylase step in the synthesis of catecholamines (RANDRUP and MUNKVAD, 1966). The potentiation of stereotyped behavior by mono-amineoxidase inhibitors might be explained by their actions on extra- and intraneuronal degradation of monoamines. However, these results do not contribute to a differentiation of which of the monoamines are involved. Drugs affecting CA storage (reserpine, tetrabenazine, α-methyl-m-tyrosine) do not antagonize amphetamine-induced stereotyped behavior, probably since they do not inhibit dopamine synthesis. Interestingly, the amphetamine stereotyped behavior is more intense, occurs with a shorter latency of onset and has a shorter duration in reserpine-treated animals. The mechanism for this reserpine-induced potentiation of amphetamine is not entirely known (see LEWANDER and JONSSON, 1973; ÄNGGÅRD, this volume). It is difficult to draw any conclusions from the results with inhibitors of the NA uptake mechanisms, since these drugs do interfere with the p-hydroxylation of amphetamine in rats and possibly other species. However, the results with these drug combinations show that amphetamine itself, rather than is p-hydroxylated metabolites, is responsible for the elicited stereotyped behavior (COSTA and GROPPETTI, 1972).

There is very little support for an involvement of 5HT neuronal systems in stereotyped behavior, except for a few studies claiming that stereotypies in guinea pigs might be influenced in opposing directions by the 5HT receptor blocker methysergide (potentiation) and 5HTP (inhibition) (WEINER et al., 1973; GOETZ and KLAWANS, 1974).

Anticholinergic drugs that are active in the CNS generally potentiate, while cholinergic drugs inhibit amphetamine-induced stereotypies. These results favor the hypothesis by ARNFRED and RANDRUP (1968) of a dopaminergic/cholinergic balance regulating various motor activities. In agreement with this hypothesis a few studies have shown that the amphetamine antagonism exerted by trifluoperazine or α-methyltyrosine is partially counteracted by anticholinergic drugs injected into the neostriatum (FOG et al., 1967, see also FOG, 1972) or systematically (COSTALL et al., 1972c; SAYERS, 1972).

δ) Neuroanatomical Site of Action of Amphetamine with Regard to Stereotyped Behavior

Intracerebral Injections of Amphetamine. Injections of microgram quantities of amphetamine into the basal ganglia induce stereotyped behavior in rats (MUNKVAD et al., 1968) and cats (COOLS and VAN ROSSUM, 1970; COOLS, 1971) more readily after reserpine pretreatment (FUXE and UNGERSTEDT, 1970). Still more localized injections, i.e., into the caudate, putamen, substantia nigra, or globus pallidus were effective, while the same amounts injected to cortex, thalamus, or hippocampus were ineffective (FOG and PAKKENBERG, 1971; COSTALL et al., 1972a). Unilateral injections of amphetamine into the striatum causes contralateral turning of the rat (MCKENZIE et al., 1972). Antagonists of amphetamine-induced stereotyped behavior such as chlorpromazine (FUXE and UNGERSTEDT, 1970), haloperidol (COOLS, 1971; COSTALL et al., 1972c), and the cholinergic agent arecoline (COSTALL et al., 1972c) are effective also when injected intraneostriatally. Anticholinergics are effective after intrastriatal injection (FOG et al., 1967; COSTALL and NAYLOR, 1972).

Although p-hydroxyamphetamine does not induce stereotyped behavior when given systematically (WALLACH and GERSHON, 1971), it may do so when injected into the striatum (FOG and PAKKENBERG, 1971a).

Intracerebroventricular infusion of amphetamine induces stereotyped behavior, although more drug is needed than for induction of locomotor activity (SEGAL et al., 1974).

Effects of Brain Lesions. Dopamine-containing neurons in the rat CNS have their cell bodies mainly in the pars compacta of the substantia nigra and their terminals in several neostriatal, paleostriatal, and mesolimbic brain areas (FUXE et al., 1969, see also COSTALL and NAYLOR, 1974). The short hypothalamic (and retinal) dopamine neurons are not considered in this context. Bilateral electrolytic lesions within the nucleus caudatus putamen were found to reduce amphetamine-induced stereotyped behavior (FOG et al., 1967, 1970; FUXE and UNGERSTEDT, 1970; NEILL et al., 1974); the reduction was approximately proportional to the size of the lesion. This could not be confirmed by DIVAC (1972). NAYLOR and OLLEY (1972), using more restricted lesions, could demonstrate that destruction of the globus pallidus was most efficient, whereas lesions in the caudate nucleus and putamen were less efficient in reducing amphetamine-induced stereotyped behavior. A detailed analysis of the effects of small bilateral lesions in various dopaminergically innervated areas on various components of the stereotyped behavior showed that destruction of the globus pallidus abolished all kinds of stereotyped behavior in spite of a persistent general hyperactivity; that destruction of tuberculum olfactorium abolished stereotyped sniffing and head movements, and destruction of the amygdala (central nucleus) abolished biting, gnawing, and licking, while other mesolimbic or neostriatal (caudatus putamen) were without effect on these behaviors (COSTALL and NAYLOR, 1974).

Lesions of DA nerve terminals by 6-hydroxydopamine injected into the caudate nucleus reduced stereotyped behavior of amphetamine, while similar lesions in the tuberculum olfactorium or nucleus accumbens did not (ASHER and AGHAJANIAN, 1974; CREESE and IVERSEN, 1974).

Bilateral electrolytic lesions of the main group of dopamine cell bodies in the zona compacta of the substantia nigra or the main nigroneostriatal bundle (COSTALL and NAYLOR, 1974) did not antagonize the stereotyped behavior after amphetamine (SIMPSON and IVERSEN, 1971; COSTALL et al., 1972b). FIBIGER et al. (1973) have reported a decrease in stereotyped behavior after 6-hydroxydopamine lesions in the substantia nigra bilaterally, but not after intraventricular injection of 6-hydroxydopamine. However, when 6-hydroxydopamine was given to rats treated with a monoamineoxidase inhibitor, which ascertained a more complete degeneration of brain DA neurons, the stereotyped behavior was reduced (BREESE et al., 1973; FIBIGER et al., 1973; FIBIGER, 1973). Administration of 6-hydroxydopamine neonatally in rats prevented amphetamine-induced stereotypies when tested as adults (CREESE and IVERSEN, 1972). Selective destruction of rat brain NA neurons had no effect on amphetamine-induced stereotyped activity, while selective DA neuron degeneration abolished the behavior (HOLLISTER et al., 1974).

Lesions in the pars reticulata of the substantia nigra, if anything, seem to potentiate stereotyped behavior (BAUM et al., 1972). Effects of amphetamine in rats with unilateral lesions of central dopamine neurons have been described (see Subsect. B.II.4).

Lesions in other regions, such as the frontal cortex in rats, enhance stereotyped behavior (IVERSEN et al., 1971). Bilateral diencephalic lesions in the medial forebrain bundle in rats causes enhanced gnawing but other stereotype movements are reduced (BOISSIER et al., 1971). Catalepsy caused by posterior hypothalamic lesions was antagonized by amphetamine (DELINI-STULA and MORPURGO, 1968). The chronic thalamic preparation (ablation of all brain structures above and in front of the thalamus including the striatum did not prevent rats from displaying stereotyped behavior (BORBÉLY et al., 1973; HUSTON and BORBÉLY, 1974). Even spinalized cats displayed movements after amphetamine, which were characterized as stereotyped and were antagonized by chlorpromazine (VALDECASASet al., 1958). Although there are contradictions in the literature on the effect of brain lesions on stereotyped behavior (probably due to failure in some studies to verify the extent and selectivity of the lesions), evidence is accumulating to show that the nigroneostriatal DA neurons are of major importance for this effect of amphetamine. Certain mesolimbic dopaminergically innervated structures might also be involved in components of this behavioral state.

ε) Correlations Between Stereotyped Behavior and Brain Catecholamines

The interaction of amphetamine with storage, release, uptake, and metabolism of brain NA or DA has been dealt with in Subsection B.I.1 above. A few studies have tried to correlate the behavioral effects with the biochemical measures. Thus, RANDRUP and JONAS (1967) and JONAS and SCHEEL-KRÜGER (1969) have shown that stereotyped behavior is related to the amphetamine-induced release of brain DA measured as increased levels of MTA after MAO-inhibition in reserpine-treated rats, rather than release of brain NA. Later, SCHEEL-KRÜGER (1972a) showed that release of brain CA rather than inhibition of their uptake was related to the appearance of stereotype motor activities. D-amphetamine, which is about 2–3 times more potent than *l*-amphetamine as inducer of stereotyped behavior, was also 2–3 times more effective as releaser of brain DA (SCHEEL-KRÜGER, 1972a).

COOK and SCHANBERG (1970) and SCHANBERG and COOK (1972) found good correlations between methamphetamine-induced stereotyped behavior and biochemical effects on release and uptake of brain NA in rats. In their studies brain dopamine was not included for examination, however. Therefore, their conclusion that stereotyped behavior is dependent on brain NA release does not appear to be well founded.

One strain of mice does not display stereotyped behavior and no increase in brain HVA was found, while in another strain both stereotyped behavior and an increase in brain HVA occurred (JORI and GARATTINI, 1973).

ζ) Other Factors Affecting Amphetamine-Induced Stereotyped Behavior

In rats previously isolated from social interaction for some time (muricidal rats) the stereotyped behavior was prolonged as compared with controls (KOSTOWSKI and CZLONKOWSKI, 1973). BORBERG (1974) reported that rats could be conditioned to show or not to show the usual stereotyped behavior pattern after amphetamine injections using electrical foot shocks.

A series of seven electroconvulsive shocks before administration of amphetamine in rats did not modify the stereotyped behavior (PAPESCHI et al., 1974).

Previous treatment with amphetamine, 2–5 weeks before, was followed by a more intense stereotyped behavior after the second amphetamine injection in rats (MAGOS 1969) and dogs (WALLACH and GERSHON, 1970). Thiamine deficient rats are more sensitive to amphetamine than controls (IWATA et al., 1969).

η) Stereotyped Behavior During Chronic Amphetamine Treatment

No definite signs of tolerance to the amphetamine-induced stereotyped behavior have been observed in mice (RICHARDSON et al., 1972; HITZEMAN et al., 1973), rats

(LEWANDER, 1968b, 1971d; SCHANBERG and COOK, 1972) and guinea pigs (LEWANDER, 1971c). UTENA et al. (1959) and UTENA (1966), however, reported decreased responsiveness with regard to stereotyped behavior in mice and guinea pigs.

No tolerance development for stereotyped behavior of amphetamine in cats has been reported (ELLISON et al., 1968; ELLINWOOD and ESCALANTE, 1970a, b; KNOLL et al., 1970; ELLINWOOD 1971a, b, 1972). However, ELLINWOOD (1971a) reported that cats treated chronically with methamphetamine showed more and more restricted behavioral repertoires, which might be regarded as tolerance for some manifestations of stereotyped behavior. At the same time other items became dominating. In the same study it was also shown that the methamphetamine injections conditioned the display of certain stereotyped behaviors; after substitution of saline stereotyped behavior was recorded for a short period after each injection.

From the studies in cats referred to above (see ELLINWOOD et al., 1972) and in monkeys (MACHIYAMA et al., 1970; ELLINWOOD, 1971b; ELLINWOOD and DUARTE-ESCALANTE, 1972; ELLINWOOD et al., 1974), it appears that chronic treatment with amphetamine induces more and more bizarre behavior and dyskinesias (hallucinatory psychotic state) differing from the acute effect and reminiscent of certain psychotic manifestations in humans. In this connection it might be mentioned that the inhibition of amphetamine-induced stereotyped behavior in rats has been used as a screening method for antipsychotic (neuroleptic) drugs (see JANSSEN et al., 1965).

θ) Summary and Conclusions

Amphetamine-induced stereotyped behavior has been observed in all species studied; however, its characteristics vary between species. This effect of amphetamine is reduced or abolished by pretreatment of animals with neuroleptic drugs and tyrosine-hydroxylase inhibitors, while it is potentiated by reserpine and anticholinergic agents. These and other findings indicate that the stereotypies are dependent on the amphetamine-induced release of newly synthesized DA and is under a dopaminergic cholinergic balance. Neuroanatomical studies have also shown that the stereotyped behavior is dependent on intact nigrostriatal and possibly mesolimbic DA neuron projections. Tolerance does not develop to amphetamine-induced stereotyped behavior. Instead, chronic amphetamine administration in certain species (cat, monkey) seem to produce increasingly bizarre behavior, reminiscent of a psychotic state in humans.

b) Motor Activity

α) Comments on Methodology

Several devices for measurement of amphetamine-induced increase in motor activity have been used. Some devices record mainly coordinated locomotion as with photoelectric cages (interruption of light-beams), leaping wheels, jiggle cages, pressing of micro-switches connected to floor squares, etc. Others types (e.g., Animex; SVENSSON and THIEME, 1969) are based on discrimination of changes in electromagnetic fields and are in addition able to pick up fine uncoordinated movements, such as various types of stereotypies. A comparison between the Animex and a photoelectric device by MAJ et al. (1971) clearly showed striking differences in results, which were of importance for the interpretation of the results.

β) Dose and Time Response Relationships

The dose-response curve in mice and rats almost (irrespective of the apparatus used) has the form of an inverted U (TAINTER, 1943; DEWS, 1953; BROWN, 1960; ADLER, 1961; KNOLL and VAJNOVSZKY, 1961; LAT, 1965; RECH and STOLK, 1970; TAYLOR and SNYDER, 1970, 1971). The explanation for this is the appearance of intense stereotyped behavior with little locomotion at higher dose levels of amphetamine (BAINBRIDGE, 1970; RANDRUP and MUNKVAD, 1970; and others).

The time response curve usually shows a peak value at 45–60 min after the injection (ADLER, 1961; DINGELL et al., 1967; LEWANDER, 1971d). Motor activity seems to be affected at comparatively lower doses of amphetamine as compared with other effects of the drug (VAN ROSSUM and SIMMONS, 1969; MAICKEL et al., 1970; MCKENZIE and SOROKO, 1972).

γ) The Relative Potency of *d*- and *l*-Amphetamine

SCHULTE et al. (1941) and NOVELLI and TAINTER (1943) seem to be the first to have compared the relative efficiency of *dl*-, *d*-, and *l*-amphetamine on motor activity. Different ratios were obtained depending on which measures were used (threshold dose $d/l \doteq 20$, maximal effect $d/l = 4$). The racemic compound was slightly less potent than the *d*-isomer. Recently, several studies have included *d*- versus *l*-comparisons. TAYLOR and SNYDER (1970, 1971) and SNYDER et al. (1972) reported *d*-amphetamine to be 10 times more potent than *l*-amphetamine with regard to motor activity in rats pretreated with MAO inhibitors and correlated this with a similar difference for NA uptake into brain NA neurons (see Subsect. B.I.1) while a 1–2:1 ratio was found for stereotyped behavior, which should be dependent on inhibition of DA uptake. A useful test for the relative importance of the involvement of NA or DA neurons in various effects of amphetamine thus seemed to be available. However, several subsequent studies have yielded varying results, with *d*/*l*-ratios of 1–2–4–10. The dose of the isomers (SVENSSON, 1971) and the species (mice, rats or cats; MAJ et al., 1972b; THORNBURG and MOORE, 1972, 1973a; NORTH et al., 1974) seem to be important variables. A rather low ratio (1–2) was obtained when amphetamine was infused intracerebroventricularly (SEGAL et al., 1974). Thus, there is general agreement that *d*-amphetamine is more potent than *l*-amphetamine but interpretations of a particular ratio to indicate the involvement of some specific neuronal mechanism seem unwarranted (cf. Subsect. B.I.1 and C.I.1.α).

δ) Effects of Other Drugs on Amphetamine Induced Increase in Motor Activity

The studies relevant to this issue are compiled in Table 2.

The importance of catecholamine neuronal mechanisms for the effect of amphetamine on motor activity is clearly evident. The amphetamine antagonism exerted by α-methyl-tyrosine, the catecholamine synthesis inhibitor, and the reversal of this antagonism by *l*-dopa (OFFERMEIER and POTGIETER, 1972), the antagonism caused by neuroleptic drugs and the lack of effect of amphetamine after selective degeneration of brain catecholamine neurons by intracerebroventricularly administered 6-hydroxydopamine, all favor this contention.

Two important issues have been much debated: (1) Whether amphetamine acts directly on brain catecholamine receptors or indirectly via release of brain catecholamines and, (2) whether dopamine or noradrenaline is responsible for the locomotor component of the increased motor activity. The finding of no effect or a potentiation

Table 2. Effects of other drugs on intensity and duration of amphetamine-induced increase in motor activity (photoelectric cages, leaping wheels, Animex motimeter, jiggle cage, Knolls apparatus, and others). Symbols: No effect: 0. Intensity, increased: ↑, decreased: ↓, Duration, reduced: ←, prolonged: →

Drug	Intensity	Duration	References: Rat	Mouse	Others
Neuroleptic drugs					
Chlorpromazine[a]	↑	→	8[c], 10[c], 72, 97[b]		
	↑			86	
	↓	→	9, 10, 97[b]		
	↓		43	3, 4, 11, 33, 34, 65, 84, 87, 100	
Thioridazine	↓		54	54	
Chloraciazine		→		47	
Fluphenazine	0 or ↓		78[i]		
Trifluoperazine	↓			79, 65	
	0			37[m]	
Perphenazine	0 or ↑		76[b], 77		
	↓		79	65	
Haloperidol	↓		79, 103	67, 103	
Aceperone	↓		79	79	
	0			37	
Pimozide	↓		53, 79	53, 83	
Spiramide	↓		79	79	
Penfluridol	↑			37[m]	
	↓			61	
Clozapine	↑		54		
	↓			54	
α-receptor blocking agents					
Phenoxybenzamine	↓		32, 53, 79	33, 34, 53, 79	
	↓ or 0		78[i]		
	0		54	4, 37	
Phentolamine	↓			84	
Dihydroergothamine	↓		79	113	
Dibozane	↓			34	
Dibenamine	0			34, 113	
Others	↓			113	
β-receptor blocking agents					
Pronethanol	↑			34	
Propranolol	↓		36, 108[i]	23, 24, 25, 55[b], 56[b]	
	↑	→	88[b], 108[i]		
	0			42	
Prindolol	↑	→	88[b]		
Practolol[k]	↓		108[i]		
	0		88[b]		
Alprenolol	↑ or ↓[g]		38		
Others	0 or ↓		36, 108[i]	25, 112	
5HT-receptor blocking agents					
Methysergide	0		87	46	
	↑			33	
	↓			89[b]	
Cyproheptadine	↓			31	
BOL 148	0			46	
Cinanserin	↓			41[b]	

Table 2 (continued)

Drug	Intensity	Duration	References: Rat	Mouse	Others
Drugs affecting storage of monoamines					
Reserpine	↑	←	13, 14, 22, 78[b,g], 93[g], 94[g], 95[g]	4	
	↑			3[b], 5, 33, 47, 69, 80[b], 86, 90, 91, 100	
	↓		32, 78[b], 85[g,j]	3[b], 81, 91[d]	
	0		69, 104	71, 84, 107, 112	16[r]
Reserpine + FLA 63	↑		14		
	↓		32		
Reserpine + αMT	↓		52, 96[b]	18, 99	
Tetrabenazine	↓		10, 13	112	
	↓ or 0		78[i]	106	
DMI + tetrabenazine	↓			106[q]	
Prenylamine	↓			33, 71	
Catecholamine-synthesis inhibitors					
Tyrosinehydroxylase inhibitors					
α-methyltyrosine (αMT)	↓		13, 17, 22, 39, 52, 58, 63[h], 78, 79, 85[g,i], 87, 96[b], 98	4, 18[b], 19[h], 31[b], 62, 64, 79, 87, 89, 99, 101, 102[s], 112, 103, 107[b], 109	
	0			37[m], 105[m]	
αMT + DMI	↓		17		
αMT + pargyline	↓		69	69	
αMT-pargyline + *l*-dopa	0		69	69	
Pargyline + αMT	0			18	
Pheniprazine + αMT	↓		96[b]		
	0			18	
αMT + *l*-dopa	0		58	4	
αMT + *l*-3-O-methyl-tyrosine	↓		58		
Dopa-decarboxylase inhibitors					
α-methyldopa	0			91	
Dopamine-β-hydroxylase inhibitors					
Disulfiram[a]	↓[d]		21		
	↓			51	
Diethyldithiocarbamate (DEDC)	↓		70	75[b]	
DEDC + nialamide			70		
FLA-63	↓			79	
	0			37, 99, 102[s]	
FLA-63 + dops	0			99	
U 14, 264[n]	0			64, 102[s]	
Tryptophanhydroxylase inhibitors					
Parachlorophenylalanine (PCPA)[l]	0		87, 98	31, 87, 89	
	↓			46	
	↑		50	67	
PCPA + 5HTP	0		50		

Table 2 (continued)

Drug	Inten-sity	Dura-tion	References		
			Rat	Mouse	Others
Aminoacid precursors to monoamines					
l-dopa	↑			57	
αMMT	0		98	4	
Monoamineoxidase inhibitors (MAOI)					
Pargyline	0			48, 89, 107	
Iproniazid	0		82		
	↑		92	12, 48, 90	
Nialamide	0		82		
Pheniprazine	↑			48	
Others	↑		78, 92		
Inhibitors of uptake of monoamines					
Tricyclic antidepressants					
Imipramine[a]	↑	→	8	47	
	0			48[b]	
	↓			68, 84	
Desmethylimipramine (DMI)[a]	↑	→	17, 110		
	↓			34, 84	
	0			48, 106	
Iprindole[a]	↑	→	29, 30, 62		
Others	↑	→	8		
Cholinergics					
Tremorine	↓			33	
Pilocarpine	↓			74[o]	
Methacholine	0			74	
Anticholinergics					
Atropine	↑			33, 34, 113	
Benactycine	↑			33	
Scopolamine	↑			113	
Others	↑			113	
Sedatives and hypnotics					
Barbiturates	0		78[i]		
	↓		78[i]	34[b]	
	↑			1, 3, 86[p]	
Meprobamate	↓			34	
	0			86[p]	
Chlordiazepoxide[a]		←	2[b]		
	↓	→	10	34[b]	
	↑			66, 86[p], 100	
	0			1	
Diazepam[a]		←	2[b]		
	↓			1, 3[b]	
	↑			66	
Nitrazepam		←	2[a, b]		
	↑			66	
Oxazepam		←	2[b]		
Others		←	2		

Table 2 (continued)

Drug	Intensity	Duration	References: Rat	Mouse	Others
Miscellaneous					
6-hydroxy-dopamine (6-OH-DA)	↓		24, 27, 111[t]		
	0 or ↑		26[f]		
MAOI + 6-OH-DA[e]	↓		27, 28		
DMI + 6-OH-DA[e]	↓		39		
p-hydroxy-amphetamine	0		49		
Amantadine	↓			59	
	0			101	
Fenfluramine	↓			6	
LSD-25	0			46	
Li^+	↑		20		
Naloxone	↓		40		
	0			40	
Lynestrol	↓			7	
Mestranol	↑			7	
Δ^9-tetrahydrocannabinol (THC)	↑		44	15, 35	
	↓		73		
Red oil (hashish extract)	↑			35	
Progesterone	0		45		
Mecamylamine	↓			34	
Guanethidine	↓			71	
SKF 525 A[a]		→	2[b]		
NH_4Cl	↓	←	8		
Ethanol	↑			3	
Urethane	↓ or 0		78[i]		
Bulbocapnine	↓		103	103	

[a] Shown to inhibit amphetamine metabolism in rats (see Sect. I, Chap. 1.A.; LEWANDER and JONSSON, 1973).
[b] Dose-, or time-response relationship investigated.
[c] In low dose.
[d] Initial depression, late prolongation.
[e] Administered intracerebroventricularly.
[f] Depending on dose.
[g] Given chronically.
[h] Tolerance to α-MT on chronic administration.
[i] Depending on dose of amphetamine.
[j] Amphetamine given intraventricularly.
[k] Does not penetrate blood brain barrier.
[l] 300 mg/kg × 1–3.
[m] 15 mg/kg amphetamine; test drug caused a reduction of stereotyped behavior and reappearance of locomotor activity
[n] 1-phenyl-3-(2-thiazolyl)-2-thiourea.
[o] Atropine and JB318 reverses pilocarpine effect, while methylatropine and JB323 do not.
[p] Chronic treatment with test drug in combination with amphetamine for 10 days.
[q] < 3.2 mg/kg of *d*-amphetamine.
[r] Monkey.
[s] CA synthesis inhibitor added in food.
[t] 6-OH-DA given intracerebroventricularly during neonatal period.

References

1. ABDALLAH et al., 1974
2. BABBINI et al., 1971
3. BANARJEE and GEH, 1973
4. BANARJEE and LIN, 1973
5. BEAUVALLET et al., 1966a
6. BERGER et al., 1973
7. BLACKHAM and SPENCER, 1969
8. BORELLA and HERR, 1971
9. BORELLA et al., 1969a
10. BORELLA et al., 1969b
11. BREESE et al., 1973
12. BROWN, 1960
13. BUUS LASSEN, 1973
14. CORRODI et al., 1970
15. CRAIGMILL et al., 1974
16. DAVIS, 1957
17. DINGELL et al., 1967
18. DOMINIC and MOORE, 1969a
19. DOMINIC and MOORE, 1969b
20. D'ENCARNACAO and ANDERSON, 1970
21. D'ENCARNACAO et al., 1969
22. ERNST, 1969
23. ESTLER and AMMON, 1967
24. ESTLER et al., 1970
25. ESTLER and AMMON, 1970
26. EVETTS et al., 1970
27. FIBIGER et al., 1973
28. FIBIGER, 1973
29. FREEMAN et al., 1970
30. FREEMAN and SULSER, 1972
31. FREY and MAGNUSSEN, 1968
32. FUXE and UNGERSTEDT, 1970
33. GALAMBOS et al., 1967
34. GARDOCKI et al., 1966b
35. GARRIOTT et al., 1967
36. GRANA and SOSSI, 1967
37. HASSELAGER et al., 1972
38. HERMAN et al., 1971a
39. HOLLISTER et al., 1974
40. HOLTZMAN, 1974
41. HUANG and HO, 1973
42. HUTCHINS and ROGERS, 1971
43. IRWIN et al., 1958
44. KUBENA and BARRY, 1970
45. LADISCH and BAUMANN, 1971
46. LAPIN et al., 1972
47. LAPIN and SCHELKUNOV, 1965
48. LEW et al., 1971
49. LEWANDER, 1971d
50. MABRY and CAMPBELL, 1973
51. MAJ and PRZEGALINSKI, 1967
52. MAJ et al., 1973
53. MAJ et al., 1972b
54. MAJ et al., 1974
55. MANTEGAZZA et al., 1968a
56. MANTEGAZZA et al., 1970
57. McGRATH and KETTELER, 1963
58. McLEAN et al., 1973
59. MENON et al., 1973
60. MENON et al., 1967
61. MIELE et al., 1972
62. MILLER et al., 1970
63. MOORE et al., 1970
64. MOORE and THORNBURG, 1973
65. MORPURGO and THEOBALD, 1964
66. MÜLLER and SCHOETENSACK, 1969
67. NEUBURG and THUT, 1974
68. OFFERMEIER and POTGIETER, 1969
69. OFFERMEIER and POTGIETER, 1972
70. PFEIFFER et al., 1966
71. PFEIFFER et al., 1968
72. PHILLIPS and BRADLEY, 1969
73. PIRCH et al., 1973
74. PROCTOR et al., 1967
75. PRZEGALINSKI and KLEINROK, 1970
76. RANDRUP and MUNKVAD, 1965a
77. RANDRUP and MUNKVAD, 1965b
78. RECH and STOLK, 1970
79. ROLINSKI and SCHEEL-KRÜGER, 1973
80. VAN ROSSUM et al., 1962
81. VAN ROSSUM and HURKMANS, 1964
82. SCHEEL-KRÜGER, 1972a
83. SCHLECHTER and BUTCHER, 1972
84. VAN DER SCHOOT et al., 1962
85. SEGAL et al., 1974
86. SETHY et al., 1970
87. SIMON et al., 1970b
88. SIMON et al., 1972
89. SIMON et al., 1970a
90. SMITH, 1963
91. SMITH, 1965
92. SPENGLER, 1962
93. STOLK and RECH, 1967
94. STOLK and RECH, 1968b
95. STOLK and RECH, 1969
96. STOLK and RECH, 1970
97. SULSER and DINGELL, 1968a, b
98. SULSER et al., 1968
99. SVENSSON, 1970
100. TEDESCHI et al., 1959
101. THORNBURG and MOORE, 1972
102. THORNBURG and MOORE, 1973b
103. TSENG and LOH, 1974
104. TRIPOD et al., 1954
105. TÖRÖK, 1972
106. WEISSMAN, 1972
107. VILLARREAL et al., 1973
108. WEINSTOCK and SPEISER, 1974
109. ZIEGLER et al., 1972
110. SULSER et al., 1966
111. CREESE and IVERSEN, 1973
112. STRUBELT et al., 1970
113. TRIPOD, 1952

of amphetamine-induced motor activity, rather than an inhibition, after depletion of the catecholamines by reserpine pretreatment, was interpreted in favor of a direct effect of amphetamine on brain catecholamine receptors (VAN DER SCHOOT et al., 1962; VAN ROSSUM et al., 1962; SMITH, 1963, 1965; VAN ROSSUM, 1967). For a detailed discussion on the reserpine-amphetamine interaction, which seems to depend on an acceleration of CA synthesis and postsynaptic receptor supersensitivity, see RECH and STOLK (1970) and SCHEEL-KRÜGER (1972b). Direct proof that amphetamine acts indirectly, i.e., via release of brain CA, was provided by the use of αMT and 6-hydroxydopamine. Tolerance develops to the amphetamine antagonistic effect of αMT (DOMINIC and MOORE, 1969b). Development of postsynaptic receptor supersensitivity seems to be one mechanism for this tolerance phenomenon (MOORE and DOMINIC, 1971; BEUTHIN et al., 1972; GEYER and SEGAL, 1973).

The antagonistic effect of catecholamine synthesis inhibition with α-MT added support to the contention that amphetamine acts via preferential release of newly synthesized rather than stored catecholamines (SULSER et al., 1968; review by SCHEEL-KRÜGER, 1972b, cf. Subsect. B.I.1 above). In studies on the relative importance of NA or DA neurons for the locomotor effect of amphetamine, the dopamine-β-hydroxylase inhibitors, disulfiram, diethyldithiocarbamate, and FLA-63, and centrally active α-receptor blockers (e.g., phenoxybenzamine) have been used as tools. The effects of these drugs have been variable (Table 2); still most studies show a slight reduction. It is clear from these studies that none of these drugs can completely abolish the motor effect of amphetamine. FLA-63 + dops (=dihydroxyphenylserine, an NA precursor) caused a potentiation of the amphetamine effect (SVENSSON, 1970).

Neuroleptic drugs, particularly those having a high specificity for brain dopamine receptors, can completely abolish the amphetamine-induced motor activity. A competitive reciprocal antagonism between some neuroleptics and amphetamine seems to exist at short pretreatment intervals (RANDRUP and MUNKVAD, 1965a; SULSER and DINGELL, 1968a, b; BORELLA et al., 1969a; BORELLA and HERR, 1971; HASSELAGER et al., 1972), whereas a metabolic interaction between some phenothiazines and thiaxanthenes causes a prolongation of the amphetamine effect (Table 2, footnote a).

Selective degeneration of brain dopamine neurons caused by injection of 6-hydroxydopamine in DMI pretreated rats (DMI prevents the entry of 6-hydroxydopamine into noradrenaline but not into dopamine neurons) also reduced all motor effects of amphetamine including locomotion (BREESE et al., 1973; HOLLISTER et al., 1974).

Taken together, the efficiency of α-methyltyrosine, neuroleptics, and 6-hydroxydopamine to completely abolish all motor activities of amphetamine and the comparatively small effect of DBH inhibitors and α-receptor blocking agents indicate that an intact brain dopamine synthesis is necessary for amphetamine to produce its motor effects, while brain noradrenaline neurons might modify the locomotor component.

Centrally active adrenergic β-receptor blocking agents may also modify motor activity (Table 2). Drugs affecting brain 5HT receptors have little or no effect on the amphetamine response. Two studies, however, have shown that pretreatment of rats or mice with PCPA, a 5HT-synthesis inhibitor, potentiated amphetamine; 5HTP administration reversed this effect. Some MAO-inhibitors seem to potentiate amphetamine, while others have no effect. Tricyclic antidepressants potentiate and prolong the effect of amphetamine in rats due to both pharmacodynamic and metabolic interactions (Sect. I, Chap. 1.A) and inhibit amphetamine in mice, where generally higher, receptor blocking doses of the drugs have been tried; see also MONTANARO and BABBINI (1965). Anticholinergics seem to potentiate and cholinergics to antagonize amphetamine-induced motor activity, which adds support to the hypothesis of a dopaminergic-cholinergic balance in the CNS. For sedative and hypnotic drugs no specific conclusions can be drawn. Interestingly, barbiturates and sometimes benzodiazepines may potentiate amphetamine under certain conditions; other studies claim the contrary, however. It is noticeable that amantadine, in itself a dopaminergic drug, and fenfluramine, an amphetamine analogue, seem to antagonize amphetamine. Ammonium chloride accelerates the renal elimination of amphetamine and thereby reduces and shortens its effects.

ε) Effects of Brain Lesions on Amphetamine-Induced Motor Activity

Trials to inhibit amphetamine-induced motor activity by lesions in the brain DA systems have been met with controversial results. Electrolytic and 6-hydroxydopamine lesions in the substantia nigra or in the caudate nucleus, putamen, or globus pallidus have been used. The electrolytic lesions suffer from the disadvantage that they disrupt both in- and outgoing neurons in an area and destroy all types of cell bodies which precludes a selective DA neuronal lesion. In this respect 6-hydroxydopamine lesions have a clear advantage. In both cases careful confirmation of the extent and site of lesions is necessary, preferably by histofluorescence methods for catecholamines and/or chemical estimations of the degree of reduction in transmitter content. These criteria have not always been met. Furthermore, the test time point after the lesion is important since, in dopamine neurons, there is an increase in amphetamine-releasable DA during a short interval following the lesion. Later there may be development of supersensitivity of DA receptors (cf. GLICK and MARSANICO, 1974a, b), increased turnover in remaining terminals, or collateral sprouting and reinnervation of receptors after longer time intervals.

Experimental electrolytic lesions of the substantia nigra (zona compacta, an area rich in DA cell bodies) caused a decrease in amphetamine-induced motor activity (SIMPSON and IVERSEN, 1971), preferentially after low doses of amphetamine (IVERSEN, 1971), the results of which could be partially reproduced in rats with low spontaneous activity (COSTALL and NAYLOR, 1973). However, COSTALL and NAYLOR (1973) found no effect of such lesions in highly active rats and CREESE and IVERSEN (1972), using localized 6-hydroxydopamine lesions, could not repeat the 1971 study. Intracerebroventricular administration of 6-hydroxydopamine was also ineffective (EVETTS et al., 1970), while other reports showed a reduction (HOLLISTER et al., 1974, see Table 2). This problem is apparently still controversial and needs clarification.

Electrolytic lesions in dopamine terminal areas showed that motor activity was not affected by lesions in the caudate nucleus and the caudate putamen, and sometimes potentiated (ASHER and AGHAJANIAN, 1974; CREESE and IVERSEN, 1974) by lesions in the globus pallidus (NAYLOR and OLLEY, 1972, cf. Stereotyped Behavior, Subsect. C.I.1.a above). GLICK and MARSANICO (1974a) found a decrease in amphetamine-induced hyperactivity 1 day after caudate lesions in mice, but no change at 1 month after the lesion.

Lesions in the frontal cortical areas in rats have repeatedly been shown to make the animals hypersensitive to amphetamine (ADLER, 1961; LYNCH et al., 1969; GLICK, 1970; GLICK et al. 1971; IVERSEN, 1971; IVERSEN et al., 1971). Similar results were obtained for both *d*- and *l*-amphetamine in rats and mice (GLICK 1971a). An issue for discussion has been whether this potentiation increases with time after the frontal lesion (ADLER, 1961; GLICK, 1970, 1972; GLICK and MARSANICO, 1974a) or decreases with time (LYNCH et al., 1969, 1971). The frontal lesions do not affect brain CA levels and α-methyl-tyrosine is an efficient antagonist of amphetamine in this preparation (IVERSEN, 1971). In monkeys prefrontal leukotomy has been reported to potentiate amphetamine (COLE and GLEES, 1957), while ambiguous results were obtained by BLUM et al. (1950). Lesions in occipital cortex have been shown to potentiate (ADLER, 1961) or to have no effect (ISAAC, 1971) on amphetamine-elicited hyperactivity in rats.

Septal lesions modify some types of behavior in rats depending on the dose of amphetamine (NOVICK and PIHL, 1969; PIHL and GREENBERG, 1969; CAREY and SALIM, 1970). An attenuation of amphetamine-induced hyperactivity has been observed after central amygdaloid lesions in rats (COLE and GAY, 1974). Lesions in the superior colliculus (ISAAC, 1971), hippocampus (GLICK and MARSANICO, 1974a), or the dorsal and median raphé region (5HT cell bodies) (NEILL et al., 1972) potentiate amphetamine-induced motor activity in rats.

In the chronic thalamic preparation, where all structures above and in front of the thalamus (including the striatum) were ablated in rats, increased motor activity was still present after amphetamine (HUSTON and BORBÉLY, 1974). This finding suggests that motor effects of amphetamine might also be mediated by mechanisms other than release of brain dopamine in the striatum.

ζ) Correlation Between Enhanced Motor Activity and Brain CA

Amphetamine given in threshold doses with regard to increased motor activity caused an acceleration of the turnover of brain DA but not of brain NA (COSTA et al., 1972; GROPPETTI et al., 1972a).

η) Other Factors Affecting Amphetamine-Induced Increase in Motor Activity

The number of mice or rats in the groups seems to be positively correlated to the amount of activity, at least using certain measuring devices (WENZEL and BROADIE, 1966; BAINBRIDGE, 1970). Different strains of mice react quite differently to amphetamine (YEN and ACTON, 1973). In some strains (C_3H, Balb/cJ) no change in motor activity was noticed (JORI and GARATTINI, 1973; MOISSET and WELCH, 1973). Age is an important factor since young rats are less reactive than adult animals (HEIMSTRA and MCDONALD, 1962; SOFIA, 1969). Individual differences between rats have been observed and generally there is a positive correlation between spontaneous activity and the drug-induced enhancement of it (IRWIN et al., 1958). Both rats (MONTANARO and BABBINI, 1965; ISAAC, 1971; SEEGAL and ISAAC, 1971) and monkeys (ALEXANDER and ISAAC, 1965) are more active after amphetamine in light, than in darkness. Mouse-killing rats are more excited by amphetamine, than nonkillers (SALAMA and GOLDBERG, 1973). Also diurnal and nocturnal monkeys react differently to amphetamine under similar lighting conditions (ISAAC and TROELSTRUP, 1969). Starvation, imposed or drug-induced, causes rats to be more sensitive to the motor stimulant effect of the drug (CAMPBELL and FIBIGER, 1971; FIBIGER et al., 1972; FIBIGER, 1973). Hypothyroid (methylthiouracil-treated) rats are clearly less sensitive to amphetamine than euthyroid animals (MANTEGAZZA et al., 1968b). Gamma radiation of the thyroid has similar effects (McDOWELL et al., 1967). Repeated electroconvulsive shocks also reduced sensitivity to amphetamine (PAPESCHI et al., 1974).

θ) Effect of Chronic Administration of Amphetamine on Motor Activity

In a previous review (KOSMAN and UNNA, 1968) several references were cited in support of the statement that tolerance to amphetamine-induced motor activity does not develop. The literature on this subject, however, reveals some controversial reports. Variations in methods of assessment, dosages, mode of administration, and duration of treatment are common. In Table 3 10 studies in rats are given. In 5 instances signs of tolerance were observed. In 2 of these (TAINTER, 1943; MAGOUR et al., 1974a) the dosages were very high, indicating that the rats showed mainly stereotyped behavior and no increase in locomotion as pointed out by MAGOUR et al. A similar mechanism might explain the results of FANELLI (1973). In the studies by HERMAN et al. (1971a) and BELL et al. (1974) moderate doses were given and tolerance started to develop after 1 month of treatment and was maximal after 4 months. The long duration of treatment in this study might be the explanation. However, a similar dose of amphetamine, for the same duration but with a different method of measurement of motor activity, did not induce tolerance (UEHLING, 1969). In addition, SHILLITO (1966) observed a weaker response to amphetamine after 3 days of repeated administration.

SEGAL and MANDELL (1974) reported a progressive augmentation of motor activity and stereotype behavior on chronic administration. At certain dose levels of amphetamine, the stereotyped behavior became increasingly prominent, which revealed itself in the recordings as a decrease of motor activity. Stereotypies and locomotion are antagonistic. These results clearly show that recordings of motor activity alone in such experiments may be misleading. Interestingly, the dominance of stereotyped behavior over locomotion remained when the animals were tested with amphetamine after 8 days of amphetamine withdrawal (SEGAL and MANDELL, 1974). In mice, pellet implantation of amphetamine for 3 days induced tolerance to exploratory behavior (locomotion and rearing) observed in an open field, while other activities were not attenuated (HITZEMAN et al., 1973). This study is the only one where dose-response curves have been plotted for acutely and chronically treated animals.

UTENA (1966) treated mice and rats (revolving drum), guinea pigs (observation), and cats (scoring), chronically with methamphetamine and found attenuated responses to the drug. Similar findings were reported by YAGI (1963) in rats.

In summary, a definite statement with regard to tolerance to amphetamine-induced increase in motor activity cannot yet be made.

Withdrawal effects, mainly nonspecific sedation or reduced motor activity compared to controls, have been reported in rats after chronic amphetamine treatment (TAINTER, 1943; YAGI, 1963; HERMANN et al., 1971a) and after methamphetamine in rats, mice, and guinea pigs (YAGI, 1963; UTENA, 1966). Abnormal motor behavior measured by a sophisticated apparatus was described in mice chronically treated

Table 3. Effects of chronic amphetamine administration on motor activity in rats

Reference	Method	Dose (parenteral administration unless otherwise stated)	Finding
ZIEVE, 1937	Revolving drum	2 mg/kg in food for 15 days	No tolerance
TAINTER, 1943	Revolving drum	3.9 or 3.6 mg/kg per day in food for 7 days	No tolerance
TAINTER, 1943	Revolving drum	37.8 mg/kg per day in food for 7 days	Tolerance
TORMEY and LASAGNA, 1960	Fotoelectric cages	20 mg/kg per day for 26 days	No tolerance
FANELLI, 1973	Fotoelectric cages	10 mg/kg per day orally for 15 days	Tolerance
LU et al., 1972	Fotoelectric cages	5 mg/kg per day for 15 days	No tolerance
HERMAN et al., 1971b	Microswitches under floor plates (Knolls motimeter)	3 mg/kg per day in food for 4–9 months	Tolerance
TILSON and RECH, 1973b	Microswitches under floor plates (Knolls motimeter)	<2 mg/kg per day for 15 days	No tolerance
SEGAL and MANDELL, 1974	Microswitches under floor plates (Knolls motimeter)	0.5–7.5 mg/kg per day for 36 days	No tolerance (progressive augmentation)
WRAY and COWAN, 1971	Y-maze	2 mg/kg, 4 times, 48 h intervals	No tolerance (progressive augmentation)
MAGOUR et al., 1974a	Animex	15–95 mg/kg per day in drinking water; stepwise increase for 5 weeks	No tolerance (progressive augmentation) (except at highest dose level)
UEHLING, 1969	Tilt cage	2 mg/kg per day in drinking water for 6.5 months	No tolerance
BELL et al., 1974	Tilt cage	1 mg/kg daily for 6 weeks	Tolerance
RECH and THUT, 1970	?	2 mg/kg every 3 h for 9 h	No tolerance
SCHUSTER and ZIMMERMAN, 1961	?	1–1.5 mg/kg daily	No tolerance

with methamphetamine (TSUYUSAKI and KOBAYASHI, 1966). This abnormal behavior persisted for several weeks after withdrawal and could be normalized by chlorpromazine.

ι) Summary and Conclusions

The increase in motor activity induced by amphetamine is a complicated behavior. The inverted, U-formed dose-response curve shows that high doses of the drug actually decrease motor activity due to increased amounts of stereotyped behaviors, which are not recorded by most measuring devices. Therefore, competing behaviors should be rated as a complement to the measurements of motor activity so that the results will not be misinterpreted.

Most studies with drugs used as pharmacologic tools and with brain lesions seem to indicate that this effect of amphetamine, like stereotyped behavior, is dependent on intact brain dopaminergic functions. In addition, release of brain NA might be of importance for the locomotor component of hyperactivity. Cholinergic mechanisms appear to antagonize the amphetamine-induced hyperactivity, and a few studies indicate a similar role for serotonergic processes.

According to most studies, tolerance does not seem to develop to this effect of amphetamine; on the contrary, chronic amphetamine treatment may increase motor activity.

A definitive statement with regard to tolerance should await the results of further investigations.

c) Behaviors Observed in the Open Field

Items of behavior, such as locomotion (square crossings), rearing, grooming, sniffing, circling, and backing, are most easily observed in open field and recorded at intervals after drug administration or by time sampling. By this procedure amphetamine has been found to affect behavioral items in different directions with very different dose- and time-response relationships in rats (SILVERMAN, 1966; JENNINGS, 1968; GUPTA et al., 1971; MAICKEL and JOHNSON, 1973), mice (WIMER and FULLER, 1965), and cats (NORTON, 1967). Rearing and locomotion, each or both together, sometimes described as "exploration," are generally increased by amphetamine while grooming spells are decreased. As pointed out previously, these behaviors are typically displayed at doses of amphetamine below 5 mg/kg. At higher dose levels stereotyped behavior takes over and locomotion, rearings, etc., are not displayed (response incompatibility; see Subsect. C.I.1.a).

Several studies have shown that the effects of amphetamine on these behavioral items are modified by pretreatment with other drugs. Thus, amphetamine-induced rearing and locomotion are generally antagonized by neuroleptic drugs and α-adrenergic blocking agents (PHILLIPS and BRADLEY, 1969; KULKARNI and DANDIYA, 1972; ROLINSKI and SCHEEL-KRÜGER, 1973; CAHN and HEROLD, 1970) or DEDC in rats (MAYER and EYBL, 1971). Imipramine potentiates amphetamine (ROBICHAUD et al., 1970). In mice, HASSELAGER et al. (1972) have made an extensive study on the effects of a series of pharmacologic tools on behavioral items of amphetamine generally showing that neuroleptic drugs and α-methyltyrosine completely inhibit amphetamine, while α-receptor blocking agents and DBH inhibitors are less effective. Similarly, α-MT antagonized amphetamine in rats, while disulfiram (a DBH inhibitor) did not (KULKARNI and DANDIYA, 1972). 5HT antagonists (LDS-25, methysergide, BOL-148, and PCPA) also seem to antagonize amphetamine in mice (LAPIN et al., 1972) while PCPA potentiated amphetamine in rats (KULKARNI and DANDIYA, 1972). Propranolol potentiates amphetamine in low doses (< 6 mg/kg) and antagonizes higher doses (CAHN and HEROLD, 1970). In conclusion, effects of drugs on most behavioral items seem to be dependent on both DA, NA, and possibly 5HT neuronal systems.

Grooming, rearing, and locomotor activity, which are abolished in rats treated with doses of amphetamine that induce stereotyped behavior, reappear after treatment with neuroleptics that reduce the stereotypies (RANDRUP and MUNKVAD, 1965a, b; DELRIO and FUENTES, 1969) but not by α-receptor blocking agents that do not antagonize stereotypies (DELRIO and FUENTES, 1969). Spiramide, a potent DA receptor blocker, antagonized stereotypies and partially restored grooming in monkeys (KJELLBERG and RANDRUP, 1974). These studies show the competitive nature of various behavioral items and the amphetamine antagonism exerted by most neuroleptic drugs.

d) Exploratory Behavior

Exploratory activity refers to a complicated investigative behavior displayed when rodents are placed in a novel environment. The most significant items in this behavior are locomotion, rearing, and sniffing. Sometimes the amphetamine effect seems to be due to an increase in locomotion rather than other exploratory behavior (BÄTTIG, 1963b; MARIOTT, 1968; KUMAR, 1969; ROBBINS and IVERSEN, 1973b). Methodologically, several techniques have been used which put emphasis on one or all of the items. Thus, besides observation, sometimes with the open field technique (described above) investigation of a changed environment, bar pressing for light (WIMER and FULLER, 1965), the Y-maze (STEINBERG et al., 1961), hole-board (see BRADLEY et al., 1968), and several automatic devices based on the motion of the animal through gates to previously closed parts of the cages (LAT, 1965; OLIVIERO et al., 1973; ROBBINS and IVERSEN, 1973b) have been used to study the effects of amphetamine. Increases in exploratory activity produced by amphetamine are dose- and time-dependent except bar pressing for light which is not affected (WIMER and FULLER, 1965).

Furthermore, there are genetic differences in mice with regard to the response to amphetamine (OLIVIERO et al., 1973). Rats previously experimented on in the test situation seemed to have a reduced response to amphetamine (STEINBERG et al., 1961; RUSHTON et al., 1963; RUSHTON and STEINBERG, 1964; review) but this was not so in the study by WRAY and COWAN (1971), however.

Barbiturates given in combination with amphetamine have repeatedly been shown to potentiate the effect of amphetamine in the Y-maze (RUSHTON and STEINBERG, 1963a, b; KUMAR, 1969) and hole-board (BRADLEY et al., 1968). No tolerance to the effect of the amphetamine-barbiturate mixture could be observed (PORSOLT et al., 1969). Also, chlordiazepoxide (RUSHTON and STEINBERG, 1966; ARMITAGE et al., 1968; RUSHTON and STEINBERG, 1967), ethanol (LEONARD and WISEMAN, 1970), and chlorpromazine in low doses (2 mg/kg) (PHILLIPS and BRADLEY, 1969) potentiated amphetamine in the Y-maze. Scopolamine (MULAS et al., 1970) antagonized amphetamine in this test. Amphetamine could not prevent the inhibitory effect of intracerebroventricularly administered 6-hydroxydopamine on Y-maze behavior (TAYLOR and LAVERTY, 1972). There is no explanation for the potentiation of amphetamine by barbiturates. Inhibition of amphetamine metabolism by chlordiazepoxide, ethanol, and chlorpromazine may well explain the reported interactions. Lesions of the caudate nucleus did not reduce the amphetamine-induced increase in exploratory activity (NEILL et al., 1974).

e) Gross Behavioral Effects in Birds

The typical behavioral items elicited by amphetamine (2–10 mg/kg) in *chicken*, referred to above (Subsect. C.I.1), have been studied with a series of pharmacologic tools.

The initial sleeping-eating behavior can be replicated by NA administration and has therefore been suggested to be due to amphetamine-induced NA release (SPOONER and FLAMING, 1972). The wing-droop, postural (crouching head retraction) position, and twittering, as well as concomitant EEG changes, are antagonized by a 5HT-receptor blocking agent, methysergide (DEWHURST and MARLEY, 1964, 1965a), a 5HT synthesis inhibitor, p-chlorophenylalanine (SCHROLD and SQUIRES, 1971), and a series of imipraminelike antidepressants, cocaine, and a few antihistaminic drugs (chlorpheniramine, diphenhydramine) (SCHROLD, 1972). In this context the tricyclic antidepressants and possibly the antihistaminic drugs are thought to inhibit amphetamine-induced 5HT release. Blocking agents of α-(phenoxybenzamine) or β-(pronethanol, dichlorisoprenaline) adrenergic receptors or anticholinergics (hyoscine, benztropine) were without amphetamine-antagonizing effects (DEWHURST and MARLEY, 1964, 1965a, b). In *pigeons* amphetamine induces chickenlike postural changes, which are also reduced by 5 HT-receptor blockers: methysergide, cyproheptadine, methergoline (RANDRUP and MUNKVAD, 1972a). These authors also showed that two DA-receptor blockers, spiramide and pimozide, reduced the locomotor activity and fast head movements in amphetamine-treated pigeons. The above evidence favors the view that in birds, amphetamine causes behavioral changes by acting through release of NA, DA, and 5HT.

In chickens repeatedly treated with amphetamine, there was a slight reduction in the postural effects and there was an increase in excitatory (stereotyped) effects (SPOONER and FLAMING, 1972).

f) General Summary and Conclusions

Amphetamine-induced changes in gross behavior are characterized by increasing motor activity and the appearance of species-typical stereotypies at the expense of reduction of other behavior, such as locomotion, rearing, and grooming. Most of these effects of amphetamine seem to be dependent on intact nigroneostriatal DA neuronal pathways and are antagonized by drugs that interfere with the functional effects of DA released by amphetamine or with synthesis of this transmitter. Brain NA neurons might modulate, e.g., amphetamine-induced locomotion; but this issue is still debated. It seems, however, that without DA neurons there are no effects of amphetamine on motor activity whatsoever. Cholinergic and possibly, serotonergic neurons seem to antagonize the DA-mediated effects of amphetamine and, thus, a dopaminergic/cholinergic-serotonergic balance may regulate various items of gross motor behaviors. During chronic administration of amphetamine the gross behavioral changes seem to increase and, in cats and monkeys, bizarre psychosis like behaviors appear. Thus, tolerance does not develop to these effects of amphetamine.

2. Social Behaviors

Amphetamine has been observed to cause changes in certain behaviors in social situations, i.e., when animals of the same species or different species are put together and allowed to interact. Small doses of the drug seem to increase and large doses to disrupt social behavior.

a) Social Interaction

Social interaction was reported to be more frequent in rats after 0.5–1 mg/kg of amphetamine (HEIMSTRA, 1962b), at least in adult age (75 days old; HEIMSTRA and McDONALD, 1962; TIKAL and BENESOVA, 1972). This effect was potentiated by atropine (RUMELHART and MUELLER, 1963). A disruption of all social activities occurred after 2–5 mg/kg in rats (HEIMSTRA, 1962a; SCHIØRRING and RANDRUP, 1971; TIKAL and BENESOVA, 1972; SYME and SYME, 1973), cats (FUNATOGAWA, 1964), dogs (ACCORNERO, 1947), and monkeys (FITZ-GERALD, 1967; SCHIØRRING, 1972; KJELLBERG and RANDRUP, 1971, 1973). Maternal behaviors in rats (FRANKOVA, 1972) and monkeys (SCHIØRRING, 1972) are also disrupted.

Pimozide, a neuroleptic drug, antagonized the amphetamine-induced disruption of social behavior in rats (SCHIØRRING and RANDRUP, 1971) while it was without effect on this behavior after amphetamine in monkeys (KJELLBERG and RANDRUP, 1971). Low doses (0.25 mg/kg) of amphetamine facilitated nonverbal communication between monkeys in a cooperative conditioning situation (MILLER et al., 1973).

b) Aggressive Behavior

The effect of amphetamine on aggressive behavior has been studied in fish, chicken, mice, rats, and cats. The Siamese fighting fish *(Betta Splendens)* is made more aggressive by amphetamine (WEISCHER, 1966). Aggressive pecking among amphetamine-treated chicken kept in flocks has been reported (DEWHURST and MARLEY, 1965b; SCHROLD and SQUIRES, 1971; SPOONER and FLAMING, 1972). An increased probability of aggressive behavior such as fighting or aggressive postures (sparring position) has been reported after low doses of amphetamine in normal mice (CHANCE, 1946; BANARJEE and GEH, 1973; ROLINSKI, 1973) or in mice aggregated after some time of isolation (WELCH and WELCH, 1969), after pretreatment of mice with *l*-dopa, reserpine, or *l*-dopa + reserpine (TEDESCHI et al., 1969; LAL et al., 1970; BANARJEE and GEH, 1973). A dose-dependent facilitation (<1 mg/kg) of aggressive behavior in grid-shocked rats after amphetamine has been reported (TEDESCHI et al., 1969; CROWLEY, 1972), in rats treated with reserpine (MORPURGO and THEOBALD, 1966) or 6-OH-DA (EVETTS et al., 1970) or in a contest situation (TERADA and MASUR, 1973). Other conditions that seem to make rats more prone to aggressive behavior are chronic amphetamine treatment (EHRICH and KRUMBHAAR, 1937; RANDRUP and MUNKVAD, 1967b), REM-sleep deprivation (FERGUSON and DEMENT, 1969), and morphine abstinence (LAL et al., 1971; THOR, 1971; CARLINI and GONZALEZ, 1972; CRABTREE and MOYER, 1973). Dextroamphetamine was more potent than the levo isomer and p-hydroxyamphetamine was without effect on aggression during morphine abstinence (LAL et al., 1971). Male rats become more aggressive against amphetamine-treated intruder rats (SHEARD, 1973). The common factor between the above pretreatments that increase the sensitivity for amphetamine to induce aggressive behavior, is not immediately evident. In an experimental ethologic study, MICZEK (1974) showed that amphetamine increased aggressivity in dominant male rats. Administration of amphetamine to the submissive rat made them more submissive and more prone to provoke aggressiveness in the untreated dominant rat. Small doses (cf. high doses, below) facilitates muricidal behavior in rats (VERGNES and CHAURAND, 1972).

Amphetamine-induced aggressive behavior was blocked by spiramide, trifluoperazine, α-methyltyrosine, and pimozide in grouped mice, and by chlorpromazine (and chlordiazepoxide) in foot-shocked mice (TEDESCHI et al., 1969), while FLA-63, phenoxybenzamine, and aceperone were without effect (ROLINSKI, 1973), which support the view that a dopaminergic mechanism may also be involved in this behavior.

Aggressive behavior has also been reported in experiments with cats given methamphetamine chronically (ELLINWOOD and ESCALANTE, 1970a). SHEARD (1967) showed that aggression in cats elicited by electrical stimulation of certain brain areas was facilitated by amphetamine. In addition, however, the attack pattern is changed simultaneously with the facilitation by amphetamine in such cat experiments (MAC DONNELL and FESSOCK, 1972). BELOZERTSEV (1974) reported, however, that amphetamine (1–3 mg/kg) suppressed the defense reaction in cats elicited by lateral hypothalamic stimulation.

Amphetamine in doses of 2–5 mg/kg or more, which induces stereotyped behavior, decreases shock-induced fighting both in mice (WELCH and WELCH, 1969) and rats (LAL et al., 1968; CROWLEY, 1972; POWELL et al., 1973), and inhibits aggressive behavior in rats paired after isolation (SILVERMAN, 1966) and muricidal behavior (KARLI, 1960; HOROVITZ et al., 1966; LEAF et al., 1969; VERGNES and CHAURAND, 1972; COLE and GAY, 1974). A decrease in fighting among mice and rats has also been reported by HART et al. (1969) and HEIMSTRA (1962b). It has not been possible to correlate inhibition of mouse-killing behavior in rats with any specific changes in brain catecholamine or serotonin turnover (SALAMA and GOLDBERG, 1970). However, the inhibition of muricidal behavior in rats by amphetamine was abolished by chlorpromazine (KARLI, 1960), but there was no inhibition of aggressive behavior in mice (VALZELLI et al., 1967a).

Amphetamine applied directly into the amygdaloid nucleus also inhibits muricidal activity in rats. Ablation of the amygdala antagonizes the muricidal blockade of systemically administered amphetamine (LEAF et al., 1969; see COLE and GAY, 1974).

Previously isolated rats, whether muricidal or not, are more sensitive to elicitation of increased motor activity and stereotyped behavior by amphetamine (KOSTOWSKI and CZLONKOWSKI, 1973; SALAMA and GOLDBERG, 1973). Similarly, mice made aggressive by previous isolation are more sensitive to the lethal effect of amphetamine (CONSOLO et al., 1965b); this does not depend on changes in the brain concentrations of amphetamine (CONSOLO et al., 1965b).

c) *Sexual Behavior*

Amphetamine has been reported to induce increased spontaneous (i.e., without specific experimental eliciting stimuli) sexual behaviors, such as the lordosis response in females or mounting behavior in both males and females, in grouped rats pretreated with reserpine (MORPURGO and THEOBALD, 1966) or in REM-sleep-deprived rats (FERGUSON and DEMENT, 1969). A facilitation of male sexual behavior, i.e., decreased intromission latency, intromission interval, ejaculation latency, and increased intromission frequency, has been shown after a dose of 0.2–2 mg/kg of amphetamine in rats (SOULAIRAC, 1963; BIGNAMI, 1966; BUTCHER et al., 1969). In dogs there was a delayed latency to erection after sexual stimulation (ALPERN et al., 1941). Amphetamine also counteracted the complete inhibition of sexual and other behaviors in male rats treated with tetrabenazine (BUTCHER et al., 1969) and the reduced sexual activity in male rats with frontal cortical lesions (SOULAIRAC and SOULAIRAC, 1972) and female rats with anterior hypothalamic lesions (HERNODON and NEILL, 1974). The effects of amphetamine on sexual behavior seem to be dose dependent since, in female rats, the lordosis response is inhibited by 2 mg/kg or more of amphetamine. This effect is potentiated by imipramine and pargyline (MEYERSON, 1968). Sexual relations between pairs of monkeys are also decreased or disrupted (KJELLBERG and RANDRUP, 1973).

Pretreatment of female rats with α-MT, a CA synthesis inhibitor, did not block the effect of amphetamine on the lordosis response while completely blocking its effects on motor activity, including stereotyped behavior (MEYERSON, 1968). H 22/54 (α-propyldopacetamide), an inhibitor of both catecholamine and serotonin synthesis, did not antagonize inhibition of the lordosis response (MEYERSON, 1968), while p-chlorophenylalanine, which inhibits only serotonin synthesis, did antagonize this effect of amphetamine (ELIASSON et al., 1972). Thus, the amphetamine-induced inhibition of the lordosis response in female rats is probably mediated through some other mechanism than release of brain catecholamines. A serotonergic mechanism may be involved. Amphetamine is reported also to prolong the estrus phase in female rats (GONZALES-BARÓN et al., 1971).

3. Conditioned and Operant Behaviors

Amphetamine and methamphetamine have been extensively investigated with respect to their effects on instrumental behaviors. The present account of the characteristics of the amphetamine-induced changes in conditioned and operant behaviors will emphasize pharmacologic aspects. However, summaries of the behavioral techniques seem necessary for the understanding of the changes induced by the stimulant. Reviews of this field have appeared previously (DEWS and MORSE, 1961; HERZ, 1961; GOLLUB and BRADY, 1965; BARRY and BUCKLEY, 1966; KELLEHER and MORSE, 1968; GROSSMAN and SCLAFANI, 1971).

a) The Conditioned Avoidance Response

Negatively reinforced (electrical shocks or intense noise; HARRISON and ABELSON, 1959; NIEMEGEERS et al., 1972) operant behavior has been studied in three different situations, namely passive avoidance, continuous avoidance, and discriminant avoidance.

α) Passive Avoidance

Rats having experienced electrical shocks on a grid floor are placed on a platform slightly elevated from the grid and the latency for stepping down on the grid is used as the measure of avoidance behavior.

Amphetamine has been reported not to change the step-down latency (BOHDANECKY and JARVIK, 1967; SOUTHGATE et al., 1971; KUMAR, 1971).

β) Continuous Avoidance

The animal is given continuous shocks from a grid floor (or intense noise) unless it presses a lever for postponement of the next shock for a limited time (e.g., 20 s). Ideally the animal presses the lever (response rate) once at the end of each postponement period in order to keep the number of shocks (reinforcements) received per unit time at a minimum.

D-amphetamine, in doses below 4 mg/kg, dose- and time-dependently increases the response rate in rats, irrespective of the operandum of the task, i.e., lever pressing or water licking (TEITELBAUM and DERKS, 1958; VERHAVE, 1958; HARRISON and ABELSON, 1959; WEISSMAN, 1959; CARLTON, 1961a, b; CARLTON and DIDAMO, 1961; WEISSMAN, 1963; STONE, 1964, 1965; SCHUSTER et al., 1966; SEGAL et al., 1967; BERNSTEIN and LATIMER, 1968; DALLEMAGNE, 1970; EVANS, 1971; HEISE and BOFF, 1971). In fact, the increase in continuous avoidance responding seems to be a very sensitive sign of amphetamine action; relative to other behavioral effects, very low doses and low brain concentrations of the drug are sufficient for elicitation of an effect (MAICKEL et al., 1969, 1970). The increased response rate is usually followed by a reduced shock rate. However, the number of responses far exceeds what is necessary for maximal shock avoidance (TEITELBAUM and DERKS, 1958; STONE, 1965; LYON, 1971).

There are marked individual differences between rats with regard to the increase in response rate (WEISSMAN, 1963; STONE, 1964, 1965; SCHUSTER et al., 1966). According to WEISSMAN (1963) a low baseline for received shocks rather than a low baseline response rate might be used to predict the individual response to the drug. Similar findings were reported by STONE (1965), who also showed that rats displaying "freezing" behavior in an open field were most effectively affected by amphetamine in the continuous avoidance situation. Among environmental factors, the size of the test compartment might influence the dose-response curve for amphetamine (HEISE and BOFF, 1971). Higher doses of amphetamine (>3–5 mg/kg) have quite opposite effects on the response and shock rate. The response rate usually drops to zero and, accordingly, the number of shocks received increases (TEITELBAUM and DERKS, 1958; VERHAVE, 1958; WEISSMAN, 1959; STONE, 1964; DALLEMAGNE, 1970; LYON, 1971; LYON and RANDRUP, 1972). This breakdown of the avoidance behavior is probably due to increased stereotypies, which eventually make responding impossible for a certain time interval (e.g., VERHAVE, 1958). With doses of amphetamine in the intermediate range, the stereotyped behavior might also express itself as very high rates of responding ("wild licking", TEITELBAUM and DERKS, 1958), or licking and biting the lever (LYON and RANDRUP, 1972), since the stereotyped animals' repertoire might be restricted to this particular behavior. Other indications of stereotype responding is the resistance to extinction, i.e., prolonged responding when the reinforcements are withdrawn (STONE, 1964; RANDRUP and MUNKVAD, 1970; LYON, 1971; LYON and RANDRUP, 1972). In another type of experiment designed to disclose stereotyped responding, it was found that rats, required to lick a water spout not more than 3 licks/s in order to avoid shocks, totally failed and continued to lick at a high rate in spite of frequent shocks (TEITELBAUM and DERKS, 1958). Stereotyped responding might also explain the change in the dose-response curve (increased response rate with rather high doses of amphetamine) observed in small test compartments (HEISE and BOFF, 1971), since the small area might have reduced the possibilities for other activities than bar pressing.

Low doses (0.25–3.0 mg/kg) increase the response rate in a continuous avoidance response situation in monkeys (HODOS et al., 1962; COOK and CATANIA, 1964; DALRYMPLE and STRETCH, 1971; HOUSER, 1973). In Houser's study the shock rate as well as the response rate increased, which was interpreted as a disturbance in stimulus control of the behavior.

γ) Discriminate Avoidance Behavior. One-way Avoidance

In the one-way avoidance situation the rat's task is to jump onto a safe platform on presentation of a light or sound signal in order to avoid electrical shocks from the grid floor. If the animal is too late, an escape response is possible after the electrical shocks are turned on. The measures are the number of correct avoidances out of a stipulated number of trials (percentage avoidances), the number of escapes per trial, and/or the latencies of avoidance or escape. KELEMAN and BOVET (1961) showed that amphetamine (0.3–1.0 mg/kg) reduced the escape latency in rats. A similar dose of amphetamine did not, however, change the number of correct avoidances in another study (GRIFFITHS and WAHLSTEN, 1974). Rats bred for low CAR performance were found to perform better under the influence of amphetamine, while rats inbred for a high CAR performance were not affected by the drug (SATINDER and PETRYSHYN, 1974). Obviously, rats with a high baseline performance do not increase the avoidance percentage, while those with a low baseline performance do.

The one-way avoidance method has also been used for studying the effects of amphetamine on learning and memory (see Subsect. C.I.4.a).

δ) Discriminate Avoidance Behavior. Two-way Avoidance

The shuttlebox is the usual arrangement of the two-way avoidance response task. A grid floor is subdivided into two equal compartments by a hurdle. On signal the animal can avoid the negative reinforcements or escape from them by jumping to the other side of the hurdle. In addition to the measures used in the one-way situation, the number of intertrial hurdle crossings also might be regarded as a measure of general activity. Since the animal has to return to an area where shocks might have been previously received, a component of passive avoidance is built into this technique.

The electrified Y-maze used by BARRETT et al. (1974) further has the advantage of making rats discriminate between the two arms and as well as deciding when to jump.

Amphetamine improves the performance of rats in the shuttle box both by reducing latencies (BÄTTIG, 1963b; NIEMEGEERS et al., 1972; BARRETT et al., 1974; BARRETT and STERANKA, 1974) and by increasing the number of avoidances (most studies referred to in this subsection). Intertrial crossings also increase after amphetamine (GUPTA and HOLLAND, 1969; SATINDER, 1971; BARRETT et al., 1974; BARRETT and STERANKA, 1974). Effective doses of amphetamine and methamphetamine are in the range of 0.5–2.0 mg/kg. Higher doses reduced the avoidance performance or disrupted the behavior (GUPTA and HOLLAND, 1969; SATINDER, 1971, 1972; ROSEN and LAFLORE, 1973). Escape responses are not, however, disrupted by high but nonlethal doses of the drugs.

An increased number of avoidance responses per trial is not always observed. Very well-trained rats may not perform better (CARDO, 1959; ROSEN and LAFLORE, 1973) and rats lesioned in the septal region, which improves avoidance performance itself, are not further facilitated by amphetamine (TONINI et al., 1963; ANISMAN and

KOKKINIDIS, 1974). So-called poor avoidance rats might be stimulated by amphetamine (CARDO, 1959; KRIECKHAUS, 1965; RECH, 1966) as well as rats being poor avoiders due to lesions in the mamillothalamic tracts (KRIECKHAUS, 1965). However, a subgroup of the poor avoiders are not stimulated by amphetamine (RECH, 1966). Similarly, a few other rat strains selectively inbred for low avoidance performance were only slightly improved by amphetamine (GUPTA and HOLLAND, 1969; SATINDER, 1971, 1972). Rats trained to discriminate whether to respond or not depending on the frequency of the tone used as the conditioned stimulus did not perform correctly under the influence of amphetamine (CARDO, 1959).

Amphetamine and methamphetamine also improve the two-way conditioned avoidance response in golden hamsters (SANSONE and RENZI, 1970, 1971; RENZI, 1971; CASTELLANO et al., 1973) and cats (WADA et al., 1963; HANSON, 1966, 1967). Mice and guinea pigs, particularly if their performance is declining after prolonged sessions (1000–1500 trials) in the shuttle box, are also stimulated by amphetamine (BOVET and OLIVERIO, 1967; OLIVERIO, 1967; SANSONE and BOVET, 1969).

A discussion of the explanations for the improved performance in the shuttle box will follow under "Acquisition of the Conditioned Avoidance Response" (Subsect. C.I.3.a.ζ).

ε) Discriminate Avoidance Response in the Skinner Box

A third modification of the discriminant CAR is the use of bar- or lever-pressing as the operandum, similar to the continuous avoidance situation. The task of the animal is to press the bar within a limited time period after a signal in order to avoid the negative reinforcements. This paradigm makes it easy to record all responses on the lever (overall response rate) and to precisely determine their occurrence in relation to the conditioned and unconditioned stimuli. A series of measures, e.g., premature responses, efficient responses, and interresponse time distribution can thus be used to describe the drug effect in addition to the number of avoidance responses and the avoidance and escape latencies. By comparison of four such measures in drug-treated versus control animals, profiles characteristic for groups of pharmacologic agents can be constructed (Bovet-Gatti profile: cf. SMYTHIES et al., 1969).

Amphetamine in doses of 0.25–3 mg/kg increases the response rate, reduces the signal response latency, and the number of shocks received (LYNCH et al., 1960; VERHAVE, 1961; HEARST and WHALEN, 1963; MOLINENGO and RICCI-GAMALERO, 1970; NIEMEGEERS et al., 1970, 1972). Methamphetamine was more potent than d-amphetamine in this test, while p-hydroxyamphetamine was without effect (COX and MAICKEL, 1972). In one study, amphetamine (1.5 mg/kg) blocked the behavior transiently (RAY, 1965). In very well-trained rats amphetamine reduced the response rate (LATZ et al., 1967). Amphetamine has a characteristic Bovet-Gatti profile in doses below 5 mg/kg with an increase in the number of premature responses, a decrease in efficient and late responses, and a reduced reaction time in addition to shorter interresponse times (SMYTHIES et al., 1969; BEATON et al., 1974). With higher doses a profile typical for hallucinogenic agents emerged (SMYTHIES et al., 1969).

Both guinea pigs (BEATON et al., 1974) and golden hamsters (SANSONE et al., 1974) show changes similar to those in rats after amphetamine administration.

ζ) Acquisition of the Conditioned Avoidance Response

In a number of studies it has been concluded that (meth-)amphetamine injected before, during, or after training sessions of continuous or signaled avoidance responding facilitates the acquisition of the behavior as compared with saline-treated control rats (VERHAVE, 1958; DOMINO et al., 1965; KRIECKHAUS et al., 1965; DOTY and DOTY, 1966; RECH, 1966; BOHDANECKY and JARVIK, 1967; KULKARNI and JOB, 1967; KULKARNI, 1968, 1972; CICALA and KREMER, 1969; EVANGELISTA et al., 1970; BAUER and DUNCAN, 1971; ORSHINGHER and FULGINITI, 1971; WOLTHUIS, 1971; ROFFMAN and LAL, 1971; BARRETT et al., 1972; GUPTA and HOLLAND, 1972a, b; DAVIES et al., 1974) and golden hamsters (SANSONE et al., 1974). This effect seems to be dose dependent in the dose range below 5 mg/kg (GUPTA and HOLLAND, 1972b; DAVIES et al., 1974); with higher doses the acquisition of the behavior was disrupted (DAVIES et al., 1974). The facilitating action of amphetamine was independent of the intensity of the delivered shocks (CICALA and KREMER, 1969).

Other studies have failed to demonstrate a facilitation of avoidance acquisition in rats (BOVET and GATTI, 1965; POWELL et al., 1965; PLOTNIKOFF, 1966) and mice (BIANCHI and MARAZZI-UBERTI, 1969). Passive avoidance acquisition was even disrupted by amphetamine in control mice, while mice lesioned in the frontal cortex were less disrupted (GLICK et al., 1971).

Interestingly, the amphetamine-treated animals seem to perform better than controls already during the first set of training trials (VERHAVE, 1958; KELEMAN and BOVET, 1961; BOHDANECKY and JARVIK, 1967; KULKARNI, 1968; 1972; ORSINGHER and FULGINITI, 1971; GUPTA and HOLLAND, 1972a; BARRETT et al., 1974; BARRETT and STERANKA, 1974) which raises the question whether or not the effect of amphetamine is on performance rather than on acquisition (BAUER and DUNCAN, 1971). Although the results of some reports showed that the maximum level of performance, which of course is higher with amphetamine, is reached after a fewer number of trials in the drug-treated animals (BAUER and DUNCAN, 1971; ROFFMAN and LAL, 1971; DAVIES et al., 1974), this was not always the case (RECH, 1966; KULKARNI, 1968; BARRETT et al., 1974; BARRETT and STERANKA, 1974). Most reports claiming an effect of amphetamine on acquisition of conditioned avoidance behavior have used the percentage of avoidances in consecutive sessions as the only measure of the level of performance. However, the effect of amphetamine on general activity is accompanied by an increased number of intertrial crossings or incorrect lever pressings (BOVET and GATTI, 1965; POWELL et al., 1965; KULKARNI, 1968; GUPTA and HOLLAND, 1972a), which casts doubt on whether or not the facilitation is real or apparent due to avoidance by chance (BOVET and GATTI, 1965).

In recent reports rats were studied in an automatic Y-maze avoidance apparatus, where the animals had to discriminate not only when to avoid but also where to jump in order to avoid or escape shocks (BARRETT et al., 1972, 1974; BARRETT and STERANKA, 1974). A series of measures in addition to the number of correct avoidances were registered in order to be able to analyze the avoidance behavior in more detail. It was confirmed that amphetamine-treated rats acquired a better performance more rapidly than controls as judged by the number of avoidances. However, during the first 5 days of training under amphetamine, the animals made fewer correct choices with regard to discrimination of the safe area and more incorrect avoidances. After 7 days, the amphetamine-treated animals had achieved a high level of performance without displaying more incorrect responses than the controls. The authors concluded that the enhanced avoidance produced by amphetamine is mediated by the increased general activity, which increased the opportunity for learning the task rather than facilitating the learning process. The fact that the avoidance responding returns to the control level or is even impaired after amphetamine withdrawal in rats (e.g., RECH, 1966; SCHUSTER et al., 1966; BARRETT et al., 1972, 1974; BARRETT and STERANKA, 1974) and hamsters (SANSONE and RENZI, 1971) again argues for an effect of the drug on performance rather than acquisition, although learning of the task obviously takes place during the amphetamine as well as the saline administration. These matters will be further discussed under "State Dependent Learning" (Subsect. C.I.4.b). Most studies on the effect of amphetamine on performance of avoidance behavior in previously trained rats (cf. previous sections, e.g. HEARST and WHALEN, 1963) have shown that amphetamine increases the avoidance behavior already after the first injection. It may be productive to the discussion of facilitation of acquisition of avoidance behavior to determine the number of predrug training sessions necessary for amphetamine to enhance the performance to the same plateau level as the one achieved after acquisition of the behavior under ongoing drug treatment. If, in saline-treated animals, a single injection of amphetamine increases the performance to the same level and after the same number of training trials that were sufficient for amphetamine-treated animals to reach their plateau, one might be able to conclude more definitely, whether or not amphetamine enhances performance rather than acquisition.

Current explanations for the enhancing effects of amphetamine on avoidance responding have been discussed by BAUER and DUNCAN (1971) and BARRETT and STERANKA (1974). The increase in general motor activity by low doses of amphetamine is obviously of importance because the animals are exposed to the contingencies of the task more frequently. Increased avoidance responding should probably not, however, be regarded purely as a measure of unspecific motor activity, but also as an increased responsiveness to environmental influences. Amphetamine's actions responsible for reduction of "freezing behavior" induced by the electrical shocks probably also contributes to better avoidance performance (HEARST and WHALEN, 1963; KRIECKHAUS et al., 1965). Changes in the threshold for foot shocks as an explanation could not be substantiated experimentally (CICALA and KREMER, 1969).

η) Effects of Other Drugs on Amphetamine-Induced Facilitation of Avoidance Responding

The continuous or signaled avoidance responses have been very useful for drug-amphetamine interaction studies. In summarizing the results of the literature compiled in detail in Table 4, a pattern of interactions emerges, which is very similar to analogous studies of other effects of amphetamine, e.g., motor activity and stereotyped behavior.

Neuroleptic drugs in sufficient dosage generally antagonize amphetamine-induced facilitation of avoidance responding. A small dose of chlorpromazine might, however, potentiate amphetamine (SCHECKEL and BOFF, 1964), probably due to its inhibition of amphetamine metabolism. The efficiency of rather specific DA receptor blocking neuroleptics, such as haloperidol and pimozide, seem to point to the involvement of a dopaminergic mechanism for this effect of amphetamine. However, NA neurons cannot be excluded from participation, since phenoxybenzamine and a few other α-adrenergic blocking agents also seem to antagonize amphetamine. The impairment of avoidance performance by reserpine and tetrabenazine is antagonized

Table 4. Effects of other drugs on amphetamine-induced changes of avoidance behavior. Asterisk (*) after reference number indicates discriminated avoidance (shuttlebox, etc.), in all other studies continuous avoidance response was used. Symbols: No effect: 0, amphetamine-induced response rate, increased: ↑ decreased: ↓. Duration, prolonged: →, reduced: ←

Drug	Response rate	Duration	References	
			Rat	Other
Neuroleptic drugs				
Chlorpromazine	↓		16*, 18, 19*[m], 23	Monkey 4[a]
Chlorpromazine	↑		30[k]	
Perphenazine	↓		18	
Trifluoperazine	↓		18	
Flufenazine	↓		24*, 26	
Haloperidol	↓		19*[m]	
Pimozide	↓		19*[m]	
Pipamperone	↓		19*[m]	
Others	↑		30	
α-receptor blocking agents				
Phenoxybenzamine	↓		26	
Dibenamine	↓		21*[c], 22*[c]	
Drugs affecting storage of monoamines				
Reserpine	0[g]		12, 23*[i], 26	Cat 8*, 9* 11*[be]
Tetrabenazine	0[g]		14	Mice 27
Guanethidine	↑		30	
CA synthesis inhibitors				
TH-inhibitor				
α-methyltyrosine (α-MT)	↓		5[b], 6*[c], 13[f], 15[f], 17*[g,h] 20, 21*[c], 22*[c], 25* 26, 34, 29*[c]	
Reserpine + α-MT	↓		26	Cat 8*, 9*
Reserpine + *l*-dopa + α-MT	0			Cat 8*, 9*
α-MT + *l*-dopa	↓		34, 35	
α-MT + *l*-dopa	(↓)		21*[c]	
MAO-inhibitor + α-MT	0		6*[c]	
5-HT-synthesis inhibitor				
p-chlorophenylalanine	0		26	
MAO-inhibitors				
Isocarboxazide	↑		30	
Iproniazid	↑		30	
Pargyline	↑	→	31	
Tricyclic antidepressants				
Imipramine		→	1, 12, 34,	
Imipramine	↑		23, 30	
Desmethylimipramine		→	1	
Desmethylimipramine	↑		30	
Amitriptyline	↑		30	
Others	↑		30	

Table 4 (continued)

Drug	Response rate	Duration	References: Rat	References: Other
Cholinergics				
Tremorine	↓		7	
Physostigmine	↓		7	
Anticholinergics				
Atropine	↑	→	3, 7, 23, 30, 33*	
	0		3	
Methylatropine	↑		30	
Scopolamine	↑		2, 30	
Others	↑		30	
Acetylcholine-synthesis inhibitor				
NVP[d]	↑		7	
Quarternary derivative of NVP	0		7	
Antihistaminics				
Promethazine	↑		30	
Promazine	↑		30	
Sedative and hypnotic drugs				
Pentobarbital	↓		16*	
Miscellaneous				
6-hydroxydopamine (6-OH-DA)	↓		32[l]	
Naloxone	↓		10*	
Naloxone	0			Mouse 10*
Urethane	↓		24*	
Adrenaline	0[g]		28	
Cocaine	↑		30	
Tyramine	↑		30	
Methylphenidate	↑		30	

[a] Competitive relation between chlorpromazine (CPZ) and amphetamine, i.e., depressant effect of CPZ was reversed by amphetamine, which was antagonized by CPZ.
[b] Methamphetamine was used.
[c] Acquisition of avoidance response during amphetamine treatment was not antagonized by α-MT.
[d] NVP=4-(*l*-naphthylvinyl)pyridine.
[e] Reserpine given intracerebrocentricularly, methamphetamine was given i.p. (no effect if methamphetamine was given intracerebrocentricularly).
[f] α-methyltyrosine was given 150 mg/kg i.p. twice daily for 2 days, amphetamine impaired avoidance response in these rats.
[g] Amphetamine antagonized the depression of avoidance responding caused by this drug.
[h] α-methyltyrosine, 50 mg/kg, was given three times at 4-h intervals.
[i] Both *d*- and *l*-amphetamine had this effect.
[k] 0.2 mg/kg
[l] Amphetamine pretreatment prevented decrease in avoidance responding caused by 6-OH-DA.
[m] The drug also restored changes in other parameters induced by amphetamine.

References

1. BERNSTEIN and LATIMER, 1968
2. CARLTON, 1961b
3. CARLTON and DIDAMO, 1961
4. DALRYMPLE and STRETCH, 1971
5. EVANS, 1971
6. FULGINITI and ORSINGHER, 1973
7. GOLDBERG and CIOFALO, 1969
8. HANSSON, 1966
9. HANSSON, 1967
10. HOLTZMAN, 1974
11. JOHN et al., 1958
12. LAPIN and SCHELKUNOV, 1965
13. MAICKEL et al., 1970
14. MCKEARNEY, 1968
15. MILLER et al., 1970
16. MOLINENGO and RICCI-GAMALERO, 1970
17. MOORE and RECH, 1967
18. MORPURGO and THEOBOLD, 1964
19. NIEMEGEERS et al., 1970
20. OLIVERIO, 1967
21. ORSINGHER and FULGINITI, 1971
22. ORSINGHER and FULGINITI, 1973
23. RAY and BIVENS, 1968
24. RECH, 1964
25. RECH and MOORE, 1968
26. RECH and STOLK, 1970
27. REITZ et al., 1972
28. REMINGTON and ANISMAN, 1974
29. ROFFMAN and LAL, 1971
30. SCHECKEL and BOFF, 1964
31. SEGAL et al., 1967
32. TAYLOR and LAVERTY, 1972
33. VASQUEZ and ISQUIERDO, 1970
34. WEISSMAN, 1961
35. WEISSMAN and KOE, 1965
36. WEISSMAN et al., 1966

by amphetamine; similar results have been obtained in studies of motor activity and stereotyped behavior.

The CA synthesis inhibitor, α-methyltyrosine, effectively antagonized amphetamine, particularly if the animals were pretreated with reserpine. In at least one cat study the α-methyltyrosine effect could be reversed by administration of *l*-dopa, a CA precursor, that bypasses the inhibited enzymatic step (HANSON, 1966, 1967). In rats this was not the case (WEISSMAN and KOE, 1965). There are a number of studies with α-methyltyrosine that are presently difficult to explain. These studies have shown that α-methyltyrosine-induced depression of avoidance behavior was reversed by amphetamine (MOORE and RECH, 1967), the conversion of amphetamine facilitation into depression of acquisition (ORSINGHER and FULGINITI, 1971, 1973) or performance of CAR (MAICKEL et al., 1970; MILLER et al., 1970) after single or repeated α-methyltyrosine injections, and the inability of α-methyltyrosine to prevent amphetamine-induced facilitation of acquisition of avoidance learning (ROFFMAN and LAL, 1971). A similar, apparently paradoxical, finding is the facilitation by amphetamine of intracranial self-stimulation, maximally depressed by previous repeated injections of α-methyltyrosine (POSCHEL and NINTEMAN, 1966). Taken together these last-mentioned findings suggest that α-methyl-tyrosine might have other, as yet undiscovered, effects in addition to being a TH inhibitor, or, that amphetamine exerts some effects via mechanisms other than CA release.

MAO-inhibitors and tricyclic antidepressants markedly potentiate and prolong the actions of amphetamine on avoidance behavior, as in most other similar interaction studies. Interaction on a neuronal level, as well as with the metabolism of amphetamine, seem to explain these findings as discussed previously (see Sect. I, Chap. 1.A).

A constant finding, as it seems, is the potentiation of amphetamine by anticholinergics and antagonism of its effects by cholinergics, which again supports the hypothesis of a catecholaminergic/cholinergic balance between the neuronal pathways in the CNS that subserves some actions of amphetamine.

Very few studies have been performed in order to evaluate the participation of serotoninergic mechanisms in the amphetamine-induced facilitation of avoidance behavior. Further, the effects of 6-hydroxydopamine, a useful tool for making selective chemical lesions of the brain CA pathways on amphetamine-induced changes of avoidance behavior has not been sufficiently explored. In the only study found in the literature, amphetamine was given before the intracerebroventricular administration of 6-hydroxydopamine, and was found to inhibit the depressant effect of this drug on the CAR.

θ) Effects of Brain Lesions on the Actions of Amphetamine on Avoidance Behavior

Septal lesions make rats perform better in CAR (TONINI et al., 1963; ANISMAN and KOKKINIDIS, 1974). These lesions do not seem to alter the effect of amphetamine on avoidance responding (TONINI et al., 1963; ANISMAN and KOKKINIDIS, 1974). Lesions in the frontal cortex in mice antagonized the disrupting effect of amphetamine on learning of a passive avoidance response (GLICK et al., 1971). Mamillothalamic tractotomy in untreated rats reduced avoidance responding, but the facilitating property of amphetamine was not abolished (KRIECKHAUS, 1965). No studies have been found where mono-aminergic pathways have been selectively lesioned in order to determine their importance for amphetamine-induced facilitation of avoidance responding.

ORSINGHER and FULGINITI (1973) have provided evidence for a role of adrenal CA in the overall effect of amphetamine on acquisition of the conditioned avoidance response. Adrenalectomy abolished the inhibitory effect of dibenamine on the amphetamine facilitation of the CAR.

ι) Effect of Repeated or Chronic Treatment with Amphetamine on Avoidance Behavior

Rats, once previously treated with amphetamine, seem to react more intensively to subsequent drug exposures when tested in the CAR situation (TILSON and RECH, 1973a). Repeated injections of amphetamine (0.5–2 mg/kg once daily) for up to 5 days did not make rats tolerant to the effects of amphetamine on avoidance responding (RECH, 1966; SCHUSTER et al., 1966; RECH and THUT, 1970; KUMAR, 1971). In another study with a longer chronic treatment period, tolerance to 1–3 mg/kg of methamphetamine was found to develop (BAUER and PICKENHAIN, 1967). In this study the rats were trained to avoid shocks only when presented one of two different tones. After an initial disruption in their discrimination of the correct tone (cf. CARDO, 1959), the chronically treated rats returned to predrug performance after 5–8 days. The animals also made increased intertrial crossings during the first few days of chronic treatment, but later these aberrant responses disappeared. The decrease in number of incorrect responses was ascribed to habituation rather than tolerance by BAUER and PICKENHAIN (1967). An initial increase and later disappearance of intertrial crossing were also found during acquisition of avoidance behavior during amphetamine treatment, (BAUER and PICKENHAIN, 1967; BARRETT and STERANKA, 1974). In monkeys chronic administration of 1–2 mg/kg of d-amphetamine for a week was accompanied by increased continuous avoidance responding without signs of tolerance development (HOUSER, 1973). However, withdrawal of amphetamine treatment was followed by a transient phase of decreased response rate and increased number of shocks as compared with the established baseline performance, indicating an abstinence reaction in these monkeys. A phase of below control performance after abrupt, but not gradual, withdrawal of chronic amphetamine treatment in rats that had learned their avoidance behavior under the influence of the drug was found by BARRETT and STERANKA (1974). In this case the subnormal performance was attributed to the state-dependent learning of the behavior, which was not transferred to the nondrug state. It presently seems difficult to differentiate tolerance–withdrawal phenomena from those ascribed to state-dependent learning in studies of the effect of chronic amphetamine treatment on learned behaviors.

ϰ) Summary and Conclusions

Amphetamine increases the response rate in a number of different conditioned avoidance tasks in all species tested. This effect is observed with low dose levels, whereas higher doses disrupt the avoidance behavior. Even the low dose effect is not necessarily commensurate with a firm stimulus control of the behavior, since, under several circumstances, the responding does not result in a reduction of the number of punishments and is not easily extinguished by turning off the reinforcements. Acquisition of the avoidance behavior is apparently facilitated by amphetamine, but there are not sufficient data to exclude the possibility that amphetamine enhances performance of this behavior in experimentally naive animals. The enhancing effect of amphetamine on avoidance responding is at least partly to be attributed to a pharmacologic action on central dopaminergic mechanisms. However, the full armamentarium of available pharmacologic and neurosurgical techniques has not been utilized, which makes it impossible at present to draw firm conclusions with regard to specific neuronal pathways mediating this effect of the drug. To the question of tolerance development to the effect of amphetamine on avoidance behavior, no definite answer can be given. The increase in response rate after amphetamine is not diminished during chronic administration, but other measures indicate an improvement of the stimulus control of the behavior. However, the apparent signs of tolerance have also been interpreted in terms of state-dependent learning and cannot be unequivocally ascribed to functional tolerance.

b) Positively Reinforced Operant Behavior

The effects of amphetamine on positively reinforced operant behavior are dependent on a series of variables including the schedule of reinforcement, baseline performance, discriminatory stimuli in the environment, the physiologic condition of the animal (e.g., food deprivation), and the type of reward. The measures of the performance are the response rate (e.g., number of lever presses per unit time), the number of reinforcements obtained per trial or session, and the temporal pattern of responding (i.e., interresponse times during various phases of the trial). The amphetamine-induced changes of these parameters will first be considered for each of a number of different schedules of reinforcement: continuous reinforcement (CRF), fixed ratio (FR), variable ratio (VR), variable interval (VI), fixed interval (FI), differential reinforcement of low rate (DRL), other more complicated schedules combined with discrimination tasks, and finally, the conditioned emotional response (CER). These schedules may be used alone or multiple schedules may be used (e.g., multiple FR–FI, where the two components are signaled by different discriminatory stimuli; mixed schedules, mixed FI–FR, no discriminatory stimuli delivered; concurrent schedules, e.g., concurrent FI, FR, where the animal is required to respond on lever A for one reinforcer on an FI basis, and on lever B for a second reinforcer on an FR basis). These combined schedules will be discussed under the respective component schedule, unless a more detailed account is warranted, or if important comparisons of the schedules or reinforcers are made. After these descriptions a few generalizations of the effects of amphetamine on operant behavior are made.

α) Continuous Reinforcement (CRF)

In a CRF schedule each response on a lever is reinforced. Amphetamine in doses below 4 mg/kg has been shown to increase the response rate in CRF responding, when rats are reinforced with heat (WEISS and LATIES, 1963; LATIES, 1971), illumination (UEHLING, 1969), water (STONE, et al., 1958; KNOWLER and UKENA, 1973), and food (WEISSMAN, 1959; OLDS, 1970, 1972; LATIES, 1971; WILL and CHECCHINATO, 1973). The dose-response curve for individual animals varies considerably (WILL and CHECCHINATO, 1973). These authors found a positive correlation between the increase in heart rate after electrical shock and the increase in response rate on the CRF schedule, indicating different levels of reactivity for each individual, which might provide one explanation for the interindividual variability. With the use of concurrent CRF schedule where rats were offered heat and food on two different levers, it could be demonstrated that under the defined experimental conditions food was a better reinforcer than heat (LATIES, 1971). In one study 0.5 mg/kg amphetamine depressed CRF responding for food (COLE, 1969). The interresponse time (IRT) distribution was found to be shifted toward shorter IRTs in the study by STONE et al. (1958), which is a natural consequense of the amphetamine-induced increase in response rate. Amphetamine-treated, food-reinforced rats show a strong resistance to extinction of CRF responding, i.e., the animals continued bar-pressing for a long period of time even after the rewards were withheld (OLDS, 1970). Possible explanations for this phenomenon will be discussed below.

Monkeys increased their response rate for light after 3 mg/kg amphetamine in a CRF schedule (FOX, 1962). The peak effect of the drug occured at 2–3 h after the injection, and the duration of action was 8 h. In the absence of light reinforcement, the amphetamine-treated monkeys displayed a below control rate of responding, indicating that the drug effect was not due to increased general motor activity.

Food-deprived dogs increased their bar-pressing for food after 1.75–4 mg/kg amphetamine without eating the delivered meat chops (ANGEL et al., 1974). This might seem paradoxical, but shows the limitations of the use of bar-pressing for food as a measure of amphetamine-induced anorexia. This issue will subsequently be more fully discussed. A selected strain of genetically nervous dogs (pointers; E-strain) only respond for food reinforcement when treated with chlordiazepoxide. Amphetamine disrupted responding (ANGEL et al., 1974).

β) Fixed Ratio (FR) Reinforcement

In a FR schedule the animal is required to respond more than once for each reinforcement. (The CRF schedule is, in fact, a special case of FR reinforcement, namely FR 1.) The control pattern of responding in the FR schedule is characterized by a steady response rate during the trial for rations up to about FR 30. With very high ratios (e.g., 100 or more) the animal usually pauses after each reinforcement before starting to bar-press again at a steady rate. For intermediate FRs the pausing may be short or variable. The FR schedule thus engenders a high overall response rate in control animals.

The characteristic effect of amphetamine on FR responding is to decrease the response rate, with exceptions mentioned below, and to increase the interresponse times, particularly the postreinforcement pause. The decrease in response rate has been shown to occur in pigeons (DEWS, 1958a; SMITH, 1965; RUTLEDGE and KELLEHER, 1965; MEGINNIS, 1967; MCMILLAN, 1968a, b, 1969; LEANDER and MCMILLAN, 1974), rats (KELLEHER and COOK, 1959; OWENS, 1960, 1963; KELLEHER et al., 1961; BROWN, 1963; CLARK and STEELE, 1966; RAY and BIVENS, 1966; GLICK and MULLER, 1971; GOLLUB and MANN, 1969; SPARBER and TILSON, 1972a, b; SPARBER and PETERSON, 1973; TILSON and RECH, 1973a; TILSON and SPARBER, 1973a, b; CAREY et al., 1974; OWENS and CAMPBELL, 1974; HEFFNER et al., 1974), and monkeys (KELLEHER and MORSE, 1964; COOK and CANTANIA, 1964; BYRD, 1973; MCKEARNEY, 1973). Further this effect of amphetamine seems to be independent of whether the reinforcer was positive, such as food (most of the studies mentioned) or water, (GLICK and MULLER, 1971; BYRD, 1973; HEFFNER et al., 1974) or negative. Electrical shocks can be used as negative (KELLEHER and MORSE, 1964) or positive (MCKEARNEY, 1973) reinforcers. Intracranial self-stimulation and intravenous self-administration of amphetamine have also been used as reinforcers and are discussed separately in later sections. However, similar results have been obtained with these rewards.

The exceptions of the generally reported decrease in FR responding after amphetamine are due to either high FRs, low dose of amphetamine, or the state of food deprivation. As mentioned previously, there is a longer postreinforcement pause when the FR is large. Amphetamine administration tends to cause a shortening of this pause and an increased response rate in the initial interval of the trial, which might lead to an increase in the overall response rate, as has been shown in pigeons (MORSE and HERRNSTEIN, 1956; FERSTER et al., 1962; MCMILLAN, 1969), rats (RAY and BIVENS, 1966), and monkeys (BYRD, 1973). Very low doses of amphetamine (less than 0.3 mg/kg) may also cause an increase in the overall response rate and particularly so at high FRs if the baseline response rate is low or if the animals are food deprived and respond for food rewards (DEWS, 1958a, b; RAY and BIVENS, 1966; MEGINNIS, 1967; GLICK and MULLER, 1971; BYRD, 1973; MCKEARNEY, 1973; OWENS and CAMPBELL, 1974).

The IRTs tend to be more constant under the influence of amphetamine in that long IRTs (>5 s) are shortened, while short IRTs (<1 s) are prolonged (MORSE and HERRNSTEIN, 1956; WEISS, 1970).

The dose response curve for amphetamine or methamphetamine in FR responding rats has the inverted U shape; very low doses of the drug increase the response rate, while doses in the range of 0.5–2.5 mg/kg progressively decrease it with doses above this level causing a complete disruption of FR responding (KELLEHER and COOK, 1959; OWENS, 1960, 1963; KELLEHER et al., 1961; BROWN, 1963; CLARK and STEELE, 1966; RAY and BIVENS, 1966; BIVENS and RAY, 1968; GLICK and MULLER, 1971; SPARBER and TILSON, 1972a, b; CAREY et al., 1974). A similarly shaped dose-response curve seems to be shifted somewhat toward higher doses in pigeons (DEWS, 1958a; SMITH, 1964; MEGINNIS, 1967; MCMILLAN, 1968a, b; LEANDER and MCMILLAN, 1974) and toward lower doses in monkeys (KELLEHER and MORSE, 1964; BYRD, 1973; MCKEARNEY, 1973). The dose-response curve in rats may show marked individual differences (GLICK and MULLER, 1971).

The duration of the disruption of FR responding is also dose dependent (OWENS, 1960; BROWN, 1963). Interestingly, the disruption may be overcome if the animal is removed from the test chamber until 10 min after the amphetamine injection (SPARBER et al., 1973). An interaction between drug and environment seems to be present. Apart from these few reports investigations on time-response relationships are lacking.

Both *d*- and *l*-amphetamine have similar effects on FR responding in rats (OWENS, 1960); the *d/l* ratio is reported to be 3–4 (SPARBER and PETERSON, 1973; TILSON and SPARBER, 1973b). Methamphetamine also has similar effects (OWENS, 1960), but there are no studies on its potency relative to amphetamine. The amphetamine metabolite p-hydroxyamphetamine reduces FR responding in a dose of 10 mg/kg (MCMILLAN, 1968a). Intracerebroventricular administration of amphetamine dose-dependently decreases FR responding and 160–900 μg so administrated is equipotent with 1.6–2.5 mg/kg given intraperitonelly (SPARBER and TILSON, 1972a).

γ) Variable Interval (VI) Reinforcement

The VI schedule is considered next, since it resembles the FR schedule in engendering a rather high response rate in control animals. This is due to the fact that reinforcements are available at random intervals ranging around a mean, e.g., 90 s, and the animal cannot predict when responding is of consequence. The effects of amphetamine on the VI schedule, according to available reports, are variable and sometimes contradictory. This seems to be dependent in part on the baseline response rate. Thus, amphetamine increases the response rate if the animal's baseline is low, and has no effect or decreases responding when the baseline is high. As for the FR schedule, the dose-response curve for amphetamine is complex. In studies on pigeons, DEWS (1958a) reported that methamphetamine has no effect on or slightly increases the response rate in doses below 1 mg/kg and decreased the response rate above this dose level. In another report, 3 mg/kg amphetamine decreased the response rate, while lower doses had no effect (TODOROV et al., 1972); these authors also reported changes in performance of a two lever concurrent VI–VI task.

Most studies in rats show that amphetamine (or methamphetamine) increases or leaves unchanged the response rate, depending on the predrug baseline rate (MILLER, 1956; KELLEHER et al., 1961; HEARST, 1961; SEGAL, 1962; RAY and BIVENS, 1966; HEARST and VANE, 1967; BIVENS and RAY, 1968; COLE, 1968; EVANS, 1971; GREEN and HARVEY, 1974. HEFFNER et al., 1974). Since the response to amphetamine is baseline dependent, the dose-response curves for different animals are dissimilar (HEARST, 1961). The same general pattern of the dose-response curve is obtained regardless of whether the reinforcer is food or water (HEARST, 1961; GREEN and HARVEY, 1974; HEFFNER et al., 1974). In one study, however, it was found that methamphetamine reduced VI responding for food, while there was an increased responding for water (POSCHEL, 1963). This observation led to the conclusion that methamphetamine reduced hunger more than thirst.

Monkeys bar-pressing for food (HANSON et al., 1967) or for electrical shocks (as a positive reinforcer) (MCKEARNEY, 1973) show, on VI schedules, also a baseline and a dose-dependent response to methamphetamine. IRTs on the VI schedule after methamphetamine in rats were found to be either shortened (SEGAL, 1962) or prolonged (HEARST, 1961), depending on whether there was an increase or a decrease, respectively, of the amphetamine-induced alteration of the response rate. In a detailed, interresponse time distribution study in pigeons, a series of peaks of different IRTs emerged under control conditions probably due to missed pecks (WEISS, 1970). Amphetamine caused shorter IRTs and reduced the number of missed pecks on the lever.

δ) Fixed Interval (FI) Reinforcement

On an FI schedule the animal is reinforced at a constant time interval (e.g., once every 2 min), provided it makes a lever response at or after this time point. During training the animal learns that responding early in the interval is of no consequence. Therefore the typical response pattern consists of a postreinforcement pause until shortly before the fixed time point, when responding at a steady rate starts and continues until reinforcement.

Thus, the basal response rate is very low or zero early in each interval and high at its end. Amphetamine (or methamphetamine) has characteristic effects on this schedule. The low response rate initially is usually increased, while the high response rate late in the interval remains unchanged or is decreased. The overall response rate might therefore be either increased, unchanged, or decreased depending on the dose of the drug. The IRTs are shortened (i.e., shifted to the left on an IRT distribution histogram), the time point in the interval at which the low response rate is shifted to a high rate is usually also shifted to the left, i.e., it occurs earlier in the time interval.

Although SKINNER and HERON (1937) and WENTINK (1938) showed that amphetamine increased the overall response rate for food on an FI schedule in rats, DEWS (1958a, b) seems to be the first to show most of the detailed (meth-)amphetamine effects described above on this schedule in pigeons. These findings were later confirmed and extended in pigeons (SMITH, 1964; RUTLEDGE and KELLEHER, 1965; LATIES and WEISS, 1966; MCMILLAN, 1968a, b, 1969, 1973; BIGNAMI and GATTI, 1969; LEANDER and MCMILLAN, 1974; BRANCH and GOLLUP, 1974), in rats, irrespective of whether the reward was food (KELLEHER et al., 1961; CLARK and STEELE, 1966; SCHUSTER et al., 1966; WUTTKE and INNIS, 1972; EVANS et al., 1973; TILSON and SPARBER, 1973a, b; WAYNER et al., 1973) or water (OLIVEIRA and GRAEFF, 1972; HEFFNER et al., 1974), cats (RICHELLE, 1969), or in monkeys rewarded with food (KELLEHER and MORSE, 1964; COOK and CATANIA, 1964; WUTTKE, 1970; MCKEARNEY, 1972, 1974; BYRD, 1973), or electric shocks as negative (KELLEHER and MORSE, 1964) or positive (MCKEARNEY, 1972, 1974) reinforcers.

The dose-response curves indicate an inverted U-shaped function with a maximal increase in the response rate with 1 mg/kg of the drug and a decreased response rate and disruption of responding at doses above 2 mg/kg both in pigeons (MCMILLAN, 1968b; BIGNAMI and GATTI, 1969; LEANDER and MCMILLAN, 1974; BRANCH and GOLLUP, 1974), rats (KELLEHER et al., 1961; RAY and BIVENS, 1966; BIVENS and RAY, 1968; OLIVEIRA and GRAEFF, 1972; CAREY and GOODALL, 1973; TILSON and SPARBER, 1973a; WAYNER et al., 1973; HEFFNER et al., 1974); and monkeys (KELLEHER and MORSE, 1964; WUTTKE, 1970; MCKEARNEY, 1972; BYRD, 1973). In a few studies stereotyped responding at doses above this limit has been reported (CLARK and STEELE, 1966; OLIVEIRA and GRAEFF, 1972). Time-response relationships are not well established for amphetamine on FI schedule responding.

The overall baseline response rate on the FI schedule is lower than, e.g., the FR or VI schedules and amphetamine, therefore, often increases the response rate on this schedule at dose levels that decrease response rates on the other schedules, as has been shown in studies using multiple or mixed FR–FI schedules in the same animals (KELLEHER et al., 1961; CLARK and STEELE, 1966; MCMILLAN, 1968b; LEANDER and MCMILLAN, 1974).

The *d*-isomer of amphetamine appears to be 3–4 times more potent than *l*-amphetamine (SPARBER and PETERSON, 1973; TILSON and SPARBER, 1973b), but the available data are not conclusive for the FI schedule. The amphetamine metabolite in rats, p-hydroxyamphetamine, also increases FI responding (MCMILLAN, 1968a). Methamphetamine increases the response rate more during darkness than during light in a 12–12 h dark-light cycle (EVANS et al., 1973).

As mentioned above, the effects of amphetamine on the temporal pattern of responding are to decrease the postreinforcement pause, which might also be expressed as a shorter quarter-life (the time period for the first 25% of the number of responses within an FI interval), and to shorten IRTs for doses of the drug that increase the response rate (DEWS, 1958a, b; LATIES and WEISS, 1966). A detailed study (BRANCH and GOLLUP, 1974) of the frequency of response rates of different magnitudes during each tenth of the FI interval in untreated pigeons showed that essentially two rates, low or high, were displayed in the records. During the initial period of the FI only low response rates were present, while in the late period only high rates occurred. The shift from low to high rate responding occurred during the 7th tenth of the FI. Thus, these results show that there is not a gradual increase in the response rate toward the end of the FI, which might have been expected from ordinary cumulative curves, but rather an alternation between high and low rates a couple of times before the animals definitely change to the typical prereinforcement high rate. Amphetamine caused the shift to occur during the 3rd to 4th tenth of the FI.

The degree of disruption of the temporal pattern of responding is under environmental influences. LATIES and WEISS (1966) were able to show that pigeons responding on an FI 5-min schedule in a chamber provided with an external clock (light-dark changes in the cage each minute) were less disrupted by amphetamine than pigeons exposed to constant light conditions. Similarly, LEANDER and MCMILLAN (1974) found that rats on multiple FR–FI schedule (different colors on light for the two components) were less disrupted than rats on a mixed FR–FI schedule with constant white light.

ε) Differential Reinforcement of Low Rates (DRL)

The DRL schedule is so arranged that the animal has to wait a certain time (e.g. 20 s) between each response in order to receive the reinforcement. Premature responses reset the equipment to zero time and a new, full-time period of nonresponding has to elapse before the next reinforced response. In untreated animals the interresponse times are rather narrowly distributed around a mean value corresponding to the requirement of the schedule. Theoretically a response rate of 3/min on a DRL 20-s schedule is enough to obtain maximal reinforcement.

In the first description of this schedule, which was designed for assessing timing behavior, SIDMAN (1955) investigated the effects of amphetamine. It was found that *dl*-amphetamine, 1.5–3 mg/kg, increased the response rate in the rat and shifted the IRT distribution toward shorter IRTs. Consequently the number of reinforcements received was decreased. These findings have been confirmed in rats (KELLEHER and COOK, 1959; CARLTON, 1961 a, b; SCHUSTER and ZIMMERMANNN, 1961; SEGAL, 1962; SCHUSTER et al., 1966; BIVENS and RAY, 1968; MORRISON, 1968; PRADHAN and DUTTA, 1970; CAREY and KRITKAUSKY, 1972; ROBBINS and IVERSEN, 1973a; CAMPBELL and SEIDEN, 1973), pigeons (HEARST and VANE, 1967) and monkeys (HODOS et al., 1962).

Dose- and time-response relationships are not well established for this schedule. Usually doses between 0.25–2.5 mg/kg have been used. Only in the study by HEARST and VANE (1967) was it found that higher doses, above 8 mg/kg, again decreased the response rate and, therefore, increased the number of reinforcements. ROBBINS and IVERSEN (1973a) tried to determine if the amphetamine-induced disruption of DRL responding was due to changes in sensory or motor functions of the animals. They concluded that the effect was due to a motor response disinhibition, rather than to an effect of their inner sense of timing.

ζ) Other Reinforcement Schedules

In a task where the rats were to continually press a lever for light, it was found that amphetamine increased the response rate but not necessarily the duration of illumination (KIERNAN, 1965; CARLTON, 1966). In another schedule the principle utilized was one in which the reward was dependent on a total of 1 min cumulative lever presses [cumulative response duration (CRD)]. Amphetamine-treated dogs were found to increase the response rate and selectively increase the proportion of short (less than 1 s) responses (WEISS and LATIES, 1964). Similarly dogs have been shown to emit short interresponse times on a differential reinforcement of long response durations and monkeys on an autoregressive reinforcement schedule (WEISS, 1970).

η) Operant Performance Combined with Discrimination Tasks

In this category of tasks, which have been applied in studies of the effects of amphetamine on various aspects of behavior, a number of discriminative situations have been utilized.

One subtype includes discriminative stimuli that are presented before and during the expected responses to be performed and indicating which of several levers to press or when to do it. Thus, it has been shown that pigeons reinforced only during the presence of a discriminative stimulus performed adequately until a dose of amphetamine was administered that interfered with the response rate (DEWS, 1955). In another study pigeons had to discriminate the duration of a visual stimulus and respond on one of two levers if the duration was short and on the other if it was long. Amphetamine dose-dependently reduced the discrimination performance in this task (STUBBS and THOMAS, 1974).

FERSTER et al. (1962) introduced a red light at regular intervals in an FR 85 schedule; responding when the light was on resulted in a 20 min time-out period (during which no reinforcements were available). Under the influence of amphetamine the pigeons were disrupted in their performance of this task.

RAY and BIVENS (1966) used three different discriminatory tasks (conditioned approach, conditioned discrimination, and stimulus self selection) and reported that the disruptive effect of amphetamine in rats was dependent on the reinforcer (water > food > sweet milk) in addition to the dose of the drug. Amphetamine given intracerebroventricularly or intraperitoneally disrupted discrimination of signaled reinforcement periods in a CRF schedule (KHAVARI, 1969). Rabbits, subjected to a similar task, i.e., pulling a ring for food reward at the sound of a buzzer, failed to discriminate the appropriate moments at doses of amphetamine that did not interfere with the ring pulling itself (SADOWSKI and LONGO, 1962; MCGAUGH et al., 1963). Also, in monkeys amphetamine disrupted discrimination in a delayed matching test (GLICK and JARVIK, 1969).

Previous behavior has been used as discriminative stimuli in tasks where the animal has to respond a certain number of times on one lever (fixed consecutive number) in order to receive a reward on operating a second lever. Pigeons subjected to such tasks after administration of amphetamine reduced the response rate on the first key and increased responding on the second, which resulted in fewer reinforcements (LATIES, 1972; BRANCH, 1974).

In the above-mentioned studies, amphetamine disrupted discrimination tasks. However, external stimuli improved the behavioral performance of pigeons and rats in other types of experiments. Thus, it has been shown that the bad performance under amphetamine can be improved if discriminatory stimuli are turned on at regular intervals or, in connection with changes in the schedule of reinforcement in, e.g., multiple schedules (LATIES and WEISS, 1966; HEISE and LILIE, 1970; CAREY and KRITKAUSKY, 1972; LATIES, 1972; LEANDER and MCMILLAN, 1974). In another study two pigeons were required to peck simultaneously on two levers in order to obtain rewards. Amphetamine in doses that disrupted performance of single pigeons did not do so if only one in the pair received amphetamine and the other was untreated (BROWN and RICHARDS, 1966).

θ) Conditioned Emotional Response (CER)

In the conditioned emotional response procedure animals are trained on various operant schedules for positive reinforcement and are then subjected to signaled periods of punishment mostly in the form of electric shocks delivered at each operant response for the positive rewards. The basic idea behind this technique is that the subject runs into a conflict whether or not to touch the lever during these periods. In trained animals the shocks may even be turned off during the signaled period of punished responding and still the animal will stop responding until the signal is gone. The measures are usually the response rate, the number of shocks received, and the number of reinforcements obtained before and during the punished periods. Under control conditions the animals may stop responding during these periods of punished responding, but this reaction is dependent on the shock level, the type of positive reinforcement, and other variables. It is important, when using this technique for drug experiments, that the shock intensity is intermediate or low, so that a certain number of responses on the lever are emitted. High intensity shocks totally depress the operant behavior (GELLER and SEIFTER, 1960; LAUENER, 1963; CAPPELL et al., 1972; FOREE et al., 1973).

The CER has been studied in amphetamine-treated pigeons (FOREE et al., 1973; MCMILLAN, 1973), rats (BRADY, 1956, 1958; TEITELBAUM and DERKS, 1958; GELLER and SEIFTER, 1960; LAUENER, 1963; BARRY et al., 1963; HENDRY and VAN TOLLER, 1964; CAPPELL et al., 1972), and in monkeys (BRADY, 1956, 1958; HANSON et al., 1967). In most rat and monkey studies (above), the basic operant schedule chosen engendered a high response rate (VI, CRF, or FR) and the effect of 0.1–3.0 mg/kg amphetamine as compared with saline was to lower the response rate during the punished period; at higher doses in some experiments it was increased (TEITELBAUM and DERKS, 1958; HILL et al., 1967; CAPPELL et al., 1972). In other words, amphetamine enhanced the effect of punishment. In two studies no effect of methamphetamine (BARRY et al., 1963) or amphetamine (HANSON et al., 1967) was observed, most probably due to the fact that responding during the punished period was already totally depressed in the control situation.

MCMILLAN (1973) and FOREE et al. (1973) have studied the effects of the baseline rate of responding on FR and FI schedules in combination with punishment on the action of amphetamine in pigeons and CAPPELL et al. (1972) in rats. In summary, it was reconfirmed that amphetamine increased low response rates and decreased high response rates, but both these effects were attenuated during periods of punished responding.

ι) Baseline Rates and the Effects of Amphetamine on Operant Behavior

In the previous descriptions of the effects of amphetamine on operant behavior, the importance of the baseline rate of responding (or baseline IRTs) for the direction and the degree of the induced changes has occasionally been mentioned. In summary, amphetamine (or methamphetamine) in doses below 2–3 mg/kg is most likely to increase the response rate when the predrug baseline is low and to decrease it when the baseline is high. The first observation of baseline rate dependency was made by MORSE and HERRNSTEIN (1956), who found that methamphetamine increased IRTs shorter than 1 s and decreased IRTs longer than 5 s in pigeons. DEWS (1958a, b) extended these observations to several schedules of reinforcement and gathered data from the literature to show that this effect was independent of the nature of the reinforcer. These suggestions were later supported in studies in rats (KELLEHER et al., 1961; SEGAL, 1962). Even responding maintained by negative reinforcement in monkeys was affected similarly by amphetamine (COOK and CATANIA, 1964). KELLEHER and MORSE (1964, 1968) have carefully reviewed the literature on this subject and showed that the critical baseline response rate in monkeys was 1/s; amphetamine increased rates below this value and decreased higher response rates. These observations have later been amply confirmed in specifically designed experiments (MCMILLAN, 1969, 1973; WUTTKE, 1970; EVANS, 1971; MCKEARNEY, 1972, 1973, 1974; BYRD, 1973; BRANCH and GOLLUP, 1974; LEANDER and MCMILLAN, 1974; HEFFNER et al., 1974).

Particularly interesting is the finding that monkeys can be shaped to work for weak electrical shocks as positive reinforcements, and the same baseline dependency for the effects of amphetamine applies for this operant behavior, also (MCKEARNEY, 1972, 1973, 1974). It should be pointed out that the failure of amphetamine to increase high rates of responding is not due to a physical impossibility to do so, since barbiturates may increase the response rate to 2–3/s, at least in pigeons (see KELLEHER and MORSE, 1968). Thus, the reduction of high response rates by amphetamine is probably due to a disruption or dampening of motor activities. There seem to be at least two conditions when the baseline dependency does not apply, namely at high doses (above 3 mg/kg) of amphetamine, which seem to disrupt responding engendered by all schedules of reinforcement, and, when the behavior is controlled by powerful discriminatory stimuli. Both these exceptions have been discussed above.

Further exceptions to the general rule have been pointed out by KELLEHER and MORSE (1968).

ϰ) Can Food and Water Reinforced Operant Behavior be Used as a Measure of Amphetamine-Induced Anorexia and Adipsia?

As pointed out in the previous section, amphetamine induced very similar changes in operant responding on various schedules of reinforcement irrespective of the reinforcer. This circumstance clearly seems to argue against the claim that the operant technique could be used for investigating the effect of amphetamine on hunger and thirst (e.g., OWENS, 1963).

Further, it has been reported in many studies that irrespective of the changes in response rates and number of reinforcements obtained, the animals do not eat the delivered food pellets. This seems to be true for pigeons (HEARST, 1961), rats (WEISSMAN, 1959; HEARST, 1961; POSCHEL, 1963; FAIDHERBE et al., 1962), rabbits (SADOWSKI and LONGO, 1962; MCGAUGH et al., 1963), and dogs (ANGEL, et al., 1974). In several studies extinction periods have been included in order to find out whether or not the availability of rewards influences the responding under amphetamine. As for avoidance behavior and self-stimulation (below), amphetamine-treated animals exhibited a prolonged responding during extinction (SKINNER and HERON, 1937; RICHELLE, 1969; OLDS, 1970, 1972; GOMER and JAKUBCZAK, 1974), which again indicates that the amphetamine-induced output of motor behavior becomes stereotypical and the animals tend to persevere once a certain behavior has been established. Alternatively, the amphetamine might enhance the rewarding value of the reinforcers and thereby influence the behavior of the animal.

However, animals deprived of food or water for 24–28 h before the experiment do show a smaller decrease in responding on an FR schedule after amphetamine than satiated animals (MEGINNIS, 1967; CAREY et al., 1974; OWENS and CAMPBELL, 1974), with one exception (GOLLUB and MANN, 1969). Other schedules have not been investigated in this respect. Thus, in animals that are not deprived of food it does not seem possible to directly translate changes in responding for food or water into degree of hunger or thirst.

λ) Spatial and Paw Preferences in Rats Responding on Levers in Operant Conditioning Experiments

Two reports are available on the choice of the right or the left lever in a two-lever situation and the preference of the right or the left paw on the FR an FI schedule (GLICK, 1973b; GLICK and JERUSSI, 1974). The results indicate that doses of amphetamine that increase the response rate decrease the side and paw preferences, while higher doses that decrease the response rate do not change or increase the side and paw preferences.

μ) Effects of Other Drugs on Amphetamine-Induced Changes in Positively Reinforced Operant Responding

Relatively few studies have been performed on the interaction of other drugs with the specific effects of amphetamine on positively reinforced operant behavior. Of the neuroleptics, only chlorpromazine has been utilized as a potential amphetamine antagonist. This drug was found to antagonize the decrease in response rate on an FI schedule caused by a high (2.5 mg/kg) dose of amphetamine in rats (BROWN, 1963). Chlorpromazine also antagonized the amphetamine-induced increase in response rate on an FI schedule in rats (RAY and BIVENS, 1968). In monkeys, chlorpromazine also antagonized the decrease both in response rate and accuracy of performance of a delayed, matching test under amphetamine (GLICK and JARVIK, 1969).

Reserpine and tetrabenzine, drugs which interfere with storage of monoamines in the nerve terminals, in relatively low doses depress operant responding on all schedules. Amphetamine has been shown to reverse the effects of these drugs on both FR and FI schedules in pigeons (SMITH, 1964; MCMILLAN, 1968b; MCKEARNEY, 1968), rats (MCKEARNEY, 1968; BUTCHER and ANDÉN, 1969; OLIVERIA and GRAEFF, 1972), and monkeys (MCMILLAN, 1968b). Interestingly, amphetamine not only restored responding to control levels but also increased the response rate early in the FI interval (SMITH, 1964; MCKEARNEY, 1968). These results have been interpreted to show that amphetamine increases low response rates (MCKEARNEY, 1968; MCMILLAN, 1968b), and that amphetamine acts directly on CA receptors in brain (SMITH, 1964). However, amphetamine has also been shown to increase *inter alia* motor activity and eating totally suppressed by these drugs, which indicates that amphetamine seems to be a general reserpine antagonist. The question whether or not amphetamine acts directly on brain CA receptors has been discussed (Subsect. B.3 and C.I).

There are only two studies on the effect of the CA synthesis inhibitor α-methyltyrosine (α-MT) on amphetamine-induced changes in positively reinforced operant behavior. Some of the changes in response rate on the FI and the VI schedules were antagonized by α-MT (EVANS et al., 1973). However, the interactions between α-MT and (meth-) amphetamine are complicated and antagonism or potentiation of certain effects of amphetamine depends on the schedule of reinforcement, the baseline response rate, the doses of α-MT, and methamphetamine, and the time point after administration of the two drugs (EVANS, 1971).

A DBH inhibitor, NSD 1034, decreased FR and FI responding; after administration of amphetamine operant responding was reinstituted (MCMILLAN, 1968b).

Two anticholinergics (atropine and scopolamine) have been shown to potentiate the response rate increasing effects of amphetamine in rats on FI responding but not the decreasing effect on a VI schedule (RAY and BIVENS, 1968). Anticholinergics also potentiated the effect of amphetamine on accuracy and response rate in the delayed matching test in monkeys (GLICK and JARVIK, 1969). The tricyclic antidepressant, imipramine, also potentiates amphetamine on FI operant behavior (RAY and BIVENS, 1968). Sedative drugs, like pentobarbital, ethanol, and diazepam potentiate amphetamine on the FI schedule in pigeons (RUTLEDGE and KELLEHER, 1965), on the cumulative response duration schedule in dogs (WEISS and LATIES, 1964), and on FI responding in cats (RICHELLE, 1969). In one rat study an antagonism between a barbiturate and amphetamine on FI responding was observed (WENTINK, 1938).

ν) Effect of Brain Lesions on Amphetamine-Induced Changes of Operant Responding

Selective lesions of rat brain CA neurons by intracerebroventricular injection of 6-OH-DA have been shown to antagonize the increase in response rate on VI and FI schedules after amphetamine, while the decreasing effect on an FR schedule was not blocked (SCHOENFIELD and ZIGMOND, 1973). SPARBER and PETERSON (1973), in a similar study, demonstrated that the actions of *d*-and *l*-amphetamine were differently affected by 6-OH-DA injections that selectively destroyed brain NA neurons. They concluded that *l*-amphetamine might be more dependent on brain NA neurons than *d*-amphetamine.

Electrolytic lesions in the medial forebrain bundle of rats, causing a decrease in both 5HT, NA, and DA levels in the forebrain, enhanced the rate increasing effect of amphetamine on a VI schedule (GREEN and HARVEY, 1974). However, lesions in other brain areas that caused selective reductions of brain NA or DA levels, i.e., probably severing NA and DA neuronal pathways, did not alter the effects of amphetamine. GREEN and HARVEY (1974) therefore concluded that the effects of amphetamine are enhanced by disruption of ascending serotonergic fibers.

Frontal cortical lesions in rats have been shown to alter the inverted U-shaped dose-response curve for amphetamine on FI responding in such a fashion that the increase in response rate after low doses was enhanced and the decrease after higher doses was accelerated. In other words the dose-response curve was both lifted and shifted to the left, which again shows the complexity of the behavioral effects of amphetamine (GLICK and MARSANICO,1974b).

ξ) Chronic Administration of Amphetamine and Operant Responding

During chronic administration of amphetamine (2–3 mg/kg daily) to rats bar-pressing for food on an FR schedule, tolerance developed within 10 days to the decrease in response rate observed after the first injections (BROWN, 1963, 1965; SPARBER and TILSON, 1972a, b; TILSON and SPARBER, 1972, 1973). However, tolerance development was observed only if the drug was given to the rats when they were allowed to perform the operant behavior after each injection; chronically treated rats that were returned to their home cages following the injections and were not retested until the 10th day did not show tolerance (BROWN, 1965, cf. below). No tolerance was observed for operant responding for illumination under amphetamine (UEHLING, 1969). Tolerance also developed more rapidly to amphetamine administered intracerebroventricularly when compared with intraperitoneal administration. Further, there was no cross tolerance between the two modes of administration (SPARBER and TILSON, 1972a). These findings suggest different mechanisms of action for amphetamine when administered centrally versus systemically.

On the FR schedule, tolerance develops to both *d*- and *l*-amphetamine but there is not a complete cross tolerance between the two isomers (TILSON and SPARBER, 1973b). Previous amphetamine experience seems to influence later responses to the drug (carry-over), since tolerance then developed faster in rats with a previous amphetamine history than in drug-naive rats (TILSON and RECH, 1973a). On the basis of finding similar degrees of release of brain NA in rats treated with a single or repeated doses of amphetamine, it has been suggested that the tolerance is due to refractoriness of CA receptors (SPARBER and TILSON, 1972a).

In rats responding on the FI schedule, it was found that tolerance develops within 8 days to the increased response rate of 0.15 mg/kg daily of amphetamine (TILSON and SPARBER, 1973a, b). However, there was not complete cross tolerance between the *d*- and *l*-isomers of the drug (TILSON and SPARBER, 1973b). Tolerance was not observed in rats during chronic administration of higher doses of amphetamine (1 mg/kg) on the FI schedule (SCHUSTER et al., 1966; SCHUSTER et al., cited by GOLLUP and BRADY, 1965).

The DRL schedule has proved to be very suitable for studies on amphetamine tolerance. The increase in overall response rate, the displacement of the IRT distribution curve toward shorter IRTs, and the decrease in number of reinforcements observed after a single dose of amphetamine, all return toward control performance during 3–4 weeks of chronic administration of 1 mg/kg amphetamine daily (SCHUSTER and ZIMMERMAN, 1961; ZIMMERMAN and SCHUSTER, 1962; SCHUSTER et al., 1966; CAMPBELL and SEIDEN, 1973). It was concluded in the study by SCHUSTER et al. (1966) that behavioral tolerance, rather than functional tolerance, develops in rats in order to meet the requirements for reinforcements. In other words, the animals learned to perform accurately although they were under the influence of the drug. This contention has been supported by CAMPBELL and SEIDEN (1973), who showed that tolerance developed only if the rats were injected before the daily DRL-session but not if treated after the sessions. These results are in agreement with the similar findings on the FR schedule referred to above and also with similar findings on tolerance to the anorexigenic effect of amphetamine (CARLTON and WOLGIN, 1971). In rats that acquired the DRL performance during chronic administration of amphetamine, the performance was disrupted when the animals were switched to saline injections (CAREY, 1973b). This result has been interpreted in terms of state-dependent learning, rather than of abstinence.

Operant techniques have also been used for studies of the possible cross tolerance between amphetamine and hallucinogenic drugs, such as mescaline and LSD. There seems to be a partial cross tolerance between mescaline and amphetamine (TILSON and SPARBER, 1973a), indicating that these drugs might share common mechanisms of action. No cross tolerance was observed between LSD and amphetamine (APPEL and FREEDMAN, 1968; TILSON and SPARBER, 1973a).

ϕ) Summary and Conclusions

There seem to be three important aspects of the varied effects of amphetamine on positively reinforced operant responding which are worth emphasizing. First, the effect of amphetamine on any schedule of reinforcement is best predicted by the baseline response rate; amphetamine increases low rates and decreases high rates. Second, these effects are independent of the type of the reinforcing agent. Third, the dose-response curve for the rate increasing effect has the shape of an inverted U and doses of amphetamine more than approx. 3 mg/kg disrupt responding on all schedules and in all species. Further, there is a complicated relationship between discriminatory stimuli and the effect of amphetamine; on one hand, amphetamine disrupts performance of tasks of discrimination, but, on the other hand, discriminatory stimuli in the environment attenuate its disruptive effects on the operant responding. Amphetamine seems to disrupt the stimulus control of operant behavior by disinhibition of motor responding rather than by interference with sensory processes.

There is some support for the view that the effects of amphetamine on operant behavior are mediated by brain monoaminergic mechanisms, but compared with other effects of the drug, the neuroanatomical and neurochemical correlates of these effects have not been fully investigated.

Tolerance develops to some of the effects of amphetamine on positively reinforced operant behaviors. There are, however, several indications that the apparent tolerance is due to relearning of the behavioral tasks in the amphetamine state, rather than to some general functional mechanism.

c) *Intracranial Self-Stimulation*

α) Amphetamine-Induced Changes of Intracranial Self-Stimulation

The first demonstration that certain brain areas in the rat electrically stimulated reinforced lever pressing (intracranial self-stimulation; OLDS and MILNER, 1954), was soon followed by the observation that amphetamine facilitates this operant behavior (OLDS, 1958; STEIN and SEIFTER, 1961).

General aspects of intracranial self-stimulation have been reviewed previously (STEIN, 1964a; WISE and STEIN, 1970; STEIN and WISE, 1973; CROW, 1972a, b; GERMAN and BOWDEN, 1974). Both neuropharmacologic and neurophysiologic experiments, as well as studies on the anatomical substrates of self-stimulation, support the view that this behavior is probably dependent on both the DA (nigrostriatal and mesolimbic) and the dorsal NA pathways (GERMAN and BOWDEN, 1974). Cholinergic mechanisms, both muscarinic and nicotinic, also appear to influence the behavior (PRADHAN and BOWLING, 1971; OLDS, 1972; NEWMAN, 1972). The present survey will be dealing exclusively with matters related to the interactions between amphetamine and self-stimulation.

Amphetamine-induced facilitation of self-stimulation has been demonstrated either as an increased rate of responding for intracerebral rewards (OLDS, 1958; STEIN, 1964a) or as a reduction of the current threshold for its elicitation (STEIN and RAY, 1960; STEIN and SEIFTER, 1961; STEIN, 1962). Recently, the latency to initiate rewarding or escape from aversive brain stimulation has been utilized as a measure (ATRENS et al., 1974). The amphetamine-induced increase in the rate of self-stimulation is largely dependent on the predrug baseline-response rate; an increased rate of self-stimulation is most easily observed if the baseline rate is low; in rats with high baseline rates a smaller increase or a decrease may be induced by amphetamine (STEIN, 1964a, b; STARK et al., 1969; OLDS, 1970, 1972; ATRENS et al., 1974; WAUQUIER and NIEMEGEERS, 1974). Therefore, the baseline level for the individual animals has often been adjusted downward by reducing the current strength or by choosing a schedule of reinforcement (variable interval, or fixed ratio rather than continous reinforcement) in order to be able to show a reliable facilitation by amphetamine (STEIN, 1962, 1964a; PHILLIPS and FIBIGER, 1973; PICK et al., 1973; and others).

It has been amply demonstrated by several different methods that amphetamine decreases the current threshold for elicitation of self-stimulation behavior (STEIN and RAY, 1960; STEIN and SEIFTER, 1961; STEIN, 1962; WANNER and BÄTTIG, 1966; STEINER and STOKELY, 1973; PICK et al., 1973). In fact, amphetamine even converts rats that are nonresponders, due to subthreshold currents, into responders (STEIN and SEIFTER, 1961). These results support the view that amphetamine, at least partly, facilitates the self-stimulation behavior through interference with specific neuronal substrates and not exclusively by general activation or other unspecific mechanisms. Amphetamine not only facilitates self-stimulation, but also makes the rats resistant to extinction of this (and other) operant behaviors (OLDS, 1970, 1972). Whether this continuation of responding in the absence of reinforcements is due to stereotype perseveration of the behavior (cf. LYON, 1971; LYON and RANDRUP, 1972) or not, to date has not been fully clarified.

Olds (1958) reported that amphetamine, in addition to its facilitation of self-stimulation, decreased escape responses to brain stimulation in aversive brain areas. Measurement of the latencies to initiate stimulation in rewarding brain areas and to escape stimulations in aversive areas (Atrens et al., 1974) has shown that amphetamine not only enhances the rewarding effects of stimulation (in, e.g., the hypothalamus), but also reduces reward in other areas (thalamus) and, further, reduces or enhances aversive brain stimulation.

Amphetamine has been shown to facilitate self-stimulation elicited within several anatomical locations in rats, i.e., the septum, the lateral, posterior, and paraventricular hypothalamus (mainly medial forebrain bundle), the ventral and dorsal midbrain tegmentum (Olds, 1958; Stein and Seifter, 1961; Stein, 1962, 1964a; Poschel and Ninteman, 1966; Wanner and Bättig, 1966; Crow, 1969; Margules, 1969; Stark et al., 1969; Atrens et al., 1974; Carey et al., 1974), as well as specific dopaminergic and noradrenergic areas like the substantia nigra (Phillips and Fibiger, 1973) and the nucleus locus ceruleus (Ritter and Stein, 1973; Stein and Wise, 1973).

A facilitation of self-stimulation in the lateral hypothalamus and the caudate nucleus has been reported after 0.1–0.5 mg/kg *d*-amphetamine in cats (Horovitz et al., 1962a, b). In another study, however, methamphetamine (0.1–1.0 mg/kg) was found to decrease the rate of self-stimulation and to increase the current threshold for elicitation of self-stimulation in the lateral hypothalamus (Umemoto and Kido, 1967). Differences in the schedule of reinforcement and the initial baseline response rate may be advanced as explanations of these contradictory findings.

A clear dose-response relationship for the amphetamine-induced enhancement of self-stimulation has been regularly observed with doses of *d*-amphetamine below 2 mg/kg. Above this dose level, no further increase and sometimes even a decrease in the rate of self-stimulation (inverted U dose-response curve) has been observed, depending on the baseline level (Horovitz et al., 1962b; Stein, 1964a; Stark et al., 1969; Olds, 1972; Domino and Olds, 1972; Stein and Wise, 1973; Phillips and Fibiger, 1973; Carey et al., 1974; Wauquier and Niemengeers, 1974). Self-stimulation of the locus ceruleus is more sensitive to amphetamine enhancement and lower doses of the drug are effective in this case (Ritter and Stein, 1973).

The time-response curves published for the effect of amphetamine on self-stimulation indicate that maximal facilitation occurs at 40–60 min after the amphetamine administration in rats (Olds, 1970; Domino and Olds, 1972; Newman, 1972; Cooper et al., 1974).

Methamphetamine seems to be as effective as amphetamine facilitating self-stimulation in rats (Poschel and Ninteman, 1966; Crow, 1969; Newman, 1972), but possibly not in cats (Umemoto and Kido, 1967). The dextroisomer of amphetamine has been reported to be 5–10 times more potent than the levo-isomer (Stein, 1964a; Stein and Wise, 1973), particularly if the stimulating electrode was located in the medial forebrain bundle, rather than in the substantia nigra, where the *d*/*l*-ratio was close to 1 (Phillips and Fibiger, 1973).

β) Effects of Other Drugs on Amphetamine-Induced Facilitation of Self-Stimulation

The response rate is decreased and/or the current threshold is increased by drugs which impair brain monoamine transmission, like chlorpromazine (Stein, 1962,

1964a; OLDS, 1972), reserpine (STEIN, 1962, 1964a), and tetrabenazine (OLDS, 1972). In all three instances amphetamine (0.5–2.0 mg/kg) reversed the effects of these drugs. If the dosages of both drugs are carefully chosen, chlorpromazine can be shown to diminish but prolong the facilitation induced by amphetamine (STEIN and SEIFTER, 1961). Repeated injections of the tyrosine hydroxylase inhibitor, α-methyltyrosine, were found to almost abolish responding for medial forebrain bundle stimulation; in rats thus treated, methamphetamine reinstated self-stimulation (POSCHEL and NINTEMAN, 1966). CROW (1969) reported that α-methyltyrosine antagonized the facilitation of tegmental self-stimulation induced by methamphetamine. Diethyldithiocarbamate, a DBH inhibitor, depressed lateral hypothalamic self-stimulation, an effect which was antagonized by amphetamine (WISE and STEIN, 1970). FLA-63, another DBH inhibitor (see Table 1), did not block the effect of amphetamine on self-stimulation (CANFIELD et al., 1973). Reserpine in combination with α-methyltyrosine completely abolished amphetamine-induced facilitation of self-stimulation in the lateral hypothalamus, while reserpine plus a DBH inhibitor did not (COOPER et al., 1974). COOPER et al. (1974) also used 6-OH-DA injected intracerebroventricularly in rats pretreated with inhibitors of the NA or DA uptake mechanisms, respectively. Thereby brain NA or DA pathways could be selectively destroyed. Amphetamine-induced facilitation of lateral hypothalamic self-stimulation was antagonized by the destruction of the DA pathways only.

Imipramine, iprindole, and cocaine potentiated the effect of amphetamine on self-stimulation (STEIN, 1962, 1964a; GLUCKMAN and BAUM, 1969), which in the light of later knowledge might probably be due to interaction of these drugs with the metabolism of amphetamine (see Sect. I, Chap. 1.A). Iproniazid, an MAO inhibitor, potentiated amphetamine (STEIN, 1964a). Methysergide, a 5HT blocking agent, was without effect on amphetamine in this respect (CROW, 1969). Amphetamine antagonized the depressant effect of physostigmine on selfstimulation (OLDS, 1972) and was itself potentiated by atropine (STEIN, 1962) and the choline acetyl transferase inhibitor, NVP (4-(1-naphtyl-vinyl)-pyridine) (MALICK and GOLDBERG, 1970). Mecamylamine, a centrally active antinicotinic drug, did not change the methamphetamine enhancement of hypothalamic self-stimulation (NEWMAN, 1972). A barbiturate, amylobarbitone, potentiated the effect of amphetamine on self-stimulation (COOPER et al., 1969) as has been shown for several other amphetamine effects.

The results obtained by the use of various pharmacologic tools for elucidation of the mechanism of action of amphetamine with respect to its effect on self-stimulation behavior are not as clear-cut as for other amphetamine effects. Particularly interesting are the findings that the effects on self-stimulation of substantial dosages of chlorpromazine and α-methyltyrosine might be reversed by methamphetamine. However, this is not direct proof against the presently held view that amphetamine enhances the release of CA induced by electrical brain stimulations since it might still be a matter of dosage (CROW, 1972b; STEIN and WISE, 1973; cf. GERMAN and BOWDEN, 1974). Brain cholinergic mechanisms seem to modulate the amphetamine-induced enhancement of stimulation in similar directions as for other effects.

γ) Effect of Repeated Injections of Amphetamine on Self-Stimulation

No tachyphylaxia to the enhancing effect on self-stimulation was observed after repeated injections of amphetamine (STEIN, 1964b). No chronic studies pointing to the possible development of tolerance to this effect of amphetamine have been found in the literature.

δ) Summary and Conclusions

Amphetamine in comparatively low doses seem to enhance intracerebral self-stimulation as evident from increasing response rates and decreased current thresholds for elicitation of the behavior. The mechanism of action for this effect of amphetamine appears to be mediated by interactions with release of central DA and NA. The effect seems to be modified by muscarinic cholinergic mechanisms. Amphetamine, thus, appears to act, at least partly, on the same neuronal substrates that subserves the acquisition and maintenance of self-stimulation under control conditions. General activation or, more probably, stereotype perseveration of motor behavior, in this case lever pressing, might contribute to some of the present findings.

d) Self-Administration of Amphetamine in Animals

α) Behavioral Patterns of Amphetamine Self-Administration

Self-administration of intravenous infusions of amphetamine and methamphetamine was shown to be rapidly initiated and maintained in rats (PICKENS et al., 1967) and in monkeys (DENEAU et al., 1969; WILSON et al., 1971; SCHLICHTING et al., 1971; STRETCH et al., 1971). Although several other amphetaminelike central stimulants and cocaine are also self-administered by animals (THOMPSON and PICKENS, 1970; WILSON et al., 1971; WILSON and SCHUSTER, 1972a, b) and are in fact interchangeable with each other (WILSON and SCHUSTER, 1972a, b; BALSTER and SCHUSTER, 1973; HOFFMEISTER and GOLDBERG, 1973), the present review will be concerned only with amphetamine and methamphetamine.

According to PICKENS (1968) and PICKENS et al. (1972), rats subjected to unlimited intravenous self-administration of the amphetamines on an FR schedule (one lever press per injection) typically spent 1–2 days pushing a lever that activated the infusion pump at irregular intervals. After 4 days they displayed a stable pattern of responding, which was interrupted by 1–2-day periods of self-selected withdrawal. Most animals died within about 14 days. Rats (YOKEL and PICKENS, 1973; DAVIS and SMITH, 1972) and monkeys (STRETCH et al., 1971) scheduled for 3–6 h of self-administration per day showed stable response patterns and survived for considerably longer periods of time. When unlimited access to self-administration was provided the food and water intakes, as well as the body weights, were reduced during the first days and the rats displayed increased motor activity and stereotyped behavior, including self-mutilations. Tolerance to the drug seems to develop after about 5 days, since the daily drug intake increased and the animals resumed feeding and drinking (PICKENS, 1968; PICKENS et al., 1972). However, at the end of a 14-day period the body weight was still reduced by 28–42%. A similar rapid initiation and maintenance of periodic self-administration was also observed in monkeys (DENEAU et al., 1969).

During the phase of stable self-administration in rats, specially designed experiments have shown that the response rate and thus, the number of infusions taken per hour, is inversely related to the amount of the drug per infusion, i.e., the daily drug intake remains fairly constant (satiation) both in rats (PICKENS et al., 1967; PICKENS and HARRIS, 1968; PICKENS, 1968; YOKEL and PICKENS, 1973) and monkeys (WILSON et al., 1971).

Interestingly, monkeys previously subjected to self-administration of phenobarbital were more rapidly initiated on amphetamine self-administration (SCHLICHTING et al., 1971). If the numbers of lever presses per injection were increased (FR 1 to FR 30), the response rate increased while the infusion rate remained constant (PICKENS et al., 1967; PICKENS and HARRIS, 1968). Sudden exchange of the drug solution for saline was followed by a short period of increased responding and then disappearance of the behaviour, i.e., a typical extinction pattern according to the operant behavior paradigm (PICKENS and HARRIS, 1968). In a two-lever choice situation, rats were able to pick the lever connected with the infusion pump and to switch levers on reversal (PICKENS, 1968); monkeys might be indiscriminate in their lever pressing at high doses of the drug, i.e., they stereotypically continue pressing a lever without drug reinforcements when intoxicated with the drug.

The intervals between infusions are longer if the dose per infusion is increased and, analogously, additional doses of the drug injected by the investigator caused a temporary dose-related pause in the self-administration (PICKENS et al., 1968). Two explanations for the dose-related response intervals have been proposed, namely that some direct effect of the drug limits the self-administration behavior or that the animals titrate themselves accurately to a certain drug concentration in the tissues (WILSON et al., 1971). Recent experiments in rats (YOKEL and PICKENS, 1974) showed that the plasma levels of amphetamine were surprisingly constant during 6 h self-administration sessions (blood samples were drawn at intervals immediately before an infusion) and, thus, favor the view that the animals are titrating themselves to a certain dose level.

Rats, self-administering both the *d*- and *l*-isomers of amphetamine and methamphetamine (YOKEL and PICKENS, 1973), demonstrated a constant ratio of approximately 2.6 between the self-titrated plasma levels of *d*- and *l*-amphetamine, respectively (YOKEL and PICKENS, 1974). This ratio was in agreement with the ratio of the self-selected daily intake of the two isomers and, thus, seems to give a measure of the relative potencies of *d*- and *l*-amphetamine to maintain this behavior. In monkeys, the *d*-/*l*-potency ratio, based on self-selected drug intake over time, is approximately 4 (BALSTER and SCHUSTER, 1973). D-amphetamine and *d*-methamphetamine seem to be approximately equipotent both in rats (YOKEL and PICKENS, 1973) and monkeys (BALSTER and SCHUSTER, 1973).

β) Effects of Other Drugs on Self-Administration of Amphetamine

A low dose of the CA synthesis inhibitor α-methyltyrosine (αMT) (5–10 mg/kg) caused an increase in the response rate at FR 1 for self-administration of amphetamine, while a high dose of αMT (20–80 mg/kg) abolished self-administration for a dose-dependent duration dependent on the dose of the inhibitor after which the animals seemed to resume the behavior at an increased response rate and then return to the baseline (PICKENS et al., 1968). This reaction to αMT has been interpreted to be due to a relative antagonism toward amphetamine at low doses or at the end of the action of a high dose, in which instances the rats respond as if a lower dose of amphetamine was received: at higher doses of αMT extinction occurred as if not amphetamine was given (vide supra). Also, the acquisition and reacquisition of amphetamine self-administration in rats as well as development of conditioned reinforcement by amphetamine can be blocked by αMT (DAVIS and SMITH, 1972, 1973).

Chlorpromazine, a CA receptor blocking agent, at 0.25 mg/kg caused an increased response rate for amphetamine in monkeys while 1 mg/kg antagonized self-administration completely (WILSON and SCHUSTER, 1972a, b). The interpretation of these effects of chlorpromazine was similar to the one given for αMT above, namely a partial amphetamine antagonism with the low dose and a total antagonism with high dose, although it was not possible to rule out the hypothesis that chlorpromazine at the low dose antagonized aversive effects of amphetamine and thereby permitted a higher amphetamine intake. The studies with α-methyltyrosine and chlorpromazine seem to indicate that central CA mechanisms are involved in this behavior.

Administration of *l*-dopa to rats did not influence self-administration of amphetamine (PICKENS et al., 1968).

A combination of a barbiturate and amphetamine was found to be self-administered by monkeys at higher daily doses than the sum of either of the components (SCHLICHTING et al., 1971). The self-administration behavior initiated and maintained by amphetamine and other dependence-producing drugs is thought to be due to their reinforcing properties (PICKENS et al., 1967; DENEAU et al., 1969; THOMPSON and PICKENS, 1970). Arguments favoring the operant conditioning paradigm include the spontaneous acquisition of the behavior, the dose-response, and the FR-response relationship and the satiation and extinctions patterns outlined above. Also, GOLDBERG (1973) has drawn attention to the parallelisms between drug-reinforced and food-reinforced operant behavior.

However, other factors may contribute to the maintenance of the self-administration behavior. In an experimental design where monkeys had been initiated on self-administration of amphetamine and subsequently subjected to extinction of the behavior, intraperitoneal injections of amphetamine elicited self-administration of saline (STRETCH et al., 1971; STRETCH and GERBER, 1973). The proposed explanations for this phenomenon were either that amphetamine induced general activation including increased lever pressing or, more likely, that the animals displayed the behavior as an expression of state-dependent learning. Further, discriminatory stimuli in the environment (e.g., buzzers) influences the rate of responding for amphetamine (STRETCH and GERBER, 1973).

γ) Summary and Conclusions

Amphetamine initiates and maintains intravenous self-administration behavior. This phenomenon is primarily regarded due to operant conditioning where the drug acts as a reinforcer. Stereotype lever pressing behavior may, in some instances, explain high rates of responding, e.g., for saline in amphetamine-pretreated animals. State-dependent learning and external conditioning may also contribute to the maintenance of the behavior. Interestingly, α-methyltyrosine and chlorpromazine, drugs which interfere with brain CA transmission and antagonize several other pharmacologic actions of amphetamine, also seem to antagonize the reinforcing properties of amphetamines in the self-administration situation. No studies on the effects of selective lesions of brain CA or other pathways have appeared in the literature on self-administration of amphetamine.

e) Amphetamine as an Aversive Stimulus

In addition to being a reinforcer in self-administration experiments, amphetamine was also shown to have aversive properties. Another amphetamine effect of this type is the induction of interoceptive discriminatory cues in state-dependent operant behaviors (Subsect. C.I.4.b). Further, the drug might serve as a conditioned stimulus for some of its own pharmacologic effects, namely motor activity (PICKENS and CROWDER, 1969) and stereotyped activity (ELLINWOOD, 1971a). Most investigations on the aversive effect of the drug have used the conditioned saccharin aversion (bait shyness) paradigm. The typical procedure is to give rats access to a saccharin solution or sweet milk (which are very palatable and are preferred over water in a two-bottle choice situation) for a limited time period and immediately afterward inject (meth-)amphetamine. After one to several such pairings, the rats are tested a few days later without amphetamine injections and the saccharin consumption is measured. In this situation it has been clearly demonstrated that amphetamine causes a reduction of the saccharin consumption, i.e., the drug probably serves as an aversive or noxious stimulus (CAPPELL and LEBLANC, 1971; BERGER, 1972; MARTIN and ELLINWOOD, 1973). It has been shown that this is a drug effect and not an artifact due to changes in the environment (test cage) (MARTIN and ELLINWOOD, 1974). It might be pointed out that several other drugs, including chlorpromazine but not saline (BERGER, 1972), have this property (CAPPELL and LEBLANC, 1973). Further, electroconvulsive shocks or footshocks were not as effective as aversive stimuli as amphetamine (BERGER, 1972). The dose of amphetamine necessary for inducing bait shyness depends on the number of pairings and varies between 1.5–2 mg/kg in a single pairing to 0.5 mg/kg if several pairings are performed (CAPPELL and LEBLANC, 1973). D- and *l*-amphetamine are both effective, but a four-fold increase in the dose of the *l*-isomer is necessary for equipotency (CAREY and GOODALL, 1974). Chlorpromazine given in combination with amphetamine during the conditioning phase did not antagonize the amphetamine effect (CAPPELL and LEBLANC, 1973). However, chlordiazepoxide given before the test trial antagonized the learned amphetamine aversion (CAPPELL and LEBLANC, 1973). Lesions in the area postrema, which reduce bait shyness for scopolamine in rats, did not antagonize the effect of amphetamine (BERGER et al., 1973).

Chronic treatment of rats with 2–4 mg/kg or 10–20 mg/kg of amphetamine daily for 15 days attenuated the development of saccharin aversion (LEBLANC and CAPPELL, 1974). This might be regarded as a sign of tolerance to the conditioned aversive properties of amphetamine, although other explanations have been discussed (LEBLANC and CAPPELL, 1974).

In another type of amphetamine aversions experiment, rats were subjected to a two-bottle choice situation: one bottle containing 0.5 mg/ml of amphetamine and the other water or a quinine solution. The animals learned to avoid the amphetamine bottle after a number of trials and drank only water or quinine solution (CAREY, 1973a). After acquisition of the amphetamine aversion in this situation, amphetamine solutions as weak as 0.001 mg/ml were recognized and avoided (CAREY, 1973a). The conditioned nature of this behavior was clearly shown in extinction experiments, where the rats continued to drink quinine even after exchange of the amphetamine solution for water (CAREY, 1973a). GLICK (1973a) showed that rats with frontal cortex lesions did not develop aversion to amphetamine solutions.

f) Maze Performance

Under this heading some studies of the effects of amphetamine on performance of maze running are summarized. The mazes may be simple runways, Y-mazes, or complicated T-mazes; sometimes in combination with swimming water mazes. This type of behavior has been included under operant behavior, since they are either negatively (water)or positively (food at the goal) reinforced. Discrimination tasks have been included in a few experiments. The measures might be running (swimming) speed and/or number of errors made.

Performance in simple runways was found to be decreased by all dose levels of amphetamine or methamphetamine (BARRY and MILLER, 1962; CARLSON et al., 1965; BREDA et al., 1969). With higher doses, increased rearing, sniffing, and other stereotypies interfere with the tasks; such rats were also resistant to extinction of the behavior (CARLSON et al., 1965). Amylobarbitone in combination with amphetamine also made rats resistant to extinction, due to potentiation of a subthreshold amphetamine dose (DUDDERIDGE and GRAY, 1974).

Rats experienced in a multiple T wading maze were unaffected by 0.25–1.0 mg/kg amphetamine at 1–2 h, while the performance was inferior to controls at 3–5 h after injection (DISPENSA and BARRETT, 1941). Reduced swimming times in a water maze after 2–4 mg/kg amphetamine was reported by MAICKEL et al. (1969). Amphetamine antagonized the deleterious effect of scopolamine in another study on maze performance (PAZZAGLI and PEPEU, 1964). Similarly, *dl*-, *d*-, and *l*-amphetamine and methamphetamine in doses above 2 mg/kg decreased the performance of rats in a multiple T maze; the relative potency of the isomers were $dl/d/l = 5/2.5/14$ on a mg/kg basis (EWING et al., 1952). Mice showed disorganized swimming and thus disrupted performance in a water T maze (LATZ et al., 1967). Discrimination in a water maze was either unaffected or disrupted by 5–10 mg/kg amphetamine in mice (CASTELLANO, 1971). In conclusion, low doses of amphetamine seldom improve performance and high doses impair maze performance in mice and rats.

g) Conditioned Reflexes

Very few studies on classical conditioned reflexes have been found in the literature. A reduction of the conditioned stimulus-response latency for conditioned salivation in dogs was reported by PLANE and BOGAERT (1953). In cats, amphetamine impaired a conditioned alimentary reflex, which was not specified further (NAVRATIL et al., 1973).

h) Miscellaneous Behavioral Studies

Since amphetamine induces a psychotic state in humans (see Sect. I, Chap. 2), it has been compared with hallucinogenic drugs on behavioral tests. Apart from the hallucinogenic profile of high doses of amphetamine in the Bovet-Gatti profile test in rats (SMYTHIES et al., 1969; BEATON and BRADLEY, 1972; see section on CAR above), amphetamine inhibits nest building in mice as do several hallucinogenic compounds (SCHNEIDER and CHENOWETH, 1970). However, psychotogenic effects of amphetamine might not show up unless the drug is administered repeatedly as is mostly the case in humans. In fact, chronically treated cats and monkeys do show gross behavioral changes that have been said to resemble certain aspects of amphetamine psychosis in man (see Subsect. C.I.1.a).

4. Effects of Amphetamine on Learning and Memory

a) Experiments on Learning and Memory

Several studies have been undertaken in order to elucidate the effects of amphetamine on acquisition, learning, and memory consolidation. The effects of amphetamine on acquisition of avoidance behavior have been discussed in a previous section (Subsect. C.I.3.a) and it was found difficult to distinguish between improved performance and a clear-cut facilitation of learning. In this section studies on learning of some other behavioral tasks are considered.

Maze learning in rats and mice was either impaired (MINKOWSKY, 1939; LATZ et al., 1967) or unaffected (EWING et al., 1952) by amphetamine. Learning of discrimination tasks in mice and rats and hamsters has been reported to be facilitated (LIBERSON et al., 1959; RENSCH and RAHMAN, 1960; GLICK, 1969, 1971 b; KRIVANEK and MCGAUGH, 1969; BÖSSER et al., 1970; RAHMAN, 1970; BAUER and DUNCAN, 1971; CASTELLANO, 1974), unaffected (MEAD, 1974), or retarded (KOSMAN, 1964; COLE, 1968; GLICK, 1971 a; BECKWITH et al., 1974; KELFER and ROSEN, 1974). Those studies reporting a facilitation of learning need some further comments. In the study of LIBERSON et al. (1959) amphetamine was used to antagonize "rigid behavior" (freezing) due to electric shocks which rats had to escape from. The golden hamsters in RENSCH and RAHMAN (1960) and RAHMAN (1970) were improved only by 0.5 mg/kg methamphetamine, while 2 mg/kg disrupted the behavior. Similar dose dependency was reported in rats (GLICK, 1971 b). BAUER and DUNCAN (1971) concluded that the better performance they observed under amphetamine was probably due to an effect on performance rather than acquisition. BÖSSER et al. (1970) treated the rats with a combination of amphetamine and amylobarbitone and observed an effect in female, but not male, rats in only one of two tasks.

Thus, four studies (one in hamsters, one in rats, and two in mice) remain which support the claim that (meth-)amphetamine facilitates learning in low to intermediate doses. Amphetamine-induced changes of the performance, of the cue value of discrimination, or of reward value of the reinforcing agents have been advanced as alternative explanations for the apparent facilitation of acquisition (BAUER and DUNCAN, 1971; GLICK, 1971 b; GOMER and JAKUBCZAK, 1974).

Although the reports on facilitation of acquisition and learning fail to be convincing, there is another category of studies, memory studies which offer more evidence for an effect of amphetamine on learning and memory consolidation. Thus, golden hamsters that had learned a discrimination task under methamphetamine were shown to remember the task in retrieval trials for approximately 100 days afterward as compared to 30 days for the controls (RENSCH and RAHMAN, 1960; RAHMAN, 1970). Mice given amphetamine before or after training in a water maze performed better than controls when tested 1 week later without amphetamine (KOSMAN, 1964). Furthermore, rats that had acquired a conditioned avoidance response (KULKARNI and JOB, 1967; EVANGELISTA et al., 1970), passive avoidance (JOHNSON and WAITE, 1971), or a conditioned emotional response (GLICK, 1969) under, after, or before administration of amphetamine, performed better than controls when retested within 1–7 days after withdrawal of the drug.

Amphetamine did not affect the retrieval and decay of stored information (WOLTHUIS, 1971). An improved performance on discrimination tasks in a recall trial was observed if amphetamine was given 30 min before the trial (BECKWITH et al., 1974). This cannot be taken as evidence for an enhancement of memory, but only as an enhanced performance of a previously learned response. Another model for memory is the spinal cord fixation time. An asymmetric posture is produced by a lesion in the cerebellum of rats, which persists if the spinal cord is not transected within 35–45 min after the first lesion. The fixation time was not changed by amphetamine in one study (MOURIVIEFF-LESUISSE and GIUREA, 1968), but was reduced, i.e., the spinal cord transection had to be performed earlier, according to another report (PALMER et al., 1970).

Taken together, the investigations on learning and memory have not provided solid evidence for an effect of amphetamine on these processes. The major difficulty seems to be to separate improved performance from facilitation of acquisition and memory consolidation. In future experiments the designs have to be improved in order to resolve the present controversies.

b) State-Dependent Learning

Although the question whether or not amphetamine improves learning is not resolved, it is clear that learning takes place under the influence of amphetamine and also that behaviors initially disrupted by amphetamine might be relearned under the influence of the drug (see sections on tolerance to operant behaviors) (Subsect. C.I.3.a and b).

The acquisition of a behavior under amphetamine has repeatedly been found to be retrieved much better if amphetamine instead of saline was injected before the recall trial (BELLEVILLE, 1964; BOHDANECKY and JARVIK, 1967; LAL, 1968, 1969; ROFFMAN and LAL, 1971; WOLTHUIS, 1971; COOPER et al., 1972). Only in one study was amphetamine found to be inactive (DAVIS et al., 1974). A clear disruption of a behavior acquired under amphetamine treatment followed on the withdrawal of the drug in rats (CAREY, 1973b; BARRETT et al., 1974; BARRETT and STERANKA, 1974). If amphetamine was withdrawn gradually, the clear dissociation between the drug and the nondrug state was attenuated (BARRETT and STERANKA, 1974). One study reported a failure of methamphetamine to produce state dependency on methamphetamine withdrawal (CROWLEY and SISEMORE, 1972).

However, the evidence that amphetamine-induced internal cues influence performance of behaviors acquired under amphetamine seems to be valid. In several experiments rats were trained to respond on one lever or to turn into one arm of a Y maze after injection of amphetamine, while the other lever or arm was the right response for saline or another drug, i.e., amphetamine serves as the discriminatory cue (HARRIS and BALSTER, 1971; WATER et al., 1972; SCHECHTER and ROSECRANS, 1973; JONES et al., 1974). Established amphetamine-dependent behaviors have been used to determine similarities and differences between amphetamine and other drugs with respect to the state they might induce. Thus, it has been shown that rats responding on two different levers for 0.3 and 2.5 mg/kg methamphetamine, respectively, could discriminate other doses of the drug as the amphetamine state (WATER et al., 1972). L-Amphetamine was recognized as *d*-amphetamine by rats trained to differentiate 4 mg/kg *d*-amphetamine from saline (SCHECHTER and ROSECRANS, 1973). Similarly, *d*- and *l*-amphetamine were not dissociated in the study by JONES et al. (1974). Parahydroxyamphetamine was not exchangeable for *d*-amphetamine, but caused the rats to respond on the saline lever (LAL and ROFFMAN, 1969; ROFFMAN and LAL, 1972; JONES et al., 1974).

The amphetamine-induced enteroceptive cues could be abolished by α-methyltyrosine or chlorpromazine (LAL and ROFFMAN, 1969; ROFFMAN and LAL, 1972). The blocking effect of α-methyltyrosine was reversed by administration of *l*-dopa (LAL and ROFFMAN, 1969). Reserpine eliminated the amphetamine state and *l*-dopa and 5-hydroxytryptophan both had to be given for a reversal of the reserpine effect (ROFFMAN and LAL, 1972). These authors also showed that a 5HT-receptor blocking agent, cyproheptadine, and the 5HT synthesis inhibitor PCPA, impaired the amphetamine state while 5-HTP reversed the PCPA effect. It was therefore concluded by ROFFMAN and LAL (1972) that the amphetamine state is dependent on both brain CA and 5HT neuronal mechanisms.

c) Summary and Conclusions

It is still questionable whether or not amphetamine facilitates learning and memory processes, since some support for both claims has been reported. It seems possible, however, to conclude that amphetamine does not impair these processes.

Amphetamine-induced state-dependent learning is presently well established and this phenomenon has proved to be of importance for elucidation of mechanisms of action of the amphetamine-induced state as well as an explanation for tolerance development to some effects of the drug and for some apparent signs of amphetamine abstinence.

II. Effects of Amphetamine on Food and Water Intake

1. Food Intake

a) Amphetamine-Induced Anorexia

Since the early observations by EHRICH and KRUMBHAAR (1937), HARRIS et al. (1947), and TAINTER (1943) that amphetamine caused a reduction of food intake in rats, this effect of the drug has been confirmed and investigated, and been utilized in the treatment of obesity.

The ED 50 for the anorexigenic effect of *d*-amphetamine in mice (SPENGLER and WASER, 1959; ABDALLAH and WHITE, 1970; ABDALLAH, 1973), rat (SPENGLER and WASER, 1959; ABDALLAH and WHITE, 1970; ALPHIN and WARD, 1969; MAICKEL et al., 1970; ABDALLAH, 1973), guinea pig (ALPHIN and WARD, 1969), cat (ABDALLAH and WHITE, 1970; ABDALLAH, 1973), and dog (ALPHIN and WARD, 1969; ABDALLAH and WHITE, 1970; CORSON et al., 1971; ABDALLAH, 1973) varies between 0.4 and 3.0 mg/kg. Chickens are an exception to the general rule of inhibition of food intake, since in this species amphetamine elicits increased pecking and eating (SPOONER and FLAMING, 1972). The ED 50 for anorexia is about one-tenth of the LD 50 for the drug (SPENGLER and WASER, 1959; ABDALLAH and WHITE, 1970; ABDALLAH, 1973) and seems to be higher than the ED 50 for several other effects of the drug (BÄTTIG, 1963b; MAICKEL et al., 1970).

Time response curves have shown that amphetamine causes an initial absolute anorexia for 0.5–3 h, depending on the dose. After this aphagic period, the animals start eating rapidly (ROSZKOWSKI and KELLEY, 1963; BIZZI et al., 1970; LEWANDER,

1971d, 1974b; LU et al., 1973). Negligence of this fact seems to be the reason for the great variability in the results obtained when the interaction of other drugs with the anorexigenic effect of amphetamine or amphetamine tolerance has been studied. Methodologically, the duration of the time period chosen for measurement of food intake and its relation to the time of injection of amphetamine are thus critical factors. It is well established that the *d*-isomer of amphetamine is more potent than the *l*-isomer (ROSZKOWSKI and KELLEY, 1963; LAWLOR et al., 1969; YELNOSKY and LAWLOR, 1970; FREY, 1970; COSTA et al., 1972; HAJOS and GARATTINI, 1973; LEWANDER, 1974b; BAEZ, 1974). Methamphetamine and several other amphetamine derivatives are also potent anorexigenics (ROSZKOWSKI and KELLEY, 1963; see review by VAN ROSSUM, 1970).

b) Effect of Other Drugs on Amphetamine-Induced Anorexia

Studies using various pharmacologic tools for the elucidation of the mechanism of the anorexigenic effect of amphetamine are summarized in Table 5. The pattern of results shows a partial similarity with what has been described for other amphetamine effects. Thus, α-methyltyrosine, a CA synthesis inhibitor, attenuates amphetamine-induced anorexia, but does not completely antagonize the initial aphagia caused by amphetamine. The attenuating effect of αMT is shown most clearly by an intermediary dose range; small doses of αMT are without effect, and large doses (>100 mg/kg) cause a decrease of food intake in rats. The NA synthesis inhibitor, disulfiram, antagonized amphetamine anorexia in one study (FREY and SCHULZ, 1973), while another DBH inhibitor, diethyldithiocarbamate, prolonged the effect of amphetamine in another (LEWANDER and JONSSON, 1973; JONSSON and LEWANDER, 1973). Neuroleptic drugs, either NA-receptor blockers, DA-receptor blockers, or both, do not completely antagonize amphetamine; at higher doses these drugs seem to potentiate or prolong the amphetamine effect due to aphagia caused by these drugs alone. It is concluded from these studies that doses of neuroleptics that completely inhibit amphetamine-induced stereotyped behavior and increased motor activity do not completely abolish the anorexigenic effect. Thus, amphetamine-induced anorexia is only partially explained by the presence or predominance of behaviors that, in themselves, decrease food intake. Further, the anorexigenic effect is at least partly caused by mechanisms other than release of brain CA.

The studies with α- and β-receptor blocking agents do not support the involvement of α- or β-receptor mechanisms for the anorexigenic effect of amphetamine. A recent finding that propranolol inhibits p-hydroxylation of amphetamine (SHOEMAN et al., 1974) prompts reinvestigation of the effect of β-receptor blocking agents on amphetamine-induced anorexia. Serotonin receptor blocking agents with assumed central activities and serotonin uptake inhibitors (e.g., chlorimipramine) were also without effect. This is in contrast to the inhibition of fenfluramine-induced anorexia shown to occur with these agents (FUNDERBURK et al., 1971; JESPERSON and SCHEEL-KRÜGER, 1973; KRUK, 1973; CLINESCHMIDT et al., 1974). Thus, it seems that amphetamine and fenfluramine might have different modes of action and that amphetamine does not utilize a serotonergic mechanism for its anorexigenic action.

An interesting feature is the antagonism by amphetamine of reserpine-induced aphagia (MENON et al., 1970; SCHMITT, 1973). This might be due to the increase in general activity caused by small to moderate doses of amphetamine in reserpine-

Table 5. Effects of other drugs on the intensity and duration of amphetamine induced anorexia. Symbols: No effect: 0. Intensity reduced (=eating): ↓, increased (more reduced food intake): ↑. Duration: prolonged: →, reduced: ←

Drug	Intensity	Duration	References		
			Rat	Mouse	Other
Neuroleptic drugs					
Chlorpromazine[e]	↑	→	4, 15		
Haloperidol	↓		5, 7		
Pimozide	↓		14[k]		
α-receptor blocking agents					
Phentolamine	↓		7		
Piperoxane	0		23		
Yohimbine	0		23		
Dibenamine	0		23		
β-receptor blocking agents					
Propranolol	0		23		
	↑			18, 19	
Prindolol	0		23		
5HT-receptor blocking agents					
Cyproheptadine	↓		7		
	0		5, 14[k]		
Methergoline[d]	0		8, 13		
	↑		5		
Methysergide	0		13, 23, 24		
	↑		5		
Cinanserin	↑		5		
Drugs affecting storage of monoamines					
Reserpine	↑		12[f], 23[i]		
	↓		21[f], 23[f]		
Tetrabenazine	0		4		
	↓		23		
Monoamine synthesis inhibitors					
Tyrosine hydroxylase inhibitor					
α-methyl-tyrosine (α-MT)	↓		3, 5, 7, 11, 17[g], 23, 26	1[g]	
α-MT + *l*-dopa	0		3		
Dopamine-β-hydroxylase inhibitor					
Disulfiram[e]	↓		7		
Tryptophane hydroxylase inhibitor					
p-chlorophenylalanine	↓		7[c]		
(PCPA)	0		7[c]		
Unnatural monoamine precursors					
α-methyl-m-tyrosine (α-MMT)	0			1	
Monoamine oxidase inhibitors					
Iproniazid	↑		25		
Pargyline	↑			1	
Others	↑		25		

Table 5 (continued)

Drug	Intensity	Duration	References: Rat	Mouse	Other
Inhibitors of monoamine uptake					
Imipramine[e]	↑		9		
Desmethylimipramine[e]	↑	→	20		
Chlorimipramine[e]	↑		13		
	0		5		
Sedatives and hypnotics					
Pentobarbital	↓			2	
Meprobamate	↓			2	
Chlordiazepoxide	↓			2	
Diazepam	↓			2	
Miscellaneous drugs					
6-hydroxydopamine[b] (6-OH-DA)	0		22[h]		
MAOI + 6-OH-DA	↓		6[a]		
p-hydroxyamphetamine	0		16		
Fenfluramine	↑		27		
Naloxone	0		10	10	
Iprindole[e]	↑		9		

[a] Tested after disappearance of aphagia caused by this drug combination.
[b] Given intracerebroventricularly.
[c] PCPA, 300 mg/kg i.p. given twice 48 and 24 h before amphetamine; a single injection had no effect.
[d] Methergoline = AHR 3009 = 8-β-carbobenzyloxy amino methyl-l-methyl-10-α-ergoline.
[e] Shown to inhibit amphetamine metabolism in vivo (see Sect. I, Chap. 1.A).
[f] Amphetamine antagonizes reserpine-induced aphagia.
[g] Depending on dose.
[h] In spite of central NA depletion to 15% of control level.
[i] Doses below 5 mg/kg i.p.
[k] Amphetamine given either i.p. or intracerebroventricularly.

References

1. ABDALLAH, 1971
2. ABDALLAH et al., 1974
3. BAEZ, 1974
4. BORELLA et al., 1969b
5. CLINESCHMIDT et al., 1974
6. FIBIGER, 1973
7. FREY and SCHULZ, 1973
8. FUNDERBURK et al., 1971
9. GLUCKMAN and BAUM, 1969
10. HOLTZMAN, 1974
11. HOLTZMAN and JEWETT, 1971
12. HULME et al., 1974
13. JESPERSEN and SCHEEL-KRÜGER, 1973
14. KRUK, 1973
15. LU et al., 1973
16. LEWANDER, 1971d
17. MAICKEL et al., 1970
18. MANTEGAZZA et al., 1970
19. MANTEGAZZA et al., 1968a
20. MENON et al., 1970
21. NEILL and GROSSMAN, 1971
22. SAMANIN et al., 1972
23. SCHMITT, 1973
24. SINGER et al., 1971
25. SPENGLER, 1962
26. WEISSMAN, et al., 1966
27. YELNOSKY and LAWLOR, 1970

treated animals, which makes feeding behavior possible. As for most effects of amphetamine, inhibitors of the uptake mechanism of monoamines and monoamine-oxidase inhibitors (which drugs also inhibit the metabolism of amphetamine in rats) potentiate amphetamine-induced anorexia.

Degeneration of brain CA neurons induced by intracerebroventricular injection of 6-hydroxydopamine in animals pretreated with an MAO inhibitor causes temporary aphagia in rats (FIBIGER, 1973). When the aphagia disappears and the animals resume eating, amphetamine anorexia is attenuated. Similar results have been obtained in rats recovered from lateral hypothalamic lesions (see below). It is not evident from this study if NA or DA neurons are involved. In another report (SAMANIN et al., 1972), where 6-hydroxydopamine was given alone, the effect of amphetamine on appetite was not altered. In this case only brain NA was completely depleted, while brain DA was reduced by 50%. These results suggest the involvement of a DA mechanism in the anorexigenic effect of amphetamine. This argument is, however, weakened by the finding of AHLSKOG and HOEBEL (1973) that lesions of the ventral NA bundle in the rat brain attenuate amphetamine anorexia. Rats chronically treated with an acetylcholine-esterase inhibitor (DFP-diisopropyl fluorophosphate) were similarly affected by amphetamine as the controls (OVERSTREET et al., 1972).

Another interesting observation is the antianorexigenic action of chlordiazepoxide and other sedative drugs (Table 5). This finding has not so far been adequately confirmed. The conditioned food or saccharin aversion caused by amphetamine (CAPPELL and LEBLANC, 1971; BERGER, 1972) is also antagonized by chlordiazepoxide (CAPPELL and LEBLANC, 1973). A punishing effect of amphetamine, thus, might contribute to their anorexigenic effect (MILLER, 1956).

In summary, drugs affecting CA functions in the brain seem to affect the anorexigenic effect of amphetamine. However, according to available evidence, amphetamine-induced release of central CA can only partially explain this effect of amphetamine; other, so far unknown, mechanisms must also be involved.

c) *Effects of Brain Lesions or Amphetamine-Induced Anorexia*

In view of the concepts of a ventromedial hypothalamic satiety center and a lateral hypothalamic hunger center, the obvious experiments to try were investigations of the effect of brain area lesions on amphetamine-induced anorexia. From Table 6 it is evident that ventromedial hypothalamic lesions do not attenuate but possibly enhance amphetamine anorexia. The only exception is the study by SHARP et al. (1962). In this study the effects of these lesions were tested at 24 h after experimentation with only four cats, and the food intake was measured for 24 h, which is too long for a disclosure of the initial absolute anorexigenic phase after amphetamine administration. These objections do not rule out a species difference, however. Taken together these studies do not support the contention that amphetamine should be dependent on the ventromedial hypothalamus for its anorexigenic effect. The lateral hypothalamic lesions, however, partially attenuated but did not abolish, amphetamine-induced anorexia. These results support the view that this area is of importance as a target area for amphetamine.

Indications for a role of the ventral NA bundle have also been proposed as a neural substrate for amphetamine anorexia (AHLSKOG and HOEBEL, 1973). Lesions of

Table 6. Effects of brain lesions on amphetamine-induced anorexia

Species	Site of lesion	Effect on amphetamine anorexia	Reference
Rat	Ventromedial Hypothalamus (VMH)	Exaggeration[a]	STOWE and MILLER, 1957
Rat	VMH	Prolongation[a]	EPSTEIN, 1959
Rat	VMH	Enhancement[a]	REYNOLDS, 1959
Cat	VMH	Attenuation[b]	SHARP et al., 1962
Mouse	VMH	Enhancement	MOGENSON et al., 1967
Rat	VMH	Enhancement[a]	BERNIER et al., 1969
Rat	VMH	No effect[a]	WISHART and WALLS, 1973
Rat	VMH	No effect[e]	PANKSEPP and BOOTH, 1973
Rat	Lateral hypothalamus	Attenuation[c]	CARLISLE, 1964
Rat	Lateral hypothalamus	Attenuation[c]	RUSSEK et al., 1973
Rat	Lateral hypothalamus	Attenuation[c]	PANKSEPP and BOOTH, 1973
Rat	Lateral hypothalamus	Attenuation[f]	BLUNDELL and LESHEM, 1974
Rat	Lateral hypothalamus	No effect[g]	BLUNDELL and LESHEM, 1974
Rat	Anterior hypothalamus	Enhancement	COLE, 1966, 1971[d]
Rat	Septum	Attenuation	CAREY and SALIM, 1970
Rat	Area postrema	Enhancement	CARLISLE and REYNOLDS, 1961
Rat	Area postrema	No effect	BERGER et al., 1973
Dog	Prefrontal Cortex	Attenuation	ANDERSSON and LARSSON, 1956
Rat	Raphé nuclei	No effect	NEILL et al., 1972
Rat	Raphé nuclei	No effect	SAMANIN et al., 1972
Rat	Ventral NA Bundle	Attenuation	AHLSKOG and HOEBEL, 1973
Rat	Dorsal NA Bundle	No effect	AHLSKOG and HOEBEL, 1973

[a] Tested in static hyperphagic phase.
[b] Tested 24 h after lesion.
[c] Tested after recovery from lateral hypothalamic aphagic phase.
[d] Supraoptic nuclei and other structures.
[e] Attenuation in some animals.
[f] Tested 8 weeks after lesion.
[g] Tested 20 weeks after lesion.

this pathway caused obesity and attenuated the anorexigenic effect of amphetamine. Lesions of the raphé nuclei in the mesencephalon, where the cell bodies of serotonin-containing neurons are located, did not affect amphetamine anorexia. However, this lesion abolishes the inhibition of food intake caused by fenfluramine (SAMANIN et al., 1972). Different modes of action for the two anorexigenics, thus, are once more suggested.

The enhancement of the amphetamine-induced anorexia by the lesions in the various other brain areas is difficult to reconcile with any relevant hypothesis for the site or mode of action of the drug.

d) Effects of Brain Stimulations on Amphetamine-Induced Anorexia

Electrical stimulation of the lateral hypothalamus elicits eating. It has been observed that higher current thresholds are required for this effect in amphetamine-treated animals (STARK and TOTTY, 1967). Furthermore, amphetamine-induced anorexia could be antagonized by lateral hypothalamic stimulation (THODE and CARLISLE, 1968). Both studies add support for an action of amphetamine within the lateral hypothalamus or on neurons with their axons passing through this area, or on nerve cells in connection with these neurons.

e) Effects of Amphetamine on the Electrical Activity of Neurons Within the Hypothalamus

Peripheral administration of amphetamine caused increased activity of the ventromedial hypothalamic neurons (BROBECK et al., 1956; KREBS et al., 1969; FOXWELL et al., 1969), while the neuronal activity in the lateral hypothalamus was unchanged (FOXWELL et al., 1969) or variably increased or decreased (KREBS et al., 1969). Iontophoretic application of amphetamine into the ventromedial or the perifornical area causes excitation of approximately one-quarter of the neurons, inhibition of one-quarter, and no effect on half of the neuronal population (KREBS and BINDRA, 1971). The finding by BROBECK et al. (1956) gave rise to the hypothesis that amphetamine mimics the ventromedial hypothalamic satiety center. However, the lesion studies described above argue against such a mechanism of action of amphetamine. The effects of amphetamine on the electric activity in the ventromedial hypothalamus must not necessarily be related to the anorexigenic effect of the drug.

f) Intracerebral Injections of Amphetamine and Food Intake

Intracerebral injection of amphetamine into the lateral hypothalamic area caused inhibition of food intake (BOOTH, 1968; LEIBOWITZ, 1970a, b; BLUNDELL and LESHEM, 1973), while injection into the ventromedial area either induced eating (LEIBOWITZ, 1970a, b) or had no effect (BLUNDELL and LESHEM, 1973) in rats. LEIBOWITZ (1970a, b) has suggested on the basis of intracerebral injections of amphetamine and a series of agents affecting α- or β-receptors in rats, that amphetamine causes anorexia by stimulating β-receptors in the lateral hypothalamus and causes eating by stimulating ventromedial α-receptors. This study needs confirmation, since peripherally administered α- or β-receptor blocking agents do not antagonize the anorexigenic effect of amphetamine as previously discussed.

Intracerebroventricular administration of NA elicits eating in amphetamine treated rats (BERGER et al., 1971). To the contrary is a recent report by BROEKKAMP et al. (1974) who showed that amphetamine in similar doses inhibits eating elicited by an NA injection into the perifornical area. The often reported finding (see BOOTH, 1968; LEIBOWITZ, 1970a) that intracerebroventricular or intracerebral NA injections elicits eating argues against NA as a mediator of amphetamine anorexia.

g) Other Factors Affecting Amphetamine-Induced Anorexia

As for several other amphetamine effects, the anorexigenic action seems to be dependent on genetic (strain) differences in dogs (CORSON et al., 1971) and mice (JORI and GARATTINI, 1973). In rats the anorexigenic effect does not become apparent until 15 days of age (LYTLE et al., 1971), and young rats are less susceptible to it than old ones (MACPHAIL and GOLLUP, 1974). Thyroid activity is of importance, since hypothyroid rats have been shown to be less sensitive to amphetamine than euthyroid rats (MANTEGAZZA et al., 1968b).

One important variable that affects the intensity of the anorexigenic effect of amphetamine is the duration of food deprivation before testing amphetamine. The longer the food deprivation period (up to 48 h) the less anorectic are rats (COLE, 1965, 1970, 1972; BERGER et al., 1971; MACPHAIL and GOLLUP, 1974). The study by SIEGEL and STERLING (1959), however, is in disagreement with this finding.

A decreased anorexigenic effect of amphetamine was also observed in fat-fed rats that had reduced stores of hepatic glycogen (RUSSEK and STEVENSON, 1972). This finding is in line with the hypothesis by RUSSEK and co-workers (RUSSEK et al., 1968; RUSSEK and BRUNI, 1970; see also review by RUSSEK, 1971) according to which amphetamine-induced glycogenolysis causes an increase in the glycose concentration in the liver cells, which in turn signals satiety via hepatic glucoreceptors and afferent neurons within the hepatic branch of the vagus nerve (NIIJIMA, 1969). Food deprivation causes depletion of liver glycogen, which might possibly explain the attenuation of the anorexigenic effect of amphetamine in deprived animals.

h) Effects of Chronic Administration of Amphetamine on Food Intake and Body Weight

Daily measurements of food intake during chronic administration of amphetamine have almost unanimously disclosed the development of tolerance to the anorexigenic effect of amphetamine in rats (SHAPIRO and FREEDMAN, 1957; HOLM et al., 1960; TORMEY and LASAGNA, 1960; ALPHIN and WARD, 1969; LAWLOR et al., 1969; LEWANDER, 1971d; GHOSH and PARVATHY, 1973; PANKSEPP and BOOTH, 1973; ABDALLAH, 1973; FANELLI, 1973; LU et al., 1973; MAGOUR et al., 1974a; SEVER and CALDWELL, 1974) and guinea pigs (ALPHIN and WARD, 1969; SEVER and CALDWELL, 1974).

Tolerance develops to *dl*- as well as *d*- and *l*-amphetamine (LAWLOR et al., 1969; LEWANDER, 1974b) and methamphetamine (YOKEL and PICKENS, 1970). Cross tolerance between *d*- and *l*-amphetamine has been reported in a preliminary form (TAGLIAMONTE et al., 1969; LEWANDER, 1974b). Several reports have drawn attention to the fact that tolerance does not develop to the initial absolute aphagic phase immediately after amphetamine administration in dogs (HARRIS et al., 1947) and rats (GHOSH and PARVATHY, 1973; LU et al., 1973; LEWANDER, 1974b). In those reports, where tolerance to the amphetamine-induced inhibition of food intake was not observed, rather small doses (<3 mg/kg) were administered (LEONARD et al., 1966; RUSSEK et al., 1968) or intermediate doses (about 5 mg/kg) for only 4–5 days (JORI and BERNARDI, 1969; ABDALLAH and WHITE, 1970). Generally higher doses, twice daily or more frequently for more than 1 week are necessary for a clear-cut tolerance development to occur. The time period chosen for assessment of the food intake in relation to the amphetamine injection should not be restricted to the first hours only after administration, since tolerance does not develop to the initial absolute anorexic phase. Food intake should not be measured for a 24-h period if high doses are not given, since only a small anorexigenic effect will be observed.

CARLTON and WOLGIN (1971) have drawn attention to the contingent tolerance to the anorexigenic effect of amphetamine. Tolerance does not develop, if the amphetamine injections were given immediately after the daily feeding period. These results should be confirmed with higher doses (> 3 mg/kg) of amphetamine.

In those studies in which body weight of the animals has been used as a measure of anorexia, tolerance has not always been found (ZIEM et al., 1970; LEWANDER, 1971d; ABDALLAH, 1973; LU et al., 1973; MAGOUR et al., 1974a), while in other studies tolerance could be demonstrated (EHRICH and KRUMBHAAR, 1937; SHAPIRO and FREEDMAN, 1957; TORMEY and LASAGNA, 1960). No single explanation for these discrepancies can be given. However, high doses (> 10 mg/kg) in combination with very short daily feeding schedules are not likely to be accompanied by tolerance development, as pointed out above. PANKSEPP and BOOTH (1973) showed that tolerance did not develop in rats with ventromedial hypothalamic lesions, but did develop in lateral hypothalamically lesioned rats. The same authors argue that tolerance development is due to accumulating food deprivation, rather than by some other mechanisms. The false transmitter (p-hydroxynorephedrine) hypothesis for amphetamine tolerance seems unlikely, since p-hydroxyamphetamine pretreatment does not attenuate the anorexigenic action of amphetamine (LEWANDER, 1971d). Further tolerance develops to the *l*-isomer of amphetamine, which is not converted to the false transmitter in rats. Finally, p-hydroxynorephedrine is not formed from any isomer in guinea pigs, in which species tolerance occurs (see above).

i) Summary of Proposed Mechanisms for the Anorexigenic Effect of Amphetamine

A series of different mechanisms for the anorexigenic action of amphetamine have been proposed and discussed in the literature (see COLE, 1967, 1973; COLE and GAY, 1974). *Behavioral incompatibility*, i.e., that amphetamine increases other behaviors such as stereotypies, exploration, motor activity (CARLSON et al., 1965; JENNINGS, 1968), and "freezing" (COLE, 1972) to the degree that feeding is made impossible, has been proposed. Against this explanation are the arguments that tolerance develops to the anorexigenic effect (see below) but not to the effects on motor activity or to stereotyped behavior that certain drugs (neuroleptics, α-methyltyrosine) completely antagonized the motor effects but not the anorexigenic action and that structure-activity relationships among several amphetamine analogues do not show a close relation between anorexigenic and motor stimulant activity (ROSZKOWSKI and KELLEY, 1963; RIVA et al., 1969; VAN ROSSUM and SIMMONS, 1969). The hypothesis that amphetamine-induced anorexia should be due simply to an increase in body temperature has been discussed and discarded by BRODIE et al. (1969). The proposal that amphetamine causes *nausea* and thereby inhibits appetite (MILLER, 1956) is hard to test. The recent hypothesis of activation by amphetamine of hepatic glucoreceptors (RUSSEK and BRUNI, 1970; RUSSEK, 1971) is interesting but needs more support.

Central mechanisms include the *ventromedial hypothalamus mimicking* and the *lateral hypothalamus blocking* hypotheses (COLE, 1973). The lack of an attenuating effect of lesions of the ventromedial hypothalamus argues against the ventromedial hypothalamic mimicking hypothesis (see above). However, the recent preliminary report of the importance of the ventral NA bundle for amphetamine anorexia has added new support to an involvement of a hypothalamic satiety center (AHLSKOG and HOEBEL, 1973). However, activation of hypothalamic α-receptors with NA and other sympathomimetics generally elicits eating (BOOTH, 1968; LEIBOWITZ, 1970a, b), which is difficult to reconcile with an amphetamine-induced release of NA as a mechanism for its anorexigenic action. Far more support for the lateral hypothalamic blocking hypothesis is available (see discussion by COLE, 1973; COLE and GAY, 1974; and above). The results presented by LEIBOWITZ (1970a, b) of an amphetamine-induced anorexia due to activation of lateral hypothalamic β-receptors are in accordance with a lateral hypothalamic mechanism, but need support. At present, the available evidence does not permit any definite conclusions with regard to the mechanism of action of the anorexigenic effect of amphetamine.

2. Hoarding Behavior

Hoarding (i.e., collection of food for future consumption) is a behavior probably related to long-term regulation of the nutritional status in certain rodents. ZUCKER and MILNER (1964) and BLUNDELL (1971) showed that hoarding behavior in rats was dose-dependently inhibited by amphetamine and other anorexigenic drugs, although low doses (<2.5 mg/kg) might increase hoarding (ZUCKER and MILNER, 1964). The time-response relationship between amphetamine and inhibition of hoarding, as well as development of tolerance to this effect of amphetamine was demonstrated by LEWANDER et al. (1973). A most interesting observation in this study was the increase of hoarding behavior after withdrawal of chronic amphetamine administration. It is not yet clear whether this is a sign of abstinence or not, since the changed behavior persisted for 3 weeks after amphetamine withdrawal and was not normalized at that time.

3. Water Intake

ANDERSSON and LARSSON (1956) reported that amphetamine inhibits drinking induced by a NaCl injection in dogs. Later, amphetamine-induced inhibition of water intake was reported in rats by several authors (EPSTEIN, 1959; MAICKEL et al., 1970; YELNOSKY and LAWLOR, 1970; KNOWLER and UKENA, 1973; WAYNER et al., 1973; WISHART and WALLS, 1973). However, low doses of the drug seem to increase water intake (RAHMAN, 1961; GLICK and MÜLLER, 1971). A dose-dependent decrease in water intake was also found in monkeys (BYRD, 1973). Drinking elicited by hypothalamic stimulation in the rat was also inhibited by amphetamine (MOGENSON, 1968). The taste of the water seems to be important, since hypodipsia is not seen after presentation of a sucrose solution to amphetamine-treated rats (MAICKEL and WEBB, 1972). Amphetamine-induced hypodipsia (like the anorexia) is slightly antagonized by pimozide (NIELSEN and LYON, 1973), α-methyltyrosine (HOLTZMAN and JEWETT, 1971) and reserpine (NEILL and GROSSMAN, 1971; HULME et al., 1974). Methysergide was without effect on amphetamine hypodipsia (SINGER et al., 1971).

Rats recovered from lateral hypothalamic lesions did not become less sensitive to the hypodipsic effect, only to the anorexigenic effect of amphetamine (CARLISLE, 1964). During chronic administration, tolerance to the hypodipsic effect of amphetamine (LAWLOR et al., 1969) and methamphetamine (YOKEL and PICKENS, 1970) was observed.

III. Autonomic, Metabolic, and Endocrine Effects of Amphetamine; Amphetamine Toxicity

1. Cardiovascular Effects of Amphetamine

The cardiovascular effects of amphetamine discussed in this section include the changes in systemic and regional blood pressure, heart rate, and peripheral resistance in the in vivo situation. Specific problems concerning the intricate details of the interaction of amphetamine with some peripheral noradrenergic neuron terminals such as in isolated atria and the aortic strip will be dealt with as well. Studies on other adrenergically innervated tissues were discussed under a previous heading (Subsect. B.II.1).

a) Arterial Blood Pressure

α) The Pressor Response

The first description of an effect of amphetamine on blood pressure can be traced back to a study by HARTUNG and MUNCH (1931) who found that the N-methylated derivative of phenylethylamine had biological effect, i.e., a pressor effect, in rats and rabbits, even when given perorally whereas phenylethylamine was active only after parenteral administration. A pressor effect with comparatively low doses (<2.5 mg/kg) of amphetamine was later demonstrated and confirmed in dogs (ALLES, 1933; DETRICK et al., 1937; PINKSTON et al., 1939), cats (DETRICK et al., 1937; ALLES, 1939; GUNN et al., 1939; WINDER et al., 1948; LOCKETT, 1950b; REINERT, 1958; HANNA, 1960; SCHMITT and SCHMITT, 1961a; DAY and RAND, 1962; HARVEY et al., 1968; WILL and BISGARD, 1972; ABDALLAH, 1974; BUYNISKI et al., 1974), rabbits (DETRICK et al., 1937; ALLES, 1939; GUNN et al., 1939; HANNA, 1960), and rats (SCHMITT and SCHMITT, 1961a; ALLAN et al., 1969; BIZZI et al., 1970).

The relative potency of amphetamine compared with adrenaline was reported to be about $^1/_{100}$–$^1/_{300}$ calculated on the doses required for certain blood pressure responses (ORZECHOWSKI, 1941; WINDER et al., 1948). The potency ratio *d*-/*l*-amphetamine has been reported to be 1–2 (ALLES, 1939; HANNA, 1960; HARVEY et al., 1968). Methamphetamine also increases the blood pressure in cats (MALORNY and ORZECHOWSKI, 1940) and dogs (WINDER et al., 1948). Parahydroxyamphetamine (HANNA, 1960; SCHMITT and SCHMITT, 1970; VALETTE et al., 1971), and p-hydroxynorephedrine (SCHMITT and SCHMITT, 1970; VALETTE et al., 1971) have been shown to have pressor effects in several animal species. However, p-hydroxyamphetamine has antihypertensive effects in addition to pressor effect under certain conditions (OHLER and WAKERLIN, 1953). It may be suggested that the delayed antihypertensive effect might be due to the depletion of NA and incorporation of p-hydroxynorephedrine into peripheral or central NA neurons.

β) The Depressor Response

Some authors have reported a depressor response after high doses (>4 mg/kg) of amphetamine in dogs (PINKSTON et al., 1939), and cats (REINERT, 1958). ALLES (1939) ascribed the depressor effect of amphetamine to deep anesthesia. LOCKETT (1950b) observed it after surgical sympathectomy and other investigators after development of tachyphylaxia to the pressor effects of amphetamine (GUNN et al., 1939; REINERT, 1958). Neither vagotomy, carotid sinus denervation, nor evisceration abolished the depressor response (PINKSTON et al., 1939). These same authors observed that ergotamine antagonized the depressor but not the pressor response. Spinalization converted the depressor to a pressor response, from which REINERT (1958) concluded that the depressor action was due to an effect of amphetamine on vasomotor centers in the CNS. Several years later HOYER and VAN ZWIETEN (1971, 1972) reported that 50–150 μg/kg of amphetamine injected into the vertebral artery of cats caused a depressor response, while intravenous injection caused a pressor response. The depressor response was abolished by centrally active α-receptor blocking agents (piperoxane, yohimbine, haloperidol) and was suggested to be due to a central noradrenergically mediated decrease in the peripheral sympathetic tone (see VAN ZWIETEN, 1973). Tachyphylaxis to the depressor response after repeated injections of amphetamine was reported by REINERT (1958).

γ) Other Effects of Amphetamine on the Systemic Blood Pressure

A sympatholytic effect of amphetamine has been reported by MALORNY and ORZECHOWSKI (1940), ORZECHOWSKI (1941), and WINDER et al. (1948). High or repeated doses of the drug antagonized the effects of NA or A on blood pressure. Other investigators have found, however, a potentiation of the pressor effects of CA after low doses of amphetamine (WINDER et al., 1948; HANNA, 1960; BOCKNIK and KULKARNI, 1973). The pressor response to carotid occlusion in cats and dogs was time-dependently decreased by 1–2 mg/kg of amphetamine (DAY and RAND, 1962; COMMARATO and LUM, 1970). Whether this finding might be explained by the central depressor action of the drug (above), its sympatholytic (see Subsect. B.II.1) or ganglionic blocking effect (REINERT, 1958; GOLD and REINERT, 1960), is not known. Similarly, the blood pressure increase after stimulation of the central stump of the cut vagus nerve was reduced by amphetamine (DAY and RAND, 1962).

δ) Heart Rate

The heart rate increases after amphetamine in most species (ALLES, 1933; DETRICK et al., 1937; GUNN et al., 1939; LOCKETT, 1950b; HARVEY et al., 1968; IWATA et al., 1969; STEPANEK, 1972; WEIS, 1973), except in thiamine-deficient rats (IWATA et al., 1969). In the conscious dog, WILL and BISGARD (1972) observed an increased heart rate, while in another study there was a drop in heart rate with 1–2 mg/kg of amphetamine, but no effect with higher doses (BUYNISKI et al., 1974). Only one study has been found on the cardiovascular effects of amphetamine in race horses (SMETZER et al., 1972). Very small, insignificant changes of the heart rate were observed after 150 mg of amphetamine. However, an increased frequency of ectopic heart beats and AV block II occurred immediately after the race in the amphetamine-treated horses.

The Isolated Heart and the Heart-Lung Preparation. Amphetamine stimulated the rate of isolated guinea pig atria in low concentrations (TRENDELENBURG et al., 1963; BENFEY and VARMA, 1964; SMITH, 1966), but reduced the rate at high (10–100 μg/ml) concentrations (TRENDELENBURG et al., 1963; SMITH, 1966) for unknown reasons. The stimulating effect was reduced by previous reserpine treatment of the animals (TRENDELENBURG et al., 1963) and potentiated by MAO inhibitors (SMITH, 1966; ANTONACCIO and SMITH, 1969). Similarly, the increase in heart rate caused by amphetamine in the dog heart-lung preparation is antagonized by previous in vivo reserpine treatment (BEJRABLAYA et al., 1958; LIEBMAN, 1961) but not by treatment with hexamethonium, a ganglionic blocking agent (BEJRABLAYA et al., 1958). The above findings support the idea that amphetamine causes the increase in heart rate by releasing endogenous noradrenaline from the NA nerve terminals; direct evidence for such an action of the drug on the heart was provided by CHIDSEY et al. (1962), who demonstrated increased concentrations of NA in the coronary sinus blood in amphetamine treated dogs.

ε) Peripheral Resistance

The total peripheral resistance increases after amphetamine (HARVEY et al., 1968; COMMARATO and LUM, 1970) and so does the local resistance in the perfused dog hindlimb (EBLE and RUDZIK, 1965), in the rat tail (ALLAN et al., 1969; BIZZI et al., 1970; BABULOVA et al., 1972) and the rabbit ear (BURN and RAND, 1958). The combination of amphetamine and tetrabenazine also increased the perfusion pressure in the rat tail and this effect was blocked by fenfluramine and some of its congeners (JESPERSEN and BONACCORSI, 1969a).

The isolated aortic strip from rabbit has been used for detailed studies ot the mechanism of action of amphetamine. Thus, *d*- and *l*-amphetamine cannot contract this preparation if the strips are taken from reserpine-treated animals, while MAO inhibitors potentiate its effects (MAENGWYN-DAVIES et al., 1966). Introduction into the in vitro system of a strip of spleen tissue from an untreated animal restored the response of the aortic strip from a reserpine-treated animal, but if the spleen tissue, as well was taken from a reserpine-treated rabbit, no restoration occurred (MAENGWYN-DAVIES et al., 1966). These experiments clearly show the importance of endogenous NA for the action of amphetamine. Cocaine inhibited the effect of low concentrations of amphetamine on the isolated aortic strip, probably by inhibiting the uptake of amphetamine into the NA neurons (MAENGWYN-DAVIES and KOPPANYI, 1966). Washing out of cocaine restored the action of amphetamine unless α-methyl-p-tyrosine or disulfiram (CA synthesis inhibitors) were introduced into the bath, again demonstrating the importance of endogenous NA for the effect of amphetamine (MAENGWYN-DAVIES and KOPPANYI, 1966).

b) Effect of Amphetamine on Pulmonary Circulation

Amphetamine seems to have little or no effects on pulmonary blood pressure in the dog (STÜHLINGER et al., 1971; WILL and BISGARD, 1972; SCHUSTER and IHLI, 1973; ABDALLAH, 1974) and the calf (BYRNE-QUINN and GROVER, 1972). Amphetamine, in contrast to some other anorexigenic agents (e.g., aminorex), did not induce pulmonary hypertension on chronic administration (BYRNE-QUINN and GROVER, 1972).

c) Effect of Amphetamine on Cerebral Circulation

On this subject, only one study has been found, showing that methamphetamine causes cerebral vasoconstriction and decreased blood flow, as visualized by angiography of the carotid artery of rhesus monkeys (RUMBAUGH et al., 1971). The importance of this finding for the central and toxic effects of the amphetamine is not known. The subject deserves further investigation.

d) Effects of Surgical Manipulations on the Pressor Response to Amphetamine

Spinalization of cats (GUNN et al., 1939; DAY and RAND, 1963b; BHAGAT, 1965a, b; DAY, 1967), pithing of rats (SCHMITT and SCHMITT, 1961a; DAY, 1967), or vagectomy of dogs (WINDER et al., 1948; COMMARATO and LUM, 1970) did not affect the pressor response to amphetamine. Adrenalectomy has been claimed to reduce the duration of the pressor action of amphetamine in cats (REINERT, 1958). Other studies have shown that adrenalectomy did not abolish the pressor effect of an amphetamine plus reserpine combination (see Subsect. C.III.e.α), however.

There is a strong correlation between the pressor response to amphetamine and increased plasma levels of CA in dogs, which supports the view that amphetamine acts by releasing NA and A (HARVEY et al., 1968). Potentiation of the pressor effect of electrical stimulation of the posterior hypothalamus also supports this view (BABULOVA et al., 1972).

e) Effects of Other Drugs on the Pressor Effect of Amphetamine

The pharmacologic analysis of the mechanisms of action of amphetamine with respect to its pressor effect (Table 7) supports the view that the drug acts indirectly through release of NA. Thus, drugs depleting stores of NA in the terminals of peripheral NA neurons, i.e., reserpine, tetrabenazine, guanethidine, and similar compounds α-methyldopa and other unnatural MA precursors and MA as well as inhibitors of CA synthesis, antagonized the effects of amphetamine. Cocaine, an NA uptake inhibitor, probably also inhibits the entrance of amphetamine into the NA terminals, thereby decreasing the pressor response. MAO-inhibitors, which increase the amounts of NA within the terminals, potentiated, and even prolonged the amphetamine-induced increase in blood pressure. In animal species, where p-hydroxylation of amphetamine occurs in the liver, the prolongation of the action of amphetamine might be due to a metabolic interaction by which the rate of metabolism of amphetamine is retarded (Sect. I, Chap. 1.A).

A series of α-receptor blocking agents inhibit the pressor response to amphetamine, while β-receptor blocking agents, anticholinergics, and 5HT/histamine receptor blockers are without influence. The lack of an inhibition of the amphetamine-induced pressor response after spinalization and pithing (above) clearly indicates that amphetamine, in causing its pressor response, acts on the noradrenergic terminals in the periphery (and the adrenomedullary cells) and not on vasomotor centers in the CNS.

Comments on the Reserpine-, Guanethidine-, Tyramine-, and Amphetamine Interactions. Some comments of qualification have to be made on the interactions between reserpine, guanethidine, or tyramine, on one hand, and amphetamine on the other.

α) The Reserpine-Amphetamine Interaction

First, the reserpine-induced inhibition of the pressor response after amphetamine seems to occur when reserpine is given in relatively high doses (>1 mg/kg) at least 3–4 h before amphetamine or in repeated doses (BURN and RAND, 1958; SCHMITT and SCHMITT, 1960). If amphetamine was administered shortly (less than 3 h) after reserpine, there was a potentiation of the pressor effect of amphetamine (SCHMITT and SCHMITT, 1960). Apparently, the NA stores in the noradrenergic terminal have to be

Table 7. Effects of other drugs on pressor effect of amphetamine. Symbols: No effect: 0, intensity increased: ↑, decreased: ↓; duration: prolonged: →, reduced: ← ( ) = weak action

Drug	Intensity	Duration	References			
			Rat	Dog	Cat	Rabbit
Neuroleptic drugs						
Chlorpromazine	0				32[g]	
α-receptor blocking agents						
Ergotamine	(↓)			9	9	9
Phentolamine	↓		16, 17		32[g]	
Phenoxybenzamine	↓		28[h]			
Dibozane	↓			34[g]		
β-receptor blocking agents						
Propranolol	(↓)		16			
5HT receptor blocking agents						
Cinanserin	0		13			
Cyproheptadine	(↓)		16, 17			
LSD-25	0				32	
Drugs affecting storage of monoamines						
Reserpine	↑		4, 27, 29	11, 22, 23, 24	32, 33	
	↓		16, 17 26[g], 28[h]	14, 24 30, 34	5[c], 31	
Tetrabenzine	↑		4			
Guanethidinelike drugs						
Guanethidine	↓		3, 26[g]	12, 15, 30		
	0[e]			8, 10	8	
TM-10[b]	↓		3			
Xylocholine	↓		2			
Unnatural monoamines and precursors						
α-methyldopa	↓			30		
α-methyl-m-tyrosine	↓		29	30		
Metaraminol	↓		29			
Tyramine	↓			20[f], 21[f]		
Octopamine	↓		29			
p-hydroxynorephedrine	↓		29			
CA synthesis inhitors						
TH-inhibitor						
α-methyltyrosine	↓		29			
DBH-inhibitor						
Disulfiram	↓		29			
DDC-inhibitor						
See α-methyldopa above						
MAO-inhibitors						
Iproniazid	↑	→		25		
Pargyline	↑	→	18		18	
Nialamide	↑	→	18		18	
Phenelzine	↑	→	18		18	
Tranylcypromine	↑	→	18		18	
Others	↑	→	18,28		18	

Table 7 (continued)

Drug	Intensity	Duration	References Rat	Dog	Cat	Rabbit
MA uptake inhibitors						
Cocaine	↓		16, 17,	9, 12	9	9
	0		26[g]		19	
Anticholinergics						
Atropine	0		1[a]	34[g]		
Ganglionic blocking agents						
Hexamethonium	↓				19	
Tetramethylammonium	0			32[g]		
Mecamylamine	0			34[g]		
Pentolinium	0			34[g]		
Miscellaneous drugs						
Fenpromex	↓				2	
Fenfluramine	↓				2	
Angiotensin	↑			7[d]		
Methylphenidate	↓		26[g]			
Hydralazine	↓		26[g]			

[a] Complicated interaction between atropine and amphetamine, suggesting atropine effect to be α-adrenergic blocking rather than antimuscarinic.
[b] 2,6-xylyl-choline ether bromide.
[c] Infusion of NA restored the response.
[d] Amphetamine potentiated pressor effect of angiotensin.
[e] Amphetamine antagonized depressor effect of guanethidine; see text.
[f] Amphetamine potentiated pressor effect of tyramine.
[g] Also antagonized pressor effect of amphetamine plus reserpine; see text.
[h] Antagonized pressor effect of harmaline plus amphetamine.

References

1. ALLAN et al., 1969
2. BERRY et al., 1971
3. BHAGAT and SHIDEMAN, 1963
4. BONACCORSI, 1968
5. BURN and RAND, 1958
6. CESSION-FOSSION, 1963a
7. COMMARATO and LUM, 1970
8. DAY and RAND, 1962
9. DETRICK et al., 1937
10. GOKHALE et al., 1966
11. HARRISON et al., 1963
12. HARVEY et al., 1968
13. HUANG and HO, 1973
14. MAXWELL et al., 1959
15. MAXWELL et al., 1960
17. NIELSEN and FREY, 1969a
17. NIELSEN and FREY, 1969b
18. RAND and TRINKER, 1968
19. REINERT, 1958
20. RUDZIK and EBLE, 1960
21. RUDZIK and EBLE, 1967
22. SCHMITT and SCHMITT, 1957
23. SCHMITT and SCHMITT, 1959
24. SCHMITT and SCHMITT, 1960
25. SCHMITT and SCHMITT, 1961a
26. SCHMITT and SCHMITT, 1961b
27. SCHMITT et al., 1962
28. SCHMITT and SCHMITT, 1964
29. SCHMITT and SCHMITT, 1970
30. STONE et al., 1962
31. TRENDELENBURG et al., 1962a
32. VALDECASAS et al., 1958
33. VIZI et al., 1968
34. YELNOSKY et al., 1966b

emptied by reserpine before a blockade of the effects of amphetamine takes place, while during the course of the reserpine-induced depletion of NA more of the transmitter is available for release by amphetamine. Secondly, if reserpine was given after amphetamine, again there was a marked potentiation of the pressor effect of amphetamine. Even after return of the increased blood pressure after amphetamine to control levels in rats, cats, or dogs, the addition of reserpine (SCHMITT and SCHMITT, 1957; VALDECASAS et al., 1958) or tetrabenazine (JESPERSEN and BONACCORSI, 1969a; BIZZI et al., 1970) caused a strong pressor response. This amphetamine plus reserpine response can be abolished by previous reserpine administration (VALDECASAS et al., 1958; YELNOSKY et al., 1966b), and by α-receptor blocking agents (e.g., hydralazine) (VALDECASAS et al., 1958; SCHMITT and SCHMITT, 1959, 1961b, 1966, 1970; YELNOSKY et al., 1966a). However, ganglionic blocking agents (VALDECASAS et al., 1958; SCHMITT and SCHMITT, 1961a; YELNOSKY et al., 1966b) or adrenalectomy (SCHMITT and SCHMITT, 1959; YELNOSKY et al., 1966b) did not alter the amphetamine plus reserpine response. MAO-inhibitors potentiated the pressor effect of reserpine in amphetamine-treated animals (SCHMITT and SCHMITT, 1961a, 1970). Interestingly, reserpine induced a pressor response even in animals that were tachyphylactic to amphetamine (YELNOSKY et al., 1966a). One interpretation of the results of the analysis of the amphetamine plus reserpine interaction is that amphetamine releases NA being displaced from its storage vesicles by reserpine and simultaneously inhibits its re-uptake (SCHMITT and SCHMITT, 1970).

β) The Guanethidine-Amphetamine Interaction

This interaction has received much attention since it was shown that amphetamine given before or after guanethidine abolished the effect of this adrenergic neuron blocking and antihypertensive agent (DAY and RAND, 1962). The effect of the first, but not of a subsequent dose of amphetamine, was inhibited, though. Although the investigation of this interaction, which apparently takes place on the NA neuron terminal level, has been performed on noradrenergically innervated tissues other than the cardiovascular system, the matter will be dealt with here. Amphetamine reversed the guanethidine-induced blockade of the pressor response to carotid occlusion, electrical stimulation of the central stump of the cut vagus nerve, or physostigmine injections in rats and cats (DAY and RAND, 1962; GOKHALE et al., 1966; SPRIGGS, 1966). Further, amphetamine antagonized the blockade induced by guanethidine, bretylium, xylocholine, and bethanidine of the electrically stimulated nictitating membrane in cat (DAY, 1962; DAY and RAND, 1963a), the innervated ileum, the ear, and the spleen in rabbits and sheep (GERKENS et al., 1969), the rat inferior eyelid (SPRIGGS, 1966; OBIANWU, 1969b; BABULOVA et al., 1972), the guinea pig vas deferens (DAY and RAND, 1963a; FOLLENFANT and ROBSON, 1970), and the rabbit perfused heart (STARKE, 1972). Biochemical studies showed that amphetamine (or methamphetamine) inhibited the uptake of guanethidine (and related drugs) into the rabbit and rat heart (MATSUMOTO and HORITA, 1963; CHANG et al., 1965) and the guinea pig vas deferens (DEAN and HUGHES, 1971, 1972). In addition, amphetamine prevented the guanethidine-induced depletion of heart CA (MATSUMOTO and HORITA, 1963; CHANG et al., 1965; MAITRE and BRUNNER, 1965), and caused a release of guanethidine from noradrenergically innervated tissues (CHANG et al., 1965; FLEGIN et al., 1970; DEAN and HUGHES, 1971, 1972).

It has thus been concluded that amphetamine inhibits the uptake of guanethidine and releases it from NA nerve terminals. It was initially thought that amphetamine and guanethidine competed for the same intraneuronal pool of NA (MATSUMOTAO and HORITA, 1963), but subsequent studies indicated that guanethidine is probably taken up and temporarily stored in at least two different pools (CHANG et al., 1965; OBIANWU et al., 1968) and that amphetamine interferes with one of them (CHANG et al., 1965; OBIANWU, 1969b).

γ) The Tyramine-Amphetamine Interaction

Tyramine given before amphetamine inhibited the pressor response of the latter substance (Table 7). However, the tyramine pressor response was potentiated if the animals (dogs, rats) were pretreated with amphetamine in a single (EBLE and RUDZIK, 1966a, b, 1970) or repeated (tachyphylaxis) injection(s) (CESSION-FOSSION, 1963a, b; EBLE and RUDZIK, 1965). The potentiation of tyramine by amphetamine seems to be due partly to inhibition of the metabolism of tyramine in the liver (EBLE and RUDZIK, 1966b, 1970).

f) Effects of Repeated Doses of Amphetamine on Blood Pressure

On repeated injections of amphetamine the pressor response becomes weaker. This tachyphylactogenic property of amphetamine is discussed in detail in Subsection B.II.1.

g) Summary and Conclusions

Amphetamine causes an increase in the systemic arterial blood pressure by increasing the peripheral resistance and accelerating the heart rate. The pressor effect is due to the release of CA from peripheral NA neurons and the adrenal medulla. All available evidence favors the view that amphetamine is an indirectly acting sympathomimetic amine in this respect, and there is no evidence against it.

In addition to the peripheral pressor effect, amphetamine seems to have a central depressor action, probably due to an indirect sympathomimetic action on central α-adrenergic receptors. Tachyphylaxis develops rapidly to the peripheral pressor effect, and possibly also to the central depressor effect.

There is some evidence for an amphetamine-induced reduction of the cerebral circulation. Such an action might be of importance for the acute and chronic toxicity of amphetamine and should therefore be further investigated.

2. Respiratory Effects of Amphetamine

Amphetamine increases respiratory rate in cats, dogs, and rabbits (ALLES, 1933; DETRICK et al., 1937; STEPANEK, 1972) and morphine does not antagonize this effect in the dog (HANDLEY and SWEENEY, 1941). Tachyphylaxis develops to this effect (DETRICK et al., 1937). Close arterial injection of amphetamine into the vascular supply of the glomus caroticum in cats showed that the drug might stimulate the glomus cells; denervation of the carotid body abolished the stimulation of the respiration (REINERT, 1958). Bronchodilatation, although poor and variable (PEDDEN et al., 1935), and nasal decongestion (WINDER et al., 1948) are other respiratory effects of amphetamine in animals. Amphetamine's effects on oxygen consumption are discussed in Subsection C.III.5.

3. Amphetamine-Induced Salivation

Amphetamine caused increased salivation in mice, probably as a secondary phenomenon (heat dissipating mechanism) to its hyperthermic effect (MALING et al., 1972; KOPPANYI and MALING, 1972; MALING and KOPPANYI, 1972). The increased salivation was reduced or abolished by α- and β-receptor blocking agents, ganglionic blocking agents, and anticholinergic drugs (idem., see Subsect. C.III.4).

4. Body Temperature

Administration of amphetamine leads to marked changes in body temperature in rats, mice, rabbits, guinea pigs, and cats. However, the dose of the drug and the mode of administration and the ambient temperature are important determinants as to whether there will be an increase or a decrease in body temperature. Thus, hypothermia occurs after intraperitoneal administration of doses below 1–5 mg/kg in rats (JELLINEK, 1971; KRUK, 1972) and mice (MCCULLOUGH et al., 1970) and intracerebroventricular injection of amounts of the drug below 100–400 μg in rats (JELLINEK, 1971; KRUK and BRITTAIN, 1972), while higher doses given peripherally in rats (JELLINEK, 1971) or mice (MCCULLOUGH et al., 1970) or intracerebroventricularly in mice (MCCULLOUGH et al., 1970) caused hyperthermia, sometimes followed by hypothermia (RIVA et al., 1969; MCCULLOUGH et al., 1970). In cats, amounts of more than 400 μg intracerebroventricularly cause a decrease, while lower doses cause an increase in body temperature (TERWELP et al., 1973).

At ambient temperatures above 20° C there is generally a hyperthermic response, while at temperatures below 15° C there is a hypothermic response in mice (DOLFINI et al., 1969a) and rats (COPER et al., 1971; YEHUDA and WURTMAN, 1972a; BORBELY et al., 1974; SCHEEL-KRÜGER and HASSELAGER, 1974) after intraperitoneal injection.

In the following presentation, amphetamine-induced hyperthermia and hypothermia will be treated separately.

a) Amphetamine-Induced Hyperthermia

An increase in body temperature induced by amphetamine was first reported in rabbits by HAFFNER (1938). This effect of amphetamine was confirmed in rabbits, guinea pigs, and mice (HAAS, 1939; KIESSIG, 1941). HAAS (1939) reported, in addition, a hypothermic response in rats. Later, the hyperthermic effect was described in rats (SIMONYI and SZENTGYÖRGI, 1949). The peak hyperthermic effect, measured in the rectum, colon, or esophagus, occurs about 30–60 min after injection of the drug and the hyperthermia persists for 2–4 h depending on the dose and the species. Dose- and time-response relationships have been repeatedly reported in rats (HAAS, 1939; BRODIE et al., 1969; GESSA et al., 1969; RIVA et al., 1969; COPER et al., 1971; MATSUMOTO and GRIFFIN, 1971; YEHUDA and WURTMAN, 1972a; MAJ et al., 1972a; HAJOS and GARATTINI, 1973; BORBÉLY et al., 1974), in mice (ASKEW, 1962; DOLFINI et al., 1969a, b; BIZZI et al., 1970; CRAIG and KUPFERBERG, 1972; MALING et al., 1972), and rabbits (HAFFNER, 1938; HAAS, 1939; HILL and HORITA, 1971; HORITA and HILL, 1973). However, BELENKY and VITOLINA (1962) showed a peak hyperthermic effect after 3 h in rabbits and cats (intraperitoneal measurement).

D-Amphetamine was found to be more potent than *l*-amphetamine in rats (HAJOS and GARATTINI, 1973; LEWANDER et al., 1975) or equally potent (MAJ et al., 1972a). On intracerebroventricular administration in the cat, *d*-, *l*-, and *dl*-amphetamine were found to be equipotent with regard to the body temperature response (TERWELP et al., 1973). Parahydroxyamphetamine, which is a metabolite of amphetamine in some species (see Sect. I, Chap. 1.A), is less potent than the parent compound, but has hyperthermic actions in rats (VALZELLI et al., 1967b; BRODIE et al., 1969) and mice (MCCULLOUGH et al., 1970). Pretreatment of rats with desmethylimipramine (DMI), an inhibitor of the monoamine uptake mechanism in NA-neurons, abolished p-hydroxyamphetamine-induced hyperthermia in rats (VALZELLI et al., 1967b). Intracerebroventricular administration of p-hydroxyamphetamine in mice causes a hypothermic response (MCCULLOUGH et al., 1970). Methamphetamine also causes hyperthermia in rats and guinea pigs (HAAS, 1939) and rabbits (THUILLIER et al., 1962).

b) Effects of Other Drugs on Amphetamine-Induced Hyperthermia

The literature on this subject has been compiled in Table 8. Neuroleptic drugs, including those with purported dopamine receptor blocking activity, antagonized amphetamine-induced hyperthermia. Generally, α- and β-receptor blocking agents were also effective in this respect. Catecholamine synthesis inhibitors, particularly α-methyltyrosine, as well as the 5HT-synthesis inhibitor, parachlorophenylalamine,

Table 8. Effect of other drugs on intensity and duration of amphetamine-induced hyperthermia. Symbols: No effect: 0, Intensity, increased: ↑, decreased: ↓. Duration, reduced: ←, prolonged: →

Drug	Intensity	Duration	References				
			Rat	Mouse	Rabbit	Cat	Guinea-pig
Neuroleptic drugs							
Chlorpromazine[a]	↓	→	8, 41, 45[c], 46, 47, 57, 61				
	0		61[l]		48		
	↓			1			
Propericiazine	↓		57				
Perphenazine	↓		45[c], 47				
Trifluoperazine	↓		45[c], 47	60[c]			
Chlorprothixene	↓		47				
Haloperidol	↓		47	60[c]			
Pimozide	↓		31, 41, 61		23		
	0		59[k]				
α-receptor blocking agents							
Phenoxybenzamine	↓			1	22		
Phentolamine	↓		16[b], 59[k]	37, 38			
Dihydroergotamine	↓		16[b]				
Dibenzyline	0		20				
Azapetin	↑		7[g]				
	↓		7[f]				
Others	↓				22		

Table 8 (continued)

Drug	Intensity	Duration	References: Rat	Mouse	Rabbit	Cat	Guinea-pig
β-receptor blocking agents							
Propranolol	↓		7[f], 12, 31, 59[k]	*37*, *38* 39, 40			
	0	→	7[g]		*22*		
KÖ 592[h]	↓		10, 20, 50	37			
INPEA	0		16				
Others	0				22		
5HT-receptor blocking agents							
BOL 148	↓			34			
LSD-25	↑				56		
Drugs affecting storage of monoamines							
Reserpine	↑		10, 47[b], 57	9[c], 28			
	↓		47[b], 61	2, 43[c]			
	0		61[l]	43[l], 51[c]	54[c]		
Reserpine + αMT	0		10				
Tetrabenazine	0			51[c], 62[c]			
p-Hydroxyamphetamine	↓		13[i], 20, 36, 52[i]				
	0						13
p-Hydroxynorephedrine	↓		10, 11, 13[i], 52				
	0						13
Norephedrine	↓		13[i], 52[i]				
Monoamine synthesis inhibitors							
Tyrosine hydroxylase inhibitors							
α-methyltyrosine (αMT)	↓		10, 14, 57	43, 44			
	0			43[l]			
	↑		4[d]				
Dopamine hydroxylase inhibitor							
Diethyldithio-carbamate[a] (DEDC)	↓			49			
Tryptophanhydroxylase inhibitor							
p-chlorophenylalamine (PCPA)	↓		32	34	33		
Aminoacid precursors to monoamines							
l-dopa	↑		5				

Table 8 (continued)

Drug	Intensity	Duration	References				
			Rat	Mouse	Rabbit	Cat	Guinea-pig
Monoamineoxidase inhibitors (MAOI)							
Tranylcypromine	↑		47				
Phenelzine	↑	→	57				
Pheniprazine	↑		5				
Pheniprazine + *l*-dopa	↑		5				
Iproniazid	↑		47				
Inhibitors of monoamine uptake							
Tricyclic antidepressants							
Imipramine[a]	↑		46, 47, 58				
Desmethylimipramine[a] (DMI)	↑	→	46, 47, 57, 58				
	↑			42			
	0			18			
	↓	→	18[j], 27	27			
Amitriptyline[a]	↑	→	46, 47, 57				
Nortryptyline[a]		→	46, 47, 57				
Chlorimipramine	↑	→	47				
Trimipramine	↑	→	47				
Others		→	47, 57				
Cocaine[a]	↑	→	47				
Cholinergics							
Tremorine	0			53[c]			
Pilocarpine	↑			60[c]			
Anticholinergics							
Atropine[a]	↑		47				
	0			30			
l-hyoscyamine	0			30			
Scopolamine	↑		47				
Methylatropine	0		47	30			
Trihexyphenidyl	↑		47				
Antihistaminics							
Promethazine	↑	→	47				
	0[c]				29		
Ganglionic blocking agents							
Chlorisondamine	0		10, 20, 50				
	↓			30			
Chlorisondamine + Atropine	0		50				
Nicotine	↓			62[h]			
Hexamethonium	0				3		
Pentolinium	0				3		
Tetramethylammonium-	↓				22		
chloride	0		14				

Table 8 (continued)

Drug	Intensity	Duration	References: Rat	Mouse	Rabbit	Cat	Guinea-pig
Miscellaneous							
6-hydroxydopamine (6-OH-DA) i.p., i.v.	↓		14, 59[k]				
	0			60			
DMI + 6-OH-DA	0		14				
Bulbocapnine	↓			62[n], 63[n]			
Fenfluramine[a]	↓	→	26[j], 61	25			
	0		61[e]				
S 992	0			25			
p-Cl-amphetamine	0		14				
Nicotinic acid	↓		10, 20				
	0				22		
Prostaglandin E_1^e (PGE_1)	↓		10, 20				
5C3MP[e]	0		5				
Curare	↓				15	22	
Toxiferin[f]	↑		7				
Decamethonium[f]	↓					3	
Diplacine[f,o]	↓					3	
Halothane	0					19[c]	
SCTZ[m]	↑			1			
Lynestrol	↓			6			
Mestranol	↑			6			
Naloxone	0		24				
SKF 525A[a]		→	47				
Acetylsalicyclic acid	0		47				
Salicylate-Ca	0			1			
Phenazetine	0		47				
Pyramidone	↑		47				
Ascorbic acid	0				21		21
Δ^9-tetrahydro-cannabinol	0			17			

[a] Shown to inhibit amphetamine metabolism in rats (see Sect. I, Chap. 1.A).
[b] Dose- and time-response relationships investigated.
[c] Amphetamine antagonizes hypothermia induced by this drug.
[d] In α-methyltyrosine tolerant animals or after αMT withdrawal.
[e] Lipolysis inhibitor; 5C3MP = 5-carboxy-3-methyl-pyrazol.
[f] Curarization; in curarized rats.
[g] Delayed rise in body temperature.
[h] 1-(3-methylphenoxy)-3-isopropylamine-2-propanol.
[i] No antagonism in female rats or guinea pigs; only antagonism in males.
[j] Initial antagonism, later prolongation.
[k] Decapitated rat.
[l] Ambient temperature > 37° C.
[m] SCTZ = 4-methyl-5-(β-chloroethyl)-thiazol.
[n] Amphetamine potentiated hypothermia induced by this drug.
[o] 1,3-bis-β-platinecium-ethoxybenzene.

References

1. ASKEW, 1962
2. BEAUVALLET et al., 1966a
3. BELENKY and VITOLINA, 1962
4. BEUTHIN et al., 1972
5. BIZZI et al., 1970
6. BLACKHAM and SPENCER, 1969
7. BORBÉLY et al., 1974
8. BORELLA et al., 1969a
9. BRITTAIN et al., 1964
10. BRODIE et al., 1969
11. BRODIE et al., 1970
12. CAHN and HEROLD, 1970
13. CALDWELL and SEVER, 1974
14. CALDWELL et al., 1974
15. CHEYMOL and LEVASSORT, 1957
16. COPER et al., 1971
17. CRAIGMILL et al., 1974
18. DOLFINI et al., 1969a
19. FELDBERG and LANG, 1970
20. GESSA et al., 1969
21. HAAS, 1939
22. HILL and HORITA, 1970
23. HILL and HORITA, 1971
24. HOLTZMAN, 1974
25. JESPERSEN and BONACCORSI, 1969b
26. JONSSON and GUNNE, 1972
27. JORI and GARATTINI, 1965
28. JORI and GARATTINI, 1973
29. KLISSIUNIS and DOSI, 1959
30. KOPPANYI and MALING, 1972
31. KRUK, 1972
32. LADEFOGED, 1973
33. LADEFOGED, 1974
34. LAPIN et al., 1972
35. LEWANDER, 1969
36. LEWANDER, 1971d
37. MALING et al., 1972
38. MALING and KOPPANYI, 1972
39. MANTEGASSA et al., 1968b
40. MANTEGAZZA et al., 1970
41. MATSUMOTO and GRIFFIN, 1971
42. MATSUMOTO and SHAW, 1971b
43. MENON and DANDIYA, 1967
44. MENON et al., 1967
45. MORPURGO and THEOBALD, 1964
46. MORPURGO and THEOBALD, 1965
47. MORPURGO and THEOBALD, 1967
48. NAKAJIMA et al., 1964
49. PRZEGALINSKI and KLEINROK, 1970
50. REID, 1970b
51. ROSS and RENYI, 1967b
52. SEVER et al., 1974
53. SPENCER, 1965
54. TEDESCHI et al., 1959
55. TERWELP et al., 1973
56. THUILLIER et al., 1962
57. VALZELLI et al., 1968
58. VALZELLI et al., 1967b
59. WEIS, 1973
60. WOLF and BUNCE, 1973
61. YEHUDA and WURTMAN, 1972a
62. ZETLER, 1970
63. ZETLER and MOOG, 1958

were antagonistic. A potentiation of the amphetamine effect was reported in α-methyltyrosine tolerant or withdrawn rats (BEUTHIN et al., 1972). Reserpine had a time-dependent action in antagonizing the increase in body temperature at certain times while potentiating the response at other times after its administration. Monoamineoxidase inhibitors potentiated amphetamine. The potentiation and prolongation of amphetamine in rats with tricyclic antidepressants is probably due to a metabolic interaction. The effects of cholinergic, anticholinergic, and ganglionic blocking agents are controversial. Among the group of miscellaneous drugs, the antagonism by 6-hydroxydopamine is most interesting, since peripheral NA-neurons are selectively degenerated by this treatment. A series of antilipolytic agents (nicotinic acid, PGE) were also antagonists of amphetamine. Drugs causing curarization generally antagonized amphetamine-induced hyperthermia. Antipyretic drugs were mainly without effect.

Species variations were frequently observed in the drug interaction studies, and it cannot be excluded that the mechanisms of action of amphetamine with regard to hyperthermia might vary between species (cf. CALDWELL et al., 1974).

c) Effects of Brain Lesions on Amphetamine-Induced Hyperthermia

BELENKY and VITOLINA (1962) showed that the amphetamine-induced hyperthermia was more pronounced in thalamic than in control cats. In cats with a section between the thalamus and the quadrigeminal bodies there was a slight potentiation of the effect of amphetamine. However, no hyperthermia was elicited by amphetamine in cats with a cut between the superior and inferior corpora quadrigemina. This led to the conclusion that the locus of action of amphetamine was in the diencephalon. These results have been partly confirmed. Thus, in chronic thalamic rats (removal by suction of the cortex, hippocampus, striatum, amygdaloid nucleus, and septum) there was a hyperthermic response to amphetamine (BORBÉLY et al., 1973; HUSTON and BORBÉLY, 1974). The diencephalic locus of action of amphetamine could not, however, be confirmed in rats. WEIS (1973) found a clear-cut hyperthermic response to amphetamine in decapitated rats, and in spinalized or pithed rats there was a reduction but no abolition of the hyperthermic response (BORBÉLY et al., 1974).

d) Other Factors Influencing Amphetamine-Induced Hyperthermia

Ambient temperature. As described in the introduction to this section, the ambient temperature is of importance for the body temperature response to amphetamine. There is a hypothermia at ambient temperatures below $+15°$ C, and a potentiation of the hyperthermic response if the ambient temperature is increased above $+20$–$+22°$ C up to $+30$–$+37°$ C in rats (YEHUDA and WURTMAN, 1972a; BORBÉLY et al., 1974; SCHEEL-KRÜGER and HASSELAGER, 1974) and mice (ASKEW, 1962; DOLFINI et al., 1969a; MALING and KOPPANYI, 1972; CRAIG and KUPFERBERG, 1972).

Isolation, Aggregation. It has repeatedly been observed that the toxicity of amphetamine is greater in aggregated than in isolated mice (see Subsect. C.III.7). A number of studies also report that the hyperthermic response to amphetamine is potentiated if mice (GREENBLATT and OSTERBERG, 1961; ASKEW, 1962; CRAIG and KUPFERBERG, 1972) and rats (DOLFINI et al., 1969c) are aggregated. Several authors have claimed a close relation between the toxicity of amphetamine in aggregated mice and lethal hyperthermia (GREENBLATT and OSTERBERG, 1961; ASKEW, 1962; DOLFINI et al., 1969a; CRAIG and KUPFERBERG, 1972; WOLF and BUNCE, 1973, see C.III.7).

Age of Animals. Young rats seem to have a smaller hyperthermic response to amphetamine than older animals (ZIEM et al., 1970).

Strain Differences. Many reports of differences in temperature response to amphetamine in mice of different strains are available (BROWN and JULIAN, 1968; DOLFINI et al., 1969a; BIZZI et al., 1970; SCOTT et al., 1971; JORI and GARATTINI, 1973; CRAIGMILL et al., 1973). The C_3H strain of mice is particularly interesting, since no hyperthermia at all develops after amphetamine at room temperature (BIZZI et al., 1970). This strain has a rather low degree of hyperthermia when ambient temperature is elevated (DOLFINI et al., 1969a) and also a low lethality toward amphetamine at 30° C ambient temperature (DOLFINI et al., 1969a). In addition, this strain of mice differs with regard to hyperactivity, stereotyped behavior, and several biochemical changes after amphetamine compared with other strains, although brain amphetamine concentrations were comparable (JORI and GARATTINI, 1973). The BD strain is another strain with a low response toward amphetamine (CRAIG and KUPFERBERG, 1972; CRAIGMILL et al., 1973).

Endocrine Influences. L-thyroxine administration potentiates the amphetamine-induced hyperthermia in mice (ASKEW, 1962) and rats (DOLFINI and KOBAYASHI, 1967; BIZZI et al., 1970). Rats made hypothyroid by thyroidectomy or α-methyl-thiouracil treatment have less hyperthermia following amphetamine than control rats (DOLFINI and KOBAYASHI, 1967; MANTEGAZZA et al., 1968b). This is not due to differences in brain amphetamine concentrations (DOLFINI and KOBAYASHI, 1967). Hypophysectomized rats do not react with hyperthermia to amphetamine; on the contrary, even at 30° C ambient temperature there is a hypothermic response (BORBÉLY et al., 1974). Adrenalectomy also reduces the hyperthermic response to amphetamine (BRODIE et al., 1969; WEIS, 1973). Whether this depends on the lack of cortical or medullary hormones has not been established.

e) *Physiologic Correlates to Amphetamine-Induced Hyperthermia*

The increase in body temperature is accompanied by an increase in the skin temperature in intact rabbits and cats (BELENKY and VITOLINA, 1962) but a decrease in skin temperature in decapitated and intact rats (WEIS, 1973; BORBÉLY et al., 1974). In mice a close correlation has been established between an increase in body temperature and in salivation, which seems to be an important heat dissipating mechanism (MALING and KOPPANYI, 1972; MALING et al., 1972). The oxygen consumption in mice and rats is increased after amphetamine, but there is no correlation between oxygen consumption and hyperthermia (CRAIGMILL et al., 1973; WEIS, 1973). Differences in heat loss through evaporation seem to be the mechanism for different sensitivity to amphetamine in Swiss Webster and BDF mice (CRAIGMILL et al., 1973).

f) *Biochemical Correlates to Amphetamine-Induced Hyperthermia*

Concomitant changes in the free fatty acid (FFA) concentration in plasma and hyperthermia after amphetamine administration in rats were found under various conditions (MANTEGAZZA et al., 1968b; GESSA et al., 1969; BRODIE et al., 1969). The β-fluoro derivatives (β-mono- and β-difluoro) of amphetamine were later shown to dissociate the two amphetamine effects in rats; these derivatives caused an increase in the FFA concentrations but not in body temperature (FULLER et al., 1972a). Similar findings have been shown for fenfluramine, the N-methyl, m-trifluoromethyl amphetamine derivative (see BIZZI et al., 1970). In mouse strains with different sensitivities to the hyperthermic action of amphetamine there is no difference with respect to the amphetamine-induced increase in FFA concentrations (CRAIGMILL et al., 1973; JORI and GARATTINI, 1973). Thus, amphetamine-induced release of FFA is not responsible per se for the increase in body temperature.

Amphetamine-induced release of peripheral NA seems to be related to the increase in body temperature. The collected evidence for this will be reviewed below. Here it is mentioned that amphetamine causes a reduction in tissue concentrations of NA in several species (see Subsect. B.I.1) and that mice that are insensitive to the hyperthermic action of amphetamine are not NA depleted (CACCIA et al., 1973; JORI and GARATTINI, 1973). Further, administration of p-hydroxynorephedrine or the immediate precursors of this substance (p-hydroxyamphetamine, norephedrine), which is a metabolite of amphetamine and acts as a false adrenergic transmitter, protects against amphetamine-induced hyperthermia in male rats (GESSA et al., 1969; BRODIE et al., 1969; LEWANDER, 1971d; CALDWELL and SEVER, 1974). Even though there is a correlation between increased brain HVA levels and hyperthermia after amphetamine in rats, the relation between increased body temperature and brain DA or NA turnover after threshold doses of amphetamine is not very striking (GROPPETTI et al., 1972a; COSTA et al., 1972). Therefore, these parameters do not seem to be causally related (JORI and BERNARDI, 1969). In amphetamine-sensitive mice the brain DA turnover seems to be more rapid than in insensitive animals (CACCIA et al., 1973). A causal relationship is not proved, however. An increased turnover rate of brain 5HT in rats seems to be secondary to amphetamine hyperthermia; both effects of amphetamine are abolished at reduced ambient temperature (REID, 1970a, b; SCHEEL-KRÜGER and HASSELAGER, 1974).

g) Effects of Chronic Amphetamine Administration on the Hyperthermic Response

HARRISON et al. (1952) were the first to report the development of tolerance to the hyperthermic effect of amphetamine and methamphetamine in rats. At a dose of 5 mg/kg daily, tolerance was present within 6 days and it had not disappeared after 3 days of amphetamine withdrawal. Tolerance development in rats has been confirmed by GESSA et al. (1969), BRODIE et al. (1969), LEWANDER (1971d), and CALDWELL and SEVER (1974). However, ZIEM et al. (1970) did not find tolerance in rats after 5 mg/kg per day for 20 days. Four injections of amphetamine in rabbits at 6-h intervals were not accompanied by tolerance development (HILL and HORITA, 1971).

The tolerance to the hyperthermic effect has been related to depletion of NA in peripheral and central NA neurons due to displacement by the false transmitter p-hydroxynorephedrine; which is formed from amphetamine (see Subsect. B.I.1). Administration of p-hydroxynorephedrine was also followed by a protection against amphetamine hyperthermia (simulation of amphetamine tolerance) in rats (BRODIE et al., 1969; GESSA et al., 1969; LEWANDER, 1971d), but only in male rats (CALDWELL and SEVER, 1974; SEVER et al., 1974). Thus, a false transmitter mechanism seemed to explain amphetamine tolerance in this respect. Later, however, it was found (LEWANDER, 1974a; LEWANDER et al., 1975) that tolerance developed to *l*-amphetamine, which is not metabolized to p-hydroxynorephedrine in rats but causes a depletion of NA stores (see Subsect. B.I.1), and to *d*-amphetamine in guinea pigs, a species in which p-hydroxylation of amphetamine does not occur (SEVER and CALDWELL, 1974). The disappearance of tolerance to the hyperthermic effect of amphetamine did not parallel the restoration of NA stores centrally or peripherally which occurred much earlier than the normal body temperature response to amphetamine (LEWANDER, 1974b; LEWANDER et al., 1975). The more rapid development of tolerance to the hyperthermic effect of amphetamine during a second chronic administration period after amphetamine withdrawal in previously tolerant rats, carry over, was not accompanied by a more rapid depletion of NA stores (LEWANDER et al., 1973; LEWANDER, 1974b).

Thus, taken together, the available data indicate that tolerance to amphetamine hyperthermia is not explained by the false transmitter mechanism alone. However, if p-hydroxynorephedrine is formed and present in the tissues it might contribute to the development of tolerance.

A more detailed analysis of the body temperature response to amphetamine after single injection and after chronic amphetamine administration demonstrated that tolerance developed more rapidly and more completely to the temperature response at 1 h after the amphetamine injection than to the 4 h temperature response (LEWANDER et al., 1975). In this study was also reported the observation that the hyperthermic response to amphetamine was converted to a hypothermic response, if the animals were fed for 7 h during daytime in contrast to animals fed for 16 h during nighttime. This finding was interpreted to indicate that the body temperature response at least partly is dependent on an exhaustible pool of calorigenic substances (LEWANDER et al., 1975).

h) Amphetamine-Induced Hypothermia

A hypothermic response to amphetamine is obtained under the following conditions (cf. above): (1) at low ambient temperature, (2) after low doses of the drug given peripherally, (3) after intracerebroventricular administration of amphetamine, and (4) in hypophysectomized rats (BORBÉLY et al., 1974). Dextroamphetamine is more potent than levoamphetamine (ratio = 3:1; YEHUDA and WURTMAN, 1972b) in inducing hypothermia. Interestingly amphetamine, given i.p. or i.c.v., prevents the decrease in body temperature after i.c.v. administration of NA in mice (BRITTAIN, 1966) and cats (FELDBERG and LANG, 1970; TERWELP et al., 1973).

A central dopaminergic mechanism for amphetamine-induced hypothermia is supported by the finding that dopamine receptor blocking agents like pimozide and haloperidol antagonize this action of amphetamine in rats (KRUK, 1972; YEHUDA and WURTMAN, 1972a, b, 1973). Intracerebroventricular administration of 6-hydroxydopamine, which causes degeneration of brain DA (and NA) neurons, also abolished the hypothermic response in rats (YEHUDA and WURTMAN, 1972b) and cats (TERWELP et al., 1973). Electrolytic lesions in brain areas rich in DA terminals like the olfactory tubercles and the nucleus accumbens also antagonized amphetamine hypothermia (but not hyperthermia) according to YEHUDA and WURTMAN (1972b, 1973), and CHIEL et al. (1974). Further, drugs that are supposed to be DA receptor agonists, such as DA and apomorphine, when given intracerebroventricularly mimic amphetamine in causing hypothermia (KRUK, 1972; YEHUDA and WURTMAN, 1972b). A series of other drugs have been tried in order to change amphetamine hypothermia and the results are given in Table 9.

An increase of the skin temperature accompanies the hypothermic reaction to amphetamine (KRUK, 1972; KRUK and BRITTAIN, 1972). Partial tolerance develops to the hypothermic effect of amphetamine at reduced ambient temperature (CHIEL et al., 1974).

i) Behavioral Thermoregulation

Rats treated with amphetamine at 30° C ambient temperature (hypothermic) orient themselves toward a heat lamp, while at +4° C (hypothermic rats) orient themselves away from the heat lamp (YEHUDA and WURTMAN, 1973, 1974a, b). Pimozide, haloperidol, and a series of α-receptor blocking agents normalized the paradoxical behavioral thermoregulation induced by amphetamine (YEHUDA and WURTMAN, 1973, 1974a, b). Agents stimulating brain DA receptors mimic amphetamine in this respect

Table 9. Effect of other drugs on intensity and duration of amphetamine-induced hypothermia. Symbols: No effect: 0. Intensity, reduced: ↓, increased: ↑

Drug	Intensity	References	
		Rat	Cat
Neuroleptic drugs			
Chlorpromazine	0	2	
Haloperidol	↓	3, 4	
Pimozide	↓	2, 3, 4	
α-receptor blocking agents			
Phenoxybenzamine	0	3, 4	
Phentolamine	↑	1	
Dihydroergotamine	↑	1	
β-receptor blocking agents			
Propranolol	0	3, 4	
INPEA	0	1	
5HT-receptor blocking agents			
Methysergide	0	3	
Drugs affecting storage of monoamine			
Reserpine	0	2	
Tyramine	↓	2	
Anticholinergics			
Ditran (JB 329)	↓	3, 4	
Miscellaneous			
6-hydroxydopamine (6-OH-DA) i.c.v.[a]	↓	3	5
6-OH-DA i.p.[a]	0	3	
5,6-dihydroxytryptamine i.c.v.[a]	0	3	
Fenfluramine	↑	2	
p-Chloroamphetamine	0	3	

[a] i.c.v. = intracerebroventricularly; i.p. = intraperitoneally.

References

1. COPER et al., 1971
2. YEHUDA and WURTMAN, 1972a
3. YEHUDA and WURTMAN, 1972b
4. YEHUDA and WURTMAN, 1973
5. TERWELP et al., 1973

(YEHUDA and WURTMAN, 1973, 1974a, b). In contrast, rats given amphetamine at +2° C were found to press a lever for hot air blows in an operant situation (LATIES, 1971).

There is no obvious explanation for the paradoxical themoregulatory behavior induced by amphetamine.

j) *Summary of Mechanisms Involved in Amphetamine-Induced Changes in Body Temperature*

According to the available evidence, amphetamine changes the body temperature by both central and peripheral mechanisms. The hypothermic action seems to be mainly dependent on a central dopaminergic mechanism, although a central cholinergic (ditran, Table 9) and a central or peripheral α-adrenergic mechanism may be involved.

Even the hyperthermic effect of amphetamine may be dependent on a central dopaminergic and cholinergic mechanism. The evidence for this is the well-documented antagonism of amphetamine hyperthermia by neuroleptic and anticholinergic drugs (Table 8) as well as the reduction of body temperature caused by spinalization and other brain lesions. The central actions might activate peripheral heat producing and/or heat dissipating mechanisms via the somatic and autonomous nervous systems.

The involvement of the peripheral sympathetic nervous system is supported by the antagonistic effects of ganglionic blocking agents, at least in mice and rabbits. However, a releasing effect directly on peripheral NA neurons seems to prevail, since amphetamine causes hyperthermia even in decapitated rats via peripheral α- and β-receptor mechanisms, but not after destruction of peripheral NA neurons by 6-hydroxydopamine (WEIS, 1973). Several lines of evidence (e.g., effect of β-receptor blocking agents and lipolysis inhibitors) indicate that the increased indirect release of FFA may take part in a heat-producing mechanism. However, since an increase in plasma FFA per se does not cause hyperthermia, this mechanism does not seem to be solely responsible for amphetamine hyperthermia, but rather acts in combination with other FFA utilizing mechanisms. The activity of the autonomic nervous system regulating skin temperature seems to cause its reduction in amphetamine-induced hyperthermia and its increase in hypothermia, which may indicate the involvement of a heat dissipating mechanism in body temperature changes. Similarly, changes in salivation in mice and the importance of the ambient temperature for the direction and intensity of the temperature response again indicate the involvement of alterations of heat dissipation as an important contribution to the overall effect. The dependence of amphetamine-induced changes in body temperature on several endocrine factors (pituitary, adrenal gland, thyroid gland) also indicates that amphetamine-induced release or inhibition of release of various hormones might interfere with heat producing or heat dissipating mechanisms.

Increased heat production via an increased skeletomuscular activity is indicated by the reduction of the hyperthermic response to amphetamine after spinalization in rats and curarization in cats (but not rats). It is most certain that amphetamine hyperthermia differs from pyrogen-induced fever, since antipyretics only antagonize the latter phenomenon. The studies on behavioral thermoregulation appear to indicate that amphetamine-hyperthermic rats behave as if they feel cold since they approach a heat lamp, while the hypothermic rats behave as if they were warm and turn away from the heat lamp. Thus, the amphetamine-induced changes in the thermosensory regulation should not be disregarded.

5. Effects of Amphetamine on General Metabolic Processes in Brain and Peripheral Tissues

a) Oxygen Consumption

An increase in oxygen consumption caused by amphetamine in morphine-treated dogs was shown by MORRISON and ABREU (1943). D-amphetamine was slightly more effective than both *l*- and *dl*-amphetamine in this respect. The oxygen consumption was increased by amphetamine also in rats (WATERMAN, 1949) and mice (HEIM et al., 1957). During the last 10 years, several studies have been devoted to understanding the mechanisms for this effect. A peripheral action of amphetamine seems to contribute, since isolated tissues in vitro are stimulated to increased oxygen uptake (ANDRES et al., 1967), and since increased oxygen consumption is also observed in decapitated rats (WEIS, 1973). Only brain tissue seems to be insensitive to amphetamine in vitro (UTENA et al., 1959) or to have a reduced respiration in the presence of high amphetamine concentrations (LEWIS and POLLACK, 1965). A decreased cerebral metabolic rate in vivo was demonstrated by NAHORSKI and ROGERS (1973).

Among the several drugs intervening with central and peripheral adrenergic mechanisms that have been tried, α-methyltyrosine, reserpine, guanethidine, and prenylamine (PFEIFFER et al., 1968), antagonized the amphetamine-induced increase in oxygen consumption in rats. Similar results with α-MT were obtained by TÖRÖK et al. (1973) who also showed, that pCPA and α-MT added in antagonizing this effect of amphetamine. The Li^+-ion also antagonized amphetamine-elicited oxygen consumption (OPITZ et al., 1971). In cats, chronic reserpine treatment sensitized the animals to this effect (VIZI et al., 1968). The β-receptor blocking agents nethalide (OPITZ et al., 1966) and propranolol (WEIS, 1973) antagonized amphetamine, as did the α-receptor blocking agent phentolamine (WEIS, 1973). However, in another study phenoxybenzamine and propranolol were without effect (PFEIFFER, 1972). This same study reported that diethyldithiocarbamate potentiated amphetamine in rats, a finding that might be due to the inhibition of p-hydroxylation of amphetamine with concomitant increase in tissue levels of the drug (see Sect. I, Chap. 1.A).

The thyroid activity seems to have a profound influence on the stimulation of oxygen consumption induced by amphetamine, since this pharmacologic effect is markedly enhanced in hyperthyroid rats while almost abolished in hypothyroid rats (CHU et al., 1969). Also, the amphetamine-induced increase in oxygen consumption of skeletal muscle in vitro was reduced in hypothyroid rats (ANDRES et al., 1967). However, amphetamine does not seem to act via release of thyroid hormones, since no signs of increased thyroid activity could be seen in histologic sections of the gland and, furthermore, thyroidectomized rats treated with triodotyronine responded well to amphetamine administration with the usual increase in oxygen consumption (CHU et al., 1969). It might be concluded from these findings that both peripheral adrenergic mechanisms and the action of thyroid hormones on target cells are necessary for the effect of amphetamine on oxygen consumption.

b) Brain Adenosine Nucleotide and Phosphate Metabolism

Amphetamine has been reported to increase brain ATP levels in rabbits (PALLADIN and RYBINA, 1953) and mice (HEIM et al., 1957). In rats, both *d*- and *l*-amphetamine, as well as methamphetamine, increase brain ATP concentrations with a concomitant decrease in ADP and AMP, whereas there was no or very minor changes in the levels of organic phosphor or creatinine phosphate (LEWIS and VAN PETTEN, 1962; ESTLER and AMMON, 1967). The increase in rat brain ATP does not seem to be caused by an inhibition of ATP degradation via ATPase (LEWIS and POLLACK, 1965). Recent studies, in which more sophisticated techniques have been applied in order to avoid spontaneous postmortem changes, no effects on rat (WILSON, 1969; NICHOLS et al.,

1970) or mouse (NAHORSKI and ROGERS, 1973) brain ATP have been observed after amphetamine, whereas brain phosphocreatine was decreased (NAHORSKI and ROGERS, 1973).

c) Carbohydrate Metabolism

MOORE et al. (1965) showed that amphetamine caused a decrease in liver, muscle, and brain glycogen concentrations in aggregated mice. The decrease in liver glycogen was confirmed by BEWSCHER et al. (1966), who also demonstrated that the glucose oxidation in whole mice was increased threefold after amphetamine. The effect on *liver glycogen* was antagonized by such drugs as phenoxybenzamine, chlorpromazine, reserpine, α-methyltyrosine (MOORE et al., 1965), and propranolol (ESTLER et al., 1970). Only aggregated mice that died after the amphetamine administration had reduced blood glucose levels (MOORE, 1966a). The decrease in liver glycogen and blood glucose was more marked in hyperthyroid mice (MOORE, 1966a). Tissue lactic acid concentrations were unchanged (MOORE, 1966b). The effects of amphetamine on liver glycogen in mice were confirmed by STOLK et al. (1970). In rats, however, an increased liver glycogen concentration was found after amphetamine (STOLK et al., 1970). Chronic, but not acute, pretreatment with reserpine reduced the effect of amphetamine on the liver glycogen stores (STOLK et al., 1970). KIRBY and TURNER (1974) found no effect of *d*-, *l*-, or *dl*-amphetamine on glucose uptake or on glycogen concentrations in a skeletal muscle preparation in vitro. Reduced blood glucose concentrations after amphetamine under certain conditions might be due to amphetamine-induced release of insulin (MCMAHON et al., 1971).

The reduced concentrations of *brain glycogen* in mice have been studied by several investigators. In methamphetamine-treated mice, the decrease in brain glycogen was inhibited by the β-blockers, propranolol, INPEA, and pronethanol (ESTLER and AMMON, 1967, 1970, 1971; HUTCHINS and ROGERS, 1970, 1971). Alpha-receptor blocking agents, anticholinergic, antiserotonergic, and antihistaminic agents were without effect (HUTCHINS and ROGERS, 1971). Pretreatment of mice with a combination of α-methyltyrosine and reserpine abolished the cerebral glycogenolytic effect of amphetamine; various inhibitors of monoamine synthesis given separately were only partially effective in this respect (HUTCHINS and ROGERS, 1973). A study of the structure activity relationships of this effect of the amphetamine showed that methamphetamine was slightly inferior to *d*- or *dl*-amphetamine, while p-hydroxyamphetamine or benzphetamine had minimal activity (ROGERS and HUTCHINS, 1972). ESTLER and MITZNEGG (1971) showed that the decrease in brain glycogen is due to increased glycogenolysis rather than to inhibition of glycogen accumulation. Recently, NAHORSKI and ROGERS (1973) found that the glycogen stores reduced by amphetamine might be located in glial cells and not in neuronal cells. In chronically methamphetamine treated rats, the initial decrease in brain glycogen and increases in brain lactate and glucose $-{}^{14}C$ incorporation had returned to control levels (MANNING et al., 1974). Interestingly, the changes observed in these parameters after acute treatment were reversed after amphetamine withdrawal (MANNING et al., 1974). Amphetamine also stimulated the incorporation of ^{14}C-atoms into brain and serum after administration of ^{14}C-mannose in isolated and aggregated mice (DE FEUDIS, 1973). Direct measurement of the activities of certain enzymes involved in glucose metabolism have shown that amphetamine reduced the phosphoglucomutase and pyruvate kinase activities in skeletal muscle of mice and rabbits, probably by reducing the availability of Mg^{2+} ions (ZWARENSTEIN and SAPEIKA, 1970; SAPEIKA et al., 1971; KAPLAN et al., 1972).

d) Lipid Metabolism

Amphetamine in doses greater than 2 mg/kg causes an increase in plasma-free fatty acids (FA) in rats (OPITZ and LOESER, 1963; KHAN et al., 1964; SANTI and FASSINA, 1964, 1965; BIZZI et al., 1965), mice (ESTLER et al., 1970; JORI and GARATTINI, 1973), and dogs (ZSOTER et al., 1966). Peak effect of the drug on plasma FFA occurs after 1 h in rats (KHAN et al., 1964). Both the amphetamine isomers are active, but *d*-amphetamine is up to 10 times more potent than *l*-amphetamine (SHAW et al., 1972; HAJOS and GARATTINI, 1973). Parahydroxyamphetamine, although effective, is considerably less active than amphetamine (BRODIE et al., 1969). Methamphetamine is also active (FALB et al., 1966). Besides the FFA, blood glycerol levels are also increased after amphetamine administration (OPITZ, 1970).

The pharmacologic analysis of the FFA-releasing effect of amphetamine has shown that the β-receptor blocking agents propranolol and Kö 592 inhibit the release in rats (FASSINA, 1966; GESSA et al., 1969; BRODIE et al., 1969; SHAW et al., 1972) and mice (ESTLER et al., 1970). In dogs propranolol and MJ 1999 had a similar effect, while some other β-receptor blockers (dichloroisoproterenol, pronethanol, butoxamine, isolpropoxamine) had no effect or even potentiated amphetamine (ZSOTER et al., 1966; OPITZ, 1970). The α-receptor blocker phenoxybenzamine was partially effective as an antagonist (FASSINA, 1966) but other α-blockers were ineffective (BRODIE et al., 1969). Reserpine pretreatment partially inhibits (FASSINA, 1966) or potentiates (BRODIE et al., 1969) amphetamine in rats and antagonizes it in dogs (OPITZ, 1970). Other depletors of tissue NA levels, like guanozane and prenylamine, were without effect on amphetamine in dogs (OPITZ, 1970). The results with ganglionic blocking agents are equivocal; chlorisondamine is without effect in rats (GESSA et al., 1969; BRODIE et al., 1969) and dogs (OPITZ, 1970), while hexamethonium inhibits and mecamylamine potentiates amphetamine in dogs (OPITZ, 1970). The CA synthesis inhibitor, α-methyltyrosine, antagonized amphetamine in rats (BRODIE et al., 1970). Chlorpromazine was ineffective in dogs (OPITZ, 1970). Three inhibitors of lipolysis, nicotinic acid, prostaglandin E_1, and 5 C 3 MP (see Table 8, Footnote 3) completely inhibited the increase in FFA after amphetamine in rats (GESSA et al., 1969; BRODIE et al., 1969; BIZZI et al., 1970). Desmethylimipramine, which potentiates most other effects of amphetamine in rats, antagonized the increase in blood FFA after amphetamine (MATSUMOTO and SHAW, 1971 a, b). In man, ergotamine and reserpine antagonized the FFA increase, as did also intravenous glucose administration (PINTER and PATTEE, 1970).

A number of other experiments have been conducted to elucidate the mechanism of amphetamines in vivo lipolytic action.

Thyroidectomy or drug-induced hypothyroidism decreases the effect of amphetamine on plasma FFA (MANTEGAZZA et al., 1968 b; SHAW et al., 1972). Adrenalectomy had no effect (BRODIE et al., 1970; SHAW et al., 1972) nor did hypophysectomy (SHAW et al., 1972). An interesting observation is that in starved rats, which have high FFA levels in plasma, amphetamine does not cause any additional increase (SHAW et al., 1972). After repeated injections with amphetamine in rats the FFA response is reduced, but is reestablished 25 h after withdrawal of amphetamine (BRODIE et al., 1969). The rapid development of tolerance to this effect of amphetamine has been related to the NA depletion in adipose tissue caused by the drug and the false transmitter mechanism (BRODIE et al., 1969; see Subsect. A.1). Pretreatment of rats with p-hydroxyamphetamine or p-hydroxynorephedrine protects rats from the FFA-increasing effect of amphetamine (GESSA et al., 1969; BRODIE et al., 1969, 1970).

It has been reported repeatedly that amphetamine does not release FFA from adipose tissue in vitro, i.e., it has no direct effect on fat cells (FASSINA, 1966; OPITZ and LOESER, 1966; FINGER et al., 1966) and the FFA concentrations within the fat tissue is not increased (BIZZI et al., 1970). In contrast to amphetamine (FASSINA, 1966), methamphetamine seems to inhibit the FFA releasing effect by NA on fat cells in vitro, but methamphetamine has no effect on FFA released by growth hormone, dexamethasone, or caffeine in vitro (DANNENBURG and KARDIAN, 1969). Fenfluramine seems to act differently on fat cells (DANNENBURG and KARDIAN, 1969). Similarly, one in vivo study (ISACHENKO et al., 1971) has reported that amphetamine antagonizes the effect of NA on lipolysis. Further, DANNENBURG and KARDIAN (1969, 1970) reported that methamphetamine inhibits the conversion of glucose into FFA in rat epididymal fat cells. Amphetamine also inhibits synthesis and release of triglycerides from liver slices (MARSH and BIZZI, 1972).

In summary, it has been suggested by several authors that amphetamine has at least two actions on adipose tissue and lipolysis: one indirect via the sympathetic nervous system and a second via release of a hypothalamic adipokinetic principle (ACTH? growth hormone?; FASSINA, 1966; OPITZ, 1970; SHAW et al., 1972). Indirect evidence for the last proposal has recently been provided by BOLT et al. (1974), who found that the increase in FFA could not be abolished by spinal transection in cats, thus excluding the possibility of a centrally elicited increase in the efferent innervation of the adipose tissue. Amphetamine does not have a direct action on fat tissue cells.

Neuronal Membrane Lipids. The turnover of phosphatidylcholine, a component of membranes in nerve endings and vesicles, has been studied under acute and chronic administration of amphetamine and found to be inhibited in several brain areas (HITZEMANN and LOH, 1973). In vitro it was found that amphetamine inhibited one specific enzymatic step in the synthesis of phosphatidylcholine (diphosphorylcholine diglyceride transferase; HITZEMANN and LOH, 1973). In rat hippocampus there was a decrease in phosphatidylcholine, while phosphatidylethanolamine increased after a single dose of amphetamine (LEONARD and NEUHOFF, 1974).

e) Protein Metabolism

Incorporation of radioactively labeled leucine into liver proteins was found to be reduced after chronic amphetamine administration in rats (MAGOUR et al., 1974b). Brain microsome protein synthesis was also inhibited by amphetamine in mice (LOH et al., 1973). Amphetamine in vitro had the same effect (LOH et al., 1973). However, DEWAR and WINTERBURN (1973) found no effect of amphetamine on rat brain protein synthesis with a similar technique.

f) Nucleic Acid Metabolism

The ribonucleic acid (RNA) content in rat brain and liver was found to be unchanged by repeated amphetamine injections (LIMA et al., 1966). Later, NASELLO and IZQUIERDO (1969) found that amphetamine increased the RNA level in the hippocampus; a decrease was observed, however, after chronic amphetamine treatment. In other brain areas, such as cortex, hypothalamus, cerebellum, the RNA content was unchanged. In mice, amphetamine has been reported to increase RNA turnover, as measured by incorporation of labeled uridine into brain RNA (DICARLO et al., 1972). Similar studies in rats treated acutely or chronically with amphetamine showed no changes in RNA synthesis or degradation in brain cortex (DEWAR and WINTERBURN, 1973). Obviously, the available studies on the brain metabolism of proteins and nucleic acids are contradictory. This should not discourage future efforts in this rather unexplored area.

g) Miscellaneous Studies on General Metabolism

In dogs amphetamine administration led to a state of compensated metabolic acidosis with increased blood lactate, hypocapnia, and reduced oxygen saturation. Sodium excretion was increased (STEPANEK, 1972) and electrolyte disturbances in brain tissue were described (ABSAVA and VYSOTSKAYA, 1970). Brain glutamate-oxaloacetate and glutamate-pyruvate transaminase activities were decreased after amphetamine (PANDA and RAY, 1973).

6. Endocrinologic Effects of Amphetamine

a) Insulin

Methamphetamine and amphetamine causes an increase of plasma insulin in rats and mice, which in mice is accompanied by a decrease in blood glucose (MCMAHON et al., 1971). This finding does not agree with the previous study by BEWSHER et al. (1966), showing amphetamine to cause hypoglycemia in mice pretreated with insulin antiserum and showing that pancreatic insulin was unchanged after amphetamine administration.

b) ACTH-Corticosteroids

Amphetamine caused a decrease in the ascorbic acid content of the adrenal gland indicative of release of corticosteroids (NASMYTH, 1949). In guinea pigs it was shown that amphetamine affects peripheral receptors that mediate release of corticosteroids via a central mechanism (NAUMENKO, 1967). In dogs amphetamine antagonized corticosteroid output induced by surgical stress (VAN LOON et al., 1971). Administration of (meth)amphetamine did not disturb the diurnal rhythm of plasma corticosteroids in rats (ONODERA et al., 1973; ASANO and MOROJI, 1974). Chronically methamphetamine-treated rats, however, had their circadian rhythm abolished and reacted more intensely than controls with an increase in plasma corticosterone after neurogenic stress (ONODERA et al., 1973).

c) Growth Hormone

Growth hormone has not been investigated after amphetamine in animals. (For studies in man see Sect. I, Chap. 1.C.)

d) Prolactin

Prolactin was increased in plasma and decreased in the pituitary gland after amphetamine in rats (LU and MEITES, 1971).

e) Pineal Gland Hormones

Amphetamine stimulates the pineal gland synthesis of N-acetylserotonin and melatonin in vitro indirectly, via release of NA that acts on pineal β-receptors (BÄCKSTRÖM and WETTERBERG, 1973).

f) Thyroid Hormones

There are several reports on the importance of thyroid hormones for different amphetamine-induced effects in the organisms, of which many have been discussed in relation to other actions of the drug. So far there are no indications that amphetamine actually releases or inhibits release of thyroid hormones (OPITZ et al., 1966).

7. Toxicity of Amphetamine

a) Lethality after Amphetamine Administration

The LD 50 for amphetamine in different species has been difficult to establish with certainty, since age and strain differences, environmental factors such as diurnal rhythm, ambient temperature, concomitant stress, noise, and illumination considerably influence the outcome of the determination. With these reservations, some LD 50 values will be quoted from the literature. Thus, the LD 50 in rats is 13 mg/kg i.p. (ALLES, 1933), in mice 27 mg/kg (ALLES, 1939), in guinea pig 40–100 mg/kg (ALLES, 1933; EHRICH et al., 1939), in rabbits 3–20 mg/kg, depending on the mode of administration (EHRICH et al., 1939; KASIRSKY et al., 1972), in dogs 5 (i.v.)–20 (i.p. or perorally) mg/kg (EHRICH et al., 1939; ZALIS and KAPLAN, 1966; ZALIS et al., 1967), in sheep 15 mg/kg (EHRICH et al., 1939), and in monkeys 25 mg/kg, according to EHRICH et al. (1939) and 4.2 mg/kg according to BLUM et al., (1950). GUNN and GURD (1940) were the first to observe increased toxicity for amphetamine in aggregated (grouped) mice as compared with isolated mice. CHANCE (1946) expanded on this observation and reported a 10-fold difference in LD 50 for isolated (117 mg/kg i.p.), as compared to grouped (14 mg/kg) mice. Similarly, the isolation/aggregation ratio was about 3–6 for methamphetamine (CHANCE, 1946; GREENBLATT and OSTERBERG, 1961). Increased toxicity to amphetamine has also been observed in aggregated versus isolated rats (STOLK and RECH, 1968a). In dogs the LD 50 for the isolated and the aggregated condition was 6.56 and 4.47 mg/kg, respectively, and not statistically significantly different (ZALIS and KAPLAN, 1966). The *d*- and *dl*-isomers of amphetamine are approximately equipotent with regard to toxicity in mice (ALLES, 1939; PROCTOR et al., 1966). The *l*-isomer, however, is less potent than *d*- or *dl*-amphetamine (ALLES, 1939; MOORE, 1963; MAJ et al., 1972a; LITTLETON, 1974). Amphetamine administration to pregnant female rats had no effect on litter size nor were teratologic effects observed in the offspring (SELIGER, 1973).

b) *Factors of Importance for Amphetamine Induced Toxicity in Aggregated and Isolated Mice and Rats*

A widespread interest in amphetamine toxicity, particularly in aggregated mice, has resulted in a large number of publications dealing with various aspects of this phenomenon.

Strain Differences. In addition to the differences in amphetamine toxicity between species (above), there are also marked differences between strains of mice (CHANCE, 1947; WEAVER and KERLEY, 1962; DOLFINI et al., 1969a, b; CRAIG and KUPFERBERG, 1972). Thus, VALZELLI et al. (1969a) reported an LD 50 for Swiss albino mice to 13 mg/kg, while the value for the C_3H strain was 113 mg/kg under identical experimental conditions. This difference could not be attributed to differences in brain amphetamine concentrations (VALZELLI et al., 1969a).

Age. Young rats seem to be less susceptible to amphetamine than adult animals (EHRICH and KRUMBHAAR, 1937) and a similar difference has been observed in aggregated, but not isolated, mice (CHANCE, 1947; CHERNOV et al., 1966; ALHAVA, 1972). Young mice have lower amphetamine concentrations in the brain compared with adults, which might partly explain the difference in susceptibility (ALHAVA and MATTILA, 1972, 1974).

Circadian Rhythm. The susceptibility to the lethal effects of amphetamine in rats varies considerably over the 24-h day, with a peak at 2 a.m., i.e., under darkness (SCHEVING et al., 1968). No diurnal rhythm was observed under continuous light conditions (SCHEVING et al., 1968). BAUER and PERLEY (1971) found a diurnal rhythm in the concentrations of amphetamine in the liver of mice, with the highest levels occurring during the dark period of the day. Whether a similar diurnal rhythm in the rate of metabolism of amphetamine occurs in rats is not known.

Stress. Various kinds of stress markedly alter the toxicity of amphetamine in both mice and rats. Thus, noise (CHANCE, 1947), electric shocks (WEISS et al., 1961; ASKEW, 1962; GOLDBERG and SALAMA, 1969), drum stress, and cold stress (GOLDBERG and SALAMA, 1969) have been reported to increase susceptibility to amphetamine.

Social stress (or aggregation), i.e., the presence of several animals in the same cage, has been most intensively studied. Under this condition both the area per animal and the total number of animals per cage seem to influence the lethality (CHANCE, 1946; HÖHN and LASAGNA, 1960; CHERNOV et al., 1966). Maximal social stress seems to occur, according to toxicity data, if all the mice in a group of 10 are treated with amphetamine; an amphetamine treated mouse was less susceptible if placed in a cage together with 9 untreated or sedated mice, and least susceptible if kept alone (WANG et al., 1969).

The configuration of the dose-response curves for amphetamine-induced lethality for aggregated mice is different from that obtained with isolated animals in that they are triphasic and monophasic, respectively (L'HUILLIER et al., 1963; GARDOCKI et al., 1966a; GEORGE and WOLF, 1966; MENNEAR and RUDZIK, 1968; STOLK and RECH, 1968a; ALHAVA, 1972). Most consistently, aggregation confers an increased susceptibility to amphetamine in the dose range 20–40 mg/kg, while the animals are again relatively resistant to amphetamine doses between 80 and 100 mg/kg i.p. Above 100 mg/kg, there is again a rapid increase in toxicity and this portion of the dose-response curve is similar to that obtained in isolated animals. For reasons discussed in detail below, different causes of death for the low dose toxicity have been suggested. Because of the triphasic dose-response curve in aggregated mice, LD 50 values cannot be determined meaningfully and the low LD 50 values for aggregated mice reported above refer to the first phase of the triphasic curve.

Aggregated mice have been reported to have slightly higher concentrations of amphetamine in the brain as compared with isolated animals given the same dose of the drug, at least after injections of doses higher than 7.5 mg/kg (CONSOLO et al., 1965b; FULLER and HINES, 1967). However, this difference does not seem to fully explain the increased toxicity of amphetamine in aggregated mice. Prolonged isolation of mice for several weeks induces aggressivity. These aggressive mice are more susceptible to amphetamine by a factor of 3–4, regardless of whether they are kept isolated or grouped after the amphetamine injection (CONSOLO et al., 1965b, c; WELCH and WELCH, 1966). There was no difference in the total brain concentration of amphetamine between aggressive and nonaggressive mice according to CONSOLO et al. (1965b). FEUDIS and MARKS (1973a, b), however, found higher levels of amphetamine in brains of aggressive mice as compared with nonaggressive animals. Previously isolated mice had also higher concentrations of amphetamine within brain synaptosomes (FEUDIS and MARKS, 1973a). Genetically aggressive mice, in contrast to those with isolation-induced aggressivity, were not more susceptible to amphetamine than nonaggressive mice (LAGERSPETZ and LAGERSPETZ, 1971). WELCH and WELCH (1972) reported that the submissive mice were less vulnerable to amphetamine than the aggressor mice in a colony, i.e., the habituation to increased social stimulation rendered the receivers of aggression more tolerant to the toxic effects of amphetamine. Obviously social stimulation might be both advantageous and disadvantageous with regard to amphetamine susceptibility.

Ambient Temperature. Most studies on amphetamine toxicity have been performed at room temperature (18–21° C). However, changes of the environmental temperature have marked effect on lethality. Thus, there is an increase in the incidence of lethality in both solitary and aggregated mice at elevated temperatures (approx. +30° C) (CHANCE, 1946, 1947; ASKEW, 1962; HÖHN and LASAGNA, 1960; CLARK et al., 1967; DOLFINI et al., 1969a; KOPPANYI and MALING, 1973). Then again, a reduction of the ambient temperature to between +10 and +15° C clearly reduced toxicity in mice (HÖHN and LASAGNA, 1960; CLARK et al., 1967; MÜLLER and VERNIKOS-DANELLIS, 1970; KOPPANYI and MALING, 1973). In fact, no group toxicity in mice was present at −10° C (HÖHN and LASAGNA, 1960), and even transfer of aggregated mice to a cold environment after the typical signs and symptoms of amphetamine intoxication had developed reduced the toxicity (CLARK et al., 1967; KOPPANYI and MALING, 1973). Further reduction of the ambient temperature to below 10° C or prolonged cold stress again increased amphetamine toxicity (HÖHN and LASAGNA, 1960).

Dehydration. Water balance seems to be of importance for the toxicity of amphetamine as well, in that dehydration increases it (MÜLLER and VERKNIKOS-DANELLIS, 1970), while hydration lessens it in mice (CHANCE, 1947).

Hormonal Influences. Hyperthyroid mice were more sensitive to the toxic effect of amphetamine (ASKEW, 1962; MOORE, 1966a). Hypothyroidism, induced by treatment with potassium perchlorate, had no effect on amphetamine toxicity in mice (MOORE, 1966a) while α-methyl-thiouracil-treated animals were more susceptible to amphetamine than controls (MANTEGAZZA et al., 1968b). Adrenalectomy seems to have little (WEISS et al., 1961; RICHARDS et al., 1966) if any (GEORGE and WOLF, 1967) effect on amphetamine toxicity in mice. However, mice subjected to both adrenalectomy and immunosympathectomy were less susceptible to amphetamine under both isolated and aggregated conditions (RICHARDS et al., 1966). Treatment of mice with dexamethasone reduced the toxicity (CLARK et al., 1967) and ACTH and cortisol administration increased it (D'ARCY and SPURLING, 1961) suggesting that the adrenocortical hormones might influence the sensitivity to amphetamine in mice.

c) *Effects of Other Drugs on Amphetamine-Induced Toxicity in Mice and Rats*

A large number of drugs, particularly those with actions on monoaminergic mechanisms, have been studied with regard to their effects on amphetamine toxicity. The outcomes of these studies are summarized in Table 10 and will be briefly commented upon. As evident, most studies deal with amphetamine toxicity in aggregated mice, only a few isolated mice and still fewer with amphetamine toxicity in rats. Even if the outcomes of the studies in aggregated and isolated mice are mostly in agreement, generalizations to toxicity under isolated conditions or to other species should be drawn with caution. It might be that the causes of death under the two conditions are different (GREENBLATT and OSTERBERG, 1961; HÖHN and LASAGNA, 1960; ALHAVA, 1972; LITTLETON, 1974).

All drugs that impair CA transmission, i.e., neuroleptics, α- and β-receptor blocking agents, depletors of CA stores, and tyrosine hydroxylase inhibitors, as well chemical sympathectomy with 6-OH-DA, reduce amphetamine toxicity. Drugs that might potentiate amphetamine's stimulant effect on CA transmission centrally and peripherally, i.e., MAO inhibitors and CA uptake inhibitors, tend to potentiate amphetamine toxicity. This clearly indicates an involvement of NA and DA neurons in the series of events leading to death of the animals.

The studies of drugs interfering with serotonergic transmission seem to point to an involvement of 5HT neurons as well, since 5HT receptor blocking agents and the inhibitor of 5HT synthesis, p-chloroamphetamine, decreases amphetamine toxicity. It is interesting to note that as for several other effects of amphetamine, cholinergic drugs, such as tremorine, decrease the effect of amphetamine, while anticholinergics potentiate the toxic effects of amphetamine. Sedative drugs and ethanol seem to interact with amphetamine differently depending on their dosage. As pointed out in many of these reports (Table 10), the sedative and hypnotic drugs may alleviate the stress factor of aggregation and thereby reduce toxicity. Among the miscellaneous drugs listed in Table 6, salicylate was without effect on toxicity as it is on the hyperthermic effect of amphetamine (Table 8). The inhibitor of glucose metabolism, 2-deoxyglucose, decreased amphetamine toxicity, which might indicate an involvement of the energy produced by glucose oxidation in the successive events leading to death. Fenfluramine, an amphetamine derivative, antagonizes amphetamine toxicity, like several other actions of amphetamine (Tables 1, 2, 8).

Table 10. Effects of other drugs on toxicity of amphetamine in mice and rats. Symbols: No effect: 0, Toxicity increased: ↑, decreased: ↓. No change: 0. aggr. = aggregated; isol. = isolated

Drug	Mice		Rats	References
	aggr.	isol.		
Neuroleptic drugs				
Chlorpromazine	↓	↓		2, 4, 12, 19, 20[d], 21, 22, 23, 30, 34, 41, 42[k], 43, 44, 51, 60, 63
	0			3
		↓		12, 20[d], 30
			↓	16, 21[o], 45
Perphenazine	↓			44, 51, 52[m]
Trifluoperazine	↓			44
Fluphenazine	↓			51
Haloperidol	↓		↓	16, 18
Others	↑			29
	↓			31, 51
α-adrenoceptor blocking agents				
Phenoxybenzamine	↓			2, 4, 14, 36, 41, 55, 60
Phentolamine	↓			34, 54
Dibenamine	↓	↓		20[d]
			↓	21[o]
Others	↓			20[d], 34, 36, 43, 54, 55
β-adrenoceptor blocking agents				
Propranolol	↓			32, 33, 42[k], 54
			↓	16
Pronethanol	↓	↓		20[d]
Dichloroisoproterenol (DCI)	0			36, 41
Nethalide	↓			36
5HT receptor blocking agents				
Methysergide	↓			59
BOL 148	↓			28
Drugs affecting storage of monoamines				
Reserpine		↑		3[a]
	↓			6, 12, 22, 30, 34, 37, 40[i], 41, 43, 47, 55, 57
	0			23[e], 42[k]
		0		30
Reserpine, chronic	↑		↑	57, 58
Tetrabenazine	↓		↓	21
Guanethidine	↓			14
Bretylium	0			41
Metaraminol	↓			37, 41
Monoamine synthesis inhibitors				
Tyrosine hydroxylase inhibitors				
α-methyl tyrosine	↓			1, 21, 37[h], 40[i]
	0			42[k]
			↓	21[o]
Dopa/5HTP decarboxylase inhibitors				
α-methyl-dopa	↓			27

Table 10 (continued)

Drug	Mice		Rats	References
	aggr.	isol.		
DA-β-hydroxylase inhibitors				
Diethyldithiocarbamate	↑			5, 53[b]
(DEDC)	0			48
Tryptophanehydroxylase inhibitors				
p-chlorophenylalanine (PCPA)	↓			28
Monoamine precursors				
5HTP	↑			47
Unnatural monoamine precursors				
α-methyl-m-tyrosine	↓			27, 37, 38, 41, 43
	0			42[k]
Monoamine oxidase inhibitors				
Iproniazid	↑			14, 22[c], 41
Pargyline	↑			46
Nialamide	↑			46
Phenelzine	↑			9, 10, 11, 46
Tranylcypromine	↑			9, 22[c], 46
Others	↑			41, 46
Inhibitors of monoamine uptake				
Imipramine	↑			17[c], 29
	↓			56[n]
Desmethylimipramine	↑ or ↓			56[n]
Others	↑ or ↓			56[n]
Cholinergics				
Tremorine	↓			35
Carbachol	0			35
Anticholinergics				
Atropine[g]	↑			20[d], 50
	0			41
Scopolamine[g]	0			34, 41
	↑			35
Ganglionic blocking agents				
Mecamylamine	↑	↑		20[d]
Pentamethazene	0	0		63
Methioplegium	0	0		63
Sedative and hypnotic drugs				
Pentobarbital	0			3
	0		0	30, 49[a]
Phenobarbital	↓		0	18, 30, 41, 45
Meprobamate	0			12
	↓			20[d], 24
Chlordiazepoxide	↑			8[b], 20[d]
	↓			8[b]
	↓		↓	21[o]
			0	45

Table 10 (continued)

Drug	Mice		Rats	References
	aggr.	isol.		
Sedative and hypnotic drugs, ctd				
Diazepam	0			3
			0	45
Others	0			12
	(↓)			34
			0	45
Miscellaneous drugs				
6-OH-DA[p]	↓	↓		62
Cinobufotenin	↑			47
LSD-25	0			34
5HT	0			34
	↑			47[l]
Fenfluramine	↓	0		25
S 992[f]	0			25
4-methyl-5-(β-chloro-ethyl)-thiazol (SCTZ)	↑			2
Diphenylhydantoin	↓			13
Ethanol	↑			3
	↓			13
Morphine	0			34, 60
Chlordane	0			49[a]
Amantadine	↓			39
Mestranol	↑			7
Lynestrol	0			7
2-deoxyglucose	↓			13
Ca-salicylate	0			2
SKF 525 A	↑			49
Δ^9-tetrahydrocannabinol	0			15
	↓			26
Procaine	0	0		63
Ouabain	0	0		63
Digitaline	0	0		63
Sparteine	0	0		63

[a] Repeated injections.
[b] Depending on dose.
[c] Reversed reserpine-induced reduction of amphetamine toxicity.
[d] Study of dose-response relationships for extensive series of drugs in aggregated and isolated mice.
[e] Amphetamine antagonized reserpine-induced reduction of toxicity of MAO inhibitors.
[f] Fenfluramine derivative.
[g] Methylscopolamine and methylatropine without effect.
[h] Dose range 30–300 mg/kg of α-methyltyrosine.
[i] less protection at elevated (+38° C) ambient temperature.
[k] In hyperthyroid mice.
[l] Given inrecerebroventricularly.
[m] Together with a choline esterase inhibitor.
[n] Detailed study, showing complicated pattern of interactions between tricyclic antidepressants and amphetamine.
[o] In stressed rats given amphetamine.
[p] Intraperitoneal administration.

References

1. ALHAVA, 1973
2. ASKEW, 1962
3. BANARJEE and GEH, 1973
4. BANARJEE and LIN, 1973
5. BARNETT and PESCHEL, 1968
6. BEAUVALLET et al., 1966a
7. BLACKHAM and SPENCER, 1969
8. BOISSIER et al., 1963
9. BRITTAIN et al., 1964
10. BROWNLEE and WILLIAMS, 1963a
11. BROWNLEE and WILLIAMS, 1963b
12. BURN and HOBBS, 1958
13. CLARK et al., 1967
14. COHEN and LAL, 1963
15. CRAIGMILL et al., 1974
16. DAVIS et al., 1974
17. DRUDI-BARACCO et al., 1963
18. FROMMEL and CHMOULIOVSKY, 1964
19. FULLER and HINES, 1967
20. GARDOCKI et al., 1966b
21. GOLDBERG and SALAMA, 1969
22. HALPERN et al., 1962a
23. HALPERN, et al., 1962b
24. L'HUILLIER et al., 1963
25. JESPERSEN and BONACCORSI, 1969b
26. KUBENA and BARRY, 1970
27. LAL et al., 1963
28. LAPIN et al., 1972
29. LAPIN and SCHELKUNOV, 1965
30. LASAGNA and MCCANN, 1957
31. LESLIE and MAXWELL, 1964
32. MANTEGAZZA et al., 1970
33. MANTEGAZZA et al., 1968a
34. MAXWELL, 1959
35. MENNEAR, 1965
36. MENNEAR and RUDZIK, 1965
37. MENNEAR and RUDZIK, 1966
38. MENNEAR and RUDZIK, 1968
39. MENON et al., 1973
40. MENON and DANDIYA, 1967
41. MOORE, 1964
42. MOORE, 1965
43. MOORE et al., 1965
44. MORPURGO and THEOBALD, 1964
45. MÜLLER and SCHOETENŞACK, 1969
46. O'DEA and RAND, 1969
47. PECZNIK and ZEI, 1961
48. PFEIFFER et al., 1966
49. PRABHU, 1972
50. PROCTOR et al., 1973
51. PROCTOR et al., 1966
52. PROCTOR et al., 1964
53. PRZEGALINSKI and KLEINROK, 1970
54. RAEVSKII and GURA, 1970
55. RUDZIK and MENNEAR, 1966
56. SOKOL and MAICKEL, 1972
57. STOLK and RECH, 1968a
58. STOLK and RECH, 1968b
59. TICE and SPOONER, 1970
60. VEDERNIKOV, 1970
61. WEIS et al., 1961
62. WOLF and BUNCE, 1973
63. DELPHAUT, 1965

d) Chronic Toxicity of Amphetamine

Repeated or chronic amphetamine administration for up to 4–6 weeks has shown that tolerance develops to otherwise lethal doses in mice (ABDALLAH, 1973), rats (LEWANDER, 1968b; MAGOUR et al., 1974a, b), cats (ELLISON et al., 1968; ELLINWOOD and ESCALANTE, 1970a), and rabbits (KASIRSKY et al., 1972). O'DEA and RAND (1969) showed that pretreatment of mice with a single nonlethal dose of amphetamine had no protective action on the lethality after a second administration, while ALHAVA (1972) presented evidence for a protective action by 5 mg/kg amphetamine given 4 h before a second challenge with the drug.

e) Pathologic-Anatomical Findings in Amphetamine-Intoxicated Animals

In the early studies by ALLES (1933, 1939), EHRICH and KRUMBHAAR (1937), and EHRICH et al. (1939), it was noted that rats and guinea pigs had small hemorrages in their lungs, livers, kidneys, and spleens, dilation of the heart, patches of necrosis in the myocardium, and increased amounts of gas and fluid in the intestines after lethal doses of amphetamine. Similar findings have been reported by others (YOKOI and AKIYAMA, 1958, cited by UTENA et al., 1959; UTENA, 1966) and have been found to occur also in dogs (ZALIS et al., 1967) and rabbits (KASIRSKY et al., 1972). Like several other sympathomimetic amines (CHAPPEL et al., 1959), amphetamine causes degeneration of the myocardium with an almost infarctionlike pathologic picture in rats (HALPERN et al., 1962c), rabbits (KASIRSKY et al., 1972), and dogs (ZALIS et al., 1967). Neuropathologic descriptions after amphetamine intoxication contain such items as hyperemia, petechial hemorrages, and glial proliferation in monkeys (DUARTE-ESCALANTE and ELLINWOOD, 1972), and enlargement and chromatolysis of cell bodies in the medulla oblongata in cats (ESCALANTE and ELLINWOOD, 1970). Further, swelling of cell bodies and their nuclei in the caudate nucleus, cortex, hypothalamus (POPOVA et al., 1972), and other changes in rat motor cortical neurons (VTYURIN and TUMANOV, 1971) have been reported.

f) On the Causes of Death in Amphetamine Intoxication

The clinical descriptions of the signs associated with death after fatal doses of amphetamine are fairly consistent. Thus, mice, rats, and dogs are described to have a reduced respiratory rate and enter into a shocklike state and fade away; if very high doses are given, convulsions might precede this moribund state (ALLES, 1933, 1939; GUNN et al., 1939; HÖHN and LASAGNA, 1960; CLARK et al., 1967; ZALIS et al., 1967; ALHAVA, 1972). Agreement on the immediate cause of death in relation to amphetamine poisoning has not been obtained. Some important contributing factors for the lethal effect of amphetamine in mice and rats that have been advanced in the literature include: hyperthermia, CA depletion, hypoglycemia, neuromuscular blockade, stress, fear, and anxiety.

Hyperthermia as the cause of death in aggregated mice was suggested by GREENBLATT and OSTERBERG (1961) and ASKEW (1962). The facts that amphetamine-treated animals kept at elevated ambient temperature have both higher body temperature and an increased susceptibility to the drug and that placement of intoxicated animals at low environmental temperature reduces the toxicity add support to hyperthermia as the cause of death. However, with certain drugs such as 6-OH-DA as well as α- and β-receptor blocking agents given as pretreatments to aggregated mice, it seems possible to dissociate the effect of amphetamine on the body temperature from its lethality (RAEVSKII and GURA, 1970; WOLF and BUNCE, 1973).

Depletion of CA has been forwarded by several authors as a cause of death in mice (BEAUVALLET et al., 1962b, 1963, 1964; MOORE, 1963; LAL and CHESSIK, 1964). The findings that hyperthyroid mice are more susceptible to the toxic effects of amphetamine, that the CA depletion in these mice was enhanced (MOORE, 1965), and that the NA depletion was more pronounced in aggregated as compared with isolated mice (LAL and CHESSIK, 1964) supported this view. A later report, however, has shown a lack of a direct correlation between the degree of CA depletion and lethality after amphetamine administration (ALHAVA and KLINGE, 1972). In addition, LITTLETON (1974) observed that aggregated mice dying after a low dose of amphetamine had a more pronounced depletion of NA than isolated mice dying after higher doses of the drug. This again shows the complexity of factors at work causing death under different conditions. The biochemical findings and the fact that most of the observations with drugs interfering with storage, synthesis, and postsynaptic effects of CA, which protect mice and rats from the lethal effects of amphetamine, are in agreement with the view that central and peripheral release of CA, rather than depletion, is of major importance in the chain of events leading to death in amphetamine poisoning.

Glycogen Depletion and Hypoglycemia. Although effects of amphetamine on tissue glycogen levels were known since 1939 (EHRICH et al., 1939), MOORE et al. (1965) were the first to consider a relation between glycogen depletion, hypoglycemia, and amphetamine toxicity in mice. This relation was later confirmed in mice and rats (STOLK et al., 1970). As described above, hypoglycemia in aggregated mice seems to be due to increased glucose utilization rather than insulin release (BEWSHER et al., 1966). Of particular interest in this connection is the finding that glucose administration only slightly protected aggregated mice from amphetamine toxicity, while 2-deoxyglucose, which inhibits glucose utilization, was far more effective in this respect (CLARK et al., 1967). This observation raises an unresolved question, whether amphetamine toxicity might be caused by glucose oxidation and the concomitant production of heat or toxic intermediate metabolites, rather than the secondary hypoglycemia and glycogen depletion.

Neuromuscular blockade is caused by amphetamine in vitro (27 μg/ml) on a phrenic nerve—diaphragm preparation (PETERSON et al., 1964). At elevated bath temperature this effect of amphetamine was enhanced about threefold and this effect may be of significance as a contributory factor to amphetamine toxicity in vivo.

Anxiety and fear as a cause of death in amphetamine poisoning in aggregated mice has been advanced on the basis of a lowered convulsive threshold to pentylenetetrazol in aggregated mice (SWINYARD et al., 1961).

The most thorough analysis of the pathophysiologic events in fatal amphetamine poisoning has been done in dogs, and include measurements of cardiovascular and respiratory functions, acidbase-balance, electrolytes in serum, blood glucose, body temperature, and a histopathologic examination of most tissues after death (ZALIS et al., 1967). The results from this study show that a hypermetabolic state with hyperpyrexia and acidosis is followed by shock, anoxia, and tissue degeneration. Thus, several simultaneous disregulations of vital functions seem to contribute to the fatal outcome. These authors have also drawn attention to the protective effect of *d*-tubocurarine on amphetamine toxicity in dogs (ZALIS et al., 1965), which has been interpreted as suggesting involvement of increased skeletal muscular activity as a contributing factor to the amphetamine-induced pathophysiology and death.

g) Summary and Conclusions

On the basis of the available literature, it seems probable that the cause of death after amphetamine administration is related to release of central and peripheral CA, which in turn causes a series of secondary effects such as increased muscular activity, increased glucose utilization, hypoglycemia, and glycogen depletion, increased body temperature, as well as changes in the cardiovascular system with organ congestion, hemorrages, etc. Various additional stresses obviously facilitate one or several of the links in this chain of events and thereby the toxicity of amphetamine. It does not seem possible to pinpoint any single factor as the only cause of death in amphetamine poisoning. On the basis of studies in animals, a regime combining neuroleptic drugs and a β-receptor blocking agent has been suggested for use in combating amphetamine intoxication in humans (DAVIS et al., 1974).

IV. Miscellaneous Pharmacologic Effects of Amphetamine

1. Electroencephalographic Changes Caused by Amphetamine

Up to 1949 amphetamine had not been observed to cause any EEG changes such as seizures or dysrythmias (TOMAN and DAVIS, 1949). A few years later the now well-known desynchronizing (increased frequency and decreased amplitude) effect on the cortical EEG was demonstrated in dogs (SCHALLEK and WALZ, 1953), cats (HIEBEL et al., 1954; ELKES et al., 1954), rabbits (RINALDI and HIMWICH, 1955; SKARINGA et al., 1969), and chickens(KEY and MARLEY, 1962). LIESIENE (1970) has drawn attention to the similarity in the EEG picture caused by amphetamine and paradoxic (REM) sleep. Racemic as well as *d*- and *l*-amphetamine had this effect, although *d*-amphetamine was more potent than *l*-amphetamine in most studies (ELKES et al., 1954; BRADLEY and ELKES, 1957; DEWHURST and MARLEY, 1965a), but not all (HANNA et al., 1963). Parahydroxyamphetamine did not cause desynchronization of the EEG in cats (WALLACH and GERSHON, 1971).

The desynchronizing effect of peripherally administered amphetamine on the EEG has been localized to an action of the drug on the brain stem (reticular formation), since brain transections at the mesencephalic level (cerveau isolé) abolished it while various pontine or spinal transections did not change this action of the drug in cats and monkeys (HIEBEL et al., 1954; ELKES et al., 1954; BRADLEY and ELKES, 1957; KILLAM et al., 1959; WHITE and DAIGNEAULT, 1959), rabbits (TAKEO and HIMWICH, 1967; FUJIMORI and HIMWICH, 1968, 1969), and chickens (MARLEY and STEPHENSON, 1971; SPOONER and FLAMING, 1972). Intracerebral microinjections of amphetamine into the reticular formation in chickens had no effect on the EEG (MARLEY and STEPHENSON, 1968), while in rabbits the cortical EEG was activated by such injections (ANICHKOV et al., 1973).

In addition, amphetamine potentiated the desynchronizing effect on EEG caused by various sensory inputs (HIEBEL et al., 1954; BRADLEY and KEY, 1958; ETEVENON, 1967) and reduced the thresholds for desynchronization of reticular or thalamic electrical stimulation (KILLAM et al., 1959). Similarly, hippocampal field responses after electrical stimulation in the septum were increased by amphetamine (STÜMPF, 1965; IZQUIERDO et al., 1968). Evoked potentials elicited by photic stimulation were changed in appearance and pattern after amphetamine in rats (SCHWARTZBAUM et al., 1971, 1974).

Drug interaction studies have shown that amphetamine does not cause desynchronization in atropine- or scopolamine-treated rabbits (LONGO and SILVESTRINI, 1957; WHITE and DAIGNEAULT, 1959), dogs (WHITE et al., 1961), cats (BRADLEY and ELKES, 1957), monkeys (WHITE et al., 1961), and rats (IZQUIERDO et al., 1968). On the contrary, the motor behavioral effects of amphetamine were not antagonized by atropine (WHITE and DAIGNEAULT, 1959; WHITE et al., 1961). Eserine, although in itself causing desynchronization of the EEG in rabbits, antagonized the effect of amphetamine (LONGO and SILVESTRINI, 1957). In the limbic system, eserine and methamphetamine seem to cause desynchronization by different mechanisms (GOGOLÁK et al., 1969). Another drug interaction that has received much attention is the antagonism by amphetamine of barbiturate sedation and EEG synchronization. Although BRADLEY and ELKES (1957) did not find it, several subsequent reports have unanimously concluded that amphetamine is a competitive antagonist of barbiturate sedation with clinical implications in barbiturate intoxication (WHITE and BOYAJY, 1959; KILLAM et al., 1959; HANNA et al., 1963; BARNES, 1966; HATCH, 1973), as well as other sedative drugs (WHITE and BOYAJY, 1959).

A pharmacologic analysis of the alerting effect of amphetamine on the EEG by aid of drugs affecting the monoaminergic mechanisms in the CNS has shown that some α-receptor blocking agents (phenoxybenzamine, dibenamine, phentolamine) and chlorpromazine block the effect (WHITE and BOYAJY, 1959; MUNOZ and GOLDSTEIN, 1961; CAHN and HEROLD, 1970). In contrast, β-receptor blocking agents (propranolol, dichlorisoprenaline) potentiate amphetamine activation of the EEG (MUNOZ and GOLDSTEIN, 1961; CAHN and HEROLD, 1970). In chickens, a 5HT receptor blocking agent, methysergide, and α-methylnoradrenaline antagonized amphetamine's EEG effect (DEWHURST and MARLEY, 1964; MARLEY and STEPHENSON, 1971) while α- or β-receptor blocking agents were without effect (DEWHURST and MARLEY, 1964). Amphetamine reduced the duration of epileptiform after-discharges in isolated cortex slabs after electrical stimulation. This effect was shared by 5HT and was antagonized by inhibition of 5HT synthesis (pCPA) or 5HT receptor blocking agents (methysergide, cyproheptadine) (VASQUEZ and KRIP, 1973).

In summary, amphetamine has various effects on EEG and brain excitability. The scarce data on mechanisms of action for these effects suggest that both cholinergic and serotonergic mechanisms might be involved. Whether the cholinergic and serotonergic events are primary or secondary to interactions with CA-ergic neuronal systems is not known.

2. Effect of Amphetamine on Seizure Susceptibility

Since the report by TAINTER et al. (1943) on *d*-amphetamine-induced elevation of the threshold for elicitation of seizures by electrical stimulation in rats, several reports on this subject have appeared with contradictory results. Thus, e.g., SWINYARD et al. (1961), LANCIAULT and WOLF (1965), and SPENCER and TURNER (1968) found a lowering of the seizure threshold in rats and mice by amphetamine. This literature has been commented on by RUDZIK and JOHNSON (1970) and GERALD and RIFFEE (1973); in both these latter studies an elevation of the threshold was found and confirmed. The differences in results seem to have been obtained depending on whether the sizures were elicited by electrical stimulation or by drugs such as pentylenetetrazol (RUDZIK and JOHNSON, 1970; GERALD and RIFFEE, 1973). Species (rats, mice) or strain differences as well as differences in the experimental conditions seem to have contributed to the controversy (see also MCKENZIE and SOROKO, 1972). Audiogenic seizures in mice were antagonized by amphetamine (GRAHAM et al., 1974). L-Amphetamine has been reported to be without effect on the seizure threshold (TAINTER et al., 1943), or to have half the potency of *d*-amphetamine (GERALD and RIFFEE, 1973).

Studies on the effect of drugs affecting brain monoamines on amphetamine-induced seizure susceptibility have shown that reserpine has no effect (PFEIFFER et al., 1968) or potentiates (WOLF et al., 1969) amphetamine's threshold increasing action. RUDZIK and JOHNSON (1970) showed that amphetamine antagonized the threshold lowering effect of reserpine itself. The CA synthesis inhibitor α-methyltyrosine antagonized amphetamine (SPENCER and TURNER, 1969; WOLF et al., 1969), while diethyldithiocarbamate (DBH inhibitor) and p-chlorphenylalanine (5HT synthesis inhibitor) had no effect (SPENCER and TURNER, 1969). It has thus been concluded that brain DA rather than NA or 5HT is involved also in this effect of amphetamine.

A separate discussion concerns the question, whether or not amphetamine potentiates the anticonvulsants diphenylhydantoin, phenobarbital, and similar agents in animals as it does in man. FREY (1964) and FREY and KAMPMANN (1966) found that amphetamine was a weak antagonist to phenobarbital and diphenylhydantoin with regard to their actions on pentylenetetrazol-induced convulsions, whereas no interaction was observed for electrically induced seizures. Later STILLE (1953) and RUDZIK and JOHNSON (1970) reported that amphetamine potentiates both these drugs as well as acetazolamide in mice and rats. In this connection it might be mentioned that amphetamine also potentiates barbiturate-induced burst suppression on the EEG (WIDERLÖV and WAHLSTRÖM, 1967).

In mice, chronically treated with amphetamine, tolerance was shown to develop to changes in various threshold measures induced by single doses of the drug (GERALD and RIFFEE, 1973).

In cats, WINTERS et al. (1969) presented evidence that high doses of amphetamine causes synchronization, rather than epileptic discharges as determined by EEG. Chronic treatment of cats with amphetamine induces characteristic changes of the electric activity in the olfactory forebrain (ELLINWOOD and ESCALANTE, 1970b; ELLINWOOD et al., 1974). Methamphetamine reportedly caused generalized seizures in cats after pretreatment with disulfiram (DBH inhibitor) (ELLINWOOD et al., 1973). These seizures seem to start with synchronization of olfactory spindles and are thought to be of importance in relation to amphetamine psychosis. In baboons, amphetamine antagonized seizures induced by photostimulation in 50% of the experiments (KILLAM, 1973).

It is obvious from this short review that amphetamine influences seizure susceptibility, although its mechanisms of action are unclear.

3. Sleep and Wakefulness

Animal experiments on the effect of amphetamine on natural sleep are rather few. In rabbits (KHAZAN and SAWYER, 1964), rats (GOLDSTEIN et al., 1967), and cats (JEWETT and NORTON, 1966; SHIMIZU and HIMWICH, 1968), amphetamine abolished paradoxic sleep (activated or REM sleep). Amphetamine also abolished neurophysiologic counterparts of REM sleep, namely the PGO-spikes, in cats (HAEFELY et al., 1973). The effects on slow wave sleep have varied (JEWETT and NORTON, 1966; SHIMIZU and HIMWICH, 1968). One study in canaries showed that in birds self-regulating their circadian rhythm, amphetamine, if given in the p.m., prolonged their waking phase and caused them to oversleep the next morning (WAHLSTRÖM and WIDERLÖV, 1968). Several studies, however, deal with the analeptic effect of amphetamine in drug-induced sleep or narcosis. HAFFNER (1938) and ALLES (1939) reported amphetamine (*d*-amphetamine was more potent than *l*-amphetamine) to reduce the sleeping time after injection of chloral hydrate in mice and rabbits. Later, pentobarbital-induced narcosis was shown to be shortened by amphetamine in cats (*d*- and *l*-amphetamine equally potent; HANNA et al., 1963) and mice (BARNES and MEYERS, 1964). Eserine, although in itself antagonizing barbiturate narcosis, antagonized this effect of amphetamine (cf. similar study on EEG, LONGO and SILVESTRINI, 1957). In contrast, LANCIAULT and WOLF (1965) did not find any change in the hexobarbital sleeping time in mice treated with amphetamine. In this species the hexobarbital metabolism is inhibited by amphetamine (LAL et al., 1970), which might obscure the outcome of studies on this drug interaction. Similarly, pentobarbital narcosis in rats was not shortened by amphetamine (BURTON et al., 1957). In dogs, cats, and calves, however, amphetamine promptly antagonized barbiturate narcosis, particularly if given in combination with yohimbine (HATCH, 1973). Amphetamine did not antagonize chlorpromazine-induced sedation, while it effectively antagonized reserpine (BURTON et al., 1957).

4. Amphetamine-Induced Analgesia

Several recent studies, referred to below, have confirmed the demonstration by KIESSIG and ORZECHOWSKI (1941) that sympathomimetic amines have analgesic properties. In mice, amphetamine has been reported to protect from writhing (OKUN et al., 1963; DAVIS and HORLINGTON, 1964) and MOORMAN and OPITZ (1967) reported that amphetamine (4 mg/kg), although without significant analgesic effect in itself, potentiated codeine in this respect. Similar findings have been reported in rats (AHMED et al., 1970). Methamphetamine-induced analgesia in rats (COLVILLE and CHAPLIN, 1964) was antagonized by diethyldithiocarbamate, p-chlorphenylalanine, and reserpine, while it was potentiated by α-methyltyrosine (VEDERNIKOV, 1969; MAJOR and PLEUVRY, 1971). The effect of α-methyltyrosine was reversible with *l*-dopa. A DA/5HT antagonizing system influencing pain sensitivity was suggested from these findings. Similar results were obtained by GÖRLITZ and FREY (1972), with the exception that α-methyltyrosine antagonized amphetamine-induced analgesia in their study, as did chlorpromazine and haloperidol. Both in mice and rats, α- and β-receptor blocking agents were without effect on amphetamine-induced analgesia (MAJOR and PLEUVRY, 1971; GÖRLITZ and FREY, 1972).

Chronic treatment of rats with methamphetamine (10 mg/kg twice daily for 5 days) or mice with amphetamine (20 mg/kg i.p. daily) was accompanied by development of tolerance to the analgesic effect and cross tolerance toward morphine (PLEUVRY, 1971; LITTLE and REES, 1974).

5. Effects of Amphetamine on Physical Performance in Animals

Tests of physical performance have mainly included measurement of the exhaustion time in a water bath or the swimming speed in a water alley. In mice, amphetamine in a wide dose range had no effect on the exhaustion time (JACOB and MICHAUD, 1961). In rats the effects on exhaustion time varied from a slight effect with doses around 4 mg/kg (BÄTTIG, 1963a, b) to a clear prolongation with 10–40 mg/kg of the drug (BHAGAT and WHEELER, 1973). In this study *dl*-, *d*-, and *l*-amphetamine were equipotent. Although swimming speed is reduced or swimming is disorganized at short, postinjection intervals in rats (KAY and BIRREN, 1958; BÄTTIG, 1963b; MAICKEL et al., 1969), there may be an improved swimming speed toward the end of long sessions of trials when control rats begin to slow down (KAY and BIRREN, 1958).

There was no significant improvement of the running speed in race horses (STEWART, 1972).

Thus, amphetamine might counteract exhaustion in prolonged testing sessions, but does not improve running or swimming speed in animals.

D. Conclusion

The increasing knowledge of the distribution and function of central and peripheral monoaminergic neurons has stimulated and made possible the intensive research into the mechanisms of action of amphetamine. It seems possible to draw a series of conclusions from the immense number of new facts reported during the last 15 years.

Biochemically, there is solid evidence to show that amphetamine causes release of central and peripheral catecholamines and interferes with the uptake, metabolism, and turnover of these transmitters. The detailed mechanism(s) of amphetamine-induced release of the catecholamines is, however, unknown. Functional evidence supports the conclusion that the main mechanism of amphetamine action is related to the release of noradrenaline and dopamine in both the peripheral and central nervous systems. The large number of reported pharmacologic effects of amphetamine is in all probability due to the fact that catecholaminergic neurons are widespread in the nervous system and involved in many biological functions.

In addition, there is some evidence that amphetamine interferes with central and peripheral serotonergic neurons in a similar fashion (i.e., release and inhibition of uptake), and some pharmacologic effects seem to be mediated via the release of this transmitter. There are no indications that amphetamine directly releases acetylcholine or any other known transmittor substance. However, since many peripheral and central catecholaminergic functions appear to be balanced by cholinergic neuronal pathways, some effects of amphetamine are modified by cholinergic and anticholinergic agents.

In spite of the surge in amphetamine research during the recent past and the fact that amphetamine may be one of the most well-understood, centrally acting drugs, there are still many unexplained findings in the literature needing further investigations.

Acknowledgement: The present review has been completed during the tenure of a position in drug addiction research supported by the Swedish Medical Research Council (B74-04P-4288-01A, B75-04P-4288-02B, B76-04P-4288-03C).

Abbreviations

CA	catecholamine (s)
MA	monoamine (s)
NA	noradrenaline
DA	dopamine
l-dopa	*l*-dihydroxyphenylalanine
NM	normetanephrine
MTA	3-methoxytyramine

Mopeg	3-methoxyphenylglycol
Dopac	dihydroxyphenylacetic acid
HVA	homovanillic acid
5HT	5-hydroxytryptamine-serotonin
5HIAA	5-hydroxyindole-3-acetic acid
5HTP	5-hydroxytryptophan
ACh	acetylcholine
GABA	γ-aminobutyric acid
TH	tyrosine hydroxylase
DBH	dopamine-β-hydroxylase
DDC	dopa/5HTP decarboxylase
MAO	monoamine oxidase
MAOI	monoamine oxidase inhibitor (s)
COMT	catechol-O-methyl-transferase
α-MD	α-methyl-dopa
α-MT	α-methyl-p-tyrosine
α-MMT	α-methyl-m-tyrosine
6-OH-DA	6-hydroxy-dopamine
PCPA	p-chlorophenylalanine
DEDC	diethyldithiocarbamate
DMI	desmethylimipramine
LSD	lysergic acid diethylamide
PGE	prostaglandin E

References

Abdallah, A. H.: On the role of norepinephrine in the anorectic effect of *d*-amphetamine in mice. Arch. int. Pharmacodyn. **192**, 72—76 (1971)

Abdallah, A. H.: Comparative study of the anorexigenic activity of 5-(3,4-dichlorophenoxymethyl)-2-amino-2-oxazoline HCl and *d*-amphetamine in different species. Toxicol. appl. Pharmacol. **25**, 344—353 (1973)

Abdallah, A. H.: Effect of APMO on pulmonary and arterial blood pressure: a comparison with aminorex. Europ. J. Pharmacol. **27**, 249—251 (1974)

Abdallah, A. H., White, H. D.: Comparative study of the anorectic activity of phenindamine, *d*-amphetamine and fenfluramine in different species. Arch. int. Pharmacodyn. **188**, 271—283 (1970)

Abdallah, A. H., White, H. D., Kulkarni, A. S.: Interaction of *d*-amphetamine with central nervous system depressants on food intake and spontaneous motor activity. Europ. J. Pharmacol. **26**, 119—121 (1974)

Absava, G. I., Vysotskaya, N. B.: Psychostimulators effect on the locomotor activity of rats, and also upon intra- and extracellular content of potassium and sodium ions in different sections of the brain. Farmakol. i Toksikol. **33**, 137 (1970)

Accornero, F.: Intossicazione sperimentale da betafenilisopropilamina. Lav. neuropsichiat. **1**, 308—327 (1947)

Adler, M. W.: Changes in sensitivity to amphetamine in rats with chronic brain lesions. J. Pharmacol. exp. Ther. **134**, 214—221 (1961)

Aghajanian, G. K., Bunney, S.: Central dopaminergic neurons: neurophysiological identification and responses to drugs. Frontiers in catecholamine research, pp. 643—648. Usdin, E., Snyder, S. (Eds.). New York: Pergamon Press 1973

Ahlskog, J. E., Hoebel, B. G.: Overeating and obesity from damage to noradrenergic system in the brain. Science **182**, 166—169 (1973)

Ahmed, S. S., Abraham, G. J. S., Ansari, M. Y.: Dose-dependent modification of codeine analgesia by *d*-amphetamine in albino rats. Arch. int. Pharmacodyn. **184**, 240—246 (1970)
Alexander, M., Isaac, W.: Effect of illumination and *d*-amphetamine on the activity of the rhesus macaque. Psychol. Rep. **16**, 311—313 (1965)
Alhava, E.: Amphetamine toxicity in adult and developing mice. Acta pharmacol. (Kbh.) **31**, 387—400 (1972)
Alhava, E.: Modification by methyltyrosine methylester (H 44/68) of the amphetamine-induced toxicity and brain catecholamine changes in developing mice. Acta pharmacol. (Kbh.) **32**, 119—128 (1973)
Alhava, E., Klinge, E.: Age and brain catecholamine content as factors influencing amphetamine toxicity in mice. Acta pharmacol. (Kbh.) **31**, 401—411 (1972)
Alhava, E., Mattila, M. J.: Amphetamine toxicity and brain amphetamine concentrations in adult and developing mice. Brit. J. Pharmacol. **44**, 352—353 (1972)
Alhava, E., Mattila, M. J.: Dose-dependent differences of amphetamine levels in brain and heart of adult and developing mice. Acta pharmacol. (Kbh.) **34**, 211—221 (1974)
Allan, D., Baird, J. R. C., Ellis, K. E. J.: Interaction between (±)-amphetamine and atropine on the rat cardiovascular system. Brit. J. Pharmacol. **37**, 367—370 (1969)
Alles, G. A.: The comparative physiological actions of *dl*-phenylisopropylamines I. Pressor effect and toxicity. J. Pharmacol. exp. Ther. **47**, 339—354 (1933)
Alles, G. A.: Comparative physiological actions of the optically isomeric phenisopropylamines. Univ. Calif. Publ. Pharmacol. **1**:**11**, 129—150 (1939)
Alpern, E. B., Finkelstein, N., Gantt, W. H.: Effect of amphetamine sulfate on the nervous activity of dogs. Amer. J. Physiol. **133**, 196 (1941)
Alphin, R. S., Ward, J. W.: Anorexigenic effects of fenfluramine hydrochloride in rats, guinea pigs and dogs. Toxicol. appl. Pharmacol. **14**, 182—191 (1969)
Andén, N.-E.: Effects of amphetamine and some other drugs on central catecholamine mechanisms. In: Costa, E., Garattini, S. (Eds.): International symposium on amphetamines and related compounds, pp. 447—462. New York: Raven Press 1970
Andén, N.-E., Carlsson, A., Häggendal, J.: Adrenergic mechanisms. Ann. Rev. Pharmacol. **9**, 119—134 (1969)
Andén, N.-E., Dahlström, A., Fuxe, K., Larsson, K.: Functional role of the nigro-neostriatal dopamine neurons. Acta pharmacol. (Kbh.) **24**, 263—274 (1966)
Andén, N.-E., Henning, M., Obianwu, H.: Effect of epsilonaminocaproic acid on adrenergic nerve function and tissue monoamine levels. Acta pharmacol (Kbh.) **26**, 113—129 (1968)
Andén, N.-E., Magnusson, T., Stock, G.: Effects of drugs influencing monoamine mechanisms on the increase in brain dopamine produced by axotomy or treatment with gammahydroxybutyric acid. Naunyn-Schmiedebergs Arch. exp. Path. Pharmak. **178**, 363—372 (1973)
Andén, N.-E., Rubenson, A., Fuxe, K., Hökfelt, T.: Evidence for dopamine receptor stimulation by apomorphine. J. Pharm. Pharmacol. **19**, 627—629 (1967)
Andén, N.-E., Svensson, T. H.: Release of dopamine from central noradrenaline nerves after treatment with reserpine plus amphetamine. J. neural Transm. **34**, 23—30 (1973)
Andersson, B., Larsson, S.: Water and food intake and the inhibitory effect of amphetamine on drinking and eating before and after "Prefrontal lobotomy" in dogs. Acta physiol. scand. **38**, 22—30 (1956)
Andres, F., Ohnesorge, F. K., DeVries, R.: Zum Mechanismus der stoffwechselstimulierenden Wirkung von Weckaminen an isolierten Organen. Naunyn-Schmiedebergs Arch. exp. Path. Pharmak. **257**, 261—262 (1967)
Angel, C., Muphree, O. D., DeLuca, D. C.: The effects of chlordiazepoxide, amphetamine and cocaine on bar-press behaviour in normal and genetically nervous dogs. Dis. nerv. Syst. **35**, 220—231 (1974)
Angelis, L. de, Soranzio, M. R., Traversa, U., Vertua, R.: 5-Hydroxytryptamine-C^{14} and dexamphetamine-C^{14} uptake by fundal sarcolemma preparations and the problem of the common receptor. Arch. int. Pharmacodyn. **210**, 99—107 (1974)
Anichkov, S. V., Borodkin, Y. S., Zaitsev, Y. V.: Direct action of amphetamine and its furane derivatives on the excitability of neuron populations in various cerebral structures. Neuropharmacology **12**, 803—812 (1973)

Anisman, H., Kokkinidis, L.: Effects of central and peripheral adrenergic and cholinergic modification on time-dependent processes in avoidance performance. Behav. Biol. **10**, 161—171 (1974)

Antonaccio, M. J., Smith, C. B.: Effects of chronic pretreatment with pargyline upon responses of the atrial pacemaker and of left atrial strips of guinea pigs to tyramine, mephentermine, *d*-amphetamine and adrenergic nerve stimulation. J. Pharmacol. exp. Ther. **170**, 97—107 (1969)

Appel, J. B., Freedman, D. X.: Tolerance and cross-tolerance among psychotomimetic drugs. Psychopharmacologia (Berl.) **13**, 267—274 (1968)

Armitage, P., Rushton, R., Steinberg, H.: Interactions of chlordiazepoxide and dexamphetamine in rats. Naunyn-Schmiedebergs Arch. exp. Path. Pharmak. **259**, 150—151 (1968)

Arnfred, T., Randrup, A.: Cholinergic mechanism in brain inhibiting amphetamine-induced stereotyped behaviour. Acta pharmacol. (Kbh.) **26**, 384—394 (1968)

Arnold, L. E., Kirilcuk, V., Corson, S. A., Corson, E. O.: Levoamphetamine and dextroamphetamine: differential effect on aggression and hyperkinesis in children and dogs, Amer. J. Psychiat. **130**, 165 (1973)

Asano, Y., Moroji, T.: Effects of methamphetamine on daily rhythms of hypothalamic norepinephrine, serotonin and plasma corticosterone levels in the rat. Life Sci. **14**, 1463—1472 (1974)

Asher, B. M., Aghajanian, G. K.: 6-Hydroxydopamine lesions of olfactory tubercles and caudate nuclei: effect on amphetamine-induced stereotyped behaviour in rats. Brain. Res. **82**, 1—12 (1974)

Askew, B. M.: Hyperpyrexia as a contributory factor in the toxicity of amphetamine to aggregated mice. Brit. J. Pharmacol. **19**, 245—257 (1962)

Åström, A.: Antisympathetic action of sympathomimetic amines. Acta physiol. scand. **18**, 295—307 (1949)

Atrens, D. M., Vietinghoff-Riesch, F. von, Karabetian, A. der, Masliya, E.: Modulation of reward and aversion processes in the rat diencephalon by amphetamine. Amer. J. Physiol. **226**, 874—880 (1974)

Axelrod, J.: Metabolism of epinephrine and other sympathomimetic amines. Physiol. Rev. **39**, 751—776 (1959)

Axelrod, J.: The metabolism, storage and release of catecholamines. In: Recent progress in hormone research, pp. 597—622. New York: Academic Press 1965

Axelrod, J.: Amphetamine: metabolism, physiological disposition and its effects on catecholamine storage. In: Costa, E., Garattini, S. (Eds.): International Symposium on Amphetamine and Related Compounds, pp. 207—216. New York: Raven Press 1970

Axelrod, J., Hertting, G., Potter, L.: Effects of drugs on the uptake and release of ^{3}-H-norepinephrine in the rat heart. Nature (Lond.) **194**, 297 (1962)

Axelrod, J., Inscoe, J. K.: The uptake and binding of circulating serotonin and the effect of drugs. J. Pharmacol. exp. Ther. **141**, 161—165 (1963)

Axelrod, J., Tomchick, R.: Increased rate of metabolism of epinephrine and norepinephrine by sympathomimetic amines. J. Pharmacol. exp. Ther. **130**, 367—369 (1960)

Axelrod, J., Whitby, L. G., Hertting, G.: Effect of psychotropic drugs on the uptake of ^{3}H-norepinephrine by tissues. Science **133**, 383—385 (1961)

Ayhan, I. H., Randrup, A.: Inhibitory effects of amphetamine, l-DOPA and apomorphine on morphine-induced behavioural excitation or rats. Arch. int. Pharmacodyn. **204**, 283—292 (1973)

Azzaro, A. J., Rutledge, C. O.: Selectivity of release of norepinephrine, dopamine and 5-hydroxytryptamine by amphetamine in various regions of rat brain. Biochem. Pharmacol. **22**, 2801—2813 (1973)

Azzaro, A. J., Ziance, R. J., Rutledge, C. O.: The importance of neuronal uptake of amines for amphetamine-induced release of ^{3}H-norepinephrine from isolated brain tissue. J. Pharmacol. exp. Ther. **189**, 110—118 (1974)

Babbini, M., Montanaro, N., Strocchi, P., Gaiardi, M.: Enhancement of amphetamine-induced stereotyped behavior by benzodiazepines. Europ. J. Pharmacol. **13**, 330—340 (1971)

Babulova, A., Cotillo, P., Bonaccorsi, A., Podvalova, I.: Effects of fenfluramine on the adrenergic system. J. Pharm. Pharmacol. **24**, 886—893 (1972)

Bäckström, M., Wetterberg, L.: Increased N-acetylserotonin and melatonin formation induced by *d*-amphetamine in rat pineal gland organ culture via a β-adrenergic receptor mechanism. Acta physiol. scand. **87**, 113—120 (1973)
Baez, L. A.: Role of catecholamines in the anorectic effects of amphetamine in rats. Psychopharmacologia (Berl.) **35**, 91—98 (1974)
Bagchi, S. P., McGeer, P. L.: Some properties of tyrosine hydroxylase from the caudate nucleus. Life Sci. **3**, 1195—1200 (1964)
Bainbridge, J. G.: The inhibitory effect of amphetamine on exploration in mice. Psychopharmacologia (Berl.) **18**, 314—319 (1970)
Baird, J. R. C.: The effects of (+)-amphetamine and (±)-phenmetrazine on the noradrenaline and dopamine levels in the hypothalamus and corpus striatum of the rat. J. Pharm. Pharmacol. **20**, 234—235 (1968)
Baird, J. R. C., Lewis, J. J.: Effects of drugs on noradrenaline and 3-hydroxytyramine (dopamine) levels and on the noradrenaline to dopamine ratio in the rat brain. Biochem. Pharmacol. **12**, 577—602 (1963)
Baird, J. R. C., Lewis, J. J.: The effects of cocaine, amphetamine and some amphetamine-like compounds on the in vivo levels of noradrenaline and dopamine in the rat brain. Biochem. Pharmacol. **13**, 1475—1482 (1964)
Baldessarini, R. J.: Release of aromatic amines from brain tissues of the rat in vitro. J. Neurochem. **18**, 2509—2518 (1971)
Baldessarini, R. J., Vogt, M.: The uptake and subcellular distribution of aromatic amines in the brain of the rat. J. Neurochem. **18**, 2519—2533 (1971)
Balster, R. L., Schuster, C. R.: A comparison of *d*-amphetamine, *l*-amphetamine, and methamphetamine self-administration in rhesus monkeys. Pharmacol. Biochem. Behav. **1**, 67—71 (1973)
Banerjee, U., Geh, S. L.: Time-related interaction patterns of amphetamine with reserpine and other central depressants. Res. Commun. chem. Path. Pharmacol. **6**, 109—122 (1973)
Banarjee, U., Lin, G. S.: On the mechanism of central action of amphetamine—the role of catecholamines. Neuropharmacology **12**, 917—931 (1973)
Barnes, C. D.: The interaction of amphetamine and eserine on the EEG. Life Sci. **5**, 1897—1902 (1966)
Barnes, C. D., Meyers, F. H.: Eserine and amphetamine: Interactive effects on sleeping time in mice. Science **144**, 1221—1222 (1964)
Barnett, A., Peschel, E.: Potentiation of methamphetamine aggregate toxicity in mice by diethyldithiocarbamate. J. Pharm. Pharmacol. **20**, 482—483 (1968)
Barrett, R. J., Leith, N. J., Ray, O. S.: Permanent facilitation of avoidance behavior by *d*-amphetamine and scopolamine. Psychopharmacologia (Berl.) **25**, 321—331 (1972)
Barrett, R. J., Leith, N. J., Ray, O. S.: An analysis of the facilitation of avoidance acquisition produced by *d*-amphetamine and scopolamine. Behav. Biol. **11**, 189—203 (1974)
Barrett, R. J., Steranka, L. R.: An analysis of *d*-amphetamine produced facilitation of avoidance acquisition in rats and performance changes subsequent to drug termination. Life Sci. **14**, 163—180 (1974)
Barry, H., Buckley, J. P.: Drug effects on animal performance and the stess syndrome. J. pharm. Sci. **55**, 1159—1183 (1966)
Barry, H., Miller, N. E.: Effects of drugs on approach avoidance conflict tested repeatedly by means of a "telescope alley." J. comp. physiol. Psychol. **55**, 201—210 (1962)
Barry, H., Wagner, S., Miller, N. E.: Effects of several drugs on performance in an approach avoidance conflict. Psychol. Rep. **12**, 215—221 (1963)
Bartolini, A., Pepeu, G.: Effect of adrenergic blockade on spontaneous and stimulated acetylcholine output from the cerebral cortex of the cat. Pharmacol. Res. Commun. **2**, 23—29 (1970)
Basso, P. del, Rusca, G., Carpi, A.: Drug-induced changes of urinary catecholamines in the rat: role of the adrenal medulla. Europ. J. Pharmacol. **13**, 83—89 (1970)
Bättig, K.: Die Wirkung von Training und Amphetamin auf Ausdauer und Geschwindigkeit der Schwimmleistung der Ratte. Psychopharmacologia (Berl.) **4**, 15—27 (1963a)
Bättig, K.: Schwimmleistung, Laufleistung, Fluchtreaktionen, Neugierde und Appetitverhalten der Ratte unter der Wirkung von Amphetamin. Pharm. Acta Helv. **39**, 427—440 (1963b)
Bauer, I., Pickenhain, L.: Untersuchungen zur Gewöhnung an Weckamine mit der Methode der bedingten Fluchtreflexe bei der Ratte. Psychopharmacologia (Berl.) **12**, 78—82 (1967)

Bauer, R. H., Duncan, N. C.: Twenty-four-hour proactive facilitation of avoidance and discrimination learning in rats by *d*-amphetamine. J. comp. physiol. Psychol. **77**, 521—527 (1971)
Bauer, W. S., Perley, J. E.: Diurnal variation of hepatic amphetamine concentrations in mice fed freely and fed single daily meals. J. Pharm. Pharmacol. **23**, 976—977 (1971)
Baum, E., North-Diehl, A., Piarroux, M.-C., Thenint, F., Boissier, J.-R.: Modifications biochimiques, histochimiques et pharmacologiques après lésion bilatérale des différentes parties de la substance noire chez le rat. J. Pharmacol (Paris) **3**, 477—486 (1972)
Beani, L., Bianchi, C., Santinoceto, L., Marchetti, P.: The cerebral acetylcholine release in conscious rabbits with semi-permanently implanted epidural cups. Int. J. Neuropharmacol. **7**, 469—481 (1968)
Beaton, J. M., Bradley, R. J.: The behavioral effects of some hallucinogenic derivatives of amphetamine. In: Ellinwood, E. H., Cohen, S. (Eds.): Current concepts on amphetamine abuse, pp. 49—57. DHEW Publ. No. (HEW) 72-9085. U.S. Government Printing Office, Washington, D.C. 1972
Beaton, J. M., LeBlanc, A. E., Webster, C. D.: The effects of *d*-amphetamine on the interresponse times of rats and guinea pigs on a modified Sidman discriminated avoidance schedule. Psychopharmacologia (Berl.) **37**, 199—203 (1974)
Beauvallet, M.: L'amphétamine et les monoamines du système nerveux central. Actualités pharmacol. **21**, 15—39 (1968)
Beauvallet, M., Fugazza, J., Godefroy, F., Solier, M.: Hyperactivité provoquée par l'amphetamine chez la souris groupées. II. Etude comparée des teneur en noradrenaline et 3,4-dihydroxytyramine (dopamine) du tissue cérébral. J. Physiol. (Paris) **56**, 532—533 (1964)
Beauvallet, M., Fugazza, J., Godefroy, F., Solier, M.: Hyperactivité provoquée par l'amphétamine chez les souris en exercice forcé et teneur en noradrénaline et 3,[3,4-]4-hydroxytyramine (dopamine) du cerveau et du coeur. J. Physiol. (Paris) **57**, 551—552 (1965)
Beauvallet, M., Fugazza, J., Godefroy, F., Solier, M.: Hyperactivité provoquée par l'amphétamine chez les souris réserpinées en exercice forcé et teneur en catécholamines du cerveau et des surrénal. J. Physiol. (Paris) **58**, 460—461 (1966 a)
Beauvallet, M., Fugazza, J., Legrand, M.: Excrétion urinaire de l'adrénaline, de la noradrénaline et de la dopamine au cours des 24 heures qui suivent l'administration d'amphétamine. C.R. Soc. Biol. (Paris) **160**, 546—550 (1966 b)
Beauvallet, M., Fugazza, J., Legrand, M.: Action de l'amphétamine sur les teneurs en noradrénaline et 5-hydroxytryptamine du cerveau de rats maintenus 4 heures à 33° C. Thérapie **22**, 1273—1276 (1967)
Beauvallet, M., Fugazza, J., Solier, M.: Variations de la teneur en adrénaline et noradrénaline des surrénales et du cerveau avant et après travail musculaire. Action de l'amphétamine. C.R. Soc. Biol. (Paris) **156**, 1258—1260 (1962 a)
Beauvallet, M., Fugazza, J., Solier, M.: Hyperexcitabilité provoquée par *l*-amphétamine chez les souris groupées et teneur en noradrénaline et adrénaline des tissus. C.R. Acad. Sci. (Paris) **257**, 3251—3253 (1963)
Beauvallet, M., Halpern, B. N., Fugazza, J., Drudi-Baracco, C.: Corrélation entre l'hyperexcitabilité provoquée par l'amphétamine chez les souris groupées et la teneur du cerveau en noradrénaline. Biochem. Pharmacol. **12**, Suppl. 8 (1962 b)
Beauvallet, M., Legrand, M., Bernard, J., Solier, M.: Actions de l'amphétamine sur la teneur en 5-hydroxytryptamine du cerveau de rat. C.R. Soc. Biol. (Paris) **163**, 855—856 (1969)
Beauvallet, M., Legrand, M., Solier, M.: Actions de l'amphétamine sur la teneur en 5-hydroxytryptamine des différentes aires du cerveau de rat. C.R. Soc. Biol. (Paris) **164**, 1462—1467 (1970)
Beauvallet, M., Solier, M.: Effect d'un exercice forcé sur la toxicité de l'amphétamine et la teneur en noradrénaline du cerveau. C.R. Soc. Biol. (Paris) **158**, 2306—2309 (1964)
Beckwith, B. E., Sandman, C. A., Alexander, W. D.: *d*-Amphetamine effects on attention and memory in the albino and hooded rat. Pharmacol. Biochem. Behav. **2**, 557—561 (1974)
Bejrablaya, D., Burn, J. H., Walter, J. M.: The action of sympathomimetic amines on heart rate in relation to the effect of reserpine. Brit. J. Pharmacol. **13**, 461—466 (1958)
Belenky, M. L., Vitolina, M.: The pharmacological analysis of the hyperthermia caused by phenamine (amphetamine). Int. J. Neuropharmacol. **1**, 1—7 (1962)

Bell, D. S., Kirkby, R. J., Preson, A. C.: Tolerance to the hyperactivating effect of methylamphetamine. Psychopharmacologia (Berl.) **36**, 41—47 (1974)
Belleville, R. E.: Control of behavior by drug-produced internal stimuli. Psychopharmacologia (Berl.) **5**, 95—105 (1964)
Belozertsev, Y. A.: Effect of psychostimulants on the avoidance reaction to stimulation of the caudate nucleus. Bull. exp. Biol. Med. **77**, 284—287 (1974)
Benfey, B. G., Varma, D. R.: Inhibition of sympathomimetic effects on the heart. Brit. J. Pharmacol. **23**, 399—404 (1964)
Benington, F., Morin, R. D.: The chemorelease of norepinephrine from mouse hearts by substituted amphetamines. J. med. Chem. **11**, 896—897 (1968)
Beretta, C., Locatelli, A.: Inhibitory activity of β-carbobenzyloxyaminomethyl-1,6-dimethyl-10α-ergoline towards stimulant effects by 5-hydroxytryptamine and amphetamine on liver fluke, *Fasciola hepatica* in vitro. J. Pharm. Pharmacol. **20**, 744—745 (1968)
Berger, B. D.: Conditioning of food aversions by injections of psychoactive drugs. J. comp. physiol. Psychol. **81**, 21—26 (1972)
Berger, B. D., Wise, C. D., Stein, L.: Norepinephrine: Reversal of anorexia in rats with lateral hypothalamic damage. Science **172**, 281—284 (1971)
Berger, B. D., Wise, C. D., Stein, L.: Area postrema damage and bait shyness. J. comp. physiol. Psychol. **82**, 475—479 (1973)
Berger, H. J., Brown, C. C., Krantz, J. C.: Fenfluramine blockade of CNS stimulant effects of amphetamine. J. pharm. Sci. **62**, 788—791 (1973)
Bernier, A., Sicot, N., LeDouarec, J. C.: Action comparée de la fenfluramine et de l'amphétamine chez les rats obèses hypothalamiques. Rev. franç. Etud. clin. biol. **14**, 762—772 (1969)
Bernstein, B. M., Latimer, C. N.: Behavioral facilitation: the interaction of imipramine and desipramine with amphetamine, alphapipradrol, methylphenidate, and thozalinone. Psychopharmacologia (Berl.) **12**, 338—345 (1968)
Berry, M. J., Poyser, R. H., Robertson, M. I.: Inhibition by appetite suppressants of the pressor response to (+)-amphetamine in anaesthetized cats. J. Pharm. Pharmacol. **23**, 140—142 (1971)
Besson, M. J., Cheramy, A., Feltz, P., Glowinski, J.: Release of newly synthesized dopamine from dopamine-containing terminals in the striatum of the rat. Proc. nat. Acad. Sci. (Wash.) **62**, 741—748 (1969a)
Besson, M. J., Cheramy, A., Feltz, P., Glowinski, J.: Dopamine: spontaneous and drug-induced release from the caudate nucleus in the cat. Brain Res. **32**, 407—424 (1971a)
Besson, M. J., Cheramy, A., Gauchy, C., Musacchio, J.: Effects of some psychotropic drugs on tyrosine hydroxylase activity in different structures of the rat brain. Europ. J. Pharmacol. **22**, 181—186 (1973)
Besson, M. J., Cheramy, A., Glowinski, J.: Effects of amphetamine and desmethylimipramine on amines synthesis and release in central catecholamine containing neurons. Europ. J. Pharmacol. **7**, 111—114 (1969b)
Besson, M. J., Cheramy, A., Glowinski, J.: Effects of some psychotropic drugs on dopamine synthesis in the rat striatum. J. Pharmacol. exp. Ther. **177**, 196—205 (1971b)
Beuthin, F. C., Miya, T. S., Blake, D. E., Bousquet, W. F.: Enhanced sensitivity to noradrenergic agonists and tolerance development to alphamethyltyrosine in the rat. J. Pharmacol. exp. Ther. **181**, 446—456 (1972)
Bewsher, P. D., Hillman, C. C., Ashmore, J.: Studies on the hypoglycemic effect of *d*-amphetamine in aggregated mice. Biochem. Pharmacol. **15**, 2079—2085 (1966)
Bhagat, B.: Amphetamine and stores of noradrenaline. J. Pharm. Pharmacol. **17**, 191—192 (1965a)
Bhagat, B.: Pressor responses to amphetamine in the spinal cat and its influence on tachyphylaxis to tyramine. J. Pharmacol. exp. Ther. **149**, 206—211 (1965b)
Bhagat, B., Shideman, F. E.: Mechanism on the inhibitory action of guanethidine on cardiovascular responses to tyramine and amphetamine. J. Pharmacol. exp. Ther. **140**, 317—323 (1963)
Bhagat, B., Wheeler, N.: Effect of amphetamine on the swimming endurance of rats. Neuropharmacology **12**, 711—713 (1973)
Bianchi, C., Marazzi-Uberti, E.: Acquisition and retention of a conditioned avoidance response in mice as influenced by Pemoline, by some of its derivatives and by some CNS stimulants. Psychopharmacologia (Berl.) **15**, 9—18 (1969)

Bignami, G.: Pharmacologic influences on mating behavior in the male rat. Effects of *d*-amphetamine, LSD-25, strychnine, nicotine and various anticholinergic agents. Psychopharmacologia (Berl.) **10**, 44—58 (1966)
Bignami, G., Gatti, G. L.: Analysis of drug effects on multiple fixed ratio 33—fixed interval 5 min in pigeons. Psychopharmacologia (Berl.) **15**, 310—332 (1969)
Biscardi, A. M., Carpi, A., Orsingher, O. A.: Urinary excretion of catecholamines in the rat after their liberation by reserpine or dexamphetamine. Brit. J. Pharmacol. **23**, 529—539 (1964)
Bivens, L. W., Ray, O. S.: Amphetamine, atropine, and meprobamate effects on operant behavior in the rat. Arch. int. Pharmacodyn. **172**, 380—392 (1968)
Bizzi, A., Bonaccorsi, A., Jespersen, S., Jori, A., Garattini, S.: Pharmacological studies on amphetamine and fenfluramine. In: Costa, E., Garattini, S. (Eds.): International Symposium on Amphetamines and Related Compounds, pp. 577—595. New York: Raven Press 1970
Bizzi, A., Garattini, S., Veneroni, E.: The action of salicylate in reducing plasma free fatty acids and its pharmacological consequences. Brit. J. Pharmacol. **25**, 187—196 (1965)
Blackham, A., Spencer, P. S. J.: Interactions of oestrogenic and progestational steroids with dexamphetamine and fencamfamin in mice. Brit. J. Pharmacol. **37**, 508 (1969)
Blaschko, H.: Amine oxidase and amine metabolism. Pharmacol. Rev. **4**, 415—458 (1952)
Blaschko, H., Richter, D., Schlossman, H.: The oxidation of adrenaline and other amines. Biochem. J. **31**, 2187—2196 (1937)
Blaschko, H., Strömblad, B. C. R.: The inhibition of human amine oxidase by the two isomers of amphetamine. Arzneimittel-Forsch. **10**, 327 (1960)
Blum, R. A., Blum, J. S., Chow, K. L.: Production of convulsion by administration of benzedrine following brain operations in monkeys. Arch. Neurol. Psychiat. (Chic.) **64**, 685—691 (1950)
Blundell, J. E.: Possible mechanism for the effect of anorexic agents on feeding and hoarding behaviour in rats. Psychopharmacologia (Berl.) **22**, 224—229 (1971)
Blundell, J. E., Leshem, M. B.: Dissociation of the anorexic effects of fenfluramine and amphetamine following intrahypothlamic injection. Brit. J. Pharmacol. **47**, 183—188 (1973)
Blundell, J. E., Leshem, M. B.: Central action of anorexigenic agents: effects of amphetamine and fenfluramine in rats with lateral hypothalamic lesions. Europ. J. Pharmacol. **28**, 81—88 (1974)
Boakes, R. J., Bradley, P. B., Candy, J. M.: Abolition of the response of brain stem neurones to iontophoretically applied *d*-amphetamine by reserpine. Nature (Lond.) **229**, 496—498 (1971)
Boakes, R. J., Bradley, P. B., Candy, J. M.: A neuronal basis for the alerting actions of (+)-amphetamine. Brit. J. Pharmacol. **45**, 391—403 (1972)
Boakes, R. J., Bradley, P. B., Candy, J. M.: Antagonism of the effects of iontophoretically applied (+)-amphetamine by chlorpromazine on single neurones. Brit. J. Pharmacol. **49**, 175—176 (1973)
Bocknik, S. E., Kulkarni, A. S.: Effects of anorectic agents on blood pressure of the dog. Arch. int. Pharmacodyn. **202**, 213—218 (1973)
Bohdanecky, S., Jarvik, M. E.: The effect of *d*-amphetamine and physostigmine upon acquisition and retrieval in a single trial learning task. Arch. int. Pharmacodyn. **170**, 58—65 (1967)
Boissier, J.-R., Etevenon, P., Piarroux, M.-C., Simon, P.: Effects of apomorphine and amphetamine in rats with a permanent catalepsy induced by diencephalic lesion. Res. Commun. chem. Path. Pharmacol. **2**, 829—836 (1971)
Boissier, J.-R., Simon, P., Guernet, M., Tillement, J.-P.: Variations des monoamines provoquées par la p-chlor-N-méthylamphétamine dans différentes parties du cerveau de rat. C.R. Soc. Biol. (Paris) **268**, 2298—2299 (1969)
Boissier, J.-R., Simon, P., Lwoff, J. M.: Sur une action paradoxale du chlordiazepoxide. Thérapie **18**, 1247—1256 (1963)
Bolt, A. G., Mulligan, B. M., Graham, G.: The lipolytic activity of amphetamine in the spinalized cat. Res. Commun. chem. Path. Pharmacol. **9**, 182—192 (1974)
Bonaccorsi, A.: Studies on the hypertensive response elicited by reserpine or tetrabenazine in rats treated with amphetamine-like drugs. Europ. J. Pharmacol. **3**, 97—105 (1968)
Bonhoff, G., Lewrenz, H.: Über Weckamine (Pervitin und Benzedrin). Berlin-Göttingen-Heidelberg: Springer 1954
Booth, D. A.: Amphetamine anorexia by direct action on the adrenergic feeding system of rat hypothalamus. Nature (Lond.) **217**, 869—870 (1968)

Borbély, A. A., Baumann, I. R., Waser, P. G.: Amphetamine and thermoregulation: studies in the unrestrained and curarized rat. Naunyn-Schmiedebergs Arch. Pharmacol. **281**, 327—340 (1974)
Borbély, A. A., Huston, J. P., Baumann, R.: Body temperature and behavior in chronic brain lesioned rats after amphetamine, chlorpromazine and γ-butyrolactone. In: Schönbaum, E., Lomax, P. (Eds.): The pharmacology of thermoregulation, pp. 447—462. Basel: Karger 1973
Borberg, S.: Conditioning of amphetamine-induced behaviour in the albino rat. Psychopharmacologia (Berl.) **34**, 191—198 (1974)
Borella, L. E., Herr, F.: Effect of ammonium chloride on the potentiation of amphetamine by psychotropic drugs in the rat. Biochem. Pharmacol. **20**, 589—595 (1971)
Borella, L., Herr, F., Wojdan, A.: Prolongation of certain effects of amphetamine by chlorpromazine. Canad. J. Physiol. Pharmacol. **47**, 7—13 (1969a)
Borella, L. E., Paquette, R., Herr, F.: The effect of some CNS depressants on the hypermotility and anorexia induced by amphetamine in rats. Canad. J. Physiol. Pharmacol. **47**, 841—847 (1969b)
Bösser, T. H., Joyce, D., Summerfield, A.: Factors affecting the acquisition of new behaviour after administration of an amphetamine barbiturate mixture. Brit. J. Pharmacol. **38**, 459—460 (1970)
Boulu, R., Rapin, J. R., Lebas, M., Jacquot, C.: Action centrale de l'amphétamine, de l'éphédrine et de leur dérive p-hydroxyl après lésion unilatérale du faisceau nigrostriatal chez le rat. Psychopharmacologia (Berl.) **26**, 54—61 (1972)
Bovet, D., Gatti, G. L.: Pharmacology of instrumental avoidance conditioning. In: Mickelson, M. Y., Longo, V. G. (Eds.): Pharmacology of conditioning, learning and retention, pp. 75—89. New York: Pergamon Press 1965
Bovet, D., Oliverio, A.: Decrement of avoidance conditioning performance in inbred mice subjected to prolonged sessions: performance recovery after rest and amphetamine. J. Psychol. **65**, 45—55 (1967)
Bradley, P. B.: Synaptic transmission in the central nervous system and its relevance for drug action. Int. Rev. Neurobiol. **11**, 1—56 (1968)
Bradley, P. B., Elkes, J.: The effects of some drugs on the electrical activity of the brain. Brain **80**, 77—117 (1957)
Bradley, P. B., Key, B. J.: The effect of drugs on arousal responses produced by electrical stimulation of the reticular formation of the brain. Electroenceph. clin. Neurophysiol. **10**, 97—110 (1958)
Bradley, P. B., Wolstencroft, J. H.: Actions of drugs on single neurones in the brain-stem. Brit. med. Bull. **21**, 15—18 (1965)
Bradley, D. W. H., Joyce, D., Murphy, E. H., Nash, B. M., Porsolt, R. D., Summerfield, A., Twyman, W. A.: Amphetamine-barbiturate mixture: Effects on the behaviour of mice. Nature (Lond.) **220**, 187—188 (1968)
Brady, J. V.: A comparative approach to the evaluation of drug effects upon affective behaviour. Ann. N.Y. Acad. Sci. **64**, 632—643 (1956)
Brady, J. V.: Animal experimental evaluation of drug effects upon behaviour. Fed. Proc. **17**, 1031—1034 (1958)
Branch, M. N.: Behavior as a stimulus: Joint effects of *d*-amphetamine and pentobarbital. J. Pharmacol. exp. Ther. **189**, 33—41 (1974)
Branch, M. N., Gollup, L. R.: A detailed analysis of the effects of *d*-amphetamine on behaviour under fixed-interval schedules. J. exp. Anal. Behav. **21**, 519—539 (1974)
Breckenridge, B. M., Norman, J. H.: The conversion of phosphorylase b to phosphorylase a in brain. J. Neurochem. **12**, 51—57 (1965)
Breda, J. B., Carlini, E. A., Sader, N. F. A.: Effects of chronic administration of (+)-amphetamine on maze performance of the rat. Brit. J. Pharmacol. **37**, 79—86 (1969)
Breese, G. R., Cooper, B. R., Smith, R. D.: Biochemical and behavioural alterations following 6-hydroxydopamine administration into brain. Frontiers in catecholamine research, pp. 701—706. Usdin, E., Snyder, S. (Eds.) New York: Pergamon Press 1973
Breese, G. R., Kopin, I. J., Weise, V. K.: Effects of amphetamine derivatives on brain dopamine and noradrenaline. Brit. J. Pharmacol. **38**, 537—545 (1970)
Brittain, R. T.: The intracerebral effects of noradrenaline and its modification by drugs in the mouse. J. Pharm. Pharmacol. **18**, 621—623 (1966)

Brittain, R.T., Jack, D., Spencer, P.S.J.: Mechanism of action of monoamine oxidase inhibitors in enhancing amphetamine toxicity. J. Pharm. Pharmacol. **16**, 565—567 (1964)
Brobeck, J.R., Larsson, S., Reyes, E.: A study of the electrical activity of the hypothalamic feeding mechanism. J. Physiol. (Lond.) **132**, 358—364 (1956)
Brodie, B.B., Shore, P.A.: A concept for a role of serotonin and norepinephrine as chemical mediators in the brain. Ann. N.Y. Acad. Sci. **66**, 631—642 (1957)
Brodie, B.B., Costa, E., Groppetti, A., Matsumoto, C.: Interaction between desipramine, tyramine and amphetamine at adrenergic neurones. Brit. J. Pharmacol. **34**, 648—658 (1968)
Brodie, B.B., Cho, A.K., Stefano, F.J.E., Gessa, G.L.: On mechanisms of norepinephrine release by amphetamine and tyramine and tolerance to their effects. Advanc. Biochem. Psychopharmacol. **1**, 219—238 (1969)
Brodie, B.B., Cho, A.K., Gessa, G.L.: Possible role of p-hydroxynorephedrine in the depletion of norepinephrine induced by *d*-amphetamine and in tolerance to this drug. In: Costa, E., Garattini, S. (Eds.): International Symposium on Amphetamines and Related Compounds, pp. 217—230. New York: Raven Press 1970
Broekkamp, C.L.E., Honig, W.M.M., Pauli, A.I., Rossum, J.M. van: Pharmacological suppression of eating behavior in relation to diencephalic noradrenergic receptors. Life Sci. **14**, 473—481 (1974)
Brown, A.M., Julian, T.: The body temperature response of two strains of mice to handling, saline and amphetamine. Int. J. Neuropharmacol. **7**, 531—541 (1968)
Brown, B.B.: CNS drug action and interaction in mice. Arch. int. Pharmacodyn. **128**, 391—414 (1960)
Brown, H.: D-amphetamine—chlorpromazine antagonism in a food reinforced operant. J. exp. Anal. Behav. **6**, 395—398 (1963)
Brown, H.: Drug-behaviour interaction affecting development of tolerance to *d*-amphetamine as observed in fixed ratio behaviour of rats. Psychol. Rep. **16**, 917—921 (1965)
Brown, H., Richards, R.K.: An interaction between drug effects and food reinforced "social" behaviour in pigeons. Arch. int. Pharmacodyn. **16**, 286—293 (1966)
Brownlee, G., Williams, G.W.: Potentiation of amphetamine and pethidine by monoamine-oxidase inhibitors, Lancet **1963a**, 669
Brownlee, G., Williams, G.W.: Monoamineoxidase inhibitors, Lancet **1963b**, 1323
Buening, M.K., Gibb, J.W.: Influence of methamphetamine and neuroleptic drugs on tyrosine hydroxylase activity. Europ. J. Pharmacol. **26**, 30—34 (1974)
Buffoni, F., Della Corte, L.: Pig plasma benzylamine oxidase. Advanc. Biochem. Psychopharmacol. **5**, 138—149 (1972)
Bunney, B.S., Aghajanian, G.K.: Electrophysiological effects of amphetamine on dopaminergic neurones. Frontiers in catecholamine research, pp. 957—962. Usdin, E., Snyder, S. (Eds.) New York: Pergamon Press 1973
Bunney, B.S., Aghajanian, G.K., Roth, R.H.: Comparison of effects of l-dopa, amphetamine and apomorphine on firing rate of rat dopaminergic neurones. Nature (Lond.) New Biol. **245**, 123—125 (1973a)
Bunney, B.S., Walters, J.R., Roth, R.H., Aghajanian, G.K.A.: Dopaminergic neurons: effect of antipsychotic drugs and amphetamine on single cell activity. J. Pharmacol. exp. Ther. **185**, 560—571 (1973b)
Burgen, A.S.V., Iversen, L.L.: The inhibition of noradrenaline uptake by sympathomimetic amines in the rat isolated heart. Brit. J. Pharmacol. **25**, 34—49 (1965)
Burn, J.H., Hobbs, R.: A test for tranquilizing drugs. Arch. int. Pharmacodyn. **113**, 290—295 (1958)
Burn, J.P., Rand, M.J.: The action of sympathomimetic amines in animals treated with reserpine. J. Physiol (Lond.) **144**, 314—336 (1958)
Burton, R.M., Sodd, M.A., Goldin, A.: The analeptic action of lysergic acid diethylamide and *d*-amphetamine on reserpine-sedated mice. Arch. int. Pharmacodyn. **112**, 188—198 (1957)
Butcher, L.L., Andén, N.-E.: Effects of apomorphine and amphetamine on schedule-controlled behavior: reversal of tetrabenazine suppression and dopaminergic correlates. Europ. J. Pharmacol. **6**, 255—264 (1969)
Butcher, L.L., Butcher, S.G., Larsson, K.: Effects of apomorphine (+)-amphetamine, and nialamide on tetrabenazine-induced suppression of sexual behavior in the male rat. Europ. J. Pharmacol. **7**, 283—288 (1969)

Buus Lassen, J.: The effect of amantadine and (+)-amphetamine on motility in rats after inhibition of monoamine synthesis and storage. Psychopharmacologia (Berl.) **29**, 55—64 (1973)
Buyniski, J. P., Smith, M. L., Bierwagen, M. E.: Cardiovascular and gross behavioural effects of amphetamine, 2-amino-1-(2,5-dimethoxy-4-methylphenyl)propane (DOM) and 2-amino-l-(2,5-dimethoxy-4-methyl-phenyl)butane (BL-3912A) in the conscious dog. Res. Commun. chem. Path. Pharmacol. **8**, 213—221 (1974)
Byrd, L. D.: Effects of *d*-amphetamine on schedule-controlled key pressing and drinking in the chimpanzee. J. Pharmacol. exp. Ther. **185**, 633—641 (1973)
Byrne-Quinn, E., Grover, R. F.: Aminorex (Menocil) and amphetamine: acute and chronic effects on pulmonary and systemic haemodynamics in the calf. Thorax **27**, 127—131 (1972)
Caccia, S., Cecchetti, G., Garattini, S., Jori, A.: Interaction of (+)-amphetamine with cerebral dopaminergic neurones in two strains of mice, that show different temperature responses to this drug. Brit. J. Pharmacol. **49**, 400—406 (1973)
Cahn, J., Herold, M.: Effects of amphetamines in rats and rabbits injected with alpha- and beta-adrenergic blocking agents. In: Costa, E., Garattini, S. (Eds.): International Symposium on Amphetamine and Related Compounds, pp. 493—510. New York: Raven Press 1970
Caldwell, J., Sever, P. S.: Evidence against a role for p-hydroxynorephinedrine in amphetamine tolerance. J. Pharmacol. (Paris) **5** suppl. 2, 14 (1974)
Caldwell, J., Sever, P. S., Trelinski, M.: On the mechanism of the hyperthermia induced by amphetamine in the rat. J. Pharm. Pharmacol. **26**, 821—823 (1974)
Campbell, B. A., Fibiger, H. C.: Potentiation of amphetamine-induced arousal by starvation. Nature (Lond.) **223**, 424—425 (1971)
Campbell, J. C., Seiden, L. S.: Performance influence on the development of tolerance to amphetamine. Pharmacol. Biochem. Behav. **1**, 703—708 (1973)
Canfield, D. R., Antelman, S. M., Lippa, A. S., Fischer, A. E.: Amphetamine-induced alteration of biogenic amines and self-stimulation. Fed. Proc. **32**, 705 (1973)
Cappell, H., Ginsberg, R., Webster, C. D.: Amphetamine and conditioned anxiety. Brit. J. Pharmacol. **45**, 525—531 (1972)
Cappell, H., Le Blanc, A. E.: Conditioned aversion to saccharin by single administration of mescaline and *d*-amphetamine. Psychopharmacologia (Berl.) **22**, 352—356 (1971)
Cappell, H., Le Blanc, A. E.: Punishment of saccharin drinking by amphetamine in rats and its reversal by chlordiazepoxide. J. comp. physiol. Psychol. **85**, 97—104 (1973)
Cardo, B.: Action de l'amphétamine dextrogyre et de l'esérine sur un conditionnement de fuite et sur les phénomènes de discrimination. J. Physiol. (Paris) **51**, 845—860 (1959)
Carey, R. J.: Acquired aversion to amphetamine solutions. Pharmacol. Biochem. Behav. **1**, 227—229 (1973a)
Carey, R. J.: Disruption of timing behavior following amphetamine withdrawal. Physiol. Psychol. **1**, 9—12 (1973b)
Carey, R. J., Goodall, E. B.: A comparison of the effects of amphetamine on fixed interval performance maintained by electrical stimulation of the brain versus food reinforcement. Pharmacol. Biochem. Behav. **1**, 237—239 (1973)
Carey, R. J., Goodall, E. B.: Amphetamine induced taste aversion—a comparison of d-versus *l*-amphetamine. Pharmacol. Biochem. Behav. **2**, 325—330 (1974)
Carey, R. J., Goodall, E. B., Procopio, G. F.: Differential effects of *d*-amphetamine on fixed ratio 30 performance maintained by food versus brain stimulation reinforcement. Pharmacol. Biochem. Behav. **2**, 193—198 (1974)
Carey, R. J., Kritkausky, R. P.: Absence of a response rate dependent effect of *d*-amphetamine on a DRL schedule when reinforcement is signalled. Psychon. Sci. **26**, 285—286 (1972)
Carey, R. J., Salim, A. P.: Changes in *d*-amphetamine reactivity resulting from septum injury. Physiol. Behav. **5**, 133—136 (1970)
Carlini, E. A., Gonzalez, C.: Aggressive behavior induced by marihuana compounds and amphetamine in rats previously made dependent on morphine. Experientia (Basel) **28**, 542—544 (1972)
Carlisle, H. J.: Differential effects of amphetamine on food and water intake in rats with lateral hypothalamic lesions. J. comp. physiol. Psychol. **58**, 47—54 (1964)
Carlisle, H. J., Reynolds, R. W.: Effect of amphetamine on food intake in rats with brain stem lesions. Amer. J. Physiol. **201**, 965—967 (1961)

Carlson, N. J., Doyle, G. A., Bidder, T. G.: The effect of *dl*-amphetamine and reserpine on runway performance. Psychopharmacologia (Berl.) **8**, 157—173 (1965)
Carlsson, A.: Physiological and pharmacological release of monoamines in the central nervous system. In: Euler, U.S. von, Rosell, S., Uvnäs, B. (Eds.): Mechanisms of release of biogenic amines, pp. 331—346. Glasgow: Pergamon Press 1966
Carlsson, A.: Pharmacology of synaptic monoamine transmission. Progr. Brain Res. **31**, 53—59 (1969)
Carlsson, A.: Amphetamine and brain catecholamine. In: Costa, E., Garattini, S. (Eds.): International Symposium on Amphetamines and Related Compounds, pp. 289—300. New York: Raven Press 1970
Carlsson, A., Corrodi, H., Fuxe, K., Hökfelt, T.: Effects of some antidepressant drugs on the depletion of intraneuronal brain catecholamine stores caused by 4, α-methyl-m-tyramine. Europ. J. Pharmacol. **5**, 367—373 (1969)
Carlsson, A., Fuxe, K., Hamberger, B., Lindqvist, M.: Biochemical and histochemical studies on the effects of imipramine like drugs and (+)-amphetamine on central and peripheral catecholamine neurons. Acta physiol. scand. **67**, 481—497 (1966 a)
Carlsson, A., Fuxe, K., Hökfelt, T.: Effect of desmethylimipramine, protriptyline and (+)-amphetamine on fluorescence of central adrenergic neurons of rats pretreated with α-methyl-DOPA and tetrabenazine or reserpine. Europ. J. Pharmacol. **2**, 196—201 (1968)
Carlsson, A., Lindqvist, M., Dahlström, A., Fuxe, K., Masuoka, D.: Effects of the amphetamine group on intraneuronal brain amines in vivo and in vitro. J. Pharm. Pharmacol. **17**, 521—523 (1965)
Carlsson, A., Lindqvist, M., Fuxe, K., Hamberger, B.: The effect of (+)-amphetamine on various central and peripheral catecholamine-containing neurones. J. Pharm. Pharmacol. **18**, 128—130 (1966 b)
Carlsson, A., Waldeck, B.: Inhibition of ^{3}H-metaraminol uptake by antidepressive and related agents. J. Pharm. Pharmacol. **17**, 243—244 (1965)
Carlsson, A., Waldeck, B.: Effects of amphetamine, tyramine and protriptyline on reserpine-resistant amine-concentrating mechanisms of adrenergic nerves. J. Pharm. Pharmacol. **18**, 252—253 (1966 a)
Carlsson, A., Waldeck, B.: Structure-activity relationships for release of ^{14}C-octopamine from adrenergic nerves by phenylethylamines. Acta pharmacol. (Kbh.) **24**, 255—262 (1966 b)
Carlsson, A., Waldeck, B.: Different mechanisms of drug induced release of noradrenaline and its congeners alfa-methylnoradrenaline and metaraminol. Europ. J. Pharmacol. **4**, 165—168 (1968)
Carlton, P. L.: Potentiation of the behavioral effects of amphetamine by imipramine. Psychopharmacologia (Berl.) **2**, 364—376 (1961 a)
Carlton, P. L.: Augmentation of the behavioral effects of amphetamine by scopolamine. Psychopharmacologia (Berl.) **2**, 377—380 (1961 b)
Carlton, P. L.: Scopolamine, amphetamine and light reinforced responding. Psychon. Sci. **5**, 347—348 (1966)
Carlton, P. L., Didamo, P.: Augmentation of the behavioral effects of amphetamine by atropine. J. Pharmacol. exp. Ther. **132**, 91—96 (1961)
Carlton, P. L., Wolgin, D. L.: Contingent tolerance to the anorexigenic effects of amphetamine. Physiol. Behav. **7**, 221—223 (1971)
Carr, L. A., Moore, K. E.: Norepinephrine: release from brain *d*-amphetamine in vivo. Science **164**, 322—323 (1969)
Carr, L. A., Moore, K. E.: Effects of amphetamine on the contents of norepinephrine and its metabolites in the effluent of perfused cerebral ventricles of the cat. Biochem. Pharmacol. **19**, 2361—2374 (1970 a)
Carr, L. A., Moore, K. E.: Release of norepinephrine and normetanephrine from cat brain by central nervous system stimulants. Biochem. Pharmacol. **19**, 2671—2675 (1970 b)
Castellano, C.: Lysergic acid diethylamide, amphetamine and chlorpromazine on water maze discrimination in mice. Psychopharmacologia (Berl.) **19**, 16—25 (1971)
Castellano, C.: Cocaine, pemoline and amphetamine on learning and retention of a discrimination test in mice. Psychopharmacologia (Berl.) **36**, 67—76 (1974)

Castellano, C., Sansone, E. M., Renzi, P., Annecker, L.: Central stimulant drugs on avoidance behaviour in hamsters. Pharmacol. Res. Commun. **3**, 287—293 (1973)

Cattabeni, F., Racagni, G., Groppetti, A.: P-hydroxynorephedrine: its selective distribution in different rat brain areas. Frontiers of catecholamine research, pp. 1035—1037. Usdin, E., Snyder, S. (Eds.) New York: Pergamon Press 1973

Cattabeni, F., Revuelta, A., Costa, E.: Effects of β-methoxy derivatives of 3-trifluoromethylphenylethylamine on food intake and brain serotonin content. Neuropharmacology **11**, 753—760 (1972)

Cession-Fossion, A.: La tyramine, à la différence de l'amphétamine, se comporte chez le rat comme une amine sympathicomimétique à action indirecte. J. Physiol. (Paris) **55**, 223—223 (1963a)

Cession-Fossion, A.: Sur les propriétés hypertensives de l'amphétamine et de la tyramine chez le rat. Arch. int. Physiol. **71**, 382—392 (1963b)

Cession-Fossion, A.: Action des amines sympathicomimétiques à action indirecte sur la médullo-surrénale du rat perfusée "in vitro". Arch. int. Physiol. **75**, 303—309 (1967)

Chance, M. R. A.: Aggregation as a factor influencing the toxicity of sympathomimetic amines in mice. J. Pharmacol. exp. Ther. **87**, 214—219 (1946)

Chance, M. R. A.: Factors influencing the toxicity of sympathomimetic amines to solitary mice. J. Physiol. (Lond.) **89**, 289—296 (1947)

Chance, M. R. A., Mansour, T. E.: A chymographic study of the action of drugs on the liver fluke *(Fasciola hepatica)*. Brit. J. Pharmacol. **4**, 7—13 (1949)

Chang, C. C.: Reversal by amphetamine of the protective effect of bretylium on reserpine-induced depletion of noradrenaline. J. Pharm. Pharmacol. **17**, 818—820 (1965)

Chang, C. C., Costa, E., Brodie, B. B.: Interaction of guanethidine with adrenergic neurons. J. Pharmacol. exp. Ther. **147**, 303—312 (1965)

Chappel, C. I., Rona, G., Balazs, T., Gaudry, R.: Comparison of cardiotoxic actions of certain sympathomimetic amines. Canad. J. Physiol. Pharmacol. **37**, 35—40 (1959)

Cheng, H. C., Long, J. P.: Effect of *d*- and *l*-amphetamine on 5-hydroxytryptamine receptors. Arch. int. Pharmacodyn. **204**, 124—131 (1973)

Chernov, H., Furness, P., Partyka, D., Plummer, A. J.: Age, confinement and aggregation as factors in amphetamine group toxicity in mice. J. Pharmacol. exp. Ther. **154**, 346—349 (1966)

Chevillard, C., Alexandre, J.-M.: In vitro effects on cardiac norepinephrine of angiotensin II and of two indirectly acting sympathomimetic amines (amphetamine and tyramine) a study of their combinations. Europ. J. Pharmacol. **19**, 223—230 (1972)

Chevillard, C., Duchène, N., Alexandre, J.-M.: Selective release of newly synthesized cardiac norepinephrine induced by angiotensin II. Europ. J. Pharmacol. **15**, 8—14 (1971)

Cheymol, M. J., Levassort, C.: Hyperthermie amphetaminique et curarisation par *d*-tubocurarine. Ann. pharm. franç. **15**, 103—106 (1957)

Chidsey, C. A., Harrison, D. C., Braunwald, E.: Release of norepinephrine from the heart by vasoactive amines. Proc. Soc. exp. Biol. (N.Y.) **109**, 488—490 (1962)

Chiel, H., Yehuda, S., Wurtman, R. J.: Development of tolerance in rats to the hypothermic effects of *d*-amphetamine and apomorphine. Life Sci. **14**, 483—488 (1974)

Chistoni, A., Beccari, E.: Richerche farmacologiche sulla fenilisopropilamina Nota III—Azione eccitante sul sistema nervoso centrale ed in particolare sul respiro. Arch. ital. Sci. farmacol. **9**, 1—40 (1940)

Chiueh, C. C., Moore, K. E.: Release of endogenously synthesized catechols from the caudate nucleus by stimulation of the nigro-striatal pathway and by the administration of *d*-amphetamine. Brain Res. **50**, 221—225 (1973)

Chiueh, C. C., Moore, K. E.: Relative potencies of *d*- and *l*-amphetamine on the release of dopamine from cat brain in vivo. Res. Commun. chem. Path. Pharmacol. **7**, 189—199 (1974a)

Chiueh, C. C., Moore, K. E.: In vivo release of endogenously synthesized catecholamines from the cat brain evoked by electrical stimulation and by *d*-amphetamine. J. Neurochem. **23**, 159—168 (1974b)

Chiueh, C. C., Moore, K. E.: Effects of alpha-methyltyrosine on *d*-amphetamine induced release of endogenously synthesized and exogenously administered catecholamines from the cat brain in vivo. J. Pharmacol. exp. Ther. **190**, 100—108 (1974c)

Christie, J. E., Crow, T. J.: Central actions of amphetamine and ephedrines after unilateral lesions of dopamine neurones. Brit. J. Pharmacol. **41**, 405 (1971a)
Christie, J. E., Crow, T. J.: Turning behaviour as an index of the action of amphetamines and ephedrines on central dopamine containing neurones. Brit. J. Pharmacol. **43**, 658—667 (1971b)
Chu, H., Opitz, K., Intemann, E.: Über die Bedeutung der Schilddrüse für die stoffwechselsteigernde Wirkung von Amphetamin. Naunyn-Schmiedebergs Arch. exp. Path. Pharmak. **263**, 358—362 (1969)
Cicala, G. A., Kremer, E.: The effect of shock intensity and *d*-amphetamine on avoidance learning. Psychon. Sci. **14**, 41—42 (1969)
Clark, F. C., Steele, B. J.: Effects of *d*-amphetamine on performance under a multiple schedule in the rat. Psychopharmacologia (Berl.) **9**, 157—169 (1966)
Clark, W. C., Blackman, H. J., Preston, J. E.: Certain factors in aggregated mice *d*-amphetamine toxicity. Arch. int. Pharmacodyn. **170**, 350—363 (1967)
Clay, G. A., Cho, A. K., Roberfroid, M.: Effect of diethylaminoethyl diphenylpropylacetate hydrochloride (SKF-525A) on the norepinephrine depleting actions of d-amphetamine. Biochem. Pharmacol. **20**, 1821—1831 (1971)
Clineschmidt, B. V., McGuffin, J. C., Werner, A. B.: Role of monoamines in the anorexigenic actions of fenfluramine, amphetamine and p-chloramphetamine. Europ. J. Pharmacol. **27**, 313—323 (1974)
Clymer, N. V., Seifter, J.: A method of screening sympathomimetic amines for stimulant action on the cerebrum. J. Pharmacol. exp. Ther. **89**, 149—152 (1947)
Cohen, M., Lal, H.: A study of the mechanism of amphetamine toxicity in aggregated mice. Pharmacologist **5**, 261 (1963)
Cole, J., Glees, P.: Some effects of methylphenidate (Ritalin) and amphetamine on normal and leucotomized monkeys. J. ment. Sci. **103**, 406—417 (1957)
Cole, S. O.: Further study of interactive effects of amphetamine and food deprivation. Psychol. Rep. **16**, 625—630 (1965)
Cole, S. O.: Increased suppression of food intake by amphetamine in rats with anterior hypothalamic lesions. J. comp. physiol. Psychol. **61**, 302—305 (1966)
Cole, S. O.: Experimental effects of amphetamine: a review. Psychol. Bull. **68**, 81—90 (1967)
Cole, S. O.: The depression of operant behaviour and retarding action on discrimination learning by amphetamine. Psychon. Sci. **10**, 19—20 (1968)
Cole, S. O.: Amphetamine depression of a multiple CRF-EXT operant schedule: a sex comparison. Psychon. Sci. **16**, 143—144 (1969)
Cole, S. O.: On the combined effect of amphetamine and food deprivation: a reply to Gollub and Mann. Psychopharmacologia (Berl.) **16**, 426—429 (1970)
Cole, S. O.: The increased effectiveness of amphetamine with anterior hypothalamic lesions. Pharmacol. Res. Commun. **3**, 255—260 (1971)
Cole, S. O.: The relationship of amphetamine-induced anorexia and freezing under free-feeding conditions. Pharmacol. Res. Commun. **4**, 71—78 (1972)
Cole, S. O.: Hypothalamic feeding mechanisms and amphetamine anorexia. Psychol. Bull. **79**, 13—20 (1973)
Cole, S. O., Gay, P. E.: Brain mechanisms underlying the effects of amphetamine on feeding and non-feeding behaviours: dissociation and overlap. Physiol. Psychol. **2**, 80—88 (1974)
Colville, K. I., Chaplin, E.: Sympathomimetics as analgesics: effects of methoxamine, methamphetamine, metaraminol and norepinephrine Life Sci. **3**, 315—322 (1964)
Commarato, M. A., Lum, B. K. B.: Potentiation of the pressor response to angiotensin by amphetamine and ephedrine. Europ. J. Pharmacol. **10**, 25—33 (1970)
Consolo, S., Garattini, S., Ghielmetti, R., Valzelli, L.: Concentrations of amphetamine in the brain in normal or aggressive mice. J. Pharm. Pharmacol. **17**, 666 (1965a)
Consolo, S., Garattini, S., Valzelli, L.: Amphetamine toxicity in aggressive mice. J. Pharm. Pharmacol. **17**, 53—54 (1965b)
Consolo, S., Garattini, S., Valzelli, L.: Sensitivity of aggressive mice to centrally acting drugs. J. Pharm. Pharmacol. **17**, 594—595 (1965c)
Consolo, S., Ladinsky, H., Garattini, S.: Effect of several dopaminergic drugs and trihexyphenidyl on cholinergic parameters in the rat striatum. J. Pharm. Pharmacol. **26**, 275—277 (1974)

Consolo, S., Ladinsky, H., Peri, G., Garattini, S.: Effect of central stimulants and depressants on mouse brain acetylcholine and choline levels. Europ. J. Pharmacol. **18**, 251—255 (1972)
Cook, J. D., Schanberg, S. M.: The effects of methamphetamine on behavior and on the uptake, release and metabolism of norepinephrine. Biochem. Pharmacol. **19**, 1165—1179 (1970)
Cook, L., Catania, A. C.: Effect of drugs on avoidance and escape behaviour. Fed. Proc. **23**, 818—835 (1964)
Cools, A. R.: The function of dopamine and its antagonism in the caudate nucleus of cats in relation to the stereotyped behaviour. Arch. int. Pharmacodyn. **194**, 259—269 (1971)
Cools, A. R., Rossum, J. M. van: Caudal dopamine and stereotyped behaviour of cats. Arch. int. Pharmacodyn. **187**, 163—173 (1970)
Cooper, B. R., Cott, J. M., Breese, G. R.: Effects of catecholmine depleting drugs and amphetamine on self-stimulation of brain following various 6-hydroxydopamine treatments. Psychopharmacologia (Berl.) **37**, 235—248 (1974)
Cooper, S. J., Joyce, D., Summerfield, A.: Self-stimulation of the brain after administration of an amphetamine-barbiturate mixture. Brit. J. Pharmacol. **36**, 192 (1969)
Cooper, S. J., Joyce, D., Summerfield, A.: Termination of treatment with an amphetamine-barbiturate mixture at different stages of training. Activ. nerv. sup. (Praha) **14**, 266—268 (1972)
Coper, H., Lison, H., Rommelspacher, H., Schulze, G., Strauss, S.: The influence of adrenergic receptor-blocking agents, amphetamine and 6-aminonicotinamide on thermoregulation. Naunyn-Schmiedebergs Arch. exp. Path. Pharmak. **270**, 378—391 (1971)
Corrodi, H., Fuxe, H., Hökfelt, T.: The effect of some psychoactive drugs on central monoamine neurons. Europ. J. Pharmacol. **1**, 363—368 (1967)
Corrodi, H., Fuxe, K., Ljungdahl, Å., Ögren, S.-O.: Studies on the action of some psychoactive drugs on central noradrenaline neurones after inhibition of dopamine-β-hydroxylase. Brain Res. **24**, 451—470 (1970)
Corson, S. A., Corson, E. O. L., Kirilcuk, V., Arnold, L. E.: Tranquilizing effects of *d*-amphetamine on hyperkinetic untrainable dogs. Fed. Proc. **30**, 206 (1971)
Costa, E., Carenzi, A., Guidotti, A., Revuelta, A.: Narcotic analgesics and the regulation of neuronal catecholamine stores. Frontiers of catecholamine research, pp. 1003—1010. Usdin, E., Snyder, S. (Eds.) New York: Pergamon Press 1973
Costa, E., Groppetti, A.: Biosynthesis and storage of catecholamines in tissues of rats injected with various doses of *d*-amphetamine. In: Costa, E., Garattini, S. (Eds.): International symposium on amphetamines and related compounds, pp. 231—355. New York: Raven Press 1970
Costa, E., Groppetti, A.: Relationships between biochemical and pharmacological responses elicited by dextroamphetamine. In: Ellinwood, E. H., Cohen, S. (Eds.): Current concepts on amphetamine abuse, pp. 117—124. DHEW Publ. No (HSM) 72-9085. Washington D.C.: U.S. Government Printing Office 1972
Costa, E., Groppetti, A., Naimzada, M. K.: Effects of amphetamine on the turn-over rate of brain catecholamines and motor activity. Brit. J. Pharmacol. **44**, 742—751 (1972)
Costa, E., Groppetti, A., Revuelta, A.: Action of fenfluramine on monoamine stores of rat tissues. Brit. J. Pharmacol. **41**, 57—64 (1971)
Costall, B., Naylor, R. J.: Modification of amphetamine effects by intracerebrally administered anticholinergic agents. Life Sci. **11**, 239—253 (1972)
Costall, B., Naylor, R. J.: The role of the substantia nigra in the locomotor stimulant action of amphetamine. Brit. J. Pharmacol. **49**, 29—36 (1973)
Costall, B., Naylor, R. J.: Extrapyramidal and mesolimbic involvement with the stereotypic activity of *d*- and *l*-amphetamine. Europ. J. Pharmacol. **25**, 121—129 (1974)
Costall, B., Naylor, R. J., Olley, J. E.: Stereotypic and anticataleptic activities of amphetamine after intracerebral injection. Europ. J. Pharmacol. **18**, 83—94 (1972a)
Costall, B., Naylor, R. J., Olley, J. E.: The substantia nigra and stereotyped behaviour. Europ. J. Pharmacol. **18**, 95—106 (1972b)
Costall, B., Naylor, R. J., Wright, T.: The use of amphetamine induced stereotyped behaviour as a model for the experimental evaluation of antiparkinson agents. Arzneimittel-Forsch. **22**, 1178—1183 (1972c)
Cowan, F. F., Cannon, C., Koppanyi, T., Maengwyn-Davis, G. D.: Reversal of phenylalkylamine tachyphylaxis by norepinephrine. Science **134**, 1069—1070 (1961)

Cox, B., Tha, S. J.: Effects of amantadine and *l*-dopa on apomorphine- and *d*-amphetamine-induced stereotyped behaviour in rats. Europ. J. Pharmacol. **24**, 96—100 (1973)
Cox, R. H., Maickel, R. P.: Comparison of anorexigenic and behavioural potency of phenylakylamines. J. Pharmacol. exp. Ther. **181**, 1—9 (1972)
Coyle, J. T., Snyder, S. H.: Antiparkinsonian drugs: inhibition of dopamine uptake in the corpus striatum as a possible mechanism of action. Science **166**, 899—901 (1969a)
Coyle, J. T., Snyder, S. H.: Catecholamine uptake by synaptosomes in homogenates of rat brain: stereospecificity in different areas. J. Pharmacol. exp. Ther. **170**, 221—231 (1969b)
Crabtree, J. M., Moyer, K. E.: Sex differences in fighting and defense induced in rats by shock and *d*-amphetamine during morphine abstinence. Physiol. Behav. **11**, 337—343 (1973)
Craig, A. L., Kupferberg, H. J.: Hyperthermia in *d*-amphetamine toxicity in aggregated mice of different strains. J. Pharmacol. exp. Ther. **180**, 616—624 (1972)
Craigmill, A. L., Canafax, D. M., Curties, F. R.: The interaction of Δ^9-tetrahydrocannabinol and *d*-amphetamine in aggregated mice. Res. Commun. chem. Path. Pharmacol. **9**, 229—241 (1974)
Craigmill, A. L., Kupferberg, H. J., Miller, J. W.: Heat production and heat loss in *d*-amphetamine hyperthermia in aggregated Swiss Webster and BDF_1 mice. Toxicol. appl. Pharmacol. **26**, 362—369 (1973)
Creese, I., Iversen, S. D.: Amphetamine response in rat after dopamine neurone destruction. Nature (Lond.) New Biol. **239**, 247—248 (1972)
Creese, I., Iversen, S. D.: Blockage of amphetamine induced motor stimulation and stereotypy in the adult rat following neonatal treatment with 6-hydroxydopamine. Brain Res. **55**, 369—382 (1973)
Creese, I., Iversen, S. D.: The role of forebrain dopamine systems in amphetamine induced stereotyped behaviour in the rat. Psychopharmacologia (Berl.). **39**, 345—357 (1974)
Crismon, C. A., Tainter, M. L.: Comparative pressor efficiency of sympathomimetic amines in the normal state and in decerebrate shock J. Pharmacol. exp. Ther. **66**, 146—170 (1938)
Crow, T. J.: Mode of enhancement of selfstimulation in rats by methamphetamine. Nature (Lond.) **224**, 709—710 (1969)
Crow, T. J.: A map of the rat mesencephalon for electrical selfstimulation. Brain Res. **36**, 265—273 (1972a)
Crow, T. J.: Catecholamine-containing neurones and electrical selfstimulation: 1. A review of some data. Psychol. Med. **2**, 414—421 (1972b)
Crow, T. J., Gillbe, C.: Methamphetamine-protryptiline interaction in rotating rats. Brit. J. Pharmacol. **38**, 458 (1970)
Crowley, T. J.: Dose-dependent facilitation or suppression of rat fighting by methamphetamine, phenobarbital or imipramine. Psychopharmacologia (Berl.) **27**, 213—222 (1972)
Crowley, T. P., Sisemore, D. D.: The failure of methamphetamine hydrochloride to produce dissociation of an escape response. Psychon. Sci. **28**, 50—52 (1972)
Dallemagne, G. H.: Action de la chlorpromazine, du méprobamate de l'amphétamine sur trois comportements conditionnés sous controle aversif. Arch. int. Pharmacodyn. **183**, 46—59 (1970)
Dalrymple, S. D., Stretch, R.: Effects of amphetamine and chlorpromazine on second-order escape behavior in squirrel monkeys. Psychopharmacologia (Berl.) **21**, 268—282 (1971)
Dannenburg, W. N., Kardian, B. C.: The effect of fenfluramine and methamphetamine on free fatty acid release in epididymal fat cells of the rat. Arch. int. Pharmacodyn. **177**, 196—210 (1969)
Dannenburg, W. N., Kardian, B. C.: Metabolic effects of fenfluramine and methamphetamine on free fatty acid release and glucose utilization in epididymal fat cells of the rat. In: Costa, E., Garattini, S. (Eds.): International Symposium on Amphetamines and Related Compounds, pp. 597—610. New York: Raven Press 1970
D'Arcy, P. F., Spurling, N. W.: The effect of cortisol and corticotrophin on amphetamine toxicity in mice under crowded and noncrowded conditions. J. Endocr. **22**, 35—36 (1961)
Davey, M. J., Farmer, J. B.: The mode of action of tyramine. J. Pharm. Pharmacol. **15**, 178—182 (1963)
Davies, J. A., Jackson, B., Redfern, P. H.: The effect of amantadine, *l*-dopa, (+)-amphetamine and apomorphine on the acquisition of the conditioned avoidance response. Neuropharmacology **13**, 199—204 (1974)

Davis, G. D.: Effects of central excitant and depressant drugs on locomotor activity in the monkey. Amer. J. Physiol. **188**, 619—623 (1957)

Davis, R. A., Horlington, M.: Effects of reserpine pre-treatment on the protective action of amphetamine and phenoxypropazine in the phenylbenzoquinone induced writhing syndrome in mice. Nature (Lond.) **201**, 306—307 (1964)

Davis, W. M., Logston, D. G., Hickenbottom, J. P.: Antagonism of acute amphetamine intoxication by haloperidol and propranolol. Toxicol. appl. Pharmacol. **29**, 397—403 (1974)

Davis, W. M., Smith, S. G.: Alpha-methyltyrosine to prevent selfadministration of morphine and amphetamine. Curr. ther. Res. **14**, 814—819 (1972)

Davis, W. M., Smith, S. G.: Blocking effect of α-methyltyrosine on amphetamine based reinforcement. J. Pharm. Pharmacol. **25**, 174—177 (1973)

Day, M. D.: Effect of sympathomimetic amines on the blocking action of guanethidine, bretylium and xylocholine. Brit. J. Pharmacol. **18**, 421—439 (1962)

Day, M. D.: The lack of crossed tachyphylaxis between tyramine and some other indirectly acting sympathomimetic amines. Brit. J. Pharmacol. **30**, 631—643 (1967)

Day, M. D., Rand, M. J.: Antagonism of guanethidine by dexamphetamine and other related sympathomimetic amines. J. Pharm. Pharmacol. **14**, 541—549 (1962)

Day, M. D., Rand, M. J.: Evidence for a competitive antagonism of guanethidine by dexamphetamine. Brit. J. Pharmacol. **20**, 17—28 (1963a)

Day, M. D., Rand, M. J.: Tachyphylaxis to some sympathomimetic amines in relation to monoamine oxidase. Brit. J. Pharmacol. **21**, 84—96 (1963b)

Dean, H. G., Hughes, I. E.: Effect of amphetamine on the uptake release and effectiveness of xylocholine in the guinea-pig vas deferens. J. Pharm. Pharmacol. **23**, 606—611 (1971)

Dean, H. G., Hughes, I. E.: Increased release of xylocholine (TM 10) from guinea-pig vas deferens treated with amphetamine sulphate. J. Pharm. Pharmacol. **24**, 183—188 (1972)

Deffenu, G., Bartolini, A., Pepeu, G.: Effect of amphetamine on cholinergic systems of the cerebral cortex of the cat. In: Costa, E., Garattini, S. (Eds.): International Symposium on Amphetamines and Related Compounds, pp. 357—368. New York: Raven Press 1970

Delini-Stula, A., Morpurgo, C.: Influence of amphetamine and scopolamine on the catelepsy induced by diencephalic lesions in rats. Int. J. Neuropharmacol. **7**, 391—394 (1968)

Delphaut, J.: Action de la digitaline, de l'ouabaine, de la sparteine, de la procaine, du pentamethazene et du metioplegium sur le hypertoxicit amph taminique de groupement chez la souris blanche. Arch. int. Pharmacodyn. **154**, 434—437 (1965)

Del Rio, J., Fuentes, J. A.: Further studies on the antagonism of stereotyped behaviour induced by amphetamine. Europ. J. Pharmacol. **8**, 73—78 (1969)

D'Encarnacao, P. S., Anderson, K.: Effects of lithium pretreatment on amphetamine and DMI tetrabenazine produced psychomotor behavior. Dis. nerv. Syst. **31**, 494—496 (1970)

D'Encarnacao, P. S., D'Encarnacao, P., Tapp, J. T.: Potentiation of amphetamine induced psychomotor activity by diethyldithiocarbamate. Arch. int. Pharmacodyn. **182**, 186—189 (1969)

Deneau, G., Yanagita, T., Seevers, M. H.: Self-administration of psychoactive substances by the monkey. A measure of psychological dependence. Psychopharmacologia (Berl.) **16**, 30—48 (1969)

Dengler, H. J., Spiegel, H. E., Titus, E. O.: Effects of drugs on uptake of isotopic norepinephrine by cat tissues. Nature (Lond.) **191**, 816—817 (1961)

Detrick, L. E., Millikan, R., Modern, F. S., Thienes, C. H.: On the pharmacology of phenylisopropylamine (benzedrine). J. Pharmacol. exp. Ther. **60**, 56—68 (1937)

Dewar, A. J., Winterburn, A. K.: Amphetamine and RNA and protein metabolism in rat brain. Brain Res. **59**, 359—370 (1973)

Dewhurst, W. G., Marley, E.,: Differential effect of sympathomimetic amines on the central nervous system. Animal behaviour and drug action, pp. 175—188. Steinberg, H. (Ed.) London: Churchill Ltd. 1964

Dewhurst, W. G., Marley, E.: The effects of alpha-methyl derivatives of noradrenaline, phenylethylamine and tryptamine on the central nervous system of the chicken. Brit. J. Pharmacol. **25**, 682—704 (1965a)

Dewhurst, W. G., Marley, E.: Action of sympathomimetic and allied amines on the central nervous system of the chicken. Brit. J. Pharmacol. **25**, 705—727 (1965b)

Dews, P. B.: The measurement of the influence of drugs on voluntary activity in mice. Brit. J. Pharmacol. **8**, 46—48 (1953)

Dews, P. B.: Studies on behaviour. II. The effects of pentobarbital, methamphetamine and scopolamine on performances in pigeons involving discriminations. J. Pharmacol. exp. Ther. **115**, 380—389 (1955)

Dews, P. B.: Studies on behavior. IV. Stimulant actions of methamphetamine. J. Pharmacol. exp. Ther. **122**, 137—147 (1958a)

Dews, P. B.: Analysis of effects of psychopharmacological agents in behavioural terms. Fed. Proc. **17**, 1024—1030 (1958b)

Dews, P. B., Morse, W. H.: Behavioral pharmacology. Ann. Rev. Pharmacol. **1**, 145—174 (1961)

Diaz, J.-L., Huttunen, M. O.: Altered metabolism of serotonin in the brain of the rat after chronic ingestion of *d*-amphetamine. Psychopharmacologia (Berl.) **23**, 365—372 (1972)

DiCarlo, R., Edel, S., Randrianarisoa, H., Mandel, P.: Amphetamine and cerebral RNA metabolism. Pharmacol. Res. Commun. **4**, 275—286 (1972)

Dingell, J. V., Owens, M. L., Norvich, M. R., Sulser, F.: On the role of norepinephrine biosynthesis in the central action of amphetamine. Life Sci. **6**, 1155—1162 (1967)

Dispensa, J., Barrett, M. E.: The effect of amphetamine (benzedrine) sulphate on maze performance of the albino rat. J. Psychol. **11**, 397—410 (1941)

Divac, I.: Drug-induced syndromes in rats with large chronic lesions in the corpus striatum. Psychopharmacologia (Berl.) **27**, 171—178 (1972)

Doda, M., György, L., Nador, K.: Influence of *d*-amphetamine on the effects of adrenergic neuron blocking agents. Arch. int. Pharmacodyn. **164**, 247—257 (1966)

Dolfini, E., Del Angel, R. A., Garattini, S., Valzelli, L.: Brain catecholamine release by dexamphetamine in three strains of mice. Europ. J. Pharmacol. **9**, 333—336 (1970)

Dolfini, E., Garattini, S., Valzelli, L.: Activity of (+)-amphetamine at different environmental temperatures in three strains of mice. J. Pharm. Pharmacol. **21**, 871—872 (1969a)

Dolfini, E., Garattini, S., Valzelli, L.: Different sensitivity to amphetamine of three strains of mice. Europ. J. Pharmacol. **7**, 220—223 (1969b)

Dolfini, E., Kobayashi, M.: Studies with amphetamine in hyper- and hypothyroid rats. Europ. J. Pharmacol. **2**, 65—66 (1967)

Dolfini, E., Tansella, M., Valzelli, L., Garattini, S.: Further studies on the interaction between desipramine and amphetamine. Europ. J. Pharmacol. **5**, 185—190 (1969c)

Dominic, J. A., Moore, K. E.: Acute effects of α-methyltyrosine on brain catecholamine levels and on spontaneous and amphetamine stimulated motor activity in mice. Arch. int. Pharmacodyn. **178**, 166—170 (1969a)

Dominic, J. A., Moore, K. E.: Supersensitivity to the central stimulant actions of adrenergic drugs following discontinuation of a chronic diet of α-methyltyrosine. Psychopharmacologia (Berl.) **15**, 96—101 (1969b)

Domino, E. F., Caldwell, D. F., Henke, R.: Effects of psychoactive agents on acquisition of conditioned pole jumping in rats. Psychopharmacologia (Berl.) **8**, 285—289 (1965)

Domino, E. F., Olds, M. E.: Effects of *d*-amphetamine, scopolamine, chlordiazepoxide and diphenylhydantoin on self-stimulation behavior and brain acetylcholine. Psychopharmacologia (Berl.) **23**, 1—16 (1972)

Domino, E. F., Wilson, A.: Psychotropic drug influences on brain acetylcholine utilization. Psychopharmacologia (Berl.) **25**, 291—298 (1972)

Dorris, R. L., Shore, P. A.: Interaction of apomorphine, neuroleptics and stimulants with α-methyl-m-tyramine, a false dopaminergic transmitter. Biochem. Pharmacol. **23**, 867—872 (1974)

Doty, B. A., Doty, L. A.: Facilitative effects of amphetamine on avoidance conditioning in relation to age and problem difficulty. Psychopharmacologia (Berl.) **9**, 234—241 (1966)

Downing, O. A.: Effect of amphetamine on the transmission of repetitive impulses through the isolated superior cervical ganglion of the rat. Brit. J. Pharmacol. **44**, 71—79 (1972)

Drudi-Baracco, C., Halpern, B. N., Bessirard, D.: Psychopharmacologie de l'imipramine étudieé à l'aide de la toxicité de groupe. C.R. Soc. Biol. (Paris) **157**, 1236—1240 (1963)

Drummond, G. I., Bellward, G.: Studies on phosphorylase b kinase from neural tissues. J. Neurochem. **17**, 475—482 (1970)

Duarte-Escalante, O., Ellinwood, E. H. Jr.: Effects of chronic amphetamine intoxication on adrenergic and cholinergic structures in the central nervous system: Histochemical observations in cats and monkeys. In: Ellinwood, E. H., Cohen, S. (Eds.): Current concepts on amphetamine abuse, pp. 97—106. DHEW Publ. No. (HSM) 72-9085. Washington D.C.: U.S. Government Printing Office 1972

Dubnick, B., Ruchi, E. W., Salama, A. I.: A comparison of 1-p-acetyldeoxyephedrine and 4-methyl-ethyl-m-tyramine as a lowering of brain serotonin and their antagonism by antidepressants. Europ. J. Pharmacol. **22**, 121—128 (1973)

Dudderidge, H. J., Gray, J. A.: Joint effects of sodium amylobarbitone and amphetamine sulphate on resistance to extinction of a rewarded running response in the rat. Psychopharmacologia **35**, 365—370 (1974)

Duhault, J., Verdavainne, C.: Modification du taux de sérotonine cérébrale chez le rat par le trifluorométhyl-phényl-2-éthyl amino propane (Fenfluramine 768 S). Arch. int. Pharmacodyn. **170**, 276—286 (1967)

Eble, J. N., Rudzik, A. D.: The potentiation of the pressor response to tyramine by amphetamine in the anesthetized dog. J. Pharmacol. exp. Ther. **150**, 375—381 (1965)

Eble, J. N., Rudzik, A. D.: The blockade of the pressor response to tyramine by amphetamine in the reserpine-treated dog. J. Pharmacol. exp. Ther. **153**, 62—69 (1966 a)

Eble, J. N., Rudzik, A. D.: Amphetamine: augmentation of pressor effects of tyramine in rats. Proc. Soc. exp. Biol. (N.Y.) **122**, 1059—1060 (1966 b)

Eble, J. N., Rudzik, A. D.: Interaction between amphetamine and sympathomimetic agents on the cardiovascular system. In: Costa, E., Garattini, S. (Eds.): International symposium on amphetamines and related compounds, pp. 513—529. New York: Raven Press 1970

Ebstein, R. P., Ebstein, B. S., Samuel, D., Berger, B. D.: Differential sensitivity of amygdala and hypothalamus to amphetamine induced release of norepinephrine. J. Neurochem. **19**, 2703—2705 (1972)

Ehrich, W. E., Krumbhaar, E. B.: The effects of large doses of benzedrine sulphate on the albino rat: functional and tissue changes. Ann. intern. Med. **10**, 1874—1888 (1937)

Ehrich, W. E., Lewy, F. H., Krumbhaar, E. B.: Experimental studies upon toxicity of benzedrine sulphate in various animals. Amer. J. med. Sci. **198**, 785—803 (1939)

El Guedri, H., Jacquot, C., Rapin, J., Cohen, Y.: Metabolism, regional distribution and pharmacokinetics of para-hydroxynorephedrine in the rat brain. J. Pharmacol. (Paris) **4**, 453—463 (1973)

Eliasson, M., Michanek, A., Meyersson, B. J.: A differential inhibitory action of LSD and amphetamine on copulatory behaviour in the female rat. "Animal Pharm." Uppsala University Biomedical Center, Uppsala, pp. 22 1972

Elkes, J., Elkes, C., Bradley, P. B.: The effect of some drugs on the electrical activity of the brain and on behaviour. J. ment. Sci. **100**, 125—128 (1954)

Ellinwood, E. H. Jr.: "Accidental Conditioning" with chronic methamphetamine intoxication: implications for a theory of drug habituation. Psychopharmacologia (Berl.) **21**, 131—138 (1971 a)

Ellinwood, E. H. Jr.: Comparative methamphetamine intoxication in experimental animals. Pharmacopsychiat. Neuropsychopharmakol. **4**, 351—361 (1971 b)

Ellinwood, E. H. Jr., Balster, R. L.: Rating the behavioural effects of amphetamine. Europ. J. Pharmacol. **28**, 35—41 (1974)

Ellinwood, E. H. Jr., Duarte-Escalante, O.: Chronic methamphetamine intoxication in three species of experimental animals. In: Ellinwood, E. H., Cohen, S. (Eds.): Current concepts on amphetamine abuse, pp. 59—68. DHEW Publ., No. (HSM) 72-9085. Washington D.C.: U.S. Government Printing Office 1972

Ellinwood, E. H. Jr., Escalante, O.: Behavior and histophathological findings during chronic methedrine intoxication. Biol. Psychiatry **2**, 27—39 (1970 a)

Ellinwood, E. H. Jr., Escalante, O.: Chronic amphetamine effect on the olfactory forebrain. Biol. Psychiatry **2**, 189—203 (1970 b)

Ellinwood, E. H. Jr., Sudilovsky, A., Grabowy, R.: Olfactory forebrain seizures induced by methamphetamine and disulfiram. Biol. Psychiatry **7**, 89—99 (1973)

Ellinwood, E. H. Jr., Sudilovsky, A., Nelson, L. M.: Behavioral analysis of chronic amphetamine intoxication. Biol. Psychiatry **4**, 215—229 (1972)

Ellinwood, E. H. Jr., Sudilovsky, A., Nelson, L. M.: Behaviour and EEG analysis of chronic amphetamine effect. Biol. Psychiatry **8**, 169—176 (1974)

Ellison, T., Siegel, M., Silverman, A. G., Okun, R.: Comparative metabolism of *dl*-^{3}H-amphetamine hydrochloride in tolerant and nontolerant cats. Proc. West. Pharmacol. Soc. **11**, 75—77 (1968)

Enna, S. J., Dorris, R. L., Shore, P. A.: Specific inhibition by alphamethyltyrosine of amphetamine-induced amine release from brain. J. Pharmacol. exp. Ther. **184**, 576—582 (1973)

Enna, S. J., Shore, P. A.: On the nature of the adrenergic neuron extragranular amine binding site. J. neural Transm. **35**, 125—135 (1974)

Epstein, A. N.: Suppression of eating and drinking by amphetamine and other drugs in normal and hyperphagic rats. J. comp. physiol. Psychol. **52**, 37—45 (1959)

Ernst, A. M.: Mode of action of apomorphine and dexamphetamine on gnawing compulsion in rats. Psychopharmacologia (Berl.) **10**, 316—323 (1967a)

Ernst, A. M.: Mode of action of apomorphine and dexamphetamine on gnawing compulsion in rats. Acta physiol. pharmacol. neerl. **14**, 341 (1967b)

Ernst, A. M.: The role of biogenic amines in the extra-pyramidal system. Acta physiol. pharmacol. neerl. **15**, 141—154 (1969)

Escalante, O. D., Ellinwood, E. H. Jr.: Central nervous system cytopathological changes in cats with chronic methedrine intoxication. Brain Res. **21**, 151—155 (1970)

Estler, C.-J., Ammon, H. P. T.: The influence of propranolol on the methamphetamine-induced changes of cerebral function and metabolism. J. Neurochem. **14**, 799—805 (1967)

Estler, C.-J., Ammon, H. P. T.: Antagonistischer Einfluß von *β*-Sympathicolytica auf die durch Methamphetamine verursachten Änderungen von Funktion und Stoffwechsel des Gehirns. Arzneimittel-Forsch. **20**, 908—909 (1970)

Estler, C.-J., Ammon, H. P. T.: Modification by two beta-adrenergic blocking drugs of the effects of methamphetamine on behavior and brain metabolism of mice. J. Neurochem. **18**, 777—779 (1971)

Estler, C.-J., Ammon, H. P. T., Fickl, W., Fröhlich, H.-N.: Substrate supply and energy metabolism of skeletal muscle of mice treated with methamphetamine and propranolol. Biochem. Pharmacol. **19**, 2957—2962 (1970)

Estler, C.-J., Mitznegg, P.: Influence of methamphetamine on incorporation of glucose into brain glycogen. Biochem. Pharmacol. **20**, 1331—1333 (1971)

Etevenon, P.: Effets de la stimulation en lumière intermittente sur les aires cortico-visuelles du Lapin. Latéralisation spécifique des résponses électroencéphalographiques et ses variations après administration d'amphétamine ou de pentobarbital. C.R. Acad. Sci. (Paris) **265**, 885—888 (1967)

Euler, U. S. von, Lishajko, F.: Effect of directly and indirectly acting sympathomimetic amines on adrenergic transmitter granules. Acta physiol. scand. **73**, 78—88 (1968)

Evangelista, A. M., Gattoni, R. C., Izquierdo, I.: Effect of amphetamine, nicotine and hexamethonium on performance of a conditioned response during acquisition and retention trials. Pharmacology (Basel) **3**, 91—96 (1970)

Evans, H. L.: Behavioral effects of methamphetamine and alphamethyltyrosine in the rat. J. Pharmacol. exp. Ther. **176**, 244—254 (1971)

Evans, H. L., Ghiselli, W. B., Patton, R. A.: Diurnal rhythm in behavioral effects of methamphetamine, p-chloromethamphetamine and scopolamine. J. Pharmacol. exp. Ther. **186**, 10—17 (1973)

Everett, G. M., Thoman, J. E. P., Smith, A. H.: Central and peripheral effects of reserpine and 11-desmethoxyreserpine (Harmonyl) on the nervous system. Fed. Proc. **16**, 295 (1957)

Everett, G. M., Yellin, T. O.: Effect of 4-chlorophenoxyamphetamine (A-6587), *d*-amphetamine and 4-chloroamphetamine on brain biogenic amines and behavior in mice. Res. Commun. chem. Path. Pharmacol. **2**, 407—414 (1971)

Evetts, K. D., Uretsky, N. J., Iversen, L. L., Iversen, S. D.: Effects of 6-hydroxydopamine on CNS catecholamines, spontaneous motor activity and amphetamine induced hyperactivity in rats. Nature (Lond.) **225**, 961—962 (1970)

Ewing, P. L., Moore, B. M., Moore, W. T.: The effect of amphetamine and related compounds on maze performance of white rats. J. Pharmacol. exp. Ther. **105**, 343—348 (1952)

Faidherbe, J., Richelle, M., Schlaug, J.: Nonconsumption of the reinforcer under drug action. J. exp. Anal. Behav. **5**, 521—524 (1962)
Falb, A., Schmidt, G., Schwabe, U.: Zusammenhänge zwischen anorexigener Wirkung und Fettsäuremobilisation durch Methamphetamine. Arzneimittel-Forsch. **16**, 733—735 (1966)
Fanelli, O.: Pharmacological and toxicological study of a new psychotropic stimulant: the 2-phenyl-5-dimethyl-tetrahydro-1,4-oxazine, in comparison with *dl*-amphetamine, phenmetrazine and pemoline-Mg. Arzneimittel-Forsch. **23**, 810—816 (1973)
Farnebo, L.-O.: On transmitter release evoked by field stimulation of monoamine neurons, pp. 1—44. Stockholm: Department of Histology, Karolinska Institutet 1971
Fassina, G.: Azioni di farmaci anoressanti sugli acidi liberi plasmatici. Arch. int. Pharmacodyn. **161**, 410—422 (1966)
Fawas, G., Simaan, J.: The tachyphylaxis caused by mephentermine and tyramine. Brit. J. Pharmacol. **24**, 526—531 (1965)
Feldberg, W., Lang, W.J.: Effects of monoamine oxidase inhibitors and amphetamine on hypothermia produced by halothane. Brit. J. Pharmacol. **38**, 181—191 (1970)
Feltz, P.: La réponse cellulaire d'excitation nigro-striatale: Test pharmacologiques par perfusion de la surface du noyau caudé. J. Physiol. (Paris) **62**, suppl. 3, 374—375 (1970)
Feltz, P., Champlain, J. de: The postsynaptic effect of amphetamine on striatal dopamine-sensitive neurones. Frontiers of catecholamine research, pp. 951—956. Usdin, E., Snyder, S. (Eds.) New York: Pergamon Press 1973
Ferguson, J., Dement, W.: The behavioral effects of amphetamine on REM deprived rats. J. psychiat. Res. **7**, 111—118 (1969)
Ferrendelli, J.A., Kinscherf, D.A., Kipnis, D.M.: Effects of amphetamine, chlorpromazine and reserpine on cyclic GMP and cyclic AMP levels in mouse cerebellum. Biochem. biophys. Res. Commun. **46**, 2114—2120 (1972)
Ferris, R.M., Maxwell, R.A.: Effects of isomers of amphetamine deoxypipradrol and methylphenidate on uptake and release of H^3-catecholamines in crude synaptosomal preparations of rat cortex, hypothalamus and striatum. Fed. Proc. **31**, abstract 2175 (1972)
Ferris, R.M., Tang, F.L.M., Maxwell, R.A.: A comparison of the capacities of isomers of amphetamine, deoxypipradrol and methylphenidate to inhibit the uptake of tritiated catecholamines into rat cerebral cortex slices. J. Pharmacol. exp. Ther. **181**, 407—416 (1972)
Ferster, C.B., Appel, J.B., Hiss, R.A.: The effects of drugs on a fixed ratio performance suppressed by a pre time-out stimulus. J. exp. Anal. Behav. **5**, 73—88 (1962)
Feudis, F.V. de: Effects of *d*-amphetamine on the incorporation of carbon atoms of D-mannose into the brains and sera of differently housed mice: short-term reversibility of these effects. Biol. Psychiatry **7**, 3—10 (1973)
Feudis, F.V. de, Marks, J.H.: Studies on the time course of entry and subcellular distribution of radioactivity of (^{3}H)-*d*-amphetamine in the brains of differentially housed mice. Experientia (Basel) **29**, 1518—1520 (1973a)
Feudis, F.V. de, Marks, J.H.: Brain to serum distribution of radioactivity of injected (^{3}H)-*d*-amphetamine in differentially housed mice. Biol. Psychiatry **6**, 85—88 (1973b)
Fibiger, H.C.: Behavioural pharmacology of *d*-amphetamine: some metabolic and pharmacological considerations. Frontiers in catecholamine research, pp. 933—937. Usdin, E., Snyder, S. (Eds.) New York: Pergamon Press 1973
Fibiger, H.C., Fibiger, H.P., Zis, A.P.: Attenuation of amphetamine-induced motor stimulation and stereotypy by 6-hydroxydopamine in the rat. Brit. J. Pharmacol. **47**, 683—692 (1973)
Fibiger, H.C., McGeer, E.G.: Effect of acute and chronic methamphetamine treatment on tyrosine hydroxylase activity in brain and adrenal medulla. Europ. J. Pharmacol. **16**, 176—180 (1971)
Fibiger, H.C., Trimbach, C., Campbell, B.A.: Enhanced stimulant properties of (+)-amphetamine after chronic reserpine treatment in the rat: mediation by hypophagia and weight loss. Neuropharmacology **11**, 57—67 (1972)
Finger, K.F., Page, J.G., Feller, D.R.: Influence of various agonists and antagonists on the release of free fatty acids from adipose tissue in vitro. Biochem. Pharmacol. **15**, 1023—1032 (1966)
Fischer, E., Heller, B.: Pharmacology of the mechanism of certain effects of reserpine in the rat. Nature (Lond.) **216**, 1221—1222 (1967)
Fitz-Gerald, F.L.: Effects of *d*-amphetamine upon behavior of young chimpanzees reared under different conditions. Neuropharmacology **5**, 1226—1227 (1967)

Fleckenstein, A., Burn, J. H.: The effect of denervation on the action of sympathomimetic amines on the nictitating membrane. Brit. J. Pharmacol. **8**, 69—78 (1953)

Fleckenstein, A., Stöckle, D.: Zum Mechanismus der Wirkungsverstärkung und Wirkungs-Abschwächung sympathomimetischer Amine durch Cocain und andere Pharmaka. Naunyn-Schmiedebergs Arch. exp. Path. Pharmak. **224**, 401—415 (1955)

Flegin, O. T., Morgan, D. H., Oates, J. A., Shand, D. G.: The mechanism of the reversal of the effect of guanethidine by amphetamines in cat and man. Brit. J. Pharmacol. **39**, 252—254 (1970)

Fog, R.: Stereotyped and non-stereotyped behaviour in rats induced by various stimulant drugs. Psychopharmacologia (Berl.) **14**, 299—304 (1969)

Fog, R.: Behavioural effects in rats of morphine and amphetamine and of a combination of the two drugs. Psychopharmacologia (Berl.) **16**, 305—312 (1970)

Fog, R.: On stereotyped and catalepsy: studies on the effect of amphetamines and neuroleptics in rats. Copenhagen: Munksgaard (Thesis) 1972

Fog, R., Pakkenberg, H.: Behavioral effects of dopamine and p-hydroxyamphetamine injected into corpus striatum of rats. Exp. Neurol. **31**, 75—86 (1971)

Fog, R., Randrup, A., Pakkenberg, H.: Lesions in corpus striatum and cortex of rat brains and the effect on pharmacologically induced stereotyped, aggressive and cataleptic behaviour. Psychopharmacologia (Berl.) **18**, 346—356 (1970)

Fog, R., Randrup, A., Pakkenberg, H.: Aminergic mechanisms in corpus striatum and amphetamine induced sterotyped behaviour. Psychopharmacologia (Berl.) **11**, 179—193 (1967)

Follenfant, M. J., Robson, R. D.: The antagonism of adrenergic neurone blockade by amphetamine and dexamphetamine in the rat and guinea-pig. Brit. J. Pharmacol. **38**, 792—801 (1970)

Foote, W. E., Sheard, M. H., Aghajanian, G. K.: LSD and amphetamine effects on midbrain raphe units. Nature (Lond.) **222**, 567—569 (1969)

Foree, D. P., Moretz, F. H., McMillan, D. E.: Drugs and punished responding. II. *d*-amphetamine induced increases in punished responding. J. exp. Anal. Behav. **20**, 291—300 (1973)

Fox, S. S.: Self-maintained sensory input and sensory deprivation in monkeys: A behavioural and neuropharmacological study. J. comp. physiol. Psychol. **55**, 438—444 (1962)

Foxwell, M. H., Funderburk, W. H., Ward, J. W.: Studies on the state of action of a new anorexigenic agent, fenfluramine. J. Pharmacol. exp. Ther. **165**, 60—70 (1969)

Frankova, S.: Effect of amphetamine and chlorpromazine on maternal behaviour of rats. Activ. nerv. şup. (Praha) **14**, 166—167 (1972)

Freeman, J. J., Miller, K. W., Sulser, F.: On the mechanism of amphetamine potentiation by iprindole. Pharmacologist **12**, 226 (1970)

Freeman, J. J., Sulser, F.: Iprindole-amphetamine interactions in the rat: the role of aromatic hydroxylation of amphetamine in its mode of action. J. Pharmacol. exp. Ther. **183**, 307—315 (1972)

Frey, H.-H.: Note on the interactions of amphetamine with anticonvulsant drugs. Acta pharmacol. (Kbh.) **21**, 290—298 (1964)

Frey, H.-H.: p-Chloroamphetamine-similarities and dissimilarities to amphetamine. In: Costa, E., Garattini, S. (Eds.): International Symposium on Amphetamines and Related Compounds, pp. 343—347. New York: Raven Press 1970

Frey, H.-H., Kampmann, E.: Interaction of amphetamine with anticonvulsant drugs. II. Effect of amphetamine on the absorption of anticonvulsant drugs. Acta pharmacol (Kbh.) **24**, 310—316 (1966)

Frey, H.-H., Magnussen, M. P.: Different central mediation of the stimulant effects of amphetamine and its p-chloro analogue. Biochem. Pharmacol. **17**, 1299—1307 (1968)

Frey, H.-H., Schulz, R.: On the central mediation of anorexigenic drug effects. Biochem. Pharmacol. **22**, 3041—3049 (1973)

Frommel, E., Chmouliovsky, M.: De l'action antidotale de l'haloperidole envers l'amphétamine chez la souris. C.R. Soc. Biol. (Paris) **158**, 48—50 (1964)

Fuentes, J. A., Del Rio, J.: Striatal homovanillic acid levels in rats after combined treatments with amphetamine and neuroleptics. Europ. J. Pharmacol. **17**, 297—300 (1972)

Fujimori, M., Himwich, H. E.: Electroencephalographic alerting sites of *d*-amphetamine and 2,5-dimethoxy-4-methyl-amphetamine. Nature (Lond.) **220**, 491—494 (1968)

Fujimori, M., Himwich, H. E.: Electroencephalographic analyses of amphetamine and its methoxy derivatives with references to their sites of EEG alerting in the rabbit brain. Int. J. Neuropharmacol. **8**, 601—613 (1969)

Fulginiti, S., Orsingher, O. A.: Effects of learning, amphetamine and nicotine on the level and synthesis of brain noradrenaline in rats. Arch. int. Pharmacodyn. **190**, 291—298 (1971)
Fulginiti, S., Orsingher, O. A.: Further evidence in support of a common adrenergic mechanism for the facilitatory action on learning of amphetamine and nicotine in rats. J. Pharm. Pharmacol. **25**, 580—581 (1973)
Fuller, R. W.: Selective inhibition of monoamineoxidase. Adv. Biochem. Psychopharmacology **5**, 339—354 (1972)
Fuller, R. W., Hines, C. W.: *d*-Amphetamine levels in brain and other tissues of isolated and aggregated mice. Biochem. Pharmacol. **16**, 11—16 (1967)
Fuller, R. W., Hines, C. W., Mills, J.: Lowering of brain serotonin level by chloramphetamines. Biochem. Pharmacol. **14**, 483—488 (1965)
Fuller, R. W., Schaffer, R. J., Roush, B. W., Molloy, B. B.: Drug disposition as a factor in the lowering of brain serotonin by chloramphetamines in the rat. Biochem. Pharmacol. **21**, 1413—1417 (1972b)
Fuller, R. W., Shaw, W. N., Molloy, B. B.: Dissociation of the lipid mobilizing and hyperthermic effects of amphetamine by beta-fluoro substitution. Arch. int. Pharmacodyn. **199**, 266—271 (1972a)
Fuller, R. W., Walters, C. P.: Inhibition of monoamine oxidase action on kynuramine by substrate amines and stereosimeric α-methyl amines. Biochem. Pharmacol. **14**, 159—163 (1964)
Funatogawa, S.: Methamphetamine-induced changes in behaviour of cats and in topographical distribution of brain serotonin. Psychiat. Neurol. jap. **66**, 743—754 (1964)
Funatogawa, S.: Methamphetamine-induced changes in behaviour of the cat and in topographical distribution of brain serotonin. Excerpta med. (Amst.) Sect. II **18**, Abstr. 4863 (1965)
Funderburk, W. H., Hazelwood, J. C., Ruckart, R. T., Ward, J. W.: Is 5-hydroxytryptamine involved in the mechanism of action of fenfluramine. J. Pharm. Pharmacol. **23**, 468—470 (1971)
Fuxe, K., Hamberger, B., Malmfors, T.: The effect of drugs on accumulation of monoamines in tubero-infundibular dopamine neurons. Europ. J. Pharmacol. **1**, 334—341 (1967)
Fuxe, K., Hökfelt, T., Ungerstedt, U.: Distribution of monoamines in the mamalian central nervous system by histochemical studies. In: Hooper, G. (Ed.): Metabolism of amines in the brain, pp. 10—22. London: McMillan 1969
Fuxe, K., Ungerstedt, U.: Histochemical studies on the effect of (+)-amphetamine, drugs of the imipramine group and tryptamine on central catecholamine and 5-hydroxytryptamine neurons after intraventricular injection of catecholamines and 5-hydroxytryptamine. Europ. J. Pharmacol. **4**, 135—144 (1968)
Fuxe, K., Ungerstedt, U.: Histochemical, biochemical and functional studies on central monoamine neurons after acute and chronic amphetamine administration. In: Costa, E., Garattini, S. (Eds.): International symposium on amphetamines and related compounds, pp. 257—288. New York: Raven Press 1970
Fuxe, K., Ungerstedt, U.: Action of caffeine and theophyllamine on supersensitive dopamine receptors: considerable enhancement of receptor response to treatment with dopa and dopamine receptor agonists. Med. Biol. **52**, 48—54 (1974)
Galambos, E., Pfeifer, A. K., György, L., Monlár, J.: Study on the excitation induced by amphetamine, cocaine and α-methyltryptamine. Psychopharmacologia (Berl.) **11**, 122—129 (1967)
Ganguly, D. K.: Effect of methylamphetamine on the brain 5-hydroxytryptamine content of isolated and aggregated rats. Experientia (Basel) **25**, 1153—1154 (1969)
Gardocki, J. F., Schuler, M. E., Goldstein, L.: Reconsideration of the central nervous system. Pharmacology of amphetamine. I. Toxicity in grouped and isolated mice. Toxicol. appl. Pharmacol. **8**, 550—557 (1966a)
Gardocki, J. F., Schuler, M. E., Goldstein, L.: Reconsideration of the central nervous system. Pharmacology of amphetamine. II. Influence of pharmacological agents on cumulative and total lethality in grouped and isolated mice. Toxicol. appl. Pharmacol. **9**, 536—554 (1966b)
Garriott, J. C., King, L. J., Forney, R. B., Hughes, F. W.: Effects of some tetrahydrocannabinols on hexobarbital sleeping time and amphetamine-induced hyperactivity in mice. Life Sci. **6**, 2119—2128 (1967)
Gauchy, C., Bioulac, B., Cheramy, A., Besson, M. J., Glowinski, J., Vincent, J. D.: Estimation of chronic dopamine release from the caudate nucleus of the macaca mulatta. Brain Res. **77**, 257—268 (1974)

Geller, I., Seifter, J.: The effects of meprobamate, barbiturates, *d*-amphetamine and promazine on experimentally induced conflict in the rat. Psychopharmacologia (Berl.) **1**, 482—492 (1960)

George, D. J., Wolf, H. H.: Dose-lethality curves for *d*-amphetamine in isolated and aggregated mice. Life Sci. **5**, 1583—1590 (1966)

George, D. J., Wolf, H. H.: Effect of adrenal demedullation and adrenalectomy on amphetamine toxicity in isolated and aggregated mice. J. Pharm. Pharmacol. **19**, 636—638 (1967)

Gerald, M. C., Riffee, W. H.: Acute and chronic effects of *d*- and *l*-amphetamine on seizure susceptibility in mice. Europ. J. Pharmacol. **21**, 323—330 (1973)

Gerkens, J. F., McCulloch, M. W., Wilson, J.: Mechanism of the antagonism between guanethidine and dexamphetamine. Brit. J. Pharmacol. **35**, 563—572 (1969)

German, D. C., Bowden, D. M.: Catecholamine systems as the neural substrate for intracranial selfstimulation. A Hypothesis. Brain Res. **73**, 381—419 (1974)

Gessa, G. L., Clay, G. A., Brodie, B. B.: Evidence that hyperthermia produced by *d*-amphetamine is caused by a peripheral action of the drug. Life Sci. **8**, 135—141 (1969)

Geyer, M. A., Segal, D. S.: Differential effects of reserpine and alpha-methyl-p-tyrosine on norepinephrine and dopamine induced behavioural activity. Psychopharmacologia (Berl.) **29**, 131—140 (1973)

Ghosh, M. N., Parvathy, S.: Tolerance pattern of the anorexigenic action of amphetamine in rats. Brit. J. Pharmacol. **49**, 658—661 (1973)

Glick, S. D.: Effects of *d*-amphetamine and frontal ablations on response suppression in rats. J. comp. physiol. Psychol. **69**, 49—54 (1969)

Glick, S. D.: Change in sensitivity to *d*-amphetamine in frontal rats as a function of time: shifting of the dose-response curve. Psychon. Sci. **19**, 57—58 (1970)

Glick, S. D.: Differential sensitivity of frontal rats to *d*-amphetamine and scopolamine. Commun. behav. Biol. **5**, 341—346 (1971 a)

Glick, S. D.: Facilitation or impairment of learning by *d*-amphetamine as a function of stimuli. Psychopharmacologia (Berl.) **21**, 353—360 (1971 b)

Glick, S. D.: Changes in amphetamine sensitivity following frontal cortical damage in rats and mice. Europ. J. Pharmacol. **20**, 351—356 (1972)

Glick, S. D.: Impaired tolerance to the effects of oral amphetamine intake in rats with frontal cortex ablations. Psychopharmacologia (Berl.) **28**, 363—371 (1973 a)

Glick, S. D.: Enhancement of spatial preferences by (+)-amphetamine. Neuropharmacology **12**, 43—47 (1973 b)

Glick, S. D., Greenstein, S.: Possible modulating influence of frontal cortex on nigro-striatal function. Brit. J. Pharmacol. **49**, 316—321 (1973)

Glick, S. D., Jarvik, M. E.: Amphetamine, scopolamine and chlorpromazine interactions on delayed matching performance in monkeys. Psychopharmacologia (Berl.) **16**, 147—155 (1969)

Glick, S. D., Jerussi, T. P.: Spatial and paw preferences in rats: their relationship to rate-dependent effects of *d*-amphetamine. J. Pharmacol. exp. Ther. **188**, 714—725 (1974)

Glick, S. D., Jerussi, T. P., Waters, D. H., Green, J. P.: Amphetamine induced changes in striatal dopamine and acetylcholine levels and relationship to rotation (circling behaviour) in rats. Biochem. Pharmacol. **23**, 3223—3225 (1974)

Glick, S. D., Marsanico, R. G.: Comparative time-dependent changes in sensitivity to locomotor effects of *d*-amphetamine in mice with caudate, hippocampal or frontal cortical lesions. Arch. int. Pharmacodyn. **209**, 80—85 (1974 a)

Glick, S. D., Marsanico, R. G.: Shifting of *d*-amphetamine dose-response curve in rats with frontal cortical ablations. Psychopharmacologia (Berl.) **36**, 109—115 (1974 b)

Glick, S. D., Muller, R. U.: Paradoxical effects of low doses of *d*-amphetamine in rats. Psychopharmacologia (Berl.) **22**, 396—402 (1971)

Glick, S. D., Nakamura, R. K., Jarvik, M. E.: Recovery of function following frontal brain damage in mice. J. comp. physiol. Psychol. **76**, 454—459 (1971)

Glick, S. D., Waters, D. H., Milloy, S.: Depletion of hypothalamic norepinephrine by food deprivation and interaction with *d*-amphetamine. Res. Commun. chem. Path. Pharmacol. **6**, 775—778 (1973)

Glowinski, J.: Effects of amphetamine on various aspects of catecholamine metabolism in the central nervous system of the rat. In: Costa, E., Garattini, S. (Eds.): International symposium on amphetamines and related compounds, pp. 301—316. New York: Raven Press 1970 a

Glowinski, J.: Release of monoamines in the central nervous system. Bayer Symposium II. New aspects of storage and release mechanism of catecholamines, pp. 237—247. Schüman, H.J., Kroneberg, G. (Eds.) Berlin-Heidelberg-New York: Springer 1970 b
Glowinski, J., Axelrod, J.: Inhibition of uptake of tritiated noradrenaline in the intact rat brain by imipramine and structually related compounds. Nature (Lond.) **204**, 1318—1319 (1964)
Glowinski, J., Axelrod, J.: Effect of drugs on the uptake, release and metabolism of H^3-norepinephrine in the rat brain. J. Pharmacol. exp. Ther. **149**, 43—49 (1965)
Glowinski, J., Axelrod, J., Iversen, L.: Regional studies of catecholamines in the rat brain. IV. Effects of drugs on the disposition and metabolism of H^3-norepinephrine and H^3-dopamine. J. Pharmacol. exp. Ther. **153**, 30—41 (1966 a)
Glowinski, J., Baldessarini, R. J.: Metabolism of norepinephrine in the central nervous system. Pharmacol. Rev. **18**, 1201—1238 (1966)
Glowinski, J., Iversen, L., Axelrod, J.: Storage and synthesis of norepinephrine in the reserpine-treated rat brain. J. Pharmacol. exp. Ther. **151**, 385—399 (1966 b)
Gluckman, M. I., Baum, T.: The pharmacology of iprindole, a new antidepressant. Psychopharmacologia (Berl.) **15**, 169—185 (1969)
Goetz, C., Klawans, H. L.: Studies on the interaction of reserpine, *d*-amphetamine, apomorphine and 5-hydroxytryptophan. Acta pharmacol (Kbh.) **34**, 119—130 (1974)
Gogolák, G., Liebeswar, G., Sumpf, Ch.: Differential action of eserine and methamphetamine on the limbic system. Arch. int. Pharmacodyn. **178**, 77—84 (1969)
Gokhale, S. D., Gulati, O. D., Udwadia, B. P.: Antagonism of the adrenergic neurone blocking action of guanethidine by certain antidepressant and antihistamine drugs. Arch. int. Pharmacodyn. **160**, 321—329 (1966)
Gold, D., Reinert, H.: The depolarizing and blocking action of some sympathomimetic amines in the cat's superior cervical ganglion. J. Physiol. (Lond.) **151**, 3—4 (1960)
Goldberg, M. E., Ciofalo, V. B.: Alteration of the behavioral effects of amphetamine by agents which modify cholinergic function. Psychopharmacologia (Berl.) **14**, 142—149 (1969)
Goldberg, M. E., Salama, A. I.: Amphetamine toxicity and brain monoamines in three models of stress. Toxicol. appl. Pharmacol. **14**, 447—456 (1969)
Goldberg, S. R.: Comparable behavior maintained under fixed ratio and second-order schedules of food presentation, cocaine injection or *d*-amphetamine injection in the squirrel monkey. J. Pharmacol. exp. Ther. **186**, 18—30 (1973)
Goldstein, L., Gardocki, J. F., Mundschenk, D. C., O'Brien, G.: The effect of psychotropic drugs on the occurrence of paradoxical sleep in rats and cats. Fed. Proc. **26**, 506 (1967)
Goldstein, M., Anagnoste, B.: The conversion in vivo of *d*-amphetamine to (+)-p-hydroxynorephedrine. Biochim. biophys. Acta (Amst.) **107**, 166—168 (1965)
Goldstein, M., McKereghan, M. R., Lauber, E.: The stereospecificity of the enzymatic amphetamine β-hydroxylation. Biochim. biophys. Acta (Amst.) **89**, 191—193 (1964)
Gollub, L. R., Brady, J. V.: Behavioural pharmacology. Ann. Rev. Pharmacol. **5**, 235—262 (1965)
Gollub, L. R., Mann, W. G. Jr.: The interaction of amphetamine and body weight on a food-reinforced operant. Psychopharmacologia (Berl.) **15**, 64—72 (1969)
Gomer, F. E., Jakubczak, L. F.: Dose-dependent selective facilitation of response-contingent light-onset behavior by *d*-amphetamine. Psychopharmacologia (Berl.) **34**, 199—208 (1974)
Gonzalez-Barón, S., Jiménez-Vargas, J., Lopez, G. G.: Cambios en el ciclo ovárico por anfetamina y reserpina. Rev. Med. Univ. Navarra **15**, 251—259 (1971)
Görlitz, B.-D., Frey, H.-H.: Central monoamines and antinociceptive drug action. Europ. J. Pharmacol. **20**, 171—180 (1972)
Graham, A. W., Aghajanian, G. K.: Amphetamine—effects on single cell activity in a catecholamine nucleus, the locus coeruleus. Nature (Lond.) **234**, 100 (1971)
Graham, J. M., Schreiber, R. A., Zemp, J. W.: Effect of *d*-amphetamine sulphate on susceptibility to audiogenic seizures in DBA-2J mice. Behav. Biol. **10**, 183—190 (1974)
Grana, E., Lilla, L.: The inhibition of amine oxidase and the central stimulating action of the stereoisomeric amphetamines and *l*-phenylethylamines. Brit. J. Pharmacol. **14**, 501—504 (1959)
Grana, E., Sossi, D.: Ricerche sull'attivita' centrale di composti ad attivita' β-simpatolitica. Farmaco, Ed. sci. **22**, 582—589 (1967)
Green, A. L.: Inhibition of rat and mouse brain monoamine oxidase by (+)-amphetamine. Biochem. J. **121**, 37 (1971)

Green, T., Harvey, J. A.: Enhancement of amphetamine action after interruption of ascending serotonergic pathways. J. Pharmacol. exp. Ther. **190**, 109—117 (1974)

Greenblatt, E. N., Osterberg, A. C.: Correlations of activating and lethal effects of excitatory drugs in grouped and isolated mice. J. Pharmacol. exp. Ther. **131**, 115—119 (1961)

Greengard, P., Kebabian, J. W.: Role of cyclic AMP in synaptic transmission in the mammalian peripheral nervous system. Fed. Proc. **33**, 1059—1067 (1974)

Griffiths, D., Wahlsten, D.: Interacting effects of handling and *d*-amphetamine on avoidance learning. Pharmacol. Biochem. Behav. **2**, 439—441 (1974)

Groppetti, A., Costa, E.: *d*-Amphetamine (A): metabolites and depletion of brain and heart norepinephrine (NE) in guinea pig and rat. Fed. Proc. **28**, 795 (1969a)

Groppetti, A., Costa, E.: Tissue concentrations of p-hydroxynorephedrine in rats injected with *d*-amphetamine: effect of pretreatment with desipramine. Life Sci. **8**, 653—665 (1969b)

Groppetti, A., Misher, A., Naimzada, M., Revuelta, A., Costa, E.: Evidence that in rat 1-benzyl-β-methoxy-3-trifluormethyl-phenethylamine (SK&F 1-39728) dissociates anorexia from central stimulation and action on brain monoamines stores. J. Pharmacol. exp. Ther. **182**, 464—473 (1972a)

Groppetti, A., Zambotti, F., Mantegazza, P.: Effect of acute and chronic *d*-amphetamine treatment on behaviour and on brain biogenic amines in guinea-pigs. Psychopharmacologia (Berl.) **26**, Suppl. 35 (1972b)

Grossman, S. P., Sclafani, A.: Sympathomimetic amines. In: Furchgott, E. (Ed.): Pharmacological and biophysical agents and behaviour, pp. 269—344. London-New York: Academic Press 1971

Groves, P. M., Rebec, G. V., Segal, D. S.: The action of *d*-amphetamine on spontaneous activity in the caudate nucleus and reticular formation of the rat. Behav. Biol. **11**, 33—47 (1974)

Gudetsky, G. A., McCall, R. B., Chiueh, C. C., Moore, K. E.: Lack of effect of amantadine on *d*-amphetamine induced release of dopamine from cat brain in vivo. Res. Commun. chem. Path. Pharmacol. **9**, 653—660 (1974)

Guilleux, H., Peterfalvi, M.: Le comportement de rotation après lésion unilatérale du striatum analysé à l'aide d'un rotomètre. J. Pharmacol. (Paris) **5**, 63—74 (1974)

Gunn, J. A., Gurd, M. R.: The action of some amines related to adrenaline. Cyclohexylalkylamines. J. Physiol. (Lond.) **97**, 453—470 (1940)

Gunn, J. A., Gurd, M. R., Sachs, I.: The action of some amines related to adrenaline: methoxyphenylisopropylamines. J. Physiol. (Lond.) **95**, 485—500 (1939)

Gunne, L.-M., Lewander, T.: Long-term effects of some dependence-producing drug on the brain monoamines. Molecular basis of some aspects of mental activity, Vol. 2, pp. 75—81. London-New York: Academic Press 1967

Gunne, L.-M., Lewander, T.: Brain catecholamines during chronic amphetamine intoxication. The Addictive States, Vol. XLVI, pp. 106—116. Baltimore: Williams and Wilkins Company 1968

Gupta, B. D., Dandiya, P. C., Gupta, M. L., Gabba, A. K.: An examination of the effect of central nervous system stimulant and antidepressant drugs on open field performance in rats. Europ. J. Pharmacol. **13**, 341—346 (1971)

Gupta, B. D., Holland, H. C.: An examination of the effects of stimulant and depressant drugs on escape/avoidance conditioning in strains of rat selectively bred for emotionality/nonemotionality intertrial activity. Int. J. Neuropharmacol. **8**, 227—234 (1969)

Gupta, B. D., Holland, H. C.: An examination of the effects of stimulant and depressant drugs on escape/avoidance conditioning in strains of rats selectively bred for emontionality/nonemotionality: a multivariate analysis of the effects of drugs on conditioned avoidance responses and intertrial activity. Neuropharmacology **11**, 23—30 (1972a)

Gupta, B. D., Holland, H. C.: Emotion as a determinant of the effects of drugs and their combination on different components of behaviour in rats. Neuropharmacology **11**, 31—38 (1972b)

Haas, H. T. A.: Ascorbinsäure und Fieber. Naunyn-Schmiedebergs Arch. exp. Path. Pharmak. **192**, 331—350 (1939)

Haefely, W., Jalfre, M., Monachon, M.-A.: NE-neurons and phasic sleep phenomena. Frontiers in catecholamine research, pp. 773—775. Usdin, E., Snyder, S. (Eds.). New York: Pergamon Press 1973

Haffner, F.: Zur Pharmakologie und Praxis der Stimulantien. Klin. Wschr. **17**, 1310—1311 (1938)

Häggendal, J., Hamberger, B.: Quantitative in vitro studies on noradrenaline uptake and its inhibition by amphetamine, desipramine and chlorpromazine. Acta physiol. scand. **70**, 277—280 (1967)

Hajos, G. J. T., Garattini, S.: A note on the effect of (+) and (−)-amphetamine on lipid metabolism. J. Pharm. Pharmacol. **25**, 418—419 (1973)

Halliwell, G., Quinton, R. H., Williams, F. E.: A comparison of imipramine, chlorpromazine and related drugs in various tests involving antonomic functions and antagonism of reserpine. Brit. J. Pharmacol. **23**, 330—350 (1964)

Halpern, B. N., Drudi-Baracco, C., Bessirard, D.: Toxicité de groupe par l'amphétamine et action de la reserpine, de la chlorpromazine et des inhibiteurs de la monoamineoxydase. C.R. Soc. Biol. (Paris) **156**, 769—773 (1962a)

Halpern, B. N., Drudi-Baracco, C., Bessirard, D.: Mise en évidence de la "toxicité de groupe" de certains inhibiteurs de la monoamineoxydase. Comparaison de leur mode d'action avec celui de l'amphétamine. C.R. Soc. Biol. (Paris) **156**, 1238—1245 (1962b)

Halpern, B. N., Morard, J. C., Drudi-Baracco, C.: Action protectrice de la réserpine sur les lésions de dégénérescence myocardique aigue provoquées par la *dl*-amphétamine et suppression de cette action par les inhibiteurs de la monoamineoxydase. C.R. Soc. Biol. (Paris) **156**, 773—779 (1962c)

Hamberger, B., Malmfors, T.: Uptake and release of alphamethyl-noradrenaline in vitro after reserpine treatment. A histochemical study. Acta physiol. scand. **70**, 412—418 (1967)

Handley, C. A., Sweeney, H. M.: The relation of central nervous system depression and stimulation to biochemical changes occurring in cerebral blood. Amer. J. Physiol. **133**, 314 (1941)

Hanna, C.: Tachyphylaxis V. Actions of optical isomers of sympathomimetic amines on rabbit aortic strips. Arch. int. Pharmacodyn. **129**, 191—200 (1960)

Hanna, C., Upton, P. D., Chambers, W. F.: Comparative effects of *d*-, *l*-amphetamine and 1-phenyl-I-(Piperidyl-2′-)-I-acetoxy-I-methane (R.P.-8228) in antagonizing barbiturate hypnosis. Arch. int. Pharmacodyn. **145**, 553—564 (1963)

Hanson, H. M., Witoslawski, J. J., Campbell, E. H.: Drug effects in squirrel monkeys trained on a multiple schedule with a punishment contingency. J. exp. Anal. Behav. **10**, 565—569 (1967)

Hanson, L. C. F.: Evidence that the central action of amphetamine is mediated via catecholamines. Psychopharmacologia (Berl.) **9**, 78—80 (1966)

Hanson, L. C. F.: Evidence that the central action of (+)-amphetamine is mediated via catecholamines. Psychopharmacologia (Berl.) **10**, 289—297 (1967)

Harris, J. E., Baldessarini, R. J.: Uptake of (^{3}H)-catecholamines by homogenates of rat corpus striatum and cerebral cortex: effect of amphetamine analogues. Neuropharmacology **12**, 669—679 (1973a)

Harris, J. E., Baldessarini, R. J.: Amphetamine-induced inhibition of tyrosine-hydroxylation in homogenates of rat corpus striatum. J. Pharm. Pharmacol. **25**, 755—757 (1973b)

Harris, R. T., Balster, R. L.: An analysis of the function of drugs in the stimulus control of operant behaviour. In: Thompson, T., Pickens, R. (Eds.): Stimulus properties of drugs, pp. 111—132. New York: Appleton Century Crofts 1971

Harris, S. C., Ivy, A. C., Searle, L. M.: The mechanism of amphetamine-induced loss of weight; a consideration of the theory of hunger and appetite. J. Amer. med. Ass. **134**, 1468—1475 (1947)

Harrison, D. C., Chidsey, C. A., Braunwald, E.: The potentiation of the cardiovascular responses to sympathomimetic amines by reserpine. J. Pharmacol. exp. Ther. **141**, 22—29 (1963)

Harrison, J. M., Abelson, R. M.: The maintenance of behaviour by the termination and onset of intense noise. J. exp. Anal. Behav. **2**, 23—42 (1959)

Harrisson, J. W. E., Ambrus, C. M., Ambrus, J. L.: Tolerance of rats towards amphetamine and methamphetamine. J. Amer. pharm. Ass. **XLI**, 539—541 (1952)

Hart, M., Dienske, H., Udo de Haes, H. A., Noach, E. L.: Ethological investigation of the effect of some psychopharmacological drugs on the behaviour of rats. Acta physiol. pharmacol. neerl. **15**, 403 (1969)

Hartung, W. H., Munch, J. C.: Amino alcohols. VI. The preparation and pharmacodynamic activity of four isomeric phenylpropylamines. J. Amer. chem. Soc. **53**, 1875—1879 (1931)

Harvey, S. C., Sulkowski, T. S., Weenig, D. J.: Effect of amphetamines on plasma catecholamines. Arch. int. Pharmacodyn. **172**, 301—322 (1968)

Hasselager, E., Rolinski, Z., Randrup, A.: Specific antagonism by dopamine inhibitors of items of amphetamine induced aggressive behaviour. Psychopharmacologia (Berl.) **24**, 485—495 (1972)

Hatch, R. C.: Experiments on antagonism of barbiturate anaesthesia with adrenergic serotonergic and cholinergic stimulants given alone and in combination. Amer. J. vet. Res. **34**, 1321—1331 (1973)

Hauschild, F.: Zur Pharmakologie des l-phenyl-2-methylaminopropane (Pervitin). Naunyn-Schmiedebergs Arch. exp. Path. Pharmak. **191**, 465—481 (1938)

Hawke, D. M., Chesher, G. B.: The effect of amphetamine and tyramine on the guinea-pig urinary bladder. Agents Actions **1/4**, 164—171 (1970)

Hearst, E.: Effects of *d*-amphetamine on behavior reinforced by food and water. Psychol. Rep. **8**, 301—309 (1961)

Hearst, E., Vane, J. R.: Some effects of *d*-amphetamine on the behaviour of pigeons under intermittent reinforcement. Psychopharmacologia (Berl.) **12**, 58—67 (1967)

Hearst, E., Whalen, R. E.: Facilitating effects of *d*-amphetamine on discriminated-avoidance performance. J. comp. physiol. Psychol. **56**, 124—128 (1963)

Heffner, T. G., Drawbaugh, R. B., Zigmond, M. J.: Amphetamine and operant behaviour in rats: Relationship between drug effect and control response rate. J. comp. physiol. Psychol. **86**, 1031—1043 (1974)

Heim, F., Leuschner, F., Estler, K. J.: Beziehungen zwischen Funktionen und Stoffwechsel des Zentralnervensystems nach Pervitin. Experientia (Basel) **13**, 462—464 (1957)

Heimstra, N. W.: Effects of amphetamine sulfate (benzedrine) on the behavior of paired rats in a competitive situation. Psychol. Rec. **12**, 25—34 (1962 a)

Heimstra, N. W.: Social influence on the response to drugs. I. Amphetamine sulfate. J. Psychol. **53**, 233—244 (1962 b)

Heimstra, N. W., McDonald, A.: Social influence on the response to drugs. III. Response to amphetamine sulfate as a function of age. Psychopharmacologia (Berl.) **3**, 212—218 (1962)

Heise, G. A., Boff, E.: Stimulant action of *d*-amphetamine in relation to test compartment dimensions and behavioural measure. Neuropharmacology **10**, 259—266 (1971)

Heise, G. A., Lilie, N. L.: Effects of scopolamine, atropine and *d*-amphetamine on internal and external control of responding on nonreinforced trials. Psychopharmacologia (Berl.) **18**, 38—49 (1970)

Hemsworth, B. A., Neal, M. J.: The effect of stimulant drugs on the release of acetylcholine from the cerebral cortex. Brit. J. Pharmacol. **32**, 416—417 (1968 a)

Hemsworth, B. A., Neal, M. J.: The effect of central stimulant drugs on acetylcholine release from rat cerebral cortex. Brit. J. Pharmacol. **34**, 543—550 (1968 b)

Hendley, E. D., Snyder, S. H.: Stereoselectivity of catecholamine uptake in noradrenergic and dopaminergic peripheral organs. Europ. J. Pharmacol. **19**, 56—66 (1972)

Hendry, D. P., Toller, C. van: Fixed ratio punishment with continuous reinforcement. J. exp. Anal. Behav. **7**, 293—310 (1964)

Herman, Z. S.: Influence of some psychotropic and adrenergic blocking agents upon amphetamine stereotyped behaviour in white rats. Psychopharmacologia (Berl.) **11**, 136—142 (1967)

Herman, Z. S., Kmieciak-Kolada, K., Drybanski, A., Sokola, A., Trzeciak, H., Chrusciel, T. L.: The influence of 1-(o.allylphenoxy)-3-isopropylamino-2-propanol hydrochloride (alprenolol) in the central nervous system of the rat. Psychopharmacologia (Berl.) **21**, 66—75 (1971 a)

Herman, Z. S., Trzeciak, H., Chrusciel, T. L., Kmieciak-Kolada, K., Drybanski, A., Sokola, A.: The influence of prolonged amphetamine treatment and amphetamine withdrawal on brain biogenic amine content and behaviour in the rat. Psychopharmacologia (Berl.) **21**, 74—81 (1971 b)

Hernodon, J. G., Neill, D. B.: Amphetamine reversal of sexual impairment following anterior hypothalamic lesions in female rats. Pharmacol. Biochem. Behav. **1**, 285—288 (1974)

Hertting, G., Axelrod, J., Whitby, G. L.: Effect of drugs on the uptake and metabolism of H^3-norepinephrine. J. Pharmacol. exp. Ther. **134**, 146—153 (1961)

Herz, A.: Drugs and the conditioned avoidance response. Int. Rev. Neurobiol. **2**, 229—277 (1961)

Hiebel, G., Bonvallet, M., Huvé, P., Dell, P.: Analyse neurophysiologique de l'action centrale de la *d*-amphétamine (maxiton). Sem. Hôp. Paris **30**, 1880—1887 (1954)

Higuchi, H., Matsuo, T., Shimamoto, K.: Effects of methamphetamine and cocaine on the depletion of catecholamine of the brain, heart and adrenal gland in rabbit by reserpine. Jap. J. Pharmacol. **12**, 48—56 (1962)

Hill, H. E., Bell, E. C., Wikler, A.: Reduction of conditioned suppression: actions of morphine compared with those of amphetamine, pentobarbital, nalorphine, cocaine, LSD-25 and chlorpromazine. Arch. int. Pharmacodyn. **165**, 212—226 (1967)

Hill, H. F., Horita, A.: Amphetamine-induced hyperthermia in rabbits. Pharmacologist **12**, 197 (1970)

Hill, H. F., Horita, A.: Inhibition of (+)-amphetamine hyperthermia by blockade of dopamine receptors in rabbits. J. Pharm. Pharmacol. **23**, 715—717 (1971)

Hitzemann, R. J., Loh, H. H.: Effect of p-hydroxynorephedrine (PONE) on the conversion of ^{3}H-tyrosine to ^{3}H-catecholamine. Pharmacology (Basel) **8**, 361—367 (1972)

Hitzemann, R. J., Loh, H. H.: Effect of *d*-amphetamine on the turnover, synthesis and metabolism of brain phosphatidylcholine. Biochem. Pharmacol. **22**, 2731—2741 (1973)

Hitzemann, R. J., Loh, H. H.: Further studies on the effect of p-hydroxynorephedrine (PONE) on norepinephrine metabolism in the rat brain. Europ. J. Pharmacol. **27**, 89—98 (1974)

Hitzemann, R. J., Loh, H. H., Carves, F. B., Domino, E. F.: The use of *d*-amphetamine pellet implantation as a model for *d*-amphetamine tolerance in the mouse. Psychopharmacologia (Berl.) **30**, 227—240 (1973)

Hitzemann, R. J., Loh, H. H., Domino, E. F.: Effects of *d*-amphetamine (A) and p-methoxyamphetamine (PNA) on adrenergic and serotonergic metabolism. Pharmacologist **12**, 197 (1970)

Ho, A. K. S., Gershon, S.: Drug-induced alterations in the activity of rat brain cholinergic enzymes. I. In vitro and in vivo effect of amphetamine. Europ. J. Pharmacol. **18**, 195—200 (1972)

Hodos, W., Ross, G. S., Brady, J. V.: Complex response patterns during temporally spaced responding. J. exp. Anal. Behav. **5**, 473—479 (1962)

Höhn, R., Lasagna, L.: Effects of aggregation and temperature on amphetamine toxicity in mice. Psychopharmacologia (Berl.) **1**, 210—220 (1960)

Hoffmeister, F., Goldberg, S. R.: A comparison of chlorpromazine, imipramine, morphine and *d*-amphetamine self-administration in cocaine-dependent rhesus monkeys. J. Pharmacol. exp. Ther. **187**, 8—14 (1973)

Hollister, A. S., Breese, G. R., Cooper, B. R.: Comparison of tyrosine hydroxylase and dopamine-β-hydroxylase inhibition and the effects of various 6-hydroxydopamine treatments on *d*-amphetamine induced motor activity. Psychopharmacologia (Berl.) **36**, 1—16 (1974)

Holm, T., Huus, I., Kopf, R., Nielsen, I. M., Petersen, D. V.: Pharmacology of a series of nuclear substituted phenyl-tertiary-butylamines with particular reference to anorexigenic and central stimulating properties. Acta pharmacol. (Kbh.) **17**, 121—136 (1960)

Holtzman, S. G.: Behavioral effects of separate and combined administration of naloxone and *d*-amphetamine. J. Pharmacol. exp. Ther. **189**, 51—60 (1974)

Holtzman, S. G., Jewett, R. E.: The role of brain norepinephrine in the anorexic effects of dextroamphetamine and monoamine oxidase inhibitors in the rat. Psychopharmacologia (Berl.) **22**, 151—161 (1971)

Horita, A., Hill, H. F.: Hallucinogens, amphetamines and temperature regulation. The pharmacology and thermoregulation, pp. 417—431. Schönbaum, S., Lomax, P. (Eds.) Basel: S. Karger 1973

Horn, A. S., Coyle, J. T., Snyder, S. H.: Catecholamine uptake by synaptosomes from rat brain. Structure-activity relationships of drugs with differential effects on dopamine and norepinephrine neurons. Molec. Pharmacol. **7**, 66—80 (1971)

Horn, A. S., Cuello, A. C., Miller, R. J.: Dopamine in the mesolimbic system of the rat brain: endogenous levels and the effects of drugs on the uptake mechanism and stimulation of adenylate cyclase activity. J. Neurochem. **22**, 265—270 (1974)

Horn, A. S.: Snyder, S. H.: Steric requirements for catecholamine uptake by rat brain synaptosomes: studies with rigid analogs of amphetamine. J. Pharmacol. exp. Ther. **180**, 523—530 (1972)

Horovitz, Z. P., Chow, M. I., Carlton, P. I.: Selfstimulation of the brain by cats. Technique and preliminary drug effects. Psychopharmacologia (Berl.) **3**, 449—454 (1962a)

Horovitz, Z. P., Chow, H. I., Carlton, P. I.: Selfstimulation of the brain by cats. Effects of imipramine, amphetamine and chlorpromazine. Psychopharmacologia (Berl.) **3**, 455—462 (1962b)

Horovitz, Z. P., Piala, J. J., High, J. P., Burke, J. C., Leaf, R. C.: Effects of drugs on the mouse-killing (Muricide) test and its relationship to amygdaloid function. Int. J. Neuropharmacol. **5**, 405—411 (1966)

Houser, V. P.: Modulation of avoidance behavior in squirrel monkeys after chronic administration and withdrawal of *d*-amphetamine of α-methyl-p-tyrosine. Psychopharmacologia (Berl.) **28**, 213—234 (1973)

Hoyer, I., Zwieten, P. A. van: The centrally induced fall in blood pressure after the infusion of amphetamine and related drugs into the vertebral artery of the cat. J. Pharm. Pharmacol. **23**, 892—893 (1971)

Hoyer, I., Zwieten, P. A. van: The central hypotensive action of amphetamine, ephedrine, phentermine, chlorphentermine and fenfluramine. J. Pharm. Pharmacol. **24**, 452—458 (1972)

Huang, J.-T., Ho, B. T.: Differentiation of pressor and behavioral effects of *d*-amphetamine and 2,5-dimethoxy-4-methylamphetamine (STP) by cinanserin. Canad. J. Physiol. Pharmacol. **51**, 976—980 (1973)

Hulme, E. C., Hill, R., North, M., Kibby, M. R.: Effects of chronic administration of drugs which modify neurotransmitter reuptake, storage and turn-over on levels of tyrosine and tryptophan hydroxylase in rat brain. Biochem. Pharmacol. **23**, 1393—1404 (1974)

Huston, J. P., Borbély, A. A.: The thalamic rat: general behaviour, operant learning with rewarding hypothalamic stimulation, and effects of amphetamine. Physiol. Behav. **12**, 433—448 (1974)

Hutchins, D. A., Rogers, K. J.: Physiological and drug-induced changes in the glycogen content of mouse brain. Brit. J. Pharmacol. **39**, 9—25 (1970)

Hutchins, D. A., Rogers, K. J.: Effect of receptor blocking drugs on the depletion of brain glycogen by amphetamine. Brit. J. Pharmacol. **43**, 504—513 (1971)

Hutchins, D. A., Rogers, K. J.: Effect of depletion of cerebral monoamines on the concentration of glycogen and on amphetamine-induced glycogenolysis in the brain. Brit. J. Pharmacol. **48**, 19—29 (1973)

Innes, I. R.: Action of dexamphetamine on 5-hydroxytryptamine receptors. Brit. J. Pharmacol. **21**, 427—455 (1963)

Innes, I. R., Kohli, J. D.: Excitatory action of sympathomimetic amines on 5-hydroxytryptamine receptors of gut. Brit. J. Pharmacol. **35**, 383—393 (1969)

Irwin, S., Slabok, K., Thomas, G.: Individual differences I. Correlation between control locomotor activity and sensitivity to stimulant and depressant drugs. J. Pharmacol. exp. Ther. **123**, 206—211 (1958)

Isaac, W.: A study of the relationship between the visual system and the effects of *d*-amphetamine. Physiol. Behav. **6**, 157—160 (1971)

Isaac, W., Troelstrup, R.: Opposite effect of illumination and *d*-amphetamine upon activity in the squirrel monkey (Saimiri) and owl monkey (Aotes). Psychopharmacologia (Berl.) **15**, 260—264 (1969)

Isachenko, V. B., Nikulcheva, N. G., Nargornev, V. A.: Action of some phenamine derivatives on the lipids metabolism. Farmakol. i Toksikol. **34**, 435 (1971)

Iversen, L. L.: The uptake and storage of noradrenaline in sympathetic nerves. London: Cambridge University Press 1967

Iversen, S. D.: The effect of surgical lesions to frontal cortex and substantia nigra on amphetamine responses in rats. Brain Res. **31**, 295—311 (1971)

Iversen, S. D., Wilkinson, S., Simpson, B.: Enhanced amphetamine response after frontal cortex lesions in the rat. Europ. J. Pharmacol. **13**, 387—390 (1971)

Ivy, A. C., Goetzl, F. R.: *d*-Desoxyephedrine a review. War Med. (Chic.) **3**, 60—77 (1943)

Iwata, H., Watanabe, K., Nishikawa, T., Ohashi, K.: Effects of drugs on behaviour, heart rate and catecholamine levels in thiamine deficient rats. Europ. J. Pharmacol. **6**, 83—89 (1969)

Izquierdo, I., Vasquez, B. J., Nasello, A. G.: Indirect effect of drugs on hippocampal field responses to commissural and to subicular stimulation. Pharmacology (Basel) **1**, 178—182 (1968)

Jacob, J., Michaud, G.: Actions de divers agents pharmacologiques sur les temps d'épuisement et le comportement de souris nageant à 20° C. Arch. int. Pharmacodyn. **133**, 101—115 (1961)

Jacquot, C., L'Hermite, P., Cohen, Y., Valette, G.: Distribution sub-cellulaire de la noradrénaline-^{3}H, de l'amphétamine-^{3}H et de l'hydroxyéphédrine-^{3}H dans le canal déférent et la rate du rat. C.R. Acad. Sci. (Paris) **273 D**, 1640—1642 (1971)

Jacquot, C., L'Hermite, P., Rapin, J.R., Cohen, Y.: Effet de la sympathectomie chimique sur la distribution tissulaire de *dl*-méthyloctopamine-^{3}H. C.R. Soc. Biol. (Paris) **167**, 221—225 (1973a)

Jacquot, C., Rapin, J., El Guedri, H., Cohen, Y.: Mécanism de fixation de l'hydroxyamphétamine-^{3}H dans des coupes de rate du rat. C.R. Acad. Sci. (Paris) **276**, 2851—2853 (1973b)

Jacquot, C., Rapin, J., Renault, H., Cohen, Y.: Effect de l'amphétamine et des ses métabolites sur l'activité de la dopamine-β-hydroxylase. C.R. Soc. Biol. (Paris) **168**, 479—484 (1974)

Janssen, P. A. J., Niemegeers, J. E., Schellekens, K. H. L.: Is it possible to predict the clinical effects of neuroleptic drugs (major tranquillizers) from animal data? Arzneimittel-Forsch. **15**, 104—117 (1965)

Janssen, P. A. J., Niemegeers, J. E., Schellekens, K. H. L., Lenaerts, F. M.: Is it possible to predict the clinical effects of neuroleptic drugs (major tranquillizers) from animal data? Arzneimittel-Forsch. **17**, 841—854 (1967)

Janssen, P. A. J., Niemegeers, J. E., Schellekens, K. H. L., Dresse, A. D., Lenaerts, F. M., Pinchard, A., Schaper, W. K. A., Nueten, J. M. van, Verbruggen, F. J.: Pimozide, a chemically novel, highly potent orally long-acting neuroleptic drug. Arzneimittel-Forsch. **18**, 261—287 (1968)

Javoy, F., Agid, Y., Bouvet, D., Glowinski, J.: In vivo estimation of tyrosine hydroxylation in the dopaminergic terminals of the rat neostriatum. J. Pharm. Pharmacol. **26**, 179—185 (1974)

Javoy, F., Hamon, M., Glowinski, J.: Disposition on newly synthesized amines in cell bodies and terminals of central catecholaminergic neurons. (1) Effect of amphetamine and thioproperazine on the metabolism of CA in the caudate nucleus, the substantia nigra and the ventromedial nucleus of the hypothalamus. Europ. J. Pharmacol. **10**, 178—188 (1970).

Javoy, F., Thierry, A. M., Kety, S. S., Glowinski, J.: The effect of amphetamine on the turnover of brain norepinephrine in normal and stressed rats. Commun. behav. Biol. **I**, 43—48 (1968)

Jellinek, P.: Dual effect of dexamphetamine on body temperature in the rat. Europ. J. Pharmacol. **15**, 389—392 (1971)

Jennings, W. A. Jr.: Effects of *dl*-amphetamines on open-field behaviours in the rat. Psychol. Rep. **22**, 345—346 (1968)

Jerussi, T. P., Glick, S. D.: Amphetamine-induced rotation in rats without leisons. Neuropharmacology **13**, 283—286 (1974)

Jespersen, J., Bonaccorsi, A.: Anti-amphetamine activity of fenfluramine and S 992 in the isolated tail artery of the rat. J. Pharm. Pharmacol. **21**, 776—777 (1969a)

Jespersen, S., Bonaccorsi, A.: Effect of fenfluramine on the *d*-amphetaminc toxicity in mice. Europ. J. Pharmacol. **8**, 364—368 (1969b)

Jespersen, S., Scheel-Krüger, J.: Evidence for a difference in mechanism of action between fenfluramine- and amphetamine-induced anorexia. J. Pharm. Pharmacol. **25**, 49—54 (1973)

Jewett, R. E., Norton, S.: Effect of some stimulant and depressant drugs on sleep cycles of cats. Exp. Neurol. **15**, 463—474 (1966)

John, E. R., Wenzel, B. M., Tschirgi, R. D.: Differential effects on various conditioned responses in cats caused by intraventricular and intramuscular injections of reserpine and other substances. J. Pharmacol. exp. Ther. **123**, 193—205 (1958)

Johnson, F. H., Waite, K.: Apparent delayed enhancement of memory following posttrial methylamphetamine hydrochloride. Experientia (Basel) **27**, 1316—1317 (1971)

Jonas, W., Scheel-Krüger, J.: Amphetamine-induced stereotyped behaviour correlated with the accumulation of O-methylated dopamine. Arch. int. Pharmacodyn. **177**, 379—389 (1969)

Jones, B. E., Guynet, P., Chéramy, A., Gauchy, C., Glowinski, J.: The in vivo release of acetylcholine from cat caudate nucleus after morphological and surgical manipulations of dopaminergic nigrostriatal neurons. Brain Res. **64**, 355—369 (1973)

Jones, C. N., Hill, H. F., Harris, R. T.: Discriminative response control by *d*-amphetamine and related compounds in the rat. Psychopharmacologia (Berl.) **36**, 347—356 (1974)

Jonsson, J., Gunne, L.-M.: Interaction of fenfluramine with *d*-amphetamine-induced excitatory behaviour and hyperthermia. Europ. J. Pharmacol. **19**, 52—55 (1972)

Jonsson, J., Lewander, T.: Effects of diethyldithiocarbamate and ethanol on the in vivo metabolism and pharmacokinetics of amphetamine in the rat. J. Pharm. Pharmacol. **25**, 589—591 (1973)

Jori, A., Bernardi, D.: Effect of amphetamine and amphetamine-like drugs on homovanillic acid concentration in the brain. J. Pharm. Pharmacol. **21**, 694—697 (1969)

Jori,A., Bernardi,D.: Further studies on the increase of striatal homovanillic acid induced by amphetamine and fenfluramine. Europ. J. Pharmacol. **19**, 276—280 (1972)

Jori,A., Dolfini,E.: On the effect of anorectic drugs on striatum homovanillic acid in rats. Pharmacol. Res. Commun. **6**, 175—178 (1974)

Jori,A., Dolfini,E., Tognoni,G. Garattini,S.: Differential effects of amphetamine, fenfluramine and norfenfluramine stereoisomers on the increase of striatum homovanillic acid in rats. J. Pharm. Pharmacol. **25**, 315—318 (1973)

Jori,A., Garattini,S.: Interaction between imipramine-like agents and catecholamine-induced hyperthermia. J. Pharm. Pharmacol. **17**, 480—488 (1965)

Jori,A., Garattini,S.: Catecholamines metabolism and amphetamine effects on sensitive and insensitive mice. Frontiers in catecholamine research, pp. 939—941. Ed. Usdin,E., Snyder,S. (Eds.) New York: Pergamon Press 1973

Kaplan,E.R., Sapeika,N., Zwarenstein,H.: Effect of fenfluramine, dexamphetamine and diethylpropion on phosphoglucomutase, 3-phosphoglycerate kinase and pyruvate kinase of mouse muscle. Res. Commun. chem. Path. Pharmacol. **3**, 47—50 (1972)

Karli,P.: Troubles du comportement induits chez le rat par l'amphétamine et le phénidylate (ritaline). Arch. int. Pharmacodyn. **122**, 344—351 (1960)

Kasirsky,G.I., Zaidi,I.H., Tansy,M.F.: LD 50 and pathologic effects of acute and chronic administration of methamphetamine HCl in rabbits. Res. Commun. chem. Path. Pharmacol. **3**, 215—231 (1972)

Kay,H., Birren,J.E.: Swimming speed of the albino rat. II. Fatigue, practice and drug effects on age and sex differences. J. Geront. **13**, 378—385 (1958)

Keleman,K., Bovet,D.: Effect of drugs upon the defensive behaviour in rats. Acta physiol. Acad. Sci. hung. **19**, 143—154 (1961)

Kelfer,D.A., Rosen,A.J.: Effects of methamphetamine, pipradrol and methylphenidate on instrumental conditioning and spontaneous motor activity in the immature rat. Psychopharmacologia (Berl.) **35**, 317—326 (1974)

Kelleher,R.T., Cook,L.: Effects of *d*-amphetamine-meprobamate, phenobarbital, mephenesin or chlorpromazine on DRL and FR schedules of reinforcement with rats. J. exp. Anal. Behav. **2**, 267—314 (1959)

Kelleher,R.T., Fry,W., Deegan,J., Cook,L.: Effects of mebrobamate on operant behaviour in rats. J. Pharmacol. exp. Ther. **133**, 271—280 (1961)

Kelleher,R.T., Morse,W.H.: Escape behavior and punished behavior. Fed. Proc. **23**, 808—817 (1964)

Kelleher,R.T., Morse,W.H.: Determinants of the specificity of behavioral effects of drugs. Ergebn. Physiol. **60**, 1—56 (1968)

Kewitz,H., Reinert,H.: Wirkung verschiedener Sympathicomimetica auf die chemisch und elektrisch ausgelöste Erregung des oberen Halsganglions. Naunyn-Schmiedeberg's Arch. exp. Path. Pharmak. **222**, 311—314 (1954)

Key,B.J., Marley,E.: The effect of the sympathomimetic amines on behaviour and electrocortical activity of the chicken. Electroenceph. clin. Neurophysiol. **14**, 90—105 (1962)

Khan,A.U., Forney,R.B., Hughes,F.W.: Plasma free fatty acids in rats after shock as modified by centrally active drugs. Arch. int. Pharmacodyn. **151**, 466—474 (1964)

Khavari,K.A.: Effects of central versus intraperitoneal *d*-amphetamine administration of learned behavior. J. comp. physiol. Psychol. **68**, 226—234 (1969)

Khazan,N., Sawyer,C.H.: Mechanisms of paradoxical sleep as revealed by neurophysiologic and pharmacologic approaches in the rabbit. Psychopharmacologia (Berl.) **5**, 457—466 (1964)

Kiernan,C.C.: Modification of the effect of amphetamine sulphate by past experience in the hooded rat. Psychopharmacologia (Berl.) **8**, 23—31 (1965)

Kiessig,H.J.: Untersuchungen über die Wirkungsweise der Sympathomimetika. VII. Über die temperatursteigernde Wirkung einiger Sympathomimetika. Naunyn-Schmiedebergs Arch. exp. Path. Pharmak. **197**, 384—397 (1941)

Kiessig,H.J., Orzechowski,G.: Untersuchungen über die Wirkungsweise der Sympathomimetica VIII. Über die Beeinflussung der Schmerzempfindlichkeit durch Sympathomimetica. Naunyn-Schmiedebergs Arch. exp. Path. Pharmak. **197**, 390—404 (1941)

Killam,E.K.: Antiepileptic actions of amphetamine in the seizure-prone baboon. Proc. West. Pharmacol. Soc. **16**, 119—122 (1973)

Killam, E. K., Gangloff, H., Konigsmark, B., Killam, K. F.: The action of pharmacologic agents on evoked cortical activity. In: Biological psychiatry, pp. 53—70. New York: Grune and Stratton 1959

Kirby, M. J., Turner, P.: The effect of amphetamine on glucose uptake into the rat isolated hemidiaphragm. J. Pharm. Pharmacol. **26**, 463 (1974)

Kjellberg, B., Randrup, A.: Various forms of stereotyped, abnormal behaviour induced by amphetamine in non-human primates. Acta physiol. scand. **330**, Suppl. 65 (1969)

Kjellberg, B., Randrup, A.: The effect of amphetamine and pimozide, a neuroleptic, on the social behaviour of vervet monkeys (Cercopithecus aethiops). In: Vinar, O., Votava, Z., Bradley, P. B. (Eds.): Advances in neuropsychopharmacology, p. 305. Amsterdam: North Holland Publishing Comp. Prague: Avicenum, Czechoslovak Medical Press 1971

Kjellberg, B., Randrup, A.: Stereotypy with selective stimulation of certain items of behaviour observed in amphetamine-treated monkeys (Ceropithecus). Pharmakopsychiat. Neuropsychopharmacol. **5**, 1—12 (1972)

Kjellberg, B., Randrup, A.: Disruption of social behaviour of vervet monkeys (Cercopithecus) by low doses of amphetamines. Pharmakopsychiat. Neuropsychopharmacol. **6**, 287—293 (1973)

Kjellberg, B., Randrup, A.: Partial restoration by a neuroleptic (spiramide) of items of grooming behaviour suppressed by amphetamine. Arch. int. Pharmacodyn. **210**, 61—66 (1974)

Klawans, H. L., Moses, H., Beaulieu, D. M.: The influence of caffeine on *d*-amphetamine and apomorphine induced stereotyped behaviour. Life Sci. **14**, 1493—1500 (1974)

Klawans, H. L., Rubowits, R., Patel, B. C., Weiner, W. J.: Cholinergic and anticholinergic influences on amphetamine-induced stereotyped behavior. J. neurol. Sci. **17**, 303—308 (1972)

Klingenstein, R. J., Wallach, M. B., Gershon, S.: A comparison of pimozide and thioridazine as antagonists of amphetamine-induced stereotyped behaviour in dogs. Arch. int. Pharmacodyn. **203**, 67—71 (1973)

Klissiunis, N., Dosi, I.: L'action de la promethazine sur la fièvre provoquée chez le lapin par la phenylisopropylamine. Thérapie **14**, 745—748 (1959)

Knapp, S., Mandell, A. J., Geyer, M. A.: Effect of amphetamines on regional tryptophan hydroxylase activity and synaptosomal conversion of tryptophan to 5-hydroxytryptamine in rat brain. J. Pharmacol. exp. Ther. **189**, 676—689 (1974)

Knoll, J., Vajnovszky, B.: Motimeter, a new sensitive apparatus for the quantitative measurement of hypermotility caused by psychostimulants. Arch. int. Pharmacodyn. **130**, 141—154 (1961)

Knoll, J., Vizi, E. S., Knoll, B.: Development of tolerance to some substituted amphetamines in rats, cats, and rabbits. Proceedings of the Twelfth Meeting of the European Society for the Study of Drug Toxicity. Excerpta Medica International Congress Series **220**, 50—55 (1970)

Knowler, W. C., Ukena, T. E.: The effects of chlorpromazine, pentobarbital, chlordiazepoxide and *d*-amphetamine on rates of licking in the rat. J. Pharmacol. exp. Ther. **184**, 385—397 (1973)

Koda, L. Y., Gibb, J. W.: Adrenal and striatal tyrosine hydroxylase activity after methamphetamine. J. Pharmacol. exp. Ther. **185**, 42—48 (1973)

Kopin, I. J.: Storage and metabolism of catecholamines: the role of monoamine oxidase. Pharmacol. Rev. **16**, 179—191 (1964)

Kopin, I. J., Fischer, J. E., Musacchio, J. M., Horst, W. D., Weise, W. K.: "False neurochemical transmitters" and the mechanism of sympathetic blockade by monoamine oxidase inhibitors. J. Pharmacol. exp. Ther. **147**, 186—193 (1965)

Kopin, I. J., Gordon, E. K.: Metabolism of administered and drug-released norepinephrine-7-H^3 in the rat. J. Pharmacol. exp. Ther. **140**, 207—216 (1963)

Koppanyi, T., Maling, H. M.: Salivation in mice as an index of adrenergic activity. II. The effects of atropines and ganglionic blocking agents on adrenergic salivation and temperature responses in mice. Arch. int. Pharmacodyn. **199**, 333—343 (1972)

Koppanyi, T., Maling, H. M.: Temperature-dependent toxicity of adrenergic agonists in mice as a basis for treating *d*-amphetamine poisoning. Proc. Soc. exp. Biol. (N.Y.) **144**, 575—580 (1973)

Korf, J., Bunney, B. S., Aghajanian, G. K.: Noradrenergic neurons morphine inhibition of spontaneous activity. Europ. J. Pharmacol. **25**, 165—169 (1974)

Kosman, M. E.: Effect of amphetamine on the learning and performance of mice in a swimming maze. Proc. Soc. exp. Biol. (N.Y.) **115**, 728—731 (1964)

Kosman, M. E., Unna, K. R.: Effects of chronic administration of the amphetamines and other stimulants on behavior. Clin. Pharmacol. Ther. **9**, 240—254 (1968)

Kostowski, W., Czlonkowski, A.: The activity of some neuroleptic drugs and amphetamine in normal and isolated rats. Pharmacology (Basel) **10**, 82—87 (1973)

Krebs, H., Bindra, D.: Noradrenaline and "chemical coding" of hypothalamic neurones. Nature (Lond.) New Biol. **229**, 178—180 (1971)

Krebs, H., Bindra, D., Campbell, J. F.: Effects of amphetamine on neuronal activity in the hypothalamus. Physiol. Behav. **4**, 685—691 (1969)

Krieckhaus, E. E.: Decrements in avoidance behaviour following mamillothalamic tractotomy in rats and subsequent recovery with *d*-amphetamine. J. comp. physiol. Psychol. **60**, 31—35 (1965)

Krieckhaus, E. E., Miller, N. E., Zimmerman, P.: Reduction of freezing behaviour and improvement of shock avoidance by *d*-amphetamine. J. comp. physiol. Psychol. **60**, 36—40 (1965)

Krivanek, J. A., McGaugh, J. L.: Facilitating effects of pre- and posttrial amphetamine administration on discrimination learning in mice. Agents Actions **1**, 36—42 (1969)

Kruk, Z. L.: The effect of drugs acting on dopamine receptors on the body temperature of the rat. Life Sci. **11**, 845—850 (1972)

Kruk, Z. L.: Dopamine and 5-hydroxytryptamine inhibit feeding in rats. Nature (Lond.) New Biol. **246**, 52—53 (1973)

Kruk, Z. L., Brittain, R. T.: Changes in body and skin temperature following intracerebroventricular injection of substances in the conscious rat: interpretation of data. J. Pharm. Pharmacol. **24**, 835—837 (1972)

Kubena, R. K., Barry, H.: Interactions of Δ^9-tetrahydrocannabinol with barbiturates and methamphetamine. J. Pharmacol. exp. Ther. **173**, 94—100 (1970)

Kulkarni, A. S.: Facilitation of instrumental avoidance learning by amphetamine: an analysis. Psychopharmacologia (Berl.) **13**, 418—425 (1968)

Kulkarni, A. S.: Selective increase in avoidance responding by methamphetamine in naive rats. Psychopharmacologia (Berl.) **24**, 449—455 (1972)

Kulkarni, S. K., Dandiya, P. C.: On the mechanism of potentiation of amphetamine induced stereotyped behavior by imipramine. Psychopharmacologia (Berl.) **27**, 367—372 (1972)

Kulkarni, A. S., Job, W. M.: Facilitation of avoidance learning by *d*-amphetamine. Life Sci. **6**, 1579—1587 (1967)

Kumar, R.: Exploration and latent learning: differential effects of dexamphetamine on components of exploratory behaviour in rats. Psychopharmacologia (Berl.) **16**, 54—72 (1969)

Kumar, R.: Extinction of fear. Effects of amylobarbitone and dexamphetamine given separately and in combination on fear and exploratory behaviour in rats. Psychopharmacologia (Berl.) **19**, 163—187 (1971)

Ladefoged, O.: The effects of LSD, psilocybin, harmaline, and amphetamine on the body temperature of para-chlorophenylalanine pretreated rats. Arch. int. Pharmacodyn. **204**, 326—332 (1973)

Ladefoged, O.: The effect of LSD, psilocybin, harmaline and amphetamine on the body temperature of para-chlorophenylalanine treated rabbits. Arch. int. Pharmacodyn. **208**, 251—254 (1974)

Ladisch, W., Baumann, P.: Influence of progesterone on norepinephrine metabolism of the rat brain in connection with amphetamine and stress. Neuroendocrinology **7**, 16—24 (1971)

Ladisch, W., Volbehr, H., Matussek, N.: Paradoxical effect of amphetamine on hyperactive states in correlation with catecholamine-metabolism in brain. In: Costa, E., Garattini, S., (Eds.): International Symposium on Amphetamines and Related Compounds, pp. 487—492. New York: Raven Press 1970a

Ladisch, W., Volbehr, H., Matussek, N.: Paradoxical amphetamine effect in hyperactive rats in relation to norepinephrine metabolism. Neuropharmacology **9**, 303—310 (1970b)

Lagerspetz, K. Y. H., Lagerspetz, K. M. J.: Amphetamine toxicity in genetically aggressive and non-aggressive mice. J. Pharm. Pharmacol. **23**, 542—543 (1971)

Lal, H.: Control of learned conditioned avoidance responses (CAR) by amphetamine and chlorpromazine. Psychopharmacologia (Berl.) **14**, 33—37 (1968)

Lal, H.: Control of learned conditioned-avoidance responses (CAR) by amphetamine and chlorpromazine. Psychopharmacologia (Berl.) **14**, 33—37 (1969)

Lal, H., Chessick, R. D.: Biochemical mechanism of amphetamine toxicity in isolated and aggregated mice. Life Sci. **3**, 381—384 (1964)
Lal, H., Defeo, J. J., Thut, P.: Effect of amphetamine on pain-induced aggression. Commun. behav. Biol. **1**, 333—336 (1968)
Lal, H., Ginocchio, S., Shefner, A.: The effect of α-methyl-3,4-dihydroxyphenylalanine (α-methyldopa) and α-methyl-meta-tyrosine (α-MMT) on amphetamine toxicity. Life Sci. **3**, 190—192 (1963)
Lal, H., Nesson, B., Smith, N.: Amphetamine-induced aggression in mice pretreated with dihydroxyphenylalanine (DOPA) and/or reserpine. Biol. Psychiatry **2**, 299—301 (1970)
Lal, H., O'Brien, J., Puri, S. K.: Morphine-withdrawal aggression Sensitization by amphetamines. Psychopharmacologia (Berl.) **22**, 217—223 (1971)
Lal, H., Puri, S. K., Fuller, G. C.: Inhibition of hepatic hexobarbital metabolism by *dextro*-amphetamine. Psychopharmacologia (Berl.) **16**, 395—398 (1970)
Lal, H., Roffman, M.: Learning associated with amphetamine state. Proc. 77th Ann. Conv. APA. 881—882 (1969)
Lal, S., Missala, K., Sourkes, T. L.: Effect of neuroleptics on brain amphetamine concentrations in the rat. J. Pharm. Pharmacol. **23**, 967—968 (1971)
Lal, S., Sourkes, T. L.: Effect of various chlorpromazine metabolites on amphetamine-induced stereotyped behaviour in the rat. Europ. J. Pharmacol. **17**, 283—286 (1972a)
Lal, S., Sourkes, T. L.: Potentiation and inhibition of the amphetamine stereotypy in rats by neuroleptics and other agents. Arch. int. Pharmacodyn. **199**, 289—301 (1972b)
Lal, S., Sourkes, T. L.: Ontogeny of stereotyped behaviour induced by ampomorphine and amphetamine in the rat. Arch. int. Pharmacodyn. **202**, 171—182 (1973)
Lal, S., Sourkes, T. L., Missala, K.: The effect of certain tranquilizers, chlorpromazine metabolites and diethyldithiocarbamate on tissue amphetamine levels in the rat. Arch. int. Pharmacodyn. **207**, 122—130 (1974)
Lanciault, G., Wolf, H. H.: Some neuropharmacological properties of the ephedrine isomers. J. pharm. Sci. **54**, 841—844 (1965)
Lapin, I. P., Oxenkrug, G. G., Azbekyan, S. G.: Involvement of brain serotonin in the stimulant action of amphetamine and of cholinolytics. Arch. int. Pharmacodyn. **197**, 350—361 (1972)
Lapin, I. P., Schelkunov, E. L.: Amphetamine-induced changes in behaviour of small laboratory animals as simple tests for evaluation of central effects of new drugs. Pharmacology of Conditioning, Learning and Retention, pp. 205—215. Mikhelson and Longo (Eds.). Oxford: Pergamon Press 1965
Laporte, J., Jané, J., Cuenca, E., Rodríguez, L., Valdecasas, F. G.: Interactions between amphetamine and reserpine in vitro. Experientia (Basel) **25**, 1161—1162 (1969)
Lasagna, L., McCann, W. P.: Effect of "tranquilizing" drugs on amphetamine toxicity in aggregated mice. Science **125**, 1241—1242 (1957)
Lat, J.: The spontaneous exploratory reactions as a tool for psychopharmacological studies. A contribution towards a theory of contradictory results in psychopharmacology. Pharmacology of conditioning, learning and retention, pp. 47—63. Mikhelson, Longo (Eds.). Oxford: Pergamon Press 1965
Laties, V. G.: Effects of *d*-amphetamine on concurrent schedules of heat and food reinforcement. J. Physiol. (Lond.) **63**, 315—318 (1971)
Laties, V. G.: The modification of drug effects on behaviour by external discriminative stimuli. J. Pharmacol. exp. Ther. **183**, 1—13 (1972)
Laties, V. G., Weiss, B.: Influence of drugs on behavior controlled by internal and external stimuli. J. Pharmacol. exp. Ther. **152**, 388—396 (1966)
Latz, A., Bain, G., Goldman, M., Kornetsky, C.: Maze learning after the administration of antidepressant drugs. J. Pharmacol. exp. Ther. **156**, 76—84 (1967)
Latz, A., Kornetsky, C., Bain, G.: Changes in conditioned behaviour under chlorpromazine and *d*-amphetamine as a function of the motivational characteristics and parametric values of the tests. Neuropharmacology **129**, 1003—1007 (1967)
Lauener, H.: Conditioned suppression in rats and the effect of pharmacological agents thereon. Psychopharmacologia (Berl.) **4**, 311—325 (1963)
Laverty, R., Sharman, D. F.: Modification by drugs of the metabolism of 3,4-dihydroxyphenylethylamine, noradrenaline and 5-hydroxytryptamine in the brain. Brit. J. Pharmacol. **24**, 759—772 (1965)

Lawlor, R.B., Trivedi, M.C., Yelnosky, J.: A determination of the anorexigenic potential of *dl*-amphetamine, *d*-amphetamine, *l*-amphetamine and phentermine. Arch. int. Pharmacodyn. **179**, 401—407 (1969)

Leaf, R.C., Learner, L., Horovitz, Z.P.: The role of the amygdala in the pharmacological and endocrinological manipulation of the aggression. In: Garattini, S., Sigg, E.G. (Eds.): Aggressive Behaviour, pp. 120—131. Amsterdam: Excerpta Medica Foundation 1969

Leake, C.D.: The amphetamines—their actions and uses. Springfield: Charles C. Thomas 1958

Leander, J.D., McMillan, D.E.: Rate dependent effects of drug. I. Comparisons of *d*-amphetamine, pentobarbital and chlorpromazine on multiple and mixed schedules. J. Pharmacol. exp. Ther. **188**, 726—739 (1974)

LeBlanc, A.E., Cappell, H.: Attenuation of punishing effects of morphine and amphetamine by chronic prior treatment. J. comp. physiol. Psychol. **87**, 691—698 (1974)

Leibowitz, S.F.: Hypothalamic beta-adrenergic "satiety" system antagonizes an alpha-adrenergic "hunger" system in the rat. Nature (Lond.) **226**, 963—964 (1970a)

Leibowitz, S.F.: Reciprocal hunger-regulating circuits involving alpha- and beta-adrenergic receptors located, respectively, in the ventromedial and lateral hypothalamus. Proc. nat. Acad. Sci. (Wash.) **67**, 1063—1070 (1970b)

Leitz, F.H.: Mechanisms by which amphetamine and desipramine inhibit the metaraminol-induced release of norepinephrine from sympathetic nerve endings in rat heart. J. Pharmacol. exp. Ther. **173**, 152—157 (1970)

Leitz, F.H., Stefano, F.J.E.: The effect of tyramine, amphetamine and metaraminol on the metabolic disposition of ^{3}H-norepinephrine released from the adrenergic neuron. J. Pharmacol. exp. Ther. **178**, 464—473 (1971)

Lemberger, L., Witt, E.D., Davis, J.M., Kopin, I.J.: The effect of haloperidol and chlorpromazine on amphetamine metabolism and amphetamine stereotyped behavior in the rat. J. Pharmacol. exp. Ther. **174**, 428—433 (1970)

Lemmer, B.: Effects of amantadine and amphetamine on serotonin uptake and release by human blood platelets. Europ. J. Pharmacol. **21**, 183—188 (1973)

Leonard, B.E.: Effect of four amphetamines on brain biogenic amines and their metabolites. Biochem. Pharmacol. **21**, 1289—1297 (1972)

Leonard, B.E., Neuhoff, V.: The effect of *d*-amphetamine on the concentration of phospholipids in the rat hippocampus. Z. Naturforsch. **29**, 187—188 (1974)

Leonard, B.E., Shallice, S.A.: Some neurochemical effects of amphetamine, methylamphetamine and p-bromomethylamphetamine in the rat. Brit. J. Pharmacol. **41**, 198—212 (1971)

Leonard, B.E., Wiseman, B.D.: The effect of ethanol and amphetamine mixtures on the activity of rats in a Y-maze. J. Pharm. Pharmacol. **22**, 967—968 (1970)

Leonard, D.A., Heil, G.C., Fellows, E.J.: Effects of anorectics in rats during chronic administration. Fed. Proc. **25**, 385 (1966)

Leslie, G.B., Maxwell, D.R.: Some pharmacological properties of thioproperazine and their modification by anti-parkinsonian drugs. Brit. J. Pharmacol. **22**, 301—317 (1964)

Lew, C., Iversen, S.D., Iversen, L.L.: Effects of imipramine, desipramine and monoamine oxidase inhibitors on the metabolism and psychomotor stimulant actions of *d*-amphetamine in mice. Europ. J. Pharmacol. **14**, 351—359 (1971)

Lewander, T.: Effects of amphetamine on urinary and tissue catecholamines in rats after inhibition of its metabolism with desmethylimipramine. Europ. J. Pharmacol. **5**, 1—9 (1968a)

Lewander, T.: Urinary excretion and tissue levels of catecholamines during amphetamine intoxication. Psychopharmacologia (Berl.) **13**, 394—407 (1968b)

Lewander, T.: Influence of various psychoactive drugs on the in vivo metabolism of *d*-amphetamine in the rat. Europ. J. Pharmacol. **6**, 38—44 (1969)

Lewander, T.: Catecholamine turn-over studies in chronic amphetamine intoxication. In: Costa, E., Garattini, S. (Eds.): International Symposium on Amphetamines and Related Compounds, pp. 317—329. New York: Raven Press 1970a

Lewander, T.: Relationship between amphetamine metabolism and brain catecholamines in the rat and guinea-pig. Acta pharmacol. (Kbh.) **44**, 62 (1970b)

Lewander, T.: On the presence of p-hydroxynorephedrine in the rat brain and heart in relation to changes in catecholamine levels after administration of amphetamine. Acta pharmacol. (Kbh.) **29**, 33—48 (1971a)

Lewander, T.: Displacement of brain and heart noradrenaline by p-hydroxynorephedrine after administration of p-hydroxyamphetamine. Acta pharmacol. (Kbh.) **29**, 20—32 (1971b)

Lewander, T.: Effects of acute and chronic amphetamine intoxication on brain catecholamines in the guinea-pig. Acta pharmacol. (Kbh.) **29**, 209—225 (1971c)

Lewander, T.: A mechanism for the development of tolerance to amphetamine in rats. Psychopharmacologia (Berl.) **21**, 17—31 (1971d)

Lewander, T.: Effects of chronic amphetamine intoxication on the accumulation in the rat brain of labelled catecholamines synthesized from circulating tyrosine-^{14}C and dopa-^{3}H. Naunyn-Schmiedebergs Arch. Pharmak. **271**, 211—233 (1971e)

Lewander, T.: Experimental and clinical studies on amphetamine dependence. In: van Praag, H. M. (Ed.): Biochemical and pharmacological aspects of dependence and Reports on marihuana research, pp. 69—84. Haarlem, N. V.: De Erven F. Bohn 1972

Lewander, T.: Mechanisms of tolerance and dependence for central stimulants. J. Pharmacol. (Paris) **5**, 44—45 (1974a)

Lewander, T.: Effect of chronic treatment with central stimulants on brain monoamines and some behavioural and physiological functions in rats guinea pigs and rabbits. In: Usdin, E. (Ed.): Neuropharmacology of monoamines and their regulatory enzymes, pp. 221—239. New York: Raven Press 1974b

Lewander, T., Jonsson, J.: Drugs and certain conditions interfering with the metabolism and excretion of amphetamine in the rat. In: Ban, T. A., Boissier, J. R., Gessa, G. J., Heimann, H., Hollister, L., Lehman, H. E., Munkvad, I., Steinberg, H., Sulser, F., Sundwall, A., Vinar, O. (Eds.): Psychopharmacology, sexual disorders and drug abuse, pp. 625—635. Amsterdam-London: North Holland Publishing Comp. Prague: Avicenum, Czechoslovak Medical Press 1973

Lewander, T., Moliis, G., Brus, I.: On amphetamine tolerance and abstinence in rats. Usdin, E., Snyder, S. (Eds.): Frontiers in catecholamine research, pp. 987—989. New York: Pergamon Press 1973

Lewander, T., Moliis, G., Brus, I.: On mechanisms of tolerance to the hyperthermic effect of *d*- and *l*-amphetamine in rats. In: Boissier, J. R., Hippius, H., Pichot, P. (Eds.): Neuropsychopharmacology, pp. 323—334. Amsterdam-New York: Excerpta Medica, American Elsevier 1975

Lewis, J. J., van Petten, G. R.: The effect of amphetamine and related compounds on the concentration of adenine nucleotides, inorganic phosphate and creatine phosphate in the rat brain. J. Pharmacol. exp. Ther. **136**, 372—377 (1962)

Lewis, J. J., Pollack, D.: Effects of drugs on rat brain in vitro respiration and adenosinetriphosphatase activity. Life Sci. **4**, 21—26 (1965)

L'Huillier, J., Grandjean, J. L., Thuillier, J.: Annulation par le procalmadiol des effets secondaires lethaux de la methamphetamine chez la souris; application à l'étude des psycholeptiques. C.R. Soc. Biol. (Paris) **157**, 2183—2185 (1963)

Liberson, W. T., Ellen, P., Feldman, R. S.: Effect of "somatic" vs "guidance" therapies on behaviour rigidity in rats. J. Neuropsychiat. **1**, 17—19 (1959)

Liebman, J.: Modification of the chronotropic action of sympathomimetic amines by reserpine in the heart-lung preparation on the dog. J. Pharmacol. exp. Ther. **133**, 63—69 (1961)

Liesiene, V. A.: Comparative study of natural sleep and the action of phenamine and chlorpromazine evaluation of the mean extrema's frequency of ECoG in cats. Farmakol. i Toksikol. **33**, 669 (1970)

Lima, I. M., Luiz, R., Carlini, E. A.: Effects of strychnine and amphetamine on ribonucleic acid content in brain and liver of rats. Med. Pharmacol. exp. **15**, 1—6 (1966)

Little, H. J., Rees, J. M.: Tolerance development to the antinociceptive actions of morphine, amphetamine, physostigmine and 2-aminoindane in the mouse. Experientia (Basel) **30**, 930—931 (1974)

Littleton, J. M.: The interaction of dexamphetamine with inhibitors of noradrenaline biosynthesis in rat brain in vivo. J. Pharm. Pharmacol. **19**, 414—415 (1967)

Littleton, J. M.: Dual mechanism of amphetamine toxicity in mice: relationship to brain catecholamines. Acta pharmacol. (Kbh.) **34**, 92—96 (1974)

Lockett, M. F.: The effect of denervation of the response of the cat's nictitating membrane to sympathomimetic amines. Brit. J. Pharmacol. **5**, 485—496 (1950a)

Lockett,M.F.: The responses of the heart rate and the systolic blood pressure to a series of sympathomimetic amines in the unanaesthetized, atropinized bitch. J. Physiol. (Lond.) **111**, 18—42 (1950b)

Löffelholz,K., Muscholl,E.: Der Einfluß von *d*-Amphetamin auf die Noradrenalinabgabe aus dem isolierten Kaninchenherzen. Naunyn-Schmiedebergs Arch. Pharmak. exp. Path. **266**, 393—394 (1970)

Loh,H., Hitzemann,R., Stolman,S.: Effect of *d*-amphetamine on brain protein synthesis. Toxicol. appl. Pharmacol. **26**, 476—481 (1973)

Longo,V.G., Silvestrini,B.: Action of eserine and amphetamine on the electrical activity of the rabbit brain. J. Pharmacol. exp. Ther. **120**, 160—170 (1957)

Loon,G.R. van, Scapagnini,U., Cohen,R., Ganong,W.F.: Effect of intraventricular administration of adrenergic drugs on the adrenal venous 17-hydroxycorticosteroid response to surgical stress in the dog. Neuroendocrinology **8**, 257—272 (1971)

Lu,T.-C., Claghorn,J.L., Schoolar,J.C.: Chronic administration of *d*-amphetamine and chlorpromazine in rats. Europ. J. Pharmacol. **21**, 61—65 (1973)

Lu,T.-C., Ho,B.T., McIsaac,W.M.: Effects of repeated administration of *dl*-amphetamine and methamphetamine on tolerance to hyperactivity. Experientia (Basel) **28**, 1461 (1972)

Lu,K.-H., Meites,J.: Inhibition by L-dopa and monoamine oxidase inhibitors of pituitary prolactin release; stimulation by methyldopa and *d*-amphetamine (35604). Proc. Soc. exp. Biol. (N.Y.) **137**, 480—483 (1971)

Lundberg,P.: Amphetamine-induced release (^{3}H)metaraminol from subcellular fractions of the mouse heart. J. Pharm. Pharmacol. **21**, 266—268 (1969)

Lundborg,P., Waldeck,B.: On the mechanism of amphetamine induced release of reserpine-resistant ^{3}H-noradrenaline and ^{3}H-α-methylnoradrenaline. Acta pharmacol. (Kbh.) **30**, 339—347 (1972)

Lynch,G.S., Ballantine II,P., Campbell,B.A.: Potentiation of behavioral arousal after cortical damage and subsequent recovery. Exp. Neurol. **23**, 195—206 (1969)

Lynch,G.S., Ballantine II,P., Campbell,B.A.: Differential rates of recovery following frontal cortical lesions in rats. Physiol. Behav. **7**, 737—741 (1971)

Lynch,V.D., Aceto,M.D., Thomas,R.K.: The effects of certain pharmacologic drugs on conditioning in the rat. I. J. Amer. pharm. Ass. sci. Ed. **49**, 205—210 (1960)

Lyon,M.: The effect of no-shock or continous shock upon avoidance behaviour in rats under *d*-amphetamine. Activ. nerv. sup. (Praha) **13**, 78—81 (1971)

Lyon,M., Randrup,A.: The dose-response effect of amphetamine upon avoidance behaviour in the rat seen as a function of increasing stereotypy. Psychopharmacologia (Berl.) **23**, 334—347 (1972)

Lytle,L.D., Moorcroft,W.H., Campbell,B.A.: Ontogeny of amphetamine anorexia and insulin hyperphagia in the rat. J. comp. physiol. Psychol. **77**, 388—393 (1971)

Mabry,P.D., Campbell,B.A.: Serotoninergic inhibition of catecholamine induced behavioral arousal. Brain Res. **49**, 381—391 (1973)

MacDonnell,M.F., Fessock,L.: Some effects of ethanol, amphetamine, disulfiram and p-CPA on seizing of prey in feline predatory attack and on associated motor pathways. Quart. J. Stud. Alcohol. **33**, 437—450 (1972)

Machiyama,Y., Utena,H., Kikuchi,M.: Behavioural disorders in Japanese monkeys produced by the long-term administration of methamphetamine. Proc. Jap. Acad. **46**, 738—743 (1970)

MacPhail,R.C., Gollub,L.R.: Independence of the effects of *d*-amphetamine and food deprivation or body weight on the food consumption of rats. Psychopharmacologia (Berl.) **34**, 163—172 (1974)

Maengwyn-Davies,G.D., Cowan,F.F., Koppanyi,T.: Tachyphylaxis. VI. Amphetamine in the isolated rabbit aortic strip. Arch. int. Pharmacodyn. **160**, 454—472 (1966)

Maengwyn-Davies,G.D., Koppanyi,T.: Cocaine tachyphylaxis and effects on indirectly-acting sympathomimetic drugs in the rabbit aortic strip and in splenic tissue. J. Pharmacol. exp. Ther. **154**, 481—492 (1966)

Magos,L.: Persistence of the effect of amphetamine on stereotyped activity in rats. Europ. J. Pharmacol. **6**, 200—201 (1969)

Magour, S., Coper, H., Fähndrich, CH.: The effects of chronic treatment with *d*-amphetamine on food intake, body weight, locomotor activity and subcellular distribution of the drug in rat brain. Psychopharmacologia (Berl.) **34**, 45—54 (1974a)

Magour, S., Coper, H., Fähndrich, CH.: Effect of chronic intoxication with (+)-amphetamine on its concentration in liver and brain and on (^{14}C)-leucine incorporation into microsomal and cytoplasmic proteins of rat liver. J. Pharm. Pharmacol. **26**, 105—108 (1974b)

Maickel, R. P., Cox Jr., R. H., Ksir, C. J., Snodgrass, W. R., Miller, F. P.: Some aspects of the behavioral pharmacology of the amphetamine. In: Costa, E., Garattini, S. (Eds.): International Symposium on Amphetamines and Related Compounds, pp. 747—759. New York: Raven Press 1970

Maickel, R. P., Cox, Jr., R. H., Miller, F. P., Segal, D. S., Russell, R. W.: Correlation of brain levels of drugs with behavioral effects. J. Pharmacol. exp. Ther. **165**, 216—224 (1969)

Maickel, R. P., Johnson, S. A.: Effects of various anorexigenic agents on open field behavior. Res. Commun. chem. Path. Pharmacol. **6**, 733—740 (1973)

Maickel, R. P., Webb, R. W.: Taste phenomena and drug effects on thirst-induced fluid consumption by rats. Neuropharmacology **11**, 283—290 (1972)

Maitre, L. von, Brunner, H.: Influence de la d'amphétamine sur la déplétion de la noradrénaline tissulaire produite par la guanéthidine (Ismeline) chez le rat. Helv. physiol. pharmacol. Acta **23**, C33—36 (1965)

Maitre, L. von, Brunner, H.: Antihypertensive und Katecholamin entleerende Wirkungen von α-Methyl-tyramin und α-Methyl-octopamin. Naunyn-Schmiedebergs Arch. exp. Path. Pharmak. **257**, 40 (1967)

Maj, J., Durek, B., Palider, W.: Differences in the locomotor activity of mice as measured by an Animex and photoresistor actometer. J. Pharm. Pharmacol. **23**, 979—981 (1971)

Maj, J., Grabowska, M., Gajda, L., Michaluk, J.: Pharmacologic action of *d*- and *l*-amphetamine. Diss. pharm. pharmacol. **24**, 7—15 (1972a)

Maj, J., Przegalinski, E.: Disulfiram and some effects of amphetamine in mice and rats. J. Pharm. Pharmacol. **19**, 341—342 (1967)

Maj, J., Sowinska, H., Baran, L.: Effects of amantadine, amphetamine and apomorphine on the locomotor activity in rats. Life Sci. **12**, 511—518 (1973)

Maj, J., Sowinska, H., Baran, L., Kapturkiewicz, Z.: The effects of clozapine, thioridazine and phenoxybenzamine on the action of drugs stimulating the central catecholamine receptors. Pol. J. Pharmacol. **20**, 437—448 (1974)

Maj, J., Sowinska, H., Kapturkiewicz, Z., Sarnek, J.: The effect of l-dopa and (+)-amphetamine on the locomotor activity after pimozide and phenoxybenzamine. J. Pharm. Pharmacol. **24**, 412—413 (1972b)

Major, C. T., Pleuvry, B. J.: Effects of α-methyl-p-tyrosine, p-chlorophenylalanine, L-β-(3,4-dihydroxyphenyl)alanine, 5-hydroxytryptophan and diethyldithiocarbamate on the analgesic activity or morphine and methylamphetamine in the mouse. Brit. J. Pharmacol. **42**, 512—521 (1971)

Malick, J. B., Goldberg, M. E.: Effects of a choline acetyltransferase inhibitor on self-stimulatory behavior in the rat. Arch. int. Pharmacodyn. **184**, 252—256 (1970)

Maling, H. M., Koppanyi, T.: Salivation in mice as an index of adrenergic activity. III. Partial temperature-dependence of adrenergic drug effects. Arch. int. Pharmacodyn. **199**, 344—357 (1972)

Maling, H. M., Williams, M. A., Koppanyi, T.: Salivation in mice as an index of adrenergic activity. I. Salivation and temperature responses to *d*-amphetamine and other sialogogues and the effects of adrenergic blocking agents. Arch. int. Pharmacodyn. **199**, 318—332 (1972)

Malorny, G., Orzechowski, G.: Untersuchungen über die Wirkungsweise der Sympathomimetica. II. Naunyn-Schmiedebergs Arch. exp. Path. Pharmak. **196**, 245—252 (1940)

Mandell, A. J., Knapp, S.: Cholinergic adaptation in the brain to chronic administration of amphetamine. In: Ellinwood, E. H., Cohen, S. (Eds.): Current concepts on amphetamine abuse, pp. 77—86. DHEW Publ. No. (HSM) 72-9085. Washington D.C.: U.S. Government Printing Office 1972

Mandell, A. J., Knapp, S., Kuczenski, R. T., Segal, D. S.: Methamphetamine-induced alteration in the physical state of rat caudate tyrosine hydroxylase. Biochem. Pharmacol. **21**, 2737—2750 (1972)

Mandell, A. J., Morgan, M.: Amphetamine induced increase in tyrosine hydroxylase activity. Nature (Lond.) **227**, 75—76 (1970)
Mann, P. J. G., Quastel, J. H.: Benzedrine (β-phenylisopropylamine) and brain metabolism. Biochem. J. **34**, 414—431 (1940)
Manning, D. H., Strang, R. H., Bachelard, H. S.: Changes in cerebral carbohydrate metabolism in the rat after acute and chronic treatment with, and withdrawal of methamphetamine. Biochem. Pharmacol. **23**, 1205—1209 (1974)
Mansour, T. E.: The effect of lysergic acid diethylamide, 5-hydroxytryptamine and related compounds on the liver fluke, *Fasciola Hepatica*. Brit. J. Pharmacol. **12**, 406—409 (1957)
Mantegazza, P., Müller, E. E., Naimzada, M. K., Riva, M.: Studies on the lack of correlation between hyperthermia, hyperactivity and anorexia induced by amphetamine. In: Costa, E., Garattini, S. (Eds.): International Symposium on Amphetamines and Related Compounds, pp. 559—575. New York: Raven Press 1970
Mantegazza, P., Naimzada, K. M., Riva, M.: Effects of propranolol on some activities of amphetamine. Europ. J. Pharmacol. **4**, 25—30 (1968a)
Mantegazza, P., Naimzada, K. M., Riva, M.: Activity of amphetamine in hypothyroid rats. Europ. J. Pharmacol. **5**, 10—16 (1968b)
Margules, D. L.: Noradrenergic rather than serotoninergic basis of reward in the dorsal tegmentum. J. comp. physiol. Psychol. **67**, 32—35 (1969)
Marley, E.: Action of some sympathomimetic amines on the cat's iris, in situ or isolated. J. Physiol (Lond.) **162**, 193—211 (1962)
Marley, E., Stephenson, J. D.: Intracerebral microinfusions of amines in young chickens. J. Physiol. (Lond.) **196**, 116—117 P (1968)
Marley, E., Stephenson, J. D.: Actions of dexamphetamine and amphetamine-like amines in chickens with brain transections. Brit. J. Pharmacol. **42**, 522—542 (1971)
Marriott, A. S.: The effects of amphetamine, caffeine and methylphenidate on the locomotor activity of rats in an unfamiliar environment. Int. J. Neuropharmacol. **7**, 487—491 (1968)
Marsden, C. A., Guldberg, H. C.: The role of monoamines in rotation induced or potentiated by amphetamine after nigral, raphé and mesencephalic reticular lesions in the rat brain. Neuropharmacology **12**, 195—211 (1973)
Marsh, J. B., Bizzi, A.: Effects of amphetamine and fenfluramine on the net release of triglycerides of very low density lipoproteins by slices of rat liver. Biochem. Pharmacol. **21**, 1143—1150 (1972)
Martin, J. C., Ellinwood Jr., E. H.: Conditioned aversion to a preferred solution following methamphetamine injections. Psychopharmacologia (Berl.) **29**, 253—261 (1973)
Martin, J. C., Ellinwood, Jr., E. H.: Conditioned aversion in spatial paradigms following methamphetamine injection. Psychopharmacologia (Berl.) **36**, 323—335 (1974)
Matsumoto, C., Griffin, W.: Antagonism of (+)-amphetamine-induced hyperthermia in rats by pimozide. J. Pharm. Pharmacol. **23**, 710 (1971)
Matsumoto, C., Horita, A.: Studies of the antagonism of guanethidine by methamphetamine. Biochem. Pharmacol. **12**, 295—297 (1963)
Matsumoto, C., Shaw, W. N.: Investigations on the mechanism of action of amphetamine-induced hyperthermia in rats. Fed. Proc. **30**, 206 (1971a)
Matsumoto, C., Shaw, W. N.: The involvement of plasma free fatty acids in amphetamine-induced hyperthermia in rats. J. Pharm. Pharmacol. **23**, 387—388 (1971b)
Matussek, N., Linsmayer, M.: The effect of lithium and amphetamine on desmethylimipramine-RO 4-1284 induced motor hyperactivity. Life Sci. **7**, 371—375 (1968)
Maxwell, D. R.: The activity of some adrenaline antagonists in tests for tranquillizing action. Neuropsychopharmacology **1**, 365—371 (1959)
Maxwell, R. A., Plummer, A. J., Povalski, H., Schneider, F.: Concerning a possible action of guanethidine (SU-5864) in smooth muscle. J. Pharmacol. exp. Ther. **129**, 24—30 (1960)
Maxwell, R. A., Povalski, H., Plummer, A. J.: A differential effect of reserpine on pressor amine activity and its relationship to other agents producing this effect. J. Pharmacol. exp. Ther. **125**, 178—183 (1959)
Mayer, O., Eybl, V.: The effect of diethyldithiocarbamate on amphetamine-induced behaviour in rats. J. Pharm. Pharmacol. **23**, 894—896 (1971)

McCullough, D. O., Milberg, J. N., Robison, S. M.: A central site for the hypothermic effects of (+)-amphetamine sulphate and p-hydroxyamphetamine hydrobromide in mice. Brit. J. Pharmacol. **40**, 219—226 (1970)
McDowell, A. A., Ziller, H. H., Krise, G. M.: Effects of previous radiation exposure on the activity response to *d*-amphetamine hydrochloride. J. genet. Psychol. **111**, 241—243 (1967)
McGaugh, J. L., de Baran, L., Longo, V. G.: Electroencephalographic and behavioural analysis of drug effects on an instrumental reward discrimination in rabbits. Psychopharmacologia (Berl.) **4**, 126—138 (1963)
McGeer, E. G., Fibiger, H. C., Wickson, V.: Differential development of caudate enzymes in the neonatal rat. Brain Res. **32**, 433—440 (1971)
McGeer, E. G., McGeer, P. L.: In vitro screen of inhibitors of rat brain tyrosine hydroxylase. Canad. J. Biochem. **45**, 115—131 (1967)
McGrath, W. R., Ketteler, H. J.: Potentiation of the antireserpine effects of dihydroxyphenylalanine by antidepressants and stimulants. Nature (Lond.) **199**, 917—918 (1963)
McKearney, J. W.: The relative effects of *d*-amphetamine, imipramine and harmaline on tetrabenazine suppression of schedule-controlled behavior in the rat. J. Pharmacol. exp. Ther. **159**, 429—440 (1968)
McKearney, J. W.: Schedule dependent effects: effects of drugs and maintenance of response produced by electric shocks. In: Gilbert, R. M., Keehn, J. D. (Eds.): Schedule effects; drugs, drinking and aggression, pp. 3—25. Toronto: University of Toronto Press 1972
McKearney, J. W.: Methamphetamine effects of responding under a multiple schedule of shock presentation. Pharmacol. Biochem. Behav. **1**, 547—550 (1973)
McKearney, J. W.: Effects of *d*-amphetamine, morphine and chlorpromazine on responding under fixed presentation. J. Pharmacol. exp. Ther. **190**, 141—153 (1974)
McKenzie, G. M., Gordon, R. J., Viik, K.: Some biochemical and behavioural correlates of a possible animal model of human hyperkinetic syndromes. Brain Res. **47**, 439—456 (1972)
McKenzie, G. M., Soroko, F. E.: The effects of apomorphine, (+)-amphetamine and l-dopa on maximal electrochock convulsions—a comparative study in the rat and mouse. J. Pharm. Pharmacol. **24**, 696—701 (1972)
McKenzie, G. M., Szerb, J. C.: The effect of dihydroxyphenylalanine, pheniprazine and dextroamphetamine on the in vivo release of dopamine from the caudate nucleus. J. Pharmacol. exp. Ther. **162**, 302—308 (1968)
McLean, J. R., Ensor, C. R., McCarthy, D. A., Bohner, B., Potoczak, D.: Effects of l-dopa, 1-3-methyltyrosine on *d*-amphetamine-induced activity and seizures induced by electroshock. Proc. Soc. exp. Biol. (N.Y.) **143**, 1083—1087 (1973)
McLean, J. R., McCartney, M.: Effect of *d*-amphetamine on rat brain noradrenaline and serotonin. Proc. Soc. exp. Biol. (N.Y.) **107**, 77—79 (1961)
McMahon, E. M., Andersen, D. K., Feldman, J. M., Schanberg, S. M.: Methamphetamine-induced insulin release. Science **174**, 66—68 (1971)
McMillan, D. E.: The effects of sympathomimetic amines on schedule-controlled behavior in the pigeon. J. Pharmacol. exp. Ther. **160**, 315—325 (1968 a)
McMillan, D. E.: Some interactions between sympathomimetic amines and amine depleting agents on the schedule controlled behaviour of the pigeon and the squirrel monkey. J. Pharmacol. exp. Ther. **163**, 172—187 (1968 b)
McMillan, D. E.: Effects of *d*-amphetamine on performance under several parameters of multiple fixed-ratio, fixed-interval schedules. J. Pharmacol. exp. Ther. **167**, 26—33 (1969)
McMillan, D. E.: Drugs and punished responding I: Rate-dependent effects under multiple schedules. J. exp. Anal. Behav. **19**, 133—145 (1973)
McNeill, J. H., Muschek, L. D.: The interaction between amphetamine and catecholamines on cardiac phosphorylase and adenyl cyclase. Arch. int. Pharmacodyn. **197**, 236—244 (1972)
Mead, P. G.: Effects of overtraining and pretrial administration of dextroamphetamine on reversal learning in rats. Percept. Mot. Skills **38**, 566 (1974)
Meginnis, R. F.: Interaction of *d*-amphetamine and food deprivation on fixed ratio behaviour of pigeons. Psychol. Rep. **20**, 355—358 (1967)
Menge, H. G., Brand, U.: Untersuchungen über die Stereotypien nach Amphetamin und Apomorphin sowie deren pharmakologische Beeinflussung. Psychopharmacologia (Berl.) **21**, 212—228 (1971)

Mennear, J. H.: Interaction between central cholinergic agents and amphetamine in mice. Psychopharmacologia (Berl.) **7**, 107—114 (1965)
Mennear, J. H., Rudzik, A. D.: The effects of alpha- and beta-adrenergic blockade on the lethality of amphetamine in aggregated mice. Life Sci. **4**, 1425—1432 (1965)
Mennear, J. H., Rudzik, A. D.: The effects of amine depleting agents on the toxicity of amphetamine in aggregated mice. Life Sci. **5**, 349—356 (1966)
Mennear, J. H., Rudzik, A.: The effect of alpha-methyl-tyrosine on the dose-lethality curve of *d*-amphetamine in aggregated mice. Life Sci. **7**, 213—218 (1968)
Menon, M. K., Clark, W. G., Fleming, R. M.: Blockade of the central effects of *d*-amphetamine sulfate by amantadine hydrochloride. Europ. J. Pharmacol. **21**, 311—317 (1973)
Menon, M. K., Dandiya, P. C.: Mechanism of the protective effect of reserpine on aggregated mice treated with (+)-amphetamine. J. Pharm. Pharmacol. **19**, 596—602 (1967)
Menon, M. K., Dandiya, P. C., Bapna, J. S.: Modification of the effect of some central stimulants in mice pretreated with α-methyl-p-tyrosine. Psychopharmacologia (Berl.) **10**, 437—444 (1967)
Menon, M. K., Fleming, R. M., Clark, W. G.: Studies on the biochemical mechanisms of the central effects of gammahydroxybutyric acid. Biochem. Pharmacol. **23**, 879—885 (1974)
Menon, M. K., Subramanian, K., Menon, N. K., Sharma, S. N.: Influence of desmethylimipramine on the appetite-lowering effects of *d*-amphetamine and other anorectics. Europ. J. Pharmacol. **12**, 156—160 (1970)
Meyerson, B. J.: Amphetamine and 5-hydroxytryptamine inhibition of copulatory behaviour in the female rat. Ann. Med. exp. Fenn. **46**, 394—398 (1968)
Miczek, K. A.: Intraspecies aggression in rats: effects of *d*-amphetamine and chlordiazepoxide. Psychopharmacologia (Berl.) **39**, 275—301 (1974)
Miele, E., Satta, M., Anania, W.: The analysis of blockade by penfluridol of *d*-amphetamine-induced hyperkinesia in mice. Pharmacol. Res. Commun. **4**, 305—310 (1972)
Miller, D. D., Fowble, J., Patil, P. N.: Inhibition of catecholamine uptake by conformationally restricted phenethylamine derivatives. J. med. Chem. **16**, 177—178 (1973)
Miller, F. P., Cox, R. H., Maickel, R. P.: The effects of altered brain norepinephrine levels on continuous avoidance responding and the action of amphetamines. Neuropharmacology **9**, 511—517 (1970)
Miller, K. W., Freeman, J. J., Dingell, J. V., Sulser, F.: On the mechanism of amphetamine potentiation by iprindole. Experientia (Basel) **26**, 863—864 (1970)
Miller, N. E.: Effects of drugs on motivation: the value of using a variety of measures. Ann. N.Y. Acad. Sci. **65**, 318—333 (1956)
Miller, R. E., Levine, J. M., Mirsky, I. A.: Effects of psychoactive drugs on nonverbal communication and group social behaviour of monkeys. J. Pers. soc. Psychol. **28**, 396—405 (1973)
Minkowsky, W. L.: The effect of benzedrine sulphate on learning. J. comp. Psychol. **28**, 349—360 (1939)
Møller-Nielsen, I., Dubnick, B.: Pharmacology of chlorphentermine. In: Costa, E., Garattini, S. (Eds.): International Symposium on Amphetamines and Related Compounds, pp. 63—74. New York: Raven Press 1970
Mogenson, G. J.: Effects of amphetamine on self-stimulation and induced drinking. Physiol. Behav. **3**, 133—136 (1968)
Mogenson, G. J., Cinnamon, A. D., Stevenson, J. A.: The anorexigenic effects of amphetamine in mice treated with aurothioglucose. Canad. J. Physiol. Pharmacol. **45**, 564—567 (1967)
Moisset, B., Welch, B. L.: Effects of *d*-amphetamine upon open field behaviour in two inbred strains of mice. Experientia (Basel) **29**, 625—626 (1973)
Molinengo, L., Ricci-Gamalero, S.: Behavioural action of the association of pentobarbital and chlorpromazine with amphetamine. Pharmacol. Res. Commun. **2**, 327—333 (1970)
Montanaro, N., Babbini, M.: Interaction between iminodibenzylderivatives or phenothiazine-derivatives and methamphetamine in the rat. Psychopharmacologia (Berl.) **7**, 321—338 (1965)
Moore, K. E.: Toxicity and catecholamine releasing actions of *d*- and *l*-amphetamine in isolated and aggregated mice. J. Pharmacol. exp. Ther. **142**, 6—12 (1963)
Moore, K. E.: The role of endogenous norepinephrine in the toxicity of *d*-amphetamine in aggregated mice. J. Pharmacol. exp. Ther. **144**, 45—51 (1964)
Moore, K. E.: Amphetamine toxicity in hyperthyroid mice: effect of endogenous catecholamines. Biochem. Pharmacol. **14**, 1831—1837 (1965)

Moore, K. E.: Amphetamine toxicity in hyperthyroid mice: effects on blood glucose and liver glycogen. Biochem. Pharmacol. **15**, 353—360 (1966 a)

Moore, K. E.: Effects of (+)-amphetamine on lactic acid concentrations in the tissues of aggregated mice. J. Pharm. Pharmacol. **18**, 196—197 (1966 b)

Moore, K. E., Carr, L. A., Dominic, J. A.: Functional significance of amphetamine-induced release of brain catecholamines. In: Costa, E., Garattini, S. (Eds.): International Symposium on Amphetamines and Related Compounds, pp. 371—384. New York: Raven Press 1970

Moore, K. E., Dominic, J. A.: Tyrosine hydroxylase inhibitors. Fed. Proc. **30**, 859—870 (1971)

Moore, K. E., Lariviere, E. W.: Effects of *d*-amphetamine and restraint on the content of norepinephrine and dopamine in rat brain. Biochem. Pharmacol. **12**, 1283—1288 (1963)

Moore, K. E., Lariviere, E. W.: Effects of stress and *d*-amphetamine on rat brain catecholamines. Biochem. Pharmacol. **13**, 1098—1100 (1964)

Moore, K. E., Rech, R. H.: Reversal of α-methyltyrosine-induced behavioural depression with dihydroxyphenylalanine and amphetamine. J. Pharm. Pharmacol. **19**, 405—407 (1967)

Moore, K. E., Sawdy, L. C., Shaul, S. R.: Effects of *d*-amphetamine on blood glucose and tissue glycogen levels of isolated and aggregated mice. Biochem. Pharmacol. **14**, 197—204 (1965)

Moore, K. E., Thornburg, J. E.: Importance of brain dopamine for the stimulant actions of amphetamine. Frontiers in catecholamine research, pp. 1031—1034. Usdin, E., Snyder, S. (Eds.). New York: Pergamon Press 1973

Moormann, J., Opitz, K.: Zur analgetischen Wirkung anorexigener Substanzen. Arzneimittel-Forsch. **17**, 343—345 (1967)

Moraes, S. de, Bussato, P. A., Carvalho, F. V.: A comparison of the effects of (+)-amphetamine, cocaine and denervation on the responses of the isolated vas deferens of the guinea pig to norepinephrine and acetylcholine. Arch. int. Pharmacodyn. **206**, 352—362 (1973 a)

Moraes, S. de, Carmo, M. M., Carvalho, V. F.: The effects of amphetamine and cocaine on the response of the isolated guinea-pig vas deferens to noradrenaline, tyramine and nerve stimulation. Arch. int. Pharmacodyn. **202**, 199—208 (1973 b)

Moraes, S. de, Carvalho de, V. F.: Reciprocal pressor effects between amphetamine and some sympathomimetic amines. Pharmacology (Basel) **1**, 129—134 (1968)

Moraes, S. de, Carvalho de, V. F.: Analysis of the supersensitivity to noradrenaline by amphetamine in the isolated vas deferens of the rat. J. Pharm. Pharmacol. **23**, 798—800 (1971)

Moraes, S. de, Carvalho de, V. F., Aragao, J. A.: Crossed tachyphylaxis between tyramine and some alpha-methylated sympathomimetic amines. Pharmacology (Basel) **3**, 168—176 (1970 a)

Moraes, S. de, Carvalho de, V. F., Wehrle, R. D.: Sensitivity changes to noradrenaline in the guinea-pig vas deferens induced by amphetamine, cocaine and denervation. J. Pharm. Pharmacol. **22**, 717—719 (1970 b)

Morgan, C. D., Cattabeni, F., Costa, E.: Methamphetamine, fenfluramine and their N-dealkylated metabolites: effect of monoamine concentrations in rat tissues. J. Pharmacol. exp. Ther. **180**, 127—135 (1972 a)

Morgan, C. D., Löfstrandh, S., Costa, E.: Amphetamine analogues and brain amines. Life Sci. **11**, 83—96 (1972 b)

Morpurgo, C., Theobald, W.: Influence of antiparkinson drugs and amphetamine on some pharmacological effects of phenothiazine derivatives used as neuroleptics. Psychopharmacologia (Berl.) **6**, 178—191 (1964)

Morpurgo, C., Theobald, W.: Influence of imipramine-like compounds and chlorpromazine on the reserpine-hypothermia in mice and the amphetamine-hyperthermia in rats. Med. Pharmacol. exp. **12**, 226—232 (1965)

Morpurgo, C., Theobald, W.: Behavioral reactions to amphetamine in reserpinized rats. Int. J. Neuropharmacol. **5**, 375—377 (1966)

Morpurgo, C., Theobald, W.: Pharmacological modifications of the amphetamine-induced hyperthermia in rats. Europ. J. Pharmacol. **2**, 287—294 (1967)

Morrison, C. F.: A comparison of the effects of nicotine and amphetamine on DRL performance in the rat. Psychopharmacologia (Berl.) **12**, 176—180 (1968)

Morrison, J. L., Abreu, E.: Effects of *d*-, *l*- and *r*-amphetamine sulfate on the oxygen consumption of morphinized dogs. Fed. Proc. **2**, 88 (1943)

Morse, W. H., Herrnstein, R. J.: Effects of drugs on characteristics of behaviour maintained by complex schedules of intermittent positive reinforcement. Ann. N.Y. Acad. Sci. **65**, 303—317 (1956)

Mourevieff-Lesuisse, F., Giurea, C.: Pharmacological reactivity of an experimental model of memory: the spinal cord fixation. Arch. int. Pharmacodyn. **176**, 471—472 (1968)

Mulas, A., Crabai, F., Pepeu, G.: The influence of repeated experience on the effects of scopolamine and of amphetamine on exploratory behaviour in the rat. Pharmacol. Res. Commun. **2**, 169—176 (1970)

Müller, H., Schoetensack, W.: Förderung und Hemmung stimulierender Amphetamineffekte durch Pharmaka mit zentraldämpfenden Eigenschaften. Naunyn-Schmiedebergs Arch. Pharmak. exp. Path. **264**, 285—287 (1969)

Müller, P. J., Vernikos-Danellis, J.: Effect of environmental temperature on the toxicity of caffeine and dextroamphetamine in mice. J. Pharmacol. exp. Ther. **171**, 153—158 (1970)

Munkvad, I., Pakkenberg, H., Randrup, A.: Aminergic systems in basal ganglia associated with stereotyped hyperactive behavior and catalepsy. Brain Behav. Evol. **1**, 89—100 (1968)

Munkvad, I., Randrup, A.: The persistance of amphetamine stereotypies of rats in spite of strong sedation. Acta psychiat. scand. **42**, 178—187 (1966)

Munoz, C., Goldstein, L.: Influence of adrenergic blocking drugs upon the EEG analeptic effects of *dl*-amphetamine in conscious unrestrained rabbits. J. Pharmacol. exp. Ther. **132**, 354—359 (1961)

Murphree, O. D., DeLuca, D. C., Angel, C.: Psychopharmacologic facilitation of operant conditioning of generically nervous catahoula and pointer dogs. Pavlov. J. biol. Sci. **9**, 17—24 (1974)

Muscholl, E.: Indirectly acting sympathomimetic amines. Pharmacol. Rev. **18**, 551—559 (1966)

Nahorski, S. R., Rogers, K. J.: In vivo effects of amphetamine on metabolites and metabolic rate in brain. J. Neurochem. **21**, 679—686 (1973)

Nakajima, H., Grandjean, J. L., L'Huillier, J., Thuillier, J.: Neurostimulants hallucinogènes et non hallucinogènes: confrontation expérimentale de leurs effets sur les températures centrale et périphiphérique. Proc. 2nd. int. Pharmacol. Meeting **2**, 263—278 (1964)

Nakamura, K., Schimamoto, K.: The effects of reserpine on the responses of the nictitating membrane in the rat. Jap. J. Pharmacol. **9**, 150—158 (1960)

Nasello, A. G., Izquierdo, I.: Effect of learning and of drugs on the ribonucleic acid concentration of brain structures of the rat. Exp. Neurol. **23**, 521—528 (1969)

Nasmyth, P. A.: The effect of some sympathomimetic amines on the ascorbic acid content of rats adrenal glands. J. Physiol. (Lond.) **110**, 294—300 (1949)

Naumenko, E. V.: Role of adrenergic and cholinergic structures in the control of the pituitary-adrenal system. Endocrinology **80**, 69—76 (1967)

Navratil, J., Medek, A., Hrbek, J., Komenda, S.: Comparison of pemoline, amphetamine and methylphenidate effects upon the conditioned alimentary motor reflexes in cats. Activ. nerv. sup. (Praha) **15**, 142—143 (1973)

Naylor, R. J., Costall, B.: The relationship between the inhibition of dopamine uptake and the enhancement of amphetamine stereotypy. Life Sci. **10**, 909—915 (1971)

Naylor, R. J., Olley, J. E.: Modification of the behavioural changes induced by amphetamine in the rat by lesions in the caudate nucleus, the caudate putamen and globus pallidus. Neuropharmacology **11**, 91—99 (1972)

Neill, D. B., Boggan, W. O., Grossman, S. P.: Behavioural effects of amphetamine in rats with lesions in the corpus striatum. J. comp. physiol. Psychol. **86**, 1019—1030 (1974)

Neill, D. B., Grant, L. D., Grossman, S. P.: Selective potentiation of locomotor effects of amphetamine by midbrain raphé lesions. Physiol. Behav. **9**, 655—657 (1972)

Neill, D. B., Grossman, S. P.: Interaction of the effects of reserpine and amphetamine on food and water intake. J. comp. physiol. Psychol. **76**, 327—336 (1971)

Neuburg, J., Thut, P. D.: Comparison of the locomotor stimulant mechanisms of action of *d*-amphetamine and *d*-amphetamine plus l-dopa possible involvement of serotonin. Biol. Psychiatry **8**, 139—150 (1974)

Newman, L. M.: Effects of cholinergic agonists and antagonists on self-stimulation behavior in the rat. J. comp. physiol. Psychol. **79**, 394—413 (1972)

Ng, K. Y., Chase, T. N., Kopin, I. J.: Drug-induced release of ^{3}H-norepinephrine and ^{3}H-serotonin from brain slices. Nature (Lond.) **228**, 468—469 (1970)
Nichols, R. E., Patterson, C. S., Walaszek, E. J.: The effects of some psychoactive drugs on rat brain adenosine triphosphate levels. Arch. int. Pharmacodyn. **184**, 19—26 (1970)
Nielsen, C. K., Frey, H.-H.: Actions of amphetamine and chloroamphetamine on peripheral adrenergic and tryptaminergic receptors. Arch. int. Pharmacodyn. **182**, 459—470 (1969a)
Nielsen, C. K., Frey, H.-H.: Zur Wirkung von Amphetamin und p-Chloramphetamin auf periphere adrenerge und tryptaminerge Receptoren. Naunyn-Schmiedebergs Arch. Pharmak. exp. Path. **263**, 240—241 (1969b)
Nielsen, E. B., Lyon, M.: Drinking behaviour and brain dopamine: antagonistic effect of two neuroleptic drugs (pimozide and spiramide) upon amphetamine—or apomorphine—induced hypodipsia. Psychopharmacologia (Berl.) **33**, 299—308 (1973)
Nielsen, C. K., Magnussen, M. P., Kampmann, E., Frey, H.-H.: Pharmacological properties of racemic and optically active p-chloroamphetamine. Arch. int. Pharmacodyn. **170**, 428—442 (1967)
Niemegeers, C. J. E., Verbruggen, F. J., Janssen, P. A. J.: The influence of various neuroleptic drugs on shock avoidance responding in rats. Psychopharmacologia (Berl.) **17**, 151—159 (1970)
Niemegeers, C. J. E., Verbruggen, F. J., Wauqier, A., Janssen, P. A. J.: The influence of haloperidol and amphetamine on two different noise-escape situations in rats. Psychopharmacologia (Berl.) **25**, 22—31 (1972)
Niijima, A.: Afferent impulse discharges from glucoreceptors in the liver of the guinea pig. Ann. N.Y. Acad. Sci. **157**, 690—700 (1969)
Nistri, A., Bartolini, A., Deffenu, G., Pepeu, G.: Investigations into the release of acetylcholine from the cerebral cortex of the cat: effects of amphetamine of scopolamine and of septal lesions. Neuropharmacology **11**, 665—674 (1972)
North, R. B., Harik, S. I., Snyder, S. H.: Amphetamine isomers: influences on locomotor and stereotyped behavior of cats. Pharmacol. Biochem. Behav. **2**, 115—119 (1974)
Norton, S.: An analysis of cat behavior using chlorpromazine and amphetamine. Int. J. Neuropharmacol. **6**, 307—316 (1967)
Norton, S.: Amphetamine as a model for hyperactivity in the rat. Physiol. Behav. **11**, 181—186 (1973)
Novelli, A. N., Tainter, M. L.: The stimulant power of secondary and tertiary phenylisopropylamines. J. Pharmacol. exp. Ther. **77**, 324—331 (1943)
Novick, I., Pihl, R.: Effect of amphetamine on the septal syndrome in rats. J. comp. physiol. Psychol. **68**, 220—225 (1969)
Obianwu, H. O.: Possible functional differentiation between the stores from which adrenergic nerve stimulation, tyramine and amphetamine release noradrenaline. Acta physiol. scand. **75**, 92—101 (1969a)
Obianwu, H. O.: Some studies on the mechanism by which *d*-amphetamine antagonizes guanethidine induced adrenergic neurone blockade. Acta physiol. scand. **75**, 102—110 (1969b)
Obianwu, H. O., Stitzel, R., Lundborg, P.: Subcellular distribution of (^{3}H)-amphetamine and (^{3}H)-guanethidine and their interaction with adrenergic neurons. J. Pharm. Pharmacol. **20**, 585—594 (1968)
O'Dea, K., Rand, M. J.: Interaction between amphetamine and monoamine oxidase inhibitors. Europ. J. Pharmacol. **6**, 115—120 (1969)
Offermeier, J., Potgieter, B.: Tricyclic antidepressives. A possible mode of action. S. Afr. med. J. **43**, 1366—1370 (1969)
Offermeier, J., Potgieter, B.: The possible central stimulant mechanisms of some appetite suppressants. S. Afr. med. J. **46**, 72—77 (1972)
Ohler, E. A., Wakerlin, G. E.: Amines in experimental hypertensions. Circulat. Res. **1**, 122—130 (1953)
Okun, R., Liddon, S. C., Lasagna, L.: The effects of aggregation, electric shock and adrenergic blocking drugs on inhibition of the writhing syndrome. J. Pharmacol. exp. Ther. **139**, 107—109 (1963)
Olds, J.: Studies of neuropharmacologicals by electrical and chemical manipulation of the brain in animals with chronically implanted electrodes. Neuropsychopharmacology **1**, 20—32 (1958)

Olds, J., Milner, P.: Positive reinforcement produced by electrical stimulation of the septal area and other regions of rat brain. J. comp. physiol. Psychol. **47**, 419—427 (1954)

Olds, M. E.: Comparative effects of amphetamine, scopolamine, chlordiazepoxide, and diphenylhydantonin on operant and extinction with brain stimulation and food reward. Neuropharmacology **9**, 519—532 (1970)

Olds, M. E.: Comparative effects of amphetamine, scopolamine and chlordiazepoxide on self-stimulation behavior. Rev. canad. Biol. **31**, 25—47 (1972)

Oliveira, L. de, Graeff, F. G.: Comparison between the effects of apomorphine and amphetamine on operant behavior. Europ. J. Pharmacol. **18**, 159—165 (1972)

Oliviero, A.: Analysis of the "anti-fatigue" activity of amphetamine. Role of central adrenergic mechanisms. Farmaco **22**, 441—449 (1967)

Oliviero, A., Eleftheriou, B. E., Bailey, D. W.: Exploratory activity: genetic analysis of this modification by scopolamine and amphetamine. Physiol. Behav. **10**, 893—899 (1973)

Onodera, I., Ito, K., Moroji, T., Yamashita, I.: Neuroendocrinological study of chronically methamphetamine-treated guinea pigs and rats. Psychiat. Neurol. jap. **75**, 409—423 (1973)

Opitz, K.: Anorexigene Phenylalkylamine und Serotoninstoffwechsel. Naunyn-Schmiedebergs Arch. Pharmak. exp. Path. **259**, 56—71 (1967)

Opitz, K.: Adipokinetic action of amphetamine—a study in the beagle dog. In: Costa, E., Garattini, S. (Eds.): International Symposium on Amphetamines and Related Coumpounds, pp. 627—639. New York: Raven Press 1970

Opitz, K., Chu, H., Grüter, H.: Beeinflussung der calorigenen Wirkung von sympathomimetischen Aminen durch Lithium. Naunyn-Schmiedebergs Arch. Pharmak. exp. Path. **268**, 379—391 (1971)

Opitz, K., Grüter, H., Loeser, A.: Zur Stoffwechselwirkung von Amphetaminen und anderen Phenylalkylaminen. Arch. int. Pharmacodyn. **161**, 183—195 (1966)

Opitz, K., Loeser, A.: Über den Einfluß appetithemmender Substanzen auf das Fettgewebe. Klin. Wschr. **41**, 193—196 (1963)

Orsingher, O. A., Fulginiti, S.: Effects of alpha-methyltyrosine and adrenergic blocking agents on the facilitating action of amphetamine and nicotine on learning in rats. Psychopharmacologia (Berl.) **19**, 231—240 (1971)

Orsingher, O. A., Fulginiti, S.: Influence of peripheral mechanisms on the facilitatory learning action of amphetamine and nicotine in rats. Pharmacology (Basel) **9**, 138—144 (1973)

Orzechowski, G.: Über die Wirkungsweise der Sympathicomimetica. IX. Naunyn-Schmiedebergs Arch. exp. Path. Pharmak. **198**, 27—33 (1941)

Osborne, N. N., Neuhoff, V., Leonard, B. E.: The effect of fenfluramine, 2,5-dimethoxy-4-methylamphetamine (DOM) and *d*-amphetamine on the concentration of serotonin and some free aminoacids in the oesophageal ganglion of Helix pomatia. Z. Naturforsch. **29**, 189—190 (1974)

Ott, T., Schmitt, M., Pohle, W., Matthies, H.: The effect of methamphetamine on the norepinephrine and 5-hydroxytryptamine contents in eleven rat brain regions. Brain Res. **25**, 171—178 (1971)

Overstreet, D. H., Hadick, D. G., Russell, R. W.: Effects of amphetamine and pilocarpine on eating behaviour in rats with chronically low acetylcholinesterase levels. Behav. Biol. **7**, 217—226 (1972)

Owens, J. E.: The influence of *dl*-, *d*- and *l*-amphetamine and *d*-meth-amphetamine on fixed-ratio schedule. J. exp. Anal. Behav. **4**, 293—310 (1960)

Owens, J. E.: Psychopharmacological studies of some 1-(chlorophenyl)-2-aminopropanes. I. Effects on appetitive-controlled behavior. J. pharm. Sci. **52**, 679—683 (1963)

Owens, J. E., Campbell, J. T.: Effects of methamphetamine on fixed ratio responding of food satiated and food deprived rats. Psychopharmacologia (Berl.) **34**, 223—232 (1974)

Paasonen, M. K., Vogt, M.: The effect of drugs on the amounts of substance P and 5-dyroxytryptamine in mammalian brain. J. Physiol. (Lond.) **131**, 617—626 (1956)

Palladin, A. V., Rybina, A. A.: Phosphorus metabolism in the brain upon stimulation of the higher nervous activity. Dokl. Akad. Nauk. SSSR, **91**, 903—905 (1953)

Palmer, G. C.: Influence of amphetamine, protriptyline and pargyline on the time course of the norepinephrine-induced accumulation of cyclic AMP in the rat brain. Life Sci. **12**, 345—355 (1973)

Palmer, G. C., Davenport, G. R., Ward, J. W.: The effect of neurohumoral drugs on the fixation of spinal reflexes and the incorporation of uridine into the spinal cord. Psychopharmacologia (Berl.) **17**, 59—69 (1970)
Panda, J. N., Ray, A. K.: Effect of chronic use of *d*-amphetamine on transaminase activity of brain. Indian J. exp. Biol. **11**, 345—347 (1973)
Panksepp, J., Booth, D. A.: Tolerance in the depression of intake when amphetamine is added to the rat's food. Psychopharmacologia (Berl.) **29**, 45—54 (1973)
Papeschi, R., Randrup, A.: Catalepsy, sedation and hypothermia induced by alphamethyl-p-tyrosine in the rat. An ideal tool for screening of drugs active on central catecholaminergic receptors. Pharmakopsychiat. Neuropsychopharmacol. **6**, 137—157 (1973)
Papeschi, R., Randrup, A., Lal, S.: Effect of ECT on dopaminergic and noradrenergic mechanisms. Psychopharmacologia (Berl.) **35**, 149—158 (1974)
Parmar, S. S.: Antagonism between stereoisomeric amphetamines, amphetamine derivatives and other monoamine oxidase inhibitors. Biochem. Pharmacol. **15**, 1497—1505 (1966)
Patil, P. N., LaPidus, J. B., Campbell, D., Tye, A.: Steric aspects of adrenergic drugs. II. Effects of *dl*-isomers and desoxy derivatives on the reserpine-pretreated vas deferens. J. Pharmacol. exp. Ther. **155**, 13—23 (1967b)
Patil, P. N., LaPidus, J. B., Tye, A.: Steric aspects of adrenergic drugs. I. Comparative effects of *dl*-isomers and desoxy derivatives. J. Pharmacol. exp. Ther. **155**, 1—12 (1967a)
Patil, P. N., LaPidus, J. B., Tye, A.: Steric aspects of adrenergic drugs. J. pharm. Sci. **59**, 1205—1234 (1970)
Paul, M. I., Pauk, G. L., Ditzion, B. R.: The effect of centrally acting drugs on the concentration of brain adenosine 3′,5′-monophosphate. Pharmacology (Basel) **3**, 148—154 (1970)
Pazzagli, A., Pepeu, G.: Amnesic properties of scopolamine and brain acetylcholine in the rat. Int. J. Neuropharmacol. **4**, 291—299 (1964)
Peczenik, O., Zei, M.: A comparative investigation of actions of reserpine, 5-hydroxytryptamine and cinobufotenine. Confin. neurol. (Basel) **21**, 488—500 (1961)
Pedden, J. R., Tainter, M. L., Cameron, W. M.: Comparative actions of sympathomimetic compounds. Bronchodilator actions in experimental bronchial spasm of parasympathetic origin. J. Pharmacol. exp. Ther. **55**, 242—256 (1935)
Pepeu, G., Bartolini, A.: Effetto di alcuni psicofarmaci sulla liberazione di acetilcolina dalla corteccia cerebrale di gatto. Boll. Soc. ital. Biol. sper. **43**, 1409—1413 (1967)
Pepeu, G., Bartolini, A.: Effect of psychoactive drugs on the output of acctylcholine from the cerebral cortex of the cat. Europ. J. Pharmacol. **4**, 254—263 (1968)
Pepeu, G. C., Bartolini, A., Deffenu, G.: Investigations into the increase of acetylcholine output from the cerebral cortex of the cat caused by amphetamine. In: Heilbronn, E., Winters, S. (Eds.): Drugs and cholinergic mechanisms, pp. 387—396. Stockholm: Res. Inst. Natl. Def. 1970
Persson, T.: Drug induced changes in ^{3}H-catecholamine accumulation after ^{3}H-tyrosine. Acta pharmacol. (Kbh). **28**, (preprint) 378—390 (1970)
Peterson, D. I., Hardinge, M. G., Tilton, B. E.: Neuromuscular block as a possible mechanism of death in amphetamine poisoning. J. Pharmacol. exp. Ther. **146**, 175—179 (1964)
Pfeiffer, A. K.: Study on the mechanism of the metabolism increase by amphetamine and by the elevation of brain catecholamines. Activ. nerv. sup. (Praha) **14**, 295—297 (1972)
Pfeiffer, A. K., Csáki, L., Fodor, M., György, L., Ökrös, I.: The subcellular distribution of (+)-amphetamine and (±)-p-chloroamphetamine in the rat brain as influenced by reserpine. J. Pharm. Pharmacol. **21**, 687—689 (1969)
Pfeiffer, A. K., Galambos, E., György, L.: Some central nervous properties of diethyldithiocarbamate. J. Pharm. Pharmacol. **18**, 254 (1966)
Pfeiffer, A. K., György, L., Fodor, M.: Role of catecholamines in the central effects of amphetamine. Acta med. Acad. Sci. hung. **25**, 441—450 (1968)
Philippu, A., Beyer, J.: Dopamine and noradrenaline transport into subcellular vesicles of the striatum. Naunyn-Schmiedebergs Arch. Pharmak. exp. Path. **278**, 387—402 (1973)
Philippu, A.; Glowinski, J., Besson, M. J.: In vivo release of newly synthesized catecholamines from the hypothalamus by amphetamine. Naunyn-Schmiedebergs Arch.-Pharmak. exp. Path. **258**, 238—250 (1974)
Phillips, A. G., Fibiger, H. C.: Dopaminergic and noradrenergic substrates of positive reinforcement: differential effects of *d*- and *l*-amphetamine. Science **179**, 575—577 (1973)

Phillips, M. I., Bradley, P. B.: The effect of chlorpromazine and *d*-amphetamine mixtures on spontaneous behaviour. Int. J. Neuropharmacol. **8**, 169—176 (1969)
Pick Cassens, G., Mills, A. W.: Lithium and amphetamine: opposite effects on threshold of intracranial reinforcement. Psychopharmacologia (Berl.) **30**, 283—290 (1973)
Pickens, R.: Self-administration of stimulants by rats. Int. J. Addict. **3**, 215—221 (1968)
Pickens, R. W., Crowder, W. F.: Effects of CS-UCS interval on conditioning of drug response with assessment of speed of conditioning. Psychopharmacologia (Berl.) **11**, 88—94 (1969)
Pickens, R., Harris, W. C.: Self-administration of *d*-amphetamine by rats. Psychopharmacologia (Berl.) **12**, 158—163 (1968)
Pickens, R., Meisch, R. A., Dougharty, J. A.: Chemical interactions in methamphetamine reinforcement. Psychol. Rep. **23**, 1267—1270 (1968)
Pickens, R., Meisch, R., McGuire, L. E.: Methamphetamine reinforcement in rats. Psychon. Sci. **8**, 371—372 (1967)
Pickens, R., Thompson, T., Yokel, R. A.: Characteristics of amphetamine self-administration by rats. In: Ellinwood, E. H., Cohen, S. (Eds.): Current concepts on amphetamine abuse, pp. 43—48. DHEW Publ. No. (HSM) 72-9085. Washington D.C.: U.S. Government Printing Office 1972
Pihl, R. O., Greenberg, I.: The interactive effect of amphetamine on the activity levels of rats with lesions in the septum. Psychon. Sci. **14**, 216—217 (1969)
Pinkston, J. O., Pinkston, J. L., Donabedian, A. S.: Vasomotor responses to phenyl-l-amino-2-propane sulphate (benzedrine sulphate) in dogs. J. Pharmacol. exp. Ther. **66**, 28—29 (1939)
Pinter, E. J., Pattee, C. J.: Fat-mobilizing action of amphetamine. In: Costa, E., Garattini, S. (Eds.): International Symposium on Amphetamines and Related Compounds, pp. 653—672. New York: Raven Press 1970
Pirch, H., Cohn, R. A., Osterholm, K. C., Barrat, E. S.: Antagonism of amphetamine locomotor stimulation in rats by single doses of marijuana extract administered orally. Neuropharmacology **12**, 485—493 (1973)
Plane, P., Bogaert, B.: Influence du maxiton (tartrate de phényl-l-amino-2-propane dextrogyre) sur les reflexes conditionels. C.R. Soc. Biol. (Paris) **147**, 108—110 (1953)
Pletscher, A., Bartholini, G., Bruderer, H., Burkhard, W. P., Gey, K. F.: Chlorinated arylalkylamines affecting the cerebral metabolism of 5-hydroxytryptamine. J. Pharmacol. exp. Ther. **145**, 344—350 (1964)
Pletscher, A., Burkhard, W. P., Bruderer, H., Gey, K. F.: Decrease of cerebral 5-hydroxytryptamine and 5-hydroxyindolacetic acid by an arylalkylamine. Life Sci. **2**, 828—833 (1963)
Pletscher, A., Da Prada, M., Burkard, W. P.: The effect of substituted phenylethylamines on the metabolism of biogenic monoamines. In: Costa, E., Garattini, S. (Eds.): International Symposium on Amphetamines and Related Compounds, pp. 331—341. New York: Raven Press 1970
Pleuvry, B. J.: Cross tolerance between methylamphetamine and morphine in the mouse. J. Pharm. Pharmacol. **23**, 969—970 (1971)
Plotnikoff, N.: Magnesium pemoline: Enhancement of learning and memory of a conditioned avoidance response. Sience **151**, 703—704 (1966)
Pohle, W., Matthies, H.: Untersuchungen über die Bedeutung der Amin-Bindung für die Entstehung der Blutdrucktoleranz. Acta biol. med. germ. **19**, 99—110 (1967)
Popova, E. N., Vavilov, A. M., Krivitskaya, G. N., Tumanov, V. P.: Changes in neurons in certain parts of the rat brain during amphetamine induced motor excitation. Bull. exp. Biol. Med. **73**, 340—343 (1972)
Porsolt, R. D., Joyce, D., Summerfield, A.: Lack of tolerance to an amphetamine-barbiturate mixture. Nature (Lond.) **223**, 1277—1278 (1969)
Poschel, B. P. H.: Effects of methamphetamine on hunger and thirst motivated variable-interval performance. J. comp. physiol. Psychol. **56**, 968—973 (1963)
Poschel, B. P. H., Ninteman, F. W.: Hypothalamic self-stimulation: its suppression by blockade of norepinephrine biosynthesis and reinstatement by methamphetamine. Life Sci. **5**, 11—16 (1966)
Potter, L. T., Axelrod, J.: Studies on the storage of norepinephrine and the effect of drugs. J. Pharmacol. exp. Ther. **140**, 199—206 (1963)
Powell, B. J., Martin, L. K., Kamano, D. K.: Failure to find improved shuttlebox avoidance performance using *d*-amphetamine sulphate. Psychol. Rep. **17**, 330 (1965)

Powell, D. A., Walters, K., Duncan, S., Holley, J. R.: The effects of chlorpromazine and *d*-amphetamine upon shock-elicited aggression. Psychopharmacologia (Berl.) **30**, 303—314 (1973)
Prabhu, V. G.: Amphetamine aggregation effect in mice under conditions of altered microsomal enzymes. Arch. int. Pharmacodyn. **195**, 81—86 (1972)
Prada, M. Da., Bartholini, G., Pletscher, A.: Effect of monoamine liberators on the metabolism of 5-hydroxytryptamine in blood platelets. Biochem. Pharmacol. **14**, 1721—1726 (1965)
Pradhan, S. N., Bowling, C.: Effects of nicotine on self-stimulation in rats. J. Pharmacol. exp. Ther. **176**, 229—243 (1971)
Pradhan, S. N., Dutta, S. N.: Comparative effects of nicotine and amphetamine on timing behavior in rats. Neuropharmacology **9**, 9—16 (1970)
Pratesi, P., Blaschko, H.: Specificity of amine oxidase for optically active substrates and inhibitors. Brit. J. Pharmacol. **14**, 256—260 (1959)
Proctor, C. D., Cho, J. B., Lundy, R. O., Mills, S. L., Mitchell, K. J.: Effect of atropine sulfate and atropine methyl bromide on *dl*-amphetamine toxicity for aggregated mice. Brain Res. **60**, 533—535 (1973)
Proctor, C. D., Potts, J. L., Ashley, L. G., Denefield, B. A.: Pilocarpine reversal of *d*-amphetamine induced increase in mouse exploratory locomotor activity. Arch. int. Pharmacodyn. **167**, 61—68 (1967)
Proctor, C. D., Potts, J. L., Lundy, R. O., Greenfield, E. J.: Phenothiazine tranquilizer reversal of amphetamine toxicity for aggregated mice. Arch. int. Pharmacodyn. **163**, 79—86 (1966)
Proctor, C. D., Ridlon, S. A., Fudema, J. J., Prabhu, V. G.: Extension of tranquilizer action by anticholinesterases. Toxicol. appl. Pharmacol. **6**, 1—8 (1964)
Przegalinski, E., Kleinrok Z.: Effect of diethyldithiocarbamate on the toxicity of amphetamine in aggregated mice. Psychopharmacologia (Berl.) **16**, 409—418 (1970)
Pugh, C. E. M., Quastel, J. H.: Oxidation of amines by animal tissues. Biochem. J. **31**, 2306—2321 (1937)
Quinton, R. M., Halliwell, G.: Effects of α-methyl-dopa and dopa on the amphetamine excitatory response in reserpinized rats. Nature (Lond.) **200**, 178—179 (1963)
Raevskii, K. S., Gura, S. J.: Effect of adrenergic blocking agents on toxicity of amphetamine in mice kept in groups or in isolation. Bull. exp. Biol. Med. **69**, 539—541 (1970)
Rahman, H.: Einfluß des Pervitins auf Gedächtnisleistungen, Verhaltensweisen und einige physiologische Funktionen von Goldhamstern. Pflügers Arch. ges. Physiol. **273**, 247—263 (1961)
Rahmann, H.: The influence of methamphetamine on learning, long-term memory, and transposition ability in golden hamsters. International Symposium on Amphetamines and Related Compounds, pp. 813—817. Costa, E., Garattini, S. (Eds.). New York: Raven Press 1970
Raiteri, M., Levi, G., Federico, R.: *d*-Amphetamine and the release of ^{3}H-norepinephrine from synaptosomes. Europ. J. Pharmacol. **28**, 237—240 (1974)
Rand, M. J., Trinker, F. R.: The mechanism of the augmentation of responses to indirectly acting sympathommetic amines by monoamine oxidase inhibitors. Brit. J. Pharmacol. **33**, 287—303 (1968)
Randrup, A., Jonas, W.: Brain dopamine and the amphetamine-reserpine interaction. J. Pharm. Pharmacol. **19**, 483—484 (1967)
Randrup, A., Munkvad, I.: On the relation of tryptaminic and serotonergic mechanisms to amphetamine induced abnormal behaviour. Acta pharmacol. (Kbh.) **21**, 272—282 (1964)
Randrup, A., Munkvad, I.: Pharmacological and biochemical investigations of amphetamine-induced abnormal behaviour. Neuropsychopharmacology **4**, 301—304 (1965a)
Randrup, A., Munkvad, I.: Special antagonism of amphetamine induced abnormal behaviour. Psychopharmacologia (Berl.) **7**, 416—422 (1965b)
Randrup, A., Munkvad, I.: Role of catecholamines in the amphetamine excitatory response. Nature (Lond.) **211**, 540 (1966)
Randrup, A., Munkvad, I.: Stereotype activities produced by amphetamine in several animal species and man. Psychopharmacologia (Berl.) **11**, 300—310 (1967a)
Randrup, A., Munkvad, I.: Various forms of stereotype activity produced by amphetamine in certain animal species and man. Neuropsychopharmacology **127**, 1224—1225 (1967b)
Randrup, A., Munkvad, I.: Brain dopamine and amphetamine-induced stereotyped behaviour. Acta Pharmacol. (Kbh.) **25**, Suppl. 4, 62 (1967c)
Randrup, A., Munkvad, I.: Behavioral stereotypies induced by pharmacological agents. Pharmakopsychiat. Neuropsychopharmacol. **1**, 18—26 (1968)

Randrup, A., Munkvad, I.: Pharmacological studies on the brain mechanisms underlying two forms of behavioural excitation: stereotyped hyperactivity and "rage." Ann. N.Y. Acad. Sci. **159**, 928—938 (1969)

Randrup, A., Munkvad, I.: Biochemical, anatomical and psychological investigations of stereotyped behavior induced by amphetamines. Costa, E., Garattini, S. (Eds.): International Symposium on Amphetamines and Related Compounds, pp. 965—713. New York: Raven Press 1970

Randrup, A., Munkvad, I.: Correlation between specific effects of amphetamines on the brain and on behaviour. In: Ellinwood, E. H., Cohen, S. (Eds.): Current concepts on amphetamine abuse, pp. 17—25. DHEW Publ. No 72-9085. Washington D.C.: U.S. Government Printing Office 1972a

Randrup, A., Munkvad, I.: Influence of amphetamines on animal behaviour: stereotypy, functional impairment and possible animal-human correlations. Psychiat. Neurol. Neurochir. (Amst.) **75**, 193—202 (1972b)

Randrup, A., Munkvad, I., Udsen, P.: Adrenergic mechanisms and amphetamine induced abnormal behaviour. Acta pharmacol. (Kbh.) **20**, 145—157 (1963)

Randrup, A., Scheel-Kruger, J.: Diethyldithiocarbamate and amphetamine stereotype behaviour. J. Pharm. Pharmacol. **18**, 752 (1966)

Rapin, J., Hamar, C., Cohen, Y.: Preferential release of newly synthesized norepinephrine induced by amphetamine and its metabolites in rat: study in vitro. J. Pharmacol. (Paris) **3**, 537—538 (1972)

Rapin, J., Hamar, C., Jacquot, Ch., Cohen, Y., Valette, G.: Libération préférentielle de la noradrénaline nouvellement synthétisée par l'éphédrine et l'hydroxyéphédrine. C.R. Acad. Sci. (Paris) **276**, 2989—2991 (1973a)

Rapin, J. R., Jacquet, C., Cohen, Y., Valette, G.: Effects of amphetamine, ephedrine and parahydroxylated derivate upon the distribution of tyrosine-3H and dihydroxyphenylalanine-3H. C.R. Soc. Biol. (Paris) **165**, 1576—1581 (1971)

Rapin, J., Jacquot, C., Hamar, C., Cohen, Y.: Mechanism of noradrenergic effects of amphetamine and its hydroxylated metabolites. Study in vitro. J. Pharmacol. (Paris) **4**, 519—526 (1973b)

Ray, O. S.: The effect of psychotogens on approach and avoidance behaviour. Int. J. Neuropsychiat. **1**, 98—103 (1965)

Ray, O. S., Bivens, L. W.: Chlorpromazine and amphetamine effects on three operant and on four discrete trial reinforcement schedules. Psychopharmacologia (Berl.) **10**, 32—43 (1966)

Ray, O. S., Bivens, L. W.: Anticholinergic potentiation of amphetamines: Reinforcement schedule effects. Arch. int. Pharmacodyn. **172**, 414—420 (1968)

Rech, R. H.: Antagonism of reserpine behavioral depression by *d*-amphetamine. J. Pharmacol. exp. Ther. **146**, 369—376 (1964)

Rech, R. H.: Amphetamine effects on poor performance of rats in a shuttle-box. Psychopharmacologia (Berl.) **9**, 110—117 (1966)

Rech, R. H., Moore, K. E.: Interactions between *d*-amphetamine and α-methyltyrosine in rat shuttle-box behavior. Brain Res. **8**, 398—400 (1968)

Rech, R. H., Stolk, J. M.: Amphetamine-drug interactions that relate brain catecholamines to behavior. Costa, E., Garattini, S. (Eds.): International Symposium on Amphetamines and Related Compounds, pp. 385—413. New York: Raven Press 1970

Rech, R. H., Thut, P. D.: Lack of tolerance to the central stimulant effects of *d*-amphetamine (dA) in rats. Pharmacologist **12**, 226 (1970)

Reid, W. D.: Hyperthermia and increased turnover of brain serotonin produced by amphetamine. Fed. Proc. **29**, 2832 (1970a)

Reid, W. D.: Turnover rate of brain 5-hydroxytryptamine increased by *d*-amphetamine. Brit. J. Pharmacol. **40**, 483—491 (1970b)

Reinert, H.: The effect of amphetamine on peripheral synaptic structures. Neuropsychopharmacology **1**, 399—404 (1958)

Reitz, R. H., Bocknik, S. E., Kulkarni, A. S.: A comparison of Mg pemoline and (+)-amphetamine effects upon avoidance behaviour and the amine pump. J. Pharm. Pharmacol. **24**, 830—832 (1972)

Remington, G., Anisman, H.: Disruptive effects of epinephrine on active avoidance behaviour—alteration by scopolamine and *d*-amphetamine. Pharmacol. Biochem. Behav. **2**, 427—430 (1974)

Rensch, B., Rahman, H.: Einfluß des Pervitins auf das Gedächtnis von Goldhamstern. Pflügers Arch. ges. Physiol. **271**, 693—704 (1960)
Renzi, P.: Effects of amphetamine on shuttle-box performance of hamsters. Pharmacol. Res. Commun. **2**, 355—361 (1971)
Reynolds, R. W.: The effect of amphetamine on food intake in normal and hypothalamic hyperphagic rats. J. comp. physiol. Psychol. **52**, 682—684 (1959)
Richards, R., Nicol, E., Young, P.: Effect of adrenalectomy and immune serum sympathectomy on the toxicity of *dl*-desoxyephedrin in single and aggregated mice. Life Sci. **5**, 853—864 (1966)
Richardson, D., Karczmar, A. G., Scudder, C. L.: Intergeneric behavioral differences among methamphetamine treated mice. Psychopharmacologia **25**, 347—375 (1972)
Richelle, M.: Combined action of diazepam and *d*-amphetamine on fixed-interval performance in cats. J. Pharmacol. exp. Ther. **12**, 989—998 (1969)
Riddell, D., Szerb, J. C.: The release in vivo of dopamine synthesized from labelled precursors in the caudate nucleus of the cat. J. Neurochem. **18**, 989—1006 (1971)
Rinaldi, F., Himwich, H. E.: Drugs effecting psychotic behavior and the function of the mesodiencephalic activating system. Dis. nerv. Syst. **16**, 133—141 (1955)
Ritter, S., Stein, L.: Self-stimulation of noradrenergic cell group (A6) in locus coeruleus of rats. J. comp. physiol. Psychol. **85**, 443—452 (1973)
Riva, M., Naimzada, M. K., Pirola, C., Mantegazza, P.: Rapporti tra attivita anoressigena, ipertermizzante ed eccitomotoria di composti strutturalmente correlati all'amfetamina. Farmaco **24**, 238—248 (1969)
Robichaud, R. C., Goldberg, M. E., Dubnick, B.: The effect of imipramine on centrally or peripherally administered amphetamine on exploratory behaviour of rats. Pharmacologist **12**, 226 (1970)
Robbins, T., Iversen, S. D.: Amphetamine-induced disruption of temporal discrimination by response disinhibition. Nature (Lond.) **245**, 191—192 (1973a)
Robbins, T., Iversen, S. D.: Dissociation of the effects of *d*-amphetamine on locomotor activity and exploration in rats. Psychopharmacologia (Berl.) **28**, 155—164 (1973b)
Roffler-Tarlov, S., Sharman, D. F., Tegerdine, P.: 3,4-Dihydroxyphenylacetic acid and 4-hydroxy-3-methoxyphenylacetic acid in the mouse striatum: a reflection of intra- and extra-neuronal metabolism of dopamine? Brit. J. Pharmacol. **42**, 343—351 (1971)
Roffman, M., Lal, H.: Facilitatory effect of amphetamine on learning and recall of an avoidance response in rats. Arch. int. Pharmacodyn. **193**, 87—91 (1971)
Roffman, M., Lal, H.: Role of brain amines in learning associated with "Amphetamine-State". Psychopharmacologia (Berl.) **25**, 195—204 (1972)
Rogers, K. J., Hutchins, D. A.: Studies on the relation of chemical structure to glycogenolytic activity in the brain. Europ. J. Pharmacol. **20**, 97—103 (1972)
Roliński, Z.: Analysis of aggressiveness-stereotypy complex induced in mice by amphetamine or nialamide and l-dopa. Pol. J. Pharmacol. Pharm. **25**, 551—558 (1973)
Roliński, Z., Scheel-Krüger, J.: The effect of dopamine and noradrenaline antagonists on amphetamine induced locomotor activity in mice and rats. Acta. pharmacol. (Kbh.) **33**, 385—399 (1973)
Rosen, A. J., La Flore, J. E.: Effects of intraperitoneal and intraventricular *d*-amphetamine administration on active avoidance performance in the rat. Life Sci. **13**, 1573—1580 (1973)
Ross, S. B., Renyi, A. L.: Blocking action of sympathomimetic amines on the uptake of tritiated noradrenaline by mouse cerebral cortex tissues in vitro. Acta pharmacol. (Kbh.) **21**, 226—239 (1964)
Ross, S. B., Renyi, A. L.: In vivo inhibition of ^{3}H-noradrenaline uptake by mouse brain slices in vitro. J. Pharm. Pharmacol. **18**, 322—323 (1966a)
Ross, S. B., Renyi, A. L.: Uptake of tritiated tyramine and (+)-amphetamine by mouse heart slices. J. Pharm. Pharmacol. **18**, 756—757 (1966b)
Ross, S. B., Renyi, A. L.: Uptake of some tritiated sympathomimetic amines by mouse brain cortex slices in vitro. Acta pharmacol. (Kbh.) **24**, 297—309 (1966c)
Ross, S. B., Renyi, A. L.: Accumulation of tritiated 5-hydroxytryptamine in brain slices. Life Sci. **6**, 1407—1415 (1967a)
Ross, S. B., Renyi, A. L.: Inhibition of the uptake of tritiated catecholamines by antidepressant and related agents. Europ. J. Pharmacol. **2**, 181—186 (1967b)

Ross, S. B., Renyi, A. L., Brunfelter, B.: Cocaine-sensitive uptake of sympathomimetic amines in nerve tissue. J. Pharm. Pharmacol. **20**, 283—288 (1968)

Rossum van, J. M.: The significance of brain dopamine psychomotor stimulant action. Biochem. Pharmacol. **12**, Suppl. 210 (1963)

Rossum van, J. M.: The significance of dopamine-receptor blockade for the action of neuroleptic drugs. Neuropsychopharmacology **129**, 321—329 (1967)

Rossum van, J. M.: Mode of action of psychomotor stimulant drugs. Int. Rev. Neurobiol. **12**, 307—383 (1970)

Rossum van, J. M., Hurkmans, J. A. Th.: Mechanism of action of psychomotor stimulant drugs. Int. J. Neuropharmacol. **3**, 227—239 (1964)

Rossum van, J. M., Schoot, van der, J. B., Hurkmans, J. A.: Mechanism of action of cocaine and amphetamine in the brain. Experientia (Basel) **15**, 229—231 (1962)

Rossum van, J. M., Simmons, F.: Locomotor activity and anorexigenic action. Psychopharmacologia (Berl.) **14**, 248—254 (1969)

Roszkowski, A. P., Kelley, N. M.: A rapid method for assessing drug inhibition of feeding behaviour. J. Pharmacol. exp. Ther. **140**, 367—374 (1963)

Roth, R. H., Walters, J. R., Aghajanian, G. K.: Effect of impulse flow on the release and synthesis of dopamine in the rat striatum. In: Usdin, E., Snyder, S. (Eds.): Frontiers in catecholamine research, pp. 567—574. New York: Pergamon Press 1973

Rubin, R. P.: The role of calcium in the release of neurotransmitter substances and hormones. Pharmacol. Rev. **22**, 389—428 (1970)

Rubin, R. P., Jaanus, S. D.: A study of the release of catecholamines from the adrenal medulla by indirectly acting sympathominetic amines. Naunyn-Schmiedebergs Arch. Pharmak. exp. Path. **254**, 125—137 (1966)

Rubin, R. P., Jaanus, S. D.: The release of nucleotide from the adrenal medulla by indirectly acting sympathomimetic amines. Biochem. Pharmacol. **16**, 1007—1012 (1967)

Rudzik, A. D., Eble, J. N.: Blockade of the pressor response to tyramine by amphetamine in the reserpine treated dog. Fed. Proc. **21**, 689 (1960)

Rudzik, A. D., Eble, J. N.: The potentiation of pressor response to tyramine by a number of amphetamine-like compounds. Proc. Soc. exp. Biol. (N.Y.) **124**, 655—657 (1967)

Rudzik, A. D., Johnson, G. A.: Effect of amphetamine and amphetamine analogues on convulsive thresholds. In: Costa, E., Garattini, S. (Eds.): International Symposium on Amphetamines and Related Compounds, pp. 715—728. New York: Raven Press 1970

Rudzik, A. D., Mennear, J.: The effect of reserpine on amphetamine toxicity in aggregated mice. J. Pharm. Pharmacol. **18**, 258 (1966)

Rumbaugh, C. L., Bergeron, R. T., Scanlan, R. L., Teal, J. S., Segall, H. D., Fang, H. C. H., McCormick, R.: Cerebral vascular changes secondary to amphetamine abuse in the experimental animal. Neuroradiology **101**, 345—351 (1971)

Rumelhart, D. E., Mueller, M. R.: Enhancement of amphetamine sulphate effects by atropine in a social situation. Psychol. Rep. **12**, 251—254 (1963)

Rushton, R., Steinberg, H.: Dose-response relations of amphetamine-barbiturate mixtures. Nature (Lond.) **197**, 1017—1018 (1963a)

Rushton, R., Steinberg, H.: Mutual potentiation of amphetamine and amylobarbitone measures by activity in rats. Brit. J. Pharmacol. **21**, 295—305 (1963b)

Rushton, R., Steinberg, H.: Modification of behavioural effects of drugs by past experience. In: Steinberg, H. (Ed.): Animal behaviour and drug action, pp. 207—219. London: Churchill Ltd. 1964

Rushton, R., Steinberg, H.: Combined effects of chlordiazepoxide and dexamphetamine on activity of rats in an unfamiliar environment. Nature (Lond.) **211**, 1312—1313 (1966)

Rushton, R., Steinberg, H.: Drug combinations and their analysis by means of exploratory activity in rats. Neuropsychopharmacology **129**, 464—470 (1967)

Rushton, R., Steinberg, H., Tinson, Ch,: Effects of a single experience on subsequent reactions to drugs. Brit. J. Pharmacol. **20**, 99—105 (1963)

Russek, M.: Hepatic receptors and the neurophysiological mechanisms controlling feeding behaviour. In: Ehrenpreis, S. (Ed.): Neurosciences research, Vol. 4, pp. 214—282. New York: Academic Press 1971

Russek, M., Bruni, E.: On the mechanism of the anorexigenic effect of adrenaline and amphetamine. An. Inst. Biol. Univ. Nal. Aution. Mexico. 41, Ser. Biol. exp. **1**, 17—24 (1970)
Russek, M., Rodriques-Zendejas, A. M., Teitelbaum, P.: The action of adrenergic anorexigenic substances on rats recovered from lateral hypothalamic lesions. Physiol. Behav. **10**, 329—333 (1973)
Russek, M., Stevenson, J. A. F.: Correlation between the effects of several substances on food intake and on the hepatic concentration of reducing sugars. Physiol. Behav. **8**, 245—249 (1972)
Russek, M., Stevenson, J. A. F., Mogenson, G. J.: Anorexigenic effects of adrenaline, amphetamine, and FMS 1A. Canad. J. Physiol. Pharmacol. **46**, 635—638 (1968)
Rutledge, C. D.: The mechanisms by which amphetamine inhibits oxidative deamination of norepinephrine in brain. J. Pharmacol. exp. Ther. **171**, 188—195 (1970)
Rutledge, C. D., Azzaro, A. J., Ziance, R. J.: The importance of neuronal uptake for the release of ^{3}H-norepinephrine by *d*-amphetamine Fed. Proc. **31**, 601 (1972a)
Rutledge, C. D., Azzaro, A. J., Ziance, R. J.: The role of monoamine oxidase in determining the amount of monoamines released by drugs in the central nervous system. Biochem. Psychopharmacol. **5**, 379—392 (1972b)
Rutledge, C. D., Azzaro, A. J., Ziançe, R. J.: Dissociation of amphetamine-induced release of norepinephrine from inhibition of neuronal uptake in isolated brain tissue. In: Usdin, E., Snyder, S. (Eds.): Frontiers in catecholamine research, pp. 973—975. New York: Pergamon Press 1973
Rutledge, C. O., Kelleher, R. T.: Interaction between the effects of methamphetamine and pentobarbital on operant behaviour in the pigeon. Psychopharmacologia (Berl.) **7**, 400—408 (1965)
Ryall, R. W.: Effects of cocaine and antidepressant drugs on the nictitating membrane of the cat. Brit. J. Pharmacol. **17**, 339—357 (1961)
Sadowski, B., Longo, V. G.: Electroencephalographic and behavioural correlates of an instrumental reward conditioned response in rabbits. A physiological and pharmacological study. Electroenceph. clin. Neurophysiol. **14**, 465-476 (1962)
Salama, A. I., Goldberg, M. E.: Effect of several models of stress and amphetamine on brain levels of amphetamine and certain monoamines. Arch. int. Pharmacodyn. **181**, 474—483 (1969)
Salama, A. I., Goldberg, M. E.: Neurochemical effects of imipramine and amphetamine in aggressive mouse-killing (Muricidal) rats. Biochem. Pharmacol. **19**, 2023—2032 (1970)
Salama, A. I., Goldberg, M. E.: Enhanced locomotor activity following amphetamine in mouse-killing rats. Arch. int. Pharmacodyn. **204**, 162—169 (1973)
Samanin, R., Ghezzi, D., Valzelli, L., Garattini, S.: The effects of selective lesioning of brain serotonin or catecholamines containing neurons on the anorectic activity of fenfluramine and amphetamine. Europ. J. Pharmacol. **19**, 318—322 (1972)
Sanan, S., Vogt, M.: Effect of drugs on the noradrenaline content of brain and peripheral tissues and its significance. Brit. J. Pharmacol. **18**, 109—127 (1962)
Sanders-Bush, E., Sulser, F.: Biochemical considerations on the mode of action of *p*-chloroamphetamine. In: Costa, E., Garattini, Ş. (Eds.): International Symposium on Amphetamines and Related Compounds, pp. 349—355. New York: Raven Press 1970
Sanghvi, I., Gershon, S.: The valuation of central nervous system stimulants in a new laboratory test for antidepressants. Life Sci. **8**, 449—457 (1969)
Sansone, M., Bovet, D.: Effects of amphetamine on the decrement of performance in avoidance conditioning of guinea pigs. Psychopharmacologia (Berl.) **16**, 234—239 (1969)
Sansone, M., Renzi, P.: Effects of amphetamine on shuttle-box performance of hamsters. Pharmacol. Res. Commun. **2**, 355—360 (1970)
Sansone, M., Renzi, P.: Shuttle-box avoidance behaviour of hamsters: a further investigation on the effects of amphetamine and methylphenidate. Pharmacol. Res. Commun. **3**, 113—119 (1971)
Sansone, M., Renzi, P., Castellano, C.: Effect of methamphetamine on discriminated lever-press avoidance behaviour in hamsters. Pharmacol. Res. Commun. **6**, 187—192 (1974)
Santi, R., Fassina, G.: Dexamphetamine and lipid mobilization in obesity. J. Pharm. Pharmacol. **16**, 130—134 (1964)
Santi, R., Fassina, G.: Anorexigenic drugs and lipid mobilization. J. Pharm. Pharmacol. **17**, 596—597 (1965)

Sapeika, N., Zwarenstein, H., Kaplan, E. R.: Effect or fenfluramine and dexamphetamine on pyruvate kinase. Res. Commun. chem. Path. Pharmacol. **2**, 125—127 (1971)
Satinder, K. P.: Genotype dependent effects of *d*-amphetamine sulphate and caffeine on escape avoidance behaviour of rats. J. comp. physiol. Psychol. **76**, 359—364 (1971)
Satinder, K. P.: Effects of intertrial crossing punishment and *d*-amphetamine sulphate on avoidance and activity in four selectively bred rat strains. Psychon. Sci. **29**, 291—293 (1972)
Satinder, K. P., Petryshyn, W. R.: Interaction among genotype unconditioned stimulus, *d*-amphetamine and one-way avoidance behaviour in rats. J. comp. physiol. Phsychol. **86**, 1059—1073 (1974)
Satinder, K. P., Royce, J. R., Yeudall, L. T.: Effects of electric shocks, *d*-amphetamine sulphate and chlorpromazine on factors of emotionality in inbred mice. J. comp. physiol. Psychol. **71**, 443—447 (1970)
Sayers, A. C.: Interaction of anticholinergic agents with alpha-methyl-p-tyrosine and *d*-amphetamine. Brit. J. Pharmacol. **45**, 170—171 (1972)
Sayers, A. C., Handley, S. L.: A study of the role of catecholamines in the response to various central stimulants. Europ. J. Pharmacol. **23**, 47—55 (1973a)
Sayers, A. C., Handley, S. L.: The role of amphetamine metabolism in the alpha-methyl-*p*-tyrosine-*d*-amphetamine catalepsy. Europ. J. Pharmacol. **24**, 306—309 (1973b)
Sayers, A. C., Spencer, P. S. J.: Effect of some amphetamine analogues on alpha-methyl-p-tyrosine-induced catalepsy in rats. Brit. J. Pharmacol. **43**, 877—880 (1971)
Schallek, W., Walz, D.: Effects of *d*-amphetamine on the electroencephalogram of the dog. Proc. Soc. exp. Biol. (N.Y.) **82**, 715—719 (1953)
Schanberg, S. M., Cook, J. D.: Effects of acute and chronic methamphetamine on brain norepinephrine metabolism. In: Ellinwood, E. H., Cohen, S. (Eds.): Current concepts on amphetamine abuse, pp. 87—95. DHEW Publ. No (HSM) 72-9085. Washington D.C.: U.S. Government Printing Office 1972
Schechter, M. D., Rosecrans, J. A.: *d*-Amphetamine as a discriminative cue: drugs with similar stimulus properties. Europ. J. Pharmacol. **21**, 212—216 (1973)
Scheckel, C. L., Boff, E.: Behavioral effects of interacting imipramine and other drugs with *d*-amphetamine, cocaine, and tetrabenazine. Psychopharmacologia (Berl.) **5**, 198—208 (1964)
Scheel-Krüger, J.: Comparative studies of various amphetamine analogues demonstrating different interactions with the metabolism of the catecholamines in the brain. Europ. J. Pharmacol. **14**, 47—59 (1971)
Scheel-Krüger, J.: Behavioral and biochemical comparison of amphetamine derivatives cocaine, benzotropine and tricyclic antidepressant drugs. Europ. J. Pharmacol. **18**, 63—73 (1972a)
Scheel-Krüger, J.: Some aspects of the mechanism of action of various stimulant amphetamine analogues. Psychiat. Neurol. Neurochir. (Amst.) **75**, 179—192 (1972b)
Scheel-Krüger, J.: On the possible interrelationship in mechanism of action between morphine, amphetamine and neuroleptic drugs. Usdin, E., Snyder, S. (Eds.): Frontiers in catecholamine research, pp. 1027—1029. New York: Pergamon Press 1973
Scheel-Krüger, J., Hasselager, E.: Studies of various amphetamines, apomorphine and clonidine on body temperature and brain 5-hydroxytryptamine metabolism in rats. Psychopharmacologia (Berl.) **36**, 189—202 (1974)
Scheel-Krüger, J., Jonas, W.: Pharmacological studies on tetrabenazine-induced excited behaviour of rats pretreated with amphetamine or nialamide. Arch. int. Pharmacodyn. **206**, 47—65 (1974)
Schelkunov, E. L.: The technique of "phenamine stereotypy" for evaluating the effect produced by remedial agents on the central adrenergic processes. Farmakol. i Toksikol. **27**, 633 (1964)
Scheving, L. E., Vedral, D. F., Pauly, J. E.: Daily circadian rhythm in rats to *d*-amphetamine sulphate: Effect of blinding and continuous illumination on the rhythm. Nature (Lond.) **219**, 621—622 (1968)
Schiørring, E.: Amphetamine induced selective stimulation of certain behaviour items with concurrent inhibition of others in an open-field test with rats. Behaviour **39**, 1—17 (1971)
Schiørring, E.: Social isolation and other behavioural changes in a group of three vervet monkeys (ceropithecus) produced by single low doses of amphetamine. Psychopharmacologia (Berl.) **26**, Suppl. 117 (1972)

Schiørring, E., Randrup, A.: Social isolation and changes in the formation of groups induced by amphetamine in an open-field test with rats. Pharmakopsychiat. Neuropsychopharmacol. **4**, 2—11 (1971)
Schlechter, J. M., Butcher, L. L.: Blockade by Pimozide of (+)-amphetamine-induced hyperkinesia in mice. J. Pharm. Pharmacol. **24**, 408—409 (1972)
Schlichting, U. U., Goldberg, S. R., Wuttke, W., Hoffmeister, F.: *d*-Amphetamine self-administration by rhesus monkeys with different self-administration histories. Excerpta med. int. Congr. Ser. **220**, 62—69 (1971)
Schmidt, M. J., Hopkins, J. T., Schmidt, D. E., Robinson, G. A.: Cyclic AMP in brain areas: Effects of amphetamine and norepinephrine assessed through the use of microwave radiation as a means of tissue fixation. Brain Res. **42**, 465—477 (1972)
Schmitt, H.: Influence of agents interfering with catecholamines and 5-hydroxytryptamine on the food suppressant effect of amphetamine and fenfluramine. J. Pharmacol. (Paris) **4**, 285—294 (1973)
Schmitt, H., Schmitt, H.: Interaction de la reserpine et de quelques amines sympathicomimétiques. C.R. Soc. Biol. (Paris) **151**, 2030—2032 (1957)
Schmitt, H., Schmitt, H.: Sur les effects hypertenseur de la reserpine après certain sympathicomimetiques. Arch. int. Pharmacodyn. **120**, 251—262 (1959)
Schmitt, H., Schmitt, H.: Modifications des effets des amines sympathicomimetiques sur la pression artérielle et la membrane nictitante par la reserpine. Arch. int. Pharmacodyn. **125**, 30—47 (1960)
Schmitt, H., Schmitt, H.: Mechanisms de la potentialisation de l'hypertension reserpinique chez le rat sans moelle. Arch. int. Pharmacodyn. **132**, 97—105 (1961 a)
Schmitt, H., Schmitt, H.: Inhibiteurs non sympathicolytiques de l'hypertension réserpinique post-amphétaminique. Arch. int. Pharmacodyn. **132**, 106—115 (1961 b)
Schmitt, H., Schmitt, H.: Interrelationships between amphetamine and harmala alkaloids. Nature (Lond.) **203**, 878—879 (1964)
Schmitt, H., Schmitt, H.: Inhibition de l'hypertension réserpinique postamphetaminique par les substances antidepressives du groupe de l'imipramine. C.R. Soc. Biol. (Paris) **160**, 303—306 (1966)
Schmitt, H., Schmitt, H.: Interactions between reserpine and amphetamine on blood pressure. In: Costa, E., Garattini, S. (Eds.): International Symposium on Amphetamines and Related Compounds, pp. 531—550. New York: Raven Press 1970
Schmitt, H., Schmitt, H., Depoilly, F.: Potentiation of the effects of indirect sympathicomimetic amines by reserpine and bretylium. Nature (Lond.) **196**, 73 (1962)
Schneider, C. W., Chenoweth, M. B.: Effects of hallucinogenic and other drugs on the nest building behaviour of mice. Nature (Lond.) **225**, 1262—1263 (1970)
Schneider, F. H.: Amphetamine-induced exocytosis of catecholamines from the cow adrenal medulla. J. Pharmacol. exp. Ther. **183**, 80—89 (1972)
Schoenfeld, R. I., Zigmond, M. J.: Behavioural pharmacology of 6-hydroxydopamine. In: Usdin, E., Snyder, S. (Eds.): Frontiers in catecholamine research, pp. 695—700. New York: Pergamon Press 1973
Schoot, van der, J. B., Ariens, E. J., Rossum, van, J. M., Hurkmans, J. A. Th.: Phenylisopropylamine derivatives, structure and action. Arzneimittel-Forsch. **12**, 902—907 (1962)
Schrold, J.: Behavioural effects of *d*-amphetamine alone and in combination with antidepressants, antihistamines or other psychotropic drugs in young chicks. Psychopharmacologia (Berl.) **23**, 115—124 (1972)
Schrold, J., Squires, R. F.: Behavioural effects of *d*-amphetamine in young chicks treated with p-Cl-phenylalanine. Psychopharmacologia (Berl.) **20**, 85—90 (1971)
Schuberth, J., Fyrö, B., Nybäck, H., Sedvall, C.: Effects of cocaine and amphetamine on the metabolism of tryptophan and 5-hydroxytryptamine in mouse brain in vivo. J. Pharm. Pharmacol. **22**, 860—862 (1970)
Schuberth, J., Sedvall, G.: Effect of amphetamines on tryptophan concentrations in mice and rats. J. Pharm. Pharmacol. **24**, 53—62 (1972)
Schulte, J. W., Reif, E. C., Bacher Jr., J. A., Lawrence, W. S., Tainter, M. L.: Further study of central stimulation from sympathomimetic amines. J. Pharmacol. exp. Ther. **71**, 62—74 (1941)

Schumann, H. J., Philippü, A.: The mechanism of catecholamine release by tyramine. Int. J. Neuropharmacol. **1**, 179—182 (1962)
Schuster, C. R., Dockens, W. S., Woods, J. H.: Behavioral variables affecting the development of amphetamine tolerance. Psychopharmacologia (Berl.) **9**, 170—182 (1966)
Schuster, C. R., Zimmerman, J.: Timing behavior during prolonged treatment with *dl*-amphetamine. J. exp. Anal. Behav. **4**, 427—330 (1961)
Schuster, J., Ihli, M.: The mechanism of some sympathomimetic amines. Pharmacology (Basel) **9**, 240—251 (1973)
Schwartzbaum, J. S., Ide-Johanson, L., Belgrade, J.: Comparative effects of scopolamine and amphetamine upon behavioural reactivity and visual evoked potentials to flashes in rats. J. comp. physiol. Psychol. **86**, 1044—1051 (1974)
Schwartzbaum, J. S., Kreinik, C. J., Gustavson, J. W.: Cortical evoked potentials and behavioural reactivity to photic stimuli in freely moving rats. Brain Res. **27**, 295—307 (1971)
Scott, J. P., Lee, C. T., Ho, J. E.: Effects of fighting, genotype, and amphetamine sulfate on body temperature of mice. J. comp. physiol. Psychol. **76**, 349—352 (1971)
Searle, L. V., Brown, C. W.: The effect of subcutaneous injections of benzedrine sulfate on the activity of white rats. J. exp. Psychol. **22**, 480—490 (1938)
Seegal, R. F., Isaac, W.: Sensory influences upon amphetamine tolerance. Physiol. Behav. **7**, 877—879 (1971)
Segal, D. S., Cox Jr., R. H., Stern, W. G., Maickel, R. P.: Stimulatory effects of pemoline and cyclopropylpemoline on continuous avoidance behavior: similarity to effects of *d*-amphetamine. Life Sci. **6**, 2567—2572 (1967)
Segal, D. S., McAllister, C., Geyer, M. A.: Ventricular infusion of norepinephrine and amphetamine; direct versus indirect action? Pharmacol. Biochem. Behav. **2**, 79—87 (1974)
Segal, D. S., Mandell, A. J.: Long-term administration of *d*-amphetamine-progressive augmentation of locomotor activity and stereotypy. Pharmacol. Biochem. Behav. **2**, 249—255 (1974)
Segal, E.: Effects of *dl*-amphetamine under concurrent VI DRL reinforcement. J. exp. Anal. Behav. **5**, 105—112 (1962)
Seliger, D. L.: Effect of prenatal maternal administration of *d*-amphetamine on rat offspring and passive avoidance learning. Physiol. Psychol. **1**, 273—280 (1973)
Sethy, V. H., Naik, P. Y., Sheth, U. K.: Effect of *d*-amphetamine sulphate in combination with CNS depressants on spontaneous motor activity of mice. Psychopharmacologia (Berl.) **18**, 19—25 (1970)
Sever, P. S., Caldwell, J.: Species differences in amphetamine tolerance. J. Pharmacol. (Paris) **5**, Suppl. 2, 91 (1974)
Sever, P. S., Caldwell, J., Williams, R. T.: Evidence against the involvement of false neurotransmitters in tolerance to amphetamine induced hyperthermia in the rat. J. Pharm. Pharmacol. **26**, 823—826 (1974)
Shapiro, S. L., Freedman, L.: Effect of dosage level of amphetamine tartronate on weights of castrate rats. Arch. int. Pharmacodyn. **112**, 419—426 (1957)
Sharp, J. C., Nielson, H. C., Porter, P. B.: The effect of amphetamine upon cats with lesions in the ventromedial hypothalamus. J. comp. physiol. Psychol. **55**, 198—200 (1962)
Shaw, W. N., Fuller, R. W., Matsumoto, C.: Studies on the mechanism of amphetamine-induced lipolysis in the rat. Europ. J. Pharmacol. **19**, 98—103 (1972)
Sheard, M. H.: The effects of amphetamine on attack behavior in the cat. Brain Res. **5**, 330—338 (1967)
Sheard, M. H.: Aggressive behaviour: modification by amphetamine, p-chlorophenylalanine and lithium in rats. Aggressologie **14**, 323—326 (1973)
Shillito, E. E.: A method for recording the effect of drugs on the activity of small animals over long periods of time. Brit. J. Pharmacol. **26**, 248—255 (1966)
Shimizu, A., Himwich, H. E.: The effects of amphetamine on the sleep-wake-fulness cycle of developing kittens. Psychopharmacologia (Berl.) **13**, 161—169 (1968)
Shoeman, D. W., Sirtori, C. R., Azarnoff, D. L.: Inhibition of amphetamine tolerance and metabolism by propranolol. J. Pharmacol. exp. Ther. **191**, 68—71 (1974)
Shore, P. A., Dorris, R. L.: Drug actions on α-methyl-m-tyramine, a false dopaminergic transmitter in the rat striatum. Frontiers in catecholamine research, pp. 523—526. Usdin, E., Snyder, S. (Eds.). New York: Pergamon Press 1973

Sidman, M.: Technique for assessing the effects of drugs on timing behaviour. Science **122**, 925 (1955)
Siegel, P., Sterling, T. P.: The anorexigenic action of dextroamphetamine sulphate upon feeding responses of different strength. J. comp. physiol. Psychol. **52**, 179—182 (1959)
Silverman, A. P.: The social behaviour of laboratory rats and the action of chlorpromazine and other drugs. Behaviour **26**, 1—38 (1966)
Simon, P., Cheimat, R., Fosset, M., Boissier, J.-R.: Inhibiteurs—adrenergiques et stereotypies provoqués par *l*-amphétamine ou l'apomorphine chez le Rat. Psychopharmacologia. (Berl.) **23**, 357—364 (1972)
Simon, P., Larousse, C., Boissier, J.-R.: Behavioral pharmacology of p-chloromethyl-amphetamine, comparison with amphetamine. In: Costa, E., Garattini, S. (Eds.): International Symposium on Amphetamines and Related Compounds, pp. 729—746. New York: Raven Press 1970a
Simon, P., Tillement, J.-P., Larousse, C., Breteau, M., Guernet, M., Boissier, J.-R.: Pharmacological and biochemical effects of amphetamine and paracloromethamphetamine. J. Pharmacol. (Paris) **1**, 95—108 (1970b)
Simonyi, J., Szentgyörgyi, D.: The effect of actedron (phenylisopropylamine) and thyroxine on the body temperature. Arch. int. Pharmacodyn. **80**, 1—14 (1949)
Simpson, B. A., Iversen, S. D.: Effects of substantia nigra lesions on the locomotor and stereotypy responses to amphetamine. Nature (Lond.) New Biol. **230**, 30—32 (1971)
Singer, G., Sanghvi, I., Gershon, S.: Exploration of certain behavioural patterns induced by psychoactive agents in the rat. Commun. behav. Biol. **6**, 307—314 (1971)
Skaringa, A. P., Milyukas, V. I., Mitskis, A. M.: Correlation between impulse activity of single neurons and high frequency electrocorticogram in physostigmine- and phenamine-induced desynchronization. Farmakol. i Toksikol. **32, 6**, 649—650 (1969)
Skinner, B. F., Heron, W. T.: Effects of caffeine and benzedrine upon conditioning and extinction. Psychol. Rep. **1**, 340—346 (1937)
Skolnik, J., Pfeiffer, A. K., Kaldor, A.: The effects of amphetamine on GABA (γ-aminobutyric acid) concentrations of mouse brain. Biochem. Pharmacol. **12**, Suppl. 8 (1962)
Smetzer, D. L., Senta, T., Hensel, D.: Cardiovascular effects of amphetamine in the horse. Canad. J. comp. Med. **36**, 185—194 (1972)
Smith, C. B.: Enhancement by reserpine and α-methyl-dopa of the effects of *d*-amphetamine upon the locomotor activity of mice. J. Pharmacol. exp. Ther. **142**, 343—350 (1963)
Smith, C. B.: Effects of *d*-amphetamine upon operant behavior of pigeons enhancement by reserpine. J. Pharmacol. exp. Ther. **146**, 167—174 (1964)
Smith, C. B.: Effects of *d*-amphetamine upon brain amine content and locomotor activity of mice. J. Pharmacol. exp. Ther. **147**, 96—102 (1965)
Smith, C. B.: The role of monoamine oxidase in the intraneuronal metabolism of norepinephrine released by indirectly acting sympathomimetic amines or by adrenergic nerve stimulation. J. Pharmacol. exp. Ther. **151**, 207—220 (1966)
Smythies, J. R., Johnson, V. S., Bradley, R. J.: Behavioural models for psychosis. Brit. J. Psychiat. **115**, 55—68 (1969)
Snyder, S. H., Taylor, K. M., Coyle, J. T., Meyerhoff, J. L.: The role of brain dopamine in behavioral regulation and actions of psychotropic drugs. In: Ellinwood, E. H., Cohen, S. (Eds.): Current concepts on amphetamine abuse, pp. 3—16. DHEW Publ. No. (HSM) 72-9085. Washington D.C.: U.S. Government Printing Office 1972
Sofia, R. D.: Effects of chlorpromazine and *d*-amphetamine in Long Evans rats of different age, body weight and brain weight. Arch. int. Pharmacodyn. **182**, 139—146 (1969)
Sokol, G. H., Maickel, R. P.: Toxic interactions of *d*-amphetamine and tricyclic antidepressants in mice. Res. Commun. chem. Path. Pharmacol. **3**, 513—521 (1972)
Soulairac, M. L.: Etudė experimentale des regulations hormonerveuses du comportement sexuel du rat male. Ann. Endocr. (Paris) **24**, Suppl. 1—98 (1963)
Soulairac, M.-L., Soulairac, A.: Action des substances neurostimulantes sur le comportement sexuel du rat mâle après lésious du cortex cérébral. J. Physiol. (Paris) **65**, 3:504 A (1972)
Southgate, P. J., Mayer, S. R., Boxhall, E., Wilson, A.: Some 5-hydroxytryptamine-like actions of fenfluramine: a comparison with (+)-amphetamine and diethylpropion. J. Pharm. Pharmacol. **23**, 600—605 (1971)

Sparber, S. B., Peterson, D. W.: Operant behavioural demonstration of qualitative differences between the *d*- and *l*-isomers of amphetamine. Frontiers in catecholamine research, pp. 969—972. Usdin, E., Snyder, S. (Eds.). New York: Pergamon Press 1973
Sparber, S. B., Tilson, H. A.: Tolerance and cross-tolerance to mescaline and amphetamine as a function of central and peripheral administration. Psychopharmacologia (Berl.) **23**, 220—230 (1972a)
Sparber, S. B., Tilson, H. A.: The releasability of central norepinephrine and serotonin by peripherally administered *d*-amphetamine before and after tolerance. Life Sci. **11**, 1059—1067 (1972b)
Sparber, S. B., Tilson, H. A., Peterson, D. W.: Environmental influences upon morphine or *d*-amphetamine induced suppression of operant behaviour. Pharmacol. Biochem. Behav. **1**, 133—136 (1973)
Spencer, P. S. J.: Activity of centrally acting and other drugs against tremor and hypothermia induced in mice by tremorine. Brit. J. Pharmacol. **25**, 442—455 (1965)
Spencer, P. S. J., Turner, T. A. R.: Effect of pretreatment with monoamine oxidase inhibitors of (+)-amphetamine on leptazol convulsions in mice and rats. J. Pharm. Pharmacol. **20**, Suppl. 122—125 (1968)
Spencer, P. S. J., Turner, T. A. R.: Blockade of biogenic amine synthesis: its effect on the responses to leptazol and dexamphetamine in rats. Brit. J. Pharmacol. **37**, 94—103 (1969)
Spengler, J.: Potenzierung der anorexigenen und psychomotorischen Wirkung von *d*-Amphetamine durch die Monoaminoxydasehemmer Iproniazid und Pivaloylbenzylhydrazin. Naunyn-Schmiedebergs Arch. exp. Path. Pharmak. **244**, 153—160 (1962)
Spengler, J., Waser, P.: Der Einfluß verschiedener Pharmaka auf den Futterkonsum von Albino-Ratten im akuten Versuch. Naunyn-Schmiedebergs Arch. exp. Path. Pharmak. **237**, 171—185 (1959)
Spooner, C. E., Flaming, D.: Monoamines, multiple units and various amphetamine actions. In: Ellinwood, E. H., Cohen, S. (Eds.): Current concepts on amphetamine abuse, pp. 27—36. DHEW Publ. No (HSM) 72-9085. Washington D.C.: U.S. Government Printing Office 1972
Spriggs, T. L. B.: Peripheral noradrenaline and adrenergic transmission in the rat. Brit. J. Pharmacol. **26**, 271—281 (1966)
Srimal, R. C., Dhawan, B. N.: An analysis of methylphenidate induced gnawing in guinea pigs. Psychopharmacologia (Berl.) **18**, 99—107 (1970)
Stark, P., Totty, C. W.: Effects of amphetamines on eating elicited by hypothalamic stimulation. J. Pharmacol. exp. Ther. **158**, 272—278 (1967)
Stark, P., Turk, J. A., Redman, C. E., Henderson, J. K.: Sensitivity and specificity of positive reinforcing areas to neurosedatives, antidepressants and stimulants. J. Pharmacol. exp. Ther. **166**, 163—169 (1969)
Starke, K.: Interactions of guanethidine and indirect-acting sympathomimetic amines. Arch. int. Pharmacodyn. **195**, 309—314 (1972)
Stein, L.: Effects and interactions of imipramine, chlorpromazine, reserpine, and amphetamine on self-stimulation: possible neurophysiological basis of depression. In: Wortis, J. (Ed.): Recent advances in biological psychiatry, Vol. 4, pp. 288—309. New York: Plenum Press 1962
Stein, L.: Self-stimulation of the brain and the central stimulant action of amphetamine. Fed. Proc. **23**, 836—850 (1964a)
Stein, L.: Amphetamine and neural reward mechanisms. In: Steinberg, H., De Reucht, A. V. S., Knight, J. (Eds.): Animal behaviour and drug action. CIBA Foundation Symposium, pp. 91—113. London: J. and A. Churchill Ltd. 1964b
Stein, L., Ray, O. S.: Brain stimulation reward "thresholds" self-determined in rat. Psychopharmacologia (Berl.) **1**, 251—256 (1960)
Stein, L., Seifter, J.: Possible mode of antidepressive action of imipramine. Science **134**, 286—287 (1961)
Stein, L., Wise, C. D.: Release of hypothalamic norepinephrine by rewarding electrical stimulation of amphetamine in the unanesthetized rat. Fed. Proc. **26**, 651 (1967)
Stein, L., Wise, C. D.: Release of norepinephrine from hypothalamus and amygdala by rewarding medial forebrain bundle stimulation and amphetamine. J. comp. physiol. Psychol. **67**, 189—198 (1969)

Stein, L., Wise, C. D.: Amphetamine and noradrenergic reward pathways. In: Usdin, E., Snyder, S. (Eds.): Frontiers in catecholamine research, pp. 963—968. New York: Pergamon Press 1973
Steinberg, H., Rushton, R., Tinson, C.: Modification of the effects of an amphetamine-barbiturate mixture by the past experience of rats. Nature (Lond.) **192**, 533—535 (1961)
Steiner, S. S., Stokely, S. N.: Methamphetamine lowers self-stimulation thresholds. Physiol. Psychol. **1**, 161—164 (1973)
Stepanek, J.: Veränderungen des Säuren-Basen-Haushalts durch Aminorex und Amphetamin beim Hund. Schweiz. med. Wschr. **102**, 411—414 (1972)
Stewart, G. A.: Drugs, performance and responses to exercise in the racehorse. Aust. vet. J. **48**, 544—547 (1972)
Stille, G.: Die Bedeutung zentralerregender Wirkungen für die Hemmung des Elektrokrampfes. Naunyn-Schmiedebergs Arch. exp. Path. Pharmak. **217**, 57—63 (1953)
Stolk, J. M., Burnett, L. S., Rech, R. H.: Association of amphetamine toxicity and tissue glycogen depletion in mice and rats. Arch. int. Pharmacodyn. **184**, 395—404 (1970)
Stolk, J. M., Rech, R. H.: Enhanced stimulant effects of *d*-amphetamine on the spontaneous locomotor activity of rats treated with reserpine. J. Pharmacol. exp. Ther. **158**, 140—149 (1967)
Stolk, J. M., Rech, R. H.: Species differences in amphetamine toxicity: effects of aggregation, acute and chronic reserpine pretreatment in mice and rats. Life Sci. **7**, 1299—1309 (1968a)
Stolk, J. M., Rech, R. H.: Enhanced stimulant effects of *d*-amphetamine in rats treated chronically with reserpine J. Pharmacol. exp. Ther. **163**, 75—83 (1968b)
Stolk, J. M., Rech, R. H.: Effect of reserpine on accumulation and removal of *d*-amphetamine-^{3}H. Biochem. Pharmacol. **18**, 2786—2788 (1969)
Stolk, J. M., Rech, R. H.: Antagonism of *d*-amphetamine by alpha-methyl-p-tyrosine. Behavioral evidence for the participation of catecholamine stores and synthesis in the amphetamine stimulant response. Neuropharmacology **9**, 247—263 (1970)
Stone, G. C.: Effects of drugs on nondiscriminated avoidance behaviour. I. Individual differences in dose response relationships. Psychopharmacologia (Berl.) **6**, 245—255 (1964)
Stone, G. C.: Effects of drug on avoidance behaviour II. Individual differences in susceptibilities. Psychopharmacologia (Berl.) **7**, 283—302 (1965)
Stone, G. C., Calhoun, D. W., Klopfenstein, M. H.: The interresponse interval as a measure of bar pressing behaviour in normal and drugged rats. J. comp. physiol. Psychol. **51**, 315—319 (1958)
Stone, C. A., Ross, C. A., Wenger, H. C., Ludden, C. T., Blessing, J. A., Totaro, J. A., Porter, C. C.: Effect of α-methyl-3,4-dihydroxyphenylalanine (methyldopa), reserpine and related agents on some vascular responses in the dog. J. Pharmacol. exp. Ther. **136**, 80—88 (1962)
Stowe, F. R., Miller, A. T.: The effect of amphetamine on food intake in rats with hypothalamic hyperphagia. Experientia (Basel) **13**, 114—115 (1957)
Strada, S. J., Sanders-Bush, E., Sulser, F.: p-Chloroamphetamine. Temporal relationship between psychomotor stimulation and metabolism of brain norepinephrine. Biochem. Pharmacol. **19**, 2621—2629 (1970)
Straschill, M., Hoffmann, K. P.: Effects of *d*-amphetamine on the activity of single neurons of the cats tectum opticum. Experientia (Basel) **25**, 373 (1969)
Stretch, R., Gerber, G. J.: Drug-induced reinstatement of amphetamine self-administration behaviour in monkeys. Canad. J. Psychol. **27**, 168—177 (1973)
Stretch, S., Gerber, J., Wood, M.: Factors affecting behavior maintained by response-contingent intravenous infusions of amphetamine in squirrel monkeys. Canad. J. Physiol. Pharmacol. **49**, 581—589 (1971)
Strubelt, O., Vogelberg, W., Zetler, G.: Hemmung der stimulierenden Wirkung von Methamphetamine und Morphin auf die Maus durch Beeinflussung endogener Katecholamine. Arzneimittel-Forsch. **20**, 32—37 (1970)
Stubbs, D. A., Thomas, J. R.: Discrimination of stimulus duration and *d*-amphetamine in pigeons—a psychological analysis. Psychopharmacologia (Berl.) **36**, 313—322 (1974)
Stühlinger, von W., Turnheim, K., Kraupp, O.: Der Einfluß einer N-alkylierung auf die Wirkung des Amphetamin auf die Hämodynamik des großen und kleinen Kreislaufes. Arzneimittel-Forsch. **21**, 342—346 (1971)
Stümpf, C.: Drug action on the electrical activity of the hippocampus. Int. Rev. Neurobiol. **8**, 77—183 (1965)

Sulser, F., Dingell, J. V.: Adrenergic mechanisms in the central action of tricyclic antidepressants and substituted phenothiazines. Agressologie **9**, 1—6 (1968a)

Sulser, F., Dingell, J. V.: Potentiation and blockade of the central action of amphetamine by chlorpromazine. Biochem. Pharmacol. **17**, 634—636 (1968b)

Sulser, F., Owens, L., Dingell, J.: On the mechanism of amphetamine potentiation by desipramine (DMI). Life Sci. **5**, 2005—2010 (1966)

Sulser, F., Owens, M., Norvich, M. R., Dingell, J. V.: The relative role of storage and synthesis of brain norepinephrine in the psychomotor stimulation evoked by amphetamine or by desipramine and tetrabenazine. Psychopharmacologia (Berl.) **12**, 322—332 (1968)

Sulser, F., Sanders-Bush, E.: Effects of drugs on amines in the CNS. Ann. Rev. Pharmacol. **11**, 209—230 (1971)

Svensson, T. H.: Effect of inhibition of catecholamine synthesis on dexamphetamine induced central stimulation. Europ. J. Pharmacol. **12**, 161—166 (1970)

Svensson, T. H.: Functional and biochemical effects of *d*- and *l*-amphetamine on central catecholamine neurons. Naunyn-Schmiedebergs Arch. Pharmak. exp. Path. **271**, 170—180 (1971)

Svensson, T. H., Thieme, G.: An investigation of a new instrument to measure motor activity of small animals. Psychopharmacologia (Berl.) **14**, 151—163 (1969)

Swinyard, E. A., Clark, L. D., Miyahara, J. T., Wolf, H. H.: Studies on the mechanism of amphetamine toxicity in aggregated mice. J. Pharmacol. exp. Ther. **132**, 97—102 (1961)

Syme, L. A., Syme, G. J.: Effects of chlorpromazine and methamphetamine on sociability in rats. Psychopharmacologia (Berl.) **32**, 81—84 (1973)

Tagliamonte, A., Tagliamonte, P., Gessa, G. L.: Cross tolerance to the anorexigenic effect of *d*-amphetamine, *l*-amphetamine, fenfluramine and desmethylimipramine in rats. Pharmacologist **11**, 264 (1969)

Tagliamonte, A., Tagliamonte, P., Perez-Cruet, J., Gessa, G.: Increase of brain tryptophan caused by drugs which stimulate serotonin synthesis. Nature (Lond.) New Biol. **229**, 125—126 (1971a)

Tagliamonte, A., Tagliamonte, P., Perez-Cruet, J., Stern, S., Gessa, G.: Effect of psychotropic drugs on tryptophan concentration in the rat brain. J. Pharmacol. exp. Ther. **177**, 475—480 (1971b)

Tainter, M. L.: Effects of certain analeptic drugs on spontaneous running activity of the white rat. J. comp. physiol. Psychol. **36**, 143—145 (1943)

Tainter, M. L., Tainter, E. G., Sherwood Lawrence, W., Neuru, E. N., Lackey, R. W., Laduena, F. P., Kirtland, H. B., Gonzalez, R. I.: Influence of various drugs on threshold for electrical convulsions. J. Pharmacol. exp. Ther. **79**, 42—51 (1943)

Takasaki, K., Nishizaea, T., Igisu, T., Kaneko, M.: Relationship between the changes on blood pressure response and the cardiac catecholamine level after single and repeated administrations of tyramine and methamphetamine in rats. Arch. int. Pharmacodyn. **206**, 150—160 (1973)

Takeo, Y., Himwich, H. E.: Comparative electroencephalographic study of phenylethylamines and their methoxy derivatives. Arch. int. Pharmacodyn. **166**, 47—59 (1967)

Taylor, B.: Dopamine-β-hydroxylase—stereochemical course of the reaction. J. biol. Chem. **249**, 454—458 (1974)

Taylor, K. M., Laverty, R.: The effects of drugs on the behavioural and biochemical action of intraventricular 6-hydroxydopamine. Europ. J. Pharmacol. **17**, 16—24 (1972)

Taylor, K. M., Snyder, S. H.: Amphetamine: Differentiation by *d* and *l* isomers of behavior involving brain norepinephrine of dopamine. Science **168**, 1487—1489 (1970)

Taylor, K. M., Snyder, S. H.: Differential effects of *d*- and *l*-amphetamine on behavior and on catecholamine disposition in dopamine and norepinephrine containing neurons of rat brain. Brain Res. **28**, 295—310 (1971)

Taylor, W. A., Sulser, F.: Effects of amphetamine and its hydroxylated metabolites on central noradrenergic mechanisms. J. Pharmacol. exp. Ther. **185**, 620—632 (1973)

Tedeschi, D. H., Fowler, P. F., Miller, R. B., Macko, E.: Pharmacological analysis of footshock-induced fighting behaviour. Aggressive behaviour, pp. 245—252. Eds. Garattini, S., Sigg, E. B. (Eds.). Amsterdam: Excerpta Medica Foundation 1969

Tedeschi, R. E., Tedeschi, D. H., Ames, P. L., Cook, L., Mattis, P. A., Fellows, E. J.: Some pharmacological observations on tranylcypromine, (SKF trans-385): a potent inhibitor of monoamine oxidase. Proc. Soc. exp. Biol. (N.Y.) **102**, 380—381 (1959)

Teitelbaum,P., Derks,P.: The effect of amphetamine on forced drinking in the rat. J. comp. physiol. Psychol. **51**, 801—810 (1958)
Terada,C.W., Masur,J.: Amphetamine- and apomorphine-induced alteration of the behavior of rats submitted to a competitive situation in a straight runway. Europ. J. Pharmacol. **24**, 375—380 (1973)
Terwelp,D.R., Kennedy,M.S., Burks,T.F.: Temperature responses to intraventricular amphetamine in unanesthetized cats. Res. Commun. chem. Path. Pharmacol. **6**, 795—804 (1973)
Thode,W.F., Carlisle,H.J.: Effect of lateral hypothalamic stimulation on amphetamine-induced anorexia. J. comp. physiol. Psychol. **66**, 517—548 (1968)
Thoenen,H., Hürlimann,A., Gey,K.F., Haefely,W.: Liberation of p-hydroxynorepinephrine from cat spleen by sympathetic stimulation after pretreatment with amphetamine. Life Sci. **5**, 1715—1722 (1966)
Thoenen,H., Hürlimann,A., Haefely,W.: Mechanism of amphetamine accumulation in the isolated perfused heart of the rat. J. Pharm. Pharmacol. **20**, 1—11 (1968)
Thompson,T., Pickens,R.: Stimulant self-administration by animals: some comparisons with opiate self-administration. Fed. Proc. **29**, 6—12 (1970)
Thor,D.H.: Amphetamine induced fighting during morphine withdrawal. J. gen. Psychol. **84**, 245—250 (1971)
Thornburg,J.E., Moore,K.E.: A comparison of the locomotor stimulant properties of amantadine and *l*- and *d*-amphetamine in mice. Neuropharmacology **11**, 675—682 (1972)
Thornburg,J.E., Moore,K.E.: Dopamine and norepinephrine uptake by rat brain synaptosomes: Relative inhibitory of *l*- and *d*-amphetamine amantadine. Res. Commun. chem. Path. Pharmacol. **5**, 81—89 (1973a)
Thornburg,J.E., Moore,K.E.: The relative importance of dopaminergic and noradrenergic neuronal system for the stimulation of locomotor activity induced by amphetamine and other drugs. Neuropharmacology **12**, 853—866 (1973b)
Thuillier,J., Nakajima,H., Grandjean,J.L., Kurihara,K.: Discrimination de psychoanaleptiques par l'hyperthermie provoquée par l'amide lysergique chez le lapin. C.R. Soc. Biol. (Paris) **156**, 445—446 (1962)
Tice,P., Spooner,C.E.: Antagonism of *d*-amphetamine by methysergide. Pharmacologist **12**, 226 (1970)
Tikal,K., Benesova,O.: The effect of some psychotropic drugs on contact behavior in a group of rats. Activ. nerv. sup. (Praha) **14**, 168—169 (1972)
Tilson,H.A., Rech,R.H.: Prior drug experience and effects of amphetamine on schedule controlled behavior. Pharmacol. Biochem. Behav. **1**, 129—132 (1973a)
Tilson,H.A., Rech,R.H.: Conditioned drug effects and absence of tolerance to *d*-amphetamine induced motor activity. Pharmacol. Biochem. Behav. **1**, 149—153 (1973b)
Tilson,H.A., Sparber,S.B.: Studies on the concurrent behavioral and neurochemical effects of psychoactive drugs using the push-pull cannula. J. Pharmacol. exp. Ther. **181**, 387—398 (1972)
Tilson,H.A., Sparber,S.B.: Similarities and differences between mescaline, lysergic acid diethylamid-25 (LSD) and *d*-amphetamine on various components of fixed interval responding in the rat. J. Pharmacol. exp. Ther. **184**, 376—384 (1973a)
Tilson,H.A., Sparber,S.B.: The effects of *d*- and *l*-amphetamine on fixed-interval and fixed-ratio behaviour in tolerant and nontolerant rats. J. Pharmacol. exp. Ther. **187**, 372—379 (1973b)
Todorov,J.C., Gorayeb,S.R.P., Correa,D.L., Graeff,F.G.: Effects of amphetamine on choice behavior of pigeons. Psychopharmacologia (Berl.) **26**, 395—400 (1972)
Török,T.: Amphetamines and locomotor activity. Activ. nerv. sup. (Praha) **14**, 302—303 (1972)
Török,T., Vizi,E.S., Knoll,J.: Különbözö amfetaminok szerkezete valamint az alapanyagcseret es hömersékletet fo ozó hatása közötti összefügges viszgálata. Acta pharm. hung. **43**, 126—136 (1973)
Toman,J.E.P., Davis,J.P.: The effects of drugs upon the electrical activity of the brain. Pharmacol. Rev. **1**, 245—292 (1949)
Tonini,G., Riccioni,M.L., Missere,G.: Evaluation of central pharmacological actions in rats with septal lesions. In: Votava,Z., Horvath,M., Vinar,O. (Eds.): Psychopharmacological methods, pp. 106—114. Oxford: Pergamon Press 1963
Tormey,J., Lasagna,L.: Relation of thyroid function to acute and chronic effects of amphetamine in the rat. J. Pharmacol. exp. Ther. **128**, 201—209 (1960)

Trendelenburg, U.: Supersensitivity and subsensitivity to sympathomimetic amines. Pharmacol. Rev. **15**, 225—276 (1963)

Trendelenburg, U., Gomez, B. A. S., Muskus, A.: Modification by reserpine of the response of the atrial pacemaker to sympathomimetic amines. J. Pharmacol. exp. Ther. **141**, 301—309 (1963)

Trendelenburg, U., Muskus, A., Fleming, W. W., Gomez, A. B.: Modification by reserpine of the action of sympathomimetic amines in spinal cats: a classification of sympathomimetic amines. J. Pharmacol. exp. Ther. **138**, 170—180 (1962a)

Trendelenburg, U., Muskus, A., Fleming, W. W., Gomez, A. B.: Effect of cocaine, denervation and decentralization on the response of the nictitating membrane to various sympathomimetic amines. J. Pharmacol. exp. Ther. **138**, 181—193 (1962b)

Tripod, J.: Beeinflussung der zentral-erregenden Wirkung von Weckaminen durch Pharmaka mit spezifischer Wirkung auf das autonome Nervensystem. Helv. physiol. pharmacol. Acta **10**, 403—412 (1952)

Tripod, J., Benn, H. J., Meier, R.: Characterization of central effects of serpasil (Reserpine) a new alcaloid of Rauwolfia serpentina B and their antagonistic reaction. Arch. int. Pharmacodyn. **96**, 406—425 (1954)

Tseng, L. F., Hitzeman, R. J., Loh, H. H.: Comparative effects of *dl*-methoxyamphetamine and *d*-amphetamine on catecholamine release and reuptake in vitro. J. Pharmacol. exp. Ther. **189**, 708—716 (1974)

Tseng, L. F., Loh, H. H.: Significance of dopamine receptor activity in *dl*-p-methoxyamphetamine and *d*-amphetamine induced locomotor activity. J. Pharmacol. exp. Ther. **189**, 717—724 (1974)

Tseng, L. F., Walaszek, E. J.: Blockade of *d*-amphetamine-induced stereotyped behaviour by bulbocapnine. Pharmacology (Basel) **7**, 255—258 (1972)

Tsuyussaki, T., Kobayshi, T.: Pharmacological analysis of voluntary movement of mouse. J. Chiba Med. Soc. (Japan) **42**, 291—304 (1966)

Uehling, B. S.: Effects of chronic *d*-amphetamine sulfate administration during development in rats. Int. J. Neuropharmacol. **8**, 43—48 (1969)

Umemoto, M., Kido, R.: Depressing effect of methamphetamine on self-stimulation in the cat. Nature (Lond.) **216**, 1333—1334 (1967)

Ungerstedt, U.: Striatal dopamine release after amphetamine or nerve degeneration revealed by rotational behaviour. Acta physiol. scand. **367**, 49—68 (1971a)

Ungerstedt, U.: On the anatomy, pharmacology and function of the nigrostriatal dopamine system. Stockholm: Kungl. Boktryckeriet, P.A. Norstedt & Söner 1971b

Ungerstedt, U., Arbuthnott, G. W.: Quantitative recording of rotational behaviour in rats after 6-hydroxydopamine lesions of the nigrostriatal dopamine system. Brain Res. **24**, 485—493 (1970)

Utena, H.: Behavioral aberrations in methamphetamine-intoxicated animals and chemical correlates in the brain. Progr. Brain Res. **21**, 192—207 (1966)

Utena, H., Ezoe, T., Kato, N., Hada, H.: Effects of chronic administration of methamphetamine in enzymatic patterns in brain tissue. J. Neurochem. **4**, 161—169 (1959)

Vaccari, A.: Hyposensitivity to 5-hydroxytryptamine in the isolated stomach fundus of the newborn rat. II. Amphetamine and receptor deficit. Biochem. Pharmacol. **17**, 2493—2495 (1968)

Vaccari, A., Cugurra, F.: Relationships between N-acetylneuraminic acid, serotonin and amphetamines on the isolated rat stomach preparation. Biochem. Pharmacol. **17**, 2399—2406 (1968)

Vaccari, A., Vertua, R.: ^{14}C-5-hydroxytryptamine and ^{3}H-D-amphetamine: uptake and contraction by the rat stomach fundus in vitro. Biochem. Pharmacol. **19**, 2105—2115 (1970)

Vaccari, A., Vertua, R., Furlani, A.: Decreased calcium uptake by rat fundal strips after pretreatment with neuraminidase or LSD in vitro. Effect of serotonin, *d*-amphetamine and eledoisin on this uptake. Biochem. Pharmacol. **20**, 2603—2612 (1971)

Valdecasas, F. G., Salva, J. A., Cuenca, E.: Action de la Réserpine sur le chat spinal traité avec des inhibiteurs de l'aminoxydase. Neuropsychopharmacology **1**, 421—426 (1958)

Valette, G., Jacquot, C., Heimburger, M.: Crossed tachyphylaxis: ephedrine, amphetamine and their own metabolites. J. Pharmacol. (Paris) **2**, 349—358 (1971)

Valzelli, L., Consolo, S., Morpurgo, C.: Influence of imipramine-like drugs on the metabolism of amphetamine. Garattini, S. (Ed.): Antidepressant drugs, pp. 61—69. Amsterdam: Excerpta Medica 1967b

Valzelli,L., Dolfini,E., Tansella,M., Garattini,S.: Activity of centrally acting drugs on amphetamine metabolism. J. Pharm. Pharmacol. **20**, 595—599 (1968)
Valzelli,L., Giacalone,E., Garattini,S.: Pharmacological control of aggressive behavior in mice. Europ. J. Pharmacol. **2**, 144—146 (1967a)
Vane,J.R.: The actions of sympathomimetic amines on tryptaminergic receptors. CIBA Foundation Symposium on Adrenergic Mechanisms, pp. 356—372. Vane,J.R., Wolstenholme,G.E.W., O'Connor,M. (Eds.). London: Churchill 1960
Vasko,M.R., Domino,L.E., Domino,E.F.: Differential effects of *d*-amphetamine on brain acetylcholine in young adult, or geriatric rats. Europ. J. Pharmacol. **27**, 145—147 (1974)
Vasquez,AJ., Izquierdo,I.: The effect of amphetamine, nicotine and atrophine on pseudoconditioned responses of rats. Pharmacology (Basel) **3**, 21—24 (1970)
Vasquez,A.J., Krip,G.: Interaction between amphetamine and serotoninergic agents on cat's isolated cerebral cortex. Biol. Psychiatry **7**, 11—22 (1973)
Vedernikov,Y.P.: Interaction of amphetamine, apomorphine, disulfiram with morphine and the role played by catecholamines in morphine analgesic action. Arch. int. Pharmacodyn. **182**, 59—64 (1969)
Vedernikov,Y.P.: The influence of single and chronic morphine administration on some central effects of amphetamine and apomorphine. Psychopharmacologia (Berl.) **17**, 283—288 (1970)
Vergnes,M., Chaurand,J.P.: Activation amphétaminique, rythme hippocampoque et comportement d'aggression interspécifique rat-souris. C.R. Soc. Biol. (Paris) **166**, 936—941 (1972)
Verhave,T.: The effect of methamphetamine on operant level and avoidance behaviour. J. exp. Anal. Behav. **1**, 207—219 (1958)
Verhave,T.: A depressant effect of methamphetamine on avoidance behaviour. Fed. Proc. **20**, 395 (1961)
Viala,D., Buser,P.: Modalités d'obtention de rythmes locomoteures chez le lapin spinal par traitements pharmacocogiques (dopa, 5-HTP, *d*-amphétamine). Brain Res. **35**, 151—165 (1971)
Villarreal,J.E., Guzman,M., Smith,C.B.: A comparison of the effects of *d*-amphetamine and morphine upon the locomotor activity of mice, treated with drugs which alter brain catecholamine content. J. Pharmacol. exp. Ther. **187**, 1—7 (1973)
Vizi,E.S., Knoll,B., Knoll,J.: Interaction between phenylalkylamine derivatives and reserpine. Int. J. Neuropharmacol. **7**, 503—509 (1968)
Voigtländer,P.F. von: Axonal transport—reciprocal effects of neuroleptics and *d*-amphetamine. Brain Res. **72**, 370—374 (1974)
Voigtländer,P.F. von, Moore,K.E.: Turning behavior of mice with unilateral 6-hydroxydopamine lesions in the striatum: effects of apomorphine, L-DOPA, amantadine, amphetamine and other psychomotor stimulants. Neuropharmacology **12**, 451—462 (1973a)
Voigtländer,P.F. von, Moore,K.E.: Involvement of nigrostriatal neurons in the in vivo release of dopamine by amphetamine, amantadine and tyramine. J. Pharmacol. exp. Ther. **184**, 542—552 (1973b)
Vtyurin,B.V., Tumanov,V.P.: The ultrastructure of the synaptic apparatus after the injection of phenamine and haloperidol. Bull. exp. Biol. Med. **72**, 110 (1971)
Vycklicky,L., Tabin,T.: Effect of amphetamine on spinal reflexes. Nature (Lond.) **204**, 384 (1964)
Wada,J.A., Wrinch,J., Hill,D., McGeer,P.L., McGeer,E.G.: Central aromatic amine levels and behaviour. Arch. Neurol. (Chic.) **9**, 69—80 (1963)
Wagner,L.A., Koerker,R.L., Schneider,F.H.: Influence of different anions on catecholamine release: induced by tyramine and amphetamine. Biochem. Pharmacol. **22**, 430 (1973)
Wagner,L.A., Koerker,R.L., Schneider,F.H.: Influence of different anions on tyramine and amphetamine uptake by cow adrenal medulla chromaffin vesicles. J. Pharmacol. **26**, 464—467 (1974)
Wahlström,G., Widerlöv,E.: The effects of (±)-amphetamine sulphate on the self-selected circacian rhythm of activity and rest in the canary. Brit. J. Pharmacol. **34**, 218P (1968)
Wallach,M.B., Angrist,B.M., Gershon,S.: The comparison of the stereotyped behavior-induced effects of *d*- and *l*-amphetamine in dogs. Commun. behav. Biol. **6**, 93—96 (1971)
Wallach,M.B., Gershon,S.: Sensitization to amphetamines, pp. 30—31. Puerto Rico: ACNP Meeting 1970
Wallach,M.B., Gershon,S.: A neuropsychopharmacological comparison of *d*-amphetamine, L-dopa and cocaine. Neuropharmacology **10**, 743—752 (1971)

Wallach, M. B., Gershon, S.: The induction and antagonism of central nervous system stimulant-induced stereotyped behaviour in the cat. Europ. J. Pharmacol. **18**, 22—26 (1972)
Wang, R. I. H., Hasegawa, A. T., Peters, N. J., Rimm, A.: Amphetamine toxicity in isolated and aggregated mice. Influence of aggregation with nonamphetamine treated, sedated mice. Psychopharmacologia (Berl.) **15**, 102—108 (1969)
Wanner, H. U., Bättig, K.: Wirkung von Nikotin und Amphetamin auf die Selbstreizung bei der Ratte. Helv. physiol. pharmacol. Acta **24**, 122—124 (1966)
Water, W. H., Richards, D. W., Harris, R. T.: Discriminative control and generalization of the stimulus properties of *d,l*-amphetamine in the rat. In: Singh, J. M., Miller, L. H., Lal, H. (Eds.): Drug addiction: experimental pharmacology, pp. 87—98. New York: Futura Publ. Company 1972
Waterman, F. A.: Relationship between spontaneous activity and metabolic rate as influenced by certain sympathomimetic compounds. Proc. Soc. exp. Biol. (N.Y.) **71**, 473—475 (1949)
Wauquier, A., Niemegeers, C. J.: Intracranial selfstimulation in rats as a function of various stimulus parameters. IV. Influence of amphetamine on medial forebrain bundle stimulation with monopolar electrodes. Psychopharmacologia (Berl.) **34**, 265—274 (1974)
Wayner, M. J., Greenberg, I., Trowbridge, J.: Effects of *d*-amphetamine on schedule induced polydipsia. Pharmacol. Biochem. Behav. **1**, 109—111 (1973)
Weaver, L. C., Kerley, T. L.: Strain difference in response of mice to *d*-amphetamine. J. Pharmacol. exp. Ther. **135**, 240—244 (1962)
Weiner, N.: Pharmacology of central nervous system stimulants. Zarafonetis, C. J. D. (Ed.): Proceedings of the international conference Drug abuse, pp. 243—251. Philadelphia: Lea and Febiger 1972
Weiner, W. J., Goetz, L., Westheimer, R., Klawans, H. L.: Serotoninergic and antiserotoninergic influences on amphetamine induced stereotyped behaviour. J. neurol. Sci. **20**, 373—379 (1973)
Weinstock, M., Speiser, Z.: Modification by propranolol and related compounds of motor activity and stereotype behaviour induced in the rat by amphetamine. Europ. J. Pharmacol. **25**, 29—35 (1974)
Weis, J.: On the hyperthermic response to *d*-amphetamine in the decapitated rat. Life Sci. **13**, 475—484 (1973)
Weischer, M.-L.: Einfluß von Anorektica der Amphetamin-Reihe auf das Verhalten des Siamesischen Kampffisches *Betta splendens*. Arzneimittel-Forsch. **16**, 1310—1311 (1966)
Weiss, B.: Amphetamine and the temporal structure of behaviour. In: Costa, E., Garattini, S. (Eds.): International Symposium on Amphetamines and Related Compounds, pp. 797—812. New York: Raven Press 1970
Weiss, B., Laties, V. G.: Effects of amphetamine, chlorpromazine and pentobarbital on behavioural thermoregulation. J. Pharmacol. exp. Ther. **140**, 1—7 (1963)
Weiss, B., Laties, V. G.: Effects of amphetamine, chlorpromazine, pentobarbital and ethanol on operant response duration. J. Pharmacol. exp. Ther. **144**, 17—23 (1964)
Weiss, B., Laties, V. G., Blanton, F. L.: Amphetamine toxicity in rats and mice subjected to stress. J. Pharmacol. exp. Ther. **132**, 366—371 (1961)
Weissman, A.: Differential drug effects upon a three-ply multiple schedule of reinforcement. J. exp. Anal. Behav. **2**, 271—284 (1959)
Weissman, A.: Interaction effects of imipramine and *d*-amphetamine on non-discriminated avoidance. Pharmacologist **8**, 60 (1961)
Weissman, A.: Correlation between baseline non-discriminated avoidance behaviour in rats and amphetamine induced stimulation. Psychopharmacologia (Berl.) **4**, 294—297 (1963)
Weissman, A.: Effects of amphetamine on locomotion in mice pretreated with desmethylimipramine, tetrabenazine or both. Psychopharmacologia (Berl.) **23**, 152—159 (1972)
Weissman, A., Koe, B. K.: Behavioral effects of L-α-methyltyrosine, an inhibitor of tyrosine hydroxylase. Life Sci. **4**, 1037—1040 (1965)
Weissman, A., Koe, B. K., Tenen, S.: Antiamphetamine effects following inhibition of tyrosine hydroxylase. J. Pharmacol. exp. Ther. **151**, 339—352 (1966)
Welch, B., Welch, A.-M.: Graded effect of social stimulation upon *d*-amphetamine toxicity, aggressiveness and heart and adrenal weight. J. Pharmacol. exp. Ther. **151**, 331—338 (1966)
Welch, B. L., Welch, A. S.: Aggression and the biogenic amine neurohumors. Garattini, S., Sigg, E. B. (Eds.): Aggressive behaviour, pp. 188—202. Amsterdam: Excerpta Medica Foundation 1969

Welch, B. L., Welch, A. S.: Control of brain catecholamines and serotonin during acute stress and after *d*-amphetamine by natural inhibition of monoamine oxidase: an hypothesis. In: Costa, E., Garattini, S. (Eds.): International Symposium on Amphetamines and Related Compounds, pp. 415—445. New York: Raven Press 1970

Welch, B. L., Welch, A. S.: Chronic social stimulation and tolerance to amphetamine: Interacting effects of amphetamine and natural nervous stimulation upon brain amines and behavior. In: Ellinwood, E. H., Cohen, S. (Eds.): Current concepts on amphetamine abuse, pp. 107—115. DHEW Publ. No 72-9085. Washington D.C.: U.S. Government Printing Office 1972

Wenger, G. R., Rutledge, C. O.: A comparison of the effects of amphetamine and its metabolites p-hydroxyamphetamine and p-hydroxynorephedrine on uptake, release and catabolism of ^3H-norepinephrine in cerebral cortex of rat brain. J. Pharmacol. exp. Ther. **189**, 725—732 (1974)

Wentink, E. A.: The effects of certain drugs and hormones upon conditioning. J. exp. Psychol. **22**, 150—163 (1938)

Wenzel, D., Broadie, L.: Effect of group size and cage design on voluntary running activity of amphetamine-treated mice. Arch. int. Pharmacodyn. **159**, 154—160 (1966)

White, R. P., Boyajy, L. D.: Comparison of physostigmine and amphetamine in antagonizing the EEG effects of CNS depressants. Proc. Soc. exp. Biol. (N.Y.) **102**, 479—483 (1959)

White, R. P., Daigneault, E. A.: The antagonism of atropine to the EEG effects of adrenergic drugs. J. Pharmacol. exp. Ther. **125**, 339—346 (1959)

White, R. P., Nash, C. B., Westerbeke, E. J., Possanza, G. J.: Phylogenetic comparison of central actions produced by different doses of atropine and hyoscine. Arch. int. Pharmacodyn. **132**, 349—363 (1961)

Widerlöv, E., Wahlström, G.: Synergism between hexobarbitone and amphetamine on the EEG burst-suppression threshold. Acta pharmacol. (Kbh.) **25**, 64 (1967)

Will, A., Bisgard, E.: Haemodynamic effects of oral aminorex and amphetamine in unanaesthetized beagle dogs. Thorax **27**, 120—126 (1972)

Will, B., Checchinato, D.: Effects of *d*-amphetamine on operant behaviour in the rat. Psychopharmacologia (Berl.) **29**, 141—149 (1973)

Willner, J. H., Samach, M., Angrist, B. M., Wallach, M. B., Gershon, S.: Drug-induced stereotyped behavior and its antagonism in dogs. Commun. behav. Biol. **5**, 135—141 (1970)

Wilson, M. C., Hitomi, M., Schuster, C. R.: Psychomotor stimulant self-administration as a function of dosage per injection in the rhesus monkey. Psychopharmacologia (Berl.) **22**, 271—281 (1971)

Wilson, M. C., Schuster, C. R.: The effects of chlorpromazine stimulant self-administration in the rhesus monkey. Psychopharmacologia (Berl.) **26**, 115—126 (1972a)

Wilson, M. C., Schuster, C. R.: The effect of various pharmacological agents on cocaine self-administration by Rhesus Monkeys. In: Ellinwood, E. H., Cohen, S. (Eds.): Current concepts on amphetamine abuse, pp. 37—41. DHEW Publ. No. (HSM) 72-9085. Washington D.C.: U.S. Government Printing Office 1972b

Wilson, W. S.: The effects of pentobarbitone, leptazol, dexamphetamine, iproniazid, imipramine, LSD, chlorpromazine, reserpine and hydroxazine on the in vivo levels of adenenine nucleotides and phosphocreatine in the rat brain. Brit. J. Pharmacol. **36**, 448—457 (1969)

Wimer, R. E., Fuller, J. L.: The effects of *d*-amphetamine sulphate on three exploratory behaviours. Canad. J. Psychol. **19**, 94—103 (1965)

Winder, C. V., Anderson, M. M., Parke, H. C.: Comparative properties of six phenylethylamines with observations on the nature of tachyphylaxis. J. Pharmacol. exp. Ther. **93**, 63—71 (1948)

Winters, W. D., Moi, K., Wallach, M. B., Marcus, R. J., Spooner, C. E.: Reticular multiple unit activity during a progression of states induced by CNS excitants. III. Electroencenceph. clin. Neurophysiol. **27**, 514—522 (1969)

Wise, C. D., Stein, L.: Amphetamine: facilitation of behavior by augmented release of norepinephrine from the medial forebrain bundle. In: Costa, E., Garattini, S. (Eds.): International Symposium on Amphetamines and Related Compounds, pp. 463—485. New York: Raven Press 1970

Wishart, T. B., Walls, E. K.: The effects of anorexic doses of dextroamphetamine on the ventromedial-hypothalamic hyperphagic rat. Canad. J. Physiol. Pharmacol. **51**, 354—359 (1973)

Wolf, H. H., Bunce, M. E.: Hyperthermia and the amphetamine aggregation phenomenon: absence of a causal relation. J. Pharm. Pharmacol. **25**, 425—427 (1973)

Wolf, H.H., Rollins, D.E., Rowland, C.R., Reigle, R.G.: The importance of endogenous catecholamines in the activity of some CNS stimulants. Int. J. Neuropharmacol. **8**, 319—328 (1969)

Wolthuis, O.L.: Experiments with UCB 6215, a drug which enhances acquisition in rats: its effects compared with those of methamphetamine. Europ. J. Pharmacol. **16**, 283—297 (1971)

Wong, D.T., Van Frank, R.M., Horng, J.S., Fuller, R.W.: Accumulation of amphetamine and p-chloroamphetamine into synaptosomes of rat brain. J. Pharm. Pharmacol. **24**, 171—173 (1972)

Wong, D.T., Horng, J.S., Fuller, R.W.: Kinetics of serotonin accumulation into synaptosomes of rat brain—effects of amphetamine and chloroamphetamines. Biochem. Pharmacol. **22**, 311 (1973)

Wray, R., Cowan, A.: The behavioral effects of levallorphan, cyprenorphine (M 285) and amphetamine on repeated Y-maze performance in rats. Psychopharmacologia (Berl.) **21**, 257—267 (1971)

Wuttke, W.: The effect of *d*-amphetamine on schedule controlled licking in the squirrel monkey. Psychopharmacologia (Berl.) **17**, 70—82 (1970)

Wuttke, W., Innis, N.K.: Drug effects upon behaviour induced by second order schedules of reinforcement; the relevance of ethological analysis. In: Gilbert, R.M., Keehn, J.D. (Eds.): Schedule effects; Drugs, drinking and aggression. Toronto: University of Toronto Press 1972

Yagi, B.: Studies in general activity. The effect of methamphetamine. Ann. Anim. Psychol. **13**, 17—47 (1963)

Yasunobu, K.T., Oi, S.: Mechanistic aspects of the bovine hepatic monoamineoxidase reaction. Advanc. Biochem. Psychopharmacol. **5**, 91—105 (1972)

Yehuda, S., Wurtman, R.J.: The effects of *d*-amphetamine and related drugs on colonic temperatures of rats kept at various ambient temperatures. Life Sci. **11**, 851—859 (1972a).

Yehuda, S., Wurtman, R.J.: Release of brain dopamine as the probable mechanism for the hypothermic effect of *d*-amphetamine. Nature (Lond.) **240**, 477—478 (1972b)

Yehuda, S., Wurtman, R.J.: Effects of *d*-amphetamine on body temperature and on behavioral thermoregulation in rats. Usdin, E., Snyder, S. (Eds.): Frontiers in catecholamine research, p. 948. New York: Pergamon Press 1973

Yehuda, S., Wurtman, R.J.: Paradoxical thermoregulatory behaviour in rats induced by (+)-amphetamine: blockade by a-adrenoceptor or dopamine receptor blocking agents. J. Pharm. Pharmacol. **26**, 210—212 (1974a)

Yehuda, S., Wurtman, R.J.: Paradoxical effects of *d*-amphetamine on behavioral thermoregulation: possible mediation by brain dopamine. J. Pharmacol. exp. Ther. **190**, 118—122 (1974b)

Yelnosky, J., Hewson, R., Mundy, J., Mitchell, J.: The cardiovascular effects of aminorex, a new anorexigenic agent. Arch. int. Pharmacodyn. **164**, 412—418 (1966a)

Yelnosky, J., Lawlor, R.B.: A comparative study of the pharmacologic actions of amphetamine and fenfluramine. Arch. int. Pharmacodyn. **184**, 374—388 (1970)

Yelnosky, J., McGill, J.S., Mastrangelo, A.S.: A comparison of the blood pressure effects of reserpine in dogs pretreated with amphetamine or tyramine. Arch. int. Pharmacodyn. **159**, 416—423 (1966b)

Yen, T.T., Acton, J.M.: Stimulation of locomotor activity of genetically obese mice by amphetamine. Experientia (Basel) **29**, 1297—1298 (1973)

Yokel, R.A., Pickens, R.: Chronic methamphetamine selfadministration by rats, Pharmacologist **12**, 226 (1970)

Yokel, R.A., Pickens, R.: Self-administration of optical isomers of amphetamine and methylamphetamine by rats. J. Pharmacol. exp. Ther. **187**, 27—33 (1973)

Yokel, R.A., Pickens, R.: Drug level of *d*- and *l*-amphetamine during intravenous self-administration. Psychoparmacologia (Berl.) **34**, 255—264 (1974)

Zalis, E.G., Kaplan, G.: The effect of aggregation on amphetamine toxicity in the dog. Arch. int. Pharmacodyn. **159**, 196—199 (1966)

Zalis, E.G., Kaplan, G., Lundberg, G.D., Knutsson, R.A.: Acute lethality of the amphetamines in dogs and its antagonism by curare. Proc. Soc. Exp. Biol. (N.Y.) **118**, 557—561 (1965)

Zalis, E.G., Lundberg, G.D., Knutsson, R.A.: The pathophysiology of acute amphetamine poisoning with pathologic correlation. J. Pharmacol. exp. Ther. **158**, 115—127 (1967)

Zetler, G.: Anticataleptic actions of amantadine hydrochloride. Naunyn-Schmiedebergs Arch. Pharmak. exp. Path. **266**, 276—278 (1970)

Zetler, G., Mahler, K., Daniel, F.: Versuche zu einer pharmakologischen Differenzierung kataleptischer Wirkungen. Naunyn-Schmiedebergs Arch. exp. Path. Pharmak. **238**, 486—501 (1960)
Zetler, G., Moog, E.: Die Bulbocapnin-Katatonie, ihre Synergisten und Antagonisten. Naunyn-Schmiedebergs Arch. exp. Path. Pharmak. **232**, 442—458 (1958)
Zetler, G., Thörner, R.: Drug-induced catalepsy as influenced by psychostimulants, apomorphine, L-dopa and yohimbine. Pharmacology (Basel) **10**, 238 (1973)
Ziance, R.J., Azzaro, A.J., Rutledge, C.O.: Characteristics of amphetamine induced release of norepinephrine from rat cerebral cortex in vitro. J. Pharmacol. exp. Ther. **182**, 284—294 (1972)
Ziance, R.J., Rutledge, C.O.: A comparison of the effects of fenfluramine and amphetamine on uptake release and catabolism of norepinephrine in rat brain. J. Pharmacol. exp. Ther. **180**, 118—126 (1972)
Ziegler, H., del Basso, P., Longo, V.G.: Influence of 6-hydroxydopamine and of α-methyl-p-tyrosine on the effects of some centrally acting agents. Physiol. Behav. **8**, 391—396 (1972)
Ziem, M.: Beziehungen zwischen zentralnervösen Wirkungen und dem Gehalt von Amphetamin in Leber und Gehirn bei jungen und alten Ratten. Naunyn-Schmiedebergs Arch. Pharmak. **266**, 480 (1970)
Ziem, M., Coper, H., Broermann, I., Strauss, S.: A comparison of some effects of amphetamine in rats of different ages (German). Naunyn-Schmiedebergs Arch. Pharmak. **267**, 208—223 (1970)
Zieve, L.: Effect of benzedrine on activity. Psychol. Rec. **1**, 393—396 (1937)
Zimmerman, J., Schuster, C.R.: Spaced responding in multiple drl schedules. J. exp. Anal. Behav. **5**, 497—504 (1962)
Zsoter, T., Tom, H., Kraml, M., Dvornik, D.: Adrenergic beta receptors and plasma free fatty acids. J. Pharmacol. exp. Ther. **152**, 425—431 (1966)
Zucker, I., Milner, P.: The effect of amphetamine on established hoarding in the rat. Psychon. Sci. **1**, 367 (1964)
Zwarenstein, H., Sapeika, N.: Effects of fenfluramine and dexamphetamine on phosphoglucomutase in vitro. Europ. J. Pharmacology **12**, 243—245 (1970)
Zwieten van, P.A.: The central action of antihypertensive drugs, mediated via central α-receptors. J. Pharm. Pharmacol. **25**, 89—95 (1973)

C. Effects of Amphetamines in Humans

L.-M. GUNNE

A. General Considerations

Due to various untoward effects and above all the development of widespread abuse and dependence on amphetamines, the medical reasons for prescribing this category of drugs have become increasingly sparse. In Scandinavian countries they are no longer prescribed for weight-reducing purposes. The main remaining indications are narcolepsy (see below under Wakefulness B.III.), the minimal brain damage syndrome in children (see below under Performance B.II.) and possibly as an adjunct in the treatment of parkinsonism (MILLER and NIEBURG, 1973).

The early literature on the action of amphetamine in humans contains reports of highly variable effects, e.g., blood-pressure rise and occasionally blood-pressure fall have resulted from amphetamine administration (FRANKE, 1938; LOMAN et al., 1939; PÜLLEN, 1939). Furthermore, highly variable effects on heart-rate, body temperature, and basal metabolism have been reported (REIFENSTEIN and DAVIDOFF, 1939a, b; IVY and KRASNO, 1941; IVY and GOETZL, 1943; BONHOFF and LEWRENZ, 1954). The subjective effects of amphetamine have been described either as an elevation of mood (in most cases) or, rarely, as a depression or sedation (NATHANSON, 1937; GWYNN and YATER, 1937; BAHNSEN et al., 1938; JACOBSEN et al., 1938; ANDERSON, 1938; HUNDLEY et al., 1939; TURNER and CARL, 1939; REIFENSTEIN and DAVIDOFF, 1939a, b; EVANS and SMITH, 1963; CAMERON et al., 1965; LADER, 1969; HOLLISTER and GILLESPIE, 1970). In individual cases the rise in blood pressure and the psychological phenomena do not necessarily coincide. Patients with little alteration in blood pressure may exhibit profound stimulation and vice versa (GUTTMANN and SARGANT, 1937).

The reason for these inconsistencies in both cardiovascular and other actions of amphetamine has not been convincingly accounted for. According to one view, the net effect of amphetamine administration is an activation of the sympathetic nervous system, followed by a feedback to the central nervous system from proprioceptive impulses, with a concomitant reduction of the effect (LEAKE, 1958). However, recently it has become clear that a central activation of catecholamine-mediated effects may in itself act in an opposite direction compared with peripheral adrenergic effects. For instance, a stimulation of noradrenergic function within the brain, by the noradrenaline receptor-stimulating compound clonidine, elicits a blood-pressure fall (SCHMITT et al., 1971). Since amphetamine exerts its action through catecholamine-containing neurons both in the brain and in the periphery, this double action may precipitate a spectrum of opposite effects. HOYER and VAN ZWIETEN (1971) reported that amphetamine injected into the vertebral arteries of cats lowers the blood pressure, whereas i.v. injections of the same doses cause an increase of blood pressure. Such an antagonism between central and peripheral sympathetic activity may be of particular importance in explaining some features of amphetamine intoxication syndromes, where cardiovascular effects may be absent or where there are even indications of a reduced peripheral sympathetic tone [cf. dysautonomic syndrome (Subsect. D.V.)].

B. Effects of Single Doses

I. Mood

In the majority of cases, amphetamine in moderate doses (5–30 mg) induces elation and elevation of mood, a sense of well-being, increased talkativeness, and an ability to concentrate, sometimes resulting in improved psychomotor skill in various tests, where fatigue may have reduced motivation. Occasionally, however, the most conspicuous effects are drowsiness, mental depression, and reduced ability to concentrate. NATHANSON (1937), BAHNSEN et al. (1938), ANDERSON (1938), HUNDLEY et al. (1939), LASAGNA et al. (1955), and HUBIN and SERVAIS (1968) have reported on a total of 276 subjects, 34 of which (12%) became depressed and dysphoric or drowsy and sedated from 20 mg of amphetamine. The time when the observation is made may be of importance, since TECCE and COLE (1974) reported that 13 out of 20 subjects given 10 mg of *d*-amphetamine experienced drowsiness, but only during the first hour after administration. At 1 h there was also a lowered electrical brain activity (contingent negative variation). Anxiety or aggressiveness may sometimes be prominent after amphetamine administration, particularly when the experimental setting or the expectancy set of the experimental subject has a structure that creates fear and suspicion. Anxiety may occur in combination with both euphoric and dysphoric reactions, and is more common when high doses have been employed. Attempts to treat mental depression with amphetamine administration (MYERSON, 1936) sometimes tend to aggravate the anxiety common in this mental state (DAVIDOFF and REIFENSTEIN, 1937; REITER, 1937; WARSTADT, 1938).

Even when the initially dominating effect is one of elation and euphoria, there are often feelings of depression and fatigue afterwards. Thus, NATHANSON (1937) reported that in a few patients given 20 mg amphetamine daily "there was a secondary depression following the initial stimulation." ROSENBERG et al. (1963) noticed that after 0.6 mg/kg of dexamphetamine given i.m. to detoxified heroin addicts, anxiety and euphoria lasted for about 4–6 h and were followed by dysphoria lasting throughout the night.

This succession of signs and symptoms, an initial stage of euphoria followed by dysphoria and depression, has proved to be difficult to demonstrate after low doses (JACOBSEN et al., 1938; DUREMAN and SCHOLANDER, 1960; HURST and BAGLEY, 1971). However, it is more easily recognized when large parenteral doses have been employed. According to GUNNE and ÄNGGÅRD (1973) 200 mg i.v. of *dl*-amphetamine given to detoxified amphetamine addicts induced a stage of euphoria, recorded by questionnaire technique, for 10–12 h (with a maximum 15 min after administration of amphetamine), followed by signs of dysphoria which were noticeable at first, but reached a maximum 14 h after the injection and persisted for 48 h (Fig. 1). Thus, with regard to mood changes, both euphoria and dysphoria may develop after

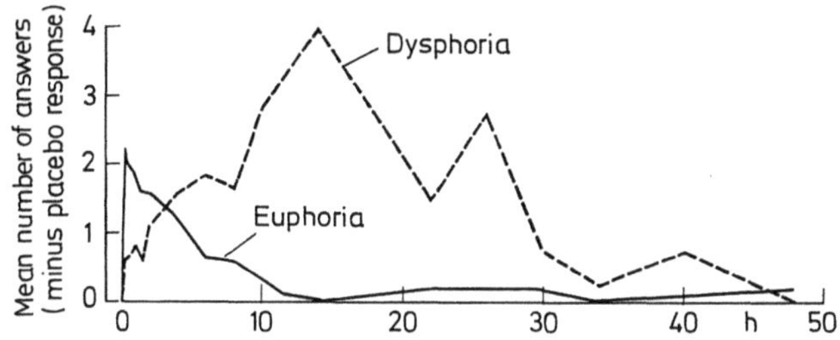

Fig. 1. Time course for development and disappearance of euphoric and dysphoric effects in 12 human subjects following 200 mg of amphetamine sulfate i.v. at zero time

amphetamine and which of these effects dominates may be at least partly dependent on dose and level of tolerance. It is tempting to speculate that a change of proportions, in the direction of increasing dysphoria during the development of tolerance, may be a premonitory sign of an oncoming amphetamine psychosis.

With regard to subjective euphoria-inducing effects, *d*-amphetamine is 4 times more potent than *l*-amphetamine (INNES and NICKERSON, 1965).

Fenfluramine (N-ethyl-3-trifluoromethylamphetamine), the most recent contribution to the series of commercially available anorexigenic amphetamines, has been found to be devoid of marked mood-elevating effects (LE DOUAREC and NEVEAU, 1970). Among experienced amphetamine addicts it was not recognized as having attractive subjective effects and 80 mg of fenfluramine could not be distinguished from placebo (GÖTESTAM and GUNNE, 1971). However, at higher doses (240 mg) fenfluramine induced a psychedelic state resembling LSD (GRIFFITH et al., 1974).

II. Performance

Amphetamine has been shown to enhance performance in humans in various tests (with the exception of intellectual tasks), particularly at the end of a test-period when boredom and irritation would have otherwise decreased performance (TROUTON and EYSENCK, 1960; WEISS and LATIES, 1962). In situations requiring subjects to maintain continuous alertness, amphetamine markedly retards the usual fall in performance with time. It does so even under conditions that would seem to mitigate its effects, such as lack of oxygen (ADLER et al., 1950; PAYNE and HAUTY, 1955; HAUTY and PAYNE, 1957). In sleep-deprived subjects, amphetamine restores only some performances to the control level (KORNETSKY et al., 1959).

In non-fatigued subjects, amphetamine generally fails to improve performance. An exception to this is demonstrated by certain studies of athletic performance, where amphetamine seems to have contributed to an improvement, for instance performance of college swimmers (SMITH and BEECHER, 1959, 1960). Thus, in addition to reducing boredom, inducing increased attentiveness, and a favorable attitude, amphetamine may also improve athletic performance (WEISS, 1969). Due to the risk of development of drug dependence, the use of amphetamine to manipulate performance in various tests is becoming increasingly rare.

In therapy, the performance-improving effect has been utilized in the treatment of hyperkinetic children. Probably there is a small group of such patients who may benefit from the effects of amphetamine (LASAGNA and EPSTEIN, 1970; FINNERTY et al., 1971; FISH, 1971; ARNOLD et al., 1972, 1973; EISENBERG, 1972). Methylphenidate has also been used with reported success (COHEN et al., 1971; FISCHER and WILSON, 1971; SYKES et al., 1972). However, suitable cases for such therapy are not easy to predict (CONNERS, 1972) and the indiscriminate use of amphetamine-like drugs in order to improve classroom performance in normal children is seriously warned against (BAZELL, 1971; DUVERNOY, 1971).

III. Wakefulness

It has been known for a long time that amphetamine causes an increased degree of wakefulness, attentiveness, awareness, and ability to concentrate, together with reduced fatigue and concomitant insomnia. Due to these effects, amphetamine was introduced in the treatment of narcolepsy (PRINZMETAL and BLOOMBERG, 1935), where it was soon found to be superior to ephedrine. It has been claimed that the *d*-isomer would be more effective as an analeptic than the *l*-isomer. ALLES (1939) and PRINZMETAL and ALLES (1939) reported that the central stimulant action of *d*-amphetamine was 2–4 times as high as that of the *l*-isomer. Recently, however, *l*-amphetamine has been used with reported success in the treatment of narcolepsy (KUHLO, 1971). Amphetamine is very effective against the sleep attack in this syndrome, but less effective in combating cataplexy, hypnagogic hallucinations, and sleep paralysis, where certain tricyclic antidepressants have been found to be superior. Amphetamine reduces both REM sleep and non-REM sleep (see below), whereas imipramine reduces only the former (ROTH et al., 1971).

The use of amphetamines as analeptics in barbiturate intoxication has been abandoned and replaced by other more effective measures, such as artificial ventilation and circulation support (EDISON, 1968).

IV. Anorexia

The appetite-depressant and weight-reducing effects have been known for a long time (DAVIDOFF and REIFENSTEIN, 1937; HILL, 1937; NATHANSON, 1937; LESSES and MYERSON, 1938; ROSENBERG, 1939). SILVERSTONE and STUNKARD (1968) conducted a controlled study of the effect on self-rated appetite, and found a demonstrable effect after 5–20 mg of dexamphetamine. However, the effect is small and there is a marked interindividual variation (HOLLISTER, 1971). HARRIS et al. (1947) found that amphetamine had only a slight effect on metabolism and ascribed most of the weight-reducing activity to decreased food intake. In psychiatric patients, who had undergone frontal lobotomy, amphetamine did not reduce the intake of food. These findings, together with an observed antagonism between amphetamine and chlorpromazine with regard to effects on appetite (SLETTEN et al., 1967), suggest that the anorexigenic effect is centrally mediated (cf. COLE, 1967).

Several compounds belonging to the amphetamine series have been used for their anorexigenic properties for weight-reducing purposes (WELSH, 1962). In fact, weight-reducing has been one of the most common reasons for prescribing amphetamine, although the rationale of this therapy has been repeatedly questioned (MODELL, 1960; BARTHOLOMEW, 1970) and the risk of creating drug dependence has been pointed out. Fenfluramine is an amphetamine-like anorexigen, which lacks excitatory effects (BIZZI et al., 1970, Seminar on Fenfluramine and Obesity, 1971). In addition to its appetite-reducing effect, amphetamine also reduces water intake in animals (ANDERSON and LARSSON, 1956) and humans (ÄNGGÅRD et al., 1970).

V. Analgesia

A low degree of analgesia has been demonstrated following amphetamine and methamphetamine administration (GOETZL et al., 1944; GRUBER and DE WEESE, 1958). Although the analgesic action of morphine is potentiated, the drowsiness induced by morphine is counteracted by amphetamine (INNES and NICKERSON, 1965). Dexamphetamine is more analgesic than the racemic mixture (LEAKE, 1958).

VI. Libido

Amphetamine has been reported to increase libido and sexual pleasure, particularly in women. In men, high doses may increase libido, but sometimes it causes impotency or delayed ejaculation (BONHOFF and LEWRENZ, 1954; CONNELL, 1958; HAMPTON, 1961; COHEN, 1969). The drug is without marked effect on sexual function when used in ordinary therapeutic doses.

VII. EEG

From qualitative EEG studies it has been known for a long time that stimulant drugs produce an EEG referred to as "alerting" or "activation." This desynchronized pattern consists of a decrease in amplitude and a concomitant increase in frequency. In the quantified EEG, this effect appears as a decrease in energy, as registered by a cumulative measurement of the area under the successive waves, with an associated decrease in variability of the energy (IVY and KRASNO, 1941; IVY and GOETZL, 1943; LEAKE, 1958; GOLDSTEIN et al., 1963; PFEIFFER et al., 1965).

It should be pointed out, however, that on the whole the amphetamines have unspectacular and possibly inconsistent EEG effects in man. In one investigation, i.v. doses of methamphetamine as large as 40–100 mg produced no observable change in half of the subjects (GASTAUT et al., 1960). Some investigators have reported that the alpha rhythm, when changed, was decreased in frequency following amphetamine (SERCL et al., 1952; GASTAUT et al., 1960), others have observed a shift in the opposite direction (GUARESCHI, 1953; BENTE and ITIL, 1957; PFLANZ and SCHRADER, 1959; KORENYI et al., 1961; FINK et al., 1971). The delta, or sleep waves, are diminished in amplitude and frequency (IVY and KRASNO, 1941; LEAKE, 1958).

Quantitative studies of the effects on the EEG spectrum have also yielded conflicting results. GIBBS and MALTBY (1943) found a shift of the spectrum toward higher frequencies and a slight increase in total voltage, whereas MONTAGU (1968) demonstrated increases at all frequencies with a shift of the spectrum toward the lower frequencies.

There are great species differences with regard to the EEG response to amphetamine. For instance the human EEG does not mimic changes found in cats or rabbits (BRAZIER, 1964).

VIII. Evoked Responses

Methamphetamine 15 mg/70 kg i.v. had no effect on visual evoked responses in man (SPILKER and CALLAWAY, 1969).

IX. REM Sleep

Paradoxical or REM (rapid eye movement) sleep is known to occupy 20–25% of human sleep and recurs about 5 times nightly (OSWALD, 1970). The EEG is of low voltage with 2–3 cps. The eyes make jerky movements and there are frequent brief bodily movements, but otherwiese profound muscular atonia. Respiration, heart rate, and blood pressure are irregular, the penis is erect, and dreaming is reported on awakening. Ordinarily the first REM sleep attack occurs after a time lag of 45–120 min. RECHTSCHAFFEN and MARON (1964) and LEWIS (1970) reported that amphetamine reduced REM sleep compared with a control when no drug was given and they found that the onset of REM sleep was delayed. This amphetamine-induced reduction of REM sleep was not influenced by pretreatment with propranolol (DUNLEAVY et al., 1971).

FEINBERG et al. (1974) questioned the depressant effect of amphetamine on REM sleep and pointed out that most earlier experiments were carried out with combinations of amphetamine and barbiturates.

X. Cerebral Blood Flow

It has been suggested by RUMBAUGH et al. (1971 a, b), mainly based on experiments with monkeys, that amphetamine may induce a major vasoconstriction within the cerebral arteries, causing a degree of cerebral anoxia which might explain the euphoric effect of the drug. However, the moderate but still euphoria-inducing dose of 25 mg daily for several days did not cause changes in cerebral blood flow in man (ANDREW and SENSENBACH, 1955).

XI. Auditory Thresholds

BRADLEY and KEY (1958) found that amphetamine lowered the thresholds for behavioral and EEG arousal produced by auditory stimulation, whereas WASSON and UNCAPHER (1966) found no effect on thresholds at any frequency (after administration of 5 mg of dexamphetamine). BENNETT (1960) reported reduced auditory acuity for 2 or 3 days after administration of 5 and 10 mg of dextroamphetamine. EYSENCK and EASTERBROOK (1960), on the other hand, suggested that dexamphetamine has no effect on auditory flutter fusion.

At high amphetamine doses an intensification of the subjective experience of various kinds of signals has been reported. Thus light, sounds, tastes, and odors have a high degree of intensity. Swedish amphetamine addicts like to gather in dark and quiet rooms, where only whispering is tolerated as long as the participants are intoxicated (GOLDBERG, 1968). GUNNE et al. (unpublished) noticed a lowered threshold for discomfort-inducing sound in amphetamine addicts given 200 mg amphetamine i.v., whereas thresholds for distinguishable sound and the stapedic reflex were unaffected.

XII. Critical Flicker Frequency

TURNER (1965) and SMART and TURNER (1966) have shown that amphetamine significantly increases critical flicker frequency at 3 h and dexamphetamine reverses the depressant effect of amylobarbitone. Similar findings have been obtained with phenmetrazine (25 mg) and diethylpropion (25 mg) (SMART et al., 1967).

XIII. Pupil Size

Dilatation of the pupils is a common sign in intoxications where high doses have been employed (KALANT, 1973). However, administration of a moderate dose (0.6 mg/kg) of dexamphetamine in subjects not dependent on amphetamine (detoxified heroin addicts) failed to cause significant changes in the size of the pupils (ROSENBERG et al., 1963). Fenfluramine, on the other hand, caused a dose-dependent increase in pupillary size (GRIFFITH et al., 1974). In a group of sleep-deprived subjects, MEYER et al. (1970) noticed pupillary dilatation even after oral administration of 10 mg of dexamphetamine.

Amphetamine instilled locally into the eye in a 0.25% solution causes mydriasis within a few minutes, which reaches a maximum after 1 h and disappears in about 8 h (IVY and KRASNO, 1941). TURNER (1970) reported a marked reduction of mydriasis when various amphetamine derivatives were instilled into the affected eye of patients who had developed a unilateral Horner's syndrome after cervical sympathectomy. When fenfluramine (1% solution) was instilled into the eyes of normal subjects, no change in pupil diameter occurred.

XIV. Cardiovascular Effects

Although earlier studies indicated similar effects on blood pressure and heart rate in man as had been noticed in animals, few data were presented to substantiate these observations statistically. The technique of registration prevalent in the 1930s and 1940s (when the early clinical studies on amphetamine and its congeners were made) allowed demonstration only of major aberrations in physiologic parameters induced by the drug. Figures for maximal rise of blood pressure measured at interindividually different time intervals after oral or parenteral administration were often given, without placebo control experiments. (For reviews see IVY and KRASNO 1941; IVY and GOETZL 1943; and HALEY 1947.)

More recent reports have confirmed a significant blood-pressure rise following injection of dexamphetamine or its various analogues (ROSENBERG et al., 1963; MARTIN et al., 1970). The observation that amphetamine raises systolic and to a lesser degree diastolic blood pressure and increases the pulse pressure, indicates that both α- and β-receptors are activated (INNES and NICKERSON, 1965). The pulse rate, by reflex, remains low during the first few hours, but tachycardia is common when, after the initial rise, the blood pressure begins to fall (PEOPLES and GUTTMANN, 1936). DEROUAUX (1942) and MATTHES (1944) have shown that *d*-amphetamine exerts its hypertensive action due to peripheral vasoconstriction. The *l*-isomer has a slightly greater effect on blood pressure than *d*-amphetamine (INNES and NICKERSON, 1965), possibly due to differences in distribution or to a lower rate of metabolism of *l*-amphetamine (GUNNE, 1967).

Cardiac output increased by 39% following i.m. injection of 15 mg of methamphetamine in normal volunteers (BERSEUS, 1943). There is a suspicion that some amphetamine-related compounds (particularly the phentermines and aminorex) may cause pulmonary hypertension, but there does not seem to be complete evidence for this effect in man (NAESS, 1970). One cause of pulmonary hypertension may be talc granuloma within the pulmonary vessels due to injection of poorly dissolved tablets (LEWMAN, 1972).

Various electrocardiographic changes have been reported, particularly cardiac arrhythmias which may occur with high doses. Chest pain, lasting for minutes or hours, has been observed following high i.v. doses (KRAMER, 1972) and in eight fatal cases, where a careful microscopic study of the myocardium was performed, signs indicative of recent myocardial necrosis were present in three young amphetamine addicts (25–32 years) without evidence of coronary atherosclerosis (GUNNE et al., 1972).

XV. Respiration

The rate and depth of respiration are increased with moderate and large amphetamine doses (INNES and NICKERSON, 1965; MARTIN et al., 1970; JÖNSSON, 1972), but there is little or no effect from normal doses of 5–20 mg (IVY and KRASNO, 1941). There are reports of a decreased rate, with a compensatory increase in depth, from low doses of amphetamine (BONHOFF and LEWRENZ, 1954) and methamphetamine (IVY and GOETZL, 1943).

XVI. Metabolic Changes

Conventional therapeutic doses of amphetamine or methamphetamine are known to cause either no change, a small fall, or a modest rise in oxygen consumption (DAVIDOFF and REIFENSTEIN, 1938; BEYER, 1939; DILL et al., 1939). The respiratory quotient was unchanged in most experiments (IVY and KRASNO, 1941; IVY and GOETZL, 1943).

The body temperature has likewise been reported to be unaffected by small doses of amphetamine. Occasionally, however, sensitive individuals may develop hyperthermia. In 14 subjects given an oral dose of 20 mg of amphetamine, three had a temperature rise not exceeding 1° C, whereas in the remaining 11 there was "no change" in body temperature (STORZ and KIRK, 1937). ROSENBERG et al. (1963) did not obtain significant effects on body temperature (maximal effect) following 0.6 mg/kg dexamphetamine i.m., whereas MARTIN et al. (1970) recorded a temperature rise from various doses of amphetamine, methamphetamine, phenmetrazine, methylphenidate, and *l*-ephedrine.

Administration of amphetamine or methamphetamine led to an increase in plasma free fatty-acid concentration (BALASSE and THYS, 1964; PINTER and PATTEE, 1968). In the latter study, where 0.2 mg/kg of amphetamine sulfate was infused i.v., there were no significant changes in blood glucose, serum total fat, or triglyceride concentration.

XVII. Blood Cells

Polycythemia may occur as a result of constriction of the spleen and (during periods of abuse) due to desiccation.

XVIII. Various Smooth-Muscle Effects

There are varying effects of amphetamine on gastrointestinal function. Both diarrhea and obstipation may occur. In the urinary bladder, the drug has a marked contractile effect on the sphincter, which has been used in the treatment of enuresis nocturna (IVY and KRASNO, 1941).

XIX. Hormones and Neurotransmitters

BESSER et al. (1969) described a rise in plasma levels of corticosteroids and growth hormone after i.v. administration of 15 mg of methamphetamine. Injection of 7.5 mg caused a rise in corticosteroids but had no effect on growth hormone. There was a rise in immunoreactive ACTH, which slightly preceded the corticosteroid response. Fenfluramine has been claimed to counteract the stress-induced rise in serum growth hormone levels (SULAIMAN and JOHNSON, 1973). This finding was questioned by DUNLEAVY et al. (1973) and still remains an open question.

In a single amphetamine abuser, 250 mg of *d*-amphetamine administered over 21 h induced a rise in the cerebrospinal-fluid content of homovanillic acid (HVA) but not of 3-methoxy-4-hydroxyphenylethylglycol, MHPG (ANGRIST et al., 1974). This finding was interpreted as indicative of an increased turnover of dopamine in the brain. Oral administration of 10 mg of methamphetamine failed to induce significant changes in the urinary excretion of adrenaline and noradrenaline (FRANKENHÄUSER and POST, 1962).

C. Effects of Chronic Administration

On repeated administration some effects of amphetamine are reduced or lost, whereas others seem to be unaffected. The previously held view (VOGEL et al., 1948; SEEVERS, 1955), that tolerance does not develop to amphetamine, or that amphetamine tolerance should be confined to vascular and other peripheral effects, can no longer be maintained. For instance, tolerance to certain subjective effects and to effects on EEG must be classified as centrally mediated.

I. Mood

In early studies there was much anecdotal material for the view that long-term administration of oral amphetamine doses causes development of tolerance to the euphoric effect. Thus, WILBUR et al. (1937), who treated 23 cases of chronic exhaustion and 21 cases of mental depression, noticed that the initial signs of exhilaration and the antifatigue effects wore off within a week with daily doses between 5 and 40 mg. GUTTMANN and SARGANT (1937), DAVIDOFF and REIFENSTEIN (1937), JAMES (1938), BAHNSEN et al. (1938); HANSEN (1944), and KNAPP (1952) also noticed a

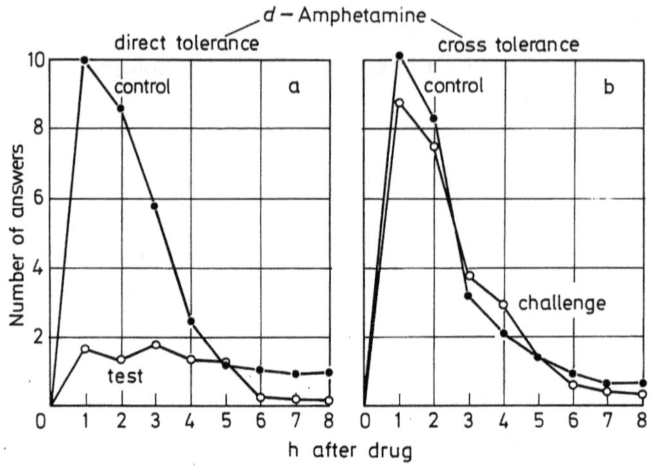

Fig. 2. Time course of euphoric response to 0.6 mg/kg i.m. of *d*-amphetamine before (●) and after (○) the chronic administration of *d*-amphetamine (*left*) and LSD (*right*) (ROSENBERG et al., Psychopharmacologia **5**, 1—15, 1963)

decrement of the subjective euphoria-like effects within a few days of chronic administration of amphetamine. In long-term experiments BAHNSEN et al. (1938) noticed that experimental subjects found it increasingly difficult to recognize the drug. However, after discontinuation for a few days the subjective effects returned on renewed administration.

ROSENBERG et al. (1963) quantitated the euphoria-like effects in a study demonstrating both direct tolerance to dexamphetamine and the absence of cross tolerance between LSD and dexamphetamine. They showed (in detoxified heroin addicts) that the subjects' response to a questionnaire, designed to study subjective amphetamine effects, changed dramatically during 14 days of daily i.m. administration of 0.6 mg/kg dexamphetamine. There was a virtual disappearance of the initial euphoric response within 14 days (Fig. 2, *left*). When LSD had been substituted for amphetamine during chronic treatment, the subjective response to amphetamine was unchanged (Fig. 2, *right*).

A corresponding study of the development of amphetamine-induced dysphoria (see p. 248) during long-term administration has not been undertaken. It is known from the studies of GRIFFITH et al. (1970), however, that amphetamine administered at hourly intervals in human volunteers leads to certain changes in mood, where dysphoric traits seem to dominate. The subjects became depressed, had an irritable, fault-finding attitude and would relate to the personnel in a "clinging and dependent manner." GRIFFITH et al. (1970 and 1972) have reported on eight such experimentally induced psychoses, whereas ANGRIST and GERSHON (1970) have studied four. The latter group of subjects seem to have displayed a mixture of elation and suspiciousness together with irritability and anxiety. One subject felt "that critical disparaging attitudes were being taken toward him by the nursing staff." Particularly in GRIFFITH's experiments the onset of psychosis seems to have been abrupt following the described period of dysphoria. BELL (1973) administered between 55 and 640 mg of methamphetamine i.v. for 60–75 min in 14 alleged amphetamine-dependent subjects. He noticed an outbreak of psychosis in 12 cases, preceded by a brief reaction of pleasure and relaxation combined with tears and emotional monologues. The time lag from injection to the first observable signs of paranoia was sometimes short, within an hour, which according to BELL would favor the view that amphetamine has a direct psychotogenic action. No psychosis was observed in two cases, who were later found not to have abused amphetamine.

In an early study from Lexington, Ky. (Publ. Health Serv. Rep. 1944, p. 57) seven patients received amphetamine daily in increasing doses up to a maximum of 1 g per day. It was noticed that the initial feelings of energy and sociability were only temporary and were replaced by "restlessness, agitation, and fear," followed by psychotic states in six subjects.

ANGRIST and GERSHON (1971) and ANGRIST et al. (1971) administered *d*- and *l*-amphetamine in separate experiments to three subjects and found the psychotogenic and mood-disturbing power of each isomer to be approximately the same (no precautions for standarizing urinary pH were taken). The limited material collected so far seems to indicate that the mediation of amphetamine psychosis is not stereospecific and is thus probably not mediated by noradrenergic brain neurons, which are preferentially activated by *d*-amphetamine (TAYLOR and SNYDER, 1970).

Withdrawal of high doses of amphetamine precipitates characteristic changes in mood, in the direction of mental depression, apathy, listlessness, and various degrees of exhaustion (KRAMER et al., 1967; GOLDBERG, 1968; SMITH, 1969a, b, c). Similarly, LEWIS et al. (1971), OSWALD et al. (1971) and STEEL and BRIGGS (1972) observed brief depressive mood changes following abrupt withdrawal of therapeutic doses of fenfluramine. Thus a depressive reaction seems to be part of the withdrawal syndrome following long-term administration of various amphetamine-related compounds.

In summary, long-term administration of amphetamine gives rise to tolerance to the euphoric effect, whereas dysphoria appears to increase gradually and is then followed by an abrupt change into psychosis. Withdrawal of various amphetamines induces a depressive reaction in most individuals.

II. Abuse Potential

Most amphetamine-like compounds, which induce abuse and drug dependence of the amphetamine type (EDDY et al., 1965), do so due to their mood-elevating properties (EDISON, 1971). Although some authors seem to rank all marketed weight-reducing amphetamines as equally dangerous in this respect (KALANT, 1973; INGHE, 1969) others have reported differences. Thus ALLMARK and RYLANDER (1968) stated that diethylpropion, although a drug which creates amphetamine-like mood elevation, is still widely regarded as a poor substitute among experienced amphetamine abusers, due to its lower effectiveness with regard to mood elevation, whereas stimulation of peripheral sympathetic tone and psychotomimetic properties are prominent. When the dose of diethylpropion was increased, tolerance to the euphoric action developed rapidly and for that reason each period of abuse was soon interrupted, generally after 2–3 days. A phenmetrazine "binge," on the other hand, could be protracted for 2–3 weeks. Phentermine has not yet been reported to create abuse and up till recently this was also true of fenfluramine. The view held by MODELL (1960) that amphetamine-like drugs owe their anorexigenic properties to the mood-elevating effects is hardly tenable after the introduction of these last-mentioned compounds. Thus, according to clinical experience, amphetamine, methamphetamine, methylphenidate, and phenmetrazine have a high abuse potential, whereas fenfluramine and phentermine have a low abuse potential. Diethylpropion appears to take an intermediate position between the two.

However, LEVIN recently (1972, 1974) reported on abuse of fenfluramine among LSD and cannabis abusers in South Africa. This group of abusers seems to have appreciated the hallucinogenic LSD-like effects, which fenfluramine exerts when applied in high doses (200–600 mg). At this dose level, the fenfluramine abusers (a total of 115) experienced euphoria with laughing attacks, followed some hours later by depressive symptoms. They reported visual and olfactory hallucinations, anxiety, sometimes with attacks of panic, nausea, and diarrhea.

III. Performance

There have been no investigations aimed specifically at the study of long-term effects of amphetamine on performance. In hyperkinetic children, the amphetamine dose has been kept constant for a considerable time with no reported diminishing effect. In high-dose abusers it has been stated that brain damage may be a complication resulting from taking amphetamine, as indicated by impairment of memory and ability to concentrate (KRAMER et al., 1967).

IV. Wakefulness

KOSMAN and UNNA (1968) stated that tolerance does not develop to the awakening action of amphetamine, as evidenced from long-term treatment of narcolepsy, where constant doses seem to deliver constant effects (PRINZMETAL and BLOOMBERG, 1935; GUTTMANN and SARGANT, 1937; BLOOMBERG, 1940; LEAKE, 1958). Furthermore, drug addicts seldom sleep during periods of abuse (KRAMER et al., 1967; GOLDBERG, 1968; SMITH, 1969 a, b, c). However, as pointed out by KRAMER (1972), when drug use is well developed, large amphetamine doses are necessary to keep the drug abusers awake, indicating that some degree of tolerance may develop also to the awakening effect. On the other hand, during a "speed binge" the increasing doses may just combat fatigue during a period of successively increasing sleep deprivation and thus the interpretation of these observations in terms of tolerance development to the awakening effect of amphetamine is difficult.

This problem was elucidated in a study (GUNNE and HOLMSTRAND, unpublished results) on two amphetamine-dependent individuals who had previously been detoxified for 3 weeks.

Figure 3 shows the design of the experiment: placebo or amphetamine tablets (0.5 mg/kg) were administered twice daily and saline or amphetamine sulfate (200 mg) was injected (single blind) at four different time points. The sleeping time was measured by twice hourly observations throughout the night and the quality of sleep was rated each morning by the staff and the experimental subjects, using the linear rating technique of AITKEN and GEYDE (1968). When 200 mg of amphetamine were injected during the first placebo tablet period, both sleeping time and ratings of sleep quality went down to zero, whereas after one week of daily oral amphet-

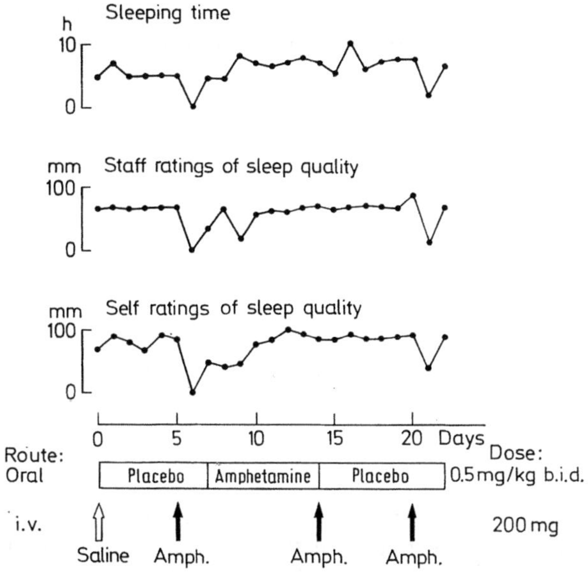

Fig. 3. Tolerance to awakening effect of 200 mg amphetamine i.v. in one detoxified amphetamine-dependent subject. Note that after first amphetamine injection on day 5, sleeping time and sleep quality are reduced, whereas this is not the case after next i.v. injection on day 14, when the patient had received 0.5 mg/kg amphetamine orally twice daily for a week.

amine, the effect of another amphetamine injection on sleep was virtually abolished. Six days later, after a second period of placebo medication, the last amphetamine injection markedly reduced sleep, although there seemed to be some degreee of remaining tolerance.

The results indicate that tolerance does develop to the awakening effect of amphetamine.

Following withdrawal after a period of abuse, amphetamine-dependent subjects often lapse into a deep sleep which lasts 24–48 h (KRAMER, 1972).

V. Anorexia

There are several reports in the literature on the development of tolerance to the anorexigenic effects of amphetamine and its congeners in man (IVY and KRASNO, 1941; BONHOFF and LEWRENZ, 1954; GALLAGHER and KNIGHT, 1958; MODELL, 1960; WELSH, 1962; INNES and NICKERSON, 1965; ROSELL, 1967). However, these statements seem to be based mainly on unsystematic clinical observations, including the "necessity" of increasing the amphetamine dose. The observation of a weight increase (following the initial reduction) in spite of a maintained constant dose (100 mg daily of diethylpropion for 12 weeks) seems to have been reported in only one single study (SEATON et al., 1961). MCKAY (1973), on the other hand, found no evidence for tolerance development to the weight-reducing effect of diethylpropion in a 24-week trial.

After withdrawal of amphetamine following a period of abuse, the dependent subject may eat ravenously. As pointed out by KRAMER (1972) this is in contrast with non-drug-assisted starvation, where the appetite is found to be diminished for many days. In summary, there is a unanimous impression that tolerance to the appetite-reducing effects of the drug develops within a few days, but there are as yet few data to support this view.

ÄNGGÅRD et al. (1970) reported on extreme degrees of negative water balance in amphetamine-dependent subjects admitted to hospital after a period of i.v. abuse. It is not known whether tolerance develops to the thirst-reducing effect of amphetamine.

VI. Analgesia

This effect has not been studied in long-term experiments.

VII. Libido

There are various reports to the effect that amphetamine increases libido in both sexes but that after long-term treatment there is a gradual reduction of sexual potency in males (BONHOFF and LEWRENZ, 1954; LADEWIG et al., 1969). KRAMER et al. (1967) stated that since ejaculation is delayed, sexual activity can be carried on for hours. When the orgasm is finally achieved, it is described as more pleasurable than without the drug. It should be pointed out, however, that as with other mind-changing drugs the perception of time is altered, which may compromise the accuracy of such statements. BELL and TRETHOWAN (1961) summarized that, similar to cocaine, amphetamine produces sexual hyperexcitability in women and in men increased sexual desire with reduced sexual powers.

VIII. EEG

EEG, with the exception of REM sleep and *evoked responses*, have not been studied in long-term experiments.

IX. REM Sleep

OSWALD and THACORE (1963) studied six experimental subjects (all dependent on amphetamine and barbiturates) who were given dexamphetamine (or phenmetrazine). Due to tolerance to amphetamine these subjects had an almost normal amount of REM sleep at night. When the drug was withdrawn they tended to sleep longer, their first REM sleep appeared abnormally early, it was increased in duration and the relative amount of REM sleep was high. When amphetamine was reinstituted the EEG sleep pattern became normal, indicating that the increase in REM sleep had been a withdrawal phenomenon. Not until 3–8 weeks after withdrawal did the sleep recordings finally return to normal. Similar observations were reported for diethylpropion but not for fenfluramine (OSWALD et al., 1968). FEINBERG et al. (1974) failed to confirm withdrawal elevations of REM sleep and raised some doubts on the data obtained by OSWALD's group.

X. Cerebral Blood Flow, Auditory Thresholds, Critical Flicker Frequency, and Pupil Size

do not seem to have been studied in relation to chronic administration of amphetamine.

XI. Cardiovascular Effects

ROSENBERG et al. (1963) noticed a significant reduction of the blood-pressure rise when amphetamine had been administered once daily for 2 weeks (0.6 mg/kg i.m.). In amphetamine addicts who had been detoxified for 1–2 weeks, 200 mg amphetamine caused a significant rise in blood pressure (mean 60 mm Hg) lasting for 8 h (JÖNSSON, 1972). The pulse rate was also significantly increased for several hours.

CITRON et al. (1970) have reported signs of periarteritis in i.v. abusers of amphetamine. However, the study suffers the deficiencies of any street drug research since the effects of multiple drug use, infection, etc. cannot be ruled out.

XII. Metabolic Effects

In the study of GRIFFITH et al. (1970), a moderate degree of hyperthermia was noticed in four out of six subjects who received dexamphetamine (about 10 mg/h) for several days. ROSENBERG et al. (1963) reported a decrease in temperature response when measured by time-action curve area, though the maximal temperature deviation was not significantly altered. JÖNSSON (1972) found no significant effect on body temperature when 200 mg amphetamine was administered i.v. in 15 amphetamine-dependent subjects who had been detoxified for 1–2 weeks prior to the experiment. This absence of effect was interpreted as indicating remaining amphetamine tolerance.

XIII. Respiration, Blood Cells, and Smooth-Muscle Effects

have not been studied in long-term experiments.

XIV. Hormones and Neurotransmitters

An increased urinary catecholamine excretion during amphetamine psychosis was reported by JÖNSSON et al. (1971a) and SCHILDKRAUT et al. (1971) reported an increased urinary level of MHPG in amphetamine intoxication, followed by a sharp decline after withdrawal, together with signs of mental depression. ANGRIST et al. (1972) found a similar urinary pattern of MHPG in only one out of three subjects given amphetamine (400 mg/24 h). The drug was without effect on CSF levels of the brain metabolites of dopamine (HVA), serotonin (5HIAA), and noradrenaline (MHPG) and it has thus not been possible to demonstrate any long-term effects of the drug on these brain monoamines in man (JÖNSSON et al., 1971a; ANGRIST et al., 1972). JÖNSSON et al. (1969a) reported an increased CSF level of choline during amphetamine intoxication with psychosis, which might be interpreted as indicating an increased liberation of brain acetylcholine. This interpretation has since been questioned, as there is no simple relationship between CSF choline and the metabolism of brain acetylcholine (SCHUBERTH and JENDEN, 1975).

D. Acute and Subacute Intoxication

When amphetamine is ingested in amounts exceeding the individual tolerance, the ensuing intoxication symptoms are very varied. In the present paper, the various clinical pictures have been tentatively subdivided into six different syndromes. Five of these (paranoid psychosis being the exception) seem to have fairly well-defined correlates in animal experiments and have been extensively studied with regard to underlying mechanisms. The intoxication syndromes generally appear isolated, but elements from two or even three syndromes may occasionally be combined. They are listed below and a selection of earlier published cases have been classified according to this system. Only the last three syndromes (4–6) represent a risk of fatal complications, whereas 1–3 do not.

I. Paranoid Psychosis

This syndrome is described in detail in Section I, Chapter 2. It is often strikingly devoid of accompanying physical symptoms as pointed out by CONNELL (1968) and generally appears as a syndrome containing purely psychic manifestations, hallucinations (mainly auditory and visual), paranoid delusions, generally without disorientation or confusion. The onset may be insidious, particularly when the route of administration of amphetamine has been oral, or occurs acutely sometimes, in amphetamine-dependent subjects, within minutes following i.v. administration. Oral abuse of amphetamine often leads to a schizophrenia-like psychotic state with a domination of auditory hallucinations and lack of insight, whereas i.v. abuse seems to create a more intense psychosis with vivid visual hallucinosis, anxiety, impulsiveness, and with a fluctuating degree of insight. The paranoid reaction may sometimes lead to sudden aggressiveness and even homicide (ELLINWOOD, 1971).

Several cases of amphetamine-induced psychosis are unrelated to long-term or periodic abuse and seem to have been precipitated by a single injection. According to the extensive literature compilation by KALANT (1973) who reported on 54 cases of acute or subacute amphetamine intoxication, more than half of the cases had transitory psychotic reactions. Certain ring-substituted amphetamines, including fenfluramine, have more pronounced hallucinogenic properties than the parent compound (NIEFORTH, 1971). Occasionally it has been claimed that methamphetamine creates chronic schizophrenia-like psychoses (TATETSU, 1960). This occurs more frequently in subjects with a heredity for schizophrenia (TATETSU, 1966).

As to the mechanisms of the amphetamine-induced paranoid psychosis, it has been speculated that an activation of dopamine neurons in the brain might evoke a model schizophrenia, while the superimposition of noradrenergic activity may elicit paranoid symptoms (SNYDER, 1972). Comparison of the capacity of the two amphetamine isomers to induce psychosis would favor a dopaminergic mediation of the syndrome (ANGRIST et al., 1971). The question whether amphetamine itself or any one of its metabolites is the psychotogenic principle has been studied by GUNNE and ÄNGGÅRD (1973). No correlation was found between amphetamine plasma levels and the intensity of the psychotic symptoms, whereas there was evidence that suggested ($n=7$) a relation between the percentage excretion of basic metabolites and degree of psychosis. Amphetamine-induced psychosis has no well-defined correlate in animal experiments.

II. Stereotyped Behavior

is another syndrome induced by high i.v. doses of amphetamine. A stereotyped behavior was first reported in animal experiments (HAUSCHILD, 1939; BONHOFF and LEWRENZ, 1954, p. 55) and its mediation via dopaminergic activity within the brain has been established (RANDRUP and MUNKVAD, 1965). In animals it consists of purposeless movements, repeated for hours in a stereotyped manner, e.g., licking and gnawing in rodents, strange head movements in cats, etc. In man, a compulsive perseverative behavior has been described as "a compulsive perfectionist need," e.g., for shoe-shining or nail polishing for hours (KRAMER et al., 1967). It has also been described as a "mechanical type of hyperactivity." "They would spend many hours trying to dismantle and repair gadgets which were in perfect order" (ibid.). Probably the tooth-grinding described by ASHCROFT et al. (1965) and the continuous rubbing of the tongue against the teeth belong to the same category of behavior. The latter behavior results in the characteristic ulcerated tongue described by CONNELL (1958).

Attempts to interrupt the stereotyped behavior generally create various degrees of anxiety. In spite of the compulsive nature of the syndrome it is often described as pleasurable by amphetamine abusers, probably since it takes place during the first 8 h after injection, when euphoria is prominent. Earlier descriptions of stereotyped behavior in man have been reported by RANDRUP and MUNKVAD (1967) who cite BONHOFF and LEWRENZ (1954, p. 104), TATETSU (1966) and CONNELL (1958, case 20) for similar case reports. GREVING (1941) had two case reports illustrating the same kind of behavior. RYLANDER (1966) reported on an investigation of phenmetrazine addicts, where a thorough study revealed these symptoms in the case histories of 29 out of 43 patients. He pointed out that the patients seldom report them spontaneously. Thus, in i.v. abusers of amphetamine and its congeners, stereotyped behavior may be a common, though often overlooked, symptom of intoxication.

III. The Choreic Syndrome

represents another frequently overlooked effect of amphetamine. There are only few case reports in the literature (BONHOFF and LEWRENZ, 1954; MATTSON and CALVERLEY, 1968; RYLANDER, 1972; CASE and MCANDREW, 1974; PALATUCCI, 1974). Although it is probably rare as an isolated phenomenon, elements of this pattern of movements can often be discovered in connection with the other intoxication syndromes. As an isolated syndrome it seems to occur only in predisposed individuals, who tend to develop the same pattern after each amphetamine administration. In children, this syndrome may develop even from low doses (5–15 mg of dextroamphetamine). In high-dose i.v. amphetamine abusers it occurs infrequently, and the symptoms can be avoided by dose reduction (GUNNE, unpublished). Among abusers the precipitation of this pattern of movements is often ascribed to minor irregularities with regard to set and setting at the moment of amphetamine injection, but this view is hardly supported by critical evidence.

The symptoms have the clinical elements of a Sydenham's chorea, i.e., hypotonicity of the limb muscles, quick jerky involuntary movements consisting of rotation, flexion, and extension of wrist and fingers, elevation of shoulders, rotations of the head, and asymmetrical movements of the facial musculature. Often the involuntary movements are superimposed upon the voluntary, rendering them incoordinate to various degrees. There are no signs of psychosis in typical cases and the movements do not seem to be ordinary restlessness and hyperactivity, since there is no evidence of stimulation and the patients are able to discuss their distressing hyperkinesia in a rather objective manner. The symptoms last for many hours and the movements can be terminated by administration of neuroleptics.

With regard to the underlying mechanism it may be pointed out that a domination of dopamine over acetylcholine is a probable underlying cause in hereditary Huntington's chorea (AQUILONIUS and SJÖSTRÖM, 1971; KLAWANS and RUBOVITS, 1972) which thus neurochemically represents the opposite of Parkinson's disease (where the dopamine system is lesioned and where akinesia dominates the picture). Accordingly, the choreic syndrome in amphetamine intoxication might be explained by an increased dopaminergic brain activity, probably combined with some as yet undefined defect within the acetylcholine system.

IV. Excitation Syndrome

States of excitation induced by amphetamine intoxication are generally combined with signs of autonomic hyperfunction: increased blood pressure, tachycardia, hyperthermia, pupillary dilation, and various degrees of motor unrest (KRISKO et al., 1969; GINSBERG et al., 1970). The skin is pale, cold, and clammy and muscular twitchings are frequent. This state may become critical and after a few hours of hyperthermia beyond 41° C and tachycardia above 180 beats/min, circulatory collapse and death may ensue. Fatal cases were earlier regarded as rare: for instance KALANT (1973) in her extensive compilation of the literature reported only six deaths in adults. From 1966–1971 no less than 18 additional deaths from amphetamine (or phenmetrazine) intoxication were reported from the area around Stockholm (ORRENIUS and MAEHLY, 1970; GUNNE et al., 1972). In California five deaths have been reported (ZALIS and PARMLEY, 1963; CRAVEY and BASELT, 1968; CRAVEY and REED, 1970). Thus i.v. abuse of high doses has greatly increased the death rates from amphetamine intoxication. Most of these cases have been excitation syndromes. At autopsy three out of eight cases studied showed microscopic lesions of ischemic nature, similar to recent myocardial infarctions. These cases (25–32 years of age) had no atherosclerosis and no history of heart disease, but all had intensive tachycardia for several hours before death.

In animal experiments an amphetamine-induced increase in motor activity has been related to an increased dopaminergic and noradrenergic brain activity (ANDÉN, 1970). Some elements of hyperactivity in rats can be eliminated by administration of dopamine-β-hydroxylase inhibitors, which reduce the noradrenaline synthesis selectively (D'ENCARNACAO et al., 1969; CORRODI et al., 1970), but a total abolition of excitation is achieved by tyrosine-hydroxylase inhibitors (RANDRUP and MUNKVAD, 1966; WEISSMAN et al., 1966).

V. Dysautonomic Syndrome

In some instances a high dose of amphetamine elicits anxiety and increasing motor unrest, but with either slight or no signs of autonomic involvement. In these cases there may be no tachycardia, normal body temperature, and normal or low blood pressure. The skin is usually pale and the patient may complain of nausea and later start vomiting. Generalized convulsions and coma may develop and at the same time there may be a fall of blood pressure. The cardiovascular shock together with repeated convulsions may soon lead to a critical state and should be regarded as very serious.

The mechanism of this syndrome is obscure, but one is reminded of the theory of RUMBAUGH et al. (1971 a, b) who suggested that constriction of the brain arteries is of importance for the development of amphetamine intoxication effects. If this is correct, the dysautonomic syndrome is a likely candidate for such a mechanism. On the other hand, as mentioned above, a vigorous stimulation of catecholamine neurons in the brain may counteract peripheral signs of sympathetic stimulation and thus account for the apparent absence of autonomic symptoms. Out of 32 non-psychotic cases of amphetamine intoxication described by KALANT (1973) only three seemed to belong to this category (ANDERSON and SCOTT, 1936; APFELBERG, 1938; SMITH, 1939) whereas another three had some elements of the syndrome (DAVIES, 1937; CURRY, 1949; WALLIS, 1957). Three lethal cases have been reported (SMITH, 1939; PONTRELLI, 1942; HARVEY et al., 1949).

VI. Cerebrovascular Accidents

Since amphetamine is known to cause a long-lasting increase in blood pressure, it is perhaps not surprising that cerebral hemorrhages may occur. Even in patients with a high degree of amphetamine tolerance the systolic blood pressure rise following 200 mg of racemic i.v. amphetamine may be as much as 100 mm Hg. A mean systolic rise of 60 mm Hg in 15 subjects was reported by JÖNSSON, 1972. Sometimes abusers may report a throbbing headache soon after an amphetamine injection, occasionally followed by hemiparesis or hemiparesthesia. Fifteen such cases have been reported, ten of which have survived (MONROE and DRELL, 1947; POTELIAKHOFF and ROUGHTON, 1956; KANE et al., 1969; GOODMAN and BECKER, 1970; WEISS et al., 1970; DINNEN, 1971). Out of the five fatal cases (GERICKE, 1945; LLOYD and WALKER, 1965; LUNDBERG cited by GOODMAN and BECKER, 1970; WEISS et al., 1970) one had also taken an MAO-inhibitor, phenelzine, in addition to amphetamine (LLOYD and WALKER, 1965).

E. Blockade of Amphetamine Effects

JÖNSSON et al. (1969), GUNNE et al. (1970), and JÖNSSON et al. (1971 b) described the blocking of amphetamine-induced euphoria and the reduction of amphetamine-induced blood-pressure rise by pretreatment with the catecholamine synthesis inhibitor, alpha-methyl-*p*-tyrosine (αMT). The self-rated euphoria was reduced to about 50% by 2 g αMT and 4 g almost eliminated the subjective effects of 200 mg of amphetamine sulfate (Fig. 4). The daily administration of 4 g/day of αMT rendered the subjects insensitive to the amphetamine blocking effect within a week and withdrawal of αMT after one week gave an enhancement of the amphetamine effect. When αMT was reinstituted 3 days after withdrawal there was another almost complete blockade. The duration of action of 4 g αMT is 48 h and for a maximal

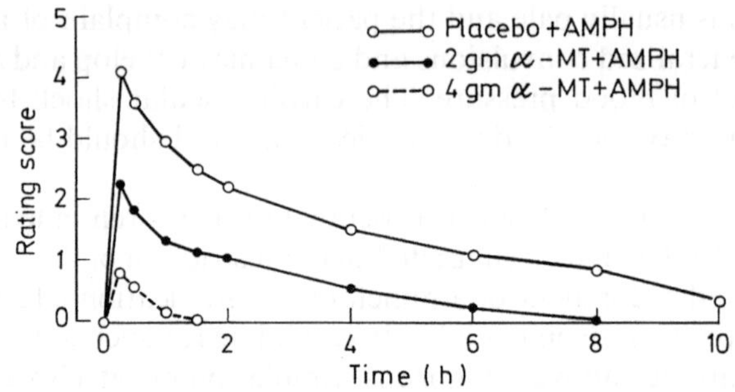

Fig. 4. Self-rated euphoria (means of 9 patients) after 200 mg amphetamine at zero time. Before amphetamine subjects were pretreated with placebo ○—○, α-methyltyrosine 2 g ●—● or 4 g ○---○ (GUNNE and ÄNGGÅRD, J. Pharmacokin., Biopharm. **1**, 481—495, 1973)

amphetamine blocking effect during long-term administration it should be given at 2-day intervals (ÄNGGÅRD and GUNNE, 1975).

Other blockers of the subjective effects of amphetamine were less effective. Some neuroleptics, chlorpromazine and pimozide, gave a reduction corresponding to about 2 g αMT, but this blockade could not be improved by increasing the dose. Chlorpromazine has been used with reported success in the treatment of amphetamine intoxication (ESPELIN and DONE, 1968). Propranolol (a sympathetic β-receptor blocker) and phenoxybenzamine (an α-receptor blocker) had no effect on amphetamine-induced euphoria.

The experiments cited indicate a possible mediation of amphetamine-induced euphoria by catecholamine-containing neurons in the brain, but do not clarify which neuron systems are of greatest importance.

Preliminary studies with *d*- and *l*-isomers administered in separate experiments, with precautions being taken to maintain an acidic urine, indicate an approximate potency for euphoria induction of 3–4:1 for the *d*- vs. the *l*-isomer. Such a quotient would favor the view that noradrenaline is of some importance, since *d*-amphetamine has greater effect on noradrenaline neurons (TAYLOR and SNYDER, 1970), but the fact that part of the euphoria is eliminated by a specific dopamine receptor blocking agent, pimozide (ANDÉN, 1970), probably indicates that dopamine also plays a role in the development of this subjective effect.

Recently, preliminary evidence for an amphetamine blocking effect of lithium has been presented in an uncontrolled study on three subjects (FLEMENBAUM, 1974).

References

Adler, H. F., Burkhardt, W. L., Ivy, A. C., Atkinson, A. J.: Effect of various drugs on psychomotor performance at ground level and at simulated attitudes of 18000 feet in a low pressure chamber. J. Aviat. Med. **21**, 221—236 (1950)

Änggård, E., Gunne, L. M.: Clinical and pharmacological studies on amphetamine dependent subjects. Neuropsychopharmacology, pp. 253—261. Boissier, J. R., Hippius, H., Pichot, P. (Eds.) Amsterdam: Excerpta Medica 1975.

Änggård, E., Gunne, L. M., Jönsson, L. E., Niklasson, F.: Pharmacokinetic and clinical studies on amphetamine dependent subjects. Europ. J. clin. Pharmacol. **3**, 3—11 (1970)

Aitken, R. C. B., Gedye, J. L.: A study of two factors which affect arousal level and the apparent duration of a ten-minute interval. Brit. J. Psychol. **59**, 253—263 (1968)

Alles, G. A.: Comparative actions of optically isomeric phenisopropylamines. J. Pharmacol. exp. Ther. **66**, 1 (1939)

Allmark, J., Rylander, G.: Ännu en form av narkomani. Läkartidningen **65**, 1530—1538 (1968)

Andén, N. E.: Effects of amphetamine and some other drugs on central catecholamine mechanisms. In: Costa, E., Garattini, S. (Eds.): Amphetamine and Related Compounds, pp. 447—462. New York: Raven Press 1970

Anderson, E. W.: Further observations on benzedrine. Brit. med. J. **1938 II**, 60—64

Anderson, E. W., Scott, W. C.: Cardiovascular effects of benzedrine. Lancet **1936 II**, 1461—1462

Andersson, B., Larsson, S.: Water and food intake and the inhibitory effect of amphetamine on drinking and eating before and after "prefrontal lobotomy" in dogs. Acta physiol. scand. **38**, 22—29 (1956)

Andrew, J. G., Sensenbach, W.: The effect of benzedrine sulfate upon cerebral blood flow and metabolism in man. J. nerv. ment. Dis. **122**, 61—64 (1955)

Angrist, B. M., Gershon, S.: The phenomenology of experimentally induced amphetamine psychosis—preliminary observations. Biol. Psychiat. **2**, 95—107 (1970)

Angrist, B. M., Gershon, S.: A pilot study of pathogenic mechanisms in amphetamine psychosis utilizing differential effects of D and L amphetamine. Pharmakopsychiat. Neuropsychopharmakol. **4**, 64—75 (1971)

Angrist, B. M., Shopsin, B., Gershon, S.: Comparative psychotomimetic effects of stereoisomers of amphetamine. Nature (Lond.) **234**, 152—153 (1971)

Angrist, B. M., Shopsin, B., Gershon, S., Wilk, S.: Metabolites of monoamines in urine and cerebrospinal fluid, after large dose amphetamine administration. Psychopharmacologia (Berl.) **26**, 1—9 (1972)

Angrist, B. M., Wilk, S., Gershon, S.: The effect of probenecid and large dose amphetamine administration on cerebrospinal fluid homovanillic acid. Biol. Psychiat. **8**, 113—114 (1974)

Apfelberg, B.: A case of benzedrine sulfate poisoning. J. Amer. Med. Ass. **110**, 575—576 (1938)

Aquilonius, S. M., Sjöström, R.: Cholinergic and dopaminergic mechanisms in Huntington's chorea. Life Sci. **10**, 405—414 (1971)

Arnold, L. E., Kirilcuk, V., Corson, S. A., Corson, E. O'L.: Levo-amphetamine and dextro-amphetamine: Different effect on aggression and hyperkinesis in children and dogs; diagnostic and biochemical implications. Amer. J. Psychiat. **130**, 165—170 (1973)

Arnold, L. E., Wender, P. H., McCloskey, K., Snyder, S. H.: Levoamphetamine and dextroamphetamine: Comparative efficacy in the hyperkinetic syndrome. Arch. gen. Psychiat. **27**, 816—822 (1972)

Ashcroft, G. W., Eccleston, D., Waddell, J. L.: Recognition of amphetamine addicts. Brit. med. J. **1965 I**, 57

Bahnsen, P., Jacobsen, E., Thesleff, H.: Den subjektive Virkning af Betafenylisopropylaminsulfat hos normale Voksne. Bibl. Læger **130**, 123—169 (1938)

Balasse, E., Thys, O.: Action adipocintétique de l'amphetamine. Path. Biol. **12**, 988—989 (1964)

Bartholomew, A. A.: Amphetamine addiction. Med. J. Aust. **1**, 1209—1214 (1970)

Bazell, R.: Panel sanctions amphetamines for hyperkinetic children. Science **171**, 1223 (1971)

Bell, D. S.: The experimental reproduction of amphetamine psychosis. Arch. gen. Psychiat. **29**, 35—40 (1973)

Bell, D. S., Trethowan, W. H.: Amphetamine addiction and disturbed sexuality. Arch. gen. Psychiat. **4**, 74—75 (1961)

Bennett, C. C.: The drugs and I. In: Uhr, L., Miller, J. G.: Drugs and Behavior. New York: Wiley 1960

Bente, D., Itil, T.: Vergleichende klinisch-elektroencephalographische Untersuchungen mit Pervitin und LSD 25. Psychotropic Drugs, pp. 284—285. Garattini, A., Ghetti, D. (Eds.) Amsterdam: Elsevier 1957

Berseus, S.: Die Wirkung von Pervitin auf den Kreislauf und die Atmung beim Menschen. Acta physiol. scand. **5**, 230—236 (1943)

Besser, G. M., Butler, P. W. P., Landon, J., Rees, L.: Influence of amphetamines on plasma corticosteroid and growth hormone levels in man. Brit. med. J. **1969 IV**, 528—530

Beyer, K. H.: The effect of benzedrine sulfate (beta-phenylisopropylamine) on metabolism and the cardiovascular system in man. J. Pharmacol. exp. Ther. **66**, 318—325 (1939)

Bizzi, A.; Bonaccorsi, A., Jespersen, J., Jori, A., Garattini, S.: Pharmacological studies on amphetamine and fenfluramine. In: Costa, E., Garattini, S. (Eds.): International Symposium on Amphetamines and Related Compounds, pp. 577—595. New York: Raven Press 1970
Bloomberg, W.: End results of use of large doses of amphetamine sulfate over prolonged periods. New Engl. J. Med. **222**, 946—948 (1940)
Bonhoff, G., Lewrenz, H.: Über Weckamine (Pervitin und Benzedrin). Berlin-Göttingen-Heidelberg: Springer 1954
Bradley, C., Key, B. J.: The effect of drugs on the arousal responses produced by electrical stimulation of the reticular formation of the brain. Electroenceph. clin. Neurophysiol. **10**, 97—100 (1958)
Brazier, M. A. B.: The effect of drugs on the electroencephalogram of man. Clin. Pharmacol. Ther. **5**, 102—116 (1964)
Cameron, J. S., Sprecht, P. G., Wendt, G. R.: Effects of amphetamines on moods, emotions and motivations. J. Psychol. (Provincetown) **61**, 93—121 (1965)
Case, Q., McAndrew, J. B.: Dexedrine dyskinesia: An unusual iatrogenic tic. Clin. Pediat. **13**, 69—72 (1974)
Citron, B. P., Halpern, M., McCarron, M., Lundberg, G. D., McCormick, R., Pincus, I. J., Tatter, D., Haverback, B. J.: Necrotizing angitis associated with drug abuse. New Engl. J. Med. **283**, 1003—1011 (1970)
Cohen, N. J., Douglas, V. I., Morgenstern, G.: The effect of methylphenidate on attentive behavior and autonomic activity in hyperactive children. Psychopharmacologia (Berl.) **22**, 282—294 (1971)
Cohen, S.: Abuse of centrally stimulating agents among juveniles in California. In: Sjöqvist, F., Tottie, M. (Eds.): Abuse of Central Stimulants, pp. 165—185. Stockholm: Almqvist and Wiksell 1969
Cole, S. O.: Experimental, effects of amphetamine: Rev. Psych. Bull. **68**, 81—90 (1967)
Connell, P. H.: Amphetamine psychosis. London: Chapman and Hall Ltd. 1958
Connell, P. H.: Amphetamine dependence. Proc. roy. Soc. Med. **61**, 178—181 (1968)
Conners, C. K.: Symposium: Behavior modification by drugs. II. Psychological effects of stimulant drugs in children with minimal brain dysfunction. Pediatrics **49**, 702—708 (1972)
Corrodi, H., Fuxe, K., Ljungdahl, Å., Ögren, S.-O.: Studies on the action of some psychoactive drugs on central noradrenaline neurons after inhibition of dopamine-β-hydroxylase. Brain Res. **24**, 451—470 (1970)
Cravey, R. H., Baselt, R. C.: Methamphetamine poisoning. J. forensic Sci. Soc. **8**, 118—120 (1968)
Cravey, R. H., Reed, D.: Intravenous amphetamine poisoning, report of three cases. J. forens. Sci. Soc. **10**, 109—112 (1970)
Curry, G. A.: Amphetamine poisoning. J. Amer. med. Ass. **140**, 850 (1949)
Davidoff, E., Reifenstein, E. C.: The stimulating action of benzedrine sulfate. J. Amer. med. Ass. **108**, 1770—1776 (1937)
Davidoff, E., Reifenstein, E. C.: A Method of studying some of the physiologic actions of benzedrine sulfate. J. Lab. clin. Med. **23**, 700—711 (1938)
Davies, I. J.: Benzedrine: a review of its toxic effects with a report of a severe case of anemia following its use. Brit. med. J. **1937 II**, 615—617
D'Encarnacao, P. S., D'Encarnacao, P. D., Tapp, J. T.: Potentiation of amphetamine induced psychomotor activity by diethyldithiocarbamate. Arch. int. Pharmacodyn. **182**, 186—189 (1969)
Derouaux, G.: Nouvelles recherches sur l'action hemostatique des amines. Arch. int. Pharmacodyn. **68**, 311—327 (1942)
Dill, D. B., Johnson, R. E., Daly, C.: Metabolic and cardiovascular effects of intramuscular injections of adrenaline and of amphetamine. Amer. J. med. Sci. **198**, 702—713 (1939)
Dinnen, A.: Cerebral haemorrhage due to stimulants. Med. J. Aust. **2**, 101—102 (1971)
Dunleavy, D. L. F., Mac Lean, A. W., Oswald, I.: Debrisoquine, guanethidine, propranolol and human sleep. Psychopharmacologia (Berl.) **21**, 101—110 (1971)
Dunleavy, D. L. F., Oswald, I., Strong, J. A.: Fenfluramine and growth hormone release. Brit. med. J. **1973 III**, 48
Dureman, I., Scholander, T.: Subjective estimation of drug-induced changes in activation level. Scand. J. Psychol. **1**, 36—40 (1960)
Duvernoy, W. F.: Amphetamine effects. J. Amer. med. Ass. **215**, 1987—1988 (1971)

Eddy, N. B., Halbach, H., Isbell, H., Seevers, M. H.: Drug dependence: its significance and characteristics. Bull. Wld Hlth Org. **32**, 721—733 (1965)
Edison, G.: Abuse of amphetamines. J. Amer. med. Ass. **205**, 174—175 (1968)
Edison, G. R.: Amphetamines: A dangerous illusion. Ann. intern. Med. **74**, 605—610 (1971)
Eisenberg, L.: Symposium: Behavior modification by drugs. III. The clinical use of stimulant drugs in children. Pediatrics **49**, 709—715 (1972)
Ellinwood, E. H.: Assault and homicide associated with amphetamine abuse. Amer. J. Psychiat. **127**, 90—95 (1971)
Espelin, D. E., Done, A. K.: Amphetamine poisoning. Effectiveness of chlorpromazine. New Engl. J. Med. **278**, 1361—1365 (1968)
Evans, W. O., Smith, R. P.: Some effects of morphine and amphetamine on intellectual functions and mood. U.S. MRL Rep. 598 (1963)
Eysenck, H. J., Easterbrook, J. A.: Drugs and personality. XI. The effects of stimulant and depressant drugs upon auditory flutter fusion. J. ment. Sci. **106**, 855—857 (1960)
Feinberg, I., Hibi, S., Braun, M., Cavness, C., Westerman, G., Small, A.: Sleep amphetamine effects in MBDS and normal subjects. Arch. gen. Psychiat. **31**, 723—731 (1974)
Fink, M., Shapiro, D. M., Itil, T. M.: EEG profiles of fenfluramine, amobarbital and dextroamphetamine in normal volunteers. Psychopharmacologia (Berl.) **22**, 362—383 (1971)
Finnerty, R. J., Soltys, J. J., Cole, J. O.: The use of D-amphetamine with hyperkinetic children. Psychopharmacologia (Berl.) **21**, 302—308 (1971)
Fischer, K. C., Wilson, W. P.: Methylphenidate and the hyperkinetic state. Dis. nerv. Syst. **32**, 695—698 (1971)
Fish, B.: The "one child, one drug" myth of stimulants in hyperkinesis. Arch. gen. Psychiat. **25**, 193—203 (1971)
Flemenbaum, A.: Does lithium block the effects of amphetamine? A report of three cases. Amer. J. Psychiat. **131**, 820—821 (1974)
Franke, H.: Klinisch vergleichende Untersuchungen über die Kreislaufmittel der Adrenalingruppe. Klin. Wschr. **17**, 1695—1700 (1938)
Frankenhauser, M., Post, B.: Catecholamine excretion during mental work as modified by centrally acting drugs. Acta physiol. scand. **55**, 74—81 (1962)
Gallagher, N. I., Knight, W. A.: Evaluation of amphetamine as an appetite depressant. J. chron. Dis. **8**, 244—252 (1958)
Gastaut, H., Plas, R., Bert, J.: Effects chez l'homme de l'administration per voie veineuse d'amphetamine. Rev. Neurol. **103**, 271 (1960)
Gericke, O. L.: Suicide by ingestion of amphetamine sulfate. J. Amer. med. Ass. **128**, 1098—1099 (1945)
Gibbs, F. A., Maltby, G. L.: Effect on the electrical activity of the cortex of certain depressant and stimulant drugs—barbiturates, morphine, caffeine, benzedrine and adrenaline. J. Pharmacol. exp. Ther. **78**, 1—10 (1943)
Ginsberg, M. D., Hertzmann, M., Schmidt-Nowara, W. W.: Amphetamine intoxication with coagulopathy, hyperthermia, and reversible renal failure. A syndrome resembling heatstroke. Ann. intern. Med. **73**, 81—85 (1970)
Götestam, K. G., Gunne, L. M.: Clinical evaluation of central stimulant effects of fenfluramine and AN 448. Brit. J. Addict. **67**, 39—44 (1971)
Goetzl, F. R., Burril, D. Y., Ivy, A. C.: The analgesic effects of sympathomimetics and the combination of morphine with amphetamines. Proc. Inst. Med. Chic. **15**, 87—88 (1944)
Goldberg, L.: Drug abuse in Sweden. Bull. Narcot. **20**, No. 1, 1—31, No. 2, 9—36 (1968)
Goldstein, L., Murphree, H. B., Pfeiffer, C. C.: Quantitative electroencephalography in man as a measure of CNS stimulation. Ann. N.Y. Acad. Sci **107**, 1045—1056 (1963)
Goodman, S. J., Becker, D. P.: Intracranial hemorrhage associated with amphetamine abuse. J. Amer. med. Ass. **212**, 480 (1970)
Greving, H.: Psychopathologische und körperliche Vorgänge bei jahrelangem Pervitinmißbrauch. Nervenarzt **14**, 395—405 (1941)
Griffith, J. D., Cavanuagh, J., Held, J., Oates, J. A.: Experimental psychosis induced by the administration of *d*-amphetamine. International Symposium on amphetamines and related compounds, pp. 897—904. Costa, E., Garattini, S. (Eds.) New York: Raven Press 1970

Griffith, J. D., Cavanaugh, J., Held, J., Oates, J. A.: Dextroamphetamine. Evaluation of psychomimetic properties in man. Arch. gen. Psychiat. **26**, 97—100 (1972)
Griffith, J. D., Nutt, J. G., Jasinski, D. R.: A comparison of fenfluramine and *d*-amphetamine in man. Clin. Pharmacol. Ther. **15**, 207 (1974)
Gruber, C. M., Weese, M. de: Methamphetamine and analgesia. Fed. Proc. **17**, 374 (1958)
Guareschi, A.: L'azione della beta-fenil-iso-propil-amina sull'attività elettrica dell'encefalo. Sist. Nerv. **5**, 129 (1953)
Gunne, L. M.: The urinary output of *d*- and *l*-amphetamine in man. Biochem. Pharmacol. **16**, 863 (1967)
Gunne, L. M., Änggård, E.: Pharmacokinetic studies with amphetamines—relationship to neuropsychiatric disorders. J. Pharmacokinet. Biopharm. **1**, 481—495 (1973)
Gunne, L.-M., Änggård, E., Jönsson, L. E.: Blockade of amphetamine effects in human subjects, pp. 249—255. ICAA. Lausanne: Tongue and Tongue, 1970
Gunne, L.-M., Holmgren, P., Lindquist, O., Saldeen, T.: Amfetamintoxikationer. Läkartidningen **69**, 4373—4376 (1972)
Guttmann, E., Sargant, W.: Observations on benzedrine. Brit. med. J. **1937 I**, 1013—1015
Gwynn, H. B., Yater, W. M.: A study of the temporary use of therapeutic doses of benzedrine sulfate in 147 supposedly normal young men (medical students). Med. Ann. D. C. **6**, 356 (1937)
Haley, T. J.: Desoxyephedrine—A review of the literature: J. Amer. pharm. Ass. **36**, 161—169 (1947)
Hampton, W. H.: Observed psychiatric reactions following use of amphetamine and amphetamine-like substances. Bull. N.Y. Acad. Med. **37**, 167—175 (1961)
Hansen, P. H.: Forfaldenhed til amphetamin (Benzedrin, Mecodrin) og Pervitin (Euforin). Ugeskr. Laeg. **106**, 95—99 (1944)
Harris, S. C., Ivy, A. C., Searle, L. M.: The mechanism of amphetamine-induced loss of weight; a consideration of the theory of hunger and appetite. J. Amer. med. Ass. **134**, 1468—1475 (1947)
Harvey, J. K., Todd, C. W., Howard, J. W.: Fatality associated with benzedrine ingestion. Delaware St. med. J. **21**, 111—115 (1949)
Hauschild, F.: Zur Pharmakologie des 1-phenyl-2-methyl-aminopropans (Pervitin). Naunyn-Schmiedebergs Arch. exp. Path. Pharmak. **191**, 465—481 (1939)
Hauty, G. T., Payne, R. B.: Effects of dextro-amphetamine upon judgment. J. Pharmacol. exp. Ther. **120**, 33—37 (1957)
Hill, J.: Benzedrine in sea-sickness. Brit. med. J. **1973 II**, 1109—1113
Hollister, L. E.: Hunger and appetite after single doses of marihuana, alcohol, and dextroamphetamine. Clin. Pharm. Ther. **12**, 44—49 (1971)
Hollister, L. E., Gillespie, H. K.: Marihuana, ethanol, and dextroamphetamine. Mood and mental function alterations. Arch. gen. Psychiat. **23**, 199—203 (1970)
Hoyer, I., Zwieten, P. A. van: The centrally induced fall in blood pressure after the infusion of amphetamine and related drugs into the vertebral artery of the cat. J. Pharm. (Lond.) **23**, 892—893 (1971)
Hubin, P., Servais, J.: Etude des effets subjectifs de l'amphetamine chez l'homme en fonction de la personnalité. Psychopharmacologia (Berl.) **12**, 239—249 (1968)
Hundley, J. M., Krautz, J. C., Hibbitts, J. T.: Dysmenorrhea—including clinical and pharmacological studies on benzedrine sulfate. Med. Clin. N. Amer. **23**, 273—293 (1939)
Hurst, P. M., Bagley, S. K.: Rebound from D-amphetamine. Psychol. Rep. **29**, 1023—1033 (1971)
Inghe, G.: The present state of abuse and addiction to stimulant drugs in Sweden. In: Sjöqvist, F, Tottie, M. (Eds.): Abuse of Central Stimulants, pp. 187—214. Stockholm: Almqvist and Wiksell 1969
Innes, I. R., Nickerson, M.: Drugs actings on postganglionic adrenergic nerve endings and structures innervated by them (sympathomimetic drugs), The Pharmacologic Basis of Therapeutics, 3. Ed., pp. 477—520. Goodman, L. S., Gilman, A. (Eds.). New York: Macmillan 1965
Ivy, A. C., Goetzl, F. R.: *d*-Desoxyephedrine: A review. War. Med. (Chic.) **3**, 60—77 (1943)
Ivy, A. C., Krasno, L. R.: Amphetamine (Benzedrine) sulfate: A review of its pharmacology. War. Med. (Chic.) **1**, 15—42 (1941)
Jacobsen, E., Bahnsen, P., Wollstein, A.: Normale Menneskers Reaktion efter Indgift av Beta-Fenylisopropylaminsulfat (Mecodrin "Meo"). Nord. med. T. **16**, 1209—1215 (1938)
James, G. W. B.: Addiction and benzedrine. Brit. J. Inebr. **36**, 13—14 (1938)

Jönsson, L.-E.: Pharmacological blockade of amphetamine effects in amphetamine dependent subjects. Europ. J. clin. Pharmacol. **4**, 206—211 (1972)

Jönsson, L.-E., Änggård, E., Gunne, L.-M.: Blockade of intravenous amphetamine euphoria in man. Clin. Pharmacol. Ther. **12**, 889—896 (1971b)

Jönsson, L.-E., Gunne, L.-M., Änggård, E.: Effects of alpha-methyltyrosine in amphetamine-dependent subjects. Europ. J. clin. Pharmacol. **2**, 27—29 (1969b)

Jönsson, L.-E., Lewander, T., Gunne, L.-M.: Amphetamine psychosis: Urinary excretion of catecholamines and concentrations of HVA and 5-HIAA in cerebrospinal fluid. Res. Commun. chem. Path. Pharmacol. **2**, 355—369 (1971a)

Jönsson, L.-E., Schuberth, J., Sundwall, A.: Amphetamine effect on the choline concentration of human cerebrospinal fluid. Life Sci. **8**, Part I, 977—981 (1969a)

Kalant, O. J.: The amphetamines. Toronto: University of Toronto Press 1973

Kane, F. J., Keeler, M. H., Reifler, C. B.: Neurological crises following methamphetamine. J. Amer. med. Ass. **210**, 556—557 (1969)

Klawans, H. L., Rubovits, R.: Central cholinergic–anticholinergic antagonism in Huntington's chorea. Neurology **22**, 107—116 (1972)

Knapp, P. H.: Amphetamine and addiction. J. nerv. ment. Dis. **115**, 406—432 (1952)

Korenyi, C., Perry, G. F., Whittier, J. R.: Placebo-controlled electroencephalographic comparison of benzphetamine with three anorexigenics. Curr. ther. Res. **3**, 207—211 (1961)

Kornetsky, C., Mirsky, A. F., Kessler, E. K., Dorff, J. E.: The effect of dextro-amphetamine on behavioral deficits produced by sleep loss in humans. J. Pharmacol. exp. Ther. **127**, 46—50 (1959)

Kosman, M. E., Unna, K. R.: Effects of chronic administration of the amphetamines and other stimulants on behavior. Clin. Pharmacol. Ther. **9**, 240—254 (1968)

Kramer, J. C.: Some observations on and a review of the effects of high-dose use of amphetamines. In: Zarafonetis, C. J. D. (Ed.): Drug abuse. Proceedings of the International Conference, pp. 253—261. Philadelphia, Pa.: Lea and Febiger 1972

Kramer, J. C., Fischman, V. S., Littlefield, D. C.: Amphetamine abuse. Pattern and effects of high doses taken intravenously. J. Amer. med. Ass. **201**, 305—309 (1967)

Kriskó, I., Lewis, E., Johnson, J. E.: Severe hyperpyrexia due to tranylcypromin-amphetamine toxicity. Ann. intern. Med. **70**, 559—564 (1969)

Kuhlo, W.: Die Narkolepsie und ihre Behandlung. Dtsch. med. Wschr. **96**, 1284—1286 (1971)

Lader, M.: Comparison of amphetamine sulphate and caffeine citrate in man. Psychopharmacologia (Berl.) **14**, 83—94 (1969)

Ladewig, D., Battegay, R., Labhardt, F.: Stimulantien: Abhängigkeit und Psychosen. Dtsch. med. Wschr. **94**, 101—107 (1969)

Lasagna, L., Epstein, L. C.: The use of amphetamines in the treatment of hyperkinetic children. International Symposium on Amphetamines and Related Compounds, pp. 849—864. Costa, E. and Garattini, S. (Eds.). New York: Raven Press 1970

Lasagna, L., Felsinger, J. M. von, Beecher, H. K.: Drug-induced mood changes in man. J. Amer. med. Ass. **157**, 1006—1020 (1955)

Leake, C. D.: The Amphetamines, Their Actions and Uses, Springfield, Ill: Charles C. Thomas 1958

Le Douarec, J. C., Neveau, C. E.: Pharmacology and biochemistry of fenfluramine. In: Costa, E., Garattini, S. (Eds.): Amphetamine and Related Compounds, pp. 75—105. New York: Raven Press 1970

Lesses, M. F., Myerson, A.: Benzedrine sulfate as an aid in the treatment of obesity. New Engl. J. Med. **218**, 119—124 (1938)

Levin, A.: The pattern of drug-taking among drug-dependent South African national servicemen. S. Afr. med. J. **46**, 1690—1694 (1972)

Levin, A.: Abuse of fenfluramine. Symposium on Fenfluramine and Derivatives Marbella, Spain 11—16 March 1974

Lewis, S. A.: Comparative Effects of some amphetamine derivatives on human sleep. In: Costa, E., Garattini, S. (Eds.): International Symposium on Amphetamines and Related Compounds, pp. 873—888. New York: Raven Press 1970

Lewis, S. A., Oswald, I., Dunleavy, D. L. F.: Chronic fenfluramine administration: Some cerebral effects. Brit. med. J. **1971 III**, 67—70

Lewman, L. V.: Fatal pulmonary hypertension from intravenous injection of methylphenidate (Ritalin) tablets. Hum. Path. **3**, 67—70 (1972)
Lloyd, J. T., Walker, D. R. H.: Death after combined dexamphetamine and phenelzine. Brit. med. J. **1965 II**, 168—169
Loman, J., Rinkel, M., Myerson, A.: Comparative effects of amphetamine sulfate (benzedrine sulfate), paradrine and propadrine on the blood pressure. Amer. Heart J. **18**, 89—93 (1939)
Martin, W. R., Sloan, J. W., Sapira, J. D., Jasinski, D. R.: Psychologic, subjective, and behavioral effects of amphetamine, methamphetamine, ephedrine, phenmetrazine, and methylphenidate in man. Clin. Pharmacol. Ther. **12**, 245—258 (1970)
Matthes, K.: Untersuchungen über das Wirkungsbild gefäßaktiver Pharmaka beim Menschen. Naunyn-Schmiedebergs Arch. exp. Path. Pharmak. **203**, 194—205 (1944)
Mattson, R., Calverley, J. R.: Dextroamphetamine-sulfate-induced dyskinesias. J. Amer. med. Ass. **204**, 108—110 (1968)
McKay, R. H. G.: Long-term use of diethylpropion in obesity. Curr. med. Res. Opin. **1**, 489—493 (1973)
Meyer, R. E., DiMascio, A., Stifler, L.: Personality differences in the response to stimulant drugs administered during a sleep-deprived state. J. nerv. ment. Dis. **150**, 91—101 (1970)
Miller, E., Nieburg, H. A.: Amphetamines. Valuable adjunct in treatment of Parkinsonism. N.Y. St. J. Med. **73**, 2657—2661 (1973)
Modell, W.: Status and prospect of drugs for overeating. J. Amer. med. Ass. **173**, 1131—1136 (1960)
Monroe, R. R., Drell, H. J.: Oral use of stimulants obtained from inhalers. J. Amer. med. Ass. **135**, 909—915 (1947)
Montagu, J. D.: The effect of d-amphetamine on the EEG response to flicker in man. Europ. J. Pharmacol. **2**, 295—300 (1968)
Myerson, A.: Effect of benzedrine sulfate on mood and fatigue in normal and in neurotic persons. Arch. Neurol. Psychiat (Chic.) **36**, 816—822 (1936)
Naess, K.: Medikamentell terapi i praksis. T. norske Laegeforen. **90**, 320—321 (1970)
Nathanson, M. H.: The central action of beta-aminopropylbenzene (Benzedrine). J. Amer. med. Ass. **108**, 528—531 (1937)
Nieforth, K. A.: Psychotomimetic phenethylamines. J. Pharm. Sci. **60**, 655—665 (1971)
Orrenius, S., Maehly, A. C.: Kasuistik—Casuistry. Lethal amphetamine intoxication. A report of three cases. Z. Rechtsmed. **67**, 184—189 (1970)
Oswald, I.: Effects on sleep of amphetamine and its derivatives. In: Costa, E., Garattini, S. (Eds.): International Symposium on Amphetamines and Related Compounds, pp. 865—871. New York: Raven Press 1970
Oswald, I., Jones, H. S., Mannerheim, J. E.: Effects of two slimming drugs on sleep. Brit. med. J. **1968 I**, 796—799
Oswald, I., Lewis, S. A., Dunleavy, D. L. F., Brezinova, V., Briggs, M.: Drugs of dependence though not of abuse: Fenfluramine and imipramine. Brit. med. J. **1971 III**, 70—73
Oswald, I., Thacore, V. R.: Amphetamine and phenmetrazine addiction: physiological abnormalities in the abstinence syndrome. Brit. med. J. **1963 II**, 427—431
Palatucci, D. M.: Single case study. Iatrogenic dyskinesia. A unique reaction to parenteral methylphenidate. J. nerv. ment. Dis. **159**, 73—76 (1974)
Payne, R. B., Hauty, G. T.: Factors affecting the endurance of psychomotor skill. Aviat. Med. **25**, 382—389 (1955)
Peoples, S. A., Guttmann, E.: Hypertension produced with Benzedrine. Lancet **1936 II, 1**, 1107—1109
Pfeiffer, C., Goldstein, L., Murphree, H. B., Sugerman, A.: Time-series, frequency analysis, and electrogenesis of the EEGs of normals and psychotics before and after drugs. Amer. J. Psychiat. **121**, 1147—1155 (1965)
Pflanz, M., Schrader, A.: Experimentelle Untersuchung über den Einfluß zentraler Stimulantien auf das Elektroencephalogramm des Menschen. Klin. Wschr. **37**, 483—485 (1959)
Pinter, E. J., Pattee, C. J.: Fat-mobilizing action of amphetamine. J. clin. Invest. **47**, 394—402 (1968)
Pontrelli, E.: Sopra un caso di avvelenamento mortale da solfato di beta-fenilisopropilamina (Simpamina). Gior clin. med. **23**, 591—600 (1942)

Poteliakhoff, A., Roughton, B. C.: Two cases of amphetamine poisoning. Brit. med. J. **1956 I**, 26—27
Prinzmetal, M., Alles, G. A.: The central nervous system stimulant effects of dextro-amphetamine sulfate. Proc. Soc. exp. Biol. (N.Y.) **42**, 206—207 (1939)
Prinzmetal, M., Bloomberg, W.: The use of Benzedrine for the treatment of narcolepsy. J. Amer. med. Ass. **105**, 2051—2054 (1935)
Public Health Service: Annual Report: Studies of drug addiction. pp. 56—57. U.S. Government Printing Office, Washington, D.C. 1944
Püllen, C.: Klinischer Beitrag zur Kreislaufwirkung des Pervitins und zur Pharmakologie des vegetativen Nervensystems. Z. Kreisl.-Forsch. **31**, 448—457 (1939)
Randrup, A., Munkvad, I.: Special antagonism of amphetamine-induced abnormal behaviour. Psychopharmacologia (Berl.) **7**, 416—422 (1965)
Randrup, A., Munkvad, I.: Role of catecholamines in the amphetamine excitatory response. Nature (Lond.) **211**, 540 (1966)
Randrup, A., Munkvad, I.: Stereotyped activities produced by amphetamine in several animal species and man. Psychopharmacologia (Berl.) **11**, 300—310 (1967)
Rechtschaffen, A., Maron, L.: The effect of amphetamine on the sleep cycle. Electroenceph. clin. Neurophysiol. **16**, 438 (1964)
Reifenstein, E. C., Davidoff, E.: The psychological effects of benzedrine sulfate. Amer. J. Psychol. **52**, 56—64 (1939a)
Reifenstein, E. C., Davidoff, E.: Benzedrine sulfate therapy. N.Y. St. J. Med., **1939b**, 42—57
Reiter, P.: Erfaringer om Benzedrin, Ugeskr. Læg. **1937**, 459—460
Rosell, S.: Aptitnedsättande medel—en litteraturgenomgång. Läkartidningen **64**, 49—56 (Suppl.) (1967)
Rosenberg, D. E., Wolbach, A. B., Miner, E. J., Isbell, H.: Observations on direct and cross tolerance with LSD and d-amphetamine in man. Psychopharmacologia (Berl.) **5**, 1—15 (1963)
Rosenberg, P.: Clinical use of benzedrine sulfate (amphetamine) in obesity. Med. World **57**, 656—659 (1939)
Roth, B., Faber, J., Nevisimalova, S., Tosovsky, J.: The influence of imipramine, dexphenmetrazine and amphetamine sulphate upon the clinical and polygraphic picture of narcolepsy–cataplexy. Schweiz. Arch. Neurol. Neurochir. Psychiat. **108**, 251—260 (1971)
Rumbaugh, C. L., Bergeron, R. T., Fang, H. C. H., McCormick, R.: Cerebral angiographic changes in the drug abuse patient. Neuroradiology **101**, 335—344 (1971a)
Rumbaugh, C. L., Bergeron, R. T., Scanlan, R. L., Teal, J. S., Segall, H. D., Fang, H. D., McCormick, R.: Cerebral vascular changes secondary to amphetamine abuse in the experimental animal. Neuroradiology **101**, 345—351 (1971b)
Rylander, G.: Preludin-narkomaner från klinisk och medicinsk-kriminologisk synpunkt. Läkartidningen **63**, 4973—4979 (1966)
Rylander, G.: Psychoses and the punding and choreiform syndromes in addiction to central stimulant drugs. Psychiat. Neurol. Neurochir. (Amst.) **75**, 203—212 (1972)
Schildkraut, J. J., Watson, R., Draskoczy, P. R., Hartmann, E.: Amphetamine withdrawal: Depression and M.H.P.G. excretion. Lancet **1971 I**, 485—486
Schmitt, H., Schmitt, H., Fenard, S.: Evidence for an alpha-sympathomimetic component in the effect of catapresan on vasomotor centres: Antagonism by piperoxane. Europ. J. Pharmacol. **14**, 98—100 (1971)
Schuberth, J., Jenden, D. J.: Transport of choline from plasma to cerebrospinal fluid in the rabbit with reference to the origin of choline and to acetylcholine metabolism in brain. Brain Res. **84**, 245—256 (1975)
Seaton, D. A., Duncan, L. J. P., Rose, K., Scott, A. M.: Diethylpropion in the treatment of "refractory obesity". Brit. med. J. **1961 I**, 1009—1011
Seevers, M. H.: Possible mechanisms of physical dependence to narcotics. In: Sevag, M. G., Reid, R. D., Reynolds, O. E. (Eds.): Origins of Resistance to Toxic Agents. New York: Academic Press 1955
Seminar on Fenfluramine and Obesity (1971), Bahamas. Ed.: Biljon, P. J. van. S. Afr. med. J. **19**, Suppl. (1971)
Sercl, M., Zdrahal, L., Jaros, O.: Electroencephalographic changes after hyperventilation and psychoton (benzedrine sulfate) in epileptics. Electroenceph. clin. Neurophysiol. **4**, 69—72 (1952)

Silverstone, J.T., Stunkard, A.J.: The anorectic effect of dexamphetamine sulphate. Brit. J. Pharmacol. **33**, 513—522 (1968)

Sletten, I.W., Ognjanov, V., Menendez, S., Sundland, D., El-Toumi, A.: Weight reduction with chlorphentermine and phenmetrazine in obese psychiatric patients during chlorpromazine therapy. Curr. ther. Res. **9**, 570—575 (1967)

Smart, J.V., Sneddon, J.M., Turner, P.: A comparison of the effects of chlorphentermine, diethylpropion and phenmetrazine on critical flicker frequency. Brit. J. Pharmacol. **30**, 307—316 (1967)

Smart, J.V., Turner, P.: Influence of urinary pH on critical flicker fusion frequency in man. Brit. J. Pharmacol. **26**, 468—472 (1966)

Smith, D.E.: Speed freaks vs. acid heads. Clin. Pediat. **8**, 185—192 (1969a)

Smith, D.E.: The characteristics of dependency in high-dose methamphetamine abuse. Int. J. Addict. **4**, 453—459 (1969b)

Smith, D.E.: Physical vs. psychological dependence and tolerance in high-dose methamphetamine abuse. Clin. Toxicol. **2**, 99—103 (1969c)

Smith, G.M., Beecher, H.K.: Amphetamine sulfate and athletic performance. J. Amer. med. Ass. **170**, 542 (1959)

Smith, G.M., Beecher, H.K.: Amphetamine, secobarbital and athletic performance. J. Amer. med. Ass. **172**, 1502—1514 (1960)

Smith, L.C.: Collapse with death following the use of amphetamine sulfate. J. Amer. med. Ass. **113**, 1022—1023 (1939)

Snyder, S.H.: Catecholamines in the brain as mediators of amphetamine psychosis. Arch. gen. Psychiat. **27**, 169—179 (1972)

Spilker, B., Callaway, E.: Effects of drugs on "augmenting/reducing" in averaged visual evoked responses in man. Psychopharmacologia (Berl.) **15**, 116—124 (1969)

Steel, J.M., Briggs, M.: Withdrawal depression in obese patients after fenfluramine treatment. Brit. med. J. **1972**, 26—27

Storz, H., Kirk, R.: Über eine neue blutdrucksteigernde Substanz. Dtsch. med. Wschr. **63**, 593 (1937)

Sulaiman, W.R., Johnson, R.H.: Effect of fenfluramine on human growth hormone release. Brit. med. J. **1973 II**, 329—332

Sykes, D.H., Douglas, V.I., Morgenstern, G.: The effect of methylphenidate (Ritalin) on sustained attention in hyperactive children. Psychopharmacologia (Berl.) **25**, 262—274 (1972)

Tatetsu, S.: Pervitin-psychosen. Folia psychiat. neurol. jap. Suppl. **6**, 25—33 (1960)

Tatetsu, S.: Langdauernde psychische Störungen infolge chronischen intravenösen Pervitinmißbrauches. Excerpta med. Int. Congr. Ser. **117**, 55 (1966)

Taylor, K.M., Snyder, S.H.: Amphetamine differentiation by *d*- and *l*-isomers of behavior involving brain norepinephrine and dopamine. Science **169**, 147 (1970)

Tecce, J.J., Cole, J.O.: Amphetamine effects in man: Paradoxical drowsiness and lowered electrical brain activity. Science **185**, 451—453 (1974)

Trouton, D., Eysenck, H.J.: The effects of drugs on behaviour. In: Eysenck, H.J. (Ed.): Handbook of Abnormal Psychology, pp. 634—696. London: Pitman Medical 1960

Turner, P.: The changes in critical flicker fusion frequency induced by some physiological procedures and by drugs, together with observations on the reliability of olfactory testing. M.D. thesis, University of London (1965)

Turner, P.: Pharmacological studies on amphetamines and related compounds in the human eye. In: Costa, E., Garattini, S. (Eds.): International Symposium on Amphetamines and Related compounds, pp. 841—848. New York: Raven Press 1970

Turner, W.D., Carl, G.P.: Temporary changes in affect and attitude following ingestion of various amounts of benzedrine sulfate. J. Psychol. **8**, 415—482 (1939)

Vogel, V.H., Isbell, H., Chapman, K.W.: Present status of narcotic addiction. With particular reference to medical indications and comparative addiction liability of the newer and older analgesic drugs. J. Amer. med. Ass. **138**, 1019—1026 (1948)

Wallis, G.G.: Amphetamine psychosis. Brit. med. J. **1957 I**, 1121

Warstadt, D.: Über therapeutische Versuche mit einem zentral wirksamen Analeptikum bei psychischen Erkrankungen (Pervitin). Wien. med. Wschr. **1938**, 1227—1230

Wasson, H. W., Uncapher, B.: Effects of dextro-amphetamine on auditory threshold in man. J. aud. Res. **6**, 351—355 (1966)
Weiss, B.: Enhancement of performance by amphetamine-like drugs. In: Sjöqvist, F., Tottie, M. (Eds.): Abuse of Central Stimulants, pp. 31—60. Stockholm: Almqvist and Wiksell 1966
Weiss, B., Laties, V. G.: Enhancement of human performance by caffeine und the amphetamines. Pharmacol. Rev. **14**, 1—36 (1962)
Weiss, S. R., Raskind, R., Morgenstern, N. L., Pytlyk, P. J., Baiz, T. C.: Intracerebral and subarachnoid hemorrhage following use of methamphetamine ("Speed"). Int. Surg. **53**, 123—127 (1970)
Weissman, A., Koe, B. K., Tenen, S.: Antiamphetamine effects following inhibition of tyrosine hydroxylase. J. Pharmacol. exp. Ther. **151**, 339—352 (1966)
Welsh, A. L.: Side effects of anti-obesity drugs. Springfield, Ill.: Charles C. Thomas 1962
Wilbur, D. L., MacLean, A. R., Allen, E. V.: Clinical observations on the effect of benzedrine sulfate. A study of patients with states of chronic exhaustion, depression and psychoneurosis. J. Amer. med. Ass. **109**, 549—554 (1937)
Zalis, E., Parmley, L. F.: Fatal amphetamine poisoning. Arch. intern. Med. **112**, 822—826 (1963)

CHAPTER 2

Amphetamine Dependence; Clinical Features

J. D. GRIFFITH

A. Naturally-Occurring Psychostimulants

Alkaloid substances are metabolic end-products of plant metabolism and therefore their function as incorporated in the bark, leaves and fruits of many plant species is difficult to deduce (ALBERT, 1965). Several have psychoactive effects similar to those of amphetamine and have long been employed as psychostimulants:

I. Khat

Khat (DREISER, 1918; RIZZOTTI, 1939; TRELLEE, 1959; LAURENT, 1962; KALANT, 1966) (one of the presumed active ingredients is *d*-norpseudoephedrine) mimics the subjective and psychotogenic effects of amphetamines (World Health Organization: Expert Committee on Addiction-Producing Drugs, 1964b; HOLLISTER, 1968) and its use is endemic among the male population in and near its areas of cultivation, principally, Ethiopia and Yemen. Wider dissemination of khat may be limited because the leaf loses its potency in about 3 days. The cultivation and use of khat has been described (LAURENT, 1962) as a somewhat unfortunate example of a marginal economy largely committed to the cultivation of a potent natural stimulant whose use has become endemic in a culture.

II. Xanthines

Xanthine derivatives, especially caffeine, as found in coffee, tea and cola beverages, are important items of international commerce. Their use is considered benign although coffee mania (MCMANAMY and SCHUBE, 1936), caffeine anxiety (REIMANN, 1967; GREDEN, 1974) and withdrawal headache (DREISBACH and PFEIFFER, 1942; HARRIE, 1970) are occasionally reported. Nevertheless, its use has in the past prompted some controversy. Eighteenth Century public coffee houses were considered somewhat disreputable. Their clientele included the unwashed but distinguished Dr. SAMUEL JOHNSON. LEWIS (1969) uses coffee as an example of the changing opinions on the dangers of psychostimulants. He quotes from the writing of CLIFFORD ALLBUTT, the Regius Professor at Cambridge, and W. E. DIXON, the noted pharmacologist: Referring to coffee,

> "the sufferer is tremulous and loses his self-command; he is subject to fits of agitation and depression. He has a haggard appearance... As with other such agents, a renewed dose of the poison gives temporary relief, but at the cost of future misery.

Tea is no better: Tea has appeared to us to be especially efficient in producing nightmares with ... hallucination which may be alarming in their intensity. Another peculiar quality of tea is to produce a strange and extreme degree of physical depression. An hour or two after breakfast at which tea has been taken ... a grievous sinking ... may seize upon a sufferer, so that to speak is an effort ... The speech may become weak and vague ... By miseries such as these, the best years of life may be spoilt."

III. Ephedrine

Ephedrine (from Ephedra plant; MA HUNG; the "yellow astringent" of Chinese herb doctors) was introduced into Western medicine by CHEN in 1923. Although amphetamine-like (MARTIN, et al., 1971) and easily available and widely used, abuse, until quite recently, has been rare (PANSE and KLAGES, 1964; ALBERT, 1965; HERRIDGE and A'BROOK, 1968; PROKOP, 1969).

IV. Absinthe

Absinthe, a cordial with a high content of alcohol, once contained an extract of wormwood. Thujune, an isomer of camphor, is present in oil of wormwood. Habitués were said to become maniacal from prolonged and excessive use. Metrazole, another camphor congener, was used as an early form of convulsive therapy for depression.

V. Cocaine

Cocaine has also been cited as an example of endemic stimulant use among Peruvian Indians. NIESCHULZ (1969), however, points out that the alkaloid is for the most part inactivated by lime, usually added before the leaf is chewed, and amounts absorbed from the buccal mucosa are rapidly inactivated. Introduced as a local anesthetic by KOHLER in 1884, cocaine was also endorsed by a number of medical authorities, including Sigmund Freud, as a safe and "nonaddicting" drug which could be used to treat alcoholism and opiumism. Both cocaine and morphine were sold openly in drug stores at the turn of this century and included deceptively in beverages and patent medicines sold as universal cure-alls. Cocaine came to be regarded as a drug which "made people crazy," and it did much to promote the passage of the Pure Food and Drug Act of 1906 (YOUNG, 1961). It is often said, but not documented (MUSTO, 1973), that the early popularity of Coca Cola was related to its cocaine content rather than to flavoring agents obtained from coca leaves. This is an exaggeration in the sense that trace amounts were discovered from improperly decocainized leaves but significant amounts were not. Cocaine has since become an important drug in illicit commerce, particularly during the 1920's and now. Cocaine has recently been demonstrated to produce amphetamine-like subjective effects in man (FISCHMAN et al., 1976) and is psychotogenic. The National Institute on Drug Abuse estimates that cocaine is now used at least on an occasional basis by 9 million people in the United States.

B. Exogenous Phenethylamine Congeners

A variety of compounds with a phenethylamine moiety are end-products of metabolism of intestinal bacteria. Tyramine and congeners are found in certain foods, such as cheese. These are readily degraded by liver enzymes and have no significant psychoactive or sympathomimetic effects under normal circumstances. It is of historic interest to note that studies of sympathomimetic drugs were initiated by BARGER and DALE (1910) who suspected that ptomaine poisoning caused by putrid meat was a result of aliphatic amines generated by bacteria. In the course of their work, they demonstrated the sructure-activity of a number of sympathomimetics, including tyramine and amphetamine.

In certain diseases, foreign alkylamines arising from gut bacteria enter the circulation, are associated with endotoxic shock (CUEVAS and FINE, 1973), and may interfere with brain and kidney function (FISCHER and BALDESSARINI, 1971). Interestingly, some patients in massive liver failure are restored to consciousness when given levodopa (FISCHER and JAMES, 1972). This may be because catecholamines synthesized from levodopa displace exogenous amines which have entered adrenergic neurons and act as "false transmitters" (see Sect. I, Chap. 1, A and B). A monoamine oxidase enzyme defect has also been postulated as a cause of migraine headache (HANINGTON, 1967; SANDLER et al., 1974).

MODELL and READER (1970) have recently taken exception to the erroneous but widespread impression that amphetamine sulfate was the first in a series of anorectic drugs. They point out that tetrahydro-β-naphthylamine, used as an anorexigenic between 1890-1910 and later discarded because of its toxicity, had both sympathomimetic and CNS properties. Better recognized is the prior use of ephedrine in the treatment of narcolepsy, nasal congestion and allergic states (CHEN and SCHMIDT, 1930).

C. Synthetic Psychostimulants

I. Amphetamine and Methamphetamine

Amphetamine itself was first prepared by EDELEANO (1887) and later by JONES and WALLIS (1926). Dr. GORDON ALLES, however, is recognized as having both synthesized several phenethylamines, including amphetamine, and with his associates (PINESS et al., 1930) demonstrated their efficacy as nasal astringents. Biographical details of the discovery of amphetamine are furnished by LEAKE (1958 and 1970). As Chairman of the Department of Pharmacology at the University of California at Los Angeles (UCLA), Leake made available to ALLES facilities for testing the pharmacology and toxicology of this drug in animals. ALLES inadvertently discovered the profound effect of amphetamine on mood while using himself as a test subject to learn of the effects of amphetamine on blood pressure.

A worldwide shortage of ephedrine, then obtained from natural sources, has been cited as the pragmatic impetus for ALLES' work. This may seem contradictory because ephedrine could be synthesized by the late 1920's; however, the efficacy of synthetic ephedrine was questioned by clinicians who favored the "natural" product (CHEN and SCHMIDT, 1930).

Amphetamine was originally introduced for use as a nasal astringent. The first such proprietary was the Benzedrine inhaler offered as an over-the-counter item in U.S. drug stores beginning in 1932. Eventually, and especially after World War II, similar preparations containing 150 to 320 mg each of amphetamine, methamphetamine or mephentermine were sold under a variety of brand names in many parts of the world, particularly North and South America, the United Kingdom and Australia. The abuse of inhalers containing the more potent amphetamines became a significant problem in the United States and, for this reason, they were withdrawn or banned between 1944-1967.

Amphetamine was later introduced into clinical practice in 1937, following several studies initiated by MYRON PRINZMETAL and ANDERSON PEOPLES, both previously associated with LEAKE at UCLA. Although the efficacy of amphetamine in the treatment of narcolepsy (PRINZMETAL and BLOOMBERG, 1935) was reported and soon confirmed (ULRICH et al., 1936), the drug was of no use in the treatment of severe emotional depression and catatonia (PEOPLES and GUTTMAN, 1936; GUTTMAN, 1936). Weight loss observed among psychiatric patients suggested that amphetamine was an anorexigenic (NATHANSON, 1937; LESSES and MYERSON, 1938; ROSENBERG, 1939, and many others); however, this property was not firmly established for several years (HARRIS et al., 1947). With reservations, amphetamine was considered useful in the treatment of mild depression and asthenia (MYERSON and RITVO, 1936; GUTTMAN, 1936; MYERSON, 1936; DAVIDOFF; 1936). The calming effect of amphetamine on hyperkinetic children was discovered by BRADLEY (1937), who administered the drug to improve the academic performance of institutionalized and brain-damaged children. Although amphetamines were usually prescribed for the treatment of obesity or mild depression, the drug was also investigated in 39 other conditions, particularly before 1941. These includes epilepsy, postencephalitic parkinsonism, schizophrenia, alcoholism, barbiturate intoxication, anesthesia, morphine and codeine addiction, tobacco habit, enuresis, migraine, heart block, disseminated cerebral sclerosis, myasthenia gravis, myotonia, head injury, polio, urticaria, Meniere's syndrome, dysmenorrhea, colic, colitis, radiation sickness, hypotension, as an adjunct to electroconvulsive treatment, carotid sinus syndrome, analgesia, premature climacteric, night blindness, and "caffeine mania" (BETT, 1946; LEAKE, 1958).

Amphetamine and methamphetamine were also used by most of the major combatants during World War II to extend the effectiveness of military personnel under certain conditions. Originally introduced for this purpose by the psychiatrist EMILIO MIRO during the Spanish Civil War, amphetamines were used principally to maintain alertness of pilots during landings following long-range bombing missions and drivers of transport and armored vehicles who had to maneuver at night to elude enemy action. Amphetamines were also put in air and naval "survival kits" during World War II and for many years afterwards. After World War II, the effectiveness of amphetamines was tested in simulated military exercises and was anything but impressive. This application apparently caused few cases of dependency; however, it did much to enhance the reputation of amphetamine as a "pep pill."

Methamphetamine was first prepared by OGATA (1919) and was demonstrated to be the desoxy derivative of ephedrine by EMDE (1929). Methamphetamine was introduced into clinical practice in Europe and Japan following the work of HAUSCHILD (1939) and others. After World War II, methamphetamine was introduced in North

America. Early investigators claimed that methamphetamine was more effective than amphetamine, probably because the *d*-isomer of methamphetamine is approximately twice as potent as racemic amphetamine, but the subjective effects and potency of methamphetamine and *d*-amphetamine in man are quite similar (MARTIN et al., 1971). Illicit amphetamines are usually methamphetamine, probably because it is relatively easy to synthesize from ephedrine, which is uncontrolled.

II. Other Amphetamine Congeners

Considerable effort has been devoted to developing drugs which have the therapeutic actions of amphetamines but less abuse liability. The most notable successes have been drugs which are nasal astringents and vasopressors. A number of side-chain amphetamine congeners have been offered as anorexigenics and antidepressants—several (phenmetrazine, methylphenidate, and to a lesser extent, diethylpropion) have been recognized as having abuse potential. Others have a lesser potency and are not commonly abused. Four ring-substituted amphetamine congeners with anorexigenic properties have been recently introduced into the pharmacopeia (chlorphentermine, clortermine, fenfluramine, and mazindol). These drugs have dysphoric and sedative effects, especially when taken in larger than recommended doses and are, therefore, less likely to be abused than amphetamine.

III. Psychedelic Amphetamines

Some ring-substituted phenethylamines and phenylisopropylamines have hallucinogenic properties resembling those of LSD (see Sect. I, Chap. 3). The effects of these drugs suggest that some hypothetical derivative may be useful in therapy. This matter occupied ALLES during his later years, and many others, but no useful compounds have evolved so far. These substituted amphetamines also have amphetamine-like activity in animals (MARTIN et al., 1977).

D. Changing Medical Views on Amphetamine Abuse and Psychotoxicity

Since the 1950's, amphetamines have become generally recognized, not only as drugs which are likely to be used as "pep pills" but also as agents which produce dependency syndromes of varying degrees of severity, often complicated by social deterioration and psychosis and the likelihood of involvement with other drugs such as alcohol, barbiturates, and minor tranquilizers, taken to increase the euphorigenic properties of amphetamine and to nullify certain of its unpleasant effects such as insomnia. Another hazard is acute overdose among neophytes, who may have a psychosis or alarming somatic symptoms. Death, however, is rare, even among indiscriminate users (BONHOFF and LAWRENZ, 1954; KALANT and KALANT, 1975).

The concept of amphetamine as a drug of dependency or one that produces psychosis was often disputed but has become better recognized as a result of a number of clinical reports. Of particular influence were KNAPP'S (1952) description of amphetamine dependency among patients "living in the shadow of strict conscience," MASAKI's (1956) report to the World Health Organization on widespread amphetamine abuse in Japan, CONNELL's (1958) landmark monograph on the amphetamine psychosis, complimented by BELL's (BELL and TRETHOWAN, 1961; BELL, 1965) work in Australia; and, finally KALANT's (1966 and 1973) meticulous synthesis of the world literature. Although amphetamines were described as relatively safe and benign drugs in the American and British medical literature as late as 1961, reports of widespread amphetamine and methamphetamine abuse in both countries (BEAMISH and KILOH, 1960; MCCONNELL and MCILWAINE, 1961; KILOH and BRANDON, 1962; GRIFFITH, 1966; ANGRIST and GERSHON, 1969a), including vivid accounts of intravenous amphetamine abuse (KRAMER et al., 1967; CAREY and MANDEL, 1968; HAWKS et al., 1969; SMITH, 1969; SMITH, 1972), dispelled this belief. In controlled experiments, amphetamine has been shown to be a psychotogenic substance (GRIFFITH et al., 1968, 1970a, b and 1972; ANGRIST and GERSHON, 1969b, 1970 and 1971) and to have subjective effects similar to those of cocaine, a similarity sometimes mentioned (STAEHELIN, 1941) but recently emphasized by BEJEROT (1970) and POST (1975).

Amphetamine-like drugs have long been considered potentially toxic substances. Amphetamine was first mentioned as a drug of abuse among students in a critical editorial in the Journal of the American Medical Association (1937), and the possibility of cardiovascular complications from unsupervised medication was a matter of medical concern. When it was accepted into the New and Nonofficial Remedies, the American Medical Association's Council on Pharmacy and Chemistry (1937) described amphetamine as specifically not recommended for developing a sense of increased energy or capacity. The same year, amphetamine was connected with the suicide of a mentally ill person who leaped to his death from the 17th floor of a hotel (SCHUBE et al., 1937). These circumstances were sufficient to cause the issue of public warnings. GALDSTON (1938 and 1940), for example, in articles for lay readers, interpreted the introduction of amphetamine, sulfonilamide, protamine zinc insulin and synthetic vitamin C as evidence that 1937 would be a year remembered as the "rebirth of chemotherapy, the science founded by PAUL EHRLICH ..." But amphetamine, he warned, was a "powerful stimulant" and self-medication was a "dangerous practice."

Initial clinical experience did little to dispel these concerns. The following year, YOUNG and SCOVILLE (1938) reported a paranoid psychosis in two patients receiving amphetamine for the treatment of narcolepsy. WAUD (1938) cited the development of tolerance to amphetamine euphoria and spoke to the possibility that a patient might increase this dose and become addicted. REIFENSTEIN and DAVIDOFF (1938) observed that patients addicted to alcohol and narcotics were able to substitute amphetamines, and were further concerned, as was GUTTMAN (1938), that the drug was becoming a problem among emotionally unstable patients and persons working under strain, e.g., actors, students, nurses, and physicians. Case histories suggesting amphetamine dependency were reported by ANDERSON (1938) and FRIEDENBERG (1940), and acute toxicity following impulsive use by a number of clinicians (NIELSEN, 1937; APFELBERG, 1938; GOTTSEGEN, 1938; EHRICH et al., 1939; SMITH, 1939; BLOOMBERG, 1940).

A further concern to medical authorities and regulatory agencies during the late 1930's and early 1940's was the spread of what might be called the amphetamine myth or illusion, i.e., that amphetamines were drugs which were not only of benefit to persons with disease but also to those who wanted to "get places," "do things," and "be somebody." The essence of the myth as it was received at an individual level is captured by JOSEPH MANKIEWICS, the noted Hollywood playwright and a subject in one of the early studies of amphetamine. After being given a tablet to take in the morning, he reported: "I felt like God ... It was unbelievable ... I wrote fifty pages of dialogue, I went into a story conference, I was absolutely brilliant" (FRANK, 1975).

BETT (1946) commenting on the "miracle drug" response to amphetamine said:

> 'It may be doubted whether, with the possible exception of the sulfonamides, penicillin, streptomycin, any therapeutic agent of modern scientific medicine ... aroused such vivid interest in professional circles and in the lay press as benzedrine sulphate. A veritable deluge of literature [was] let loose concerning its uses and abuses, its virtues, vices, and potentialities. In the salad days of its therapeutic career, benzedrine not so much achieved, as had thrust upon it an immense press publicity, even notoriety: the "Confidence Drug"—"Pep Pill"—"Reach for a Pill instead of a Cocktail." The wave of sensationalism engulfed apprehensive students, tired city magnates and housewives bored with the monotony of their unexciting lives. The wave of sensationalism spread, engulfing psychologists and philosophers, who were intrigued by the vast possibilities of a drug which promised to open unexplored fields of consciousness.'

MONROE and DRELL (1947) mentioned a newspaper advertisement for a charm-bracelet and pill box "for aspirin when you have a headache; for Benzedrine when you are blue," and the popular ditty, "Who put the Benzedrine in Mrs. Murphy's Ovaltine." For a time amphetamine and methamphetamine were not especially attractive to narcotic addicts nor was illicit traffic a problem. HIMMELSBACH et al. (1977) made a survey of drug preferences among U.S. addicts in 1944, and found that amphetamine ranked fifth in preference—after narcotics, alcohol, coffee, and Coca Cola. Nor, when given an opportunity to self-administer the drug to test its abuse potential were there many volunteers. After 1944, a dramatic change occurred.

Widespread amphetamine abuse (nasal inhalers) was reported among a sizable percentage of inmates of U.S. prisons (MONROE and DRELL, 1947; JACKSON, 1971). Moreover, some of the "New York Benzedrine Set," so called by the columnist EARL WILSON, which by then included "less sophisticated segments of society," had discovered that amphetamines and barbiturates in combination produced what was called a "bolt and a jolt." Spree use of amphetamines by teenagers was an object of satire in the Broadway play "New Faces of 1952" and had evolved into epidemic intravenous use in this age group as well as among adults by 1957, in and around Oklahoma City, St. Louis (RAWLIN, 1968; JACKSON, 1971), and other U.S. cities. By 1962, it was estimated that the annual U.S. production of amphetamine and methamphetamine was 100000 lbs, or enough to supply every man, woman and child with 250 mg a year. Moreover, half was diverted into illicit channels. In Sweden, where amphetamine abuse was first recognized as a serious social problem, the introduction and intravenous abuse of phenmetrazine created a similar problem. In Japan, where methamphetamine was introduced and advertised as a drug for "the elimination of drowsiness and repletion of the spirit," its ability to produce dependence was recognized by 1947, and its widespread abuse followed shortly thereafter.

As a result of these problems, amphetamine and methamphetamine were placed under ordinary prescription controls in North America and most European countries by 1939, and under narcotic controls in Germany in 1941 and in Sweden in 1944. Similar measures were considered in the United States in 1949 but were withdrawn when the Benzedrine inhaler was replaced by a preparation containing propylhexidrine.

On the other hand, there was a conflicting medical opinion that the abuse potential and psychotoxic properties of amphetamine were perhaps overstated. Many authors pointed out that amphetamine did not produce a "craving" or withdrawal reactions. Tolerance to the hypertensive effects of amphetamine and the ability of some mentally ill patients to tolerate the drug were emphasized by SHORVON (1945). MYERSON (1940) stated that he had not seen a single case of addiction; GRAHN (1958) that he had seen only one case (and the probability of another) in 16 years of practice. Nor were the possibilities of amphetamine dependency considered an important risk as mentioned in standard pharmacological texts or reviews.

Federal legislation in the United States restricting the availability of amphetamines has been gradual and piecemeal. The Benzedrine inhaler was withdrawn by its manufacturer in 1949 and similar preparations containing amphetamine and dextroamphetamine were banned by Federal agency ruling in 1957. Methamphetamine inhalers, however, were exempted until 1965 and had by that time become included in widespread patterns of intravenous use (GRIFFITH, 1966). Problems with mephentermine inhalers were similarly identified (ANGRIST et al., 1970). Amphetamine tablets and injectable amphetamines were placed under prescription control in 1938 and regulation of these items was given to the Food and Drug Administration, which reported that during the 1950's, illicit traffic in amphetamines had become the greatest single reason for arrests under the Pure Food and Drug Laws. The Drug Abuse Control Amendments of 1965 required that detailed records be kept on the manufacture, shipment and distribution (including pharmacy prescription records) of stimulants. Moreover, the life of amphetamine prescriptions was limited to six months or five refills. Injectable methamphetamine preparations were withdrawn. As a result of continued illicit traffic, particularly that involving international smuggling from South America, the more potent amphetamines were scheduled with narcotics under the Comprehensive Drug Abuse Prevention and Control Act of 1970. While wholesale diversion of amphetamines manufactured for legitimate use is not the problem it once was in the United States, importation of cocaine has increased. It is estimated that $ 1.2 billion worth of cocaine is smuggled into North America annually.

E. Contemporary Medical Indications

The risks and medical benefits resulting from the use of amphetamines have been subjects of continuing debate by both medical authorities and regulatory agencies.

I. Obesity

Although the ability of amphetamines to curb appetite of certain obese patients has been well demonstrated, their effectiveness in treating obesity is controversial. Only about one-third of patients treated with amphetamines have significant weight loss and tolerance develops to amphetamine's ability to curb appetite. After six weeks of continuing administration, amphetamines are no more effective than a placebo. Moreover, when amphetamines are discontinued, appetite returns and few patients maintain their loss of weight (STUNKARD and MCLAREN-HUME, 1959; MODELL, 1960; FEINSTEIN, 1960; SILVERSTONE and STUNKARD, 1968; MUNRO et al., 1968). During treatment a diet pill can become a pep pill, with an associated risk of the patient becoming dependent (KLINGER, 1965; SEATON and DUNCAN, 1965; PRAGG, 1968; LADEWIG and BATTEGAY, 1971; SWANSON et al., 1973). Thus, as LEWIS (A.J., 1970) says, amphetamines can only offer a prop and have no demonstrated role in the long-term management of obesity. The rationale of increased longevity from weight reduction (MARKS, 1960) resulting from the administration of amphetamines has been weakened by a recent critical review (MANN, 1974).

Since the 1950's, several amphetamine-like drugs have been introduced as anorexigenics and described as being as effective as dextroamphetamine or methamphetamine in curbing appetite but possessing less activity as psychostimulants. These are commonly prescribed by U.S. physicians (LASAGNA, 1973). Recent studies of their mood-altering effects (MARTIN et al., 1971; JASINSKI and NUTT, 1972; JASINSKI et al., 1974; GRIFFITH et al., 1975) as well as epidemiologic data of abuse suggest that their abuse potential has been underestimated. Possible exceptions are several of the newer ring-substituted amphetamines which are dysphoric and sedative in therapeutic doses.

Conversely, while the psychotoxicity and effectiveness of amphetamine-like anorexigenics leave much to be desired, it is a gross oversimplification to dismiss obesity as a trivial problem which must be endured by those who lack will power. If anything, it is a problem that has more important social and medical ramifications than does drug addiction (CLAYTON et al., 1973; FOX, 1973 and 1974; BRADHAM, 1974; KEESEY and POWLEY, 1975; MILLER and PARSONAGE, 1975; ROBINSON et al., 1975). While current research in the area raises more questions than answers, there is little doubt that the problem is one that deserves continued investigation. An estimated $ 100 million is spent each year for worthless obesity cures by persons seeking to lose weight (Committee on Government Operations of the U.S. House of Representatives, 1958).

II. Narcolepsy

Narcolepsy is a sleep disorder characterized by incapacitating daytime sleep attacks (narcolepsy proper), cataplexy, hypnogogic hallucinations and/or sleep paralysis (YOSS and DALY, 1957; ZARCONE, 1973). The disorder is considered to be of organic rather than of psychogenic origin. Differential diagnosis, however, can be difficult since pathological drowsiness can be a component of several functional mental illnesses. For this reason, short latency of onset of rapid eye movement (REM) sleep (RECHTSCHAFFEN et al., 1963; DEMENT et al., 1966) is of aid in making a diagnosis. Onset of narcolepsy usually begins during the second decade of life. Often described as rare, it is probably more common than generally recognized. Prevalency rates have been roughly estimated at 0.03 to 0.05 percent. It is an incapacitating illness in that a patient is unable to work in certain occupations or drive an automobile without the danger of abruptly losing consciousness. Paradoxically, nocturnal sleep is often troubled.

Dextroamphetamine and methylphenidate are partially effective in the treatment of this condition. YOSS (YOSS and DALY, 1957 and 1958; YOSS, 1969; YOSS et al., 1969) has developed a pupillographic method for estimating the optimum dose. Unfortunately, these drugs cause nocturnal insomnia and their efficacy disappears with the development of tolerance. This may be offset by increasing the dose, but at the risk of causing an amphetamine psychosis or socially deviant behavior not in keeping with the patient's personality. Tricyclic antidepressants (imipramine and desmethylimipramine) have also been shown to prevent cataplexy (HISHIKAWA et al., 1966; ROTH et al., 1971) but do not eliminate daytime sleep attacks. ZARCONE (1973) points out that while a combination of methylphenidate and imipramine is theoretically dangerous, these drugs can be used in combination to control both sleep attacks and cataplexy.

III. Hyperkinetic Syndromes of Children

Amphetamines are also used to treat certain behavioral disorders of children. Clinical interest in these problems developed during the 1918 epidemic of encephalitis in the United States (EBAUGH, 1923; KAHN and COHEN, 1934). Among the children who survived, some showed an intractable change in personality, characterized by a short attention span, distractability, irritability, impulsivity, explosiveness, destructiveness, erratic school performance and poor social adjustment (BRADLEY and BOWEN, 1941; BENDER and COTTINGTON, 1942; LAUFER and DENHOFF, 1957; WEISS et al., 1968). The abnormality is usually seen before the child enters the first grade of school and is more often diagnosed among boys than girls (6:1). The more seriously affected children may require special classroom or institutional care. Intelligence is usually normal. An organic etiology is often assumed, particularly in children with clear-cut neurological deficits and histories of brain injury; however, many children manifest only the behavioral deficit while others have borderline EEGs and other "soft" neurological signs (LAUFER et al., 1957; BUCHSBAUM and WENDER, 1973; FEINBERG et al., 1974; SATTERFIELD et al., 1974; RAPOPORT et al., 1974a; ZAHN et al., 1975).

A number of amphetamines have been demonstrated as effective adjuncts in hyperkinetic children, and these include: amphetamine, dextroamphetamine, methamphetamine (BRADLEY, 1937 and 1950; SCHILDER, 1938; BENDER and COTTINGTON, 1942; WEISS et al., 1968; EPSTEIN et al., 1968); methylphenidate (CONNERS and EISENBERG, 1963; RAPOPORT et al., 1974b); levoamphetamine (ARNOLD et al., 1972) and pemoline. A report of caffeine's effectiveness (SCHNACKENBERG, 1973) was not confirmed (GARFINKEL et al., 1975; HUESTIS et al., 1975). Response to amphetamines is usually determined by clinical trial, and it is not unusual for hyperkinetic children to show a dramatic improvement in their ability to concentrate and control their impulsive behavior—usually within 1–7 days after treatment is started. About one-half do not respond and of these, 10 percent are made worse. Treatment with minor tranquilizers is deleterious (ZRULL et al., 1964) but imipramine is effective as a second-line drug (WINSBERG et al., 1972; RAPOPORT et al., 1974b).

Treatment with amphetamine must be weighed against the possibility of certain side effects. Retardation of growth (weight and height) has been reported which is not easily recognized nor always compensated for when the drug is withdrawn (SAFER and ALLEN, 1973; RAPOPORT et al., 1974b; QUINN and RAPOPORT, 1975). Two cases of amphetamine psychosis among children being treated for hyperkinesis have been reported (NEY, 1967), but psychosis is more common among children of less than 3 years of age who take amphetamine accidentally (KALANT, 1966).

The use of amphetamines for the treatment of hyperkinetic children has stimulated such emotional articles as "Chaining children with chemicals" (VONDER HAAR, 1975), as well as critical reviews (FISH, 1971; SROUFE and STEWART, 1973) that point out problems of diagnosis and of substituting drug therapy for other forms of management. Judicious use of amphetamines in hyperkinetic children has been endorsed by experts (FREEDMAN, 1971). Amphetamine abuse among pre-adolescents has not been a problem. Many children find amphetamines unpleasant and some refuse to take them.

Some hyperkinetic children have persistent behavioral symptoms well into the third decade of life. The evidence for this has recently been reviewed in a paper by WOOD et al. (1976). They suggest, as did KNAPP (1952), that some forms of amphetamine abuse may be an attempt to control antisocial impulses originating from minimal brain dysfunction and not an expression of antisocial impulses *per se*. Interestingly, a prolonged attention span and fascination with the trivial ("punding") is a toxic effect of high-dose amphetamine abuse among adults.

IV. Emotional Depression

Amphetamines have long been recognized as contraindicated in severe emotional depression. Most investigators have had either negative or undesirable results in serious depression, especially those accompanied by anxiety or mania, since the stimulating actions of amphetamine may make such patients suicidal or combative. A similar activation of accessory symptoms was also noted among patients with an active schizophrenic illness. The response of both types of patients to single doses of dextroamphetamine and methylphenidate has been reinvestigated and reviewed by several investigators (JANOWSKY et al., 1973; POST et al., 1974; POST, 1975; VAN KAMMEN and MURPHY, 1975).

On the other hand, amphetamines were once quite popular as a means of treating mild depressions (disappointment, neurasthenia, psychoneurosis, the so-called "tired housewife syndrome," work-related fatigue, and alcoholic hangover). Recent studies of the effectiveness of amphetamines in the management of mild depression suggest that these drugs are little better than a placebo. For this reason, the British and American Medical Associations (Committee on Alcoholism and Addiction and Council on Mental Health, 1966; Committee on Medical Science, Education and Research, 1968) have recommended that amphetamines no longer be used for this purpose. The U.S. Food and Drug Administration now recognizes only methylphenidate for this use and categorizes the drug as of "questionable effectiveness pending further proof of efficacy." Nevertheless, amphetamines are still defended for this use by some physicians for well-supervised cases. For certain patients, this type of use may be justified, for example, among women with dysmenorrhea who are prescribed 3 tablets a month. The use of amphetamines for the treatment of depression also presents a circumstance for diversion and dependency.

V. Analgesia

Amphetamine in combination with a potent narcotic has been proven to have considerable utility in the sense that the mixture produces profound analgesia without respiratory depression (IVY et al., 1944; HANDLEY and ENSBERG, 1945; EVANS, 1962 and 1967; EVANS and SMITH, 1964; EVANS and BERGNER, 1964). There is little doubt that such combinations have a significant abuse potential (JASINSKI and NUTT, 1972), and they are known on the street as "BAM" (Benzedrine-and-methadone). However, this risk would not seem to be significantly greater than for narcotics alone. Despite their usefulness, amphetamine-narcotic analgesic preparations are not much used.

VI. Miscellaneous Applications

Amphetamine and methylphenidate are sometimes prescribed to counteract the sedative effects of phenothiazines in schizophrenic patients. The utility of these drugs in this application is questionable. Amphetamine's use as an analeptic in the treatment of acute barbiturate poisoning has not endured with the demonstration of the effectiveness of supportive care. Amphetamines are sometimes used as a second drug in the treatment of petit mal epilepsy or to reverse the dystonic effects of phenothiazine drugs. A combination of amphetamine and atropine remains a reference drug against which drugs for motion sickness are compared (WOOD and GRAYBIEL, 1968). It is said that cocaine is used by Peruvian Indians to overcome air hunger while working at high altitudes; however, amphetamine worsens the symptoms of altitude sickness. Its effectiveness in stuttering is marginal (GINN and HOFMAN, 1953; FISH and BOWLING, 1962).

F. Psychotoxicity

I. Acute Intoxication

Acute amphetamine intoxication is commonly reported in the clinical literature as a problem in management. Some of these cases represent patients who have an idiosyncratic hypersensitivity to therapeutic doses, while other patients use amphetamines to attempt suicide. There are reports, however, which suggest that some patients ingest large amounts of amphetamines over a short period of time to enhance the subjective effects experienced from a lesser dose (CONNELL, 1958; KALANT, 1966). Weekend "spree" use and "over-amping" is also common among both neophytes and experienced members of the "speed scene" (CAREY and MANDEL, 1968; SMITH, 1969; SMITH, 1972).Spree use is also common among cocaine abusers who will "snort" or inject it until they become psychotic. The consequences of acute and subacute amphetamine intoxication are detailed in Section I, Chapter 1, C. Patients who acutely overdose with amphetamines often present with a psychosis and show little in the way of sympathomimetic hyperstimulation. In others, autonomic effects may be alarming and occasionally fatal (BONHOFF and LAWRENZ, 1954; ONG, 1962; ZALIS and PARMLEY, 1963; KALANT, 1966; KRISKO et al., 1969; GINSBERG et al., 1970; KALANT and KALANT, 1975). Accidental poisoning from amphetamines is also encountered in pediatric practice. ESPELIN and DONE (1968) have demonstrated the efficacy of chlorpromazine in the treatment of such cases.

II. Dependency Syndromes

The World Health Organization Expert Committee on Addiction-Producing Drugs (1964a) redefined drug dependence of the amphetamine type as

> "A state arising from repeated administration of amphetamine or an agent with amphetamine-like effects on a periodic or continuous basis. Its characteristics include (1) a desire or need to continue taking the drug; (2) consumption of increasing amounts to obtain greater excitatory and euphoric effects or to combat more effectively depression and fatigue, accompanied in some measure by the development of tolerance; (3) a psychic dependence on the effects of the drug related to a subjective and individual appreciation of the drug's effects; and (4) general absence of physical dependency so that there is no characteristic abstinence syndrome when the drug is discontinued."

KALANT (1966 and 1973), among others, emphasizes the fact that amphetamine-related psychosocial impairment involves a continuum as broad and varied as that associated with the use of beverage alcohol. Signs and behavior associated with mild degrees of amphetamine dependency and intoxication are difficult to separate from ordinary acts of misjudgment which in a non-drug using person would be explained as lack of foresight or enthusiasm. More obvious states of dependency are frequently associated with mental illness, alcoholism or antisocial behavior. KALANT also predicted that because many people can use amphetamines without difficulty, control of amphetamines might prove to be as difficult as the enforcement of National Prohibition. This has proven to be the case.

Relatively benign forms of amphetamine abuse and iatrogenically induced dependency have been described (FRIEDENBERG, 1940; MYERSON, 1940; BRANDON and SMITH, 1962; KILOH and BRANDON, 1962; SCOTT and WILLCOX, 1965; KALANT, 1966; COLE, 1972; LLOYD, 1973) among persons from a variety of subcultures. These include patients who were once commonly prescribed amphetamines on a chronic basis, most of whom, it would appear, do not increase their dose or suffer important somatic or behavioral toxicity as long as they are supplied with drugs. Tolerance to the euphorigenic effects of amphetamine might seem to make continued use unrewarding; however, some patients are convinced amphetamines are useful, and if an attempt is made to reduce their dose or withdraw drugs, they complain of depression and inability to work. As FRIEDENBERG (1940) said in the first report of such cases, "the problem is not a minor one." Continued prescribing may result in frank dependency which may not become evident for years. On the other hand, discontinuing amphetamines may encourage the patient to seek a physician who is a more liberal prescriber or to make contact with an illicit supplier. Both circumstances are described in ROBINSON'S (1974) autobiographical work. Nor is the physician comfortable. "Frequently they are faced with the decision whether to give in to the demands of patients for these substances or to face a reduction in their income. As one doctor said, 'I feel my self-respect is being nibbled away'" (KILOH and BRANDON, 1962). The British Medical Association (Committee on Medical Science, Education and Research, 1968) gives reputable physicians some latitude in the management of such cases, but in the United States maintaining a patient on amphetamines invites the risk of loss of license, or if the patient encounters difficulties, tort action.

Frank states of amphetamine dependency accompanied by important behavioral and social impairment may develop slowly over a period of time ranging from months to years. Such cases were rarely described prior to KNAPP'S (1952) classic description of patients using amphetamines. Amphetamine dependency is particularly tragic in persons with a strong commitment to family, work and reputation. The insidious nature of amphetamine dependency has also been captured in FRANK's (1975) sympathetic biography of the actress, Judy Garland.

Typically, such patients begin using amphetamines during their late teens or early twenties, usually to cope with study, work, obesity or personal difficulties such as depression or shyness. He (or she, since women are equally vulnerable) may use modest amounts for years during which time the drug becomes firmly ingrained as a "problem solver." Such use was easily rationalized as harmless before amphetamines became known as dangerous drugs; indeed, the use of "pep pills" was sometimes portrayed in motion pictures and books as an expression of dedication to work (BERNSTEIN, 1958). An early problem with chronic amphetamine use is insomnia which leads to the use of alcohol, barbiturates and minor tranquilizers.

Social and occupational performance may be quite adequate when modest doses are used, but, in the presence of family or occupational stresses, the dose is frequently increased. This leads to impairment of critical judgment which is expressed as garrulousness, feverish activity, start-and-stop projects, suspiciousness, unrealistic goals, debts and gambling. DAVIDSON's (1964) articulate patient explains this best:

> "The trouble is that one gets very indiscriminating in one's output of the tremendous energy which results. I'm just as likely to clean out the attic and all the closets (which nobody wants me to do) as to do the day's work which has been assigned to me. I used to write one piece of copy 8 times in 8 beautiful versions—just as a *tour de force* rather than to write 8 different copy jobs, which the boss would have preferred."

Because amphetamines rapidly induce tolerance, such persons must withdraw or "crash," usually on weekends. Another writer explained her symptoms when crashing as follows:

> 'Off speed I felt the way people say they feel when they are "coming down with something—a cold." My eyes did not see, it seemed, as sharply. My head felt full of lint. My limbs felt heavy and dull ... "Sure," I said to my editor [wanting a rewrite], "I've got the time. I'll call you right back and let you know." To dial the Time Signal was a challenge for my ingenuity. I had to avoid telling him I can't write without speed' (ROBINSON, 1974).

Remarkably, many patients can disguise the extent of amphetamine abuse even from their spouses and explain their behavior as overwork, temporary irritation, and the like. Many apparently discontinue the drug spontaneously; others present ultimately with an amphetamine psychosis, as suicides, or are forced into treatment by legal action or threat of divorce. Their prognosis is not good. Occasionally, these patients experience panic episodes, feel they are having a "heart attack," and seek treatment in hospital emergency rooms.

After World War II, a sharp increase in the non-medical use of amphetamine-like drugs occurred in Japan, North America, and several European countries. This phenomenon includes a substantial number of persons who used amphetamines intravenously, a practice encouraged by membership in amphetamine-using cults and despite the obvious disadvantages of repeated intoxication with amphetamines obtained from illicit suppliers of dubious reputation. Repeated bouts of psychoses attended such use, and prepsychotic syndromes having a mystical, concrete, and obsessive-compulsive quality known as "punding" or "knick-knacking" were identified by several investigators (RYLANDER, 1966, 1967, and 1972; KRAMER et al., 1967;

ELLINWOOD, 1967 and 1972). Commonplace objects were endowed with special meanings by amphetamine-intoxicated persons, and the user would become fascinated by certain activities such as repetitive polishing and cleaning, taking appliances apart and sorting parts, or arranging colored pebbles into a primitive "speed art."

The sudden appearance of large numbers of intravenous users of amphetamines in places where drug addiction was relatively unknown prompted restrictive legislation in a number of countries, some making provision for the compulsory treatment and segregation of addicts. In the United States, cocaine is rapidly replacing amphetamine as a drug of abuse and is being rediscovered by certain medical and judicial authorities as a "non-addicting" drug suitable for "recreational use" which promotes "social interaction."

III. Amphetamine Psychosis

1. Clinical Syndrome

The amphetamine psychosis is an amorphous clinical syndrome characterized by well-systematized *paranoid delusions* with ideas of reference (the amphetamine psychosis proper), with or without *altered affect* (mood), *illusions*, and *hallucinations* of all the senses. Usually absent are manifestations of *organic delirium* such as disorientation, memory impairment, and clouding of consciousness. Most patients show little intellectual or psychomotor impairment, and the direction of the paranoia is usually consistent with the patient's social identification. As an example, a housewife on anorexigenics may develop the delusion that she is an object of criminal attention and seek police protection; or a person with an antisocial background may barricade himself in his room to avoid what he believes to be police surveillance. The best descriptions of the amphetamine psychosis are to be found in CONNELL's monograph (1958), the works of BELL (BELL and TRETHOWAN, 1961; BELL, 1965 and 1971), and KALANT's (1966) scholarly analysis of the clinical literature.

The amphetamine psychosis was first described by YOUNG and SCOVILLE (1938) as an intensification of latent paranoid coping mechanisms in two patients receiving amphetamine for the treatment of narcolepsy. An attempt on their part to reproduce the psychosis in one patient was unsuccessful. For two decades, it was considered a rare complication. KALANT (1966) mentions that only 71 cases were reported before 1958. Further, the psychosis was often believed to be a consequence of an underlying mental instability rather than a drug effect *per se*. Following CONNELL's (1958) detailed description of the syndrome which included descriptions of the psychosis in persons who were otherwise normal, the syndrome has been identified with depressing regularity in a number of countries, particularly during epochs of wide-spread amthetamine abuse (SANO and NAGASAKA, 1956; MASAKI, 1956; BEAMISH and KILOH, 1960; TATESU, 1960, 1972; BELL and TRETHOWAN, 1961; HAMPTON, 1961; MCCONNELL and MCILWAINE, 1961; KILOH and BRANDON, 1962; MCCONNELL, 1963; BREITNER, 1963 and 1965; BELL, 1965; TAKACS, 1965; GRIFFITH, 1966; JOHNSON and MILNER, 1966; RYLANDER, 1967; KRAMER et al., 1967; ELLINWOOD, 1967; GOLDBERG, 1968; BEJEROT, 1968; HAWKS et al., 1969). The true incidence is unknown. There is reason to believe that many episodes of amphetamine psychosis are not seen by physicians.

The amphetamine psychosis can be induced by a number of amphetamines. These include amphetamine and its isomers, methamphetamine, methylphenidate, phenmetrazine, and ephedrine (see KALANT, 1966). The psychosis resembles syndromes produced by cocaine and khat. It also resembles mental abnormalities sometimes produced by disulfiram (Antabuse) and corticosteroids (ROME and BRACELAND, 1950; HEATH et al., 1965; ADAMS and HOPE, 1966; LIDDON and SATRAN, 1967; KNEE and RAZANI, 1974). It tends to differ from LSD-like hallucinogenic states in that persons using these drugs can often experience vivid visual hallucinations but "know" that their perceptions are drug-induced and ego-alien and/or can be persuaded that this is the case. Conversely, a person experiencing amphetamine intoxication can present with true paranoia in the absence of hallucinatory phenomena.

The amphetamine psychosis also mimics variants of acute paranoid schizophrenia, and differentiating between these two syndromes can be difficult. Many patients who abuse amphetamine conceal this fact on admission and some continue to use the drug surreptitiously (BELL, 1965; ROCKWELL and OSTWALD, 1968) while in the hospital. Moreover, physical signs of amphetamine intoxication are often absent in adult patients and behavior can be easily interpreted as manifestations of anxiety or agitation. A urine positive for amphetamine and the disappearance of the psychosis within 7 days of drug withdrawal are considered satisfactory criteria for a diagnosis. Patients who remain psychotic for longer intervals are usually assumed to be true schizophrenics in whom the use of amphetamine is either incidental or has precipitated a psychosis (CONNELL, 1958; BELL, 1965; LEWIS, E., JR., 1970; ELLINWOOD, 1972; ANGRIST and GERSHON, 1972). As a rule, patients who abuse amphetamines chronically also use other drugs, including barbiturates, minor tranquilizers and alcohol. The intoxicating effects of these drugs and their withdrawal symptoms further complicate making a firm diagnosis and initiating proper treatment.

The drug histories obtained from such patients are extremely varied. A psychosis may follow a single dose; however, chronic use is more common. KALANT (1966), for example, reports that of 231 cases of amphetamine psychosis, 201 had histories of chronic use of from one to five years. The amount of drug involved also varies. Some patients have experienced a psychosis following an acute dose well within the therapeutic dose range, while others use substantially larger amounts. The reasons for this difference in dose susceptibility, also observed in controlled human studies, are not known. As previously mentioned, it was thought that a susceptibility to the psychogenic effects of amphetamine occurred in persons with a tendency toward mental illness, particularly schizophrenia, sociopathy, or paranoid tendencies. The psychosis has been observed in persons who are psychiatrically normal. Further, others who are indisputably schizoid or have paranoid tendencies use and abuse amphetamines without becoming psychotic. Thus, it is not possible to predict whether amphetamine ingestion will produce a psychosis in a patient.

2. Clinical Features

The majority of patients with an amphetamine psychosis present with a *mood disturbance.* A few are quite euphoric, approaching grandiosity in some instances, with logorrhea and flight of ideas (BELL, 1965). Most, however, are either emotionally "blunted," which gives them a schizoid-like appearance, or are frankly depressed.

Some have attempted suicide prior to admission. Emotional *depression* becomes more obvious shortly after the patient is withdrawn from amphetamines. Restlessness and agitation may accompany any mood state. Irritability and aggressiveness may be a considerable problem in management, especially among those who have also ingested alcohol or barbiturates. Because the affective component may be relatively more impressive than the paranoid states, which can be transitory or denied by the patient, it is not unusual for some patients to be diagnosed as having a manic-depressive illness. This type of syndrome has been most frequently reported in the Japanese literature.

A few patients manifest an *organic delirium.* This is often interpreted as alcohol, barbiturate, or minor tranquilizer abuse and withdrawal. Some cases of severe intoxication suggest that amphetamine may alone produce a toxic delirium or amnesia (BROWN, 1949; PETERSON and SOMERVILLE, 1949; DELAY et al., 1954; ABELY et al., 1960). Thus, some of the manifestations of amphetamine intoxication are difficult to distinguish from others of the more common forms of drug intoxication.

The *paranoid component* may consist of delusions of persecution, grandiosity, and influence with ideas of reference. Experimental studies in humans suggest that the onset of the paranoid component may occur abruptly whether the drug is administered acutely (BELL, 1973) or chronically over a period of days (GRIFFITH et al., 1970a, b and 1972; ANGRIST and GERSHON, 1970 and 1972). It has also been shown that frank paranoia may exist in the absence of hallucinations, excitability, apprehension or other mood states (GRIFFITH et al., 1972). Patients retain a considerable degree of ego functioning and social orientation, while at the same time describing well-systematized delusions which range from plausible to fantastic. Amphetamine-intoxicated patients may also attach special and personal meanings to various objects, sounds and behavior. These ideas of reference are likely to produce searching behavior. As an example, patients attempt to discover hidden cameras, microphones and the like. While an examination of these objects may convince the patient that they are ordinary, he will then develop other delusions. A substantial proportion of hospitalized amphetamine abusers experience visual and auditory *hallucinations,* as well as other sensory distortions. These range from transitory illusions, which a patient can often dismiss as unreal, to concrete visions. Gustatory and tactile hallucinations are also common. Thus, a patient may believe that he is being served decomposed or poisoned food or that his skin is infested by parasites. ELLINWOOD (1972) states that psychedelic distortion of body images may occur such as distortion of the face, hands "melting" or sensations of body transparency which allow thoughts to be "read."

BELL (1965) explains that many patients are able to recognize that their vivid hallucinations are not real. On the other hand, there may be no insight into paranoid delusions. Even among subjects in controlled experiments, an explanation that the paranoid delusion is drug-induced will be rejected by subjects as a subterfuge rather than a reasonable explanation.

Some patients maintain a *retrospective belief* in the validity of their delusions (KRAMER et al., 1967). While most soon set such thoughts aside, this is less true of those who live in antisocial cultures. KRAMER et al., (1967) for example, describes "speed freaks" who become increasingly bolder in their confrontations with the police. This type of behavior has also been observed in cocaine abusers. ELLINWOOD (1971) has described acts of violence among persons who have sustained a psychosis.

Group violence committed by amphetamine abusers has become known as "sharking." Most authorities believe, however, that most crime and violence associated with amphetamines is a consequence of disputes which arise during illicit distribution.

3. Withdrawal, Hospital Course and Treatment

Patients hospitalized for the treatment of an amphetamine psychosis are best managed on a closed psychiatric ward where personnel are experienced in coping with paranoid and depressed patients. Typically, such patients exhibit a withdrawal syndrome characterized by emotional depression, seclusiveness, hyperphagia and hypersomnia (KNAPP, 1952; CONNELL, 1958; BELL and TRETHOWAN, 1961; SCHILDKRAUT et al., 1971; WATSON et al., 1972). Unlike patients addicted to narcotics, however, drug "craving" is rarely a problem, at least initially, and, while the depression can be distressing, it is not easily relieved by less than psychotogenic doses of amphetamine. For these reasons, most authorities recommend that amphetamines be abruptly withdrawn and the patient's urine acidified to hasten the excretion of amphetamine and amphetamine metabolites. In about three days, a patient usually becomes much less depressed but can remain paranoid for 7-10 days. When the paranoid component persists beyond this time, the possibility of a functional psychosis in which amphetamine is only contributory or incidental must be considered.

Patients who are paranoid but not agitated may not require pharmacologic intervention. It has not been shown that major tranquilizers shorten the duration of the paranoid component. Patients who are agitated, particularly because of frightening hallucinations, improve dramatically if given major tranquilizers such as chlorpromazine or haloperidol. Because patients who abuse amphetamines have impaired postural vasomotor compensatory mechanisms, they are particularly sensitive to sympatholytic actions of major tranquilizers. Thus, a test dose in the order of 10 mg of chlorpromazine given subcutaneously is recommended before proceeding to larger doses.

Although amphetamine intoxication and withdrawal are rarely life-threatening, some patients present with signs of sympathomimetic and central stimulation (hypertension, hyperpyrexia, delirium and seizures). These constitute a medical emergency and deserve prompt supportive treatment and pharmacological intervention (ESPELIN and DONE, 1968; CHAPEL, 1973). The diagnosis of the etiology of delirium tremens is difficult, and sedative-hypnotic or alcohol withdrawal as well as other causes of acute brain syndrome should be considered (see Part I, Sect. III, Chaps. 2 and 3). The use of barbiturates in treating other features of amphetamine toxicity is not recommended otherwise. The amphetamine-intoxicated patient is notoriously difficult to sedate with barbiturates and is often made belligerent and abusive.

Symptoms of emotional depression may persist for months in some patients, many of whom return to the use of amphetamines upon discharge from the hospital. The possibility that these patients may have an underlying affective disorder should not be discounted (Committee on Alcoholism and Addiction and Council on Mental Health, 1966), and it may be appropriate to treat them in conventional ways for their depressions. The prognosis of amphetamine abusers is said to be poor despite best efforts.

G. Experimentally-Induced Amphetamine Psychoses

The amphetamine psychosis has been reproduced under controlled conditions by several investigators. The first such study was conducted by HIMMELSBACH and his associates who attempted to determine the addiction liability of amphetamine by allowing narcotic addicts to self-administer the drug. They observed that subjects tended to "lose control of their intake" and experienced personality changes such as withdrawal and a loss of interest in their appearance and neatness of their rooms. One subject exhibited a frank paranoid psychosis. The tendency of amphetamine abusers to escalate their dose and discontinue it spontaneously would later be rediscovered and called a "run" and "crashing." Unfortunately, this work has only recently been published (HIMMELSBACH et al., 1977) although brief notice was made at the time (Public Health Service Reports, 1944).

The psychotogenic properties of amphetamine were later studied in a number of experiments reported after 1968 (GRIFFITH et al., 1968, 1970a, b, 1972; ANGRIST and GERSHON, 1969b, 1970, 1972; JÖNSSON and GUNNE, 1970; ANGGÅRD et al., 1970, ANGGÅRD et al., 1973; BELL, 1973; ANGRIST et al., 1974). The findings of these studies can be briefly summarized as follows: During gradual administration of amphetamine over a period of up to five days, subjects experience transitory euphoria followed by a long interval during which time they are mildly depressed, hypochondriacal, spend most of their time in bed, and lose interest in their surroundings and activities such as watching television, conversing with attractive nurses, and completing hobby projects initiated during the placebo period. A prepsychotic phase lasting several hours then develops. Subjects are taciturn and deny abnormal thoughts. Subjects explain in retrospect that they experience transitory paranoid delusions lasting for several minutes, which they are able to dismiss as unreal later. Most confine themselves to their rooms at this time, and while previously open and trusting, become quite reserved and negativistic. If directly questioned, they give oblique answers and specifically deny paranoid ideation. The onset of frank paranoia is abrupt. Subjects begin to talk freely about their paranoid ideas as if they had gained some sudden "insight" into the "real" matters which explain their circumstances. One subject, after five days of dextroamphetamine, turned off the room lights and sat lotus-fashion in his bed. When questioned, he explained that "light travels faster in the dark" and he was aware of a giant oscillator in the ceiling which influenced his thoughts and the behavior of passers-by in the hall. When told that his ideas were drug-induced, he rejected this explanation in favor of the idea that the drug was a subterfuge to conceal a secret experiment of the influence of radio waves on the mind. Two subjects believed they were being discussed on television; another that his ex-wife had hired an assassin to dispatch him. He was most put out when the investigator did not guard the door while he (the subject) guarded the window. Still another subject believed that he was being photographed through a mirror. When encouraged to examine the mirror, he concluded that the camera "must be in the ceiling." Psychotic behavior ceased within 1—8 hours after the drug was withdrawn except in one subject who manifested paranoid thinking at intervals for about three days. Visual and auditory hallucinations were not reported (GRIFFITH et al., 1972).

Further studies of amphetamine intoxication conducted by ANGRIST and GERSHON (1970) indicate that a rapid escalation of dose of amphetamine or dextroamphetamine produced mood alterations more consistent with the borderline states of grandiosity reported by BELL (1967). Moreover, paranoid states with visual and auditory hallucinations were produced. They have also shown the effectiveness of haloperidol in the treatment of this condition. BELL (1973) has also reported that acute doses of methamphetamine given intravenously will produce a paranoid psychosis with hallucinatory phenomena which resembles that produced by chronic amphetamine with a short latency of onset. As noted in Section I, Chapter 1, of the volume, studies conducted by Swedish investigators suggest that administration of amphetamine prolongs the duration of the psychosis.

It would thus appear that amphetamines produce a dose-dependent response which is characterized by: (1) an initial dose-dependent alteration of mood to which (2) tolerance develops except when the dose is rapidly escalated, followed by (3) a paranoid psychosis which, in the face of a large dose, may be accompanied by hallucinations. Predrug personality does not appear to be a critical factor in the development of the psychoses, in that subjects without tendencies toward mental illness develop a psychosis which does not appear to be different from that of patients who have histories of mental abnormality. The minimum psychotogen dose observed was 50 mg of *d*-amphetamine/day for three days.

References

Abely, P., Bobin, P., Geier, S.: Toxicomanie mixte aux amphétamines et aux dérivés de l'oxazine chez de jeunes sujets. Ann. méd.-psychol. **118**, 167—172 (1960)

Adams, R.D., Hope, J.: Schizophrenia, paranoia, puerperal and endocrine psychoses. In: Harrison, T.R., Adams, R.D., Bemmett, I.L., Resnick, W.H., Thorn, G.W., Wintrobe, M.M. (Eds.): Principles of Internal Medicine, 5th ed. New York: McGraw-Hill 1966

Änggård, E., Gunne, L.-M., Jönsson, L.E.: Relationships between pharmacokinetic and clinical parameters in chronic amphetamine abuse. Acta pharmacol. (Kbh.) **28** (Suppl. 1) 92 (1970)

Änggård, E., Jönsson, L.E., Hogmark, A.L., Gunne, L.-M.: Amphetamine metabolism in amphetamine psychosis. Clin. Pharmacol. Ther. **14**, 870—880 (1973)

Albert, A.: Selective toxicity. New York: Wiley 1965

Anderson, E.W.: Further observations on Benzedrine. Brit. med. J. **1938 II**, 60—64

Angrist, B.M., Gershon, S.: Amphetamine abuse in New York City—1966 to 1968. Semin. Psychiat. **1**, 195—207 (1969a)

Angrist, B.M., Gershon, S.: Amphetamine-induced schizophreniform psychosis. In: Siva Sankar, D.V. (Ed.): Schizophrenia: Current Concepts and Research. New York: PJD Press 1969b

Angrist, B.M., Gershon, S.: The phenomenology of experimentally induced amphetamine psychosis. Preliminary observations. Biol. Psychiat. **2**, 95 (1970)

Angrist, B.M., Gershon, S.: A pilot study of pathogenic mechanisms in amphetamine psychosis utilizing differential effects of d and 1-amphetamine. Pharmakopsychiat. Neuro-Psychopharmakol. **4**, 64—75 (1971)

Angrist, B.M., Gershon, S.: Some recent studies on amphetamine psychosis-unresolved issues. In: Ellinwood, E.H., Cohen, S. (Eds.): Current concepts on amphetamine abuse. DHEW Publication No. (HSM) 72-9085. Washington, D.C.: U.S. Govt. Printing Office 1972

Angrist, B.M., Lee, H.K., Gershon, S.: The antagonism of amphetamine-induced symtomatology by a neuroleptic. Amer. J. Psychiat. **131**, 817—819 (1974)

Angrist, B.M., Schweitzer, J.W., Gershon, S., Friedhoff, A.J. Mephentermine psychosis: Misuse of the Wyamine inhaler. Amer. J. Psychiat. **126**, 1315—1317 (1970)

Apfelberg, B.: A case of Benzedrine sulfate poisoning. J. Amer. med. Ass. **110**, 575—576 (1938)

Arnold, L.E., Wender, P.H., McCloskey, K., Synder, S.H.: Levoamphetamine and dextroamphetamine: Comparative efficacy in the hyperkinetic syndrome. Assessment of target symptoms. Arch. gen. Psychiat. **27**, 816—822 (1972)

Barger, G., Dale, H. H.: Chemical structure and sympathomimetic action of amines. J. Physiol. (Lond.) **41**, 19—59 (1910)
Beamish, P., Kiloh, L. G.: Psychoses due to amphetamine consumption. J. ment. Sci. **106**, 337—343 (1960)
Bejerot, N.: An epidemic of phenmetrazine dependence. Epidemiological and clinical aspects. In: Wilson, C. W. M. (Ed.): The Pharmacological and Epidemiological Aspects of Adolescent Drug Dependence. Oxford: Pergamon 1968
Bejerot, N.: A comparison of the effects of cocaine and synthetic central stimulants. Brit. J. Addict. **65**, 35—37 (1970)
Bell, D. S.: Comparison of amphetamine psychosis and schizophrenia. Brit. J. Psychiat. **111**, 701—707 (1965)
Bell, D. S.: Addiction to stimulants. Med. J. Aust. **1**, 41—45 (1967)
Bell, D. S.: The precipitants of amphetamine addiction. Amer. J. Psychiat. **119**, 171—177 (1971)
Bell, D. S.: The experimental reproduction of amphetamine psychosis. Arch. gen. Psychiat. **29**, 35—40 (1973)
Bell, D. S., Trethowan, W. H.: Amphetamine addiction. J. nerv. ment. Dis. **133**, 489—496 (1961)
Bender, L., Cottington, F.: The use of amphetamine sulfate (Benzedrine) in child psychiatry. Amer. J. Psychiat. **99**, 116—121 (1942)
Bernstein, T. M.: Watch your Language. Great Neck, N. Y.: Channel Press 1958, 183
Bett, W. R.: Benzedrine sulphate in clinical medicine: a survey of the literature. Postgrad. med. J. **22**, 205—218 (1946)
Bloomberg, W.: End results of use of large doses of amphetamine sulfate over prolonged periods. New Engl. J. Med. **222**, 946—948 (1940)
Bonhoff, G., Lawrenz, H.: Über Weckamine (Pervitin and Benzedrin). Berlin-Göttingen-Heidelberg: Springer 1954
Bradham, G. B.: The energy requirements of obesity. Sth. med. J. (Bgham, Ala.) **67**, 967—969 (1974)
Bradley, C.: The behavior of children receiving Benzedrine. Amer. J. Psychiat. **94**, 577—585 (1937)
Bradley, C.: Benzedrine and Dexedrine in the treatment of children's behaviour disorders. Paediatrics **5**, 24—37 (1950)
Bradley, C., Bowen, M.: Amphetamine (Benzedrine) therapy of children's behavior disorders. Amer. J. Orthopsychiat. **11**, 92—103 (1941)
Brandon, S., Smith, D.: Amphetamines in general practice. J. Coll. gen. Practit. **5**, 603—606 (1962)
Breitner, C.: Appetite suppressing drugs as an etiologic factor in mental illness. Psychosomatics **4**, 327—333 (1963)
Breitner, C.: The hazard of amphetamine medication. Psychosomatics **6**, 217—219 (1965)
Brown, C. T.: Benzedrine habituation. Milit. Surg. **104**, 365—370 (1949)
Buchsbaum, M., Wender, P.: Average evoked reponses in normal and minimally brain dysfunctioned children treated with amphetamine. Arch. gen. Psychiat. **29**, 764—770 (1973)
Carey, J. T., Mandel, J.: A San Francisco Bay area "speed" scene. J. Hlth. Soc. Behav. **9**, 164—174 (1968)
Chapel, J. L.: Emergency room treatment of the drug-abusing patient. Amer. J. Psychiat. **130**, 257—259 (1973)
Chen, K. K., Schmidt, C. F.: Ephedrine and related substances. Medicine (Baltimore) **9**, 1—117 (1930)
Clayton, R. N., Grossman, L., Mayeda, T. K.: Hypothalamic obesity: The myth of the centromedial nucleus. Science **182**, 488—489 (1973)
Cole, J. O.: Clinical uses of the amphetamines. In: Ellinwood, E. H., Cohen, S. (Eds.): Current Concepts on Amphetamine Abuse. DHEW Publication No. (HSM) 72-9085. Washington, D. C.: U. S. Government Printing Office 1972
Committee on Alcoholism and Addiction and Council on Mental Health: Dependence on amphetamines and other stimulant drugs. J. Amer. med. Ass. **197**, 1023—1027 (1966)
Committee on Government Operations of the U.S. House of Representatives: False and misleading advertising (weight reducing remedies). Report No. 33, 1958
Committee on Medical Science, Education and Research (Council of the British Medical Association): Control of amphetamine preparations. Brit. med. J. **1968 IV**, 572—573

Connell, P. H.: Amphetamine psychosis. Maudsley Monograph No. 5 London: Institute of Psychiatry 1958
Conners, C. K., Eisenberg, L.: The effects of methylphenidate on symptomatology and learning in disturbed children. Amer. J. Psychiat. **120**, 458—464 (1963)
Council on Pharmacy and Chemistry: Present status of Benzedrine sulphate. J. Amer. med. Ass. **109**, 2064—2069 (1937)
Cuevas, P., Fine, J.: Production of fatal endotoxin shock by vasoactive substances. Gastroenterology **64**, 285—291 (1973)
Davidoff, E.: A clinical study of the effect of Benzedrine therapy on self-absorbed patients. Psychiat. Quart. **10**, 652—659 (1936)
Davidson, H. A.: Confessions of a goof ball addict. Amer. J. Psychiat. **120**, 750—756 (1964)
Delay, J., Pichot, P., Lemperière, T., Sadoun, R.: Psychoses amphétaminiques et pseudo-psychoses amphétaminiques. Ann. méd.-psychol. **112**, 51—57 (1954).
Dement, W., Rechtschaffen, A., Gulevich, G.: The nature of the narcoleptic sleep attack. Neurology (Minneap.) **16**, 18—33 (1966)
Dreisbach, R. H., Pfeiffer, C.: Caffeine withdrawal headache. J. Lab. clin. Med. **28**, 1212—1219 (1942)
Dreiser, T.: Khat. In: The Best Short Stories of Theodore Dreiser. New York: World Publishing Company 1918
Ebaugh, F. G.: Neuropsychiatric sequelae of acute epidemic encephalitis in children. Amer. J. Dis. Child. **25**, 89—97 (1923)
Edeleano, L.: Über einige Derivate der Phenylmethacrylsäure und der Phenylisobuttersäure. Verh. dtsch. chem. Ges. **20**, 616—622 (1887)
Ehrich, W. E., Lewy, F. H., Krumbhaar, E. G.: Experimental studies upon the toxicity of Benzedrine sulphate in various animals. Amer. J. med. Sci. **198**, 785—803 (1939)
Ellinwood, E. H., Jr.: Amphetamine psychosis: I. Description of the individuals and process. J. nerv. ment. Dis. **144**, 273—283 (1967)
Ellinwood, E. H., Jr.: Assault and homicide associated with amphetamine abuse. Amer. J. Psychiat. **127**, 1170—1175 (1971)
Ellinwood, E. H., Jr.: Amphetamine psychosis. Individuals, settings and sequences. In: Ellinwood, E. H., Cohen, S. (Eds.): Current Concepts on Amphetamine Abuse. DHEW Publication No. (HSM) 72-9085. Washington, D. C.: U. S. Government Printing Office 1972
Emde, H.: Über Diastereomerie. I. Konfiguration des Ephedrins. Helv. chim. Acta **12**, 365—376 (1929)
Epstein, L. C., Lasagna, L., Connors, C. K., Rodriguez, A.: Correlation of dextroamphetamine excretion and drug response in hyperkinetic children. J. nerv. ment. Dis. **146**, 136—146 (1968)
Espelin, D. E., Done, A. K.: Amphetamine poisoning. Effectiveness of chlorpromazine. New Engl. J. Med. **278**, 1361—1365 (1968)
Evans, A. O.: The synergism of autonomic drugs on opiate or opioid-induced analgesia: A discussion of its potential utility. Milit. Med. **127**, 1000—1003 (1962)
Evans, W. O.: The effect of stimulant drugs on opiate induced analgesia. Arch. Biol. Med. Exp. **4**, 144—149 (1967)
Evans, W. O., Bergner, B. A.: A comparison of analgesic potencies of morphine, pentazocine, and a mixture of methamphetamine and pentazocine in the rat. J. New Drugs **4**, 82—85 (1964)
Evans, W. O., Smith, R. P.: Some effects of morphine and amphetamine on intellectual functions and mood. Psychopharmacologia (Berl.) **6**, 49—56 (1964)
Feinberg, I., Hibi, S., Braun, M., Cavness, C., Westerman, G., Small, A.: Sleep amphetamine effects in MBDS and normal subjects. Arch. gen. Psychiat. **31**, 723—731 (1974)
Feinstein, A. R.: The treatment of obesity: An analysis of methods, results, and factors which influence success. J. chron. Dis. **11**, 349—393 (1960)
Fischer, J. E., Baldessarini, R. J.: False neurotransmitters and hepatic failure. Lancet **1971 II**, 75—79
Fischer, J. E., James, J. H.: Treatment of hepatic coma and hepatorenal syndrome. Amer. J. Surg. **123**, 222—230 (1972)
Fischman, M. W., Schuster, C. R., Resnekov, L., Shick, F. E., Krasnegor, N. A., Fennel, W., Freedman, D. X.: Cardiovascular and subjective effects of intravenous cocaine administration in humans. Arch. gen. Psychiat. **33**, 983—989 (1976)

Fish, B.: The "one child, one drug" myth of stimulants in hyperkinesis. Importance of diagnostic categories in evaluating treatment. Arch. gen. Psychiat. **25**, 193—203 (1971)
Fish, C. H., Bowling, E.: Effect of amphetamines on speech defects in the mentally retarded. Calif. Med. **96**, 109—111 (1962)
Fox, F. W.: The enigma of obesity. Lancet **1973 II**, 1487—1488
Fox, F. W.: The enigma of mass regulation with special reference to obesity. S. Afr. med. J. **48**, 287—301 (1974)
Frank, G.: Judy. New York: Harper & Row 1975
Freedman, D. X.: Report of the Conference on the use of stimulant drugs in the treatment of behaviorally disturbed young school children. Psychopharmacol. Bull. **7**, 23—29 (1971)
Friedenberg, S.: Addiction to amphetamine sulfate. J. Amer. med. Ass. **114**, 956 (1940)
Galdston, I.: One year's gains against disease. The progress of medicine in 1937. Reader's Digest **32**, 88—92 (1938)
Galdston, I.: Pep-teasers. Hygenia **18**, 878—880, 918—919 (1940)
Garfinkel, B. D., Webster, C. D., Sloman, L.: Methylphenidate and caffeine in the treatment of children with minimal brain dysfunction. Amer. J. Psychiat. **132**, 723—727 (1975)
Ginn, S. A., Hofman, L. B.: The use of *d*-amphetamine in severe behavior problems in children. Sth. med. J. (Bgham, Ala.) **46**, 1124—1127 (1953)
Ginsberg, M. D., Hertzman, M., Schmidt-Nowara, W. W.: Amphetamine intoxication with coagulopathy, hyperthermia, and reversible renal failure. Ann. intern. Med. **73**, 81—85 (1970)
Goldberg, L.: Drug abuse in Sweden. Bull. Narcot. **20** (1), 1—3, (2) 9—36 (1968)
Gottsegen, G.: Intoxicatio *β*-phenylisopropylamin (Aktedron). Orv. Hetil. **82**, 1139—1141 (1938)
Grahn, H. V.: Amphetamine addiction and habituation. Amer. Practit. **91**, 387—389 (1958)
Greden, J. F.: Anxiety or caffeinism: A diagnostic dilemma. Amer. J. Psychiat. **131**, 1089—1092 (1974)
Griffith, J. D.: A study of illicit amphetamine drug traffic in Oklahoma City. Amer. J. Psychiat. **123**, 560—569 (1966)
Griffith, J. D., Carr, C. B., Jasinski, D. R.: Relative effects of benzphetamine and *d*-amphetamine on mood, blood pressure and appetite in man. Clin. Pharmacol. Ther. **17**, 235 (1975)
Griffith, J. D., Cavanaugh, J. H., Held, J., Oates, J. A.: Experimental psychosis induced by the administration of *d*-amphetamine. In: Costa, E., Garattini, S. (Eds.): International Symposium on Amphetamines and Related Compounds. New York: Raven Press 1970a
Griffith, J. D., Cavanaugh, J., Held, J., Oates, J. A.: Dextroamphetamine: Evaluation of psychotomimetic properties in man. Arch. gen. Psychiat. **26**, 97—100 (1972)
Griffith, J. D., Cavanaugh, J., Oates, J. A.: Paranoid psychosis in man induced by administration of *d*-amphetamine. Pharmacologist **10**, 180 (1968)
Griffith, J. D., Cavanaugh, J., Oates, J. A.: Psychosis induced by the administration of *d*-amphetamine to human volunteers. In: Efron, D. H. (Ed.): Psychotomimetic Drugs. New York: Raven 1970b
Guttman, E.: The effect of Benzedrine on depressive states. J. ment. Sci. **82**, 618—620 (1936)
Guttman, E.: Benzedrine: Uses and abuses. Proc. roy. Soc. Med. **32**, 388—391 (1938)
Hampton, W. H.: Observed psychiatric reactions following use of amphetamine and amphetamine-like substances. Bull. N.Y. Acad. Med. **37**, 167—175 (1961)
Handley, C. A., Ensberg, D. L.: A comparison of amphetamine sulfate with other stimulants of the central nervous system in morphine respiratory depression. Anesthesiology **6**, 561—564 (1945)
Hanington, E.: Preliminary report on tyramine headache. Brit. med. J. **1967 II**, 550—551
Harrie, J. R.: Caffeine and headache. J. Amer. med. Ass. **213**, 628 (1970)
Harris, S. C., Ivy, A. C., Searle, L. M.: The mechanism of amphetamine-induced loss of weight: A consideration of the theory of hunger and appetite. J. Amer. med. Ass. **134**, 1468—1472 (1947)
Hauschild, F.: Zur Pharmakologie des 1-phenyl-2-methyl-amino-propans (Pervitin). Naunyn-Schmiedebergs Arch. exp. Path. Pharmak. **191**, 465—481 (1939)
Hawks, D., Mitcheson, M., Ogborne, A., Edwards, G.: Abuse of methylamphetamine. Brit. med. J. **1969 II**, 715—721
Heath, R. G., Nessehoff, W., Bishop, M. P., Byers, L. W.: Behavioral and metabolic changes associated with administration of tetraethylthiuram disulfide (Antabuse). Dis. nerv. Syst. **26**, 99—105 (1965)

Herridge, C. F., A'Brook, M. F.: Ephedrine psychosis. Brit. med. J. **1968 II**, 160
Himmelsbach, C. K., Cleek, D. M., Lloyd, B. J., Jr.: Studies on the addiction liability of Benzedrine sulfate. In: Martin, W. R., Isbell, H. (Eds.): Drug addiction and the U.S. Public Health Service. Washington, D.C.: U.S. Government Printing Office 1977 (in press)
Hishikawa, Y., Ida, H., Nakai, K., Kaneko, Z.: Treatment of narcolepsy with imipramine (Tofranil) and desmethylimipramine (Pertofran). J. neurol. Sci. **3**, 453—461 (1966)
Hollister, L. E.: Chemical psychoses. LSD and related drugs. Springfield, Ill.: Thomas 1968
Huestis, R. D., Arnold, L. E., Smeltzer, D. J.: Caffeine versus methylphenidate and *d*-amphetamine in minimal brain dysfunction: A double-blind comparison. Amer. J. Psychiat. **132**, 868—870 (1975)
Ivy, A. C., Goetzl, F. R., Burrill, D. Y.: Morphine-dextroamphetamine analgesia. War Med. **6**, 67—71 (1944)
Jackson, C. O.: The amphetamine inhaler: A case study of medical abuse. J. Hist. Med. allied Sci. **26**, 187—196 (1971)
Janowsky, D. S., El-Yousef, M. K., Davis, J. M., Sekerke, J. H.: Antagonistic effects of physostigmine and methylphenidate in man. Amer. J. Psychiat. **130**, 1370—1376 (1973)
Jasinski, D. R., Nutt, J. G.: Progress report on the assessment program of the NIMH Addiction Research Center. Presented to Committee on Problems of Drug Dependence, National Academy of Sciences—National Research Council 1972
Jasinski, D. R., Nutt, J. G., Griffith, J. D.: Effects of diethylpropion and *d*-amphetamine after subcutaneous and oral administration. Clin. Pharmacol. Ther. **16**, 645—652 (1974)
Jönsson, L. E., Gunne, L.-M.: Clinical studies of amphetamine psychosis. In: Costa, E., Garattini, S. (Eds.): International Symposium on Amphetamines and Related Compounds. New York: Raven 1970
Johnson, J., Milner, G.: Amphetamine intoxication and dependence in admissions to a psychiatric unit. Brit. J. Psychiat. **112**, 617—619 (1966)
Jones, L. W., Wallis, E. S.: Beckmann rearrangement involving optically active radicals. J. Amer. chem. Soc. **48**, 169—181 (1926)
Journal of the American Medical Association: Editorial: Benzedrine sulfate "pep pills." J. Amer. med. Ass. **108**, 1973—1974 (1937)
Kahn, E., Cohen, L.: Organic driveness—brain stem syndrome and an experience. New Engl. J. Med. **210**, 748—756 (1934)
Kalant, H., Kalant, O. J.: Death in amphetamine users: Causes and rates. Canad. med. Ass. J. **112**, 299—304 (1975)
Kalant, O. J.: The amphetamines. Toxicity and addiction. Springfield, Ill.: Thomas 1966
Kalant, O. J.: The amphetamines. Toxicity and addiction, 2nd ed. Springfield, Ill.: Thomas 1973
Keesey, R. E., Powley, T. L.: Hypothalamic regulation of body weight. Amer. Scient. **63**, 558—565 (1975)
Kiloh, L. G., Brandon, S.: Habituation and addiction to amphetamine. Brit. med. J. **1962 II**, 40—43
Klinger, W.: Zur Frage der Suchtentstehung durch Appetitzügler. Z. ärztl. Fortbild. **59**, 49—50 (1965)
Knapp, P. H.: Amphetamine and addiction. J. nerv. ment. Dis. **115**, 406—432 (1952)
Knee, S. T., Razani, J.: Acute organic brain syndrome: A complication of disulfiram therapy. Amer. J. Psychiat. **131**, 1281—1282 (1974)
Kramer, J. C., Fishman, V. S., Littlefield, D. C.: Amphetamine abuse. Pattern and effects of high doses taken intravenously. J. Amer. med. Ass. **201**, 305—309 (1967)
Krisko, I., Lewis, E., Johnson, J. E.: Severe hyperpyrexia due to tranylcypromine-amphetamine toxicity. Ann. intern. Med. **70**, 559—564 (1969)
Ladewig, D., Battegay, R.: Abuse of anorexics with special reference to newer substances. Int. J. Addict. **6**, 167—172 (1971)
Lasagna, L.: Attitudes toward appetite suppressants. A survey of U.S. physicians. J. Amer. med. Ass. **225**, 44—48 (1973)
Laufer, M. W., Denhoff, E.: Hyperkinetic behavior syndrome in children. J. Pediat. **50**, 463—474 (1957)
Laufer, M. W., Denhoff, E., Solomans, G.: Hyperkinetic impulse disorder in children's behavior problems. Psychosom. Med. **19**, 38—49 (1957)

Laurent, J. M.: Toxique et toxicomanie peu connus, "le cath." Ann. méd.-psychol. **120**, 649—657 (1962)
Leake, C. D.: The Amphetamines. Their Actions and Uses. Springfield, Ill.: Thomas 1958
Leake, C. D.: The long road for a drug from idea to use. In: Ayd, F. J., Blackwell, B. (Eds.): Discoveries in Biological Psychiatry. Philadelphia: Lippincott 1970
Lesses, M. F., Myerson, A.: Human autonomic pharmacology. XVI. Benzedrine sulfate as an aid in the treatment of obesity. New Engl. J. Med. **218**, 119—124 (1938)
Lewis, A.: Definitions and perspectives. In: Steinberg, H. (Ed.): Scientific Basis of Drug Dependence. New York: Grune & Stratton 1969
Lewis, A.: Amphetamines, Barbiturates, LSD and Cannabis: Their Use ans Misuse. London: Her Majesty's Stationery Office 1970
Lewis, E., Jr.: Drug abuse. The problem in the United States. Appl. Ther. **12**, 11—15 (1970)
Liddon, S. C., Satran, R.: Disulfiram (Antabuse) psychosis. Amer. J. Psychiat. **123**, 1284—1289 (1967)
Lloyd, G.: Withdrawal of prescriptions for amphetamines. Brit. med. J. **1973 I**, 101—103
Mann, G. V.: The influence of obesity on health. New Engl. J. Med. **291**, 1: 178—185; 2: 226—232 (1974)
Marks, H. H.: Influence of obesity on morbidity and mortality. Bull. N.Y. Acad. Med. **36**, 296—312 (1960)
Martin, W. R., Sloan, J. W., Sapira, J. D., Jasinski, D. R.: Physiologic, subjective, and behavioral effects of amphetamine, methamphetamine, ephedrine, phenmetrazine and methylphenidate in man. Clin. Pharmacol. Ther. **12**, 245—258 (1971)
Martin, W. R., Vaupel, D. B., Sloan, J. W., Bell, J. A., Nozaki, M., Bright, L. D.: The mode of action of LSD-like hallucinogens and their identification. In: Psychopharmacology of Hallucinogens. New York: Pergamon Press 1977 (in press)
Masaki, T.: The amphetamine problem in Japan. Wld Hlth Org. techn. Rep. Ser. **102**, 14—21 (1956)
McConnell, W. B.: Amphetamine substances in mental illnesses in Northern Ireland. Brit. J. Psychiat. **109**, 218—224 (1963)
McConnell, W. B., McIlwaine, R. J.: Amphetamine substances and mental illness in Northern Ireland. Ulster med. J. **30**, 31—34 (1961)
McManamy, M. C., Schube, P. G.: Caffeine intoxication: Report of a case the symptoms of which amounted to a psychosis. New Engl. J. Med. **215**, 616—620 (1936)
Miller, D. S., Parsonage, S.: Resistance to slimming. Adaptation or illusion? Lancet **1975 I**, 773—775
Modell, W.: Status and prospect of drugs for overeating. J. Amer. med. Ass. **173**, 1131—1136 (1960)
Modell, W., Reader, G. G.: Anorexiants. In: Modell, W. (Ed.): Drugs of Choice 1970—1971. St. Louis: C. V. Mosby Company 1970
Monroe, R. R., Drell, H. J.: Oral use of stimulants obtained from inhalers. J. Amer. med. Ass. **135**, 909—915 (1947)
Munro, J. F., MacCuish, A. C., Wilson, E. M., Duncan, L. J. P.: Comparison of continuous and intermittent anorectic therapy in obesity. Brit. med. J. **1968 I**, 352—354
Musto, D. F.: The American Disease. New Haven: Yale University Press 1973
Myerson, A.: Effect of Benzedrine sulfate on mood and fatigue in normal and in neurotic persons. Arch. Neurol. Psychiat. (Chic.) **36**, 816—822 (1936)
Myerson, A.: Addiction to amphetamine (Benzedrine) sulfate. J. Amer. med. Ass. **115**, 2202 (1940)
Myerson, A., Ritvo, M.: Benzedrine sulfate and its value in spasm of the gastro-intestinal tract. J. Amer. med. Ass. **107**, 24—26 (1936)
Nathanson, M. H.: The central action of beta-aminopropylbenzene (Benzedrine). J. Amer. med. Ass. **108**, 528—531 (1937)
Ney, P. G.: Psychosis in a child, associated with amphetamine administration. Canad. med. Ass. J. **97**, 1026—1029 (1967)
Nielsen, E. G.: Selvmordsforsg med Mecordin. Ugeskr. Laeg. **99**, 601—602 (1937)
Nieschulz, O.: Cocaine and cocaineism. Münch. med. Wschr. **111**, 2270—2281 (1969)
Ogata, A.: Desoxyephedrine—a contribution to the structure of ephedrine. J. Pharm. Soc. Japan **451**, 751—764 (1919)

Ong,B.H.: Dextroamphetamine poisoning. New Engl. J. Med. **266**, 1321—1322 (1962)
Panse,F., Klages,W.: Clinico-psychopathological observations on chronic abuse of ephedrine and related substances. Arch. Psychiat. Nervenkr. **206**, 69—85 (1964)
Peoples,S.A., Guttman,E.: Hypertension produced with Benzedrine—its psychological accompaniments. Lancet **1936 I**, 1107—1109
Peterson,B.H., Somerville,D.M.: Excessive use of "Benzedrine" by a psychopath. Med. J. Aust. **2**, 948—949 (1949)
Piness,G., Miller,H., Alles,G.A.: Clinical observations of phenylaminoethanol sulphate. J. Amer. med. Ass. **94**, 790—791 (1930)
Post,R.M.: Cocaine psychoses: A continuum model. Amer. J. Psychiat. **132**, 225—231 (1975)
Post,R.M., Kotin,J.K., Goodwin,F.K.: Effects of cocaine in depressed patients. Amer. J. Psychiat. **131**, 511—517 (1974)
Pragg,H.M.van: Abuse of, dependence on and psychoses from anorexigenic drugs. In: Meyler,L., Peck,H.M. (Eds.): Drug-Induced Diseases, Vol. 3. New York: Excerpta Medica Foundation 1968
Prinzmetal,M., Bloomberg,W.: The use of Benzedrine for the treatment of narcolepsy. J. Amer. med. Ass. **105**, 2051—2054 (1935)
Prokop,H.: On cases of ephedrine and Preludin addiction. Wien. klin. Wschr. **81**, 269—271 (1969)
Public Health Service Reports: Sutdies of drug addiction, 1944
Quinn,P.O., Rapoport,J.L.: One-year follow-up of hyperactive boys treated with imipramine or methylphenidate. Amer. J. Psychiat. **132**, 241—245 (1975)
Rapoport,J.L., Quinn,P.O., Bradbard,G., Riddle,K.D., Brooks,E.: Imipramine and methylphenidate treatments of hyperactive boys. Arch. gen. Psychiat. **30**, 789—793 (1974b)
Rapoport,J.L., Quinn,P.O., Lamprecht,F.: Minor physical anomalies and plasma dopamine-beta-hydroxylase activity in hyperactive boys. Amer. J. Psychiat. **131**, 386—390 (1974a)
Rawlin,J.W.: Street level abusage of amphetamines. In: Russo,J.R. (Ed.): Amphetamine abuse. Springfield, Ill.: Thomas 1968
Rechtschaffen,A., Wolpert,E.A., Dement,W.C., Mitchell,S.A., Fisher,C.: Nocturnal sleep of narcoleptics. Electroenceph. clin. Neurophysiol. **15**, 599—609 (1963)
Reifenstein,E.C.,Jr., Davidoff,E.: The treatment of alcoholic psychoses with Benzedrine sulfate. J. Amer. med. Ass. **110**, 1811—1812 (1938)
Reimann,H.A.: Caffeinism: A cause of long-continued, low-grade fever. J. Amer. med. Ass. **202**, 1105—1106 (1967)
Rizzotti,G.: The known facts about khat. Diffusion of use of this drug in South Ethiopia. Arch. ital. Sci. farmacol. **3**, 1—19 (1939)
Robinson,J.: Bed/Time/Story. New York: Random House 1974
Robinson,R.G., McHugh,P.R., Folstein,M.F.: Measurement of appetite disturbances in psychiatric disorders. J. psychiat. Res. **12**, 59—68 (1975)
Rockwell,D., Ostwald,P.: Amphetamine use and abuse in psychiatric patients. Arch. gen. Psychiat. **18**, 612—616 (1968)
Rome,H.P., Braceland,F.J.: Use of cortisone and ACTH in certain diseases: psychiatric aspects. Proc. Mayo Clin. **25**, 495—497 (1950)
Rosenberg,P.: Clinical use of Benzedrine sulfate (amphetamine) in obesity. Med. Wld (N.Y.) **57**, 656—659 (1939)
Roth,B., Faber,J., Nevsimalova,S., Tosovsky,J.: The influence of imipramine, dexphenmetrazine and amphetamine sulfate upon the clinical and polygraphic picture of narcolepsy-cataplexy. Schweiz. Arch. Neurol. Neurochir. Psychiat. **108**, 251—260 (1971)
Rylander,G.: Preludin-narkomaner fran klinish och medicinsk-kriminogisk synpunkt. Läkartidningen **63**, 4973—4980 (1966)
Rylander,G.: Addiction to Preludin intravenously injected. Excerpta Medica, Int. Congr. Series 150. Proc. 4th World Congr. Psychiat. (1967)
Rylander,G.: Psychoses and the punding and choreiform syndromes in addiction to central stimulant drugs. Psychiat. Neurol. Neurochir. (Amst.) **75**, 203—212 (1972)
Safer,D.J., Allen,R.P.: Factors influencing the suppressant effects of two stimulant drugs on the growth of hyperactive children. Pediatrics **51**, 660—667 (1973)

Sandler, M., Youdin, M. B. H., Hanington, E.: A phenethylamine oxidising defect in migraine. Nature (Lond.) **250**, 335—337 (1974)
Sano, I., Nagasaka: Über chronische Weckaminsucht in Japan. Fortschr. Neurol. Psychiat. **24**, 391—394 (1956)
Satterfield, J. H., Cantwell, D. P., Saul, R. E., Yusin, A.: Intelligence, academic achievement, and EEG abnormalities in hyperactive children. Amer. J. Psychiat. **131**, 391—395 (1974)
Schilder, P.: Psychological effect of Benzedrine sulfate. J. nerv. ment. Dis. **87**, 584—587 (1938)
Schildkraut, J. J., Watson, R., Draskoczy, P. R., Hartman, E.: Amphetamine withdrawal depression and MHPG excretion. Lancet **1971 II**, 485—486
Schnackenberg, R. C.: Caffeine as a substitute for Schedule II stimulants in hyperkinetic children. Amer. J. Psychiat. **130**, 796—798 (1973)
Schube, P. G., McManamy, M. C., Trapp, C. E., Myerson, A.: Effect of Benzedrine sulfate on certain abnormal mental states. Amer. J. Psychiat. **94**, 27—32 (1937)
Scott, P. D., Willcox, D. R. C.: Delinquency and the amphetamines. Brit. J. Psychiat. **111**, 865—875 (1965)
Seaton, D. A., Duncan, L. J. P.: The uses and abuses of appetite suppressants. Brit. J. clin. Pract. **19**, 89—93 (1965)
Shorvon, H. J.: Use of Benzedrine sulphate by psychopaths. The problem of addiction. Brit. med. J. **1945 II**, 285—286
Silverstone, J. T., Stunkard, A. J.: The anorectic effect of dexamphetamine sulphate. Brit. J. Pharmacol. **33**, 513—522 (1968)
Smith, D. E.: Speed freaks vs. acid heads—conflict between drug subcultures. Clin. Pediat. **8**, 185—188 (1969)
Smith, L. C.: Collapse with death following the use of amphetamine sulfate. J. Amer. med. Ass. 113, 1022—1023 (1939)
Smith, R. C.: Compulsive methamphetamine abuse and violence in the Haight-Ashbury district. In: Ellinwood, E. H., Cohen, S (Eds.): Current Concepts on Amphetamine Abuse. DHEW Publ. No. (HSM) 72—9085. Washington, D.C.: U.S. Government Printing Office 1972
Sroufe, L. A., Stewart, M. A.: Medical progress. Treating problem children with stimulant drugs. New Engl. J. Med. **289**, 407—413 (1973)
Staehelin, J. E.: Pervitin-psychose. Z. ges. Neurol. Psychiat. **173**, 598—620 (1941)
Stunkard, A., McLaren-Hume, M.: The results of treatment for obesity. Arch. intern. Med. **103**, 79—85 (1959)
Swanson, D. W., Weddige, R. L., Morse, R. M.: Abuse of prescription drugs. Mayo Clinic Proc. **48**, 359—367 (1973)
Takacs, L.: Über Psychosen nach Abusus von Anorexigenen. Psychiat. Neurol. med. Psychol. (Lpz.) **17**, 183—185 (1965)
Tatetsu, S.: Pervitin-psychosen. Folia psychiat. neurol. jap. Suppl. **6**, 25—33 (1960)
Tatetsu, S.: Methamphetamine psychosis. In: Ellinwood, E. H., Cohen, S. (Eds.): Current concepts on amphetamine abuse. DHEW Publication No. (HSM) 72-9085. Washington, D.C.: U.S. Government Printing Office 1972
Trellee, M.: The pharmacodynamics of khat. Bull. Narcot. **11**, 43—44 (1959)
Ulrich, H., Trapp, C. E., Vidgoff, B.: The treatment of narcolepsy with Benzedrine sulphate. Ann. intern. Med. **9**, 1213—1221 (1936)
van Kammen, D. P., Murphy, D. L.: Attenuation of the euphoriant and activating effects of *d*- and *l*-amphetamine by lithium carbonate treatment. Psychopharmacologia (Berl.) **44**, 215—224 (1975)
Vonder Haar, T. A.: Chaining children with chemicals. Progressive **39**, 13—17 (1975)
Watson, R., Hartmann, E., Schildkraut, J. J.: Amphetamine withdrawal: Affective state, sleep patterns, and MHPG excretion. Amer. J. Psychiat. **129**, 263—269 (1972)
Waud, S. P.: The effects of toxic doses of benzyl methyl carbinamine (Benzedrine) in man. J. Amer. med. Ass. **110**, 206—207 (1938)
Weiss, G., Werry, J., Klaus, M., Douglas, V., Sykes, D.: Studies on the hyperactive child. V. The effects of dextroamphetamine and chlorpromazine on behaviour and intellectual functioning. J. Child Psychol. **9**, 145—156 (1968)
Winsberg, B. G., Bialer, I., Kupietz, S., Tobias, J.: Effects of imipramine and detroamphetamine on behavior of neuropsychiatrically impaired children. Amer. J. Psychiat. **128**, 1425—1431 (1972)

Wood, C. D., Graybiel, A.: Evaluation of sixteen anti-motion sickness drugs under controlled laboratory conditions. Aerospace Med. **39**, 1341—1344 (1968)
Wood, D. R., Reimherr, F. W., Winder, P. H., Johnson, G. E.: Diagnosis and treatment of minimal brain dysfunction in adults. Arch. gen. Psychiat. **33**, 1453—1460 (1976)
World Health Organization: Expert Committee on Addiction-Producing Drugs: Terminology in regard to drug abuse. Wld Hlth Org. techn. Rep. Ser. **273**, 9—10 (1964a)
World Health Organization: Expert Committee on Addiction-Producing Drugs: Khat (*Catha edulis*). Wld Hlth Org. techn. Rep. Ser. **273**, 10 (1964b)
Yoss, R. E.: Treatment of narcolepsy. Mod. Treat. **6**, 1263—1274 (1969)
Yoss, R. E., Daly, D. D.: Criteria for the diagnosis of the narcoleptic syndrome. Proc. Mayo Clin. **32**, 320—328 (1957)
Yoss, R. E., Daly, D. D.: On the treatment of narcolepsy. Med. Clin. N. Amer. **52**, 781—787 (1958)
Yoss, R. E., Meyer, N. J., Ogle, K. N.: The pupillogram and narcolepsy: a method to measure decreased levels of wakefulness. Neurology (Minneap.) **19**, 921—923 (1969)
Young, D., Scoville, W. B.: Paranoid psychosis in narcolepsy and the possible danger of Benzedrine treatment. Med. Clin. N. Amer. **22**, 637—645 (1938)
Young, J. H.: The toadstool millionaires. Princeton, N. J.: Princeton University Press 1961
Zahn, T. P., Abate, F., Little, B. C., Wender, P. H.: Minimal brain dysfunction, stimulant drugs, and autonomic nervous system activity. Arch. gen. Psychiat. **32**, 381—387 (1975)
Zalis, E. G., Parmley, L. F., Jr.: Fatal amphetamine poisoning. Arch. intern. Med. **112**, 822—826 (1963)
Zarcone, V.: Narcolepsy. New Engl. J. Med. **288**, 1156—1166 (1973)
Zrull, J. P., Westman, J. C., Arthur, B., Rice, D. L.: A comparison of diazepam, *d*-amphetamine and placebo in the treatment of the hyperkinetic syndrome in children. Amer. J. Psychiat. **121**, 388—389 (1964)

CHAPTER 3

Pharmacology and Classification of LSD-like Hallucinogens

W. R. MARTIN and J. W. SLOAN

A. Introduction

Hallucinogens of the LSD-type have been used since antiquity (SCHULTES, 1969, 1970a, b). Peyote and mushrooms containing psilocin (HEIM and WASSON, 1958; WASSON, 1959) were used by Indians of the southwestern United States and Central America in religious rites. LEWIN (1964) and others conducted experiments with peyote buttons and mescaline at the turn of this century and described their major subjective effects including visual and auditory perceptual changes, some autonomic changes such as perspiration and shivering as well as elevation of mood, which LEWIN described as follows: "The individual is usually in a state of extremely good humor and full of feelings of intellectual and physical energy."

Although the subjective changes produced by these drugs have been known in Western World culture since the beginning of the 20th century, there was little interest in these types of drugs until the late 1940s and early 1950s except by a few scientists and psychiatrists. A number of factors were responsible for their subsequent increased popularity and eventual abuse.

The discovery of the hallucinogenic actions of LSD by HOFFMAN and its systematic investigation by STOLL (1947) rekindled the interest in the effects of hallucinogens as model psychoses. The elucidation of the pharmacology of LSD excited the imagination of scientists who felt that it might be a key in unlocking the cause of mental illness in which perceptual disorders and hallucinations play a role. LSD also excited the imagination of mystics, journalists, and other writers, and its effects on perception were given extensive publicity.

In the 1960s the use of LSD became widespread in the United States. Probably several factors contributed to the emergence of this problem. Some have felt that the affluence, increased leisure time, and hedonistic orientation that prevailed particularly during the 1960s may have played a role in the spread of hallucinogen abuse. In this circumstance, the extensive treatment and publicizing of the subjective effects of LSD by the press may well have catalyzed lay experimentation with these drugs. The emphasis of the exotic nature of these drug-induced changes on the psyche and the relating of them to mystical experiences (HUXLEY, 1954) came at a time when the rejection of materialism was in fad. Finally, LSD-like hallucinogens are euphoriants, a property that is not generally recognized but which may be of critical importance in their use by drug abusers. The relative importance of these various influences on the abuse of LSD-like hallucinogens cannot be ascertained with any certainty. Neither can the impact, cost, and harm that hallucinogen abuse causes be precisely measured (see Part I, Sect. I, Chap. 1). Regardless, the abuse of these drugs became sufficiently widespread that they became subject to legal control in the United States and a protocol has been formulated which would place them under international control.

The primary purpose of this chapter will be to delineate the major pharmacologic properties of this group of drugs with several ends in view: (1) An attempt will be made to sharpen the criteria that can be used in identifying LSD-like hallucinogens. (2) The sites and modes of action of these drugs will be examined in some detail from the perspective that they may give insight into the basic organic psychopathology of addicts, drug abusers, and psychopaths as well as patients who have perceptual and thinking disorders. (3) The problems associated with the classification and scheduling of these drugs will be discussed.

This monograph is not intended to be a comprehensive review of all the literature that is related to LSD-like hallucinogens.

The interested reader is referred to other related monographs, review, and symposia (BRADLEY et al., 1964; BRAWLEY and DUFFIELD, 1972; BURGER, 1968; CHOLDEN, 1956; COHEN, 1967; COLE et al., 1973; EFRON, 1968, 1970; EFRON et al., 1967; GARATTINI and GHETTI, 1957; HIMWICH, 1971; HOFFER and OSMOND, 1967; HOLLISTER, 1968; JACOBSEN, 1963; KEUP, 1970; KILLAM, 1962; KOELLE, 1958; SMYTHIES, 1970; SMYTHIES et al., 1970; WELLS, 1974; WIKLER, 1957; WOOLLEY, 1962).

B. Classification of LSD-like Hallucinogens

I. Problem

In the reviews of the sites and modes of action of LSD, it is quite apparent that LSD and related drugs have many sites of action in the brain, modes of action, and effects. LSD acts at all levels of the brain from the spinal cord to the cerebral cortex. It acts as both an agonist and antagonist. Some of its effects can be mimicked by such diverse agents as 5-HT, tryptamine, and other indoleamines as well as a variety of phenethylamines and catecholamines. It is thus not surprising that with these diverse characteristics various LSD-like hallucinogens differ in some of their pharmacologic activity. In the end the judgment as to which hallucinogens should be considered LSD-like must be based on a consensus of data.

II. Effects of LSD-like Hallucinogens, Mescaline, and Other Hallucinogens

The effects of LSD-like hallucinogens on autonomic and somatomotor behavior for various species are summarized in Table 1.

1. Man

a) Behavioral and Subjective Changes

Even the earliest studies with peyote and mescaline revealed that these agents produced changes in the functioning of the autonomic and somatomotor nervous systems as well as changes in perception and mood. The most frequent early change in mood is the emergence of feelings of euphoria (LEWIN, 1964; BERINGER, 1928) and is most reproducibly experienced after small doses. With larger doses feelings of ner-

Table 1. Summary of the autonomic, behavioral, and somatomotor effects of LSD in various species

	Mouse	Rat	Rabbit	Cat	Dog	Monkey	Man
Autonomic							
Mydriasis	+	+	+	+	+	+	+
Blood pressure				$\downarrow$ or $\uparrow$			$\uparrow$
Pulse rate			$\uparrow$	$\downarrow$ or 0	$\uparrow$		$\uparrow$
Respiration			$\uparrow$		$\uparrow$		$\uparrow$
Blood sugar			$\uparrow$				$\uparrow$
Salivation		+		+	+		+
Nausea							+
Vomiting					+		+
Lacrimation				+			
Piloerection	+	+	+				
Body temperature		$\uparrow$	$\uparrow$	$\uparrow$	$\uparrow$	$\uparrow$	$\uparrow$
Behavioral							
Startle response		$\uparrow$	$\uparrow$	$\uparrow$	$\uparrow$		
Activity	$\uparrow$	$\uparrow$	$\uparrow$			$\downarrow$	
Analgesia	—	—		+	+	+	+
Aggression	$\downarrow$ or $\uparrow$					$\downarrow$	
Head twitching	+						
Somatomotor							
Hyperreflexia			+	+	+		+
Ataxia				+	+	+	
Spasticity				+	+		

+ present; – absent; $\uparrow$ increased; $\downarrow$ decreased

In addition to appropriate references in the text, the following papers were sources of data for the construction of this table: ROTHLIN et al. (1956). ROTHLIN and CERLETTI (1956). ROTHLIN (1957a, b). HOLLISTER and SJOBERG (1964). DIXON (1968). MILLER and MILLER (1970).

vousness and anxiety may predominate, and with still larger doses, fear and terror may emerge. Some patients experience feelings of depersonalization (GUTTMAN and MACLAY, 1936) and may become catatonic. Changes in visual perception have been extensively analyzed and reviewed by KLUVER (1942) who has discovered fundamental "form constants" which are (1) "grating, lattice, fretwork, filigree, honeycombed, or chessboard patterns, (2) cobweb, (3) tunnel, funnel, alley, cone or vessel, and (4) spirals." In addition, colors may be brighter and objects have an iridescent quality. There are, as subsequently discussed, some changes in feeling states that may be regarded as somasthetic in nature. Auditory hallucinations are rare, but hyperacusis has been reported. Perception of time is also altered.

ABRAMSON et al. (1955a) investigated in a systematic way those subjective changes that were reported under placebo conditions and under graded doses of orally administered LSD. Changes produced by the 50 μg dose included visual disturbance such as difficulty in focusing, somasthetic changes, and paresthesias such as things moving, unsteadiness, weakness, dizziness, changes in sensitivity of skin, and peculiar feelings or lightness of limb, and mood changes such as dreamlike state, drowsiness, inner trembling. With the 100 μg dose level additional subjective changes

Table 2. Clinical grades of the LSD reaction

Grade 1:	Anxiety and nervousness without perceptual distortion or hallucinations.
Grade 2:	Anxiety and nervousness and visual perceptual distortions without true hallucinations.
Grade 3:	Anxiety, nervousness, perceptual distortions, and true hallucinations, but with insight maintained.
Grade 4:	Same as grade 3 except that insight (realization that the effects are due to the drug) is lost.

were reported: Blurred eyesight, objects seeming far away, things moving, and autonomic changes such as sweating, nausea, and feeling ill. ISBELL et al. (1956) constructed a questionnaire based on the findings of ABRAMSON et al. (1955a). The scores on these questionnaires increased in a dose-related manner following the administration of LSD. In addition, a clinical grading system was devised (Table 2) and it was found that clinical grade of the reaction was also dose-related. KLEE et al. (1961) also using a clinical rate scale found a dose-related increase in LSD effects. LSD elevates the Psychasthenia, Schizophrenia, Paranoia, and Taylor Anxiety scales of the MMPI (BELLEVILLE, 1956). HILL et al. (1963a, b) administered the ARCI to patients who had received LSD (1.0 and 1.5 μg/kg) and derived the LSD Significant scale which contained those items that differentiated the LSD treatment condition from the placebo treatment. This scale contains items that are related to feelings of euphoria and sedation, anxiety, restlessness and dysphoria, disruption of thought processes, paresthesia and perceptual changes, changes in self-image and autonomic changes (Table 3). Both HILL et al. (1963b) and ISBELL et al. (1956) observed that lower doses of LSD produced feelings of euphoria while larger doses produced feelings of nervousness and anxiety. DITMAN and WHITTLESEY (1959) compared the remembrance of subjective effects of subjects who had experienced delirium tremens and subjects who had received LSD. Delirium tremens had the attributes of anxiety, horror, depression, irritation, and paranoia while the LSD experiences had the characteristics of euphoria, good humor, relaxation, and a sense of wonderment. HAERTZEN (1966) subsequently developed an LSD Group Variability scale which contained items that differentiated the effects of LSD from other treatment conditions which included no-drug, placebo, morphine, pentobarbital, chlorpromazine, pyrahexyl, amphetamine, and alcohol. This scale contains many items from the LSD significant scale as well as others except for those related to euphoria. Other aspects of the LSD psychosis have also been reported such as loss of control (ARONSON and KLEE, 1960; LINTON and LANGS, 1962), anxiety, hostility (SALVATORE and HYDE, 1956), and disturbance of thought and speech (RINKEL et al., 1952).

LSD decreases the ability of subjects to judge time (BOARDMAN et al., 1957), reduces scores on the Wechsler-Bellvue Intelligence Quotient (LEVINE et al., 1955; ISBELL et al., 1956; SILVERSTEIN and KLEE, 1958), disrupts motor performance (LANDIS and CLAUSEN, 1954; ABRAMSON et al., 1955c), increases reaction time (ABRAMSON et al., 1955d; EDWARDS and COHEN, 1961), impairs arithmetic calculation (JARVIK et al., 1955c), color perception (HOLLISTER and HARTMAN, 1962), simple discriminations (JARVIK et al., 1955a), recognition and recall (JARVIK et al., 1955b) possibly set (WIKLER et al., 1965), and the ability to judge spatial relationship and size

Table 3. True items of the LSD significant scale

Euphoria

2. I have a pleasant feeling in my stomach.
91. I feel as if something pleasant had just happened to me.
98. I am in the mood to talk about the feeling I have.
243. Everything looks dreamlike.
265. A thrill has gone through me one or more times since I started the test.
279. I feel a very pleasant emptiness.
345. I feel so good that I know other people can tell it.
354. I feel like joking with someone.
398. This is one day that I'll put my troubles on the shelf.
429. I feel more excited than dreamy.
509. Everybody seems unusually cheerful today.

Sedation

86. It seems harder than usual to move around.
243. Everything looks dreamlike.

Anxiety, restlessness, and dysphoria

6. It makes me impatient to have to sit still.
29. I am unusually restless.
100. I cannot bring myself to be content.
182. I feel so miserable that other people must be aware of it.
201. I feel anxious and upset.
410. I often feel that I must get up and walk around.

Disruption of thought processes

153. It would be hard for me to concentrate on arithmetic problems right now.
164. It seems I'm spending longer than I should on each of these questions.
225. I don't feel like reading anything right now.
361. During this test I have had difficulty in keeping the questions and answers in order.
408. I have to force myself to read these questions.

Paresthesia and perceptual changes

2. I have a pleasant feeling in my stomach.
79. My face feels hot.
294. My skin seems to be unusually sensitive.
364. I have a metallic taste in my mouth.
376. My hands feel light.
387. My hands or feet feel heavy.
458. Parts of my body feel numb.
464. I have difficulty in focusing my eyes.
495. My clothes feel different to the touch.
499. I feel an increasing awareness of bodily sensations.
534. My sight seems to come and go.

Body feeling

247. I seem to be a changed person.
267. I have a weird feeling.
293. My lips feel drawn back as if in a smile.
389. I feel like gritting my teeth.
400. I am thirsty for something hot like coffee.
401. My head feels light.
442. It makes me uncomfortable to lie down.
504. I feel now as I have felt after a very exciting experience.
529. My thoughts seem to come and go.
540. I have a floating feeling.
541. My speech is not as loud as usual.

Numbers are the items of the Addiction Research Center Inventory (HILL et al., 1963b).

(ABRAMSON et al., 1955b, e; LIEBERT et al., 1958). Mescaline also disrupts motor performance and color perception (SALOMON et al., 1949; LANDIS and CLAUSEN, 1954; HOLLISTER and HARTMAN, 1962; HARTMAN and HOLLISTER, 1963).

b) Physiologic Changes

The overt physiologic changes produced by LSD-like hallucinogens in man include an increase in blood pressure, pulse rate, respiratory rate, body temperature, pupillary diameter and respiratory rate, as well as sweating, salivation, lacrimation, nausea and vomiting, tremor, ataxia, and hyperreflexia (LEWIN, 1964; BERINGER, 1928; STOLL, 1947; FORRER and GOLDNER, 1951; ISBELL et al., 1956; DIMASCIO et al., 1957; FREEMAN, 1958; GREINER et al., 1958). The increase in body temperature may reach pathologic levels (FRIEDMAN and HIRSCH, 1971). LSD also has an antidiuretic effect (KIES et al., 1957). This syndrome has been called diencephalic but as will subsequently be emphasized (see Subsect. D) LSD-like hallucinogens affect all levels of the central nervous system.

c) Effects of Related Drugs

α) Indoleamines and LSD Derivatives

Psilocybin and psilocin also induce a typical LSD-like syndrome in man, producing hyperthermia, tachycardia, increased respiratory rate and blood pressure, mydriasis, and facilitation of the patellar reflex as well as changes in perception and mood (ISBELL, 1959; MALITZ et al., 1960; HOLLISTER, 1961; WOLBACH et al., 1962b; KEELER, 1965a). It also increases scores on the Schizophrenia and Hypochondriasis scales of the MMPI (KELLER, 1965b). ISBELL et al. (1959b) studied a series of LSD congeners and reported their autonomic, somatomotor, and subjective effects. Some of these compounds produced LSD-like changes and others did not. Since ISBELL et al. (1959b) studied only the effects of single doses and not whether they produced either direct or exhibited cross tolerance to LSD, those that produced LSD-like effects are classified as "probably LSD-like" in Table 4. ABRAMSON (1959) studied the subjective effects of several LSD congeners, and the results of these studies are also incorporated into Table 4. 2,3-Dihydro-LSD also produces autonomic, subjective, and perceptual changes that are similar to LSD (GORODETZKY and ISBELL, 1964). Ololiuqui seeds of the morning glory *Ipomea violacea* which contain *d*-lysergic acid amide, *d*-iso-lysergic acid amide, chanoclavine, elymoclavine, and ergometrine did not produce LSD-like effects in man (ISBELL and GORODETZKY, 1966). BOL seems to be devoid of LSD-like activity in man (JARVIK et al., 1955d; ISBELL et al., 1959c); however, RICHARDS et al. (1958) reported an LSD-like reaction in one patient after the ingestion of BOL. Although FABING (1956) and FABING and HAWKINS (1956) reported that bufotenine produced LSD-like effects, ROSENBERG et al. (1963a) and TURNER and MERLIS (1959) were unable to confirm these observations. MARTIN and SLOAN (1970) found that intravenously infused tryptamine increased blood pressure, dilated pupils, enhanced the patellar reflex, and produced perceptual distortions. HALE and REED (1962) produced perceptual distortions and hallucinatory phenomena with methysergide in patients treated for migraine headache. MENDELS (1967) found that methysergide worsened the disease of 5 of 12 chronic schizophrenic patients. FISH et al. (1969) observed stimulation and apparent euphoria in schizophrenic children treated with methysergide. SZARA (1956) and SAI-HALASZ et al. (1958) found that DMT produced LSD-like subjective and autonomic changes in man and showed that it was metabolized to 3-indoleacetic acid (SZARA, 1956). ROSENBERG et al. (1963a) confirmed these observations and found that DMT significantly increased blood pressure and pupillary diameter, lowered the threshold for the patellar reflex, and produced LSD-like subjective effects. It also increased pulse and respiratory rate but these changes were not significant, and it did not produce an elevation of body temperature. SZARA (1957), SZARA et al. (1966) and BÖSZÖRMÉNYI et al. (1959) found that DET also produced LSD-like subjective and autonomic changes. SZARA et al. (1966), however, reported that some subjects had paranoid ideation. In this regard, BELLEVILLE (1956) and ISBELL et al. (1956) reported that LSD elevated the Paranoia scale of the MMPI. BÖSZÖRMENYI (1960) felt that DET was more stimulating and exciting than psilocybin and that psilocybin may have been more disruptive to integrative processes. N,N-Dipropyl tryptamine (DPT) and 6-fluorodiethyl tryptamine (6-FDET) increased heart rate and blood pressure. DPT produced significant subjective changes that were similar to those produced by LSD while the changes produced by 6-FDET were less striking (SZARA, 1970).

β) β-Phenethylamines and Amphetamines

The methods of ABRAMSON et al. (1955a) and ISBELL et al. (1956) for characterizing LSD-like drugs using quantitative methods for measuring both subjective and physiologic changes have been described in detail. ISBELL and his collaborators (see Subsect. B. III) also used the phenomena of tolerance and cross tolerance to determine if experimental drugs resembled LSD. For historical reasons the pharmacology of mescaline has been discussed with LSD. SHULGIN et al. (1969) has employed a mescaline unit which is a ratio of what has been called an effective dose of mescaline (3.75 µg) divided by the mean of the threshold and maximally effective dose of the experimental compound in producing LSD-like subjective effects. The criteria for measuring these end points have not been defined nor their validity or reproducibility established; nevertheless, more phenethylamine and amphetamine derivatives have been studied using this procedure than any other. Those compounds that are thought to have mescaline or LSD-like activity are summarized in Tables 4 and 5, and some are discussed in detail elsewhere. 3,4-Dimethoxyphenylethylamine is without mescaline-like effects in man (HOLLISTER and FRIEDHOFF, 1966). It is noteworthy that DOM (2,5-dimethoxy-4-methylamphetamine) which has an activity of 80 mescaline units has been subject to street abuse and was found by SNYDER et al. (1967), HOLLISTER et al. (1969), and FAILLACE et al. (1970) to produce LSD-like subjective and autonomic changes in man. 3,4,5-Trimethoxyamphetamine which was found to produce LSD-like autonomic changes, motor incoordination, and perceptual distortions in doses of 1–2 µg (PERETZ et al., 1955; SHULGIN et al., 1961) in man has also been abused. MDA (3,4-methylene-dioxyamphetamine) and MMDA (5-methoxy-3,4-methylene-dioxyamphetamine) which seem to have some but not all of the properties of mescaline have also had street abuse (SHULGIN, 1964a, b; SHULGIN and SARGENT, 1967). MDA probably can produce clonic and tonic seizures and increases body temperature, pulse rate, and blood pressure in man (RICHARDS and BORGSTEDT, 1971). ALLES (1959) reported that MDA produced hallucinations when self-administered. In the studies of NARANJO et al. (1967) hallucinations were not reported by subjects. TAYLOR et al. (1970) and THIESSEN and COOK (1973) have reviewed the literature concerning the abuse of MDA and other hallucinogens. DOET (2,5-dimethoxy-4-ethylamphetamine) produced feelings of euphoria, anxiety, nervousness and lightheadedness, and mydriasis but no perceptual distortions or changes in blood pressure, pulse rate, or body temperature (SNYDER et al., 1970, 1971). SHULGIN et al. (1969) found that paramethoxyamphetamine (PMA) was a hallucinogen in man and has been abused, apparently producing hallucinations, tachycardia, hyperthermia, hypertension, and convulsions (CIMBURA, 1974).

d) Interactions with LSD

A variety of factors alter the responsivity of subjects to LSD. Sleep loss enhances autonomic responsivity and disruption of performance produced by LSD (SAFER, 1970). Chlorpromazine antagonizes the subjective and autonomic changes produced by LSD and mescaline (HOCH, 1955; SCHWARZ et al., 1955; GIBERTI and GREGORETTI, 1955; ISBELL and LOGAN, 1957). Phenoxybenzamine and ergotamine antagonize some of the autonomic changes but intensify or do not markedly alter the subjective changes produced by LSD (ISBELL et al., 1959a; BERTINO et al., 1960; MATUSSEK and HALBACH, 1964).

The serotonin antagonist, benanserin (BAS), does not antagonize the subjective changes produced by LSD (ISBELL et al., 1959a; KLEE et al., 1960). Reserpine and scopolamine worsen the LSD reaction (ISBELL and LOGAN, 1957; ISBELL et al., 1959a) while reserpine in combination with iproniazid does not alter it (DEMAAR et al., 1960). Prednisone and progesterone decrease the disruptive effects of LSD on mental function (ABRAMSON and SKLAROFSKY, 1960; KRUS et al., 1961). LSD may decrease spinal fluid 5-HIAA (BOWERS, 1972). BERTINO et al. (1959) reported that nicotinic acid may have an ameliorating effect on LSD-induced subjective effects.

2. Other Species

The effects of LSD and related drugs on behavior have been studied in insects, fish, and birds, and the interested reader is referred to APPEL'S (1968) review. Only the effects of LSD-like hallucinogens in mammals will be reviewed here.

a) Mouse

Most studies of the effects of LSD in the mouse have been concerned with behavior and motor activity. LSD causes piloerection, agitation, and retropulsion (WOOLLEY and SHAW, 1957). KELLER and UMBREIT (1956) first noted that LSD caused head shaking in the mouse when the back of the head was touched. CORNE and PICKERING (1967) investigated the effects of a number of compounds on head twitching in the mouse and found that LSD, psilocin, mescaline, and DMT increased the incidence of head twitching, and their potency in inducing head twitching was related to their hallucinogenic potency. Other types of hallucinogens such as atropinelike drugs and phencyclidine also increased head twitching. Bufotenine and 5-HTP, which are not hallucinogenic, increased head twitching. Methysergide, which is probably an LSD-like hallucinogen, and nalorphine, which is yet another type of hallucinogen, did not produce head twitching. Chlorpromazine, haloperidol, pheniprazine, methysergide,

BOL, cyproheptadine, diphenhydramine, morphine, and sodium succinate antagonized the head twitch effects of LSD-like hallucinogens. Further, psilocybin, DMT, and mescaline antagonized LSD. DET and U-22, 394A increase head twitching in the mouse (TANG et al., 1968). SILVA and CALIL (1975) found that LSD, mescaline, and 5-HTP increased head twitching while Δ^9-THC, amphetamine, morphine, apomorphine, and chlorpromazine decreased it. Tryptamine, but not DMT, increases locomotor activity in the mouse, while both antagonize reserpine depression (VANE et al., 1961). LSD and LAE decrease activity in waltzing mice (ROTHLIN and CERLETTI, 1952). LSD increases spontaneous activity in the decorticated mouse (NEUHOLD et al., 1957). LSD, but not bufotenine, delays pole climbing (SMYTHIES, 1959). LSD reduces aggressiveness in mice (UYENO and BENSON, 1965). LSD, BOL, psilocybin, DMT, and DET antagonize morphine analgesia in the mouse (JACOB et al., 1962, 1964). Intracerebrally injected LSD produces tachypnea, Straub tail reaction, hyperexcitability, jerky movements, piloerection, and excitability in the mouse (HALEY, 1957a, b). Mescaline, intracerebrally, also caused tachypnea and some aggressiveness. In addition, mice flattened their ears against their heads and exhibited head scratching. LSD, but not BOL, antagonizes reserpine's potentiating effect on barbiturate sleeping time (TAESCHLER and CERLETTI, 1957; ASTON and CULLUMBINE, 1960).

b) Rat

Large doses (mg/kg) of LSD induce sniffing, side-to-side head movements, licking, biting of a wire mesh floor, increased motor activity, piloerection and mydriasis in rats (DIXON, 1968). In addition, psilocybin and to a lesser degree mescaline cause rats to walk backward and pivot on their hindfeet (SCHNEIDER, 1968). DOM increases locomotor activity, head twitching, and backward walking (YAMAMOTO and UEKI, 1975). LSD enhanced the startle response in intact rats but not raphé lesioned rats (DAVIS and SHEARD, 1974a). DMT also enhances the startle response in low doses but depresses it in larger doses (DAVIS and SHEARD, 1974b). Mescaline and psilocybin increase submissiveness in rats (UYENO, 1967a). These changes along with head shaking are shared by other hallucinogens including cyclazocine, levallorphan, amphetamine, atropine, and phencyclidine.

BRIMBLECOMBE (1963) found that the most consistent effect of LSD-like hallucinogens on open field behavior was to decrease the frequency of defecation. LSD does not exhibit analgesic activity in the rat using the hot plate method and antagonizes the analgesic action of morphine and methadone (DHAWAN and GUPTA, 1959). Intrastriatal administration of mescaline and 3,4-dimethoxyphenylethylamine in the rat caused salivation, aversive head turning, forelimb tremors, and convulsions (LITTLE and DILL, 1969). In the rat, tryptamine causes backward locomotion, Straub tail, bradypnea and dyspnea, and clonic convulsions (TEDESCHI et al., 1959). Large doses of LSD (30 and 100 μg/kg) decreased copulatory behavior in male testosterone-treated rats (MALMNÄS, 1973).

LSD and related drugs disrupt a variety of behavior. LSD increases the starting latency and slows swimming time of rats (UYENO, 1968). BOL, the three isomers of trimethoxyamphetamine and amphetamine also produced a dose related increase in the starting latency of swimming as do psilocybin, DMT, DOM, DOET, 4-methoxy-DMT, and 6-hydroxy-DMT (UYENO, 1971). LSD does not appear to markedly alter the ability of rats to make tone or temporal discriminations (DYKSTRA and APPEL, 1974).

LSD in small doses (ca. 0.05–0.1 μg/kg) may increase avoidance behavior while larger doses (0.2 μg/kg or greater) may impair it (COOK and WEIDLEY, 1957; TAESCHLER et al., 1960: HAMILTON, 1960; JARRARD, 1963). WINTER and FLATAKER (1956) found that LSD produced a dose-related increase in rope climbing time for a food reward, which was partially antagonized by 5-HT and 5-HTP and worsened by reserpine. LSD prolongs maze running time for a food reward in a dose-related manner (GLOW, 1957) and impairs maze learning (MILLER and MILLER, 1970). LSD also increases the latency and the force of conditioned avoidance jumping (PAWLOWSKI, 1962). In contrast to these observations are those of APPEL et al. (1967) who found that LSD did not markedly alter either unconditioned or conditioned escape behavior. STOFF et al. (1974) found that LSD decreased the latency of escape-avoidance behavior in a shuttle box. DMPEA was without effect. Frontal lobe lesions enhance the ability of LSD to disrupt discriminative maze performance (FRIEDMAN and CAREY, 1975). LEVIS and CALDWELL (1971) found that both mescaline and 3,4-dimethoxyphenethylamine produced only a modest alteration of avoidance behavior when a low stimulus intensity was used but found that using a strong stimulus intensity, both compunds disrupted avoidance behavior.

JARRARD (1963) found that low doses of LSD enhanced bar-pressing (variable interval schedule) for food while larger doses decreased it. These observations have been confirmed and extended by FREEDMAN et al. (1964), GARDNER (1965), and APPEL (1971). HAMILTON and WILPIZESKI (1961) found that LSD produced a dose-related decrease in food intake. FREEDMAN et al. (1964) found that LSD disrupted operant behavior for food using both a fixed ratio and a variable interval schedule. The disruptive effect was least in food-deprived rats (APPEL et al., 1968). LSD decreases light-reinforced behavior, while BOL does not (LOWE and WILLIAMS, 1972). LSD-treated rats showed more position perseveration and a greater decrement in learning than saline-treated rats (BUTTERS, 1966). Disruption of conditioned behavior is not specific for LSD-like hallucinogens (SILVA and CALIL, 1975). A closer analysis of the effects of LSD on bar-pressing using a FI schedule indicates that both small and large doses of LSD increased bar-pressing rates during the early part of the interval suggesting that it disrupted the timing of behavior (ALTMANN and APPEL, 1971, 1975). TILSON and SPARBER (1973) found that mescaline also increased bar-pressing during the early part of the FI period but decreased it during the latter half. GESSNER and PAGE (1962) studied the effects of a series of indoleamines on a conditioned avoidance response and found that 5-methoxy-DMT, DMT, LSD, and DET disrupted performance. Although LSD disrupts a variety of types of behavior, it also facilitates learning in the rat (SCHECHTER and WINTER, 1971; BIGNAMI, 1972; KING et al., 1974). BOL does not facilitate learning and mescaline inhibits it (SCHECHTER and WINTER, 1971; KING et al., 1974).

It is, of course, impossible to compare subjective and perceptual changes in man with behavioral changes in animals; however, through the use of drugs as discriminative stimuli, the internal cues that they produce can be compared (WINTER, 1973). HIRSCHHORN and WINTER (1971) found that both LSD (0.03 μmol/kg or greater) and mescaline (1–10 μmol/kg) could serve as discriminative cues in rat bar-pressing for food (variable interval schedule) and that rats could not distinguish between them.

WINTER (1973, 1974b) further found that rats could not distinguish between mescaline on the one hand and 2,3,4-TMPEA and DMPEA on the other but could distinguish between barbital and LSD or mescaline (HIRSCHHORN and WINTER, 1975). SCHECHTER and ROSECRANS (1972) using a T-maze found that mescaline and psilocybin produced a cuing effect similar to LSD and were respectively 900 and 90 times less potent. Amphetamine was less effective as a cue. 3,4-Dimethoxyphenethylamine, however, can be discriminated from mescaline (WINTER, 1974a). BECKER et al. (1967) had obtained similar results with LSD in the pigeon. CAMERON and APPEL (1972) found that a conditioned stimulus associated with LSD would suppress bar-pressing behavior. Mescaline inhibits electrical self-stimulation of the hypothalamus in the rat (BAILEY and PRADHAN, 1972).

A variety of steroids attenuate the disruption of rat pole climbing by LSD (BERGEN et al., 1960). Among the most potent are testosterone, progesterone, and cortisol. Steroids also inhibit the metabolism of LSD by rat liver but there is a poor correlation, either positive or negative, between their metabolic and behavioral actions (BERGEN et al., 1962). LSD is also less effective in disrupting pole climbing in pregnant rats (BERGEN, 1972). The ability of LSD to disrupt operant behavior in the rat is enhanced by reserpine, tetrabenazine, and tranylcypromine and antagonized by chlorpromazine (APPEL and FREEDMAN, 1964). DIXON (1968) found that the behavioral effects of LSD in the rat are antagonized by Dibenamine, phenoxybenzamine, α-MT, pronethalol, propranolol, BOL, and chlorpromazine but not by pCPA, and enhanced by reserpine and desmethylimipramine. SUGRUE (1969) also failed to find any antagonism of LSD's excitatory effects by reserpine. On the other hand, SUGRUE (1969) and APPEL et al. (1970) found that neither α-MT or *d*-amphetamine altered LSD's ability to disrupt operant behavior while pCPA pretreatment increased LSD's ability to disrupt operant behavior (APPEL et al., 1970). VOTAVA et al. (1967) found that small doses of LSD reversed the sedation and blepharospasms produced by reserpine pretreatment, while larger doses produced vocalization and aggression. Methysergide produced effects resembling doses of LSD.

c) Rabbit

For the most part, LSD-like hallucinogens increase body temperature and produce piloerection, mydriasis, and hyperreflexia in the rabbit (JACOB and LAFILLE, 1963). Mescaline produces miosis. LSD increases spontaneous activity in the decorticate rabbit (NEUHOLD et al., 1957).

d) Cat

Small to modest parenterally administered doses of LSD (up to 50 μg/kg) produce only minimal behavioral changes which superficially may resemble drowsiness (URSIN, 1962) but which are associated with increased wakefulness (KAY et al., 1976). Larger doses of LSD produce ataxia, spasticity, catatonia, head shaking, extrusion of the claws, piloerection, mydriasis, hyperthermia, and signs that could be interpreted as feline delirium (ROTHLIN, 1957a, b; ADEY et al., 1962; BARRATT and PRAY, 1965). When LSD is administered intraventricularly, mydriasis, paw sweating, salivation, urination, hypersensitivity to light and sound, sham rage, and some aggressiveness are seen (GADDUM and VOGT, 1956; HALEY and DASGUPTA, 1958). Relatively large doses are required to produce these changes. Doses that would produce marked subjective changes in man cause no gross changes when administered intracerebrally in the cat (GADDUM and VOGT, 1956). Intracerebrally administered BOL is inactive and does not antagonize the effects of LSD. LSD disrupts operant behavior in the cat (ADEY et al., 1962; LINDSLEY et al., 1968). KEY (1964) found that LSD shortened the latency of a conditioned escape response but enhanced the effect of a distracting stimulus. LSD decreased the rate of extinction and increased stimulus generalization using a conditioned avoidance procedure and depending on the experimental design produced a modest decrement in the learning of the response (KEY, 1961b).

Tryptamine produces a variety of changes in the cat causing signs of sympathetic activation including mydriasis, retraction of nictitating membrane, piloerection, motor signs such as extension of limbs and convulsions and affective changes such as hissing and snarling (LAIDLAW, 1912). Cats receiving DOM (1–8 mg/kg) exhibit hissing, growling, biting, tremors, ataxia, piloerection, arching of the back, and tonic extension of the back (WALLACH et al., 1972). These effects are generally antagonized by BOL, methysergide, cyproheptadine, chlorpromazine, and tripelennamine, but not by reserpine, α-MT, haloperidol, phenoxybenzamine, propranolol, or pCPA. Tachyphylaxis develops to these effects. ELDER and DILLE (1962) found that the behavioral and autonomic effects of LSD in the cat could be antagonized by chlorpromazine and phenoxybenzamine, but not by hydralazine, atropine, or scopolamine. Surgical sympathectomy and adrenal demedullation did not alter LSD actions. Reserpine pretreatment enhanced them.

e) Dog

The effects of LSD in the dog resemble those in man in many ways. Small doses (2–15 μg/kg) produce only modest behavioral changes including wakefulness, occasionally tracking, and extrusion of the tongue. Autonomic changes include mydriasis, hyperthermia, tachycardia, and tachypnea (MARTIN and EADES, 1970). Larger doses (0.2 mg/kg) cause ataxia and spasticity (ROTHLIN, 1957b).

Intraventricular injection of LSD in the dog produces salivation, retching, emesis, tachypnea, mydriasis, head shaking, ataxia, whining, and micturition (HALEY, 1956; HALEY and DASGUPTA, 1958). Psilocin, mescaline, and DMA also increase pupillary diameter, respiratory rate, pulse rate, and body temperature (MARTIN and EADES, 1970; VAUPEL and NOZAKI, 1975; VAUPEL et al., 1976). DMA in large doses decreases appetite in the dog. LSD and tryptamine prolong the latency of the skin twitch reflex in the dog (MARTIN et al., 1974; NOZAKI et al., 1976a).

TMA produces catatonia, swaying, decreased response to environmental stimuli, and mydriasis in the dog (PERETZ et al., 1955). MDA facilitates the flexor reflex, induces the stepping reflex in the chronic spinal dog, and in this regard resembles LSD. It is, however, more effective than LSD in dilating pupils, retracting the nictitating membrane, and producing hyperthermia, effects which are seen with amphetamines (NOZAKI et al., 1976a). Further, behavioral changes produced by MDA have elements of both LSD and amphetamine syndromes. Tracking and extrusion of the tongue, LSD-like, and head tossing and anorexia, amphetaminelike, were seen. MMDA also has mixed effects; facilitation of the flexor reflex, production of the stepping reflex and increased respiratory rate are LSD-like changes, while marked pupillary dilation is more commonly seen following amphetamine. MMDA produces barking, tracking, extrusion of the tongue, and nystagmus (NOZAKI et al., 1976b). Both MDA and MMDA increase the latency of the skin twitch reflex in the chronic spinal dog.

MARTIN and EADES (1970) found that cyproheptadine and chlorpromazine antagonized the spinal cord effects of LSD, mescaline, psilocin, and tryptamine; whereas, phenoxybenzamine did not. Similarly, cyproheptadine but not phenoxybenzamine blocked these effects of DOM. VAUPEL and NOZAKI (1975) found that cyproheptadine but not phenoxybenzamine antagonized the spinal cord facilitatory actions of 2,5-DMA in the chronic spinal dog. There was a trend for cyproheptadine, chlorpromazine, and phenoxybenzamine to antagonize the effects of LSD, mescaline, and psilocin on respiratory rate and pupillary diameter (MARTIN and EADES, 1970). Cyproheptadine but not phenoxybenzamine antagonized DMA's ability to induce the stepping reflex, to dilate pupils, and to stimulate respiration. Cyproheptadine antagonized MDA induction of stepping. Chlorpromazine not only antagonized MDA's effect on the stepping reflex but also antagonized its effect on pupils, respiration and body temperature (NOZAKI et al., 1976a). Similar results were obtained in studies of interactions between MMDA and phenoxybenzamine, cyproheptadine, and chlorpromazine (NOZAKI et al., 1976b).

f) Monkey

Following large doses of LSD (1 mg/kg) and bufotenine, the rhesus monkey assumes a prone position or becomes ataxic, but has muscle strength, is tame, and is indifferent to visual and painful stimuli (EVARTS, 1956, 1957, 1958b). LSD, psilocin, DMT and mescaline induce mydriasis and salivation and reduce aggressiveness in the baboon (WALTER et al., 1971; MELDRUM and NAQUET, 1971). Aggressiveness is also reduced in the squirrel monkey by LSD (UYENO, 1967B).

EVARTS (1958b) found that moderate to large doses of LSD did not impair visual discriminative tasks; however, they produced a moderate impairment of delayed responses (EVARTS, 1958b). SIEGEL et al. (1974) found that graded doses of LSD and DMT increased the frequency of yawning, tracking, fear, grimacing, head and body shakes, and stereotypic movements which include hand and head movements and rocking. FUSTER (1959), JARVIK and CHOROVER (1960), and UYENO et al. (1968) observed that LSD disrupts operant and discriminative behavior in the monkey. SHARPE et al. (1967) found that LSD's disruptive effect on discriminative behavior was greater for more difficult discriminative tasks. BOL, mescaline, several TMA isomers, as well as amphetamine disrupt operant behavior in the squirrel monkey (UYENO et al., 1968). DMT increased the latency of onset of bar-pressing for food using an FR schedule but not the overall response rate in the squirrel monkey (COLE and PIEPER, 1973). Tolerance did not develop with chronic administration of DMT.

g) Others

LSD disrupts conditioned avoidance behavior in the guinea pig (SIVADJIAN, 1970) and induces stereotypic walking patterns in the goat (KOELLA et al., 1964). This effect is greater in female than in male goats. Further, this stereotypic walking shows a cyclic pattern when LSD is administered chronically within a period of about 6 days.

III. Tolerance and Cross Tolerance

1. Man

ISBELL et al. (1955, 1956) showed tolerance to the subjective effects of LSD as well as to its effects on blood pressure, pupils, and the patellar reflex. Tolerance developed rapidly, seemed to be greatest for its effects on mental state and mydriasis and less for facilitation of patellar reflex and increase in blood pressure. Subjects returned to their control level of responsivity to LSD by the third day after LSD was discontinued. ABRAMSON et al. (1956) and CHOLDEN et al. (1955) also showed that tolerance developed to the effects of LSD on mood and perception.

ISBELL and his collaborators made extensive use of the phenomena of tolerance and cross tolerance to LSD-like agents as one means of establishing pharmacologic equivalence. Chronic administration of BOL attenuated the response to LSD (GINZEL and MAYER-GROSS;, 1956; ABRAMSON et al., 1957; ISBELL et al., 1959c). Although BOL is virtually devoid of LSD-like activity (JARVIK et al., 1955d; ISBELL et al., 1959c) in man and because single doses administered in combination with LSD did not alter the LSD response (CLARK and BLISS, 1957; ISBELL et al., 1959c), ISBELL et al. (1959c) concluded that chronic administration of BOL induced cross tolerance. The agonistic and antagonistic actions of BOL should be reexamined for only these data argue against the hypothesis that the agonistic effects of LSD-like agents are responsible for their behavioral, subjective, and physiologic effects as well as their ability to induce tolerance and cross tolerance.

ABRAMSON et al. (1957, 1958, 1960) found that cross tolerance could be produced between psilocybin and LSD and 1-methyl-LSD. ISBELL et al. (1961) found that direct tolerance to psilocybin was somewhat more erratic than tolerance to LSD. Although unequivocal tolerance developed to psilocybin's ability to produce subjective changes and cause mydriasis, tolerance to the facilitation of the patellar reflex and increase in body temperature were less constant, and no significant tolerance developed to its ability to produce tachycardia or increase blood pressure. Similarly, cross tolerance to LSD in psilocybin-tolerant subjects and psilocybin in LSD-tolerant subjects was less global than tolerance to LSD in LSD-tolerant subjects. BALESTRIERI (1960) failed to observe tolerance to psilocybin.

In a study of the effects of chronically adminstered mescaline, WOLBACH et al. (1962a) found that a statistically significant level of cross tolerance developed to its ability to cause subjective changes and mydriasis. Although there was a trend for the effect on body temperature, pulse rate, blood pressure, and the patellar reflex to be diminished, these changes were not statistically significant. The mescaline tolerant subjects showed a significant degree of cross tolerance to the effects of LSD on subjective effects, pupils, and blood pressure but not on temperature, pulse rate, or facilitation of the patellar reflex. Subjects tolerant to LSD are cross tolerant to the effects of mescaline (BALESTRIERI, 1957; BALESTRIERI and FONTANARI, 1959; WOLBACH et al., 1962a) and the degree of cross tolerance appears to be more complete than in mescaline tolerant subjects.

A significant degree of cross tolerance to the subjective effects of DMT but not to its effects on pulse rate, blood pressure, the patellar reflex, or pupils was observed in LSD tolerant subjects (ROSENBERG et al., 1964). BALESTRIERI (1960) observed that chronic treatment with methysergide reduced the autonomic and psychic responsivity of patients to LSD. HOLLISTER et al. (1969) and ANGRIST et al. (1974) found that chronic administration of DOM induced tolerance to its subjective, autonomic, and somatomotor effects.

ROSENBERG et al. (1963 b) studied the effects of amphetamine in LSD tolerant subjects and found that its effects on temperature, pulse rate, blood pressure, pupillary diameter, patellar reflex, and subjective state were not significantly altered. Subjects who received amphetamine chronically became tolerant to its effects on temperature, blood pressure, and subjective state. They were not cross tolerant to the effects of LSD. ISBELL et al. (1964) also studied tolerance and cross tolerance between the atropinelike hallucinogens, scopolamine and N-ethyl-3-piperidyl benzylate (JB-318), and LSD. Scopolamine and JB-318 slowed pulse rate, dilated pupils, and produced subjective changes. Scopolamine and JB-318 caused anxiety, perceptual distortions, and visual hallucinations like LSD but differed in that confusion and sedation were also a prominent part of their effects. When scopolamine and JB-318 were administered chronically, tolerance developed to their effects on pupils and pulse rate. In general, subjects tolerant to scopolamine were cross tolerant to JB-318 but not LSD; subjects tolerant to JB-318 were cross tolerant to scopolamine but not LSD; and subjects tolerant to LSD were not cross tolerant to scopolamine or JB-318. ISBELL and JASINSKI (1969) compared the effects of LSD and Δ^9-THC in nontolerant and LSD tolerant subjects. Δ^9-THC did not alter body temperature, blood

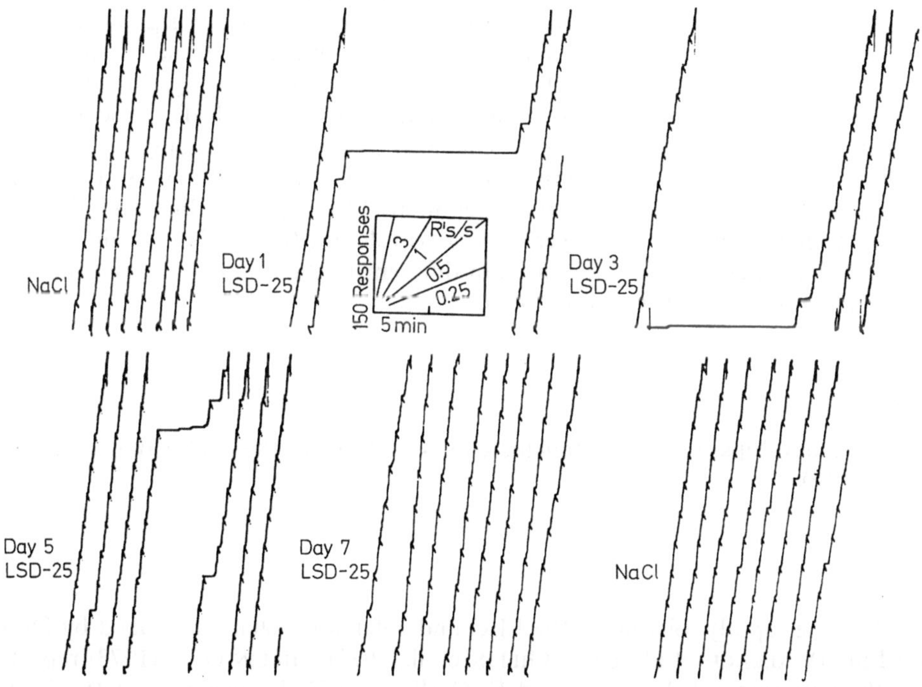

Fig. 1. Cumulative records of bar-pressing on an FR 40 (fixed-ratio 40) schedule of reinforcement for a single rat. Control sessions before and after tolerance run and 1st, 3rd, 5th, and 7th days of treatment with 130 µg of LSD-25/kg are shown. [Reproduced from FREEDMAN et al. (1964) with permission]

pressure, pupillary diameter, or the patellar reflex. It caused a much greater tachycardia than LSD. There was a similarity in subjective states produced by the two drugs. Subjects tolerant to LSD were not cross tolerant to Δ^9-THC.

2. Rat

FREEDMAN et al. (1958) found complete tolerance to LSD's ability to delay rope climbing when it was administered daily. Tolerance was more erratic when it was administered every 48h and did not develop when administered at a 72h interval. Mescaline also increased rope climbing time, and tolerance developed to this effect when it was administered chronically; however, tolerance developed less rapidly than to LSD. Chronic administration of bufotenine also induces tolerance to its ability to delay rope climbing and cross tolerance to LSD (MAHLER and HUMOLLER, 1959). Rats tolerant to LSD are also cross tolerant to bufotenine. FREEDMAN et al. (1964) found that chronic administration of LSD induced tolerance to its ability to disrupt operant behavior for food (see Fig. 1). In an extensive comparison, APPEL and FREEDMAN (1968) showed that in addition to LSD, BOL, psilocybin, and mescaline disrupted operant behavior and that when they were administered chronically tolerance developed to their disruptive effects. Further, rats tolerant to LSD were cross tolerant to psilocybin, mescaline, and to a lesser degree to BOL. They were not cross tolerant to amphetamine. Rats tolerant to BOL showed a trend toward being cross tolerant to LSD and psilocybin but this trend was not statistically significant. Similarly, animals tolerant to psilocybin and mescaline showed a trend toward being cross tolerant to LSD but again this trend was not statistically significant. Similar results have been reported by TILSON and SPARBER (1973) for LSD, mescaline, and amphetamine using a fixed interval schedule. Using both prolongation of rope climbing and inhibition of operant behavior, SILVA et al. (1968) showed that rats tolerant to Δ^9-THC were not cross tolerant to LSD or mescaline and that rats tolerant to mescaline and LSD were not cross tolerant to Δ^9-THC. Brain levels of LSD following a challenging dose are not different in the nontolerant and the LSD or mescaline-tolerant rat (WINTER, 1971).

3. Rabbit

Tolerance develops to the pyretogenic effects of LSD in the rabbit (GOGERTY and DILLE, 1956).

4. Cat

Cats become rapidly tolerant to the EEG and behavioral changes evoked by DOM and LSD (WALLACH et al., 1972; GILLIN et al., 1973). GILLIN et al. (1973) reported that the autonomic, behavioral, and EEG effects of DMT increased with chronic administration. Tolerance develops rapidly to DOM's behavioral effects in the cat and pretreatment with LSD and mescaline confers cross tolerance to DOM (WALLACH et al., 1974).

5. Dog

Tolerance develops to the ability of LSD to evoke the stepping reflex, facilitate the patellar reflex, and increase pulse rate, respiratory rate, and pupillary diameter in the chronic spinal dog (MARTIN and EADES, 1972). LSD-tolerant chronic spinal dogs are cross tolerant to tryptamine (MARTIN and EADES, 1972) and DMA (VAUPEL and NOZAKI, 1975; VAUPEL et al., 1976) but to a lesser degree to MDA (NOZAKI et al., 1976a).

IV. Criteria for LSD-like Hallucinogens

Characteristics

LSD-like hallucinogens produce mood and perceptual changes, disruption of thinking processes, and hallucinations. The mood changes are predominantly euphoric at low doses and dysphoric (anxiety, fear, apprehension) at higher doses. They alter the functioning of the autonomic nervous system. These changes, particularly those associated with low doses, are characterized by sympathetic stimulation and arousal but as will be discussed in Subsect. D, other types of changes may be produced by these drugs. Finally, these drugs have marked effects on motor function. Generally, reflex activity is facilitated but paradoxically ataxia and impairment of motor performance are also seen. These constellations of effects are all agonistic actions of LSD-like hallucinogens and constitute their pharmacologic profile.

The phenomena of direct and cross tolerance to these drugs have also assisted in their identification. Direct tolerance develops to their subjective, behavioral, autonomic and somatomotor effects and, by and large, cross tolerance can be demonstrated between several hallucinogens. Thus, mescaline, psilocin, and LSD seem to be quite similar in this regard, and there is not only a concordance between data obtained in man and animals but a good concordance between their pharmacologic profiles in man and animals and their abilities to induce tolerance.

There are, however, two drugs that have been extensively studied that raise some questions about the validity of the tolerance-cross tolerance paradigm. BOL which appears to be virtually devoid of agonistic actions appears to induce cross tolerance to LSD in both man and the rat. In contrast, DMT which appears to be a strong agonist of the LSD type showed suprisingly little attenuation of its autonomic and somatomotor effects in LSD-tolerant subjects. Before the limits of the cross tolerance technique can be clearly evaluated, additional experiments must be done particularly relating the pharmacologic profile to tolerance and cross tolerance in several species.

Yet another technique which may be of help in the identification of LSD-like hallucinogens is the use of antagonists (see Subsect. C). Most investigators have found that agonistic actions of LSD cannot be antagonized by α-adrenergic blocking agents, suggesting that noradrenergic processes are not of critical importance in the behavioral actions of LSD. On the other hand, chlorpromazine has been found to be a potent and effective antagonist of LSD. In the chronic spinal dog, cyproheptadine antagonizes most of the agonistic actions of the LSD-like hallucinogens in doses that have little α-adrenolytic (MARTIN and EADES, 1970), cholinolytic (VAUPEL and MARTIN, 1973), or serotonolytic activity (BELL et al., 1976). These observations, among others, have suggested that LSD may exert its agonistic activity through a tryptaminergic mechanism (MARTIN and EADES, 1970). The development and use of increasingly selective antagonists may also assist in the identification of LSD-like hallucinogens.

It may well be that certain congeners of LSD have several modes of action. Thus, the work of VAUPEL et al. (1976) indicates that small doses of DMA may have LSD-like activity, while larger doses also have amphetaminelike effects. Studies of interactions of LSD-like agents and amphetaminelike agents may provide information that may allow a more balanced classification particularly of the phenethylamine and amphetamine congeners that are also hallucinogens.

The classification of many drugs that have been claimed to have LSD-like activity is problematic. Table 4 presents an effort to categorize a variety of criterion drugs. Category I contains those substances for which studies of pharmacologic profiles

Table 4. Criterion drugs

I. LSD-like	
Mescaline	
D-LSD	
Psilocybin	
Psilocin	
II. Probably LSD-like	
PMA	D-lysergic acid morpholid
2,5 DMA	D-lysergic acid pyrrolidide
3,4,5 TMA	2,3 dihydro D-LSD
DOM	1-methyl D-LSD
Tryptamine	1-acetyl D-LSD
DMT	Oxymethyl D-LSD
Methysergide	
D-lysergic acid dimethylamide	
D-lysergic acid monoethylamide	
D-1-methyl-lysergic acid monoethylamide	
D-1-acetyl-lysergic acid monoethylamide	
III. Probably LSD-like but with other properties	
MDA	
MMDA	
IV. Probably not LSD-like	
BOL	
5-HTP	
V. Not LSD-like	
Amphetamine	
PEA	
DMPEA	
DOET	
Bufotenine	
1-Methyl-2-brom-D-LSD	
L-LSD	
D-Iso-lysergic acid diethylamide	
Scopolamine	
JB-318	
Δ^9-THC	

using objective and validated methods, tolerance, and cross tolerance experiments and interaction studies with antagonists have yielded concordant data. Category II contains those drugs (a) that appear to be LSD-like on the basis of existing experimental data, but for which critical data have not been evolved, or (b) that have an overall spectrum of LSD activity but for which there are some discordant data. DMT, as has been discussed, is such a drug. Category III contains those substances that have LSD properties but which also have other properties that are manifest with euphoric and hallucinogenic dose levels. There are drugs not included in this category that have other actions when dose levels several orders of magnitude larger than those required to produce LSD-like effects are administered. Category IV contains those drugs that do not appear to be LSD-like but which have some actions that resemble LSD. It is probable that drugs in this category include: (a) drugs that are weak partial agonists of the LSD type, and (b) drugs that mimic some of LSD's actions that are not involved in the production of mood and perceptual changes.

In Category V are those drugs that are structurally related to LSD-like hallucinogens but where there are hard data indicating that they do not produce LSD-like effects. The information for many of these drugs is fragmentary; thus, this classification, although the best approximation, must be regarded as tentative for agents in Categories II—V.

V. Chemical Structures of LSD-like Hallucinogens

Table 5 summarizes the structures of chemicals that have been thought to have LSD-like activity. It is problematic whether there are sufficient data for many of them to allow a critical judgment about their pharmacologic classification. They all probably change feeling states, perception, and mentation. Those chemicals that appear in Table 5 are indicated in the right column as LSD-like (I) or probably LSD-like (II). In addition, the estimates of their potency in man in mescaline units and in relation to LSD are also tabulated.

1. β-Phenethylamines

β-Phenethylamine has pharmacologic actions which resemble in part those produced by amphetamine although some effects are different (MARTIN and EADES, 1976). It does not produce LSD-like effects. Mono-methoxy ring-substituted phenethylamines do not disrupt operant behavior in rats (SMYTHIES et al., 1967a) and are probably not LSD-like agents. Disubstituted methoxy PEA's also do not disrupt operant behavior in rats and 3,4-dimethoxy-PEA is not a psychotogen in man (HOLLISTER and FRIEDHOFF, 1966). Only mescaline (3,4,5) of the trisubstituted methoxy-PEA's disrupted operant behavior in rats, and all tetra- and penta-methoxy-substituted PEA's disrupted rat operant behavior (SMYTHIES et al., 1967a). SHULGIN and CARTER (1975) have reported that 2,5-dimethoxy-4-methyl-PEA and 2,5-dimethoxy-4-bromo-PEA produce mydriasis, relaxation, and some internal harmony in human subjects. These changes can be seen with small to modest doses of both LSD and amphetamine.

Table 5. Chemical structures and potencies of drugs in man that have some LSD-like activity

No.	Name	2	3	4	5	6	Mescaline unit	Potency LSD=100	Category
β-Phenethylamines									
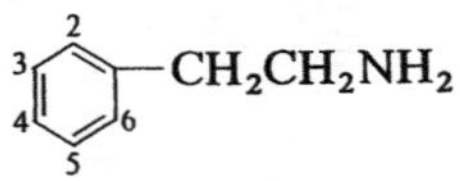									
1	Mescaline		OCH_3	OCH_3	OCH_3		1	0.03	I
2		OCH_3	OCH_3	OCH_3	OCH_3				
3		OCH_3	OCH_3	OCH_3	OCH_3	OCH_3			
4		OCH_3		BR	OCH_3				
5		OCH_3		CH_3	OCH_3				
6		OCH_3	$—O—CH_2—O—$				< 5		
7			OCH_3	$—O—CH_2—O—$			1		
8	DMPEA		OCH_3	OCH_3					
Amphetamines									
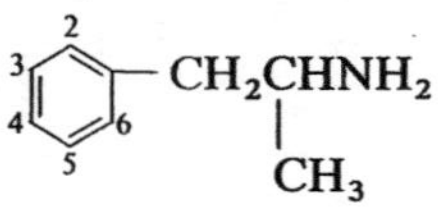									
9	PMA			OCH_3			5		II
10	DMA	OCH_3		OCH_3			5		

11		OCH_3			OCH_3		8	II
12		OCH_3	OCH_3	OCH_3			< 2	
13	TMA		OCH_3	OCH_3	OCH_3		2	II
14		OCH_3	OCH_3		OCH_3		4	
15		OCH_3	OCH_3			OCH_3	13	
16		OCH_3		OCH_3	OCH_3		17	
17		OCH_3		OCH_3		OCH_3	10	
18		OCH_3	OCH_3	OCH_3	OCH_3		6	
19		OCH_3		OCH_3	OC_2H_5		< 7	
20		OCH_3		OC_2H_5	OCH_3		15	
21		OC_2H_5		OCH_3	OCH_3		< 7	
22	DOM	OCH_3		CH_3	OCH_3		80	II
23	DOET	OCH_3		C_2H_5	OCH_3			
24		OCH_3		BR	OCH_3			
25	MDA		—O—CH_2—O—				3	
26		—O—CH_2—O—		OCH_3			3	
27		OCH_3	—O—CH_2—O—				10	
28	MMDA		—O—CH_2—O—		OCH_3		2	
29			—O—CH_2—O—			OCH_3	21	
30		OCH_3		—O—CH_2—O—			12	
31			OCH_3	—O—CH_2—O—			3	
32				—O—CH_2—O—		OCH_3	18	
33		OCH_3	—O—CH_2—O—		OCH_3		12	
34			—O—CH_2—O—		OCH_3	OCH_3	5	
35		OCH_3	OCH_3	—O—CH_2—O—			5	
36			OCH_3	—O—CH_2—CH_2—O—				

Table 5. (continued)

No.	Name	2	3	4	5	6	Mescaline unit	Potency LSD = 100	Category
Indoleamines									
	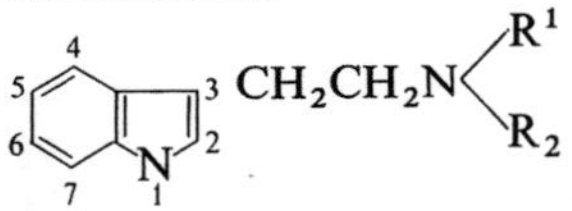								
			4	R_1	R_2				
37	Tryptamine								II
38	DMT			CH_3	CH_3			0.003	II
39	DET			C_2H_5	C_2H_5				
40	DPT			C_3H_7	C_3H_7				
41	Psilocin		OH	CH_3	CH_3			2.2	II
42	Psilocybin		H_2PO_4	CH_3	CH_3			1.5	I
Lysergic acid amides									
	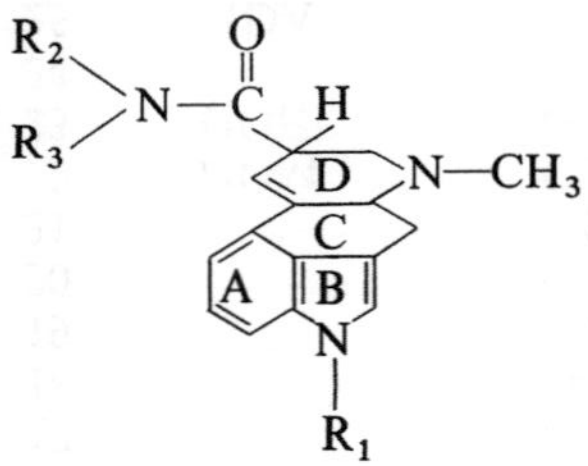								
			R_1	R_2	R_3				
43	LSD			C_2H_5	C_2H_5			100	I
44				CH_3	CH_3			10	II
45				C_2H_5				5	II
46				$-(CH_2)_2-O-(CH_2)_2-$				10	II

47			$—C_4H_8—$		5—10	II
48		CH_3	C_2H_5	C_2H_5	30	II
49		CH_3	C_2H_5		4	II
50		$—\overset{\|}{C}—CH_3$ $\|\|$ O	C_2H_5	C_2H_5	100	II
51		$—C—CH_3$ $\|\|$ O	C_2H_5		7	II
52		$—OCH_3$	C_2H_5	C_2H_5	66	II
53	Methysergide	CH_3	$CH\,CH_2CH_3$ | CH_2 | O $\|\|$			II
54	Dihydro-LSD				17	II

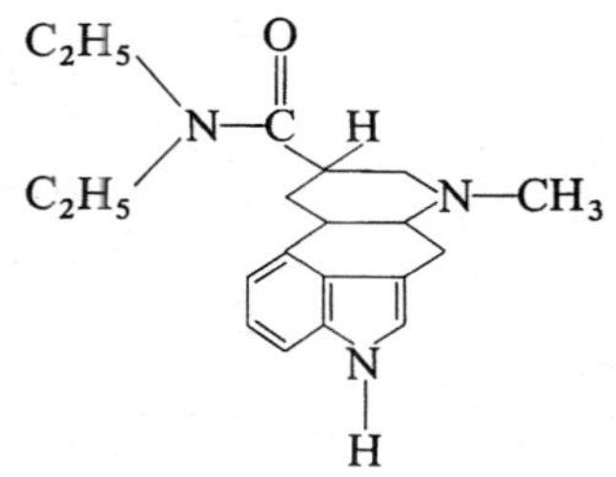

Mescaline units are discussed in the text.
Potency is expressed in terms of LSD (100%) and drugs with lesser potencies have proportionately lower percentages.
An open entry in a column indicates a hydrogen.
Categories are defined in table 4 and text.

2. Amphetamines

A large number of substituted amphetamines have been synthesized and studied in man, many by SHULGIN and his collaborators (PERETZ et al., 1955; SHULGIN et al., 1961; SHULGIN, 1964a, b; NARANJO et al., 1967; SHULGIN and SARGENT, 1967; SMYTHIES et al., 1967b; SNYDER et al., 1967; SHULGIN, 1968; SHULGIN et al., 1969, 1971; SNYDER et al., 1971). Of these PMA, DMA, and TMA appear to be LSD-like substances (see Subsect. B. II. 1. c. β). MDA and MMDA may have some LSD-like activity but also appear to have amphetaminelike effects. Data on the remaining compounds suggest that they produce definite subjective effects in man, but it has not been conclusively established that they are LSD-like hallucinogens.

3. Indoleamines

Studies of the pharmacologic profiles and of tolerance of the indoleamines have been extensively reviewed (Subsect. B. II and B. III). There is little question that psilocin and psilocybin resemble LSD in their pharmacologic actions. Tryptamine also has many LSD-like properties in man and the dog. Dimethyl, diethyl, and dipropyl-tryptamine also have agonistic actions that closely resemble those of LSD. The observation that tolerance to LSD does not confer cross tolerance to several of DMT's autonomic effects suggests that these N-disubstituted compounds may differ in some respects from LSD. The duration of action of DMT is less than that of LSD. Further, we have not been able to demonstrate tachyphylaxis to the actions of tryptamine in the dog (MARTIN and EADES, 1972). Tryptamine's duration of action is less than that of DMT. It is possible that a long duration of action is a necessary attribute of LSD-like hallucinogens for them to effectively induce tolerance (see Subsect. E. III).

4. Lysergic Acid Amines

The pharmacologic actions and cross tolerance studies of these compounds have for the most part been reviewed in Subsections B. II and B. III of this chapter. GOGERTY and DILLE (1957) studied the effects of *d*-lysergic acid morpholide (compound 46) in mice, rabbits, cats, dogs, and man, and found that its pharmacologic profile was similar to LSD. All of the lysergic acid amides listed in Table 5 produce typical LSD-like subjective effects and autonomic changes and are in all probability LSD-like agonists. Direct and cross tolerance studies in man have not been conducted on these compounds, except for LSD, and thus an important body of information is missing which is critical for their classification.

C. Mode of Action

I. Serotonin Antagonist Theory

GADDUM and HAMEED (1954) found that LSD selectively antagonized the effects of 5-HT on the rat uterus, whereas, mescaline did not. WOOLLEY and SHAW (1954a) also found that LSD as well as yohimbine and harmine antagonized the effects of 5-HT on aortic rings and rat uterus. The demonstration of the presence of 5-HT in the brain (TWAROG and PAGE, 1953; AMIN et al., 1954; BOGDANSKI et al., 1956) gave the

5-HT antagonistic action of LSD special significance as it related to its hallucinogenic effects. The relationship between 5-HT agonistic actions and 5-HT antagonists, however, has proved to be extremely complicated (see GYERMEK, 1966). Not only may there be several types of 5-HT receptors (GADDUM and PICARELLI, 1957) but agents that act as antagonists in some preparations act as agonists in other preparations. WOOLLEY and SHAW (1954b) and GADDUM and HAMEED (1954) suggested that LSD's ability to antagonize the effects of 5-HT might in some way be related to its hallucinogenic activity which in turn might be related in some way to schizophrenia. Although the initial studies demonstrating LSD and 5-HT antagonistic activity were conducted on smooth muscle preparations, subsequent studies showed that LSD also antagonizes the stimulatory effects of 5-HT on brain neurons. Further, LSD antagonizes the depressant effects of intracerebrally administered 5-HT, an action shared by morphine, methadone, and amphetamine but not by BOL (GADDUM and VOGT, 1956).

ROBERTS and STRAUGHAN (1967) found that iontophoretically administered 5-HT had both stimulatory and inhibitory actions on cortical neurons of the acutely decerebrated cat. The excitatory effects of 5-HT were antagonized by LSD, BOL, methysergide, and 2′-(3-dimethylaminopropylthio)cinnamanilide but not its inhibitory effects. BOAKES et al. (1970) studied the interactions between iontophoretically administered LSD, methysergide, and BOL, and several neurotransmitters in brain stem neurons of the decerebrate cat and also observed that LSD antagonized the excitatory but not the inhibitory effects of 5-HT. LSD also antagonized the excitatory effects of glutamate on neurons that were excited by 5-HT but not neurons that were inhibited by 5-HT. Methysergide's effects were similar to those of LSD except that methysergide was less potent. BOL antagonized the excitatory effects of 5-HT in about 20% of the neurons studied. Systemically administered LSD also antagonized the stimulatory effects of 5-HT. LSD, methysergide, and BOL did not antagonize the effects of NE, ACh, glycine, GABA, or D,L-homocysteic acid. GONZALEZ-VEGAS (1971) found that mescaline antagonized the inhibitory effect of NE and DA but not that of 5-HT, GABA, and glycine on brain stem neurons of the unanesthetized decerebrate cat. Mescaline did not antagonize the excitatory effects of NE, DA, 5-HT, or ACh. LSD and BOL antagonize the inhibitory effects of both iontophoretically applied NE and 5-HT in neurons of the olfactory bulb of the decerebrate rabbit (BLOOM et al., 1964).

II. Serotonin Agonist Theory

COSTA (1956) first reported that LSD and mescaline in low concentrations facilitated the contractile action of 5-HT on the rat uterus. GREENBERG (1960) found that LSD mimicked the actions of 5-HT on *Venus mercenaria* heart. A variety of 5-HT antagonists including BOL, chlorpromazine, and phenoxybenzamine [for an extensive treatment of 5-HT antagonists, see GYERMEK (1966)] are not LSD-like hallucinogens. Although these observations did not reject the hypothesis that some of the LSD-like actions could be a consequence of its 5-HT antagonistic property, they made it increasingly difficult to ascribe the hallucinogenic actions of LSD to this mode of action.

FREEDMAN (1961) found that large doses of LSD elevated rat brain 5-HT levels and antagonized the 5-HT depleting action of reserpine. This action was shared by *d*-1-acetyl-lysergic acid diethylamide, but not BOL, *d*-lysergic acid diethylamide, or 1-methyl-*d*-lysergic acid butanolamide. LSD increases levels of 5-HT in most brain areas of the rat and in the reserpinized rat. LSD also decreases brain 5-HIAA. The increase in brain 5-HT is in the particulate fraction while the decrease in 5-HIAA is in the supernatant fraction (FREEDMAN and GIARMAN, 1962; ROSECRANS et al., 1967). Increases in rat brain 5-HT are also seen after chronic administration of LSD (FREEDMAN and GIARMAN, 1962; PETERS, 1974). Pretreatment with pargyline may decrease or increase the changes in rat brain 5-HT levels produced by LSD (PETERS, 1974). LSD does not alter brain tryptophan levels or tryptophan hydroxylase activity (PETERS, 1974). LSD, but not BOL, decreases the turnover of 5-HT in rat brain (LIN et al., 1969) and decreases the release of 5-HT from rat brain slices (CHASE et al., 1967).

ANDEN et al. (1968) found that LSD, but not BOL or methysergide, antagonized the ability of α-propyldopacetamide, which antagonizes tryptophan hydroxylase, to lower brain and spinal cord 5-HT in the rat. In subsequent studies, ANDEN et al. (1971, 1974) found that this property was shared by psilocybin, DOM, to a lesser degree by DMT and tryptamine, but not by DMPEA, mescaline and bufotenine. Psilocybin and DMT alone did not significantly alter brain levels of 5-HT. Psilocybin and DMT, however, markedly antagonized the decrease in fluorescence intensity of 5-HT nerve terminals produced by α-propyldopacetamide. Tryptamine produced a weak antagonism and bufotenine was devoid of activity. Large doses of LSD and related hallucinogens caused hyperextension of the hindlimbs of the acute spinal rat as well as athetoid movements. 5-HTP in nialamide pretreated acute spinal rats produced similar changes suggesting that LSD-like hallucinogens were acting as serotonergic agonists (ANDEN et al., 1968, 1971). LSD decreases the release of 5-HT into the cerebral ventricles while mescaline increases release (TILSON and SPARBER, 1972). The increase in brain 5-HT in the rat evoked by increasing environmental temperature is antagonized by LSD but not BOL (AGHAJANIAN and WEISS, 1968).

AGHAJANIAN et al. (1968, 1970) studied the effects of several psychotogens on the activity of midbrain raphé neurons of chloral hydrate anesthetized rats. Dorsal and median raphé neurons were inhibited by systemically administered LSD (10–25 μg/kg). BOL also inhibited raphé units. As has been previously discussed, BOL has not been shown to cause hallucinations in man; however, when it was administered chronically, it did decrease the effect of LSD suggesting that it may have produced cross tolerance (ISBELL et al., 1959c). Mescaline and DOM inhibited the activity of some while they increased the activity of other raphé units. Neither CPZ nor pCPA antagonized the inhibitory effects of LSD on raphé units. Of interest is the fact that the firing rate of units in the ventral tegmental nucleus was increased by mescaline and DOM. LSD when applied iontophoretically to raphé units also inhibited their firing (AGHAJANIAN et al., 1972). In contrast, iontophoretically applied mescaline inhibited only 50% of raphé units. Inhibition was accompanied by a decrease in spike amplitude suggesting that mescaline may have a local anesthetic action. These results were confirmed by HAIGLER and AGHAJANIAN (1973) who also compared mescaline with NE. Mescaline and NE were applied iontophoretically to raphé units and also administered intravenously. Of the 33 units to which these drugs were applied iontophoretically, 9 responded in the same way and 24 differently. When mescaline was administered intravenously, 3 of the 11 responsive units responded the same as to iontophoretically applied NE and 8 differently.

Stimulation of the midbrain raphé of the rat inhibits cortical potentials evoked by paw stimulation and this inhibition is enhanced by LSD (SAMANIN et al., 1972). HORITA and GOGERTY (1958) found that 5-HTP caused hyperthermia in the rabbit and found that it was attenuated in the LSD-tolerant rabbit and concluded that there was a serotonergic mechanism responsible for the LSD-pyretogenic action.

BENNETT and SNYDER (1976) have studied saturable and stereospecific binding of LSD and saturable binding of 5-HT to brain homogenate. They have observed that the ability of 5-HT and proven LSD-like hallucinogens to prevent the binding of tritiated 5-HT and LSD is nearly equal. However, a variety of antagonists (cyproheptadine, mianserine and CPZ) are more potent in antagonizing the binding of LSD than 5-HT by a factor of 30 to 100. BENNETT and SNYDER (1976) have postulated that the 5-HT receptor can exist in an agonist and an antagonist configuration to account for these differences in affinity and that LSD is a mixed agonist-antagonist. An alternative interpretation is that LSD has agonistic activity at two receptors (see C. III, this chapter).

III. Tryptamine Agonist Theory

MARTIN and EADES (1970) first demonstrated the similarity of action of small doses of LSD-like hallucinogens and tryptamine in the chronic spinal dog. Tryptamine, LSD, psilocin, and mescaline all facilitated the hindlimb flexor reflex, evoked the stepping reflex, and increased respiratory rate, pulse rate, and pupillary diameter. These changes were antagonized by cyproheptadine and CPZ but not by phenoxybenzamine. Tryptamine was found to also increase blood pressure, pupillary diameter, and the patellar reflex, and produce perceptual distortions in man (MARTIN and SLOAN, 1970). Further, dogs that are tolerant to LSD are cross-tolerant to tryptamine (MARTIN and EADES, 1972). BELL and MARTIN (1974) found that low doses of LSD, tryptamine, and methysergide facilitated the C-fiber reflex in the acute spinal cat and that the facilitatory actions were antagonized by cyproheptadine.

VANE et al. (1961) first noted that tryptamine and substituted tryptamines had activity at many levels of the central nervous system and VANE (1960), GELDER and VANE (1962), and DEWHURST and MARLEY (1965a, b) proposed on the basis of EEG studies conducted in the chick and in peripheral tissues that there were tryptamine receptors in the brain which when activated by agonists had an excitatory effect. They further suggested that methoxamine and amphetamine activated these receptors. Studies of the actions of methoxamine and amphetamine on spinal cord reflex activity in the dog and cat, however, strongly suggest that they exert their actions at different receptor sites than tryptamine. Like tryptamine, methoxamine and amphetamine also facilitate the flexor reflex of the chronic spinal dog and methoxamine evokes the stepping reflex; however, methoxamine's effects are antagonized by phenoxybenzamine and to a lesser degree by cyproheptadine (MARTIN and EADES, 1967, 1970). Methoxamine and tryptamine also facilitate the mono- and polysynaptic segmental reflexes of the acute spinal cat (VAUPEL and MARTIN, 1976). The facilitatory actions of methoxamine are antagonized selectively by phenoxybenzamine while those of tryptamine are antagonized by cyproheptadine. Thus, the spinal cord facilitatory effects of LSD and tryptamine are exerted through a different mode of action than those of methoxamine and are probably not noradrenergic agonists.

Tryptamine is structurally similar to 5-HT and shares many actions in common. GADDUM (1953) on the basis of cross-tachyphylaxis studies in the guinea pig ileum thought that tryptamine and 5-HT produced their contractile effect by acting at the same receptor. VANE et al. (1961) also suggested that tryptamine and 5-HT act on the same receptor in the central nervous system. Tryptamine, as well as the racemic mixtures of α-methyltryptamine, α-ethyltryptamine, N-N dimethyltryptamine, and 5-methoxy-α-methyltryptamine in a dose of 0.5 mg/kg, facilitated the patellar and flexor reflexes. 5-HT (1 mg/kg) also facilitated these reflexes after a longer latency. These facilitatory effects were antagonized by both methysergide and BOL, 5-HT antagonists. TEDESCHI et al. (1959) found that tryptamine seizures were enhanced by monoamine oxidase inhibitors (iproniazid and β-phenylisopropyl-hydrazine) and to a lesser degree by ephedrine, pipradrol, *d*-amphetamine, methylphenidate, and caffeine, as well as the convulsants pentylenetetrazol and nikethamide. Tryptamine's convulsant effects were antagonized by the phenothiazines CPZ, prochlorperazine, and trifluoperazine, as well as by LSD and BOL. TEDESCHI concluded that tryptamine and serotonin acted at the same central receptors because: (1) Both tryptamine and 5-HTP produced seizures and their convulsant properties were enhanced by iproniazid, and (2) BOL antagonized both tryptamine and 5-HTP seizures.

On the other hand, WOOLEY and SHAW (1953, 1957) presented two lines of evidence suggesting that the tryptamine receptor was distinguishable from a 5-HT receptor: (1) The activity of 5-HT in constricting the sheep carotid ring was retained to a greater degree following storage at 4° C for 24 h than was the activity of tryptamine. (2) 1-Benzyl-2-methyl-5-hydroxytryptamine antagonized the vasopressor effect of serotonin but not that of tryptamine in the pentobarbital anesthetized dog. More recently, BELL et al. (1976) have shown that L-tryptophan also facilitates the C-fiber reflex of the acute spinal cat and its facilitatory effect is antagonized by α-methyl dopa and cyproheptadine but not by pCPA. 5-HTP did not consistently alter the amplitude of the C-fiber reflex. These studies indicate that tryptamine and LSD are not exerting their spinal cord facilitatory activity through a serotonergic mechanism.

Tryptamine has been found to be present at all levels of the central nervous system (MARTIN et al., 1972; SNODGRASS and HORN, 1973) and to be present in the brain of the rat, guinea pig, cat, dog, steer, and man (MARTIN et al., 1972; SAAVEDRA and AXELROD, 1972a, b, 1973; SNODGRASS and HORN, 1973; PHILIPS et al., 1974; SLOAN et al. 1975). MARTIN et al. (1975) have obtained evidence that there are tryptaminergic pathways descending the spinal cord from the mesencephalon and cerebellum and ascending from the spinal cord. It has been further shown that the brain releases tryptamine (MARTIN et al., 1974; SLOAN et al., 1975). The above cited observations support the hypothesis that there are tryptaminergic pathways in the brain and that the LSD hallucinogens exert some of their effects by mimicking tryptamine. Further, the mode of action of tryptamine on spinal cord reflexes is pharmacologically distinguishable from NE and 5-HT.

IV. Other Brain Amines and Polypeptides

A large body of data indicates that LSD probably interacts with noradrenergic and dopaminergic systems in the brain. Its actions on catecholaminergic systems, however, are not shared by all LSD-like hallucinogens. LSD, psilocybin, and DMT significantly decrease brain NE levels and tend to decrease DA levels in rats pretreated with α-MT (ANDEN et al., 1968, 1971) indicating that they increase the turnover of these catecholamines. SUGRUE (1969) observed that LSD and psilocybin decrease NE levels in the rat hypothalamus. PETERS (1974) did not observe any changes in rat brain levels of NE, tyrosine, or tyrosine hydroxylase activity following single doses while chronically administered LSD decreased all three. DE LA TORRE (1968) found that LSD increased catecholamine fluorescence in the rat septum. LSD decreased brain NE and DA and increased 5-HT in reserpine pretreated rats (VOTAVA et al., 1967).

LEONARD and TONGE (1969) found that LSD increased rat brain NE levels and decreased DA levels while mescaline decreased both NE and DA levels. Paradoxically, LSD and mescaline increased depletion of both brain NE and DA in rats pretreated with α-methyl-m-tyrosine but not in rats pretreated with α-MT. PIERI et al. (1974) found that a large dose of LSD (100 μg/kg) induced contralateral turning in rats with nigrostriatal lesions, suggesting that it was acting as a dopaminergic agonist. This action was antagonized by haloperidol. DMT, BOL, and methysergide were devoid of activity. LSD stimulates rat brain adenyl cyclase activity and this effect is antagonized by chlorpromazine, haloperidol, BOL, cyproheptadine and 1-methyl-LSD (von HUNGEN et al., 1974). LSD and BOL also antagonize the ability of NE and DA to stimulate rat brain adenyl cyclase activity (PALMER and BURKS, 1971; von HUNGEN et al., 1974). LSD and BOL increase DOPA brain levels of rats that have received a central decarboxylase inhibitor (HOLLUNGER and PERSSON, 1975). This effect of LSD and BOL was antagonized by apomorphine, and the LSD, but not the BOL response, was enhanced by haloperidol. PMA and MDA decrease uptake and increase release of NE from the rabbit atria, whereas, 2,5-DMA does not (PATON, 1975). The experiments of HORITA and HAMILTON (1973) also suggest that there may be a dopaminergic component to LSD's behavioral effects in the rabbit.

LSD does not alter the release of ACh from the cortex of the unanesthetized midpontine cat (DOMINO and BARTOLINI, 1972).

LSD enhances the activity of substance P on spinal cord reflex activity (KRIVOY, 1961). SMITH and WALASZEK (1962) found that LSD enhanced the activity of substance P, substance A, bradykinin, and ACh on the guinea pig ileum. KRIVOY (1957) found that LSD impeded the degradation of substance P while SMITH and WALASZEK (1962) were unable to find any evidence for inhibition of the degradation of substance P by LSD either in the rabbit brain or guinea pig ileum. NANDY and BOURNE (1964) have reported that LSD inhibits specific and nonspecific cholinesterase as well as monoamine oxidase in rat spinal cord and spinal ganglia. BERDE and CERLETTI (1956) observed that LSD expanded fish melanophores, that this activity was shared by other LSD congeners but was unrelated to either their antiserotonin activity or the hallucinogenic effect.

V. Distribution

LSD readily crosses the blood brain barrier and achieves its maximal concentration within minutes after administration. Only a small portion of the administered dose is found in the brain with higher concentrations being found in plasma and bile (STOLL et al., 1955; HALEY and RUTSCHMANN, 1957; AXELROD et al., 1957). Although its half-life varies between species (AXELROD et al., 1957) it is approximately 2 h in the monkey and cat. LSD is found throughout the brain, being present in the highest concentrations in the pituitary and pineal glands, the geniculates and colliculi, periaqueductal gray, hypothalamus, and limbic system (SNYDER and REIVICH, 1966). The gray matter has higher concentrations than the white (ARNOLD et al., 1958). Studies of the sites of binding using LSD-^{3}H and radioautography in the rat showed that LSD in the pineal and pituitary glands as well as the hippocampus was not specifically distributed while it was localized to the periphery (presumably the membrane) of certain cell bodies of the cerebellum (Purkinje cells), cerebral cortex, caudate nucleus, colliculi, olive, midbrain raphé, tegmentum, and reticular formation (DIAB et al., 1971). It had both a specific and nonspecific localization in the choroid plexus and granular cells of the cerebellum.

FARROW and VAN VUNAKIS (1972, 1973) have studied the binding of LSD to subcellular fractions of the rat brain. LSD binds to the synaptosomal and the synaptic membrane fractions of the rat cerebral cortex and to a lesser extent to the brain stem and to the myelin or mitochondrial fractions. Two binding sites with different affinities for LSD were identified. A variety of agents were found to inhibit LSD binding at the high affinity binding site including the LSD-like agonists mescaline, psilocin but not psilocybin, DMT and tryptamine, the antagonist ergonovine, and chlorpromazine, as well as 5-HT and bufotenine which are probably not LSD-like hallucinogens. L-LSD did not inhibit binding. MARCHBANKS (1967) had previously shown that D- and L-LSD, D-l-acetyl LSD, BOL, psilocybin and bufotenine but not mescaline inhibited the binding of 5-HT to nerve endings.

The distribution of intravenously administered mescaline-C^{14} in cat brain has also been studied using radioautography and regional analysis (NEFF et al., 1964). It is also widely distributed, localizing more in the gray than white matter. The highest concentrations were observed in the pituitary, pyriform cortex, and cerebral cortex.

D. Sites of Action

I. Spinal Cord

SLATER et al. (1955) found that LSD (0.1 mg/kg) enhanced the stimulatory action of 5-HT on the flexor reflex of the acute spinal cat. Low doses of LSD (10–20 μg/kg) facilitate the patellar, flexor, and crossed extensor reflexes of the unanesthetized and dial-urethane anesthetized acute spinal cat (WEIDMANN and CERLETTI, 1957; LITTLE et al., 1957; DISTEFANO and LEARY, 1958). Large doses (up to 100 μg/kg) inhibit the flexor and crossed extensor reflex (WEIDMANN and CERLETTI, 1957; DISTEFANO and LEARY, 1958). DISTEFANO and LEARY (1958) found that larger doses of LSD decreased inhibition of the flexor reflex produced by stimulating the contralateral tibial or femoral nerve but that this dysinhibition was unrelated to the facilitatory effects of LSD. KRIVOY (1961) found that LSD increased DR IV in the decerebrate spinal cat and that this increase was facilitated by substance P. Low doses of LSD (15 μg/kg) markedly facilitated the C-fiber reflex and modestly facilitated the segmental polysynaptic reflex of the acute decerebrate spinal cat (BELL and MARTIN, 1974) while large doses (200–300 μg/kg) facilitated the monosynaptic but inhibited the polysynaptic segmental reflex in the acute spinal cat (GEIGER and CERVONI, 1958). Psilocin, psilocybin, DMT, DET, tryptamine, α-ethyltryptamine, and U-22,394A also facilitated the flexor and patellar reflexes in the cat (WEIDMANN and CERLETTI, 1960; VANE et al., 1961; TANG et al., 1968). WEIDMANN and CERLETTI (1960) found that whereas psilocin and psilocybin facilitated the patellar reflex of the acute spinal cat, bufotenin and 5-HT inhibited it and the 6- and 7-hydroxy congeners were inactive. Further, 4-OH-tryptamines also decreased the amplitude of the patellar reflex. Other N-substituted 4-OH-tryptamines facilitated the patellar reflex.

TANG et al. (1968) found that DET and U-22,394A not only facilitated the flexor reflex but evoked rhythmic hindlimb movements in the acute spinal cat. LSD, psilocin, mescaline, DOM, DMA, methysergide and tryptamine facilitate the flexor reflex and evoke the stepping reflex in the chronic spinal dog (MARTIN and EADES, 1970; VAUPEL and NOZAKI, 1975). These effects were antagonized by cyproheptadine and chlorpromazine but not by phenoxybenzamine. LSD but not BOL and methysergide cause hindlimb extension in the acute spinal rat (ANDEN et al., 1968). Psilocybin, DMT, DOM, PMA, DMPEA, and to a lesser degree tryptamine enhance the extensor reflex of the acute spinal rat (ANDEN et al., 1971, 1974). Mescaline has little or no activity.

II. Pons Medulla

A number of changes caused by LSD-like hallucinogens are probably in part mediated by the brain stem including its effects on pulse rate, respiratory rate, and blood pressure. These stimulatory effects of LSD on sympathetic activity and respiration have not been well studied.

1. Cardiovascular Changes

Modest doses of LSD (ca. 25–50 μg/kg) produce a modest increase in blood pressure in the anesthetized intact rat and a larger increase in the pithed rat (SALMOIRAGHI et al., 1957). Mescaline produces bradycardia and hypoglycemia in the rat (SPECK, 1957). Psilocybin and bufotenine also increase blood pressure in the anesthetized rat and this effect is antagonized by BOL. LSD does not alter cardiac output in the rat but does increase blood flow in the frontal and parietal region of the cerebral cortex (GOLDMAN et al., 1975). In the pentobarbital anesthetized cat, modest to large doses of LSD cause a depressor response and bradycardia. The depressor response can be converted to a pressor response by ganglionic blocking drugs or pithing and the bradycardia partially decreased by vagotomy or atropine (SALMOIRAGHI et al., 1957; ROTHLIN, 1957a; CERVONI et al., 1963). LSD decreases blood pressure in the unanesthetized midpontine cat (DOMINO and BARTOLINI, 1972). In the unanesthetized acute spinal cat, LSD (7.5 μg/kg) increases blood pressure (KAY et al., 1976). In the unanesthetized cat intraventricular LSD produces a rapid decrease in blood pressure and a slowing of the heart (VOGT et al., 1957). In the unanesthetized spinal cat psilocin and psilocybin produce a modest pressor effect (WEIDMANN and CERLETTI, 1960). GINZEL (1955, 1957, 1958) studied the effects of LSD on vasomotor responses in the chloralose-vagotomized cat. In large doses LSD administered systemically and intraventricularly decreased vasopressor responses evoked by carotid occlusion, stimulation of the carotid chemoreceptor with Sebacinyl-bis-choline, potassium cyanide, and asphyxia, and sciatic nerve stimulation. LSD diminishes the baroreceptive pressor but not depressor response. LSD blocked chemoreceptor pressor responses in the decerebrate cat. GINZEL studied several other substances related to LSD on vascular reflexes and found that BOL, lysergic acid pyroledide, lysergic acid monoethylamide, and dihydroergotamine were also depressants when administered intraventricularly, whereas *l*-acetyl-LSD and mescaline were inactive. Tachycardia produced by LSD, mescaline, psilocin, and tryptamine in the chronic spinal dog is antagonized by cyproheptadine, chlorpromazine, and phenoxybenzamine (MARTIN and EADES, 1970).

2. Respiration

Although systemic administration of small doses of LSD stimulates respiratory rate in unanesthetized animals (see Subsect. B. II), it depresses respiration when administered intraventricularly to the chloralose-anesthetized vagotomized cat (GINZEL, 1957, 1958). Respiratory reflexes, however, are not affected by LSD. Cyproheptadine, chlorpromazine, and phenoxybenzamine tend to antagonize the respiratory stimulant effect of LSD, mescaline, psilocin, and tryptamine in the chronic spinal dog (MARTIN and EADES, 1970).

3. Chemoreceptive Trigger Zone

DHAWAN and GUPTA (1961) found that moderate to large doses of LSD antagonized the emetic effects of apomorphine, morphine, and Hydergine. LSD, mescaline, and tryptamine, on the other hand, are known to produce emesis in man (HOCH et al., 1952; MARTIN and SLOAN, 1970).

4. Tolerance

As discussed in Subsect. B. III, tolerance develops to the tachycardia, tachypnea, and vasopressor effects of LSD-like agents when they are administered chronically.

III. Mesencephalon

1. Effects on the Eye, Optic Tract, and Lateral Geniculate

EVARTS et al. (1955) found that LSD, but not bufotenine, in large doses, reduced the optic tract response to retinal light stimulation in the cat. Both LSD and bufotenine in modest dose levels markedly suppressed the lateral geniculate postsynaptic response but did not alter the response of the visual cortex to stimulation of the optic radiation. EVARTS (1958a) found that tryptamine and DMT but not 5-HT and BOL depressed transmission through the lateral geniculate of the cat. These findings were in part confirmed by BISHOP et al. (1958) who showed that BOL, mescaline, and 5-HT also depress short and longer latency postsynaptic lateral geniculate potentials. BISHOP et al. (1959) found that tetanic stimulation of the optic tract did not relieve the LSD depression of the lateral geniculate postsynaptic potential and concluded that LSD's action was to produce a post synaptic block of a naturally occurring excitatory neurotransmitter. In an extensive study, CURTIS and DAVIS (1962) microiontophoresed a large number of indoleamines onto neurons of the lateral geniculate of the pentobarbital anesthetized cat and found that 5-HT, as well as many other indoleamines (tryptamine, 4-HT, 6-HT, 7-HT, melatonin, psilocin, psilocybin, bufotenine and 5-HTP) and LSD congeners (LSD, BOL, methysergide, ergometrine, and methylergometrine but not ergotamine) decreased orthodromically evoked potentials. Phenethylamines were generally weak depressants; the most potent were dopamine, mescaline, NE, and tyramine. Most of the proven LSD-like hallucinogens were less potent than 5-HT, had a delayed onset of action, and their effects persisted longer than 5-HT. Many substances that were not LSD-like hallucinogens also had substantial depressant activity. The depressant effect of 5-HT on lateral geniculate transmission was not blocked by 5-HT antagonists nor enhanced by monoamine oxidase inhibitors. The above listed agents depressed orthodromic but not antidromic evoked lateral geniculate response and did not antagonize glutamate excitation of lateral geniculate cells. CURTIS and DAVIS (1962) concluded that these agents did not alter the conductance of the postsynaptic element but felt that they could be antagonizing either pre- or postsynaptically an excitatory transmitter. LSD produced a dose-related decrement of the spontaneous firing of and evoked activity in lateral geniculate units in the urethane-pentobarbital anesthetized cat (MCKAY and HORN, 1971; HORN and MCKAY, 1973). BOL had a lesser activity. LSD also increased the firing rate of some lateral geniculate units (MCKAY and HORN, 1971).

MOURIZ-GARCIA et al. (1969) found that LSD had no effect on the electroretinogram, both increased and decreased the activity of retinal ganglion cells, and had no effect on lateral geniculate units. APTER and PFEIFFER (1957) reported that LSD and mescaline increased spontaneous activity of the electroretinogram and in the optic tract of the pentobarbital-anesthetized cat. SCHWARTZ and CHENEY (1965a, b) have shown that LSD decreases spontaneous activity in the optic tract and lateral geniculate in both light and darkness in the pentobarbital-anesthetized and to a lesser degree in the unanesthetized cat. It does not markedly alter responsivity to changing stimuli.

2. Pupils

LSD-like hallucinogens dilate pupils in several species (see Subsect. B. II). This is probably a central action for pupillary dilation occurs when LSD-like drugs are administered to the ventricles. Systemically administered LSD does not appear to be effective as a mydriatic in the acutely decerebrated dog as it is in the intact dog (L.G. SHARPE and W.R. MARTIN, unpublished observations).

IV. Diencephalon-thermal Regulations

LSD-like hallucinogens increase body temperature in the rat, rabbit, cat, dog, monkey, and man. Depending on the species it increases both heat-generating (increased motor activity, hyperlipemia, and hyperglycemia) and heat-conserving mechanisms (piloerection, constriction of ear vessels) (CERLETTI, 1956; HORITA and HILL, 1972). LSD effects on body temperature are probably mediated by the diencephalon since decerebration but not decortication decreases its action (NEUHOLD et al., 1957). HORITA and DILLE (1954) first demonstrated that LSD increased body temperature in the rabbit and that antipyrine, dihydroergotamine, Hydergine, and Dibenamine did not antagonize this effect but anesthetic doses of pentobarbital did. BOL has only a modest and an inconstant pyretogenic effect and it antagonizes LSD's hyperthermic effect (HORITA and GOGERTY, 1958; JACOB and LAFILLE, 1963). JACOB and LAFILLE (1963) demonstrated that psilocybin, mescaline, DMT, and DET were also pyretogenic. In subsequent studies, JACOB (1965) found that bufotenine, psilocin, 5-HTP, and 5-methoxy-N-N′-dimethyltryptamine were also pyretogens and that tryptamine and 6-hydroxy-N-N′-dimethyltryptamine may be weak pyretogens. BRIMBLECOMBE (1967) studied a series of tryptamine derivatives and also found that intravenously injected DMT, DET, and bufotenine as well as other substituted tryptamine derivatives increased rectal temperature in the rabbit while 5-HTP did not. U-22,394A as well as DET were found to increase body temperature in the rabbit by TANG et al. (1968).

Curare and guanethidine, but curiously not hexamethonium, decrease the pyretogenic affects of LSD (ELDER and SHELLENBERGER, 1962; HORITA and HILL, 1972). BOL, chlorpromazine, phenoxybenzamine, chronic reserpine treatment, cinanserin, cyproheptadine and pCPA, but not imipramine or pimozide, also antagonize LSD's pyretogenic effect (HORITA and GOGERTY, 1958; ELDER and SHELLENBERGER, 1962; RUBIN et al., 1964; JACOB and LAFILLE, 1964). Because the pyretogenic effect of 5-HTP was attenuated in LSD-tolerant animals and by prior administration of BOL, HORITA and GOGERTY (1958) felt that a serotonergic mechanism was involved. HORITA and HAMILTON (1969) reported that α-MT antagonized the behavioral effects of LSD on behavior but not on body temperature in the rabbit. DOPA treatment restored the behavioral effects of LSD in α-MT treated rabbits and this restoration was not altered by inhibiting dopamine-β-hydroxylase (HORITA and HAMILTON, 1973). On closer examination α-MT treatment enhanced the hyperthermic effects of low doses of LSD but attenuated the effect of higher doses. pCPA enhances the pyretogenic effects of LSD and psilocybin while antagonizing the effects of amphetamine and harmaline in the rat (LADEFOGED, 1973). Haloperidol and thioridazine produce a modest antagonism of LSD's pyretogenic effect while markedly antagonizing the pyretogenic effect of apomorphine (ROSZELL and HORITA, 1975). Morphine antagonizes the pyretogenic effect of LSD in the rabbit, and with repeated administration of morphine, tolerance develops to this effect. This action is shared by some but not all morphinelike analgesics (DHAWAN, 1960).

IV. Cortex

1. Effects of LSD-like Hallucinogens on Cortical Neurons

LSD iontophoretically applied to cat cortical neurons has a weak but prolonged depressant effect. Iontophoretically applied mescaline both excites (KRNJEVIĆ and PHILLIS, 1963; ROBERTS and STRAUGHAN, 1968) and inhibits (BRADSHAW et al., 1971) cortical neurons of the pentobarbital-anesthetized cat. Sotalol, a β-blocker, and methysergide antagonize the excitatory effects. BEVAN et al. (1974) found that some cortical neurons excited by mescaline were also excited by NE and 5-HT but the response of neurons to mescaline and NE were most frequently the same. When neurons were desensitized to the excitant effect of mescaline, they showed decreased responsivity to NE and 5-HT. Sotalol antagonized the excitant but not the depressant effects of mescaline, while methysergide antagonized both the excitant and depressant effects.

2. Seizure Activity

TEDESCHI et al. (1959) studied the effects of a variety of drugs upon forelimb clonus in the rat produced by tryptamine and found that LSD, BOL, chlorpromazine, prochlorperazine, and trifluoperazine, but not mescaline, antagonized tryptamine clonus. Tryptamine clonus was enhanced by amphetaminelike agents and monoamine oxidase inhibitors. DET and U-22,394A produce tryptaminelike tremors and clonus in the rat (TANG et al., 1968). CLINESCHMIDT and LOTTI (1974), investigating the effects of several antagonists on tryptamine forelimb clonus and 5-HTP head twitch, found that methergoline and methysergide were more selective antagonists of tryptamine than were cyproheptadine and cinanserin.

3. Transcallosal Potentials

MARRAZZI and HART (1955) and MARRAZZI (1957) reported that LSD, mescaline, and bufotenine when injected close arterially decreased transcallosal potentials, as did serotonin, adrenochrome, amphetamine, and epinephrine, in the anesthetized cat. These results were partially confirmed by PURPURA (1956b) in the lightly anesthetized cat and by PICKWORTH and SHARPE (1976) in the unanesthetized decerebrate dog. PICKWORTH and SHARPE (1976) observed that LSD depressed the transcallosal potential, tryptamine had no effect, while DMT increased it. LANGFITT and FINNEY (1959b) did not observe a consistent effect of LSD on the transcallosal potential.

4. EEG, Cortical Evoked Potentials, and Sleep

a) Rat

LSD-induced head shaking in the rat is preceded by spindling in the cortex and lateral geniculate (DEPOORTERE and LOEW, 1971). LSD increases waking state and delays and decreases REM and slow wave sleep in the rat (LOEW et al., 1970; DEPOORTERE and LOEW, 1971). Mescaline produces EEG activation in the rat, and in large doses bursts of spikes associated with eye blinks (SPECK, 1958). In rats receiving mescaline chronically, an acute injection produced a slowing of the EEG. DOM desynchronizes and flattens the EEG of the rat (FLORIO et al., 1969). Topically applied DMPEA enhances cortical potentials evoked by forepaw pad stimulation (DEAR and MALCOLM, 1967).

b) Rabbit

The rabbit has been used extensively to study the effects of LSD-like hallucinogens on the EEG. Several investigators have found that LSD enhances the EEG activating effects of sensory stimuli, and large doses (50–100 μg/kg) activate the EEG (DELAY et al., 1952; VERDEAUX and MARTY, 1954; RINALDI and HIMWICH, 1955a, b; YUI and TAKEO, 1958; LONGO, 1962; KHAZAN and MCCASH, 1965). LSD also activates the EEG of encephale isolé rabbits but not rabbits whose brain stem has been transected at the postcollicular to postpontine level (SCHWEIGERDT et al., 1966), indicating a medullary site of action.

LSD desynchronizes hippocampal theta activity and inhibits hippocampal unit activity in unanesthetized curarized rabbits (BRÜCKE et al., 1961). LSD also desynchronizes hippocampal theta activity in rabbits pretreated with physostigmine but does not alter the firing pattern of driving septal units (STUMPF et al., 1962; STUMPF, 1964). The EEG activating effects of LSD are antagonized by atropine, chlorpromazine, pentobarbital (RINALDI and HIMWICH, 1955a), meprobamate, and possibly reserpine (YUI and TAKEO, 1958). Psilocybin produces EEG activation, behavioral arousal, and hyperglycemia (STEINER and SULMAN, 1963). 5-Methoxytryptamine, DMT, DET, 5-methoxydimethyltryptamine, bufotenine, *d*-lysergic acid monoethylamide, *d*-lysergic acid morpholide, mescaline, *d*-l-methyl-lysergic acid diethylamide, and *d*-lysergic acid dimethylamide also activate the rabbit EEG, while tryptamine, *d*-lysergic acid, *d*-isolysergic diethylamide, *l*-lysergic acid, *d*-Isolysergic diethylamide, *d*-lysergic acid diethylamide, BOL, and methysergide do not. On the basis of studies conducted in rabbits whose brain stems had been transected at various levels, it was concluded that LSD and 5-methoxytryptamine produced desynchronization at the mesencephalon level, mescaline, 3,4-dimethoxypenethylamine, DMT, bufotenine, and 5-methoxydimethyltryptamine at the medullary level, and the remainder at spinal level by enhanced afferent input (RINALDI and HIMWICH, 1955a; SCHWEIGERDT and HIMWICH, 1964; KHAZAN and MCCASH, 1965; TAKEO and HIMWICH, 1965, 1967; SCHWEIGERDT et al., 1966). DOMER and LONGO (1962) found that tryptamine flattened and desynchronized the EEG and produced hippocampal spiking. DE BARAN et al. (1963) found that tryptamine inhibited the firing of hippocampal units. Agroclavine and elymoclavine activate the EEG of the rabbit, while dihydroagroclavine and dihydroelymoclavine produce slow wave activity (YUI and TAKEO, 1958). FUJIMORI and HIMWICH (1968) found that DOM activated the EEG of the intact, postpontine, and spinal but not the mesencephalic rabbit and suggested that DOM inhibits a pontine synchronizing center. U-22,394A caused EEG activation in the intact and postpontine-transected rabbit and to a lesser degree in the prepontine preparation (TANG et al., 1968). Small doses of DOM activate the EEG and produce hippocampal theta activity, while larger doses produce EEG seizure activity. The activated EEG is associated with a variety of behavioral changes including movements, stupor, and catatonia, while the seizure activity is accompanied by localized tremors and clonus (FLORIO et al., 1969). There are conflicting observations on the effects of LSD on evoked cortical potentials in the rabbit. VERDEAUX and MARTY (1954) found that LSD depressed the transcallosal potential. KHAZAN and MCCASH (1965) found that LSD, DMT, and DET increased the latency and decreased the amplitute of evoked potentials in area I of the visual cortex. Further, potentials evoked in the frontal cortex by stimulation of the reticular formation were also depressed by LSD (KHAZAN and SAWYER, 1964). On the other hand, KOELLA and WELLS (1959) found that LSD decreased the latency and increased the amplitude of light-evoked potentials in the visual cortex. SMYTHIES et al. (1960) found that mescaline produced a dose-related enhancement of light-evoked potentials in the visual cortex; however, with larger doses (20–40 μg/kg) the enhancement was preceded by a short-lasting depression. ETEVENON and BOISSIER

(1972) found that LSD increased the signal-to-noise ratio for visually evoked potentials in the lateral geniculate and optic cortex of the rabbit.

c) Cat

Reports on the effects of LSD on the EEG of the cat seem to be conflicting. BRADLEY and ELKES (1957) reported that small doses of LSD desynchronized the EEG of the intact cat and the pentobarbital-anesthetized cat but not that of the encephale or cerveau isolé preparations. DOMINO and BARTOLINI (1972) observed an increase in slow wave activity in the unanesthetized midpontine cat. INGVAR and SODERBERG (1956) found that close arterially administered LSD caused an early desynchrony followed by hypersynchrony in the encephale isolé cat. TAKAGI et al. (1958) and KAY et al. (1976) found that LSD produced a modest desynchronization in the encephale isolé preparation. PASSOUANT et al. (1956) found that large doses of LSD produced cortical and hippocampal slow wave activity. LANGFITT and FINNEY (1959a) found that a decrease in frequency was the most consistent change in the EEG of the unanesthetized succinylcholine immobilized cat; whereas, a decrease in both the amplitude and frequency was seen in the pentobarbital-anesthetized cat. BARRATT and PRAY (1965) found that the predominant effect of large doses of LSD was cortical and hypothalamic desynchronization with occasional spindle and slow wave bursts. ADEY et al. (1962) found that moderate to large doses of LSD produced high voltage slow waves in the hippocampus, entorhinal cortex, amygdaloid, thalamus, midbrain reticular formation and, to a lesser extent, in the visual cortex and likened this slow wave activity to EEG seizure discharges. These EEG seizures were associated with a disruption in operant behavior. Psilocin and psilocybin also produced slow wave activity in the hippocampus, medulla and visual cortex, which was also associated with a disruption of operant behavior. WINTERS et al. (1969) found that LSD initially desynchronized the EEG, increased the firing rate of reticular units and this was followed by intermittent EEG hypersynchrony. ADEY et al. (1965) found that days after the administration of a dose of LSD there was an enhanced EEG synchrony in the hippocampus and entorhinal cortex during a discriminative performance. SCHWARZ et al. (1956a) found that small (15 µg) doses of LSD administered intraventricularly desynchronized the EEG, while VOGT et al. (1957) found that large doses (750 µg) produced slow waves in the cortex and fast waves in the putamen, pallidum, and amygdala.

LSD enhances and prolongs auditory EEG desynchronization but does not affect desynchronization evoked by stimulation of the reticular formation or peripheral nerves (KILLAM and KILLAM, 1956; BRADLEY and KEY, 1958; URSIN, 1962; KOSTOWSKI, 1971). Further, arousal produced by stimulation of the temporal lobe or amygdala is not changed by LSD (URSIN, 1962), while arousal produced by painful stimulation is shortened by LSD (KOSTOWSKI, 1971). KOSTOWSKI (1971) found that synchronization and desynchronization produced by low frequency stimulation of the reticular formation or the pontine raphé were decreased by LSD. Both behavioral and EEG arousal produced by LSD are antagonized by chlorpromazine (BRADLEY and HANCE, 1957). Most investigators have found that the EEG effects of LSD depend to some degree on the behavioral state of the cat—when behaviorally alert, a desynchronized EEG is obtained—when quiet, slow wave activity prevails (BRADLEY and ELKES, 1957; EVARTS, 1958b). In the unanesthetized or curarized cat, LSD does not markedly affect the recruiting response (KILLAM and KILLAM, 1956; EVARTS, 1958b). However, in cats that have been anesthetized with pentobarbital or ether, LSD decreases the recruiting response (PURPURA, 1956b). EEG activity evoked by a pulsed (flickering) conditioned visual stimuli was enhanced by LSD when correct responses were made (LINDSLEY et al., 1968). The spontaneous EEG had more bursts of slow wave activity. Postreinforcement EEG synchrony and the reward contingent positive variation which are seen only when the cat is subjected to diffused light after performing an operant task can be observed in complete darkness following LSD (MARCZYNSKI, 1972). LSD, mescaline, psilocin, and amphetamine decrease the cortical secondary response evoked by sciatic nerve stimulation of the dial-urethane anesthetized cat (ROTH, 1966).

DMT produces bursts of 6–10 Hz high voltage activity in the intact cat (VANE et al., 1961). DOM increases firing of reticular neurons and produces a mixture of hypersynchrony and desynchrony of the EEG of the cat (FLORIO et al., 1969; WALLACH et al., 1972). Intraperitoneally administered mescaline hypersynchronized the EEG and did not alter the firing rate of reticular units (WINTERS et al., 1969). Intraventricularly administered mescaline produces EEG seizure activity (SCHWARZ et al., 1956a) as does mescaline applied topically to the cerebral cortex (ROVETTA, 1956; CRIGHEL and STOICA, 1961, 1962; MISON-CRIGHEL et al., 1964). No marked changes in cortical glutamine, glutamic acid, or GABA were associated with mescaline seizures (MISON-CRIGHEL et al., 1964). The convulsant effects of mescaline are enhanced by stimulation of the reticular formation and antagonized by chlorpromazine (CRIGHEL and STOICA, 1961).

Small to moderate doses of LSD (2–30 μg/kg) enhance potentials evoked in the auditory and visual cortex by respectively clicks and flashes of light in both the cat and rabbit (PURPURA, 1956b; KOELLA and WELLS, 1959). Larger doses of LSD (40–60 μg/kg) decrease the amplitude of the auditory potential. The effects of LSD on evoked potentials are antagonized by reserpine. As discussed above, LSD inhibits recruiting and transcallosal responses in addition to secondary cortical discharges evoked by sciatic nerve stimulation. ROVETTA (1956), on the other hand, was unable to see any marked effect of LSD on visual evoked potentials with either systemically administered or topically applied LSD in the chloralose-urethane-anesthetized cat.

Topically applied mescaline enhances cortical and lateral geniculate potentials evoked by direct cortical stimulation as well as by visual and auditory stimuli (ROVETTA, 1956; KREINDLER et al., 1962; CRIGHEL and STOICA, 1962). LANGFITT and FINNEY (1959a) did not see any significant changes in auditory, visual, or tactile evoked potentials in either the unanesthetized or pentobarbital-anesthetized cat.

KEY (1961a) found that LSD decreased sleep in cats and habituation to sensory stimuli. HOBSON (1964), STERN et al. (1972), and KAY et al. (1976) have found that LSD decreased slow wave and REM sleep and increased waking. LSD increases PGO spikes in slow wave sleep and decreases them during REM sleep. BOL does not alter sleep in the cat (STERN et al., 1972), while tryptamine produces changes similar to LSD (KAY et al., 1976).

d) Monkey

LSD increases the frequency and decreases the amplitude of the EEG of the baboon (WALTER et al., 1971) and produces paroxysmal bursts in the hippocampus and septum of the rhesus monkey (MONROE and HEATH, 1961). Psilocybin, DMT, and methysergide produce paroxysmal slow wave activity in the EEG of the baboon, while mescaline has little effect (MELDRUM and NAQUET, 1971). LSD does not alter the electroretinogram or optic tract evoked potential but decreases visually evoked potential in the lateral geniculate and visual cortex of the baboon (WALTER et al., 1971). Psilocybin also reduces lateral geniculate and cortical evoked potentials (MELDRUM and NAQUET, 1971).

e) Man

There is general agreement that LSD increases the frequency of alpha rhythms (RINKEL et al., 1952; GASTAUT et al., 1953; SHAGASS, 1967; KOREIN and MUSACCHIO, 1968) and reduces the amplitude of the EEG (MONROE et al., 1957; RODIN and LUBY, 1966; SHAGASS, 1967). ELKES et al. (1954) found that LSD tended to decrease alpha responsivity. GOLDSTEIN et al. (1963) found that LSD (1 μg/kg) decreased electrogenesis of the EEG of prisoner patients but not of schizophrenic patients. SCHWARZ et al. (1956b) observed an increase in paroxysmal activity recorded from depth electrodes of schizophrenic patients. GASTAUT et al. (1953) and KOREIN and MUSACCHIO (1968) found that LSD increased beta activity and KOREIN and MUSACCHIO (1968) also observed an increase in slow wave focal activity in patients who had cerebral lesions. MONROE et al. (1957) found that LSD and mescaline induced paroxysmal activity in the septum, hippocampus, and amygdala. WIKLER (1954) found that hallucinogenic doses of mescaline desynchronized the EEG and accelerated alpha activity. Psilocybin also increases the frequency of alpha activity in some subjects (MALITZ et al., 1960). MERLIS and DENBER (1956) reported that mescaline decreased alpha activity and that it was restored by chlorpromazine. DET increases the frequency and diminishes the amplitude of alpha activity (BOSZÖRMÉNYI et al., 1959).

All studies of the effects of LSD on cortical evoked potentials in man have indicated a decrement. CHAPMAN and WALTER (1965) found that LSD reduced potentials evoked in the visual cortex, particularly those components that had a frequency of 6–12 Hz. RODIN and LUBY (1966) found that LSD reduced both primary evoked potential of the visual cortex and the associated rhythmic afterdischarge. Similarly, SHAGASS (1967) observed that LSD decreased the amplitude and increased the frequency (as well as decreased late-coming slow wave activity) of both visual and somasthetic evoked potentials. GREEN (1965) found that a large dose of LSD delayed the onset of REM sleep, increased the duration of REM episodes, and increased vocalization during REM episodes.

E. Conclusions and Discussion

I. Pharmacologic Characteristics of LSD-like Hallucinogens

A large body of data indicates that there is a group of hallucinogenic drugs which produce a characteristic syndrome in man consisting of changes in perception, mood, behavior, autonomic function, and somatomotor reflexes. When many of these drugs are administered chronically, tolerance is induced to their effects. The effects of these LSD-like hallucinogens on behavior are clearly pernicious for they disrupt the function of the central nervous system, decreasing the ability of the user to perform a variety of tasks, distorting his ability to perceive, and giving rise to delusions and hallucinations. The effects of LSD on performance, autonomic function, and reflex activity have also been demonstrated in several species of animals. The perceptual and mood changes produced by LSD, of course, cannot be demonstrated in animals; however, certain enteroceptive effects can be identified. There are many changes in behavior and autonomic function that are produced by LSD-like agents that are also produced by drugs that do not cause hallucinations, LSD-like or otherwise. Further, some of these effects are antagonized by drugs that do not antagonize LSD hallucinatory effects. These behavioral effects include head shakes or twitches in the mouse, retropulsion, delay of swimming and other behavioral effects in the rat. Several of the important actions of LSD in man can be investigated in animals, and with sufficiently broad-based studies, LSD-like hallucinogens can be identified in animals with probably a high degree of certainty.

LSD-like hallucinogens produce euphoria in man; however, they do not appear to have positive reinforcing properties in other animals. Most investigators have found that neither rats nor monkeys will initiate bar-pressing behavior for LSD, mescaline, or psilocin.

II. LSD Hallucinogens-Agonists

The strongest argument that LSD-like hallucinogens act as agonists is that many of their actions can be markedly and competitively antagonized by blockers that in themselves do not or minimally affect the parameters under investigation (MARTIN and EADES, 1970). Further, substantive evidence has been obtained that LSD-like hallucinogens mimic the effects of the naturally occurring active brain amine, tryptamine (Subsect. C. III).

III. Tolerance

There are no well-established theories of LSD-induced tolerance, and as a consequence, critical experiments have not been performed. At this time, speculation about the nature of tolerance must come from consideration of existing data. In the subsequent speculations, the following facts are of importance: (1) LSD and related hallucinogens are agonists; (2) tolerance to their agonistic actions develops rapidly; (3) long-acting agonists seem to be more effective than shorter-acting agonists in producing tolerance; (4) tolerance is lost rapidly following the discontinuation of the hallucinogen; and (5) no signs of abstinence have been observed following withdrawal.

Since there are no signs of abstinence in LSD-tolerant subjects, it is reasonable to assume that neither contraadaptive mechanisms are recruited or hypertrophy of redundant pathways has occurred (HIMMELSBACH, 1943; MARTIN, 1968) nor has LSD activated inhibitory pathways which initiate the development of denervation sensitization (JAFFE and SHARPLESS, 1965).

Both the development of tolerance and the absence of physical dependence would argue against the development of spare active receptors (COLLIER, 1965). One can thus assume that some sort of adaptation has occurred at the LSD receptor. Because tolerance is rapidly established and lost, it is unlikely that spare inactive receptors are induced (COLLIER, 1965). The observations that long-acting LSD agonists such as LSD are apparently more effective in inducing tolerance than short-acting agonists such as DMT and tryptamine would be consistent with both the rate theory of PATON (1961) and the potential theory of STRAUB (1905). Long-acting LSD agonists would be thought to dissociate from the receptor less rapidly than shorter-acting agonists. The fact that there are apparently competitive antagonists of LSD with a duration of action comparable to that of LSD that are devoid of agonistic activity does not support the rate theory. The very existence of competitive antagonists, of course, is not in keeping with the potential theory. This reasoning has led the authors to postulate that an unknown type of adaptation to the agonistic action of LSD- and tryptaminelike substances occurs when the agonist remains on the receptor. This adaptation may be analogous to that seen with depolarizing blockers of ACH. The presence of the agonist on the receptor not only induces adaptation but prevents another agonist from occupying the receptor and in this respect functions as a competitive antagonist. The fact that there are probably differences in the rate that tolerance develops to various effects of LSD in different species may reflect the multiple modes of action of LSD and a difference in the structure of the receptors in different species.

IV. Modes of Action

On the basis of data currently available, it would seem that LSD-like hallucinogens have multiple modes and sites of action. The mode of action of LSD and related drugs may be different at different sites in the same animal and in different species. Further, there are subtle differences in the actions of LSD-like hallucinogens of differing structures. There is a growing body of evidence that many LSD-like drugs may interact with several types of receptors. Thus, there is strong evidence that LSD interacts with both 5-HT and tryptamine receptors in the brain. Certain phenethylamines and amphetamine congeners may well interact with both tryptamine receptors and amphetamine receptors. The latter receptors are not well characterized at this time. Recently, MARTIN and EADES (1974) have presented evidence of a phenethylaminergic receptor in the dog spinal cord which could be distinguished from α-adrenergic and tryptaminergic receptors. Some, but not all, of the actions of amphetamines and some LSD-like hallucinogens may be explained by their interaction with phenethylaminergic receptors. Some of the actions of substituted amphetamines (e.g., MMDA, see Subsect. B. II. 2) are different from those of tryptamine, methoxamine, phenethylamine, and amphetamine and cannot be readily explained on the basis of interactions with tryptaminergic, α-noradrenergic, or phenethylaminergic receptors. Thus, there may be other as yet unidentified receptors which have not been characterized and which may or may not have endogenous transmitters. These receptors must be structurally related to tryptaminergic, serotonergic, and phenethylaminergic receptors and have pharmacologic significance. They may also have pathophysiologic significance.

The 5-HT antagonist explanation of the hallucinatory action of LSD seems to be untenable at this time. The principal reason for rejecting this hypothesis is that there are many serotonin antagonists which are not hallucinogens. BOL, a close congener of LSD and a 5-HT antagonist, was one of the first critical drugs which did not produce hallucinations. Further, mescaline, an LSD-like hallucinogen, does not antagonize the excitatory actions of 5-HT on brain neurons.

Although LSD has been shown to have serotonergic properties in some peripheral tissues, the most relevant observations that implicate serotonin in LSD actions are that LSD increases brain levels of 5-HT by inhibiting serotonergic neurons with a consequent decrease in its release and turnover. Two suggestions have been made to explain these observations: (1) LSD may act as a 5-HT agonist at 5-HT synapses, thus increasing negative feedback to serotonergic neurons, (2) LSD may have a direct inhibitory action on serotonergic neurons. The fact that LSD inhibits raphé units when applied directly to them favors the latter hypothesis. The behavioral significance of these observations is not clear. BOL also inhibits raphé units, while mescaline and DOM inhibit some but not all raphé units. It is probable that a serotonergic mechanism is involved in the hyperthermic actions of LSD and that LSD and related drugs are acting postsynaptically as agonists. 5-HT is probably not involved in the disruption of behavior in the rat caused by LSD.

The concordance between the action of LSD-like hallucinogens and tryptamine on spinal cord reflex activity is excellent, and it is highly probable that LSD-like hallucinogens exert their spinal cord facilitatory effects by acting as tryptaminergic agonists. The existing evidence suggests that a tryptaminergic mechanism may be responsible for the hallucinatory effects of LSD; however, the data on this issue are fragmentary. Because of tryptamine's very short duration of action, experiments comparing it with LSD and related drugs must be conducted with this property in mind. To date, appropriate experiments comparing the autonomic, behavioral, and subjective effects of tryptamine and LSD have not been conducted.

There is neurochemical evidence that LSD not only stimulates catecholaminergic neurons but may act as a dopaminergic agonist. This latter action is shared by other ergot alkaloids which are not hallucinogens. Further, DMT is not a dopaminergic agonist. LSD and BOL also antagonize the stimulatory actions of NE and DA on adenylcyclase activity. The relationship of these diverse actions of LSD on catecholaminergic brain processes and its pharmacologic actions is not clear at this time. Noradrenergic processes do not appear to be involved in LSD's effects on mood, perception, or spinal cord reflex activity since α-adrenergic antagonists do not alter these effects of LSD. A noradrenergic process may be involved in the disruption of behavior in the rat and rabbit caused by LSD. Noradrenergic processes may be involved in some autonomic effects of LSD; however, the site of action of α-adrenergic antagonists has not been identified and they may be exerting their effects centrally, peripherally or both. It is possible that there is a catecholaminergic mechanism in the pyretogenic actions of LSD. Table 6 summarizes the role that various transmitters may have in the actions of LSD.

V. Mechanisms of Action

The mechanisms responsible for the agonistic actions of LSD have not been clearly identified. LSD applied directly to raphé, lateral geniculate, and cortical neurons inhibits their activity. PURPURA (1956b) demonstrated that LSD inhibited axodendritic synapses. The inhibitory action of LSD on raphé units is only partially shared

Table 6. Possible role of neurotransmitters in the actions of LSD hallucinogens

	5-HT	NE	T	DA
Hallucinations	Doubtful	Doubtful	Possible	Doubtful
Hyperthermia	Probable	Possible	?	Possible
Spinal cord reflex facilitation	Doubtful	Doubtful	Probable	Doubtful
Disruption of behavior	Doubtful	Possible	?	Possible

by mescaline. Further, BOL which does not produce hallucinations also inhibits raphé neurons. Many indoles and lysergic acid derivatives which are not hallucinogens inhibit lateral geniculate neurons, while the hallucinogenic phenethylamine mescaline is relatively inactive. Thus, it is unlikely that the neuron inhibitory effects of LSD are related to its hallucinatory activity. Further, it does not seem probable that these inhibitory actions are responsible for the stimulatory effects of the LSD-like hallucinogens on spinal cord reflex activity or the autonomic nervous system.

Little is known about the mechanisms underlying the ability of LSD to disrupt behavior. The observation that BOL also disrupts behavior in the rat would argue that a different mechanism is involved in this effect than in hallucinatory phenomena. Alternatively, the rat LSD receptor may differ stereochemically from the human LSD receptor. Behavioral disruption is associated with changes in the electrical activity of the limbic system of the cat.

A great many neurophysiologic studies have been done attempting to define and understand the effects of LSD on perception and mood, and these have been authoritatively reviewed (ELKES, 1961; PURPURA, 1967). LSD effects on the visual system are paradoxical. It has a direct excitatory effect on the retina and in small doses facilitates transmission at the lateral geniculate, but in large doses inhibits transmission. A possible explanation of these paradoxical observations is that LSD is a partial agonist of a naturally occurring transmitter; however, one of the authors (W.R.M.) has attempted to test this hypothesis using the flexor reflex of the chronic spinal dog and has found no evidence for it. The stimulatory effects of LSD on the spinal cord, lateral geniculate body, and the autonomic nervous system may be related to FISCHER'S (1968) concept that hallucinogens cause time contraction and a related increase in data content of a stimulus as perceived by the brain. To rephrase and somewhat oversimplify this concept—hallucinogens decrease the amount of sensory input necessary for perception. Hallucinations, in this framework, may be viewed as a perception that occurs without sensory stimulation. BRADLEY and ELKES (1957) first proposed that LSD increased the input of sensory stimuli into the reticular formation. This formulation gives rise to yet another paradox relating to LSD's hallucinogenic activity. LSD-like hallucinogens activate the EEG. It is well-known that activation and sensory stimulation inhibit the development of hallucinatory phenomena in patients while under the influence of LSD-like hallucinogens, while a paucity of stimuli facilitates the development of hallucinations.

It would appear that the EEG may be activated through the stimulation of the reticular formation. Some hallucinogens stimulate the reticular formation at the mesencephalic level; others at the medullary or spinal level; while yet others may produce activation by inhibiting brain stem synchronizing centers. The activation of the EEG would not seem to be a ready explanation of LSD's hallucinatory actions as argued above. It could, however, be related to their eurphoria-producing properties. It is worthwhile to point out that amphetaminelike drugs are also euphoriants and activate the EEG. The EEG-desynchronizing effects of LSD may be related to its ability to disrupt sleep. Some presumed LSD-like hallucinogens produce overt EEG seizure. Most LSD-like hallucinogens produce slow wave activity and WINTERS (1967) and WINTERS et al. (1969) have suggested that the slow wave activity may be related to hallucinations.

There have been a number of suggestions as to how LSD-like hallucinogens act. Most of the existing evidence suggests that they produce at least their acute effects at the nerve membrane, probably at a synaptic region by interacting with receptor sites of naturally occurring transmitters, and either act as agonists or antagonists. The type of interaction is unknown. FALKENBERG (1972) feels that there is not adequate biologic data to allow the identification of molecular structures that give rise to LSD-like activity. A number of interesting hypotheses have been advanced. SMYTHIES (1970) and SMYTHIES et al. (1970) have suggested that LSD-like hallucinogens may antagonize the interaction of 5-HT with either membrane RNA or intracellular RNA or DNA. This interaction could alter protein synthesis or the intimate structure of the membrane. The importance of both lipophilicity of LSD-like hallucinogens perhaps involving π electron interactions (SNYDER and RICHELSON, 1968) and of alteration of the electron charge density of the benzene ring (KANG and GREEN, 1970a, b) have been emphasized in facilitating steric configurations (COOPER and WALTERS, 1972).

VI. Abuse Potential and Identification of LSD-like Drugs

LSD-like hallucinogens continue to be extensively abused in the United States. The incidence of serious drug reactions as indicated by the DAWN Management Report of July 1975 (U.S. Department of Justice, Drug Enforcement Administration) would indicate that LSD abuse (484 cases) is comparable in magnitude to abuse of barbiturates (513 cases) and amphetamines (694 cases) and greater than the abuse of cocaine (249 cases). There is ample evidence that LSD-like hallucinogenes disrupt behavior and distort perceptual processes. Through the use of a variety of techniques, drugs with LSD-like properties can be identified (MARTIN, 1973). Not only do these agents produce a characteristic syndrome, they in all probability produce this syndrome by interacting with specific receptors and evidence has accumulated that the actions of LSD are mediated through a tryptamine receptor. Cyproheptadine appears to be a relatively selective competitive antagonist of the agonistic actions of tryptamine and other LSD-like hallucinogens (MARTIN and SLOAN, 1974; MARTIN et al., 1976), is useful in further characterizing the mode of tryptaminergic LSD-like hallucinogens and can assist in distinguishing them from noradrenergic and phenethylaminergic agonists. Finally, LSD-tolerant animals and humans have assisted in further characterizing LSD-like hallucinogens. Additional studies of LSD-induced tolerance need to be performed to resolve some of the paradoxical data, to determine if the LSD receptors responsible for tolerance are the same in all species, and whether the ability to induce tolerance is related to the duration of action of LSD-like hallucinogens.

Most Usual Abbreviations

α-MT	α-Methyl-p-tyrosine
ACh	Acetylcholine
ARCI	Addiction Research Center Inventory
BOL	*D*-2-Bromlysergic acid diethylamide
CPZ	Chlorpromazine
DA	Dopamine
DET	Diethyltryptamine
DMA	Dimethoxyamphetamine
DMPEA	3,4-Dimethoxy-β-phenethylamine
DMT	Dimethyltryptamine
DOET	2,5-Dimethoxy-4-ethylamphetamine
DOM	2,5-Dimethoxy-4-methylamphetamine
DOPA	3-(3,4-Dihydroxyphenyl)-*L*-alanine
DPT	Dipropyltryptamine
6-FDET	6-Fluorodiethyltryptamine
FI	Fixed interval
FR	Fixed ratio
GABA	γ-Aminobutyric acid
5-HIAA	5-Hydroxyindole acetic acid
5-HTP	5-Hydroxytryptophan
5-HT	Serotonin
JB-318	N-ethyl-3-piperidyl benzylate
LAE	*D*-lysergic acid monoethylamide
LSD	*D*-lysergic acid diethylamide
MDA	3,4-methylene-dioxyamphetamine
MMDA	5-methoxy-3,4-methylene-dioxyamphetamine
MMPI	Minnesota Multiphasic Personality Inventory
pCPA	p-Chlorophenylalanine
PEA	β-phenethylamine
PGO	Pontine geniculate occipital
REM	Rapid eye movement
Δ^9-THC	Δ^9-Tetrahydrocannabinol
TMA	Trimethoxyamphetamine
TMPEA	Trimethoxyphenethylamine
U-22,394A	(a 1,2-di-substituted N-alkyltryptamine)

References

Abramson, H. A.: Lysergic acid diethylamide (LSD-25): XXIX. The response index as a measure of threshold activity of psychotropic drugs in man. J. Psychol. (Provincetown) **48**, 65—78 (1959)

Abramson, H. A., Jarvik, M. E., Gorin, M. H., Hirsch, M. W.: Lysergic acid diethylamide (LSD-25): XVII. Tolerance development and its relationship to a theory of psychosis. J. Psychol. (Provincetown) **41**, 81—105 (1956)

Abramson, H. A., Jarvik, M. E., Hirsch, M. W.: Lysergic acid diethylamide (LSD-25): VIII. Effect upon two measures of motor performance. J. Psychol. (Provincetown) **39**, 455—464 (1955c)
Abramson, H. A., Jarvik, M. E., Hirsch, M. W.: Lysergic acid diethylamide (LSD-25): X. Effect on reaction time to auditory and visual stimuli. J. Psychol. (Provincetown) **40**, 39—52 (1955d)
Abramson, H. A., Jarvik, M. E., Hirsch, M. W., Ewald, A. T.: Lysergic acid diethylamide (LSD-25): V. Effect on spatial relations abilities. J. Psychol. (Provincetown) **39**, 435—442 (1955b)
Abramson, H. A., Jarvik, M. E., Kaufman, M. R., Kornetsky, C., Levine, A., Wagner, M.: Lysergic acid diethylamide (LSD-25). I. Physiological and perceptual responses. J. Psychol. (Provincetown) **39**, 3—60 (1955a)
Abramson, H. A., Rolo, A., Sklarofsky, B., Stache, J.: Production of cross-tolerance to psychosis-producing doses of lysergic acid diethylamide and psilocybin. J. Psychol. (Provincetown) **49**, 151—154 (1960)
Abramson, H. A., Sklarofsky, B.: Lysergic acid diethylamide (LSD-25) antagonists. Arch. gen. Psychiat. **2**, 88—93 (1960)
Abramson, H. A., Sklarofsky, B., Baron, M. O., Fremont-Smith, N.: Production of tolerance to psychosis-producing doses of lysergic acid diethylamide. Science **15**, 1020 (1957)
Abramson, H. A., Sklarofsky, B., Baron, M. O., Fremont-Smith, N.: Lysergic acid diethylamide (LSD-25) antagonists. II. Development of tolerance in man to LSD-25 by prior administration of MLD-41 (1-methyl-*d*-lysergic acid diethylamide). Arch. Neurol. Psychiat. (Chic.) **79**, 201—207 (1958)
Abramson, H. A., Waxenberg, S. E., Levine, A., Kaufman, M. R., Kornetsky, C.: Lysergic acid diethylamide (LSD-25). XIII. Effect on Bender-Gestalt Test performance. J. Psychol. (Provincetown) **40**, 341—349 (1955e)
Adey, W. R., Bell, F. R., Dennis, B. J.: Effects of LSD-25, psilocybin, and psilocin on temporal lobe EEG patterns and learned behavior in the cat. Neurology (Minneap.) **12**, 591—602 (1962)
Adey, W. R., Porter, R., Walter, D. O., Brown, T. S.: Prolonged effects of LSD on EEG records during discriminative performance in cat: Evaluation by computer analysis. Electroenceph. clin. Neurophysiol. **18**, 25—35 (1965)
Aghajanian, G. K., Foote, W. E., Sheard, M. H.: Lysergic acid diethylamide: Sensitive neuronal units in the midbrain raphe. Science **161**, 706—708 (1968)
Aghajanian, G. K., Foote, W. E., Sheard, M. H.: Action of psychotogenic drugs on single midbrain raphe neurons. J. Pharmacol. exp. Ther. **171**, 178—187 (1970)
Aghajanian, G. K., Haigler, H. J., Bloom, F. E.: Lysergic acid diethylamide and serotonin: Direct actions on serotonin containing neurons in rat brain. Life Sci. **1**, 615 (1972)
Aghajanian, G. K., Weiss, B. L.: Block by LSD of the increase in brain serotonin turnover induced by elevated ambient temperature. Nature (Lond.) **220**, 759—796 (1968)
Alles, G. A.: Some relations between chemical structure and physiological action of mescaline and related compounds. In: Abramson, H. A. (Ed.): Neuropharmacology. New York: Josiah Macy Jr. Foundation 1959
Altman, J. L., Appel, J. B.: Effects of LSD on fixed-interval responding in the rat. Proceedings, 79th Annual Convention, American Psychological Association, 1971
Altman, J. L., Appel, J. B.: LSD and fixed-interval responding in the rat. Pharmacol. Biochem. Behav. **3**, 151—155 (1975)
Amin, A. H. T., Crawford, T. B. B., Gaddum, J. H.: Distribution of substance P and 5-hydroxytryptamine in central nervous system in dog. J. Physiol. (Lond.) **126**, 596—618 (1954)
Auden, N.-E., Corrodi, H., Fuxe, K.: Hallucinogenic drugs of the indolealkylamine type and central monoamine neurons. J. Pharmacol. exp. Ther. **179**, 236—249 (1971)
Auden, N.-E., Corrodi, H., Fuxe, K., Hökfelt, T.: Evidence for a central 5-hydroxytryptamine receptor stimulation by lysergic acid diethylamide. Brit. J. Pharmacol. **34**, 1—7 (1968)
Auden, N.-E., Corrodi, H., Fuxe, K., Meek, J. L.: Hallucinogenic phenylethylamines: Interactions with serotonin turnover and receptors. Europ. J. Pharmacol. **25**, 176—184 (1974)
Angrist, B., Rotrosen, J., Gershon, S.: Assessment of tolerance to the hallucinogenic effects of DOM. Psychopharmacologia (Berl.) **36**, 203—207 (1974)
Appel, J. B.: The effects of "psychotomimetic" drugs on animal behavior. In: Efron, D. H. (Ed.): Psychopharmacology; A Review of Progress, 1957—1967. PHS Publication No. 1836. Washington, D. C.: U. S. Government Printing Office 1968
Appel, J. B.: Effects of LSD on time-based schedules of reinforcement. Psychopharmacologia (Berl.) **21**, 174—186 (1971)

Appel, J. B., Freedman, D. X.: Chemically-induced alterations in the behavioral effects of LSD-25. Biochem. Pharmacol. **13**, 861—869 (1964)
Appel, J. B., Freedman, D. X.: Tolerance and cross-tolerance among psychotomimetic drugs. Psychopharmacologia (Berl.) **13**, 267—274 (1968)
Appel, J. B., Freedman, D., X., Filby, Y. M.: The effects of three psychoactive drugs on two varieties of escape behavior. Arch. int. Pharmacodyn. **167**, 179—193 (1967)
Appel, J. B., Lovell, R. A., Freedman, D. X.: Alterations in the behavioral effects of LSD by pretreatment with *p*-chlorophenylalanine and α-methyl-*p*-tyrosine. Psychopharmacologia (Berl.) **18**, 387—406 (1970)
Appel, J. B., Whitehead, W. E., Freedman, D. X.: Motivation and the behavioral effects of LSD. Psychon. Sci. **12**, 305—306 (1968)
Apter, J. T., Pfeiffer, C. C.: The effect of the hallucinogenic drugs LSD-25 and mescaline on the electroretinogram. Ann. N. Y. Acad. Sci. **66**, 508—514 (1957)
Arnold, O. H., Hofmann, G., Leupold-Löwenthal, H.: Untersuchungen zum Schizophrenieproblem. IV. Mitteilung: Die Verteilung des C-14-radioaktiven lyserg-säurediäthylamids (C-14-LSD-25) im tierischen Organismus. Wien. Z. Nervenheilk. **15**, 15—27 (1958)
Aronson, H., Klee, G. D.: Effect of lysergic acid diethylamide (LSD-25) on impulse control. J. nerv. ment. Dis. **131**, 536—539 (1960)
Aston, R., Cullumbine, H.: The effects of combinations of ataraxics with hypnotics, LSD and iproniazid in the mouse. Arch. int. Pharmacodyn. **126**, 219—227 (1960)
Axelrod, J., Brady, R. O., Witkop, B., Evarts, E. V.: The distribution and metabolism of lysergic acid diethylamide. Ann. N. Y. Acad. Sci. **66**, 435—444 (1957)
Bailey, P. T., Pradhan, S. N.: Effects of Δ^9-tetrahydrocannabinol and mescaline on self-stimulation. Neuropharmacology **11**, 831—838 (1972)
Balestrieri, A.: Crossed tolerance between LSD-25 and mescaline. In: Garattini, S., Ghetti, V. (Eds.): Psychotropic Drugs. Amsterdam: Elsevier Publishing Company 1957
Balestrieri, A.: Studies on cross tolerance with LSD-25, UML-491 and JB-336. Psychopharmacologia (Berl.) **1**, 257—269 (1960)
Balestrieri, A., Fontanari, D.: Acquired and crossed tolerance to mescaline, LSD-25, and BOL-148. Arch. gen. Psychiat. **1**, 279—282 (1959)
Barratt, E. S., Pray, S. L.: Effect of a chemically depressed amygdala on the behavioral manifestations produced in cats by LSD-25. Exp. Neurol. **12**, 173—178 (1965)
Becker, D. I., Appel, J. B., Freedman, D. X.: Some effects of lysergic acid diethylamide on visual discrimination in pigeons. Psychopharmacologia (Berl.) **11**, 354—364 (1967)
Bell, J. A., Martin, W. R.: Studies of tryptamine and lysergic acid diethylamide (LSD) on cutaneous C-fiber and polysynaptic reflexes in the cat. J. Pharmacol. exp. Ther. **190**, 492—500 (1974)
Bell, J. A., Martin, W. R., Sloan, J. W., Buchwald, W. F.: The effect of L-tryptophan on spinal cord C-fiber reflexes. J. Pharmacol. exp. Ther. **196**, 373—379 (1976)
Belleville, R. E.: MMPI score changes induced by lysergic acid diethylamide (LSD-25). J. clin. Psychol. **12**, 279—282 (1956)
Bennett, I. P. and Snyder, S. H.: Serotonin and lysergic acid diethylamide binding in rat brain membranes; relationship to postsynaptic serotonin receptors. Mol. Pharmacol. **12**, 373—389 (1976)
Berde, B., Cerletti, A.: Über den Melanophoreneffekt von d-lysergsäure-diäthylamid und verwandten Verbindungen. Helv. physiol. pharmacol. Acta **14**, 325—333 (1956)
Bergén, J. R.: Suppression of lysergic acid diethylamide (LSD) effects in pregnant rats. Res. Commun. chem. Pathol. Pharmacol. **3**, 313—319 (1972)
Bergen, J. R., Krus, D., Pincus, G.: Suppression of LSD-25 effects in rats by steroids. Proc. Soc. exp. Biol. (N.Y.) **105**, 254—256 (1960)
Bergen, J. R., Perkins, L., Hayano, M.: Action of steroids on lysergic acid diethylamide (LSD) metabolism. Proc. Soc. exp. Biol. (N.Y.) **111**, 427—428 (1962)
Beringer, K.: La intoxicación por la mescalina. Arch. argent. Neurol. **2**, 145—154 (1928)
Bertino, J. R., Klee, G. D., Collier, D., Weintraub, W.: Clinical studies with dibenzyline and lysergic acid diethylamide. J. clin. exp. Psychopath. **21**, 293—299 (1960)
Bertino, J. R., Klee, G. D., Weintraub, W.: Effects of certain vitamins and histamines on the LSD psychosis. J. ment. Sci. **105**, 1095—1099 (1959)
Bevan, P., Bradshaw, C. M., Roberts, M. H. T., Szabadi, E.: The effect of microelectrophoretically applied mescaline on cortical neurones. Neuropharmacology **13**, 1033—1045 (1974)

Bignami, G.: Facilitation of avoidance asquisition by LSD-25; possible effects on drive modulating systems. Psychopharmacologia (Berl.) **25**, 146—151 (1972)
Bishop, P. O., Burke, W., Hayhow, W. R.: Lysergic acid diethylamide block of lateral geniculate synapses and relief by repetitive stimulation. Exp. Neurol. **1**, 556—568 (1959)
Bishop, P. O., Field, G., Hennessy, B. L., Smith, J. R.: Action of d-lysergic acid diethylamide on lateral geniculate synapses. J. Neurophysiol. **21**, 529—549 (1958)
Bloom, F. E., Costa, E., Salmoiraghi, G. C.: Analysis of individual rabbit olfactory bulb neuron responses to the microelectrophoresis of acetylcholine, norepinephrine and serotonin synergists and antagonists. J. Pharmacol. exp. Ther. **146**, 16—23 (1964)
Boakes, R. J., Bradley, P. B., Briggs, I., Dary, A.: Antagonism of 5-hydroxytryptamine by LSD 25 in the central nervous system: a possible neuronal basis for the actions of LSD 25. Brit. J. Pharmacol. **40**, 202—218 (1970)
Boardman, W. K., Goldstone, S., Lhamon, W. T.: Effects of lysergic acid diethylamide (LSD) on the time sense of normals. Arch. Neurol. Psychiat. (Chicago) **78**, 321—324 (1957)
Bogdanski, D. F., Pletscher, A., Brodie, B. B., Udenfriend, S.: Identification and assay of serotonin in brain. J. Pharmacol. exp. Ther. **117**, 82—88 (1956)
Böszörményi, Z.: Psilocybin and diethyltryptamine: Two tryptamine hallucinogens. Neuro-psychopharm. **2**, 226—229 (1960)
Böszörményi, Z., Dér, P., Nagy, T.: Observations on the psychotogenic effect of N-N diethyltryptamine, a new tryptamine derivative. J. ment. Sci. **105**, 171—181 (1959)
Bowers, M. B., Jr.: Acute psychosis induced by psychotomimetic drug abuse. II. Neurochemical findings. Arch. gen. Psychiat. **27**, 440—442 (1972)
Bradley, P. B., Elkes, J.: The effects of some drugs on the electrical activity of the brain. Brain **80**, 77—117 (1957)
Bradley, P. B., Flugel, F., Hoch, P. H. (Eds.): Neuro-Psychopharmacology. New York: Elsevier Publishing Company 1964
Bradley, P. B., Hance, A. J.: The effect of chlorpromazine and methopromazine on the electrical activity of the brain in the cat. Electroenceph. clin. Neurophysiol. **9**, 191—215 (1957)
Bradley, P. B., Key, B. J.: The effect of drugs on arousal responses produced by electrical stimulation of the reticular formation of the brain. Electroenceph. clin. Neurophysiol. **10**, 97—110 (1958)
Bradshaw, C. M., Roberts, M. H. T., Szabadi, E.: Effect of mescaline on single cortical neurones. Brit. J. Pharmacol. **43**, 871—873 (1971)
Brawley, P. and Duffield, I. C.: The pharmacology of hallucinogens. Pharmacol. Rev. **24**, 31—66 (1972)
Brimblecombe, R. W.: Effects of psychotropic drugs on open-field behaviour in rats. Psychopharmacologia (Berl.) **4**, 139—147 (1963)
Brimblecombe, R. W.: Hyperthermic effects of some tryptamine derivatives in relation to their behavioural activity. Int. J. Neuropharmacol. **6**, 423—429 (1967)
Brücke, F., Gogolák, G., Stumpf, Ch.: Die Wirkung von LSD auf die Makro- und Mikrotätigkeit des Hippocampus. Naunyn-Schmiedebergs Arch. exp. Path. Pharmak. **240**, 461—468 (1961)
Burger, A. (Ed.): Drugs affecting the central nervous system. New York: Marcel Dekker, Inc. 1968
Butters, N.: The effect of LSD-25 on spatial and stimulus perseverative tendencies in rats. Psychopharmacologia (Berl.) **8**, 454—460 (1966)
Cameron, O. G., Appel, J. B.: Conditioned suppression of bar-pressing behavior by stimuli associated with drugs. J. exp. Anal. Behav. **17**, 127—137 (1972)
Cerletti, A.: Lysergic acid diethylamide (LSD) and related compounds. In: Abramson, H. A. (Ed.): Neuropharmacology. (Transactions of the Second Conference, May 25, 26, and 27, Princeton, N. J.). New York: Josiah Macy, Jr. Foundation 1956
Cervoni, P., Bertino, J. R., Geiger, L. E.: Medullary vagal effects of *d*-lysergic acid diethylamide in the decerebrate cat. Nature (Lond.) **199**, 700—701 (1963)
Chapman, L. F., Walter, R. D.: Actions of lysergic acid diethylamide on averaged human cortical evoked responses to light flash. In: Wortis, J. (Ed.): Recent Advances in Biological Psychiatry. Vol. 7. New York: Plenum Press 1965
Chase, T. N., Breese, G. R., Kopin, I. J.: Serotonin release from brain slices by electrical stimulation: Regional differences and effect of LSD. Science **157**, 1461—1463 (1967)

Cholden, L. (Ed.): Lysergic acid diethylamide and mescaline in experimental psychiatry. New York: Grune & Stratton 1956
Cholden, L. S., Kurland, A., Savage, C.: Clinical reactions and tolerance to LSD in chronic schizophrenia. J. nerv. ment. Dis. **122**, 211—221 (1955)
Cimbura, G.: PMA deaths in Ontario. Canad. med. Ass. J. **110**, 1263—1267 (1974)
Clark, L. D., Bliss, E. L.: Psychopharmacological studies of lysergic acid diethylamide (LSD-25) intoxication: Effects of premedication with BOL-128 (2-bromo-d-lysergic acid diethylamide), mescaline, atropine, amobarbital, and chlorpromazine. Arch. Neurol. Psychiat. (Chic.) **78**, 653—655 (1957)
Clineschmidt, B. V., Lotti, V. J.: Indoleamine antagonists: Relative potencies as inhibitors of tryptamine- and 5-hydroxytryptophan-evoked responses. Brit. J. Pharmacol. **50**, 311—313 (1974)
Cohen, S.: Psychotomimetic agents. Ann. Rev. Pharmacol. **7**, 301—318 (1967)
Cole, J. M., Pieper, W. A.: The effects of N,N-dimethyltryptamine on operant behavior in squirrel monkeys. Psychopharmacologia (Berl.) **29**, 107—112 (1973)
Cole, J. O., Freedman, A. M., Friedhoff, A. J. (Eds.): Psychopathology and psychopharmacology. Baltimore: Johns Hopkins University Press 1973
Collier, H. O. J.: A general theory of the genesis of drug dependence by induction of receptors. Nature (Lond.) **205**, 181—182 (1965)
Cook, L., Weidley, E.: Behavioral effects of some psychopharmacological agents. Ann. N. Y. Acad. Sci. **66**, 740—752 (1957)
Cooper, P. D., Walters, G. C.: Stereochemical requirements of the mescaline receptor. Nature (Lond.) **238**, 96—98 (1972)
Corne, S. J., Pickering, R. W.: A possible correlation between drug-induced hallucinations in man and a behavioural response in mice. Psychopharmacologia (Berl.) **11**, 65—78 (1967)
Costa, E: The effects of hallucinogenic and tranquilizing drugs on the serotonin evoked uterine contractions. Psychiat. Res. Rep. Amer. psychiat. Ass. **4**, 11—23 (1956)
Crighel, E., Stoica, E.: Cercetâri asupra focarului mescalinic sigmoidian. Stud. Cercet. Neurol. **6**, 547—555 (1961)
Crighel, E., Stoica, E.: Cercetâri asupra rolului unor sisteme aferente în declansarea activitatii unui focar convulsiv. Stud. Cercet. Neurol. **7**, 173—179 (1962)
Curtis, D. R., Davis, R.: Pharmacological studies upon neurones of the lateral geniculate nucleus of the cat. Brit. J. Pharmacol. **18**, 217—246 (1962)
Davis, M., Sheard, M. H.: Effects of lysergic acid diethylamide (LSD) on habituation and sensitization of the startle response in the rat. Pharmacol. Biochem. Behav. **2**, 675—683 (1974a)
Davis, M., Sheard, M. H.: Biphasic dose-response effects of N-N-dimethyltryptamine on the rat startle reflex. Pharmacol. Biochem. Behav. **2**, 827—829 (1974b)
Dear, E. M. A., Malcolm, J. L.: A study of the effect of 3,4-dimethoxyphenylethylamine on cortical evoked potentials in the rat. Int. J. Neuropharmacol. **6**, 529—542 (1967)
DeBaran, L., Gogolak. G., Longo, V. G., Stumpf, Ch.: The action of tryptamine upon the electrical activity of single hippocampal neurons. J. Pharmacol. exp. Ther. **139**, 337—344 (1963)
DeLaTorre, J. C.: Effect of LSD-25 on the septal region of the rat brain. Nature (Lond.) **219**, 954—955 (1968)
Delay, J., Lhermitte, F., Verdeaux, G., Verdeaux, J.: Modifications de l'electrocorticogramme du lapin par la diethylamide de l'acide *d*-lysergique (LSD 25). Rev. neurol. **86**, 81—90 (1952)
DeMaar, E. W. J., Williams, H. L., Miller, A. I., Pfeiffer, C. C.: Effects in man of single and combined oral doses of reserpine, iproniazid and *d*-lysergic acid diethylamide. Clin. Pharmacol. Ther. **1**, 23—30 (1960)
Depoortere, H., Loew, D. M.: Alterations in sleep/wakefulness cycle in rats following treatment with (+)-lysergic acid diethylamide (LSD-25). Brit. J. Pharmacol. **41**, 402P (1971)
Dewhurst, W. G., Marley, E.: The effects of α-methyl derivatives of noradrenaline, phenylethylamine and tryptamine on the central nervous system of the chicken. Brit. J. Pharmacol. **25**, 682—704 (1965a)
Dewhurst, W. G., Marley, E.: Action of sympathomimetic and allied amines on the central nervous system of the chicken. Brit. J. Pharmacol. **25**, 705—727 (1965b)
Dhawan, B. N.: Blockade of LSD-25 pyrexia by morphine. Arch. int. Pharmacodyn. **127**, 307—313 (1960)

Dhawan, B. N., Gupta, G. P.: LSD-25 antagonism of morphine analgesia. Arch. int. Pharmacodyn. **123**, 132—139 (1959)
Dhawan, B. N., Gupta, G. P.: Antiemetic activity of *d*-lysergic acid diethylamide. J. Pharmacol. exp. Ther. **133**, 137—139 (1961)
Diab, I. M., Freedman, D. X., Roth, L. J.: [^{3}H]lysergic acid diethylamide: Cellular autoradiographic localization in rat brain. Science **173**, 1022—1024 (1971)
Dimascio, A., Greenblatt, M., Hyde, R. W.: A study of the effects of LSD: Physiologic and psychological changes and their interrelations. Amer. J. Psychiat. **114**, 309—317 (1957)
DiStefano, V., Leary, D. E.: Effect of LSD on inhibitory systems in the spinal cord. Fed. Proc. **17**, 364 (1958)
Ditman, K. S., Whittlesey, J. R. B.: Comparison of the LSD-25 experience and delirium tremens. Arch. gen. Psychiat. **1**, 47—57 (1959)
Dixon, A. K.: Evidence of catecholamine mediation in the "aberrant" behaviour induced by lysergic acid diethylamide (LSD) in the rat. Experientia (Basel) **24**, 743—747 (1968)
Domer, F. R., Longo, V. G.: Effect of 5-hydroxytryptophan on the cerebral electrical activity of the rabbit. Arch. int. Pharmacodyn. **136**, 204—218 (1962)
Domino, E. F., Bartolini, A.: Effects of various psychotomimetic agents on the EEG and acetylcholine release from the cerebral cortex of brainstem transected cats. Neuropharmacology **11**, 703—713 (1972)
Dykstra, L. A., Appel, J. B.: Effects of LSD on auditory perception: A signal detection analysis. Psychopharmacologia (Berl.) **34**, 289—307 (1974)
Edwards, A. E., Cohen, S.: Visual illusion, tactile sensibility and reaction time under LSD-25. Psychopharmacologia (Berl.) **2**, 297—303 (1961)
Efron, D. H. (Ed.): Psychopharmacology; a review of progress, 1957—1967. PHS Publication No. 1836. Washington, D. C.: U. S. Government Printing Office 1968
Efron, D. H. (Ed.): Psychotomimetic drugs. New York: Raven Press 1970
Efron, D. H., Holmstedt, B., Kline, N. S. (Eds.): Ethnopharmacologic search for psychoactive drugs. PHS Publication No. 1645. Washington, D. C.: U. S. Government Printing Office 1967
Elder, J. T., Jr., Dille, J. M.: An experimental study of the participation of the sympathetic nervous system in the lysergic acid diethylamide (LSD) reaction in cats. J. Pharmacol. exp. Ther. **136**, 162—168 (1962)
Elder, J. T., Shellenberger, M. K.: Antagonism of lysergic acid diethylamide (LSD)-induced hyperthermia. J. Pharmacol. exp. Ther. **136**, 293—297 (1962)
Elkes, J.: Drugs influencing effect and behavior: Possible neural correlates in relation to mode of action. In: Smith, A., Herbert, C. C., Straus, R. (Eds.): The physiology of emotions. Springfield, Ill.: Charles C. Thomas 1961
Elkes, J., Elkes, C., Bradley, P. B.: The effect of some drugs on the electrical activity of the brain, and on behaviour. J. ment. Sci. **100**, 125—128 (1954)
Etevenon, P. R., Boissier, J. R.: LSD effects on signal-to-noise ratio and lateralization of visual cortex and lateral geniculate during photic stimulation. Experientia (Basel) **28**, 1338—1340 (1972)
Evarts, E. V.: Some effects of bufotenine and lysergic acid diethylamide on the monkey. Arch. Neurol. Psychiat. (Chic.) **75**, 49—53 (1956)
Evarts, E. V.: A review of the neurophysiological effects of lysergic acid diethylamide (LSD) and other psychotomimetic agents. Ann. N. Y. Acad. Sci. **66**, 479—495 (1957)
Evarts, E. V.: Effects of a series of indoles on synaptic transmission in the lateral geniculate nucleus of the cat. In: Pennes, H. H. (Ed.): Progress in neurobiology. 3. Psychopharmacology. New York: Hoeber-Harper 1958a
Evarts, E. V.: Neurophysiological correlates of pharmacologically-induced behavioral disturbances. In: Solomon, H. C., Cobb, S., Penfield, W. (Eds.): The brain and human behavior. Res. Publ. nerv. ment. Dis., Vol. 36. Baltimore: Williams & Wilkins 1958b
Evarts, E. V., Landau, W., Freygang, W., Jr., Marshall, W. H.: Some effects of lysergic acid diethylamide and bufotenine on electrical activity in the cat's visual system. Amer. J. Psychol. **182**, 594—598 (1955)
Fabing, H. D.: On going beserk: A neurochemical inquiry. Amer. J. Psychiat. **113**, 409—415 (1956)

Fabing, H. D., Hawkins, J. R.: Intravenous bufotenine injection in the human being. Science **123**, 886—887 (1956)
Faillace, L. A., Snyder, S. H., Weingartner, H.: 2,5-Dimethoxy-4-methylamphetamine: Clinical evaluation of a new hallucinogenic drug. J. nerv. ment. Dis. **150**, 119—126 (1970)
Falkenberg, G.: The molecular structure of some psychoactive indolealkylamines and related substances. Stockholm: Tryckeri Balder AB 1972
Farrow, J. T., van Vunakis, H.: Binding of *d*-lysergic acid diethylamide to subcellular fractions from rat brain. Nature (Lond.) **237**, 164—166 (1972)
Farrow, J. T., van Vunakis, H.: Characteristics of *d*-lysergic acid diethylamide binding to subcellular fractions derived from rat brain. Biochem. Pharmacol. **22**, 1103—1113 (1973)
Fischer, R.: Space-time coordinates of excited and tranquilized states. In: Jakob, K. (Ed.): Psychiatry and art. Proceedings IVth Int. Coll. Psychopathology of Expression. Basel-New York: S. Karger 1968
Fish, B., Campbell, M., Shapiro, T., Floyd, A., Jr.: Schizophrenic children treated with methysergide (Sansert). Dis. nerv. Syst. **30**, 534—540 (1969)
Florio, V., Lipparini, F., De Carolis, A. S., Longo, V. G.: EEG and behavioral effects of 2,5-methoxy-4-methyl-amphetamine (DOM, STP). Arch. int. Pharmacodyn. **180**, 81—88 (1969)
Forrer, G. R., Goldner, R. D.: Experimental physiological studies with lysergic acid diethylamide (LSD-25). Arch. Neurol. Psychiat. (Chic.) **65**, 581—588 (1951)
Freedman, D. X.: Effects of LSD-25 on brain serotonin. J. Pharmacol. exp. Ther. **134**, 160—166 (1961)
Freedman, D. X., Aghajanian, G. K., Ornitz, E. M., Rosner, B. S.: Patterns of tolerance to lysergic acid diethylamide and mescaline in rats. Science **127**, 1173—1174 (1958)
Freedman, D. X., Appel, J. B., Hartman, F. R., Molliver, M. E.: Tolerance to behavioral effects of LSD-25 in rat. J. Pharmacol. exp. Ther. **143**, 309—313 (1964)
Freedman, D. X., Giarman, N. J.: LSD-25 and the status and level of brain serotonin. Ann. N. Y. Acad. Sci. **96**, 98—106 (1962)
Freeman, H.: Pupil dilatation in normal and schizophrenic subjects following lysergic acid diethylamide ingestion. Arch. Neurol. Psychiat. (Chic.) **79**, 341—344 (1958)
Friedman, H., Carey, R. J.: Effects of LSD-25 on performance of a visual discrimination task in brain-damaged rats. Pharmacol. Biochem. Behav. **3**, 421—424 (1975)
Friedman, S. A., Hirsch, S. E.: Extreme hyperthermia after LSD ingestion. J. Amer. med. Ass. **217**, 1549—1550 (1971)
Fujimori, M., Himwich, H. E.: Electroencephalographic alerting sites of d-amphetamine and 2,5-dimethoxy-4-methyl-amphetamine. Nature (Lond.) **220**, 491—494 (1968)
Fuster, J. M.: Lysergic acid and its effects on visual discrimination in monkeys. J. nerv. ment. Dis. **129**, 252—256 (1959)
Gaddum J. H.: Tryptamine receptors. J. Physiol. (Lond.) **119**, 363—368 (1953)
Gaddum, J. H., Hameed, K. A.: Drugs which antagonize 5-hydroxytryptamine. Brit. J. Pharmacol. **9**, 240—248 (1954)
Gaddum, J. H., Picarelli, Z. P.: Two kinds of tryptamine receptor. Brit. J. Pharmacol. **12**, 323—328 (1957)
Gaddum, J. H., Vogt, M.: Some central actions of 5-hydroxytryptamine and various antagonists. Brit. J. Pharmacol. **11**, 175—179 (1956)
Garattini, S., Ghetti, V. (Eds.): Psychotropic drugs. New York: Elsevier Publishing Company 1957
Gardner, E. L.: Effects of LSD-25 on bar-pressing behavior in the hooded rat. Psychon. Sci. **3**, 507—508 (1965)
Gastaut, H., Ferrer, S., Castells, C.: Action de la diéthylamide de l'acide *d*-lysergique (LSD 25) sur les fonctions psychiques et l'électroencéphalogramme. Confin. neurol. (Basel) **13**, 102—120 (1953)
Geiger, L. E., Cervoni, P.: Effect of *d*-lysergic acid diethylamide on spinal reflexes. Proc. Soc. exp. Biol. (N.Y.) **99**, 179—181 (1958)
Gelder, M. G., Vane, J. R.: Interaction of the effects of tyramine, amphetamine and reserpine in man. Psychopharmacologia (Berl.) **3**, 231—241 (1962)
Gessner, P. K., Page, I. H.: Behavioral effects of 5-methoxy-N,N-dimethyltryptamine, other tryptamines, and LSD. Amer. J. Physiol. **203**, 167—172 (1962)

Giberti, F., Gegoretti, L.: Prime esperienze di antagonismo psicofarmacologico Psicosi sperimentale da LSD e trattemanto con cloropromazine e reserpina. Sist. nerv. **4**, 301—310 (1955)
Gillin, J. C., Cannon, E., Magyar, R., Schwartz, M., Wyatt, R. J.: Failure of N,N-dimethyltryptamine to evoke tolerance in cats. Biol. Psychiat. **7**, 213—220 (1973)
Ginzel, K. H.: Some central effects of lysergic acid diethylamide on vasomotor responses. J. Physiol. (Lond.) **129**, 61 P—62 P (1955)
Ginzel, K. H.: The effect of lysergic acid diethylamide on some autonomic reflex patterns: In: Garattini, S., Ghetti, V. (Eds.): Psychotropic drugs. New York: Elsevier Publishing Company 1957
Ginzel, K. H.: The effect of *d*-lysergic acid diethylamide and other drugs on the carotid sinus reflex. Brit. J. Pharmacol. **13**, 250—259 (1958)
Ginzel, K. H., Mayer-Gross, W.: Prevention of psychological effects of *d*-lysergic acid diethylamide (LSD 25) by its 2-brom derivative (BOL 148). Nature (Lond.) **178**, 210 (1956)
Glow, P. H.: Some acute effects of *d*-lysergic acid diethylamide on the rat. Quart. J. exp. Psychol. **9**, 206—214 (1957)
Gogerty, J. H., Dille, J. M.: Tolerance to the pyretogenic effects of lysergic acid diethylamide. J. Pharmacol. exp. Ther. **116**, 450—452 (1956)
Gogerty, J. H., Dille, J. M.: Pharmacology of *d*-lysergic acid morpholide (LSM). J. Pharmacol. exp. Ther. **120**, 340—348 (1957)
Goldman, H., Fischer, R., Nicolov, N., Murphy, S.: Lysergic acid diethylamide affects blood flow to specific areas of the conscious rat brain. Experientia (Basel) **31**, 328—329 (1975)
Goldstein, L., Murphree, H. B., Sugerman, A. A., Pfeiffer, C. C., Jenney, E. H.: Quantitative electroencephalographic analysis of naturally occurring (schizophrenic) and drug-induced psychotic states in human males. Clin. Pharmacol. Ther. **4**, 10—21 (1963)
Gonzalez-Vegas, J. A.: Antagonism of catecholamine inhibition of brain stem neurones by mescaline. Brain Res. **35**, 264—267 (1971)
Gorodetzky, C. W., Isbell, H.: A comparison of 2,3-dihydro-lysergic acid diethylamide with LSD-25. Psychopharmacologia (Berl.) **6**, 229—233 (1964)
Green, W. J.: The effect of LSD on the sleep-dream cycle. J. nerv. ment. Dis. **140**, 417—426 (1965)
Greenberg, M. J.: Structure-activity relationship of tryptamine analogues on the heart of *Venus mercenaria*. Brit. J. Pharmacol. **15**, 375—388 (1960)
Greiner, T., Burch, N. R., Edelberg, R.: Psychopathology and psychophysiology of minimal LSD-25 dosage; a preliminary dosage-response spectrum. Arch. Neurol. Psychiat. (Chicago) **79**, 208—210 (1958)
Guttmann, E., Maclay, W. S.: Mescalin and depersonalization: Therapeutic experiments. J. Neurol. Psychopath. **16**, 193—212 (1936)
Gyermek, L.: Drugs which antagonize 5-hydroxytryptamine and related indolealkylamines. In: Eichler, O., Farah, A. (Eds.): 6-Hydroxytryptamine and related indolealkylamines. Handbook of experimental pharmacology, Vol. 19. Berlin-Heidelberg-New York: Springer 1966
Haertzen, C. A.: Development of scales based on patterns of drug effects, using the Addiction Research Center Inventory (ARCI). Psychol. Rep. **18**, 163—194 (1966)
Haigler, H. J., Aghajanian, G. K.: Mescaline and LSD: Direct and indirect effects on serotonin-containing neurons in brain. Europ. J. Pharmacol. **21**, 53—60 (1973)
Hale, A. R., Reed, A. F.: Prophylaxis of frequent vascular headache with methysergide. Amer. J. med. Sci. **243**, 92—98 (1962)
Haley, T. J.: Pharmacological effects from drugs injected intracerebrally in unanesthetized animals. J. Amer. pharm. Ass. **45**, 604—607 (1956)
Haley, T. J.: Intracerebral injection of psychotomimetic and psychotherapeutic drugs into conscious mice. Acta pharmacol. (Kbh.) **13**, 107—112 (1957a)
Haley, T. J.: 5-Hydroxytryptamine antagonism by lysergic acid diethylamide after intracerebral injection in conscious mice. J. Amer. pharm. Ass. **46**, 428—430 (1957b)
Haley, T. J., Dasgupta, S. R.: Intracerebral injection of lysergic acid diethylamide in conscious dogs and cats. Arch. int. Pharmacodyn. **113**, 296—301 (1958)
Haley, T. J., Rutschmann, J.: Brain concentrations of LSD-25 (Delysid) after intracerebral or intravenous administration in conscious animals. Experientia (Basel) **13**, 199—202 (1957)
Hamilton, C. L.: Effects of LSD-25 and amphetamine on a running response in the rat. Arch. gen. Psychiat. **2**, 104—109 (1960)

Hamilton, C. L., Wilpizeski, C.: Effects of LSD-25 on food intake in the rat. Proc. Soc. exp. Biol. (N.Y.) **108**, 319—321 (1961)
Hartman, A. M., Hollister, L. E.: Effect of mescaline, lysergic acid diethylamide and psilocybin on color perception. Psychopharmacologia (Berl.) **4**, 441—451 (1963)
Heim, R., Wasson, R. G.: Les champignons hallucinogenes du Mexique. Etudes ethnologiques, taxinomiques, biologiques, physiologiques et chimiques. Paris: Muséum National d'Histoire Naturelle 1958
Hill, H. E., Haertzen, C. A., Wolbach, A. B., Jr., Miner, E. J.: The Addiction Research Center Inventory: Standardization of scales which evaluate subjective effects of morphine, amphetamine, pentobarbital, alcohol, LSD-25, pyrahexyl and chlorpromazine. Psychopharmacologia (Berl.) **4**, 167—183 (1963a)
Hill, H. E., Haertzen, C. A., Wolbach, A. B., Jr., Miner, E. J.: The Addiction Research Center Inventory: Appendix. I. Items comprising empirical scales for seven drugs. II. Items which do not differentiate placebo from any drug condition. Psychopharmacologia (Berl.) **4**, 184—205 (1963b)
Himmelsbach, C. K.: Morphine. IV. With reference to physical dependence. Fed. Proc. **2**, 201—203 (1943)
Himwich, H. E. (Ed.): Biochemistry, schizophrenias and affective illnesses. Baltimore: Williams & Wilkins Company 1971
Hirschhorn, I. D., Winter, J. C.: Mescaline and lysergic acid diethylamide (LSD) as discriminative stimuli. Psychopharmacologia (Berl.) **22**, 64—71 (1971)
Hirschhorn, I. D., Winter, J. C.: Differences in the stimulus properties of barbital and hallucinogens. Pharmacol. Biochem. Behav. **3**, 343—347 (1975)
Hobson, J. A.: The effect of LSD on the sleep cycle of the cat. Electroenceph. clin. Neurophysiol. **17**, 52—56 (1964)
Hoch, P. H.: Experimental psychiatry. Amer. J. Psychiat. **111**, 787—790 (1955)
Hoch, P. H., Cattell, J. P., Pennes, H. H.: Effects of mescaline and lysergic acid (d-LSD-25). Amer. J. Psychiat. **108**, 579—584 (1952)
Hoffer, A., Osmond, H. (Eds.): The hallucinogens. New York: Academic Press 1967
Hollister, L. E.: Clinical, biochemical and psychologic effects of psilocybin. Arch. int. Pharmacodyn. **130**, 42—52 (1961)
Hollister, L. E.: Chemical psychoses. LSD and related drugs. Springfield: Charles C. Thomas 1968
Hollister, L. E., Friedhoff, A. J.: Effects of 3,4-dimethoxyphenylethylamine in man. Nature (Lond.) **210**, 1377—1378 (1966)
Hollister, L. E., Hartman, A. M.: Mescaline, lysergic acid diethylamide and psilocybin: Comparison of clinical syndromes, effects on color perception and biochemical measures. Comprehens. Psychiat. **3**, 235—241 (1962)
Hollister, L. E., Macnicol, M. F., Gillespie, H. K.: An hallucinogenic amphetamine analog (DOM) in man. Psychopharmacologia (Berl.) **14**, 62—73 (1969)
Hollister, L. E., Sjoberg, B. M.: Clinical syndromes and biochemical alterations following mescaline, lysergic acid diethylamide, psilocybin and a combination of the three psychotomimetic drugs. Comprehens. Psychiat. **5**, 170—178 (1964)
Hollunger, G., Persson, S. A.: The effect of LSD and 2-bromo LSD on the DOPA accumulation after central and peripheral decarboxylase inhibition. Europ. J. Pharmacol. **31**, 156—158 (1975)
Horita, A., Dille, J. M.: Pyretogenic effect of lysergic acid diethylamide. Science **120**, 1100—1101 (1954)
Horita, A., Gogerty, J. H.: The pyretogenic effect of 5-hydroxytryptophan and its comparison with that of LSD. J. Pharmacol. exp. Ther. **122**, 195—200 (1958)
Horita, A., Hamilton, A. E.: Lysergic acid diethylamide: Dissociation of its behavioral and hyperthermic actions by DL-α-methyl-p-tyrosine. Science **164**, 78—79 (1969)
Horita, A., Hamilton, A. E.: The effects of DL-α-methyltyrosine and L-DOPA on the hyperthermic and behavioral actions of LSD in rabbits. Neuropharmacology **12**, 471—476 (1973)
Horita, A., Hill, H. F.: Hallucinogens, amphetamines and temperature regulation. In: Schoenbaum, E., Lomax, T. (Eds.): The pharmacology of thermoregulation. Symposium, San Francisco, 1972. Basel: Karger 1972

Horn,G., McKay,J.M.: Effects of lysergic acid diethylamide on the spontaneous activity and visual receptive fields of cells in the lateral geniculate nucleus of the cat. Exp. Brain Res. **17**, 271—284 (1973)

Huxley,A.: The doors of perception. New York: Harper & Row 1954

Ingvar,D.H., Soderberg,U.: The effect of LSD-25 upon the cerebral blood flow and EEG in cats. Experientia (Basel) **12**, 427—428 (1956)

Isbell,H.: Comparison of the reactions induced by psilocybin and LSD-25 in man. Psychopharmacologia (Berl.) **1**, 29—38 (1959)

Isbell,H., Belleville,R.E., Fraser,H.F., Wikler,A., Logan,C.R.: Studies on lysergic acid diethylamide (LSD-25). I. Effects in former morphine addicts and development of tolerance during chronic intoxication. Arch. Neurol. Psychiat. (Chicago) **76**, 468—478 (1956)

Isbell,H., Fraser,H.F., Wikler,A., Belleville,R.E.: Tolerance to diethylamide of lysergic acid (LSD-25). Fed. Proc. **14**, 21 (1955)

Isbell,H., Gorodetzky,C.W.: Effect of alkaloids of ololiuqui in man. Psychopharmacologia (Berl.) **8**, 331—339 (1966)

Isbell,H., Jasinski,D.R.: A comparison of LSD-25 with (-)-Δ^9-trans-tetrahydrocannabinol (THC) and attempted cross tolerance between LSD and THC. Psychopharmacologia (Berl.) **14**, 115—123 (1969)

Isbell,H., Logan,C.R.: Studies on the diethylamide of lysergic acid (LSD-25). II. Effects of chlorpromazine, azacyclonol, and reserpine on the intensity of the LSD-reaction. Arch. Neurol. Psychiat. (Chicago) **77**, 350—358 (1957)

Isbell,H., Logan,C.R., Miner,E.J.: Studies on lysergic acid diethylamide (LSD-25). III. Attempts to attenuate the LSD-reaction in man by pretreatment with neurohumoral blocking agents. Arch. Neurol. Psychiat. (Chic.) **81**, 20—27 (1959a)

Isbell,H., Miner,E.J., Logan,C.R.: Relationships of psychotomimetic to anti-serotonin potencies of congeners of lysergic acid diethylamide (LSD-25). Psychopharmacologia (Berl.) **1**, 20—28 (1959b)

Isbell,H., Miner,E.J., Logan,C.R.: Cross tolerance between *d*-2-Brom-lysergic acid diethylamide (BOL-148) and the *D*-diethylamide of lysergic acid (LSD-25). Psychopharmacologia (Berl.) **1**, 109—116 (1959c)

Isbell,H., Rosenberg,D.E., Miner,E.J., Logan,C.R.: Tolerance and cross tolerance to scopolamine, N-ethyl-3-piperidyl benzylate (JB-318) and LSD-25. Neuropsychopharmacology **3**, 440—446 (1964)

Isbell,H., Wolbach,A.B., Wikler,A., Miner,E.J.: Cross tolerance between LSD and psilocybin. Psychopharmacologia (Berl.) **2**, 147—159 (1961)

Jacob,J.: Quelques réactions enzymatiques conduisant à des substances hallucinogènes. Thérapie **20**, 252—259 (1965)

Jacob,J., Lafille,C.: Charactérisation et détection pharmacologiques des substances hallucinogénes. I. Activites hyperthermisantes chez le lapin. Arch. int. Pharmacodyn. **145**, 528—545 (1963)

Jacob,J., Lafille,C.: Antagonistes de l'action hyperthermisante du lysergamide chez de lapin. Proc. 2nd int. pharmacol. Meet. **2**, 249—261 (1964)

Jacob,J., Lafille,C., Loiseau,G., Echinard-Garin,P., Barthelemy,C.: Recherche d'équivalents expérimentaux au pouvoir hallucinogène; Action des hallucinogènès sur la température rectale du lapin et sur les effets de la morphine chez la souris. Med. exp. (Basel) **7**, 296—304 (1962)

Jacob,J., Loiseau,G., Echinard-Garin,P., Barthelemy,C., Lafille,C.: Caractérisation et détection pharmacologiques des substances hallucinogènes. II.—Antagonismes vis-a-vis de la morphine chez la souris. Arch. int. Pharmacodyn. **148**, 14—30 (1964)

Jacobsen,E.: The clinical pharmacology of the hallucinogens. Clin. Pharmacol. Ther. **4**, 480—503 (1963)

Jaffe,J.H., Sharpless,S.K.: The rapid development of physical dependence on barbiturates. J. Pharmacol. exp. Ther. **150**, 140—145 (1965)

Jarrard,L.E.: Effects of D-lysergic acid diethylamide on operant behavior in the rat. Psychopharmacologia (Berl.) **5**, 39—46 (1963)

Jarvik,M.E., Abramson,H.A., Hirsch,M.W.: Lysergic acid diethylamide (LSD-25): IV. Effect on attention and concentration. J. Psychol. (Provincetown) **39**, 373—383 (1955a)

Jarvik, M. E., Abramson, H. A., Hirsch, M. W.: Lysergic acid diethylamide (LSD-25): VI. Effect upon recall and recognition of various stimuli. J. Psychol. (Provincetown) **39**, 443—454 (1955b)
Jarvik, M. E., Abramson, H. A., Hirsch, M. W.: Comparative subjective effects of seven drugs including lysergic acid diethylamide (LSD-25). J. abnorm. soc. Psychol. **51**, 657—662 (1955d)
Jarvik, M. E., Abramson, H. A., Hirsch, M. W., Ewald, A. T.: Lysergic acid diethylamide (LSD-25): VIII. Effect on arithmetic test performance. J. Psychol. (Provincetown) **39**, 465—473 (1955c)
Jarwik, M. E., Chorover, S.: Impairment by lysergic acid diethylamide of accuracy in performance of a delayed alternation test in monkeys. Psychopharmacologia (Berl.) **1**, 221—230 (1960)
Kang, S., Green, J. P.: Correlation between activity and electronic state of hallucinogenic amphetamines. Nature (Lond.) **226**, 645 (1970a)
Kang, S., Green, J. P.: Steric and electronic relationships among some hallucinogenic compounds. Proc. nat. Acad. Sci. (Wash.) **67**, 62—67 (1970b)
Kay, D. C., Martin, W. R., Croughan, J. L.: The electrocorticographic and cardiovascular effects of tryptamine and LSD in the encephale isole cat. In preparation (1976)
Keeler, M. H.: The effects of psilocybin on a test of after-image perception. Psychopharmacologia (Berl.) **8**, 131—139 (1965a)
Keeler, M. H.: Similarity of schizophrenia and the psilocybin syndrome as determined by objective methods. Int. J. Neuropsychiat. **1**, 630—634 (1965b)
Keller, D. L., Umbreit, W. W.: "Permanent" alteration of behavior in mice by chemical and psychological means. Science **124**, 723—724 (1956)
Keup, W. (Ed.): Origin and mechanisms of hallucinations. New York: Plenum Press 1970
Key, B. J.: Effects of chlorpromazine and lysergic acid diethylamide on the rate of habituation of the arousal response. Nature (Lond.) **190**, 275—277 (1961a)
Key, B. J.: The effect of drugs on discrimination and sensory generalisation of auditory stimuli in cats. Psychopharmacologia (Berl.) **2**, 352—363 (1961b)
Key, B. J.: The effect of LSD-25 on the interaction between conditioned and non-conditioned stimuli in a simple avoidance situation. Psychopharmacologia (Berl.) **6**, 319—326 (1964)
Khazan, N., McCash, D.: Effects of LSD-25, N,N-dimethyltryptamine (DMT), and N,N-diethyltryptamine (DET) on the photic evoked responses in the unanesthetized rabbit. Arch. int. Pharmacodyn. **154**, 474—483 (1965)
Khazan, N., Sawyer, C. H.: Mechanisms of paradoxical sleep as revealed by neurophysiologic and pharmacologic approaches in the rabbit. Psychopharmacologia (Berl.) **5**, 457—466 (1964)
Kies, M. W., Horst, D., Evarts, E. V., Goldstein, N. P.: Antidiuretic effect of lysergic acid diethylamide in humans. Arch. Neurol. Psychiat. (Chic.) **77**, 267—269 (1957)
Killam, E. K.: Drug action on the brain-stem reticular formation. Pharmacol. Rev. **14**, 175—223 (1962)
Killam, K. F., Killam, E. K.: The action of lysergic acid diethylamide on central afferent and limbic pathways in the cat. J. Pharmacol. exp. Ther. **116**, 35—36 (1956)
King, A. R., Martin, I. L., Melville, K. A.: Reversal learning enhanced by lysergic acid diethylamide (LSD): Concomitant rise in brain 5-hydroxytryptamine levels. Brit. J. Pharmacol. **52**, 419—426 (1974)
Klee, G. D., Bertino, J., Callaway, E., III, Weintraub, W.: Clinical studies with LSD-25 and two substances related to serotonin. Part I. Clinical reactions of normal human subjects given LSD-25 and an antimetabolite of serotonin (1 benzyl-2,5-dimethylserotonin hydrochloride) (BAS). J. ment. Sci. **106**, 301—305 (1960)
Klee, G. D., Bertino, J., Weintraub, W., Callaway, E., III: The influence of varying dosage on the effects of lysergic acid diethylamide (LSD-25) in humans. J. nerv. ment. Dis. **132**, 404—409 (1961)
Kluver, H.: Mechanisms of hallucinations. In: Studies in personality. New York: McGraw-Hill 1942
Koella, W. P., Beaulieu, R. F., Bergen, J. R.: Stereotyped behavior and cyclic changes in responses produced by LSD in goats. Int. J. Neuropharmacol. **3**, 397—403 (1964)
Koella, W. P., Wells, C.-H.: Influence of LSD-25 on optically evoked potentials in the nonanesthetized rabbit. Amer. J. Physiol. **196**, 1181—1184 (1959)
Koelle, G. B.: The pharmacology of mescaline and *d*-lysergic acid diethylamide (LSD). New Engl. J. Med. **258**, 25—32 (1958)

Korein, J., Musacchio, J. M.: LSD and focal cerebral lesions. Behavioral and EEG effects in patients with sensory defects. Neurology (Minneap.) **18**, 147—152 (1968)
Kostowski, W.: Evoked synchronization and desynchronization in the EEG of the cat. Diss. pharm. pharmacol. **23**, 471—479 (1971)
Kreindler, A., Crighel, E., Sotirescu, N.: Modificările răspunsului cortical direct produse do mescalina si reserpina. Stud. Cercet. Neurol. **7**, 287—296 (1962)
Krivoy, W.: The preservation of substance P by lysergic acid diethylamide. Brit. J. Pharmacol. **12**, 361—364 (1957)
Krivoy, W. A.: Potentiation of substance P by lysergic acid diethylamide in vivo. Brit. J. Pharmacol. **16**, 253—256 (1961)
Krnjević, K., Phillis, J. W.: Actions of certain amines on cerebral cortical neurones. Brit. J. Pharmacol. **20**, 471—490 (1963)
Krus, D. M., Wapner, S., Bergen, J., Freeman, H.: The influence of progesterone on behavioral changes induced by lysergic acid diethylamide (LSD-25) in normal males. Psychopharmacologia (Berl.) **2**, 177—184 (1961)
Ladefoged, O.: The effects of LSD, psilocybin, harmaline and amphetamine on the body temperature of para-chlorophenylalanine pretreated rats. Arch. int. Pharmacodyn. **204**, 326—332 (1973)
Laidlaw, P. P.: The physiological action of indolethylamine. Biochem. J. **6**, 141—150 (1912)
Landis, C., Clausen, J.: Certain effects of mescaline and lysergic acid on psychological functions. J. Psychol. (Provincetown) **38**, 211—221 (1954)
Langfitt, T. W., Finney, L. A.: Effects of lysergic acid diethylamide (LSD) on cortical sensory evoked potentials in the cat. Arch. Neurol. (Chic.) **1**, 258—268 (1959a)
Langfitt, T. W., Finney, L. A.: Effects of LSD on cortex of cat correlated with changes in vital signs. Arch. Neurol. (Chic.) **1**, 269—281 (1959b)
Leonard, B. E., Tonge, S. R.: The effects of some hallucinogenic drugs upon the metabolism of noradrenaline. Life Sci. **8**, 815—825 (1969)
Levine, A., Abramson, H. A., Kaufman, M. R., Markham, S.: Lysergic acid diethylamide (LSD-25): XVI. The effect on intellectual functioning as measured by the Wechsler-Bellevue Intelligence Scale. J. Psychol. (Provincetown) **40**, 385—395 (1955)
Levis, D. J., Caldwell, D. F.: The effects of a low dosage of mescaline and 3,4-dimethoxyphenylethylamine under two levels of aversive stimulation. Biol. Psychiat. **3**, 251—257 (1971)
Lewin, L.: Phantastica: Narcotic and stimulating drugs; their use and abuse. New York: E. P. Dutton & Co. 1964
Liebert, R. S., Werner, H., Wapner, S.: Studies in the effect of lysergic acid diethylamide (LSD-25); self- and object-size perception in schizophrenics and normal adults. Arch. Neurol. Psychiat. (Chic.) **79**, 580—584 (1958)
Lin, R. C., Ngai, S. H., Costa, E.: Lysergic acid diethylamide: Role in conversion of plasma tryptophan to brain serotonin (5-hydroxytryptamine). Science **166**, 237—239 (1969)
Lindsley, D. F., Carpenter, R. S., Killam, E. K., Killam, K. F.: EEG correlates of behavior in the cat. I. Pattern discrimination and its alteration by atropine and LSD-25. Electroenceph. clin. Neurophysiol. **24**, 497—513 (1968)
Linton, H. B., Langs, R. J.: Subjective reactions to lysergic acid diethylamide (LSD-25); measured by a questionnaire. Arch. gen. Psychiat. **6**, 352—368 (1962)
Little, K. D., DiStefano, V., Leary, D. E.: LSD and serotonin effects on spinal reflexes in the cat. J. Pharmacol. exp. Ther. **119**, 161 (1957)
Little, M. D., Dill, R. E.: Mescaline and other O-methylated β-phenylethylamines: Intrastriatal induction of tremor in rats. Brain Res. **13**, 360—366 (1969)
Loew, D. M., Depoortere, H., Vigouret, J. M.: Wirkungen von d-lysergsäurediäthylamid (LSD-25) und 2-brom-d-lysergsäurediäthylamid (BOL-148) auf das Schlafverhalten der Ratte. Naunyn-Schmiedebergs Arch. exp. Path. Pharmak. **266**, 394—395 (1970)
Longo, V. G.: Epinephrine and drugs acting on adrenergic receptors; 5-hydroxytryptamine and related compounds. In: Longo, V. G. (Ed.): Electroencephalographic atlas for pharmacological research; effect of drugs on the electrical activity of the rabbit brain. Rabbit Brain Research, Vol. 2. New York: Elsevier Publishing Company 1962
Lowe, G., Williams, D. I.: The effect of LSD-25 on light-reinforced behaviour in the rat. Psychopharmacologia (Berl.) **27**, 255—263 (1972)

Mahler, D. J., Humoller, F. L.: Effect of lysergic acid diethylamide and bufotenine on performance of trained rats. Proc. Soc. exp. Biol. (N. Y.) **102**, 697—701 (1959)

Malitz, S., Esecover, H., Wilkens, B., Hoch, P. H.: Some observations on psilocybin, a new hallucinogen, in volunteer subjects. Comprehens. Psychiat. **1**, 8—17 (1960)

Malmnäs, C. O.: Monaminergic influence on testosterone-activiated copulatory behavior in the castrated male rat. Acta physiol. scand., Suppl. 395 (1973)

Marchbanks, R. M.: Inhibitory effects of lysergic acid derivatives and reserpine on 5-HT binding to nerve ending particles. Biochem. Pharmacol. **16**, 1971—1979 (1967).

Marczynski, T. J.: Lysergic acid diethylamide (LSD-25) mimicks the effect of diffuse light input on EEG correlates of conditioned operant behavior in cats. Exp. Neurol. **34**, 255—263 (1972)

Marrazzi, A. S.: The effects of certain drugs on cerebral synapses. Ann. N. Y. Acad. Sci. **66**, 496—507 (1957)

Marrazzi, A. S., Hart, E. R.: Relationship of hallucinogens to adrenergic cerebral neurohumors. Science **121**, 365—367 (1955)

Martin, W. R.: A homeostatic and redundancy theory of tolerance to and dependence on narcotic analgesics. In: Wikler, A. (Ed.): The addictive states. Res. Publ. Ass. nerv. ment. Dis., Vol. 46. Baltimore: Williams and Wilkins 1968

Martin, W. R.: Assessment of the abuse potentiality of amphetamines and LSD-like hallucinogens in man and its relationship to basic animal assessment programs. In: Goldberg, L., Hoffmeister, F. (Eds.): Psychic dependence. Definition, assessment in animals and man. Theoretical and clinical implications. Berlin-Heidelberg-New York: Springer 1973

Martin, W. R., Eades, C. G.: Pharmacological studies of spinal cord adrenergic and cholinergic mechanisms and their relation to physical dependence on morphine. Psychopharmacologia (Berl.) **11**, 195—223 (1967)

Martin, W. R., Eades, C. G.: The action of tryptamine on the dog spinal cord and its relationship to the agonistic actions of LSD-like psychotogens. Psychopharmacologia (Berl.) **17**, 242—257 (1970)

Martin, W. R., Eades, C. G.: Cross tolerance to tryptamine in the LSD tolerant dog. Psychopharmacologia (Berl.) **27**, 93—98 (1972)

Martin, W. R., Eades, C. G.: Effects of phenethylamine (PEA) in the chronic spinal dog. Pharmacologist **16**, 205 (1974)

Martin, W. R., Eades, C. G.: Effects of β-phenethylamine in the chronic spinal dog. In preparation (1976)

Martin, W. R., Sloan, J. W.: Effects of infused tryptamine in man. Psychopharmacologia (Berl.) **18**, 231—237 (1970)

Martin, W. R., Sloan, J. W.: The possible role of tryptamine in brain function and its relationship to the actions of LSD-like hallucinogens. Mt. Sinai J. Med. (N. Y.) **41**, 276—282 (1974)

Martin, W. R., Sloan, J. W., Buchwald, W. F., Bridges, S. R.: The demonstration of tryptamine in regional perfusates of the dog brain. Psychopharmacologia (Berl.) **37**, 189—198 (1974)

Martin, W. R., Sloan, J. W., Buchwald, W. F., Clements, T. H.: Neurochemical evidence for tryptaminergic ascending and descending pathways in the spinal cord of the dog. Psychopharmacologia (Berl.) **43**, 131—134 (1975)

Martin, W. R., Sloan, J. W., Christian, S. T., Clements, T. H.: Brain levels of tryptamine. Psychopharmacologia (Berl.) **24**, 331—346 (1972)

Martin, W. R., Sloan, J. W., Vaupel, D. B., Bell, J. A., Nozaki, M.: Tryptamine in the brain and spinal cord: Its role in the LSD response. In: Usdin, E., Sandler, M. (Eds.): Trace amines and the brain. New York: Marcel Dekker 1976

Matussek, N., Halbach, A.: Über den Einfluß von Ergotamin auf die LSD-"Psychose" beim Menschen. Psychopharmacologia (Berl.) **5**, 158—160 (1964)

McKay, J. M., Horn, G.: Effects of LSD on receptive fields of single cells in the lateral geniculate nucleus of the cat. Nature (Lond.) **229**, 347—349 (1971)

Meldrum, B. S., Naquet, R.: Effects of psilocybin, dimethyltryptamine, mescaline and various lysergic acid derivatives on the EEG and on photically induced epilepsy in the baboon. Electroenceph. clin. Neurophysiol. **31**, 563—572 (1971)

Mendels, J.: The effect of methysergide (an antiserotonin agent) on schizophrenia: A preliminary report. Amer. J. Psychiat. **124**, 849—852 (1967)

Merlis, S., Denber, H. C. B.: Studies on mescaline. V. Electroencephalographic evidence for the antagonism between mescaline and chlorpromazine hydrochloride. J. nerv. ment. Dis. **123**, 542—545 (1956)
Miller, R. C., Miller, E. K.: Lack of a long-term effect of LSD on Y-maze learning in mice. Nature (Lond.) **228**, 1107—1108 (1970)
Mison-Crighel, N., Luca, N., Crighel, E.: The effect of an epileptogenic focus, induced by topical application of mescaline, on glutamic acid, glutamine and GABA in the neocortex of the cat. J. Neurochem. **11**, 333—340 (1964)
Monroe, R. R., Heath, R. G.: Effects of lysergic acid and various derivatives on depth and cortical electrograms. J. Neuropsychiat. **3**, 75—82 (1961)
Monroe, R. R., Heath, R. G., Mickle, W. A., Llewellyn, R. C.: Correlation of rhinencephalic electrograms with behavior. A study on humans under the influence of LSD and mescaline. Electroenceph. clin. Neurophysiol. **9**, 623—642 (1957)
Mouriz-Garcia, A., Schmidt, R., Arlazoroff, A.: Effects of LSD on the spontaneous and evoked activity of retinal and geniculate ganglion cells. Psychopharmacologia (Berl.) **15**, 382—391 (1969)
Nandy, K., Bourne, G. H.: The effects of D-lysergic acid diethylamide tartrate (LSD-25) on the cholinesterases and monamine oxidase in the spinal cord: A possible factor in the mechanism of hallucination. J. Neurol. Neurosurg. Psychiat. **27**, 259—267 (1964)
Naranjo, C., Shulgin, A. T., Sargent, T.: Evaluation of 3,4-methylenedioxyamphetamine (MDA) as an adjunct to psychotherapy. Med. Pharmacol. exp. **17**, 359—364 (1967)
Neff, N., Rossi, G. V., Chase, G. D., Rabinowitz, J. L.: Distribution and metabolism on mescaline-C^{14} in the cat brain. J. Pharmacol. exp. Ther. **144**, 1—7 (1964)
Neuhold, K., Taeschler, M., Cerletti, A.: Beitrag zur zentralen Wirkung von LSD: Versuche über die Lokalisation von LSD-Effekten. Helv. physiol. pharmacol. Acta **15**, 1—7 (1957)
Nozaki, M., Vaupel, D. B., Martin, W. R.: A pharmacologic comparison of 3,4-methylenedioxyamphetamine and LSD in the chronic spinal dog. In preparation (1976a).
Nozaki, M., Vaupel, D. B., Martin, W. R.: A pharmacologic comparison of 2-methyl-3,4-methylenedioxyamphetamine and LSD in the chronic spinal dog. In preparation (1976b)
Palmer, G. C., Burks, T. F.: Central and peripheral adrenergic blocking actions of LSD and BOL. Europ. J. Pharmacol. **16**, 113—116 (1971).
Passouant, P., Passouant-Fontaine, Th., Cadilhac, J.: Action du L. S. D. 25 sur le comportement et les rythmes corticaux et rhinencephaliques du chat chronique. C. R. Soc. Biol. (Paris) **150**, 2237—2241 (1956)
Paton, D. M.: Effect of amphetamine, 3,4-methylenedioxyamphetamine, *p*-methoxyamphetamine and related amphetamines on uptake of metaraminol and efflux of noradrenaline in adrenergic nerves of rabbit atria. J. Pharm. Pharmacol. **27**, 361—362 (1975)
Paton, W. D. M.: A theory of drug action based on the rate of drug-receptor combination. Proc. roy. Soc. B **154**, 21—69 (1961)
Pawlowski, A. A.: Effects of lysergic acid diethylamide on the modification of a learned jumping response. J. Neuropsychiat. **4**, 81—86 (1962)
Peretz, D. I., Smythies, J. R., Gibson, W. C.: A new hallucinogen: 3,4,5-trimethoxyphenyl-β-aminopropane; with notes on the stroboscopic phenomenon. J. ment. Sci. **101**, 317—329 (1955)
Peters, D. A. V.: Comparison of the chronic and acute effects of *d*-lysergic acid diethylamide (LSD) treatment on rat brain serotonin and norepinephrine. Biochem. Pharmacol. **23**, 231—237 (1974)
Philips, S. R., Durden, D. A., Boulton, A. A.: Identification and distribution of tryptamine in the rat. Canad. J. Biochem. **52**, 447—451 (1974)
Pickworth, W. B., Sharpe, L. G.: The transcallosally evoked potential following administration of LSD, mescaline, DMT, apomorphine and methoxamine in the dog. Fed. Proc. **35**, 270 (1976)
Pieri, L., Pieri, M., Haefely, W.: LSD as an agonist of dopamine receptors in the striatum. Nature (Lond.) **252**, 586—588 (1974)
Purpura, D. P.: Electrophysiological analysis of psychotogenic drug action. I. Effect of LSD on specific afferent systems in the cat. Arch. Neurol. Psychiat. (Chic.) **75**, 122—131 (1956a)
Purpura, D. P.: Electrophysiological analysis of psychotogenic drug action. II. General nature of lysergic acid diethylamide (LSD) action on central synapses. Arch. Neurol. Psychiat. (Chic.) **75**, 132—143 (1956b)

Purpura, D. P.: Neurophysiological action of LSD. In: DeBold, R. C., Leaf, R. C. (Eds.): LSD, man & society. Middletown, Conn.: Wesleyan University Press 1967
Richards, K. C., Borgstedt, H. H.: Near fatal reaction to ingestion of the hallucinogenic drug MDA. J. Amer. med. Ass. **218**, 1826—1827 (1971)
Richards, N., Chapman, L. F., Goodell, H., Wolff, H. G.: LSD-like delirium following ingestion of a small amount of its brom analog (BOL-148). Ann. intern. Med. **48**, 1078—1083 (1958)
Rinaldi, F., Himwich, H. E.: The cerebral electrographic changes induced by LSD and mescaline are corrected by Frenquel. J. nerv. ment. Dis. **122**, 424—432 (1955a)
Rinaldi, F., Himwich, H. E.: Frenquel corrects certain cerebral electrographic changes. Science **122**, 198—199 (1955b)
Rinkel, M., DeShon, H. J., Hyde, R. W., Solomon, H. C.: Experimental schizophrenia-like symptoms. Amer. J. Psychiat. **108**, 572—578 (1952)
Roberts, M. H. T., Straughan, D. W.: Excitation and depression of cortical neurones by 5-hydroxytryptamine. J. Physiol. (Lond.) **193**, 269—294 (1967)
Roberts, M. H. T., Straughan, D. W.: Actions of noradrenaline and mescaline on cortical neurones. Naunyn-Schmiedebergs Arch. exp. Pharmak. **259**, 191—192 (1968)
Rodin, E., Luby, E.: Effects of LSD-25 on the EEG and photic evoked responses. Arch. gen. Psychiat. **14**, 435—441 (1966)
Rosecrans, J. A., Lovell, R. A., Freedman, D. X.: Effects of lysergic acid diethylamide on the metabolism of brain 5-hydroxytryptamine. Biochem. Pharmacol. **16**, 2011—2021 (1967)
Rosenberg, D. E., Isbell, H., Miner, E. J.: Comparison of a placebo, N-dimethyltryptamine and 6-hydroxy-N-dimethyltryptamine in man. Psychopharmacologia (Berl.) **4**, 39—42 (1963a)
Rosenberg, D. E., Isbell, H., Miner, E. J., Logan, C. R.: The effect of N,N-dimethyltryptamine in human subjects tolerant to lysergic acid diethylamide. Psychopharmacologia (Berl.) **5**, 217—227 (1964)
Rosenberg, D. E., Wolbach, A. B. Jr., Miner, E. J., Isbell, H.: Observations on direct and cross tolerance with LSD and D-amphetamine in man. Psychopharmacologia (Berl.) **5**, 1—15 (1963b)
Roszell, D. K., Horita, A.: The effects of haloperidol and thioridazine on apomorphine- and LSD-induced hyperthermia in the rabbit. J. psychiat. Res. **12**, 117—123 (1975)
Roth, W. T.: The effect of LSD, mescaline, and *d*-amphetamine on the evoked "secondary discharge". Psychopharmacologia (Berl.) **9**, 253—258 (1966)
Rothlin, E.: Lysergic acid diethylamide and related substances. Ann. N. Y. Acad. Sci. **66**, 668—676 (1957a)
Rothlin, E.: Pharmacology of lysergic acid diethylamide and some of its related compounds. J. Pharm. Pharmacol. **9**, 569—587 (1957b)
Rothlin, E., Cerletti, A.: Über einige pharmakologische Untersuchungen an Mäusen mit congenitaler Drehsucht. Helv. physiol. pharmacol. Acta **10** 319—327 (1952)
Rothlin, E., Cerletti, A.: Pharmacology of LSD-25. In: Cholden, L. (Ed.): Lysergic acid diethylamide and mescaline in experimental psychiatry. New York: Grune & Stratton 1956
Rothlin, E., Cerletti, A., Konzett, H., Schalch, W. R., Taeschler, M.: Zentrale vegetative LSD-Effekte. Experientia (Basel) **12**, 154 (1956)
Rovetta, P.: Effect of mescaline and L.S.D. on evoked responses especially of the optic system of the cat. Electroenceph. clin. Neurophysiol. **8**, 15—24 (1956)
Rubin, B., Piala, J. J., Burke, J. C., Craver, B. N.: A new, potent and specific serotonin inhibitor, (SQ 10,643) 2′-(3-dimethylaminopropylthio) cinnamanilide hydrochloride: Antiserotonin activity on uterus and on gastrointestinal, vascular, and respiratory systems of animals. Arch. int. Pharmacodyn. **152**, 132—143 (1964)
Saavedra, J. M., Axelrod, J.: Psychotomimetic N-methylated tryptamines: Formation in brain in vivo and in vitro. Science **175**, 1365—1366 (1972a)
Saavedra, J. M., Axelrod, J.: A specific and sensitive enzymatic assay for tryptamine in tissues. J. Pharmacol. exp. Ther. **182**, 363—369 (1972b)
Saavedra, J. M., Axelrod, J.: Effect of drugs on the tryptamine content of rat tissues. J. Pharmacol. exp. Ther. **185**, 523—529 (1973)
Safer, D. J.: The effect of LSD on sleep-deprived men. Psychopharmacologia (Berl.) **17**, 414—424 (1970)
Sai-Halász, A., Brunecker, G., Szára, S.: Dimethyltryptamin: Ein neues Psychoticum. Psychiat. et Neurol. (Basel) **135**, 285—301 (1958)

Salmoiraghi, G. C., McCubbin, J. W., Page, I. H.: Effects of *d*-lysergic acid diethylamide and its brom derivative on cardiovascular responses to serotonin and on arterial pressure. J. Pharmacol. exp. Ther. **119**, 240—247 (1957)
Salomon, K., Gabrio, B. W., Thale, T.: A study on mescaline in human subjects. J. Pharmacol. exp. Ther. **95**, 455—459 (1949)
Salvatore, S., Hyde, R. W.: Progression of effects of lysergic acid diethylamide (LSD). Arch. Neurol. Psychiat. (Chic.) **76**, 50—59 (1956)
Samanin, R., Valzelli, L., Gumulka, W.: Inhibitory effect of midbrain raphe stimulation on cortical evoked potentials in rats. Psychopharmacologia (Berl.) **24**, 373—379 (1972)
Schechter, M. D., Rosecrans, J. A.: Lysergic acid diethylamide (LSD) as a discriminative cue: Drugs with similar stimulus properties. Psychopharmacologia (Berl.) **26**, 313—316 (1972)
Schechter, M. D., Winter, J. C.: Effect of mescaline and lysergic acid diethylamide on flicker discrimination in the rat. J. Pharmacol. exp. Ther. **177**, 461—467 (1971)
Schneider, C.: Behavioural effects of some morphine antagonists and hallucinogens in the rat. Nature (Lond.) **220**, 586—587 (1968)
Schultes, R. E.: Hallucinogens of plant origin. Science **163**, 245—254 (1969)
Schultes, R. E.: The New World Indians and their hallucinogenic plants. Morris Arboretum Bull. **21**, 3—14 (1970a)
Schultes, R. E.: The botanical and chemical distribution of hallucinogens. Ann. Rev. Plant Physiol. **21**, 571—598 (1970b)
Schwartz, A. S., Cheney, C.: Effect of LSD on the tonic activity of the visual pathways of the cat. Life Sci. **4**, 771—778 (1965a)
Schwartz, A. S., Cheney, C.: Retinal effects of high doses of LSD in the cat. Exp. Neurol. **13**, 273—282 (1965b)
Schwarz, B. E., Bickford, R. G., Rome, H. P.: Reversibility of induced psychosis with chlorpromazine. Proc. Mayo Clin. **30**, 407—417 (1955)
Schwarz, B. E., Sem-Jacobsen, C. W., Petersen, M. C.: Effects of mescaline, LSD-25, and adrenochrome on depth electrograms in man. Arch. Neurol. Psychiat. (Chic.) **75**, 579—587 (1956b)
Schwarz, B. E., Wakim, K. G., Bickford, R. G., Lichtenheld, E. R.: Behavioral and electroencephalographic effects of hallucinogenic drugs. Arch. Neurol. Psychiat. (Chic.) **75**, 83—90 (1956a)
Schweigerdt, A. K., Himwich, H. E.: An electrographic study of bufotenin and 5-hydroxytryptophan. J. Pharmacol. exp. Ther. **144**, 253—259 (1964)
Schweigerdt, A. K., Stewart, A. H., Himwich, H. E.: An electrographic study of *d*-lysergic acid diethylamide and nine congeners. J. Pharmacol. exp. Ther. **151**, 353—359 (1966)
Shagass, C.: Effects of LSD on somatosensory and visual evoked responses and on the EEG in man. In: Wortis, J. (Ed.): Recent advances in biological psychiatry, Vol. 9. New York: Plenum Press 1967
Sharpe, L. G., Otis, L. S., Schusterman, R. J.: Disruption of size discrimination in squirrel monkeys (Saimiri sciureus) by LSD-25. Psychon. Sci. **7**, 103—104 (1967)
Shulgin, A. T.: Psychotomimetic amphetamines: Methoxy 3,4-dialkoxy-amphetamines. Experientia (Basel) **20**, 366—370 (1964a)
Shulgin, A. T.: 3-Methoxy-4,5-methylenedioxy amphetamine, a new psychotomimetic agent. Nature (Lond.) **201**, 1120—1121 (1964b)
Shulgin, A. T.: The ethyl homologs of 2,4,5-trimethoxyphenylisopropylamine. J. med. Chem. **11**, 186—187 (1968)
Shulgin, A. T., Bunnell, S., Sargent, T. III: The psychotomimetic properties of 3,4,5-trimethoxyamphetamine. Nature (Lond.) **189**, 1011—1012 (1961)
Shulgin, A. T., Carter, M. F.: Centrally active phenethylamines. Psychopharmacol. Commun. **1**, 93—98 (1975)
Shulgin, A. T., Sargent, T.: Psychotropic phenylisopropylamines derived from apiole and dillapiole. Nature (Lond.) **215**, 1494—1495 (1967)
Shulgin, A. T., Sargent, T., Naranjo, C.: Structure-activity relationships of one-ring psychotomimetics. Nature (Lond.) **221**, 537—541 (1969)
Shulgin, A. T., Sargent, T., Naranjo, C.: 4-Bromo-2,5-dimethoxyphenylisopropylamine, a new centrally active amphetamine analog. Pharmacology **5**, 103—107 (1971)

Siegel, R. K., Brewster, J. M., Jarvik, M. E.: An observational study of hallucinogen-induced behavior in unrestrained Macaca mulatta. Psychopharmacologia (Berl.) **40**, 211—223 (1974)

Silva, M. T. A., Calil, H. M.: Screening hallucinogenic drugs: Systematic study of three behavioral tests. Psychopharmacologia (Berl.) **42**, 163—171 (1975)

Silva, M. T. A., Carlini, E. A., Claussen, U., Korte, F.: Lack of cross-tolerance in rats among (-)Δ^9-*trans*-tetrahydrocannabinol (Δ^9-THC), cannabis extract, mescaline and lysergic acid diethylamide (LSD-25). Psychopharmacologia (Berl.) **13**, 332—340 (1968)

Silverstein, A. B., Klee, G. D.: Effects of lysergic acid diethylamide (LSD-25) on intellectual functions. Arch. Neurol. Psychiat (Chic.) **80**, 477—480 (1958)

Sivadjian, J.: L'action du diéthylamide de l'acide lysergique (LSD-25) sur le conditionnement et la sudation. Thérapie 25, 813—821 (1970)

Slater, I. H., Davis, K. H., Leary, D. E., Boyd, E. S.: The action of serotonin and lysergic acid ethylamide on spinal reflexes. J. Pharmacol. exp. Ther. **113**, 48—49 (1955)

Sloan, J. W., Martin, W. R., Clements, T. H., Buchwald, W. F., Bridges, S. R.: Factors influencing brain and tissue levels of tryptamine: Species, drugs and lesions. J. Neurochem. **24**, 523—532 (1975)

Smith, C. M., Walaszek, E. J.: Interaction of Substance P and other smooth muscle stimulants with LSD-25 on the guinea pig ileum. Arch. int. Pharmacodyn. **138**, 429—436 (1962)

Smythies, J. R.: Quantitative measurement of the effect of lysergic acid diethylamide on mice and its interactions with other drugs. Nature (Lond.) **138**, 545—546 (1959)

Smythies, J. R.: The mode of action of psychotomimetic drugs. Neurosci. Res. Prog. Bull. **8**, 1—152 (1970)

Smythies, J. R., Benington, F., Morin, R. D.: The mechanism of action of hallucinogenic drugs on a possible serotonin receptor in the brain. Int. Rev. Neurobiol. **12**, 207—233 (1970)

Smythies, J. R., Bradley, R. J., Johnston, V. S., Benington, F., Morin, R. D., Clark, L. C., Jr.: Structure-activity relationship studies on mescaline. Psychopharmacologia (Berl.) **10**, 379—387 (1967a)

Smythies, J. R., Johnston, V. S., Bradley, R. J.: Alteration by pretreatment with iproniazid and an inactive mescaline analogue of a behavior change induced by mescaline. Nature (Lond.) **216**, 196—197 (1967b)

Smythies, J. R., Koella, W. P., Levy, C. K.: The effect of mescaline on the optic evoked potentials in the unanesthetized rabbit. J. Pharmacol. exp. Ther. **129**, 462—470 (1960)

Snodgrass, S. R., Horn, A. S.: An assay procedure for tryptamine in brain and spinal cord using its (^{3}H)dansyl derivative. J. Neurochem. **21**, 687—696 (1973)

Snyder, S. H., Faillace, L., Hollister, L.: 2,5-Dimethoxy-4-methyl-amphetamine (STP): A new hallucinogenic drug. Science **158**, 669—670 (1967)

Snyder, S. H., Reivich, M.: Regional localization of lysergic acid diethylamide in monkey brain. Nature (Lond.) **209**, 1093—1095 (1966)

Snyder, S. H., Richelson, E.: Psychedelic drugs: Steric factors that predict psychotropic activity. Proc. nat. Acad. Sci. (Wash.) **60**, 206—213 (1968)

Snyder, S. H., Weingartner, H., Faillace, L. A.: DOET (2,5-dimethoxy-4-ethylamphetamine) and DOM (STP) (2,5-dimethoxy-4-methylamphetamine), new psychotropic agents: Their effects in man. In: Efron, D. H. (Ed.): Psychotomimetic drugs. New York: Raven Press 1970

Snyder, S. H., Weingartner, H., Faillace, L. A.: DOET (2,5-dimethoxy-4-ethylamphetamine), a new psychotropic drug; effects of varying doses in man. Arch. gen. Psychiat. **24**, 50—55 (1971)

Speck, L. B.: Toxicity and effects of increasing doses of mescaline. J. Pharmacol. exp. Ther. **119**, 78—84 (1957)

Speck, L. B.: Electroencephalographic changes in the rat with mescaline intoxication. J. Pharmacol. exp. Ther. **122**, 201—206 (1958)

Steiner, J. E., Sulman, F. G.: Simultaneous studies of blood sugar, behavioural changes and EEG on the wake rabbit after administration of psilocybin. Arch. int. Pharmacodyn. **145**, 301—308 (1963)

Stern, W. C., Morgane, P. J., Bronzino, J. D.: LSD: Effects on sleep patterns and spiking activity in the lateral geniculate nucleus. Brain Res. **41**, 199—204 (1972)

Stoff, D. M., Mandel, I. J., Gorelick, D. A., Bridger, W. H.: Acute and chronic effects of LSD and 3,4-dimethoxyphenylethylamine on shuttlebox escape/avoidance in rats. Psychopharmacologia (Berl.) **36**, 301—312 (1974)

Stoll, A., Rothlin, E., Rutschmann, J., Schalch, W.R.: Distribution and fate of ^{14}C-labeled lysergic acid diethylamide (LSD-25) in the animal body. Experientia (Basel) **11**, 396—397 (1955)
Stoll, W.A.: D-lysergic acid diethylamide (LSD-25), a phantasticum of the ergot group. Schweiz. Arch. Neurol. Psychiat. **60**, 1—45 (1947)
Straub, W.: Zur chemischen Kinetik Muskarinwirkung und des Antagonismus Muskarin-atropin. Pflügers Arch. ges. Physiol. **119**, 127—151 (1905)
Stumpf, Ch.: Drug action on septal and hippocampal units. In: Bradley, P.B., Flugel, F., Hoch, P. (Eds.): Neuropharmacology. Amsterdam: Elsevier Publishing Company 1964
Stumpf, Ch., Petsche, H., Gogolak, G.: The significance of the rabbit's septum as a relay station between the midbrain and the hippocampus. II. The differential influence of drugs upon both the septal cell firing pattern and the hippocampus theta activity. Electroenceph. clin. Neurophysiol. **14**, 212—219 (1962)
Sugrue, M.F.: A study of the role of noradrenaline in behavioral changes produced in the rat by psychotomimetic drugs. Brit. J. Pharmacol. **35**, 243—252 (1969)
Szara, S.: Dimethyltryptamin: Its metabolism in man; the relation of its psychotic effect to the serotonin metabolism. Experientia (Basel) **12**, 441—442 (1956)
Szara, S.: The comparison of the psychotic effect of tryptamine derivatives with the effects of mescaline and LSD-25 in self experiments. In: Garattini, S., Ghetti, V. (Eds.): Psychotropic drugs. Amsterdam: Elsevier Publishing Company 1957
Szara, S.: DMT (N,N-dimethyltryptamine) and homologues: Clinical and pharmacological considerations. In: Efron, D.H. (Ed.): Psychotomimetic drugs. New York: Raven Press 1970
Szara, S., Rockland, L.H., Rosenthal, D., Handlon, J.H.: Psychological effects and metabolism of N,N-diethyltryptamine in man. Arch. gen. Psychiat. **15**, 320—329 (1966)
Taeschler, M., Cerletti, A.: Some observations on the interaction of reserpine and LSD. J. Pharmacol. exp. Ther. **120**, 179—183 (1957)
Taeschler, M., Weidmann, H., Cerletti, A.: Die Wirkung von LSD auf die Reaktionszeiten bei einer bedingten Fluchtreaktion und im Analgesietest. Helv. physiol. Acta **18**, 43—49 (1960)
Takagi, H., Yamamoto, S., Takaori, S., Ogiu, K.: The effect of LSD and reserpine on the central nervous system of the cat. The antagonism between LSD and chlorpromazine or reserpine. Jap. J. Pharmacol. **7**, 119—134 (1958)
Takeo, Y., Himwich, H.E.: Mescaline, 3,4-dimethoxyphenylethylamine, and adrenaline: Sites of electroencephalographic arousal. Science **150**, 1309—1310 (1965)
Takeo, Y., Himwich, H.E.: The significance of methyl groups in the electroencephalographic effects of indolealkylamines in the rabbit. Biochem. Pharmacol. **16**, 1013—1022 (1967)
Tang, A.H., Uhlinger, B.J., Schroeder, L.A.: Comparison of N,N-diethyltryptamine and U-22,-394A, a new tryptamine derivative. Arch. int. Pharmacodyn. **176**, 74—85 (1968)
Taylor, R.L., Maurer, J.I., Tinklenberg, J.R.: Management of "bad trips" in an evolving drug scene. J. Amer. med. Ass. **213**, 422—425 (1970)
Tedeschi, D.H., Tedeschi, R.E., Fellows, E.J.: The effects of tryptamine on the central nervous system, including a pharmacological procedure for the evaluation of iproniazid-like drugs. J. Pharmacol. exp. Ther. **126**, 223—232 (1959)
Thiessen, P.N., Cook D.A.: The properties of 3,4-methylenedioxy-amphetamine (MDA). I. A review of the literature. Clin. Toxicol. **6**, 45—52 (1973)
Tilson, H.A., Sparber, S.B.: Studies on the concurrent behavioral and neurochemical effects of psychoactive drugs using the push-pull cannula. J. Pharmacol. exp. Ther. **181**, 387—398 (1972)
Tilson, H.A., Sparber, S.B.: Similarities and differences between mescaline, lysergic acid diethylamide-25 (LSD) and *d*-amphetamine on various components of fixed interval responding in the rat. J. Pharmacol. exp. Ther. **184**, 376—384 (1973)
Turner, W.J., Merlis, S.: Effect of some indolealkylamines on man. Arch. Neurol. Psychiat. (Chic.) **81**, 121—129 (1959)
Twarog, B.M., Page, I.H.: Serotonin content of some mammalian tissues and urine and a method for its determination. Amer. J. Physiol. **175**, 157—161 (1953)
Ursin, H.: The lack of effect of LSD-25 on amygdaloid and cortical attention responses. Psychopharmacologia (Berl.) **3**, 317—330 (1962)
U.S. Department of Justice, Drug Enforcement Administration: DAWN Management Report, July 1975

Uyeno, E. T.: Effects of mescaline and psilocybin on dominance behavior of the rat. Arch. int. Pharmacodyn. **166**, 60—64 (1967a)
Uyeno, E. T.: Lysergic acid diethylamide and dominance behavior of the squirrel monkey. Arch. int. Pharmacodyn. **169**, 66—69 (1967b)
Uyeno, E. T.: Hallucinogenic compounds and swimming response. J. Pharmacol. Exp. Ther. **159**, 216—221 (1968)
Uyeno, E. T.: Relative potency of amphetamine derivatives and N,N-dimethyltryptamines. Psychopharmacoligia (Berl.) **19**, 381—387 (1971)
Uyeno, E. T., Benson, W. M.: Effects of lysergic acid diethylamide on attack behavior of male albino mice. Psychopharmacologia (Berl.) **7**, 20—26 (1965)
Uyeno, E. T., Otis, L. S., Mitoma, C.: Behavioral evaluation of hallucinogenic trimethoxy-amphetamines in squirrel monkeys (Saimiri Sciureus). Commun. Behav. Biol. **1**, 83—90 (1968)
Vane, J. R.: The actions of sympathomimetic amines on tryptamine receptors. In: Adrenergic mechanisms. CIBA Foundation Symposium. Boston: Little, Brown and Company 1960
Vane, J. R., Collier, H. O. J., Corne, S. J., Marley, E., Bradley, P. B.: Tryptamine receptors in the central nervous system. Nature (Lond.) **191**, 1068—1069 (1961)
Vaupel, D. B., Martin, W. R.: Interaction of oxotremorine with atropine, chlorpromazine, cyproheptadine, imipramine and phenoxybenzamine on the flexor reflex of the chronic spinal dog. Psychopharmacologia (Berl.) **30**, 13—26 (1973)
Vaupel, D. B., Martin, W. R.: Actions of methoxamine and tryptamine and their interactions with cyproheptadine and phenoxybenzamine on cat spinal cord segmental reflexes. J. Pharmacol. exp. Ther. **196**, 87—96 (1976)
Vaupel, D. B., Nozaki, M.: A comparison of the effects of 2,5-dimethoxyamphetamine (DMA) and lysergic acid diethylamide (LSD) in the dog. Fed. Proc. **34**, 767 (1975)
Vaupel, D. B., Nozaki, M., Martin, W. R.: A pharmacologic comparison of 2,5-dimethoxyamphetamine and LSD in the chronic spinal dog. Drug Alc. Depend. **2**, 45—63 (1977)
Verdeaux, G., Marty, R.: Action sur l'electroencephalogramme de substances pharmacodynamiques d'intert clinique. Rev. neurol. **91**, 405—427 (1954)
Vogt, M., Gunn, C. G., Jr., Sawyer, C. H.: Electroencephalographic effects of intraventricular 5-HT and LSD in the cat. Neurology (Minneap.) **7**, 559—566 (1957)
Von Hungen, K., Roberts, S., Hill, D. F.: LSD as an agonist and antagonist at central dopamine receptors. Nature (Lond.) **252**, 588—589 (1974).
Votava, Z., Glisson, S. N., Himwich, H. E.: Behavioral reaction of rats pretreated with reserpine to LSD-25. Int. J. Neuropharmacol. **6**, 543—547 (1967)
Wallach, M. B., Friedman, E., Gershon, S.: 2,5-Dimethoxy-4-methyl-amphetamine (DOM), a neuropharmacological examination. J. Pharmacol. exp. Ther. **182**, 145—154 (1972)
Wallach, M. B., Hine, B., Gershon, S.: Cross tolerance or tachyphylaxis among various psychotomimetic agents on cats. Europ. J. Pharmacol. **29**, 89—92 (1974)
Walter, S., Balzano, E., Vuillon-Cacciuttolo, G., Naquet, R.: Effects comportementaux et electrographiques du diethylamide de l'acide d-lysergique (LSD 25) sur le papio papio photosensible. Electroenceph. clin. Neurophysiol. **30**, 294—305 (1971)
Wasson, R. G.: The hallucinogenic mushrooms of Mexico: An adventure in ethnomycological exploration. Trans. N. Y. Acad. Sci. **21**, 325—336 (1959)
Weidmann, H., Cerletti, A.: Die Wirkung von d-lysergsäure-diäthylamid und 5-hydroxytryptamin (Serotonin) auf spinale Reflexe der Katze. Helv. physiol. pharmacol. Acta **15**, 376—383 (1957)
Weidmann, H., Cerletti, A.: Studies on psilocybin and related componds. I. Communication. Structure/activity relationship of oxyindole-derivatives with regard to their effect on the knee jerk of spinal cats. Helv. physiol. pharmacol. Acta **18**, 174—182 (1960)
Wells, B.: Psychedelic drugs. New York: Jason Aronson 1974
Wikler, A.: Clinical and electroencephalographic studies on the effects of mescaline, N-allylnormorphine and morphine in man. A pharmacologic analysis of the functions of the spontaneous electrical activity of the cerebral cortex. J. nerv. ment. Dis. **120**, 157—175 (1954)
Wikler, A.: The relation of psychiatry to pharmacology. Baltimore: Williams & Wilkins Company 1957
Wikler, A., Haertzen, C. A., Chessick, R. D., Hill, H. E., Pescor, F. T.: Reaction time ("mental set") in control and chronic schizophrenic subjects and in postaddicts under placebo, LSD-25, morphine, pentobarbital and amphetamine. Psychopharmacologia (Berl.) **7**, 423—443 (1965)

Winter, C. A., Flataker, L.: Effects of lysergic acid diethylamide upon performance of trained rats. Proc. Soc. exp. Biol. (N. Y.) **92**, 285—289 (1956)
Winter, J. C.: Tolerance to a behavioral effect of lysergic acid diethylamide and cross-tolerance to mescaline in the rat: Absence of a metabolic component. J. Pharmacol. exp. Ther. **178**, 625—630 (1971)
Winter, J. C.: A comparison of the stimulus properties of mescaline and 2,3,4-trimethoxy-phenylethylamine. J. Pharmacol. exp. Ther. **185**, 101—107 (1973)
Winter, J. C.: The effects of 3,4-dimethoxyphenylethylamine in rats trained with mescaline as a discriminative stimulus. J. Pharmacol. exp. Ther. **189**, 741—747 (1974a)
Winter, J. C.: Hallucinogens as discriminative stimuli. Fed. Proc. **33**, 1825—1832 (1974b)
Winters, W. D.: Neuropharmacological studies and postulates on excitation and depression in the central nervous system. In: Wortis, J. (Ed.): Recent advances in biological psychiatry, Vol. 9. New York: Plenum Publishing Company 1967
Winters, W. D., Mori, K., Wallach, M. B., Marcus, R. J., Spooner, C. E.: Reticular multiple unit activity during a progression of states induced by CNS excitants. Electroenceph. clin. Neurophysiol. **27**, 514—522 (1969)
Wolbach, A. B., Jr., Isbell, H., Miner, E. J.: Cross tolerance between mescaline and LSD-25 with a comparison of the mescaline and LSD reactions. Psychopharmacologia (Berl.) **3**, 1—14 (1962a)
Wolbach, A. B., Jr., Miner, E. J., Isbell, H.: Comparison of psilocin with psilocybin, mescaline and LSD-25. Psychopharmacologia (Berl.) **3**, 219—223 (1962b)
Woolley, D. W.: The biochemical bases of psychoses or the serotonin hypothesis about mental diseases. New York: John Wiley and Sons, Inc. 1962
Woolley, D. W., Shaw, E.: Antimetabolites of serotonin. J. biol. Chem. **203**, 69—79 (1953)
Woolley, D. W., Shaw, E.: A biochemical and pharmacological suggestion about certain mental disorders. Proc. nat. Acad. Sci. (Wash.) **40**, 228—231 (1954a)
Woolley, D. W., Shaw, E.: Some neurophysiological aspects of serotonin. Brit. med. J. **1954 IIb**, 122—126 (1954b)
Woolley, D. W., Shaw, E. N.: Evidence for the participation of serotonin in mental processes. Ann. N. Y. Acad. Sci. **66**, 649—667 (1957)
Yamamoto, T., Ueki, S.: Behavioral effects of 2,5-dimethoxy-4-methyl-amphetamine (DOM) in rats and mice. Europ. J. Pharmacol. **31**, 156—162 (1975)
Yui, T., Takeo, Y.: Neuropharmacological studies on a new series of ergot alkaloids. The effects on electrocorticogram of rabbits. Jap. J. Pharmacol. **7**, 162—168 (1958)

SECTION II

Marihuana Dependence

CHAPTER 1

Cannabis

Its Chemistry, Pharmacology, and Toxicology

L. S. HARRIS, W. L. DEWEY, and R. K. RAZDAN

A. Introduction

For the past decade no subject, with the exception of the Vietnam War, has aroused such intense, continuous discussion and controversy in the Western World as has marihuana. This is particularly true in the United States. Some insight into the reasons for this may be obtained by perusal of the brilliant treatise by BONNIE and WHITEBREAD (1974) which traces the development of legal measures to control the use of marihuana in the United States. Additional insights may also be gained by a consideration of how other cultures view the use of this "deceptive weed" (RUBIN, 1975).

Marihuana is derived from the plant *Cannabis sativa*, which has been used for millenia as a source of hemp fiber, plant oil and drugs, both medicinal and social. Indeed, marihuana is one of our oldest drugs. It is purported to have been referred to by the Chinese emperor and physician Shen Nung in a "Pharmacopoeia" written about 2700 B.C., but this is a myth perpetuated by many modern writers. The drug was referred to in a Chinese "Treatise on Medicine" dated about the second century A.D. (LI, 1974). Perhaps the earliest documented references to the drug were found in Assyria in the eighth century B.C. The worldwide extent of the marihuana problem is difficult to ascertain. The use of the drug is endemic in certain areas and a number of United Nations reports have estimated that there are upwards of 200000000 users in the world. However, the reliability of these figures is open to some question. The latest data concerning the extent of cannabis use for a variety of countries and geographical regions may be found in the Third Annual Marihuana Report (SECRETARY, H.E.W., 1973).

The data for the United States are much better since a large number of surveys have been carried out in this country concerning the incidence and extent of the use of marihuana. These have been summarized in Appendix 1 of the Technical Papers of the First Report of the National Commission of Marihuana and Drug Abuse (SHAFER, 1972). Although there are wide differences from study to study, when the results are averaged they agree quite well with the Commission's own survey in 1971. This survey indicated that 14% of the youth (12–17 years of age) and 15% of the adults (18 years and older) reported that they had tried marihuana at least once (ABELSON et al., 1972). Further breakdown of these figures shows a clear age and education relationship. For instance, the use of marihuana in the adult population is highly concentrated in the young adults (18–24 years). This fits the finding that the incidence among college students "having ever used marihuana" was 44%. In addition to age, the survey findings indicate that a number of other factors are significantly correlated with marihuana use. These include sex, socioeconomic status, religious affiliation, political orientation, student status, region of the country and type of community.

The Commission's review also demonstrated a marked increase in use from earlier studies. A recent survey conducted by the Drug Abuse Council (1975) reveals an incidence of "having ever used" and "currently use" among teenagers of 14% and 5%, and among adults 18% and 8%, respectively. These figures appear to indicate that there has been little increase in the number of teenagers experimenting with marihuana. They also indicate that the increase in use among adults is the result of movement from the earlier teenager group into the adult age group and a confirmation of the high incidence among college students and their peers. For instance, the incidence of "having ever used" and "currently use" among the 18–25 year old group is 49% and 25%, respectively. This is contrasted with the 40–49 year old age group where the comparable figures are 9% and 5%. While there is some reason to believe that the rate of entry of new users has plateaued, the overall incidence of use in the population is growing as the older groups fade out. Thus, it was estimated in 1971 that 8 million Americans were using marihuana regularly. Three years later over 12 million were regular users. If the present trends hold, there is every reason to believe that these figures will continue to increase. There is also evidence to indicate that similar trends are occurring in other countries and this growing population of regular users must be taken into account in any consideration of the drug, be it biomedical, social, or legal.

The recent rapid increase in the use of marihuana in the United States and Europe has caused an intense upsurge of biomedical and public health interest in the drug. This, coupled with a large increase in research funding has led, in turn, to a rapid proliferation of scientific and lay literature. We will not make an attempt to review this literature in a comprehensive fashion. For instance, the number of papers published in the last 10 years has greatly exceeded the entire total previously published. Recently, a large number of books, proceedings, and reviews have been published which thoroughly cover most aspects of marihuana (WOLSTENHOLME and KNIGHT, 1965; SOLOMON, 1966; JOYCE and CURRY, 1970; SNYDER, 1971; GRINSPOON, 1971; SINGER, 1971; WAY, 1971; PATON and CROWN, 1972; LEWIS, 1972; BRECHER, 1972; MECHOULAM, 1973; NAHAS, 1973; MILLER, 1974; NAHAS, 1976; and BRAUDE and SZARA, 1976). In addition, reports have been issued by two national commissions (LEDAIN, 1970; SHAFER, 1972) and, since 1971, annual reports on "Marihuana and Health" have been presented to the U.S. Congress by the Secretary of Health, Education and Welfare as mandated by law (SECRETARY, H. E. W., 1971—1975).

This chapter will concentrate on the chemical, metabolic, pharmacological, toxicological and therapeutic aspects of marihuana. No attempt will be made to be exhaustive. However, some attempt will be made to be critical and to relate the material to the problem of abuse.

B. Chemistry

Marihuana is derived from the plant *Cannabis sativa L. Cannabis* is closely allied to *Humulus*, the genus of the hop plant and botanists now generally agree that *Canna-*

bis is a monotypic genus. Early taxonomists tended to include *Cannabis* in the *Urticaceae* family and as late as the early part of this century they favoured *Moraceae*, but the recent trend is towards the separation of *Cannabis* into a distinct family *Cannabaceae*, sometimes written as *Cannabiaceae*, *Cannabidiaceae*, or *Cannabinaceae*. In summary, the *Cannabaceae* differ from *Moraceae* in the form of the flowers and fruit, and are herbs or vines devoid of a milky latex (SCHULTES, 1970).

The plant *C. sativa* is an annual and grows to a height of 1–5 m (3–15 ft.) (STEARN, 1970). The flowers grow on the upper part of the plant. It is dioecious (i.e., the male and the female plants are separate); however, occasionally the male and female flowers are known to occur on the same plant. The male plant usually grows taller than the female and dies earlier. The main physiologically active constituents of marihuana occurs in the resinous exudate of the flowers and until very recently, it was believed that only the female plant yielded the active resin. It is now well established that both the male and female plants of the same variety contain similar amounts of cannabinoids, including the active constituents (VALLE et al., 1968; OHLSSON et al., 1971; FETTERMAN et al., 1971). The term "cannabinoid" is used for the typical C_{21}-group of compounds present in *C. sativa* and encompasses their analogs and transformation products (MECHOULAM and GAONI, 1967).

Numerous preparations of *Cannabis* (SCHULTES, 1970; MECHOULAM and GAONI, 1967), which are eaten, drunk or smoked, are known. The most potent preparation is the unadulterated resin, which is known as hashish in the Middle East and charas in India. Marihuana (pot, grass, etc.), used most in the United States, refers to the dried flowering tops and leaves of the plant which are smoked in a pipe or cigarette. The bhang of India is a concoction made in water or milk from the flowering tops of the plant and is usually drunk.

C. sativa is generally of two types—one used primarily for its drug content and the other for its fiber. The material used for drug purposes varies in its potency according to its geographical location. It is well known, for example, that the variety grown in Mexico or the United States is less potent than that from the Middle East and parts of Asia (BOUQUET, 1950, 1951; KABELIK et al., 1960).

A useful classification based on cannabinoid content of the plant has been suggested by GRLIC (1962, 1964). Thus, samples which have high cannabidiolic acid, tetrahydrocannabinoids or cannabinol content are called "unripe," "ripe," and "overripe," respectively. He has further shown that seeds from different countries yield plants with resin of different degrees of "ripeness." Conclusions along similar lines have been drawn by other workers in the field (KORTE and SIEPER, 1965; CLAUSSEN and KORTE, 1968; OHLSSON et al., 1971; FETTERMAN et al., 1971), and it appears that the mix of cannabinoids produced by the plant is controlled more by the type of seed than the quality of soil, climatic conditions, etc., but this may change after a few generations. Furthermore, it has been shown that cannabinoids are found in all parts of the plant, the content decreasing in the order bracts, flowers, leaves, smaller stems, larger stems, roots and seeds (FETTERMAN et al., 1971). Various laboratories have analyzed *Cannabis* preparations, but it is difficult to compare the results as the mode of preparation and storage conditions vary from sample to sample. However, extensive recent experience with the careful cultivation and storage of *Cannabis* has been obtained from the program at the University of Mississippi, which has been providing standardized sources of supply for research (WALLER, 1976). In general, this group has found a wide variation amoung the plant types (FETTERMAN et al., 1971; QUIMBY et al., 1973), as well as in the cannabinoid content (TURNER and HADLEY, 1973; TURNER et al., 1973), with little deterioration during controlled storage. For instance, at room temperature, in semidarkness, in air, deterioration was about 7% after one year. At lower temperatures there was even less loss (TURNER and HADLEY, 1973).

There are two main numbering systems in use at present for cannabinoids. The dibenzopyran numbering system is used in this article. The other system used quite frequently in literature is the monoterpenoid number system (Fig. 1).

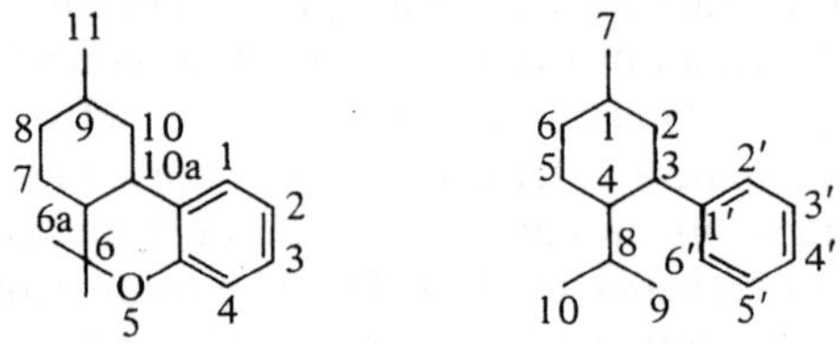

Dibenzopyran numbering

Monoterpene numbering (based on p–cymene)

Fig. 1. Cannabinoid numbering systems

Recently, various reviews have appeared on marihuana chemistry (MECHOULAM, 1970; NEUMEYER and SHAGOURY, 1971). The chemistry of cannabinoids in general was reviewed extensively in 1967 (MECHOULAM and GAONI, 1967) and was updated to cover literature until 1971 (RAZDAN, 1973). A book has now appeared on the chemistry, pharmacology, metabolism and clinical effects of marihuana, which covers the chemistry until 1972 (MECHOULAM, 1973).

Isolation

The principle activity of *Cannabis* is due to (-)-*trans*-Δ^9-tetrahydrocannabinol (Δ^9-THC, *1*), which was first isolated in pure form from hashish in 1964. For the isolation of Δ^9-THC and other constituents, the plant or hashish is extracted with petroleum ether and the extract subjected to repeated chromatography either directly or after separation into neutral and acidic fractions. In some cases, counter current distribution or its combination with chromatography has been used to separate the various constituents (MECHOULAM and GAONI, 1967; RAZDAN, 1973). WILLINSKY (1973) described the following method which has been used with good success for the isolation of cannabinoids from marihuana.

Crude marihuana yields approximately one—third of its weight in "manicured" material (i.e., marihuana from which all seeds and stalks have been removed by seiving). In this procedure, twenty grams of "manicured" plant material was extracted with two 150 ml portions of petroleum ether (b.p. 50—60° C). The two extracts were pooled and refrigerated overnight at 4° C to remove plant waxes, then filtered cold, and evaporated at 40° C *in vacuo*. For column chromatography the residue was redissolved in 10 ml of petroleum ether, and Florisil (100—120 mesh) was added and mixed by swirling until all the residue appeared to be adsorbed to it. The Florisil plus extracted residue was placed in a (3 × 40 cm) column packed with 260 g Florisil (100–120 mesh) in a pentane slurry and eluted with increasing proportions of ether in pentane. Fractions of 100 ml each were collected. Six crude fractions were isolated by this chromatographic method. Repeated chromatography yields pure fractions.

To isolate the constituents from hashish, the petroleum ether extract was separated into neutral and acidic fractions (GAONI and MECHOULAM, 1971). After repeated chromatography of the neutral fraction of Florisil and acid-washed alumina and alumina containing 12% silver nitrate, the following compounds were isolated (see Fig. 2): a mixture of waxy non-cannabinoid compounds was first obtained followed by cannabicyclol (*8a*), cannabidiol (*4a*), Δ^9-THC (*1*), cannabinol (*5a*), cannabichromene (*7a*), cannabigerol (*6a*), and polar constituents and polymers. The acidic fraction after esterification followed by column chromatography provided methyl esters of cannabinolic acid A (*6c*) but no Δ^9-THC acid A (*3a* and in some samples Δ^9-THC acid B (*3b*) was isolated (MECHOULAM et al., 1969). The cannabielsoic acids (SHANI and MECHOULAM, 1970) (*9a* and *9b*), being more polar, are only sparingly soluble in petroleum ether and were isolated (by column chromatography on silica gel) from the benzene extract of the residue remaining after petroleum ether extraction of hashish.

In addition to the column supports described above for chromatography, various other supports have been used for the separation of cannabinoids (RAZDAN, 1973; WILLINSKY, 1973). A novel solvent system, benzene-n-hexane-diethylamine (25:10:1), has also been claimed to separate cannabidiol, THC and cannabinol on a silica gel column (ARAMAKI et al., 1968).

Another technique, largely due to the work of KORTE and co-workers, which has played a major role in the isolation of many cannabinoids is the counter current distribution method. Crystalline cannabidiol (*4a*) (KORTE and SIEPER, 1960) from red oil, the first isolation of Δ^9-THC acid A (*3a*) (KORTE et al., 1965), cannabichromene (*7a*) (CLAUSSEN et al., 1966), cannabicyclol (*8a*) (CLAUSSEN et al., 1968), cannabidivarol (14) (VOLLNER et al., 1969), and cannabitriol ester of cannabidiolic acid (12) (VON SPULAK et al., 1968), were obtained utilizing this technique.

A wide variety of cannabinoids have been isolated from the plant (Fig. 2). The isolation and chemistry of most of the cannabinoids illustrated here have been reviewed (MECHOULAM and GAONI, 1967; RAZDAN, 1973; MECHOULAM, 1973). However, a number of minor constituents have since been isolated, namely *4b* (BERCHT et al., 1973), *5b* (BERCHT et al., 1973), *8b* (SHOYAMA et al., 1972), *10* (VAN GINNEKEN et al., 1972), *11*, (BERCHT et al., 1974), *13b* (PARIS et al., 1973), *16* (DEZEEUW et al., 1973), and *17* (VREE et al., 1972). In addition, four new cannabinoids were isolated in minute quantities by glass capillary chromatography (FRIEDRICH-FICHTL and SPITELLER, 1975). The isolation of cannabicitran (*11*) is interesting and suggests a search for Δ^9-*cis*-THC and *iso*-THC's in the plant, since a chemical equilibrium exists between these compounds (RAZDAN and ZITKO, 1969). Δ^8-THC (*2*) was first isolated in 1966 and is approximately just as active as Δ^9-THC (HIVELY et al., 1966). It has been found to occur only in a very few varieties of the plant although chemically it is thermodynamically more stable than Δ^9-THC. This has led to numerous chemical problems in the synthesis of THC's which will be discussed later.

Fig. 2.

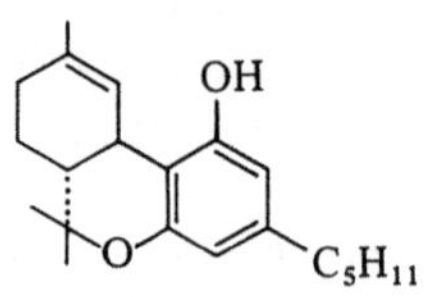

1 Δ^9–Tetrahydrocannabinol
Δ^9–THC (Δ^1–THC)

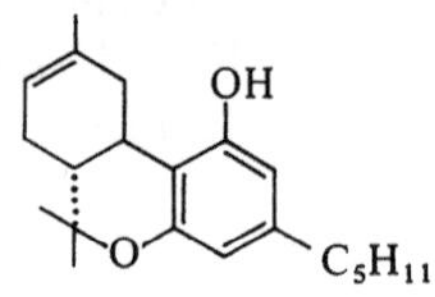

2 Δ^8–THC
($\Delta^{1(6)}$–THC, Δ^6–THC)

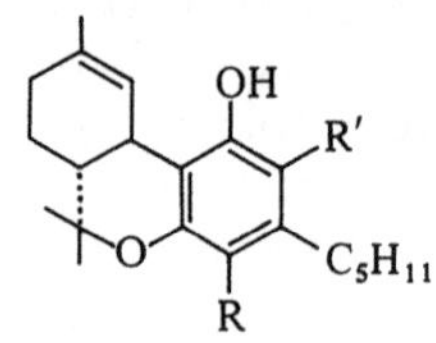

3 (a) R = H, R′ = COOH
Δ^9–THC acid A
(b) R = COOH, R′ = H
Δ^9–THC acid B

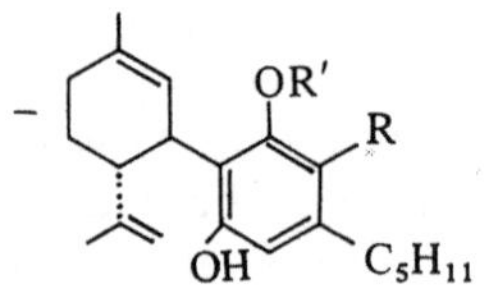

4 (a) R′ = R = H Cannabidiol (CBD)
(b) R′ = CH_3, R = H, CBD monomethyl ether
(c) R′ = H, R = COOH Cannabidiolic acid

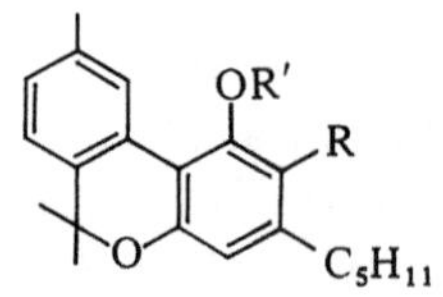

5 (a) R′ = R = H Cannabinol (CBN)
(b) R′ = CH_3, R = H CBN monomethyl ether
(c) R′ = H, R = COOH Cannabinolic acid

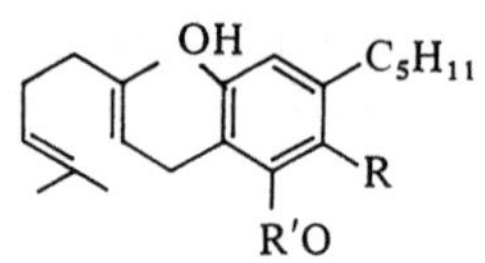

6 (a) R′ = R = H Cannabigerol
(b) R′ = CH_3, R = H
(c) R′ = H, R = COOH Cannabigerolic acid
(d) R′ = CH_3, R = COOH

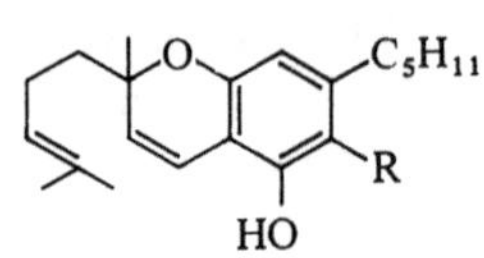

7 (a) R = H, Cannabichromene
(b) R = COOH, Cannabichromenic acid

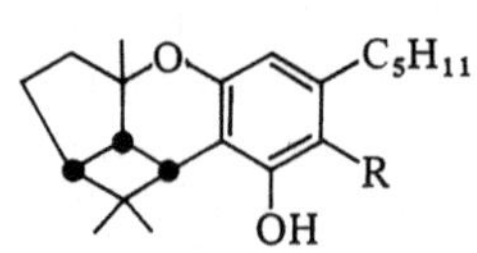

8 (a) R = H, Cannabicyclol
(b) R = COOH, Cannabicyclolic acid

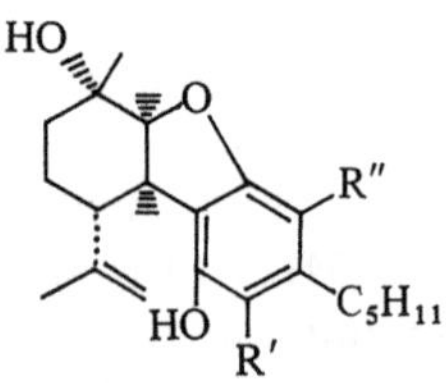

9 (a) R′ = COOH, R″ = H Cannabielsoic acid A
(b) R′ = H, R″ = COOH Cannabielsoic acid B

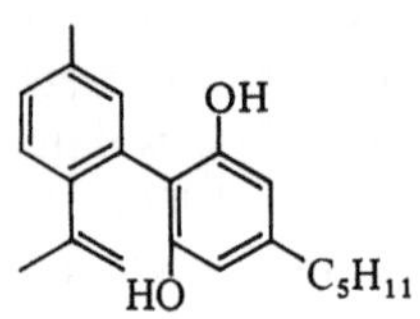

10 Cannabinodiol

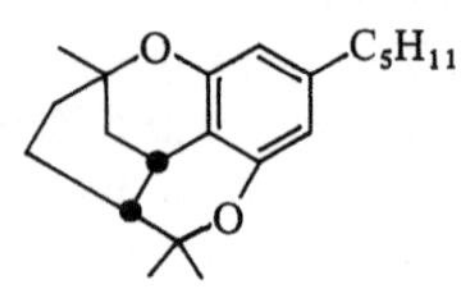

11 Cannabicitran

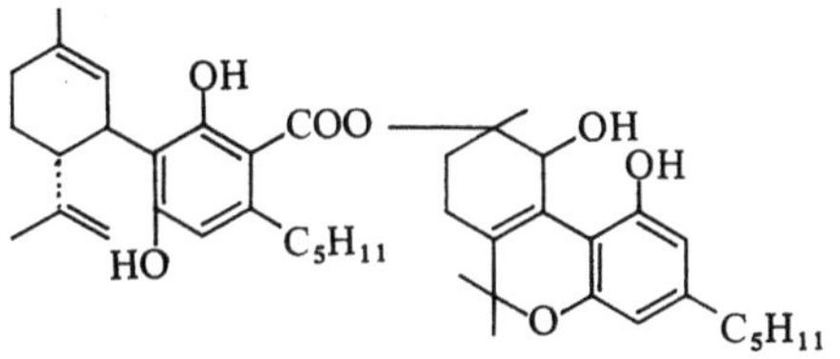

12 Cannabitriol ester of Cannabidiolic acid

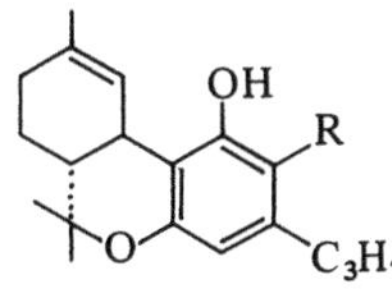

13 (a) R = H Δ^9–Tetrahydrocannabivarol (tetrahydrocannabidivarol, tetrahydrocannabivarin propyl–THC

(b) R = COOH Δ^9–Tetrahydrocannabivarol acid

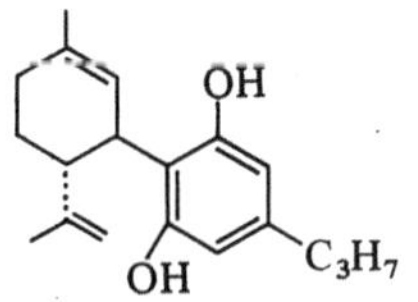

14 Cannabidivarol (Cannabidivarin, propyl cannabidiol)

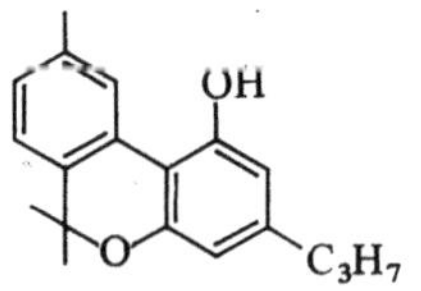

15 Cannabivarol (Cannabidivarol, Cannabivarin, propyl Cannabinol)

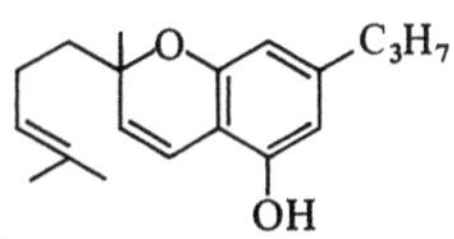

16 Cannabivarichromene

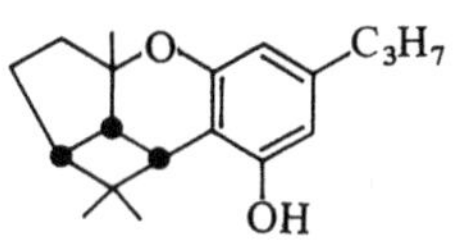

17

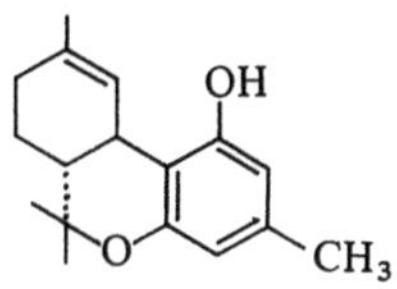

18 Δ^9–Tetrahydrocannabiorcol

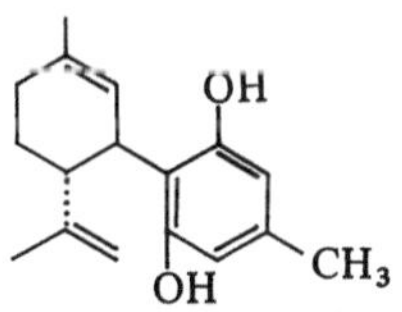

19 Cannabidiorocol

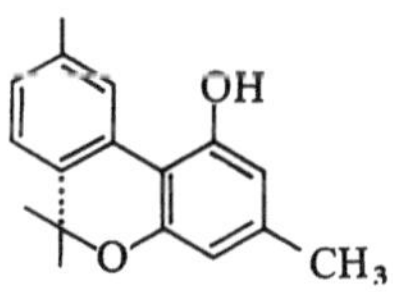

20 Cannabiorcol

The isolation of Δ^9-THC acid *3a*) (KORTE et al., 1965) was an important finding in terms of correlation between biological activity and cannabinoid content of various samples of *Cannabis* or hashish. The cannabinoid acids are biologically inactive but, on heating, they decarboxylate to furnish the corresponding neutral cannabinoids. Thus, the amount of Δ^9-THC acids should be taken into consideration when the amount of Δ^9-THC available on smoking is estimated. Similar consideration would be given to analysis of cannabinoids by gas-liquid chromatography since the acids decarboxylate to the parent compound on the column.

Table 1. Yield of cannabinoids from marihuana and hashish

Compound	Marihuana[a] Neutral and acid cannabinoids (% dry weight)	Hashish[b] Neutral cannabinoids (% dry weight)	Acid cannabinoids (% dry weight)
Cannabicyclol (CBC)	0.09	0.11	3.5
Cannabidiol (CND)	0.04 (0.07)[c]	3.74 (1.4) (2.5)	
Δ^8-Tetrahydrocannabinol (Δ^8-THC)	Not detected	Not detected	
Δ^9-Tetrahydrocannabinol (Δ^9-THC)	0.72 (1.1)	3.30 (1.4) (3.4)	2.1
Cannabinol (CBN)	0.26 (0.52)	1.30 (0.3) (1.2)	0.2
Cannabichromene	Not detected	0.19	
Cannabigerol	0.11	0.30	0.2

[a] As percentage of marihuana (street sample of unknown origin) by glc analysis of total cannabinoid content (neutral plus acid cannabinoids).
[b] As percentage of hashish (Lebanese, 1 year old, confiscated) by glc analysis of neutral cannabinoids and acid cannabinoids.
[c] Parentheses indicate partial analysis of a sample.

In addition to cannabinoids other materials have been isolated from *C. sativa*. Alkaloids trigonellin and cannabisativine, which has a novel structure (LOTTER et al., 1975), have been reported. Piperidine and choline are present and the presence of unidentified atropinic and muscarinic substances in the water-extract of the plant have also been described. In addition, other alkaloids of unknown structure and possessing some physiological activity have been found in marihuana. Since they are present only in very low concentration (0.02% dry wt basis), it is unlikely that they contribute appreciably towards activity of the plant in man.

Plant waxes like *n*-nonecasane, various phenolics and sugars and essential oils have been identified in *C. sativa*. The characteristic odor of the plant can be partly ascribed to the presence of terpenes: e.g., caryophyllene, β-farnesene, α-selinene, β-phellandrene and limonene (MECHOULAM, 1973).

A typical sample of marihuana (WILLINSKY, 1973) and hashish (GAONI and MECHOULAM, 1971) contains the cannabinoids shown in Table 1.

A detailed discussion of the analytical aspects of *Cannabis* chemistry including detection of cannabinoids in biological fluids is presented by WILLINSKY (1973). A combination of thin-layer chromatography together with gas-liquid chromatography seems adequate for the detection of various cannabinoids in the plant, but methods to obtain quantitative data still remain unreliable. However, the details of a gas-liquid chromatographic method which claims to have good quantitative reproducibility have recently been disclosed (WILLINSKY, 1973). On the other hand, the assay of THC in biological fluids in 10^{-11} -10^{-13} mol/ml range still presents many problems. The use of a recently reported immunofluorescent assay technique (GRANT et al., 1972) may be helpful.

As stated earlier, the principle active constituent of *C. sativa* is (-)-*trans*-Δ^9-THC (*1*) whereas the other active (-)-*trans*-isomer (*2*) is found only in a few varieties of the plant. It is obvious that the other two isomers of *1* and *2* are the Δ^9-10a, 6a-*cis*- and

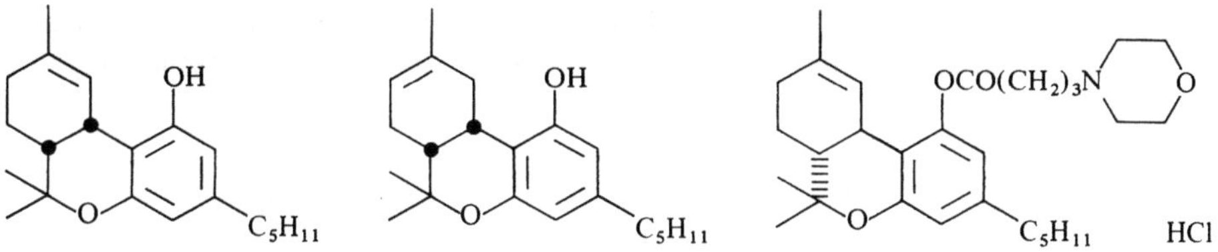

21 22 23

Fig. 3

Δ^8-10a, 6a-*cis*-THC's (*21* and *22*, Fig. 3). Both the *cis* compounds have been synthesized (ULISS et al., 1975), but have not been found in the plant. On theoretical grounds the *trans*-compounds (*1* and *2*) are expected to be more thermodynamically stable compared to the *cis* compounds *21* and *22*. Furthermore, of the two *trans*-isomers *1* and *2*, the latter has been found to be more stable since Δ^9-THC (*1*) is very easily isomerized to Δ^8-THC (*2*) on treatment with acids. The main interest pharmacologically, therefore, centers around the less stable *trans*-Δ^9-THC (*1*) its various derivatives and metabolites. This has posed many synthetic problems since it is mostly the more stable derivatives of *trans*-Δ^8-THC (*2*) that are formed during chemical reactions.

Thus far only three stereospecific syntheses of (-)-Δ^9-THC (*1*) have been described, and two of these, from (-)-verbenol (GAONI and MECHOULAM, 1971) and (+)-*p*-mentha-2, 8-dien-1-ol (PETRAZILKA et al., 1969), give (-)-Δ^8-THC (*2*), which must be transformed to *1* by addition and elimination of hydrogen chloride. These two processes, therefore, require three steps and involve at least two very tedious and careful chromatographic separations (RAZDAN et al., 1972). The third process achieved the synthesis of (-)-Δ^9-THC (*1*) in a single step from *trans*-(+)-2-carene oxide (RAZDAN and HANDRICK, 1970), but suffered from the difficult separation of *trans*-Δ^9-THC (*1*) and *cis*-Δ^9-THC (*21*) which were major products. The synthesis and availability of (-)-Δ^9-THC and its derivatives have, therefore, not been easy and straightforward.

Very recently, RAZDAN and co-workers (1974) have reported a simple one-step synthesis of (-)-Δ^9-THC (*1*) from the commercially available (+)-*p*-mentha-2,8-dien-1-ol and olivetol. By this process *1* of very high optical purity is formed in 50% yield (glc) and is isolated in 31% yield after a simple and quick column chromatography. The purity of Δ^9-THC is >96% by glc and, in contrast to the previously described process from (+)-*p*-mentha-2,8dien-1-ol, no Δ^8-THC (*2*) is formed under the new conditions. Furthermore, by a slight change in the reaction conditions of the new process, (-)-cannabidiol (4a) is obtained on a preparative scale. In addition, the same authors have reported another one-step synthesis of (-)-Δ^9-THC (*1*) from chrysanthenol (RAZDAN et al., 1975), which may have some biogenetic implications.

Extensive stability studies (RAZDAN and HANDRICK, 1970; MECHOULAM, 1973) have been carried out on Δ^8- and Δ^9-THC's. Although Δ^9-THC is easily converted to cannabinol (*5a*) on normal exposure to air and light, it can be stored without deterioration for long periods (months) in ethanol under nitrogen in a refrigerator. Δ^8-THC is relatively much more stable than *1*.

Δ^9-THC is a resinous water insoluble gum, which poses problems for pharmacological testing. However, various formulations particularly for injectables have been described and reviewed (MECHOULAM, 1973). In addition, ZITKO et al. (1972) have reported a morpholinobutyric acid ester HCL (23, SP-111, Fig. 3) of Δ^9-THC which is crystalline and water soluble. Its activity in laboratory animals is similar to that of Δ^9-THC and it appears that the principle activity of compound *23* may be due to hydrolysis in vivo to Δ^9-THC.

The biogenesis of cannabinoids has been postulated from geraniol phosphate, which is a common precursor of terpenoids (RAZDAN, 1973; MECHOULAM, 1973). In this scheme cannabidiol is the key intermediate in the formation of Δ^9-THC. However, recently various reports have appeared in the literature which indicate that certain samples of *Cannabis* and hashish contain Δ^9-THC and cannabichormene (*7a*), but no cannabidiol (*4a*) (TURNER and HADLEY, 1973). The absence of cannabidiol (*4a*) suggests that either a modification of, or a different biogenesis may be operative in certain sub-species of *Cannabis* (RAZDAN et al., 1975). To date, no experimental work on the biogenesis of cannabinoids has been reported.

C. Distribution and Fate

In man, the active principle of marihuana (Δ^9-THC) is rapidly and effectively absorbed by inhalation. It is also rather well absorbed from the gastrointestinal tract. The ratio of dose necessary to produce a "high" by oral administration compared to smoking has been estimated by ISBELL et al. (1967) to be approximately 3. When given intravenously, Δ^9-THC is approximately 10–20 times more potent than when administered orally. For instance, HOLLISTER and GILLESPIE (1973) found 1–2 mg of Δ^9-THC given intravenously approximated a 20 mg oral dose. Analagous intravenous doses were earlier reported by LEMBERGER et al. (1971) and PEREZ-REYES et al. (1972) to produce psychoactive effects. Similar oral to parenteral ratios are observed in a variety of animal species. Some discrepancies do occur, but these can usually be explained by the different formulations or vehicles used.

Pharmacokinetic studies in man using intravenously administered radiolabeled Δ^9-THC revealed a $T_{\frac{1}{2}}$ in naive subjects of 56 h with radioactivity continuing to be excreted in urine and mainly the feces for more than 8 days (LEMBERGER et al., 1970). Later studies in chronic users gave a $T_{\frac{1}{2}}$ of 28 h (LEMBERGER et al., 1971). These findings of long half-lives pointed to the possibility of cumulative effects on repeated medication. However, HOLLISTER and REAVEN (1974) using a hashish extract, found that five consecutive doses over 4 days did not lead to any significant accumulation of drug or metabolites.

One explanation of the prolonged $T_{\frac{1}{2}}$ and high fecal excretion of the cannabinoids involves the recycling through the enterohepatic circulation. KLAUSNER and DINGELL (1971) suggested this on the basis of liver perfusion studies in the rat. TURK et al. (1973) using bile cannulated rats and radiolabeled Δ^9-THC found that after administration, nearly 60% of the drug was excreted by way of the bile within 48 h. Less than 10% of the drug or its metabolites were excreted in the urine over this time period. After oral administration a much larger portion of radioactivity was excreted by way of the feces (51.5%) than the bile (21.5%) and the material in the feces was mainly unchanged Δ^9-THC. Thus, there is good evidence that the enterohepatic circulation plays a major role in the pharmacokinetics of the cannabinoids.

Another explanation for the long $T_{\frac{1}{2}}$ of the cannabinoids involves sequestration in body tissue. The high lipid solubility of these compounds lends credence to this hypothesis. Thus, AGURELL et al. (1970) found high levels of radioactivity in body fat and spleen 3 days after the i.v. administration of ^{3}H-Δ^9-THC to rabbits. After inhalation by rats, HO et al. (1970) found significant radioactivity in the brain, lung and liver 7 days later. MCISAAC et al. (1971), using radioautography, found high brain levels in the monkey 24 h after the i.v. administration of ^{3}H-Δ^9-THC, while LAYMAN and MILTON (1972) found persistent radioactivity in the brains of rats 30 days after i.p. administration. The results of this study may be complicated by the reported poor and slow absorption of the drug after this route of administration (HO et al., 1971a). KENNEDY and WADDELL (1972), using pregnant mice and radioautography, found high concentrations of radioactivity after either intravenous or subcutaneous administration in the fetal liver, intestinal contents, Harder's gland, fat, corpora lutea, and adrenal cortex. The concentration was high in maternal lung and spleen after i.v. administration but not after s.c. administration. After 24 h, there was a striking migration of radioactivity to fatty and connective tissues. Δ^9-THC crosses the placental barrier, but fetal concentrations are much less than maternal. Similarly, MARTIN et al. (1976c) consistently found lower levels of radioactivity in fetal dog tissue than in the same maternal tissues. The pregnant dogs were sacrificed 30 min after the i.v. injection of 0.5 mg/kg Δ^9THC. Little activity was found in either maternal or fetal fat in these dogs. FREUDENTHAL et al. (1972), using oral and intravenous routes of administration, found similar distribution of radioactivity in fetal and maternal mice as reported by KENNEDY and WADDELL (1972). They reported that most of the cannabinoids were present in the free rather than the conjugated form. In addition, KREUZ and AXELROD (1973) reported that after a single subcutaneous injection of Δ^9-THC, concentrations of radioactivity in fat were 10 times greater than in any other tissue examined and persisted for 2 weeks. Thus, there is good evidence to implicate redistribution as a factor in the prolonged $T_{\frac{1}{2}}$ of the cannabinoids.

However, MARTIN et al. (1976a) found low levels of radioactivity in dogs treated i.v. with 0.5 mg/kg Δ^9-THC 30 min prior to sacrifice. The concentration of radioactivity in fat was considerably less than that found in plasma at the same time period. The highest concentration of radioactivity was found in pancreas, heart, liver, kidney, and especially adrenal glands. These experimenters also confirmed previous observations from our and other laboratories that high concentrations of radioactivity were excreted in bile following the i.v. injection of radiolabeled Δ^9-THC to rats. Δ^9-THC and its metabolites were distributed throughout most organs of the body. Additionally, there was relatively equal distribution of Δ^9-THC throughout 27 different brain areas. In addition, the daily administration of 0.5 mg/kg cold Δ^9-THC for 4 days prior to the injection of radiolabeled drug did not appreciably alter the distribution of radioactivity in dogs.

One of the accepted truisms of pharmacology is that the activity of a drug is related to the plasma concentration of free drug as opposed to total concentration including the bound material. A number of studies have been done which demonstrate that Δ^9-THC is highly bound to blood constituents. DEWEY et al. (1970a) have reported that Δ^9-THC is specifically bound to lipoproteins in the blood with a high degree of selective binding to this substance. Low-affinity binding occurs to plasma albumin. These results were confirmed and extended by AGURELL et al. (1970), and WAHLQUIST et al. (1970). A few papers have appeared concerning the subcellular distribution of Δ^9-THC. COLBURN et al. (1974) reported that high concentrations of both polar metabolites and the parent compound, Δ^9-THC, are found in synaptosomal fractions of rat brain. These observations were confirmed and extended by MARTIN et al. (1976b), who showed that approximately 14% of the total radioactivity in the brain of dogs was found in the synaptosomal fraction. Further subfractionation showed that the majority of the radioactivity found in the synaptosomal fraction was associated with synaptic vesicles.

It is interesting that in the distribution studies of MARTIN et al. (1976b), higher concentrations of radioactivity were found in the gray area as opposed to heavy lipoid white matter of cerebellum and cortex of dogs. These data were confirmed in additional studies in rats and mice (DEWEY et al., 1976). END et al. (1976) found that NB2A neuroblastoma cells in culture took up much more radioactive Δ^9-THC from the medium than did glial cells. These results suggest that the higher concentration of radioactivity in the neural cells of the gray matter may be due to a preferential uptake of Δ^9-THC in neural cells as opposed to glial cells.

One of the most heavily investigated aspects of the pharmacology of marihuana, its constituents and analogues, is their metabolism. A number of metabolites of Δ^9-THC have been identified. The evidence is overwhelming that Δ^9-THC is converted to 11-hydroxy Δ^9-THC in laboratory animals and in man. It is also accepted that a portion of the parent compound is converted directly to the 8-hydroxy metabolite and that conversion of both to 8,11-dihydroxy-Δ^9-THC occurs. These metabolites have not only been found in tissue and plasma of laboratory animals and man, but have also been found to be synthesized by liver homogenates in vitro.

There are two isomers of 8-hydroxy-Δ^9-THC and 8,11-dihydroxy-Δ^9-THC. That is, the hydroxy group at the 8 position may be either equitorial or axial to the ring system. It has been reported by HOLLISTER (1974) that the 8-β-hydroxy-Δ^9-THC is approximately two-thirds as potent as the 8-α-hydroxy-Δ^9THC. AGURELL et al. (1972) identified a number of other monohydroxy metabolites of Δ^9-THC. These metabolites have been identified as alcohols of each of the 5 carbon atoms on the pentyl side chain of ring C. In addition, AGURELL et al. (1976) has shown that the 11-hydroxy-Δ^9-THC also may be hydroxylated at each of these positions in the pentyl side chain to form dihydroxy metabolites of the parent compound.

It is obvious that a number of acetyl and diacetyl metabolites of Δ^9-THC exist. These metabolites are formed by the acetylation on the phenolic hydroxyl and at the oxidated 11 position. Subsequently, or possibly, incidental with acetylation, the compounds are converted to the carboxylic acids. It is felt that these compounds may be finally conjugated with glucuronic acid and excreted as inactive conjugates.

The essence of the metabolism of Δ^8-THC is quite similar to that described above for Δ^9-THC. One exception is that the monohydroxyl substitutions on the A ring appear at the 7 rather than at the 8 position. The hydroxylations are allylic to the double bond in both isomers, that is, at the 8 and 11 carbons in Δ^9-THC and at the 7 and 11 carbons in Δ^8-THC.

Soon after the isolation and identification of Δ^9-THC, a number of studies appeared in which the pharmacological effect of the 11-hydroxy-Δ^9-THC was compared with that of the parent compound. The intent of these studies was to determine if, in fact, the 11-hydroxy metabolite was active. CHRISTENSEN et al. (1971) showed that Δ^9-THC was approximately one-half as potent as the 11-hydroxy metabolite when the two drugs were administered by the intravenous route of administration. Additionally, Δ^9-THC was approximately equally potent when given either by i.v. administration or intraventricularly in the brain. On the other hand, 11-hydroxy-Δ^9-THC was much more potent when administered intraventricularly than when given by the intravenous route. These authors concluded that the Δ^9-THC given intraventricularly had to be transported out of the brain, through the systemic circulation to the liver, where it was converted to 11-hydroxy-Δ^9-THC and then returned to the brain via the blood stream where it produced its pharmacological effect. The endpoint of the pharmacological effects in this experiment was an alteration in the overt behavior of mice. These experimental procedures do not allow strict quantification of the effects but these workers did find a similar order of potency for Δ^8-THC and its metabolites as those described above for Δ^9-THC. In later experiments, LEMBERGER et al. (1972) and PEREZ-REYES et al. (1972) showed that Δ^9-THC was less potent than 11-hydroxy-Δ^9-THC in man.

FOLTZ et al. (1970) demonstrated that the 11-hydroxy-delta-8-THC was more potent than Δ^8-THC in producing a cannabinoid like syndrome in rats. The workers postulated that Δ^8-THC also had to be converted to the 11-hydroxy metabolite before pharmacological activity was seen. Considerable controversy has existed about this point since the report that effects of 11-hydroxy-Δ^9-THC were similar to those observed with Δ^9-THC when administered intravenously to volunteers. It was obvious that 11-hydroxy-Δ^9-THC is an active agent; however, definitive experiments to rule out the pharmacological effect of the parent compound have not been done. These experiments would require the interference of hydroxylation of Δ^9-THC to the 11-hydroxy metabolite and the demonstration that no cannabinoid like effects occur.

A number of observations support the hypothesis that, in fact, 11-hydroxy-Δ^9-THC is the active substance. 11-hydroxy-Δ^9-THC is more potent than the parent compound in most all experimental procedures both in laboratory animals and man. The onset of the "high" in man following the injection of Δ^9-THC was more closely related to the blood level of the 11-hydroxy metabolite than of the parent compound. Δ^9-THC is rapidly converted to 11-hydroxy metabolite in both experimental animals and in man.

There are also a number of studies which indicate that 11-hydroxylation may not be necessary for cannabinoid like activity. MARTIN et al. (1976b) have shown that there are higher concentrations of Δ^9-THC in the brain of dogs than there are of the 11-hydroxy metabolite when the animals are sacrificed 30 min after the i.v. administration of Δ^9-THC. These data alone do not suggest that the parent compound is active but do indicate that there is enough of the parent compound in the brain to account for the central nervous system effects of this drug. Secondly, there are many synthetic compounds which have heterocyclic substitutions in the A ring which cannot be converted to an 11-hydroxy metabolite, yet these compounds have pronounced cannabinoid like activity in laboratory animals and in man. Thirdly, there are yet another number of cannabinoids which have close structural similarities to Δ^9-THC, but which cannot be converted to an 11-hydroxy metabolite, and these compounds also have pronounced pharmacological activity similar to Δ^9-THC (WILSON and MAY, 1975; MARTIN et al., 1975).

A number of studies have shown that the metabolism of other cannabinoids such as cannabinol and cannabidiol is very similar to that described above for Δ^8- and Δ^9-THC. That is, they are hydroxylated in the 11 position and they are also hydroxylated in the pentyl side chain. Additionally, a number of dihydroxy metabolites have been isolated and identified. A large number of reviews have been written on the metabolism of the cannabinoids; the reader is referred to any one of the following articles for more detail on the metabolism of these interesting agents (WALL, 1971; AGURELL et al., 1972; SECRETARY, H.E.W., 1971–1975; BRAUDE and SZARA, 1976; NAHAS, 1976).

D. Pharmacology

Much of the early scientific work with marihuana was characterized by conflicting results and a peculiar inability to replicate results from laboratory to laboratory. Much of this confusion was attributed to a lack of standardized plant material and purified extracts, and poor analytical methods. When GAONI and MECHOULAM (1964) isolated Δ^9-THC, the purported active principle of marihuana, we had hopes that the problem would be solved. The advent, due to the efforts of the National Institute on Drug Abuse, of carefully standardized plant materials and sufficient quantities of the major compounds found in the plant, have enabled careful and systematic studies. The technical problems, however, have not disappeared. Most of the cannabinoids are highly water insoluble and necessitate solubilizing and suspending agents which have often confounded the research results. There are vagaries of absorption and distribution depending on the route of administration and medication form which are difficult to control and reconcile. Thus, we are now faced with an increasing body of literature which requires careful examination and interpretation.

I. Central Nervous System

1. General Effects in Man

Perhaps the most graphic and literate description of the psychic effects of *Cannabis* can be found in the monograph *Der hachusch et de l'alienation mentale*, published in 1845 by MOREAU, which has recently become available in an English edition (MOREAU, 1973). MOREAU divided the hashish experience into eight phases of phenomena. The first a feeling of happiness; the second, excitement and dissociation of ideas; the third, errors in time and space; the fourth, enhancement of the senses, especially hearing; the fifth, fixed ideas of delusions; the sixth, damage to the emotions; the seventh, irresistible impulses; and the eighth, illusions and hallucination. He noted that the effects were dose-related and his observations are still valid. The modern works on the actions of marihuana in man have been thoroughly reviewed by PATON and PERTWEE (1973). They have proposed that the primary action of *Cannabis* is on perception and have classified the work in the field along those lines. A more detailed examination of recent work on the neurophysiological and behavioral effects of marihuana in man can be found in the chapters in MILLER (1974).

A number of human studies are of particular importance in the context of this chapter. For instance, WEIL et al. (1968) compared marihuana and a placebo in both naive and experienced marihuana smokers, in a neutral laboratory setting. They found that the naive subjects did not report as intense subjective effects as the experienced users and that the effects reported were different. This finding must be taken into account in the interpretation of most human studies. In addition, the naive subjects had an impaired performance on a number of simple intellectual and psychomotor performance tests while the users had a lesser degree of impairment. Marihuana—related tachycardia and dilation of the conjunctival blood vessels were observed in both groups of subjects. These symptoms are probably the most reproducible effects produced by *Cannabis* in man. This was, in a sense, a landmark study since it reestablished the legitimacy of studying marihuana in noninstitutionalized subjects.

The first careful demonstration that the effects of marihuana could be mimicked by Δ^9-THC was carried out by ISBELL et al. (1967). They compared the effects of marihuana and Δ^9-THC in experienced marihuana smokers and found that they could not distinguish between them. Later ISBELL and JASINSKI (1969) compared high doses of Δ^9-THC with LSD. The objective effects were markedly different although they could not readily distinguish differences in the subjective effects of the drugs by their methodology. Of interest was the observation that subjects tolerant to LSD were not tolerant to Δ^9-THC indicating a different mechanism of action.

HOLLISTER and his colleagues have systematically examined many of the constituents of marihuana, a number of the metabolites of THC, and a few totally synthetic analogues of THC. This work has recently been summarized (HOLLISTER, 1974). The position of the double bond had only quantitative effects. Thus, Δ^8-THC was only slightly less active than Δ^9-THC while the $\Delta^{6\alpha,10\alpha}$-isomer was considerably less potent. The alkyl side chain also has a strong influence on potency and does appear to alter the qualitative effects. For instance, the potency of propyl compound, tetrahydrocannabivarin, is only about one-fourth that of Δ^9-THC while synhexyl, the n-hexyl-$\Delta^{6\alpha,10\alpha}$-derivative, has about one-third the potency. The dimethyl-heptyl-$\Delta^{6\alpha,10\alpha}$-analog (DMHP) has quite potent activity in animals but has little psychic activity in man at doses which produce marked cardiovascular activity (SIDELL et al., 1973; LEMBERGER et al., 1974). These results are complicated by the large number of isomeric forms of this compound. Large doses of cannabinol and cannabidiol produced no discernable effects. As to the metabolites, both 11-hydroxy-Δ^8 and 11-hydroxy-Δ^9-THC were somewhat more potent than their parent compounds. There was no difference in onset of activity but time to peak activity was faster and the effects somewhat more severe and prolonged.

In addition to its subjective effects in man, marihuana has been found, in appropriate doses, to produce a decrement in psychomotor performance (MANNO et al., 1974), to interfere with attention and produce loss of efficiency in a variety of memory tasks (DELONG and LEVY, 1974; DARLEY and TINKLENBERG, 1974) to adversely affect motivation (CAPPELL and PLINER, 1974; MENDELSON et al., 1976) although this data should be interpreted with great care (MEYER, 1975) and to reduce physical strength or the inclination towards strenuous exertion in tasks requiring sustained effort (HOLLISTER et al., 1968; HOLLISTER, 1971). For centuries there has been a popular association between *Cannabis* use and aggressive behavior. TINKLENBERG

(1974) has recently reviewed the literature in this regard and reached the same conclusion as has nearly every official body which has studied *Cannabis* beginning with the Indian Hemp Drug Commission Report (1893–1894). "There is no convincing evidence that the pharmacological properties of marihuana incite or enhance human aggression". Indeed, there is suggestive evidence that marihuana may reduce aggressiveness in some individuals.

2. Overt Behavior

The injection of Δ^9-THC and other cannabinoids with psychoactive effects in man to laboratory animals at very high doses produces a generalized central nervous system depression. In general, larger doses of Δ^9-THC are needed to cause changes in the overt behavior of rodents than other species of laboratory animals. This depression is characterized by hypothermia, decreased activity, respiratory depression and at higher doses, prostration. There is a wide range between the dose which produces prostration and the lethal dose in mice and rats. Rodents depressed with pharmacological doses of Δ^9-THC are easily aroused and usually hyperexcitable. That is, although their spontaneous activity is depressed, when aroused they are hyperactive when compared to control animals. Low doses of psychoactive cannabinoids produce a unique syndrome in mice, which is characterized by tremors, crossing of the fore limbs and spreading of the hind limbs when the animals are held inverted by the tail.

Δ^9-THC and other cannabinoids have particularly unusual effects on the overt behavior of dogs. This phenomenon was first described by WALTON (1938) and later semiquantitated in reports from our laboratory (DEWEY et al., 1972b; SPAULDING et al., 1972). Dogs titrated with Δ^9-THC have the following characteristics. Doses as low as 0.1 or 0.2 mg/kg intravenously produce slight depression including a decrease in spontaneous activity. Although depressed, these animals are hyperexcitable (over react to a thrust with a hand). Slightly higher doses cause prance-like placement of front feet, increased hyperexcitability, and static ataxia. This static ataxia is characterized by swinging movements, side to side and forward and backward without movement of the feet. This swaying motion caused by Δ^9-THC in dogs is not produced in this species by any other group of central nervous system acting drugs known to us. We have attempted to quantitate the effects of various doses of Δ^9-THC and other psychoactive cannabinoids based on the onset and duration of the static ataxia after intravenous administration. Higher doses of Δ^9-THC produce prostration in dogs. The duration of the hyperexcitability and static ataxia is proportional to the dose but even at relatively small doses often lasts for approximately 2–4 h. Animals given a relatively high dose of Δ^9-THC (approx. 2–4 mg/kg, intravenously) will show a significant increase in the amount of time spent sleeping during the next 24 h following the disappearance of the hyperexcitability and static ataxia. Marked tolerance has been shown to develop to the static ataxia and hyperexcitability produced by Δ^9-THC in dogs, but tolerance was not observed to the sedative-hypnotic activity following chronic administration of high doses of the drug (DEWEY et al., 1972b).

Another test used to predict possible psychoactive effects of cannabinoids is the Gayer arreflexia test in rabbits (GAYER, 1928). Marihuana, Δ^9-THC and other cannabinoids which were later found to produce a psychological "high" in man decreased the blink reflex of rabbits in a dose responsive manner. The effects of Δ^9-THC in rhesus monkeys were first described by SCHECKEL et al. (1968), and more completely by GRUNFELD and EDERY (1969). Generally, Δ^9-THC produces depressant effects in rhesus monkeys. The depressive syndrome induced in these animals by cannabinoids does not differ significantly from that observed with other CNS drugs, for instance, minor and major tranquilizers and sedatives and hypnotics.

The dog static ataxia test has been found to be a reliable tool for predicting potency of cannabinoids in producing a psychological "high" in man. A comparison of the results of clinical trials of a number of synthetic compounds as well as many metabolites has verified this fact. Cannabinoids such as cannabinol and cannabidiol which do not produce a psychological "high" in man do not produce hyperexcitability nor static ataxia in dogs even at doses high enough to produce prostration. One possible exception is DMHP, which has pronounced effects on the overt behavior of dogs and produces profound cardiovascular effects in man at doses below those which are psychoactive (SIDELL et al., 1973; LEMBERGER et al., 1974).

3. Aggression

Although a large number of studies have been reported on the effects of *Cannabis* extract, Δ^9-THC and other cannabinoids on aggressive behavior in laboratory animals; considerable controversy and confusion exists. The majority of studies of the effects of cannabinoids on aggression have been carried out in rodents and can conveniently be divided into two groups: those describing the effects of cannabinoids on stress induced aggression, which includes isolation, starvation or electric shock induced aggression, and those describing the effects of cannabinoids on predatory behavior, which generally is demonstrated by the attack and killing behavior of an animal.

Shock induced aggression, that is fighting time of pairs of rodents being shocked, was shown to be decreased by doses of 2.5–50 mg/kg Δ^9-THC in the study of DUBINSKY et al. (1973). CARDER and OLSEN (1972) had previously shown that 1 and 2 mg/kg Δ^9-THC decreased and doses of 0.12–0.5 mg/kg Δ^9-THC increased the fighting time of shocked rats. CARLINI et al. (1972) and CARLINI and MASUR (1969, 1970) presented findings which suggested that starvation contributed to the development of aggression seen during the chronic administration of marihuana. Aggressiveness develops in rats maintained in isolation for prolonged periods of time and *Cannabis* extract has been shown to supress this aggressive behavior (DUBINSKY et al., 1973; SANTOS et al., 1966).

KILBEY et al. (1971) showed that Δ^9-THC caused an increase in the latency of rats to attack and kill frogs when tested immediately after drug administration. The peak drug effect was found to be less than 15 min after administration in a subsequent study (KILBEY et al., 1972). Additional reports have appeared which shown that Δ^9-THC reduces mouse-killing behavior in rats (DUBINSKY et al., 1973; KILBEY et al., 1973). The results of the study by ALVES and CARLINI (1973) suggest that Δ^9-THC has dual muricidal activity in rats. A dose of Δ^9-THC (8.8 mg/kg i.p.) which inhibited mouse killing behavior in rats when injected 4 consecutive sessions increased muricidal activity in previously non-muricidal rats when given for 40 days. Unfortunately the rats on the chronic study were food deprived for 22 h each day and, as mentioned previously, reports from the same laboratory have shown that starvation contributed to the aggression induced by Δ^9-THC in rats. However, in a more recent study MICZEK (1976) has shown that the chronic administration of Δ^9-THC for 60 days to rats given food and water ad libitum also induced mouse-killing behavior in rats that were nonmuricidal prior to the exposure to the cannabinoid.

In a recent review of the effects of marihuana on human aggression, TINKLENBERG (1974) concluded that marihuana did not increase human aggression but indicated that the pharmacological data suggested that marihuana might reduce human aggression (see previous section). This is consistent with most of the animal data.

We have recently begun to study the effects of Δ^9-THC on the social behavior of relatively free ranging rhesus monkeys. A preliminary experiment showed that the administration of Δ^9-THC to the leader and to the second ranked male monkey in a colony of 20 monkeys was followed by a complete loss of rank of the second monkey. This animal remained at the bottom of the social order long after medication was discontinued. There was no change in the rank of the lead animal. Other reports exist on the effects of Δ^9-THC on social dominance in animals but a clear picture has not emerged. The results of experiments currently in progress might shed some light on the effects of cannabinoids on this type of behavior.

4. Operant or Conditioned Behavior

The effects of cannabinoids on learned behaviors have been reviewed in the annual reports on "Marihuana and Health" presented to Congress (SECRETARY, H. E. W., 1971–1975) and by DOMINO (1971), CARLINI (1973), and more recently by BALSTER and FORD (1976) and MCMILLAN (1976). The reader is directed to these articles for a more extensive review since only some of the major findings will be described in this article. As pointed out previously in this chapter, the effects of cannabinoids on the overt behavior of most species are a combination of depressant and stimulant properties. The depressant effects of the cannabinoids have been most evident in experiments employing operant techniques. Although at low doses in certain species, the cannabinoids produced a small increase in the rate of response. The cannabinoids have been found to have generally disruptive effects on performance using avoidance procedures.

A number of papers have appeared which describe a decrease in the rate of responding induced by *Cannabis* in animals trained on a fixed ratio (FR) schedule of reinforcement (HENRIKSSON and JARBE, 1971; FRANKENHEIM et al., 1971; PETERSON et al., 1971; and PRADHAN et al., 1972). MCMILLAN et al. (1970) reported that Δ^9-THC at doses as low as 1.8 mg/kg given intramuscularly 2.5 h prior to testing abolished key-pecking in pigeons on a multiple FI, FR schedule of food reinforcement. BLACK et al. (1970) showed that 0.3 mg/kg was the minimal effective dose of Δ^9-THC in this procedure. KOSERSKY et al. (1974) reported that the effects of the primary metabolite 11-hydroxy-Δ^9-THC were indistinguishable from the parent compound using this paradigm. MASUR et al. (1972) showed that 5.0 mg/kg Δ^9-THC decreased the rate of responding of rats trained to respond under a continuous reinforcement schedule. Similarly in most studies *Cannabis* reduced the rate of responding of animals trained on a fixed interval (FI) schedule of reinforcement (FRANKENHEIM et al., 1971; MAN-

NING and ELSMORE, 1972). Responding rates were also decreased by large doses of Δ^9-THC in rats trained to respond on a variable interval schedule (VI) of reinforcement (CARLINI, 1968; GRISHAM and FERRARO, 1972). It appears therefore, that Δ^9-THC and other cannabinoids depress activity induced by many different schedules of reinforcement. As described in detail in another portion of this review, a marked tolerance has been shown to develop to these depressant effects of Δ^9-THC and other cannabinoids.

Most CNS acting drugs increase response rate at low levels and decrease high rates in responding. There is evidence that cannabinoids are somewhat unusual in this regard in that they generally depress the rate of responding regardless of the control rate of behavior. This lack of rate dependent effects of both Δ^8- and Δ^9-THC have been demonstrated in pigeons and rats on a multiple fixed ratio-fixed interval schedule (FRANKENHEIM et al., 1971). However, at times, a biphasic dose-response relationship has been observed with schedules which generate low rates of responding. An alternation between periods of no responding and periods of increased rates of responding have also been observed when cannabinoids have been given to animals working at a low response rate.

The uniqueness of the pharmacological effects of Δ^9-THC and related cannabinoids is suggested by the observations that the stimulus properties of Δ^9-THC are interchangeable with Δ^8-THC, hashish smoke, *Cannabis* extract and the primary metabolite of Δ^8-THC, 11-OH Δ^8-THC; but not with cannabinol, cannabidiol or a large number of drugs from other classes (GONZALEZ et al., 1972; BARRY and KRIMMER, 1976). FORD and BALSTER (1975) have extended these observations to show an exchange of the stimulus properties of Δ^9-THC to a large series of analogs and metabolites of Δ^9-THC which have pharmacological profiles similar to Δ^9-THC in animals and man but not to other analogs and metabolites which are devoid of "Δ^9-THC like" effects.

5. Analgesia

A number of reports have appeared which suggest that marihuana and its constituents have antinociceptive activity in a variety of rodent tests. BICHER and MECHOULAM (1968) found that 10 mg/kg of Δ^9-THC was effective in antagonizing the writhing response induced by acetic acid in mice. They also reported that 20 mg/kg Δ^8- and Δ^9-THC increased latency in the mouse tail-flick and hot-plate tests. When they increased the dose to 40 mg/kg, all pain sensations were abolished. These authors concluded that the analgesic activity of the cannabinoids was approximately half that of morphine sulfate. GILL et al. (1970) reported that 200 mg/kg of *Cannabis* extract was needed to completely abolish the writhing response induced by the i.p. injection of phenylbenzoquinone. Neither of these papers contain great detail in the experimental design, doses used or time between the injection of the cannabinoids and the injection of the antinociceptive agent. DEWEY et al. (1972a) found that relatively high doses of Δ^8- and Δ^9-THC were needed to produce inhibition of the writhing response in mice. They found that Δ^8- and Δ^9-THC were more that 150 times less potent than morphine in this procedure.

GALLAGER et al. (1972) reported that Δ^9-THC prolonged latency in the hot-plate and tail-flick tests at 15, 45, 75, and 165 min after injection. These observations were confirmed by SEGELMAN et al. (1974), who found that Δ^9-THC has an ED_{50} of 1.3 mg/kg in the mouse hot-plate test when administered by the i.p. route of administration. CHESHER et al. (1973) reported that both Δ^8- and Δ^9-THC were more potent than pethidine in the mouse hot-plate procedure. The effects of pethidine were antagonized by nalorphine while this narcotic antagonist did not antagonize the effect of the cannabinoids in this procedure. DEWEY et al. (1972a) found that the ED_{50} for Δ^8- and Δ^9-THC in the mouse hot-plate test varied between 29 and 90 mg/kg, depending on the time of administration prior to testing. The ED_{50} for morphine in this test procedure in their hands is approximately 5 mg/kg; therefore, once again Δ^9-THC was found to be a much less potent analgesic than morphine in our laboratory. BUXBAUM (1972) reported that Δ^9-THC was equipotent with morphine in the rat hot-plate test; however, he found that the cannabinoid is much less active than the opiate in mice. BUXBAUM (1972) also reported that Δ^9-THC was active in the rat tail-flick test and that it was equipotent with morphine in this procedure as well. DEWEY et al. (1972a) reported that Δ^9-THC was inactive in the mouse tail-flick procedure, and a more recent study from this laboratory (BLOOM et al., 1975) confirmed these results. It is our belief that Δ^9-THC and other cannabinoids are active in the hot-plate test due to their central nervous system depressant effects rather than to a specific antinociceptive activity. Generalized central nervous system depression does not show up in the tail-flick test, which, as we use it, is a somewhat more specific test for antinociception. However, it should be noted that the narcotic-antagonist analgesics are also inactive in our usual tail-flick procedure, yet these compounds are good analgesics in man.

In an uncontrolled study, DEWEY et al. (1972b) observed that i.v. injections of Δ^9-THC produced antinociceptive effects in dogs. These experiments were not designed to study antinociception but observations of the dogs' response to painful stimuli suggested that the cannabinoids might have analgesic activity in this species. This observation was confirmed by studies of KAYMAKCALAN et al. (1974) who observed antinociceptive effects of Δ^9-THC on the response to tooth stimulation in dogs. These workers also observed tolerance to this effect when the cannabinoid was given every 24 h for 8 days. Although tolerance has been shown to develop to many of the effects of Δ^9-THC and other cannabinoids, as we will discuss in greater detail later, we did not see signs of physical dependence as evidenced by a withdrawal syndrome when the cannabinoid was administered repeatedly at a very high dose, and abruptly withdrawn.

Recently, WILSON and MAY (1975) reported that 11-hydroxy-Δ^9-THC was a more potent antinociceptive agent in the mouse hot-plate test than the parent compound, Δ^9-THC. They also found that compounds which cannot be converted to an 11-hydroxy metabolite were inactive in this test procedure. MARTIN et al. (1976a) have recently confirmed and extended the observation by BLOOM et al. (1975) that an analogue of Δ^9-THC, 9-*nor*-9β-hydroxy-hexahydrocannabinol, is a potent analgesic in the mouse tail-flick test as well as in the other procedures used for measuring antinociception in laboratory animals.

The (-) isomer of 9-*nor*-9β-hydroxy-hexahydrocannabinol produces an ED_{50} of 1.6 mg/kg in the mouse tail-flick test in our laboratory. The ED_{50} for morphine is 5.8. This compound is the only cannabinoid tested in our laboratory which is equal to or more potent than morphine in the mouse tail-flick test. It is our opinion that

Δ^9-THC and most other cannabinoids do not produce morphine-like antinociception in laboratory animals. However, the results of all the data presented to date suggest clearly that the cannabinoids produce effects in animals which warrant additional testing in the search for a useful analgesic agent in this class of drugs.

Controversy also exists concerning the possible anti-inflammatory activity of cannabinoids. KOSERSKY et al. (1973) reported that Δ^9-THC had antipyretic and hypothermic activity but was devoid of antinociceptive activity except at doses which produced catatonia. These authors also reported that Δ^9-THC was inactive in the rat paw edema test, suggesting that Δ^9-THC does not have anti-inflammatory activity in rats. However, SOFIA et al. reported that Δ^9-THC (1973b) and other natural occurring cannabinoids (1973a) were orally effective anti-edema agents in rats. The reason for the discrepancy in these conflicting reports is unknown.

II. Biogenic Amines

We do not understand completely the mechanism of action of any drug on the central nervous system. Overwhelming evidence exists to implicate the involvement of one or another of the neurotransmitter systems of the brain in the actions of most centrally acting drugs. TRUITT and ANDERSON (1971) reviewed the effects of tetrahydrocannabinols and their metabolites on brain biogenic amine systems. They concluded at that time that Δ^9-THC increased serotonergic tone while not altering norepinephrine.

1. NE, E and DA

An updated list of the effects of Δ^8- and Δ^9-THC on brain levels and turnover rates of biogenic amines is presented in Table 2. As can be seen in the first portion of this table, Δ^9-THC causes little chance in basal levels of norepinephrine in brains of mice, rats or monkeys at doses ranging from 0.5 up to and including 50 mg/kg. Subsequent research has shown that the very high doses used by HOLTZMAN et al. (1969) increased 5-HT turnover; however, 200 and 500 mg/kg of Δ^9-THC are clearly in excess of pharmacological doses. As a matter of fact, the 24 h LD_{50} for Δ^9-THC given intraperitoneally to mice is 168 mg/kg (DEWEY et al., 1972a). Studies designed to investigate the effects of Δ^9-THC on norepinephrine levels in brain parts of rodents indicate that mouse telencephalon and brain stem and rat hypothalamus and brain stem were without change following doses of Δ^9-THC. The data in Table 2 indicate, however, that generally there is an increase in the turnover rate of NE in rats and mice following doses of Δ^9-THC between 30 and 100 mg/kg. SCHILDKRAUT and EFRON (1971) reported that the i.p. injection of 80 mg/kg Δ^9-THC increased the rate of disappearance of an intracisternally administered dose of ^{3}H-NE. HO et al. (1973) reported that Δ^9-THC caused an increase in the activity of rat brain tyrosine hydroxylase.

The essence of the data on the effects of Δ^9-THC on basal levels and turnover rate of DA are identical to those described above for Δ^9-THC on NE. That is, there is no change in the basal level of DA but the majority of the studies show that there is an increase in the rate of dopamine synthesis in mice and rats.

2. 5–HT

The picture is somewhat less clear as to the effects of Δ^9-THC on basal levels of 5–HT in rodent brain. In a number of the studies reported in Table 2, there was an increase of 5–HT following Δ^9-THC. In the majority of the papers the increase in 5–HT levels was not significantly different from control. Although there are few exceptions, one could conclude that Δ^9-THC does not alter 5–HT levels. There have been few reports of the effect of Δ^9-THC in turnover of 5–HT in rodent brain. One study has indicated that there is a 50% decrease in the turnover rate of 5–HT at 20 mg/kg Δ^9-THC. However, it has also been shown that there is no change in the turnover rate of 5–HT in either the hypothalamus or brain stem of rats treated with either 5 or 50 mg/kg Δ^9-THC by the i.p. route of administration. These data were confirmed in rat forebrain where again there was no change in the turnover rate of 5–HT following 5.5 mg/kg Δ^9-THC. These conflicting data on 5–HT turnover are also somewhat reflected in the reports that there is either a decrease, increase or no change in levels of the serotonin metabolite, hydroxyindoleacetic acid, following the injection of Δ^9-THC. Sofia et al. (1971) reported that although Δ^9-THC did not block the monoamine oxidase, there was a 66% increase in brain 5–HT with a concomitant 50% decrease in synthesis rate. These workers suggested that Δ^9-THC affected the synaptic membrane in such a way that it altered 5–HT metabolism. A number of workers have shown that Δ^9-THC blocks the pharmacological effects of reserpine and inhibits its depletion of 5–HT. Recently, Johnson et al. (1975) have reported that Δ^9-THC increases the binding of reserpine to synaptic vesicle subcellular fractions of rat brain. These reports suggest overall that there is not a clear understanding as to the effects of Δ^9-THC on serotonergic systems of the brain.

Essentially the effects of Δ^8-THC amine levels and turnover rates in the brain are identical to those observed with Δ^9-THC. There are not enough comparative studies to evaluate the potency of the two drugs on levels or turnover rates of the transmitters in the central nervous system. Generally speaking, the changes are on the same order of magnitude with Δ^8- as they are with Δ^9-THC.

There are a number of explanations for the often contradictory data contained in Table 2. A wide wide range of doses, all the way from 0.5–500 mg/kg Δ^9-THC, have been studied. Different workers have used a wide variety of vehicles in these investigations. As mentioned previously in this report, the vehicle used affects absorption, distribution and many other factors. Further research into the pharmacological effects of Δ^9-THC and other cannabinoids on brain biogenic amine levels and turnover rates are needed. We are currently pursuing this endeavor in our laboratory. Preliminary results of these studies have been included in Table 2 and have been reported by Bloom et al. (1976a).

3. ACh

Few papers have appeared concerning the effect of marihuana and its constituents on cholinergic function in the brain. Domino (1971) has shown that large doses of Δ^9-THC which cause the animals to become comatose also cause an increase in total mouse brain acetylcholine levels. Domino also reported that 10 and 32 mg/kg Δ^9-THC inhibited the depletion of brain acetylcholine induced by intraventricularly administered hemicholinium. Doses of 3, 6, and 11 mg/kg Δ^9-THC were shown to

Table 2. Effect of delta-9-THC on levels and turnover of biogenic amines in the brain

Species	Route	Dose (mg/kg)	Pretreatment	Effect	References
Norepinephrine Levels:					
Mouse	i.v.	5+10 200+500	45 min	Decrease	HOLTZMAN et al., 1969
Mouse	i.v.	0.5, 1, 2, 5, 25, and 50 10	15 min	<15% change 25% change	Ho et al., 1972
Mouse	i.v.	10	5—60 min	No change	Ho et al., 1974
Mouse	s.c.	1, 3, 10, 30, and 100	30 min	No change	BLOOM et al., 1976a
Mouse telencephalon and brain stem	i.p.	10	10 and 30 min	No change	WELCH et al., 1971
Rat	Inhalation 7, 14, and 21 days	10 min daily exposure	20 h	No change	Ho et al., 1971b
Rat	s.c. or i.p. p.o. s.c.	25 75 100	1 h 3 h	No change	MAITRE et al., 1970
Rat	i.v.	2 mg/kg daily for 7 days	18 h	No change	MAZURKIEWICZ-KWILECKI and FILCZEWSKI, 1973
Rat	i.p.	80	45 min	No change	SCHILDKRAUT and EFRON, 1971
Rat	p.o.	10	1 h	No change from vehicle	GRAHAM et al., 1974
Rat hypothalamus or brain stem Rat brain stem Rat hypothalamus	i.p.	5 50 50	30, 60 or 120 min 30 min 30, 60 or 120 min	No change Decrease	YAGIELA et al., 1974
Monkey	i.v.	0.5, 2, and 10	15 min	No change	Ho et al., 1972
Norepinephrine turnover and synthesis:					
Mouse	s.c.	1 and 3 10, 30, and 100	^{3}H-tyrosine given i.v. 10 min prior to sacrifice	No change Increase	BLOOM et al., 1976b
Rat	s.c. and i.p. p.o. s.c. (3 h prior to sacrifice)	25 75 100	^{3}H-tyrosine given i.v. 1 h prior to sacrifice	Increase Increase No change	MAITRE et al., 1970
Rat	i.v.	2 daily for 7 days	^{3}H-tyrosine given infusion immediately prior to sacrifice	Increased[a]	MAZURKIEWICZ-KWILECKI and FILCZEWSKI, 1973

Table 2 (continued)

Species	Route	Dose (mg/kg)	Pretreatment	Effect	References
Rat	i. p.	30	^{3}H-tyrosine given i. v. 1 h after Δ^9-THC and 1 h prior to sacrifice	Increased	MAITRE et al., 1973
Rat hypothalamus and brain stem	i. p.	5 or 50	Decline of N.C. after α M.T.	No change	YAGIELA et al., 1974
Dopamine levels:					
Mouse	s.c.	1, 3, 10, 30, and 100	30 min	No change	BLOOM et al., 1976a
Mouse telencephalon and brain stem	i. p.	10	10 and 30 min	No change	WELCH et al., 1971
Rat	s.c. and i. p. p.o. s.c.	25 75 100	1 h 1 h 3 h	No change	MAITRE et al., 1970
Rat	i.v.	2 daily for 7 days	18 h	No change	MAZURKIEWICZ-KWILECKI and FILCZEWSKI, 1973
Rat	p.o.	10	1 h	Decrease	GRAHAM et al., 1974
Dopamine turnover and synthesis:					
Mouse	s.c.	1 and 3	^{3}H-tyrosine given i. v. 10 min prior to sacrifice	No change	BLOOM et al., 1976b
		10, 30, and 100		Increase	
Rat	s.c. and i.p.	25	^{3}H-tyrosine given i. v. 1 h prior to sacrifice	Increase	MAITRE et al., 1970
	p.o.	75		Increase	
	s.c. (3 h prior to sacrifice)	100		No change	
Rat	i. p. (2 h prior to sacrifice)	30	^{3}H-tyrosine given 1 h after Δ^9-THC	Increase	MAITRE et al., 1973
Serotonin levels:					
Mouse	i. p.	10, 100, 200, and 500	45 min	Increase	HOLTZMAN et al., 1969
Mouse	i.v.	0.5—50 10	15 5—60 min	10% change No change	HO et al., 1972
Mouse telencephalon and brain stem	i. p.	10	30 min	Increase	WELCH et al., 1971
Rat	Inhalation 10, 14, and 21 days	10 min daily exposure	20 h	No change	HO et al., 1971b
Rat	i.v.	1 daily for 3, 7 and 12 days	24 h	No change	JOHNSON, 1974

Table 2 (continued)

Species	Route	Dose (mg/kg)	Pretreatment	Effect	References
Rat	i.p.	6, 9 or 14 20	30	No change 66% change	SOFIA et al., 1971
Rat	i.p.	80	45 min	6% increase $p<0.01$	SCHILDKRAUT and EFRON, 1971
Rat cortex, cerebellum	i.v.	1.25	2, 30, 210 min	No change	JOHNSON, 1974
Rat midbrain, medulla			2	Increase	
Rat midbrain			30 and 210	No change	
Rat medulla			30	Increase	
Rat medulla			210	No change	
Rat hypothalamus or brain stem	i.p.	5 or 50	30, 60, and 120	No change	YAGIELA et al., 1974
Rat forebrain	i.v. ivt.	1 50 µg	5 min to 4 h 30, 45, and 60 min	No change No change	GALLAGER et al., 1972
	ivt.	50 µg	15	Increase	
Rat cortex	i.v.	0.1, 1, and 10 mg/kg	45 min 3 and 24 h	Increase No change	OUELLET et al., 1973
	i.p.	5, 10, and 50 mg/kg	0.75, 3, and 24 h	No change	
Rat brain stem	i.v.	0.1, 1, and 10 mg/kg	0.75, 3, and 24 h	No change	
	i.p.	5, 10, and 50 mg/kg	45 min	No change	
	i.p.	5, 10, and 50 mg/kg	3 and 24 h	Decrease	
Monkey	i.v.	0.5 2 and 10	15	Decrease No change	Ho et al., 1972
Serotonin turnover and synthesis:					
Rat	i.p.	20	Build-up of 5-HT after pargyline	50% decrease	SOFIA et al., 1971
Rat forebrain	i.p.	5.5	Build-up of 5-HT after pargyline	No change	GALLAGER et al., 1972
Rat hypothalamus or brain stem	i.p.	5 or 50	Build-up of 5-HT after pargyline	No change	YAGIELA et al., 1974
5-HIAA levels:					
Rat	i.v. 2 × wk. for 5 wks.	10	1 h	15% decrease	Ho et al., 1974
Rat	Inhalation 14 and 21 days	10 min exposure	20 h	Decrease	Ho et al., 1971b
Rat forebrain	i.v. ivt. ivt.	1 50 µg 50 µg	5 min to 4 h 15 min 30, 45, and 60 min	No change Increase No change	GALLAGER et al., 1972

Table 2 (continued)

Species	Route	Dose (mg/kg)	Pretreatment	Effect	References
Normetanephrine levels:					
Rat	Inhalation 14 and 21 days	10 min daily exposure	20 h	Decrease	Ho et al., 1971b
Rat	i.v., 24 × wk. for 5 wks.	10	1 h	No change	Ho et al., 1974
Effect of Δ^8-THC on levels of biogenic amines in the brain					
Norepinephrine levels:					
Mouse	i.v.	0.5, 1, 2, 5, 10, 25, and 50	15 min	< 25% change	Ho et al., 1972
		30	15 min	35% decrease	
		30	30 or 60 min	No change	
Rat	i.p.	100	1, 2, 3, 5, 7, and 10 h	No change	LEONARD, 1971
Rat	i.p.	10	Not given	No change	LITTLETON and MACLEAN, 1974
Monkey	i.v.	0.5, 2 or 10	15 min	No change	Ho et al., 1972
Dopamine levels:					
Rat	i.p.	100	1, 2, 8, 5, 7, and 10 h	No change	LEONARD, 1971
Rat	i.p.	10	Not given	No change	LITTLETON and MACLEAN, 1974
Monkey	i.v.	0.5, 2, and 10	15 min	No change	Ho et al., 1972
Serotonin levels:					
Mouse	i.v.	0.5, 1, and 25	15 min	No change	Ho et al., 1972
		2, 5, 30, and 50	15 min	20% increase	
		10	15 min	40% increase	
		10	5 and 30 min	20% increase	
		10	45 and 60 min	No change	
Rat	i.p.	100	1, 2, 3, 5, 7, and 10 h	No change	LEONARD, 1971
Rat	i.p.	10	Not given	No change	LITTLETON and MACLEAN, 1974
5-HIAA metabolite levels:					
Rat	i.p.	100	1, 2, 3, 5, 7, and 10 h	No change	LEONARD, 1971
Normetanephrine levels:					
Rat	i.p.	100	1, 2, 3, 5, 7, and 10 h	No change	LEONARD, 1971

[a] Increased catecholamine synthesis (dopamine and norepinephrine not separated).

reduce release of acetylcholine from cerebral cortex of cats (DOMINO, 1971). This decrease in acetylcholine release caused by 3 mg/kg Δ^9-HTC was antagonized by 25 mg/kg *d*-amphetamine.

The suggestion has been made (BROWN, 1971) that the behavioral effects of Δ^8-THC might involve anticholinergic mechanisms. From an extensive review, DREW and MILLER (1974) have hypothesized that Δ^9-THC produces many of its behavioral effects through an anticholinergic mechanism. These authors support this view by a striking comparison of the effects of cannabinoids and anticholinergic drugs on a large number of behavioral paradigms. Furthermore, the authors point out that many of these behaviors are strongly associated with the limbic system of the brain. The hippocampal lesions influence cognitive behavior much in the same way as the cannabinoids and anticholinergic drugs. Since the main neuronal hippocampal afferents are believed to be cholinergic, the possibility that Δ^9-THC disrupts cognitive function by interfering with cholinergic neurons in the hippocampus appears quite plausible. Some evidence obtained from studies concerned with peripheral effects support the hypothesis that cannabinoids have anticholinergic properties. DEWEY et al. (1972a) have shown that Δ^9-THC and Δ^8-THC possess slight, nonspecific inhibitory activity of the isolated rat jejunum. Neither Δ^8- nor Δ^9-THC showed agonistic activity in either the rat vas deferens preparation or in the ileum. Both Δ^8- and Δ^9-THC were less active than papaverine as noncompetitive inhibitors of cholinergically induced contractions of the jejunum.

Δ^8- and Δ^9-THC were found to be weak inhibitors of peristaltic activity in the gastrointestinal tract as measured by the charcoal meal test in mice. Δ^9-THC was less than one-tenth as potent as morphine, the positive control used in this procedure. Δ^8-THC was less active than Δ^9-THC in this test system. The parasympatholegtic activity of Δ^9-THC was confirmed and extended by CAVERO et al. (1972b).

III. Cardiovascular and Autonomic Effects

Drugs of abuse like all agents with pronounced effects on the central nervous system might be expected to produce pronounced indirect, if not direct, effects on the cardiovascular system. The cannabinoids are not an exception. A number of workers have reported that marihuana smoking produced tachycardia in humans (WEIL et al., 1968; JOHNSON and DOMINO, 1971; RENAULT et al., 1971, and BEACONSFIELD et al., 1972). Tolerance does not develop to this dose-related effect and the increase in heart rate produced by epinephrine or atropine is potentiated by marihuana. The involvement of the beta adrenergic system in the cardiovascular effects of marihuana was supported by the work of MALIT et al. (1975) who reported that the i.v. injection of 135 or 201 µg/kg Δ^9-THC produced a dose-related increase in pulse rate, a slight increase in blood pressure and a decrease in total peripheral resistance.

The oral administration of 300 µg/kg Δ^9-THC produced premature ventricular contractions in one subject, T-wave and S–T segment changes in two and tachycardia in six of seven volunteers (KOCHAR and HOSKO, 1973). ARONOW and CASSIDY (1974) found that smoking one or two marihuana cigarettes decreased the exercise time prior to angina. Increased pulse rate caused by the oral administration of 30 mg Δ^9-THC was antagonized by 30 or 60 mg of cannabidiol (KARNIOL et al., 1974). The possible antagonistic effects of high doses of cannabidiol on electrocardiographic changes or angina induced by Δ^9-THC have not been investigated.

Marihuana and a number of its constituents and their analogs have pronounced effects on the cardiovascular system of many species of laboratory animals. The water insolubility of these substances requires that they be administered in an organic solvent, many of which have their own effects on the cardiovascular system. Therefore, rigorous controls are required in these experiments and one should be concerned about vehicle effects on the cardiovascular system which are in the same direction as those seen with the cannabinoid even though the vehicle at the concentration used does not cause a significant alteration. At times the cannabinoid might be potentiating the vehicle effect rather than causing a direct effect of its own.

A number of reviews have been written on the cardiovascular effects of marihuana, its extracts, constituents and their analogues. The reader is referred to these articles for a more detailed description. We have attempted to point out the more salient features of this work but the reader may consult these articles for more detail (HARDMAN et al., 1971; CLARK, 1975; and HARDMAN and HOSKO, 1976).

Δ^9-THC produces a prolonged and large decrease in heart rate and blood pressure in most experimental animals. The fall in blood pressure following the i.v. injection of Δ^9-THC in anesthetized dogs has an onset of 3 min and lasts at least 1 1/2 h, often not returning to control for up to 4 h (DEWEY et al., 1970d; CAVERO and JANDHYALA, 1971). A decrease in heart rate accompanies the fall in blood pressure and is related to it in terms of duration of action. GRAHAM and LI (1973) reported that a *Cannabis* extract, Δ^8-THC and Δ^9-THC but not cannabinol and cannabidiol produced a fall in blood pressure and heart rate in rats. WILLIAMS et al. (1973) also reported that Δ^9-THC reduced the blood pressure in non-immobilized rats and that it blocked the hypertension induced by immobilization in this species. BIRMINGHAM (1973) has shown that the i.p. injection of 3 mg/kg Δ^9-THC reduced the blood pressure in rats with adrenal regeneration hypertension, and that tolerance developed to this effect. It is important to point out that although the actual decrease in blood pressure was less after the final administration of Δ^9-THC this may be due to the fact that the resting blood pressure was decreasing thoughout the experiment. NAHAS et al. (1973) simultaneously reported that the oral administration of 5 and 25 mg/kg Δ^9-THC produced hypotension in spontaneously hypertensive rats and that tolerance developed to the hypotension but not to the loss in body weight upon daily administrations. KAYMAKCALAN and SIVIL (1974) reported that tolerance did not develop to the bradycardia produced by Δ^9-THC in rats. However, the premedication heart rate on the last day of medication approximates the lowest rate following the initial injection. LEWIS et al. (1974) reported that tolerance did not develop to the hypotensive effects of Δ^9-THC in spontaneously hypertensive rats treated i.p. for 9 successive days.

DEWEY et al. (1970d) reported that the i.v. administration of 3 mg/kg Δ^9-THC caused a decrease in cardiac output and total peripheral resistance in anesthetized dogs. Studies by CAVERO et al. (1972a) confirmed that Δ^9-THC caused a reduction in cardiac output but they did not observe a change in peripheral resistance. In a later report, CAVERO et al. (1973) suggested that the decrease in cardiac output was the result of a decrease in venous return and partially due to the decrease in heart rate.

The decrease in venous return might well be due to venous pooling. Further studies by CAVERO et al. (1974) confirmed that the decrease in cardiac output was due to a decrease in venous return. In this study, they paced the heart, found no change in contractility but still observed a decrease in cardiac ouput. When cardiac output was returned to normal by an i.v. infusion of dextran, the decrease in blood pressure persisted apparently due to reduced peripheral resistance.

A number of investigators have concluded that the depressant effects of Δ^9-THC and other cannabinoids on heart rate and blood pressure were due to a decrease in sympathetic outflow from the central nervous system. HARDMAN et al. (1957) suggested that a decrease in sympathetic tone to peripheral vessels was responsible for the hypotensive effects of dimethylheptylpyran (DMHP). CAVERO et al. (1973) indicated that the negative chronotropic effect of Δ^9-THC could be partially blocked by atropine, propranolol, vagotomy or spinal section and that the maximal degree of bradycardia was observed when both the sympathetic and parasympathetic innervations were intact.

In each of these series of experiments the blood pressure measurements were made in unanesthetized rats. DEWEY et al. (1972b) found that although Δ^9-THC produced pronounced depression of blood pressure and bradycardia in anesthetized dogs it did not significantly alter blood pressure or heart rate in unanesthetized dogs when 2 mg/kg Δ^9-THC was injected intravenously 30 min prior to the measurements. A few studies on the possibly direct effects of Δ^9-THC on the cardiac tissue and vasculature have appeared. There is little evidence that Δ^9-THC has pronounced effects on the cardiac musculature. MANNO et al. (1970) reported that the infusion of Δ^9-THC produced a decrease in the force of contraction of the isolated rat heart. High concentrations caused an initial increase in the force of contraction which was followed by a long lasting decrease. However, as mentioned above, CAVERO et al. (1974) did not find a decrease in contractility in their i.v. experiments in dogs.

The effects of Δ^9-THC on the vasculature have been less well studied. BROWN et al. (1974) found that high doses of Δ^9-THC did not have a direct effect on the vasculature of the bat wing. Recently, ADAMS et al. (1976b) reported that Δ^8- and Δ^9-THC caused a brief vasoconstriction in the rat which was followed by prolonged hypotension. The vasoconstriction was dose-related and was reduced by reserpine and the α-blocking agent phentolamine. These results suggested that the cannabinoids caused a release of NE from adrenergic neurons. ADAMS et al. (1976a) also reported that tolerance developed to the hypotension and bardycardia but not to the pressor effects of Δ^9-THC in the rat.

A number of reports have appeared on the cardiovascular effects of synthetic tetrahydrocannabinols. The cardiovascular effects of two heterocyclic synthetic cannabinoids have been reported (DEWEY et al., 1970b).

DAGIRMANJIAN and BOYD (1962) reported that DMHP produced a pronounced and prolonged fall in blood pressure of the anesthetized cat but it returned to baseline within 2 h. LEMBERGER et al. (1974) reported that an i.v. injection of 0.2 mg DMHP produced resting and postural hypotension and pronounced tachycardia in man without significant psychologic effects. Doses of Δ^9-THC and 11-OH-Δ^9-THC relatively equally effective to DMHP on heart rate produced significant psychologic effects in man. There are two asymmetric carbon atoms in the side chain of DMHP and, therefore, eight possible isomers. The information presented in the papers does not allow one to be sure that the same isomers were tested by HARDMAN et al. (1957); DAGIRMANJIAN and BOYD (1962); and LEMBERGER et al. (1974).

IV. Endocrine Effects

Little has been published concerning the effects of *Cannabis* or the cannabinoids on the endocrine system of animals or man. DEWEY and his colleagues (1970c) found that Δ^9-THC was a potent stimulator of ACTH secretion as evidenced by depletion of adrenal ascorbic acid in rats. Low doses of Δ^9-THC, which were inactive in the unanesthetized rat, stimulated ACTH secretion in the pentobarbital anesthetized animal. This makes Δ^9-THC different from other drugs reported to stimulate ACTH. Non-stimulating doses of Δ^9-THC did not block the secretion of ACTH induced by epinephrine. Tolerance to the ACTH stimulating effect of Δ^9-THC did not develop after five daily administrations. BARRY et al. (1970) simultaneously reported that Δ^9-THC activates the pituitary-adrenal axis using plasma corticosterone as an index. This work was extended (KUBENA et al., 1971) to indicate a hypothalamic or central site for this action of Δ^9-THC.

Hunger is often reported as a consequence of *Cannabis* use. It was thought that this might be due to hypoglycemia and early uncontrolled animal experiments with crude extracts seemed to confirm this. Recent studies in animals and man using plant material or pure Δ^9-THC show that fasting blood glucose levels are unaffected by marihuana (see SECRETARY, H. E. W., 1971–1975 for reviews of this material). Recently, HOLLISTER and REAVEN (1974) found elevated glucose levels as compared to control values in five of six subjects given a glucose tolerance test after receiving a 6 mg i.v. dose of Δ^9-THC. Plasma insulin levels were not affected while growth hormone levels rose following the glucose load after Δ^9-THC. These findings suggested that a resistance to insulin was responsible for the glucose intolerance. No such effects were seen when similar subjects smoked a cigarette containing 19 mg Δ^9-THC. JONES and BENOWITZ (1976) administered an insulin tolerance test to six subjects before and after a 10 day period of high dose oral ingestion of Δ^9-THC. The response to insulin was unchanged as reflected by blood glucose, ACTH and cortisol changes. However, the usual increase in growth hormone after insulin was blocked.

HARMON and ALIPOULUS (1972) reported three cases of gynecomastia in males who were long-term, high-dose marihuana users. This finding prompted KOLODNY et al. (1974) to study plasma testosterone levels in 20 adult males who used marihuana at least 4 days a week for 6 months. They found plasma testosterone levels significantly lower than a control group of age-matched men. Either abstention from marihuana or administration of chorionic gonadatropin during continued use resulted in marked increases in testosterone levels. Liver function, circulating gonadatropins, prolactin, cortisol and thyroxine were within normal limits. Six of the 17 men examined showed oligospermia. These findings implicated a central effect through the hypothalamus or pituitary. Later studies by this group (KOLODNY et al., 1976) using subjects exposed to marihuana smoking for 94 days reported a depression of plasma testosterone levels within 3 h of smoking a single marihuana cigarette. Weekly blood samples on the same subjects showed a significantly lower level after 5 weeks of daily smoking when compared to the initial abstinence period. However, none of the levels were markedly below the normal range. On the other hand, MENDELSON et al. (1974) in a carefully controlled acute and chronic marihuana smoking study using healthy male volunteers found no statistically significant difference in plasma testosterone levels during or after smoking marihuana when compared to presmoking baselines. Some question has been raised concerning the methodology employed in their study (KOLODNY, 1975 and MENDELSON et al., 1975). It should also be noted that SCHAEFER et al. (1975) reported not finding lower testosterone levels in groups of subjects exposed to Δ^9-THC in marihuana cigarettes as compared to controls, while CICERO et al. (1975) found no difference in plasma testosterone levels between control subject and heroin or methadone subjects who where also using marihuana. On the other hand, JONES and BENOWITZ (1976) did report elevated testosterone levels in their clinical study of chronic oral ingestion of Δ^9-THC, and ROSENKRANTZ and BRAUDE (1976) reported depressed testosterone levels in rats exposed to marihuana smoke.

It is apparent from the literature reviewed that marihuana or its active principles may have important albeit subtle effects on the body's neuroendocrine systems. This is an area which deserves considerably more intensive investigation.

V. Cellular Effects

Concern for the potential toxic effects of marihuana has led to a number of basic studies on a variety of cell function. For instance, MANN et al. (1971) compared alveolar macrophages from nonsmokers, regular marihuana smokers and tobacco smokers. There were some morphological differences and while fewer macrophages were recovered from marihuana smokers than nonsmokers there was no discrimination in their defensive capacity. LEUCHTENBERGER and LEUCHTENBERGER (1971), using mouse lung explants exposed to marihuana in tobacco smoke, also noted some morphological changes, and an abnormal proliferation of epithelioid cells accompanied by an increased mitotic index and a stimulation of DNA synthesis. They also noted that cigarette smoke alone was more cytotoxic than the mixed tobacco-marihuana smoke. Later studies (LEUCHTENBERGER et al., 1973) extended these studies to human lung explants and compared marihuana alone cigarettes against standard tobacco cigarettes. Again the tobacco was more cytotoxic and the same abnormalities in DNA metabolism and morphology were produced by both tobacco and marihuana smoke. A more complete resume of this elegant work is found in LEUCHTENBERGER and LEUCHTENBERGER (1976). In the same vein, MAGUS and HARRIS (1971) have reported that marihuana smoke condensates when applied to the skin of mice produced the same epidermal changes as did tobacco tar. Δ^9-THC itself had no activity.

A variety of studies have looked at the effects of marihuana on chromosomes both in vitro and in vivo. NEU et al. (1970) using human leucocyte cultures found Δ^8-THC and Δ^9-THC to have little or no effect on chromosomes although higher concentrations did decrease the mitotic index. Similar findings were reported by STENCHEVER and ALLEN (1972). This group later looked at chromosomes in chronic marihuana users (STENCHEVER et al., 1974), and reported an increase in the number of damaged or broken chromosomes. Similar results were obtained by HERHA and

OBE (1974) and by KUMAR and KUNWAR (1972) who examined direct bone marrow preparations. It should be noted, however, that their results are accounted for by only two out of seven subjects examined. On the other hand MARTIN et al. (1973), DORRANCE et al. (1970), and GILMOUR et al. (1971) did not find an increase in the frequency of cells with abnormal chromosomes. NAHAS et al. (1974) did report an increase in abnormal chromosomes in chronic marihuana smokers but the data were not statistically significant. Three recent prospective studies in man (MATSUYAMA et al., 1973; NICHOLS et al., 1974; and MATSUYAMA and JARVIK, 1976) failed to find any significant effect on chromosomes. All of the studies described have had one or another methodological problem and the question has not been answered definitively. However, with the additional failure to detect mutagenic effects with Δ^9-THC (LEGATOR et al., 1976), the weight of the evidence points to no great problem, at the present time, concerning genetic damage by marihuana or its active principles.

ZIMMERMAN and MCCLEAN (1973), studying macromolecular synthesis in tetrahymena, indicated that low concentrations of Δ^9-THC produced cytolysis and inhibited DNA, RNA and protein synthesis while NAHAS and his colleagues (1974) reported that lymphocytes from chronic marihuana users had a decreased reactivity to mitogens as measured by thymidine uptake. On the other hand, SILVERSTEIN and LESSIN (1974), using the in vivo skin response to 2,4-dinitrochlorobenzene (DCNB), found no difference between smokers and nonsmokers. These authors did note that lymphocyte reactivity to phytohemagglutinin was depressed in the marihuana smokers (SILVERSTEIN and LESSIN, 1976). These findings prompted a number of studies concerning the effects of marihuana and a variety of pure cannabinoids on immune systems and macromolecular synthesis. LEVY et al. (1974) showed that Δ^9-THC prolonged the survival of alogenic skin grafts, inhibited primary antibody production to sheep erythrocytes (SRBC) and inhibited both T- and B-cell reactivity in mice. LEFKOWITZ and YANG (1974) also demonstrated a reduced response to SRBC in Δ^9-THC treated mice using the hemolytic plaque assay. They also noted a depletion of spleen cells. Later studies (MUNSON et al., 1976) pointed to complex and subtle effects on both T- and B-cell function in mice with little effect on reticuloendothelial system activity as measured by the intravascular clearance of colloidal carbon. NAHAS and his colleagues (1976) studied a number of naturally occurring cannabinoids and their metabolites on the incorporation of thymidine in cultured lymphocytes. All of the compounds tested produced some inhibition, with cannabinol and cannabidiol being more potent than Δ^9-THC. HARRIS et al. (1974, 1976) and MUNSON et al. (1975) have reported that Δ^9-THC and other cannabinoids have antineoplastic activity in mice bearing the Lewis lung adenocarcinoma. Cannabidiol appeared to enhance tumor growth. Δ^9-THC was inactive against L-1210 leukemia but did inhibit Friend leukemia virus induced splenomegaly. Lewis lung cells incubated in vitro showed a dose dependent inhibition of thymidine and uridine uptake. CARCHMAN et al. (1975, 1976) using the same in vitro tumor cell system demonstrated that Δ^9-THC and other selected cannabinoids inhibited the incorporation of thymidine into DNA. Δ^9-THC, but not other compounds, showed some selectivity by inhibiting DNA synthesis in the tumor at concentrations significantly less than those necessary to inhibit DNA synthesis in bone marrow cells. The inhibition of DNA synthesis could not be accounted for by blockage of transport of precursor into the cell.

E. Toxicology

There have been very few reports of deaths attributed to acute overdose with *Cannabis*. Most of these have come from the older literature from India and represent large oral doses of the potent resinous material, charas (DEAKIN, 1880; Indian Hemp Drug Commission Report, 1893–94; EWENS, 1904; WALTON, 1938; BOUQUET, 1951). The recent case reports of HEYNDRICKX and his colleagues (1969) and HUGHES et al. (1970) relate deaths in which *Cannabis* may be implicated. There are, however, some doubts about these (SHAFER Marihuana Report, Appendix Vol. 1, 1972). SMITH (1968) and BRILL et al. (1970) state that they have been unable to find any reliable reports of human fatalities due purely to marihuana.

The lack of *Cannabis*-associated lethality in man, despite its widespread use, probably reflects the high therapeutic ratio of the drug. A number of reports have described the acute toxicity of *Cannabis* extracts or its active principles Δ^8- and Δ^9-

Table 3. Acute toxicity of Δ^8 and Δ^9-THC

Species	Route of administration	Acute LD_{50} mg/kg	
		Δ^8-THC	Δ^9-THC
Mouse	[a]i.v.	31	60
	[b]i.v.		43
	[a]i.p.	210	168
	[b]i.p.		455
	[a]p.o.	>2000	1900
	[b]p.o.		482
Rat	[a]i.v.	97	100
	[b]i.v.		29
	[a]i.p.	560	430
	[b]i.p.		373
	[a]p.o.	>2000	>2000
	[b]p.o.		666
	[c]p.o.	1980 males 860 females	1910 males 1040 females
	Marihuana extract[c]	3300 males	1380 males
Monkey	[c]p.o.	No deaths 131—3, 150 mg/kg Δ^8 or Δ^9-THC or 5000 mg/kg *Cannabis* extract. Three additional monkeys were given 6000, 7000 or 9000 mg/kg with no deaths reported.	
	[d]i.v.	128 mg/kg Δ^9-THC lethal, 92 mg/kg not.	
Dogs	[c]p.o.	No deaths 66—3000 mg/kg Δ^8 or Δ^9-THC except for 1/2 by drug aspiration at 353 mg/kg. One dog treated with 5000 mg/kg of the extract also died of drug aspiration.	

[a] From HARRIS (1971) and DEWEY et al. (1972a). Drugs were administered as suspensions in albumin. Male Swiss-Webster mice and Sprague-Dawley rats were used.

[b] From PHILLIPS et al. (1971b). Δ^9-THC was purified from extracted *Cannabis* and administered in 10% Tween 80-saline. Male Cox mice and Hollzman rats were used.

[c] From THOMPSON et al. (1973b). Compounds were administered in 2.5% ethanol-sesame oil, sesame oil or in some cases of high doses in absolute ethanol. The marihuana extract contained 25—32% Δ^9-THC, 2.9% cannabidiol, 3.5% cannabinol and 4% ethanol. The rats were Fisher.

[d] From THOMPSON et al. (1974a). Δ^9-THC was administered in a sesame oil-Tween 80-saline mixtures.

THC. These findings are detailed in Table 3. While there are some differences, they are not large and probably due, in part, to differences in the formulations used. As has been indicated elsewhere in this review, we do not to date have an ideal way to prepare medication forms of this highly water insoluble material. Other considerations must include differences in the strains and sex of the animals. In general, it appears that females are more susceptible to the drug. In any case, except for the intravenous route, the acute toxicity of the cannabinoids is quite low. Indeed, in the dog and monkey an acute lethal dose level could not be reached by the oral route of administration. Whether the intravenous values represent a true picture of the toxicity of the drug is open to some question. One could postulate that the highly water insoluble cannabinoids precipitate out in the blood and produce death as a result of this. Some credence for this is obtained from the data of THOMPSON et al. (1973b), who reported that the acute i.v. LD_{50} in the dog was 100 mg/kg but that the min. lethal dose ranged from 25–99 mg/kg depending on the concentration of the solution used. Similar data were obtained in the monkey. On pathological examination of the animals which succumbed after acute intravenous administration, the only consistent finding was hemorrhagic, edematous and generalized congestive changes in the lungs.

PHILLIPS et al. (1971a) reported a 30-day sub-acute toxicity study in rats with Δ^9-THC. The drug was given intraperitoneally as a water suspension in doses up to 30 mg/kg/day. Both depression and hyperexcitability were seen and tolerance was reported not to occur. There were no deaths and no evidence of inflammatory response or incomplete drug absorption was seen at the injection sites. MANNING et al. (1971) reported a similar study with Δ^9-THC in rats using small doses by both the i.p. (4 mg/kg) and intragastric route (8 mg/kg). The animals ate less and gained less weight than controls. After 30 days, medication was stopped and although the growth rate of the medicated animals returned to control levels, the medicated groups still weighed less than controls 30 days post-medication. Nonsuppurative peritonitis was noted in all rats who had received Δ^9-THC intraperitoneally.

THOMPSON et al. (1973a) reported the first of an extensive series of chronic toxicity studies of marihuana and its active principles. This paper describes the results of a 119 day chronic oral study in rats involving Δ^8-THC, Δ^9-THC, and a marihuana extract (CME). The doses used ranged from 50–500 mg/kg/day for Δ^8-THC and Δ^9-THC to 150–1500 mg/kg/day for CME. A biphasic toxicity was observed with generalized depression to which tolerance developed over the first week followed by hyperactivity, irritability and aggressiveness. These signs were progressive and convulsions began to appear about day 50. Cumulative toxicity occurred in the first few days with a 70% incidence of lethality at the acute LD_4 in males. There was a consistently greater incidence of mortality in the females and all deaths occurred in the interval from 36–72 h after the initial dose. Growth was depressed in all drug treated groups despite an increase in food and water consumption after 50 days. Pathologic changes were limited to splenic hypocellularity and vacuolization of the adrenal, while hemotalogical and biochemical parameters were unremarkable except for a slight non dose-dependent leukopenia, decreased SGOT and decreased coagulation time. A later 6-month study in Fisher rats (ROSENKRANTZ et al., 1975), using 2, 10, and 50 mg/kg Δ^9-THC, reported CNS depression, ataxia, poikilothermia and passivity, to which tolerance developed over the first 10 days. Over the next 10 days irritability, hypersensitivity, and aggression were the predominant signs noted. Tremors and clonic convulsions developed in 50% of the high dose animals after day 70 with 12% at the median dose. Tolerance also developed to the CNS stimulation but cumulative lethality was seen at the highest dose. After 180 days, there was a decreased growth rate at the two highest doses which returned to control levels 30 days after cessation of treatment. There was a 28–45% increase in SGOT and a 46–69% increase in SGPT in the male animals at the highest doses. These, and a related hyperglycemia in both sexes, were the only clinical chemistry, hemotological or urinalysis parameters which were not within normal limits. There were some elevated organ to body weights at the highest doses, but no morphological changes were seen. The brains of animals used in this study were utilized in an attempt to correlate behavioral events in time with brain protein and RNA levels and acetylcholinesterase (AChE) activity (LUTHRA et al., 1975). There appeared to be some decrease in protein levels at the higher doses, but RNA was essentially unchanged when expressed as μg/mg protein. AChE activity was elevated in the males but diminished in the females. This study corrected some of the methodological problems of an earlier report (LUTHRA and ROSENKRANTZ, 1974). However, the relationship between the often random biochemical changes observed and behavioral changes is not apparent. It would have been nice to have had non-vehicle controls to eliminate this factor.

In a 28 day sub-acute study in monkeys, THOMPSON et al. (1974a) administered Δ^9-THC intravenously (5, 15, and 45 mg/kg/day) or orally (50, 250, and 500 mg/kg/day). Two of four animals died in the intravenous study at the highest dose of acute hemorrhagic pneumonia. Two of eight animals died at the highest oral dose and one of six at the lowest dose. These animals had pathological changes characterized as atrophy of the pancreas, hemorrhagic ulcerative colitis, myeloid hyperplasia of the bone marrow, vascular nephrosis and severe serum electrolytic imbalance. No deaths were seen at 250 mg/kg. Lethargy, huddled posture, bradypnea, hypothermia, bradycardia, weight loss, anorexia and constipation were observed in both medication groups, and tolerance was seen to most of these effects. Monkeys treated intravenously had a dose-related anemia and increased sulfobromophthalein retention. No remarkable blood or enzyme changes occurred in the animals medicated orally. A later 90-day oral study in the monkey with Δ^9-THC and a marihuana extract has been completed (THOMPSON et al., 1974b), but the complete results have not as yet been published. The findings were similar to those obtained in the 28 day study with pathology associated only with the moribund or dead animals. Hematologic and blood chemistry parameters were generally not affected.

A 28-day sub-acute study was carried out in the rabbit using the subcutaneous route of administration (THOMPSON et al., 1975). The Δ^9-THC was dissolved in sesame oil and given at doses of 15.9, 45.0 or 153.4 mg/kg/day. All the rabbits survived treatment but there were severe dose-related dermal responses and depression of body weight, liver weight, and liver glycogen. Hematology and gross pathology were essentially negative but terminally there was a decrease in blood glucose and alkaline phosphatase, and an increase in serum potassium in the high dose animals only.

A variety of inhalation toxicity studies have also been undertaken. These are summarized in the paper by ROSENKRANTZ and BRAUDE (1976). Rats were exposed

Table 4. The effects of marihuana derivatives on pregnancy and fetal development in laboratory animals

Species and strain	Cannabinoid preparation	Dose (mg/kg) day[a]	Route	Days of gestation treated	Effects: Terato-genicity	Effects: Others	Reference
Rat							
Stock unstated	*Cannabis* extract	0.2%	p.o.	—[b]	None	Reduced fertility	MERAS, 1965
Inbred albino	*Cannabis* resin	4.2	i.p.	1—6	Yes	Increased fetal resorption, decreased fetal weight and size	PERSAUD and ELLINGTON, 1968
Long Evans	Δ^9-THC	0.01—200	s.c.	1—20	None	Smaller litters	BORGEN et al., 1971
Long Evans	Δ^9-THC	0.01—10	s.c.	1—20	None	Fetal hemorrhages (skin),	
		25—100	s.c.	1—20	None	Smaller litters	
				1—20[c]	None	Subdermal bruising	
Wistar	Δ^9-THC	100—250	s.c.	9—16	None	Stunted fetuses, high neonatal mortality	PACE et al., 1971
Charles River albino	Δ^9-THC or marihuana	5—50	p.o.	6—15	None	None	HALEY et al., 1973
	extract 16% Δ^9-THC	0.5—15	p.o.	6—18	None	None	
Wistar	Δ^9-THC	10	s.c.	10—12	None	None	BORGEN et al., 1973
Wistar	THC	30—120	s.c.	10—12	None	Increased stillbirths	UYENO, 1973
Charles River albino	Δ^9-THC or marihuana extract 16% Δ^9-THC	0.5—5	p.o.	15[d]	None	None	KEPLINGER et al., 1973
Sprague-Dawley	Δ^9-THC	25—100	s.c.	6—15	None	Decreased maternal weight gain	BANNERJEE et al., 1975

Mouse							
Stock unstated	*Cannabis* resin	16	i.p.	1—6	None	Complete resorption zygote	PERSAUD and ELLINGTON, 1967
				6	None	Stunted offspring	
Cox albino	Δ^9-THC	200	i.p.	8—9	None	Increased resorptions	PHILLIPS et al., 1971b
Swiss Webster	Δ^9-THC 93.2%	200	i.p.	8—13	None	In utero death, decreased fetal weight	HARBISON and MANTILLA-PLATA, 1972
	Δ^9-THC	50—300	i.p.	8—16	None	Increased resorptions	
	Δ^9-THC	300	i.p.	12 or 14	Yes	None	MANTILLA-PLATA et al., 1973
Charles River CDI	Δ^9-THC	5—150	p.o.	6—15	None	None	FLEISHMAN et al., 1975
Hamster							*Hamster*
Golden	*Cannabis* extract	25—300	s.c.	618	Yes	Runting, fetal edema	GEBER and SCHRAMM, 1969
Golden	Marihuana extract 17.1% Δ^9-THC	200—300	s.c.	6—8	None	None	PACE et al., 1971
Rabbit							
New Zealand white	*Cannabis* extract	130—500	s.c.	7—10	None	Embryocidal, fetal edema	GEBER and SCHRAMM, 1969
New Zealand white	Δ^9-THC or marihuana extract 16% Δ^9-THC	5—50	p.o.	6—15	None	None	
	Δ^9-THC	0.15—15	p.o.	6—18	None	None	HALEY et al., 1973
Nonhuman primate							
Chimpanzee	Marihuana extract or Δ^9-THC	1.4—2.1	p.o.	—[e]	None	None	GRILLY et al., 1974

[a] Unless otherwise indicated; [b] given for months; [c] given every other day for 30 days before conception; [d] males treated 60 days and female 14 days before mating, data presented in this manuscript; [e] males and females received 50—155 doses before mating; time between the last dose and impregnation was 1.5 months to 1.5 years; from FLEISHMAN et al. (1975)

to from 4–16 puffs of standard marihuana cigarettes or placebo marihuana smoke 6 times per week for up to 80 days. The inhalation toxicity paralleled that observed after oral administration except for a dose-related focal pneumonitis seen after chronic inhalation which was not so apparent after oral administration. Again, a cumulative lethal toxicity occurred after prolonged treatment by both routes of administration.

A large number of teratology studies have been carried out utilizing crude marihuana extracts, marihuana inhalation and pure Δ^8- and Δ^9-THC. These findings have been summarized in Appendix 1 of Shafer (1972), by Falek (1975), and by Fleischman et al. (1975), and are illustrated in Table 4, from Fleischman et al. (1975). Many of the studies have suffered from methodological faults, including lack of controls and the use of abnormally high doses. While one recent report (Mantilla-Plata and Harbison, 1976) presented evidence for cleft palate in mice and rabbits after high doses of Δ^9-THC, another careful study by Fleischman et al. (1975) reported no such effects in rats. In general, the results have indicated that marihuana and its active principles often produce stunted or reabsorbed fetuses, particularly after high doses. As is often the case, the findings are inconclusive.

There have been three isolated case reports of birth defects in offspring of parents who had used marihuana (Hecht et al., 1968; Carakushansky et al., 1969; Gelehrter, 1970). However, given the other drugs used by these individuals, the usual level of birth defects in the "normal" population, and the high incidence of marihuana use in the population, little conclusion can be drawn from these isolated reports. Teratogenic effects may occur, but this is uncertain and the essentially negative mutagenetic studies quoted earlier in this chapter argue against preliminary reports of serious defects in the second generation of treated animals. In any case, the Shafer Commission (1972) statement concerning the use of the U.S. Food and Drug Administration label warning on drugs to pregnant women and those of child-bearing age is relevant. "Safe use of the drug during pregnancy and lactation has not been established; therefore, in administering the drug to pregnant patients, nursing mothers, or women of child bearing potential, the potential benefits must be weighed against the possible hazards. Animal reproduction studies have yielded inconclusive results ... There have been reports of congenital malformation associated with the use of the drug, but a causal relationship has not been confirmed."

Two carefully controlled high dose chronic marihuana studies have been carried out in man. Jones and Benowitz (1976) reported a 30-day study using oral Δ^9-THC in doses up to 210 mg/24 h. They noted a rapid tolerance development to many of the behavioral and cardiovascular effects. No psychiatric, paranoid or marked anxiety effects occurred at these doses, but there was some blunting of the cardiovascular response to exercise. The intraocular pressure was reduced and plasma testosterone levels fell although there was no change in sperm counts. Definite withdrawal signs characterized by restlessness, hyperactivity, insomnia, irritability and poor appetite were noted. In another study, Cohen et al. (1976) carried out a 94 day study using marihuana smoking. These subjects smoked an average of 4.8 marihuana cigarettes each day with a range from 1.7–10. The cigarettes weighed 900 mg and analyzed as 2.2% Δ^9-THC. While no global view of this study has been published to date, a number of findings are available. For instance, lowered testosterone levels were reported (Kolodny et al., 1976) as was reduced intraocular pressure (Hepler et al.,

1976) and bronchodilation (TASHKIN et al., 1976). Some physiological and psychological effects (FRANK et al., 1976) as well as effects on motivation (LESSIN and THOMAS, 1976) have also been reported from this global study. To date, however, nothing truly remarkable from a toxicological point-of-view has emerged from these studies. Another chronic study in man (MENDELSON et al., 1976), using operant conditioning, sheds some doubt on the existence of an "amotivational syndrome" associated with marihuana use, at least in this highly motivated and intelligent group of subjects.

Two rather comprehensive retrospective studies have been carried out using existing populations of chronic, heavy *Cannabis* users in an attempt to get a better idea of the long-term effects of health. One of these studies was carried out in Jamaica and the results are presented in great detail in RUBIN and COMITAS (1975). In one facet of this overall study, 30 chronic users were compared with matched control subjects (BEAUBRUN and KNIGHT, 1973). These subjects smoked an average of seven cigarettes a day having a Δ^9-THC content of nearly 3%. There was no significant difference between the groups in the incidence of mental illness, or abnormalities of mood, thought or behavior. The second study was carried out in Greece and compared 47 chronic hashish users with controls (FINK et al., 1976; STEFANIS et al., 1976). No significant differences were found between the groups in resting EEG, echoencephalographic measures, psychological test performances, and physical, neurological and mental states with the exception of some differences in Weschler Adult Intelligence subtests and a higher incidence of personality disorders in the users. Interpretation and generalization of the findings from these studies should be approached with caution since cross-cultural factors may play an important role in the results (DORNBUSH, 1974).

F. Tolerance

MIRAS (1965) reported that following 30 days of daily treatment, tolerance developed to the diarrhea caused by the initial ingestion of *Cannabis* resin in rats. Tolerance also developed to the hypomobility and weight loss when the medications were continued for 90 or 120 days, respectively. Cross-tolerance between *Cannabis* resin and barbiturates was suggested when rats treated for six months with resin required a higher dose of barbiturate than control rats to induce sleep. CARLINI (1968) showed that tolerance developed to the disruptive effects of a fixed dose of *Cannabis* extract in laboratory animals. The initial ingestion of the *Cannabis* extract in rats caused a reduction in avoidance of an electric shock and a decrease in the rate of responding for water reinforcement. Even though tolerance developed to the effects of marihuana extract on impairment of pole climbing behavior, the rats were able to discriminate this marihuana extract from control solution (BUENO and CARLINI, 1972). SILVA et al. (1968) were the first to report that tolerance developed to the effects of repeated exposures of a fixed dose of Δ^9-THC on pole climbing and schedule-controlled behavior in rats. The first report of tolerance to Δ^9-THC in a non-rodent species was by DEWEY et al. (1969) who reported that upon chronic medication tolerance developed to the static ataxia produced by Δ^9-THC when administered intravenously to dogs. In all the above studies, tolerance developed upon chronic

Table 5. Tolerance to cannabinoids in laboratory animals

Substance administered	Species	Effect	Reference
Δ^9-THC	Frog	Loss of righting reflex and inhibition of leg withdrawal	ABEL, 1973
Δ^9-THC	Chicken	Hypothermia	ABEL et al., 1973
Δ^9-THC	Mouse	Catalepsy	GILL et al., 1970
Δ^9-THC	Mouse	Hyperthermia	LAWRENCE and PERTWEE, 1973
Δ^9-THC	Mouse	Anticonvulsant	KARLER et al., 1974
Δ^9-THC	Rat	Discriminative-Sidman avoidance schedule	WEBSTER et al., 1973
Δ^9-THC	Rat	EEG and certain non-conditioned behaviors	PIRCH et al., 1972b
Δ^9-THC	Rat	Behavioral and anticonvulsant activity	FRIED and MCINTYRE, 1973
Δ^9-THC	Rat	Hypotension	NAHAS et al., 1973
Δ^9-THC	Rat	Analgesia	KAYMAKCALAN and DENEAU, 1971
Δ^9-THC	Rat	Disruptive effects on open field behavior, balance beam and a swimming test	POTVIN and FRIED, 1972
Δ^9-THC	Rat	Decreased locomotor activity	DAVIS et al., 1972
Δ^9-THC	Rat	Ataxia and vocalization	FRIED and MCINTYRE, 1973
Δ^9-THC	Rat	Hypothermia	LOMAX, 1971
Δ^9-THC	Rat	Decreased defecation	MASUR et al., 1971
Δ^9-THC	Monkey	Spaced responding	MANNING, 1973
Δ^9-THC	Squirrel monkey	Hypothermia	FORD, 1974
Δ^9-THC	Squirrel monkey	Free operant avoidance behavior	FORD, 1974
Δ^9-THC	Rhesus monkey	Decreased response rate on a DRL schedule	HARRIS et al., 1972
Δ^9-THC	Rhesus monkey	High-voltage, slow wave EEG	HARDMAN et al., 1971
Δ^9-THC	Baboon	Anticonvulsant effects	KILLAM and KILLAM, 1972
Δ^9-THC	Speciosa monkey and chimpanzee	Decreased accuracy and amount of responding on DRL schedule for food	FERRARO, 1972
Δ^8-THC	Rat	Prolongation of hexabarbital sleeping	RATING et al., 1972
Δ^8-THC	Rat	Hypothermia	RATING et al., 1972
Δ^8-THC	Dog	Static ataxia	MARTIN et al., 1975
Δ^9-THC, hashish marihuana	Man	EEG changes, tachycardia and euphoria	FINK et al., 1972
Δ^9-THC and 11-OH-Δ^9-THC	Pigeon	Decreased key picking for food	KOSERSKY et al., 1974
Marihuana extract and Δ^9-THC	Rat	Decreased EEG voltages	PIRCH et al., 1973
9-*nor*-Δ^8-THC	Dog	Static ataxia	MARTIN et al., 1975
Δ^8, Δ^9-THC and crude marihuana extract	Rat	Depressant effect	THOMPSON et al., 1973a

Table 5 (continued)

Substance administered	Species	Effect	Reference
11-methyl-Δ^8-THC	Dog	Static ataxia	MARTIN et al., 1975
DMHP and Δ^8-THC	Rat	Conditioned avoidance	DELINI-STULA, 1973
Cannabis extract	Rat	Cardiovascular and respiratory depression	GRAHAM and LI, 1973
SP111 A	Spider	Web building	FORD, 1974
Synhexyl and DMHP	Pigeon	Schedule-controlled behavior	BLACK et al., 1970
DMHP	Rat	Prolongation of hexobarbital sleeping	DELINI-STULA, 1973
SP-1*	Pigeon	Schedule-controlled behavior	CAMPO, 1973

medication of a fixed dose. MCMILLAN et al. (1970) demonstrated the development of tolerance to the depressant effects of increasing doses of Δ^9-THC on conditioned behavior in pigeons. Many reports have appeared which clearly show that tolerance develops to many of the effects of Δ^9-THC in many species of laboratory animals. A number of these observations were included in previous reviews on tolerance to cannabinoids from our laboratory (MCMILLAN et al., 1972; MCMILLAN and DEWEY, 1972). A number of the papers dealing with tolerance to cannabinoids reported subsequent to these reviews are presented in Table 5. It is obvious from all these reports that tolerance develops to many of the effects of cannabinoids in laboratory animals.

Tolerance to the psychoactive cannabinoids develops rapidly, being apparent on the second administration of the drug (DEWEY et al., 1972b). MCMILLAN et al. (1972) reported that the duration of tolerance to Δ^9-THC in the pigeon was between 3 and 6 months. The magnitude of the tolerance is in the range of that observed with opiates. For instance, FORD and MCMILLAN (1972) reported that a dose of Δ^9-THC 6000 times the originally effective dose was inactive in the tolerant pigeon, while complete tolerance to most of the behavioral effects was observed with a dose of Δ^9-THC 320 times higher than the originally effective dose in dogs. Higher doses were not administered due to the excessive quantity of vehicle needed and the lack of justification for using such large quantities of the drug. We expect that higher doses would not have produced behavioral changes in the dogs injected daily with the cannabinoid.

Tolerance develops to many of the effects of Δ^9-THC in animals and is similar to other drugs in that tolerance does not develop to all the effects of these drugs in every species. For instance, tolerance does not develop to the miosis produced by opiates, but does develop to analgesia and respiratory depression produced by these agents. PIRCH et al. (1972b) reported that tolerance developed to certain non-conditioned behavioral changes but not to other changes in the same rats. SOFIA and BARRY (1974) did not observe tolerance to the effects of daily administrations of Δ^9-THC on body weight and food intake of rats. Tolerance was not observed to develop to daily doses of Δ^9-THC on the impairment of performance of chimpanzees on an intricate reinforcement schedule (FERRARO and GRILLY, 1973). Since they did not drastically elevate the dosage, the question of a state-dependence effect is raised.

The mechanism by which tolerance develops to Δ^9-THC has been the subject of a number of investigations. It appears that the tolerance is not due to an alteration in absorption, distribution or metabolism (MCMILLAN et al., 1973; DEWEY et al., 1973). The repeated administration of Δ^9-THC in pigeons did not produce a significant alteration in the level of total radioactivity in plasma at any time between 1 min and 14 days after administration of ^{3}H-Δ^9-THC. Chronic administration of Δ^9-THC did not cause a diminution in the amount of radioactivity in the brain following an injection of ^{3}H-Δ^9-THC. As a matter of fact, the level of the radioactivity increased in the brain during repeated administrations with ^{3}H-Δ^9-THC. Experiments in dogs (MARTIN et al., 1976b) showed that the distribution of Δ^9-THC in peripheral tissues of 27 different brain parts was not significantly different in tolerant and nontolerant animals. An investigation of the subcellular distribution of radioactivity in tolerant and nontolerant dogs demonstrated that the only significant difference in the two groups was that there was significantly less radioactivity in the synaptic vesicle fraction of the tolerant dog brains than in the nontolerant animals. This suggested that possibly the mechanism by which tolerance developed was due to a decrease in the amount of Δ^9-THC in the synaptic vesicle, the organelle of the cell involved with storage and release of the neurotransmitter. Further experimentation in mice and rats did not confirm this original observation in dogs (DEWEY et al., 1976). That is, there was no difference in the level of radioactivity in the synaptic vesicle fraction of brains from tolerant mice and rats as opposed to animals given only a single injection of ^{3}H-Δ^9-THC. Tolerance to the cannabinoids is much more difficult to demonstrate in mice or rats than dogs; however, some indication of tolerance was observed in these studies. In conclusion, it appears that the mechanism for the development of tolerance to Δ^9-THC is still to be elucidated.

Physical Dependence

Although most investigators agree that pronounced tolerance develops to the behavioral effects of Δ^9-THC in many species, there is some controversy whether this compound differs significantly from the opiates in that when the drug is abruptly withdrawn, a withdrawal syndrome is not generally observed. For instance, dogs were given high i.v. doses of Δ^9-THC daily and developed complete tolerance to many of the effects caused by acute lower doses (DEWEY et al., 1972b). Tolerance did not develop to the anorexia or sedative effects. When the medications were discontinued, the dogs continued to be depressed for some time, but did not go into a period of excitability as would be seen following the cessation of chronic injections of a barbiturate. Pigeons given chronic high doses of Δ^9-THC showed a normal rate of responding in an operant schedule indicating pronounced tolerance. When the drug was taken away, there was a decrease in the rate of responding. This decrease in response rate was not felt to be an indication of withdrawal since it was not reversed by an injection of Δ^9-THC. Opiates or other drugs which produce physical dependence and a withdrawal syndrome upon cessation of drug administration reverse the withdrawal when readministered. In addition, no overt signs which could be interpreted as an indication of a withdrawal syndrome were seen in these pigeons.

On the other hand, studies by HIRSCHHORN and ROSECRANS (1974) have shown that the injection of the pure narcotic antagonist, naloxone, to animals treated chronically with Δ^9-THC precipitates a behavioral syndrome which has many of the characteristics of opiate withdrawal. This suggests, therefore, that a withdrawal syndrome can be induced with an antagonist, but is not seen upon spontaneous withdrawal from the compound. HINE et al. (1975a, b) have reported that the injection of Δ^9-THC but not cannabidiol to morphine dependent rats reduced the intensity of withdrawal induced by naloxone given 1 h after the cannabinoid. These data support the work of HIRSCHHORN and ROSECRANS (1974) in that Δ^9-THC like morphine and other opiates given prior to opiate antagonist appear to competitively block the intensity of the withdrawal syndrome.

A number of papers have appeared which describe an alteration in electrocorticogram (PIRCH et al., 1972a), increased motor activity (DAVIS et al., 1972) and irritability (CHOPRA and CHOPRA, 1957) of laboratory animals following the cessation of chronic administration of marihuana extract.

DENEAU and KAYMAKCALAN (1971) reported that a behavioral syndrome including yawning, tremor, irritability and depression was observed following the cessation of chronic administration of Δ^9-THC to rhesus monkeys.

Recently, JONES and BENOWITZ (1976) have reported that a withdrawal syndrome similar to that seen after the cessation of chronic treatment with central nervous system depressants was observed when hospitalized volunteers were abruptly discontinued on a medication schedule of 30 mg Δ^9-THC every 4 h for 10–18 days. The withdrawal syndrome was characterized by restlessness, poor appetite, irritability, hyperactivity, sweating and increased salivation. However, a withdrawal syndrome has not been reported in studies in humans where lower doses and shorter medication periods were used.

Thus, there is still some question as to whether a true physical dependence occurs with *Cannabis* or its active constituents.

G. Therapeutic Potential

Although the therapeutic use of marihuana has a long and venerable history (MIKURIYA, 1969; SNYDER, 1971; SCHULTES, 1973; and RUBIN, 1975), until recently there have been no carefully controlled studies demonstrating the usefulness of the drug for any illness. For centuries *Cannabis* has been prescribed for the treatment of mental disorders, pain, epilepsy, asthma, and for a variety of cardiovascular and other diseases. In the past few years, there has been a re-examination of these older uses and this, combined with some new findings, have led to a resurgence of the belief that the unique pharmacological properties of *Cannabis* may be put to important therapeutic use. A recent symposium and conference have summed up much of the progress to date in this area (see PARS, A.C.S. Symposium, 1974; COHEN and STILLMAN, 1975).

In brief, Δ^9-THC has been tried as an analgesic in man and the results were equivocal. NOYES et al. (1974 and 1975) reported good analgesic activity, while REGELSON et al. (1976) could find no such effect. Two other studies (HILL et al., 1974; ROFT et al., 1976) reported hyperalgesia. Δ^9-THC has been reported to have sedative-hypnotic properties in insomniacs (COUSENS and DI MASCIO, 1973), but side effects were noted. HOLLISTER (1971), reporting on his experience, could find nothing promising about the use of Δ^9-THC for this purpose.

Promising therapeutic results, however, have been found in three areas: in the treatment of nausea and vomiting, as a bronchodilator in asthmatics, and for lowering intraocular pressure in glaucoma.

SALLEN et al. (1975), in a controlled randomized double-blind study, compared Δ^9-THC and a placebo in cancer patients receiving chemotherapeutic drugs known to cause nausea and vomiting. Δ^9-THC had a highly significant antiemetic effect. Eighty-one percent of these patients reported being "high" and 9% had disturbing psychic effects. Another study (REGELSON et al., 1976) also reported some antiemetic effects and an increased appetite and maintenance of body weight in terminal cancer patients.

VACHON et al. (1973) and TASHKIN et al. (1973) carried out the first careful studies of marihuana on respiratory dynamics. VACHON et al. (1973) reported that in controlled marihuana smoking, airway resistance fell, while functional residual capacity remained the same and specific airway conductance increased. Flow-volume loops showed an increase in flow rate and the response of the respiratory center to CO_2 was unimpaired. TASHKIN et al. (1973) obtained similar results with both marihuana smoking and orally administered Δ^9-THC. Both studies showed a marked dilation of the airways with effects after smoking lasting 60 min and after oral administration, up to 6 h. These results were similar to those seen with isoproterenol and distinctly different from the acute bronchoconstrictor effect produced by tobacco smoke. These findings also prompted both groups to examine the possible therapeutic use of marihuana or Δ^9-THC in asthmatics. TASHKIN et al. (1974) carried out a double-blind crossover study comparing smoked natural marihuana or oral Δ^9-THC against a placebo in ten subjects with stable bronchial asthma. Their findings indicated that smoked marihuana or oral Δ^9-THC produced a significant bronchodilation lasting at least two h in these subjects. VACHON et al. (1976a, b) obtained similar results with marihuana smoking and recently have reported good bronchodilator effects using an aerosol of Δ^9-THC in propylene glycol and water. Using a 1.25% Δ^9-THC aerosol, they were able to abtain a good bronchodilation effect without a subjective "high," or tachycardia. It would appear that new and useful antiasthmatic compounds or formulations may emerge from the cannabinoid field.

In a study of the effects of marihuana on the visual system, HEPLER and FRANK (1971) made the serendipitous observation that there was an impressive fall in intraocular pressure after smoking a single marihuana cigarette. This preliminary report was extended by HEPLER and FRANK (1972), who reported that marihuana decreased pupillary size, but not responsiveness to light, decreased tear secretion, and decreased intraocular pressure. Later studies from the same group (HEPLER et al., 1976) demonstrated that repeated smoking causes a 20–30% fall in intraocular pressure lasting 4–5 h. No tolerance to this effect appeared. These studies have been confirmed in animals and man for both marihuana and Δ^9-THC (GREEN and PEDERSON, 1973; GREEN and BOWMAN, 1976; PURNELL and GREGG, 1975; SHAPIRO, 1974). In a recent study in patients with proven glaucoma, HEPLER and PETRUS (1976) report that frequently, but not invariably, there was a substantial decrease in intraocular pressure following smoking of marihuana or ingestion of Δ^9-THC. The reduction in intraocular pressure was additive to the conventional glaucoma medications these patients were receiving. Topical application has been shown to be practical (GREEN and BOWMAN, 1976) and may obviate many of the difficulties and side effects associated with smoking marihuana or ingestion of Δ^9-THC. The finding that a synthetic nitrogen-containing benzopyran modeled on the cannabinoids also shares this property (DREN et al., 1975) makes this a potentially very exciting area for the development of a new and useful therapeutic agent based on marihuana.

References

Abel, E. L.: Development of tolerance to Δ^9-THC in the frog. Experientia (Basel) **29**, 1527—1528 (1973)

Abel, E. L., McMillan, D. E., Harris, L. S.: Tolerance to the hypothermic effects of Δ^9-tetrahydrocannabinol as a function of age in the chicken. Brit. J. Pharmacol. **47**, 452—463 (1973)

Abelson, H., Cohen, R., Schrayer, D.: Public attitudes toward marihuana, Part I. Princeton: Response Analysis Corp. 1972

Adams, M. D., Chait, L. D., Earnhardt, J. T.: Tolerance to the cardiovascular effects of Δ^9-tetrahydrocannabinol in the rat. Brit. J. Pharmacol. **56**, 43—48 (1976a)

Adams, M. D., Earnhardt, J. T., Dewey, W. L., Harris, L. S.: Vasoconstrictor actions of Δ^8 and Δ^9-tetrahydrocannabinol in the rat. J. Pharmacol. exp. Ther. **196**, 649—656 (1976b)

Agurell, S., Binder, M., Fonseka, K., Gustafsson, B., Lindgren, J. E., Leander, K., Martin, B., Nilsson, I. M., Nordquist, M., Ohlsson, A., Widman, M.: Cannabinoids: Metabolites hydroxylated in the pentyl side chain. In: Nahas, G. G. (Ed.): Marihuana: Chemistry, biochemistry and cellular effects. Berlin-Heidelberg-New York: Springer 1976

Agurell, S., Dahmen, J., Gustafsson, B., Johansson, I. B., Leander, K., Nilsson, I., Nilsson, J. L. G., Nordquist, M., Ramsey, C. H., Ryrfeldt, A., Sandberg, F., Widman, M.: Metabolic fate of tetrahydrocannabinol. In: Paton, W. D. M., Crown, J. (Eds.): Cannabis and its derivatives. London: Oxford University Press 1972

Agurell, S., Nilsson, I. M., Ohlsson, A., Sandberg, F.: On the metabolism of tritium-labelled Δ^1-tetrahydrocannabinol in the rabbit. Biochem. Pharmacol. **19**, 1333—1339 (1970)

Alves, C. N., Carlini, E. A.: Effects of acute and chronic administration of cannabis sativa extract on the mouse killing behavior of rats. Life Sci. **13**, 75—85 (1972)

Aramaki, H., Tomiyasu, N., Yoshimura, H., Tsukamoto, H.: Forensic chemical studies on marihuana. I. A detection method of the principle constituents by thin layer and gas chromatography. Chem. pharm. Bull. **16**, 822—826 (1968)

Aronow, W. S., Cassidy, J.: Effect of marihuana and placebo-marihuana smoking on angina pectoris. New Engl. J. Med. **291**, 65—67 (1974)

Balster, R. L., Ford, R. D.: The discriminative stimulus properties of cannabinoids: A review. In: Ho, B. T., Chute, D. L., Richards III, D. W. (Eds.): Drug discrimination and state dependent learning. New York: Academic Press (in press) 1976

Bannerjee, B. N., Galbreath, C., Sofia, R. D.: Teratologic evaluation of synthetic Δ^9-tetrahydrocannabinol in rats. Teratology **11**, 99—102 (1975)

Barry, H., Krimmer, E. C.: Discriminative delta-9-tetrahydrocannabinol stimulus tested with several doses, routes, intervals and related compounds. In: Braude, M. C., Szara, S. (Eds.): Pharmacology of marihuana. New York: Raven Press 1976

Barry, H. III, Perhoch, J. L., Kubena, R. K.: Tetrahydrocannabinol activation of pituitary-adrenal function. Pharmacologist **12**, 258 (1970)

Beaconsfield, F., Ginsburg, J., Rainsbury, R.: Marihuana smoking: Cardiovascular effects in man and possible mechanisms. New Engl. J. Med. **287**, 210—212 (1972)

Beaubrun, M. H., Knight, F.: Psychiatric assessment of 30 chronic users of cannabis and 30 matched controls. Amer. J. Psychiat. **130**, 309—311 (1973)

Bercht, C. A. L., Lousberg, R. J. J. Ch., Kuppers, F. J. E. M., Salemink, C. A.: Cannabicitran: A new naturally occurring tetracyclic diether from Lebanese Cannabis Sativa. Phytochemistry **13**, 619—621 (1974)

Bercht, C.A.L., Lousberg, R.J.J.Ch., Kuppers, F.J.E.M., Salemink, C.A., Vree, T.B., Van Rossum, J.M.: Cannabis VII. Identification of cannabinol methyl ether from hashish. J. Chromatog. **81**, 163—166 (1973)
Bicher, H.I., Mechoulam, R.: Pharmacological effects of two active constituents of marihuana. Arch. int. Pharmacodyn. **172**, 24—31 (1968)
Birmingham, M.K.: Reduction by Δ^9-tetrahydrocannabinol in the blood pressure of hypertensive rats bearing regenerated adrenal glands. Brit. J. Pharmacol. **48**, 169—171 (1973)
Black, M.D., Woods, J.H., Domino, E.F.: Some effects of (–)-Δ^9-tetrahydrocannabinol and other cannabis derivatives on schedule-controlled behavior. Pharmacologist **12**, 258 (1970)
Bloom, A.S., Dewey, W.L., Harris, L.S., Brosius, K.: Brain catecholamines and the antinociceptive activity of (±)-9-nor-9β-OH-hexahydrocannabinol. Neuroscience Abstr. **5**, 242 (1975)
Bloom, A.S., Johnson, K.M., Bowman, F.J., Dewey, W.L.: Correlation between the effects of several cannabinoids on catecholamine synthesis, body temperature and behavior. Fed. Proc. **35**, 506 (1976a)
Bloom, A.S., Johnson, K.M., Dewey, W.L., Harris, L.S.: Correlation between hypothermia and effects on catecholamine synthesis of several cannabinoids. Unpublished data (1976b)
Bonnie, R.J., Whitebread, C.H. II: The marihuana conviction. A history of marihuana prohibition in the United States. Charlottesville: University Press of Virginia 1974
Borgen, L.A., Davis, W.M., Pace, H.B.: Effects of synthetic Δ^9-tetrahydrocannabinol on pregnancy and offspring in the rat. Toxicol. appl. Pharmacol. **20**, 480—486 (1971)
Borgen, L.A., Davis, W.M., Pace, H.B.: Effect of prenatal Δ^9-tetrahydrocannabinol on the development of rat offspring. Pharmacol. Biochem. Behav. **1**, 203—206 (1973)
Bouquet, J.: Cannabis. I. Bull. Narcot. **2**, 14 (1950)
Bouquet, J.: Cannabis. III. Cannabis intoxication. IV. Problems of hashish intoxication. V. Supply. Bull. Narcot. **3**, 22—45 (1951)
Braude, M.C., Szara, S. (Eds.): The pharmacology of marihuana. New York: Raven Press 1976
Brecher, E.M. (Ed.): Licit and illicit drugs. Boston-Toronto: Little, Brown and Co. 1972
Brill, N.O., Crumpton, E., Frank, I.M., Hochman, J.S., Lomax, P., McGlothlin, W.H., West, L.J.: The marihuana problem. Ann. intern. Med. **73**, 449—465 (1970)
Brown, D.J., Miller, F.N., Longnecker, D.E., Greenwald, E.K., Harris, P.D., Forney, R.B.: The influence of Δ^9-tetrahydrocannabinol on cardiovascular and subcutaneous microcirculatory systems in the bat. J. Pharmacol. exp. Ther. **188**, 624—629 (1974)
Brown, H.: Some anticholinergic-like behavioral effects of *trans*-(–)-Δ^8-tetrahydrocannabinol. Psychopharmacologia (Berl.) **21**, 294—301 (1971)
Bueno, O.F.A., Carlini, E.A.: Dissociation of learning in marihuana tolerant rats. Psychopharmacologia (Berl.) **25**, 49—56 (1972)
Buxbaum, D.M.: Analgesic activity of Δ^9-tetrahydrocannabinol in the rat and mouse. Psychopharmacologia (Berl.) **24**, 275—280 (1972)
Campo, R.A.: Development of tolerance in pigeons to behavioral effects of a new benzopyran derivative. J. Pharmacol. exp. Ther. **184**, 521—527 (1973)
Cappell, H., Pliner, P.: Regulation of the self-administration of marijuana by psychological and pharmacological variables. Psychopharmacologia (Berl.) **40**, 65—76 (1974)
Carakushansky, G., Neu, R.L., Gardner, L.I.: Lysergide and cannabis as possible teratogens in man. Lancet **1969 I**, 150—151
Carchman, R.A., Harris, L.S., Munson, A.E.: Evaluation of cannabinoids as antineoplastics in vitro. Sixth Int. Congr. of Pharmacol., Helsinki, Finland, July 20—25, p. 554 (1975)
Carchman, R.A., Harris, L.S., Munson, A.E.: The inhibition of DNA synthesis by cannabinoids. Cancer Res. **36**, 95—100 (1976)
Carder, B., Olsen, J.: Marihuana and shock induced aggression in rats. Physiol. Behav. **8**, 599—602 (1972)
Carlini, E.A.: Tolerance to chronic administration of *Cannabis sativa* (marihuana) in rats. Pharmacology **1**, 135—142 (1968)
Carlini, E.A.: Acute and chronic behavioral effects of *Cannabis sativa*. Proc. 5th Int. Congr. Pharmacol. **1**, 31—43 (1973)
Carlini, E.A., Hamaoui, A., Martz, R.M.: Factors influencing the aggressiveness elicited by marihuana in food-deprived rats. Brit. J. Pharmacol. **44**, 794—804 (1972)

Carlini, E. A., Masur, J.: Development of aggressive behavior in rats by chronic administration of *Cannabis sativa*. Life Sci. **8**, 607—620 (1969)

Carlini, E. A., Masur, J.: Development of fighting behavior in starved rats by chronic administration of (—)-delta-9-*trans*-tetrahydrocannabinol and Cannabis extract. Commun. behav. Biol. **5**, 57—61 (1970)

Cavero, I., Buckley, J. P., Jandhyala, B. S.: Parasympatholytic activity of (—)-Δ^9-*trans*-tetrahydrocannabinol in mongrel dogs. Europ. J. Pharmacol. **19**, 301—304 (1972b)

Cavero, I., Buckley, J. P., Jandhyala, B. S.: Hemodynamic and myocardial effects of (—)-Δ^9-*trans*-tetrahydrocannabinol in anesthetized dogs. Europ. J. Pharmacol. **24**, 243—251 (1973)

Cavero, I., Ertel, R., Buckley, J. P., Jandhyala, B. S.: Effects of (—)-Δ^9-*trans*-tetrahydrocannabinol on regional blood flow in anesthetized dogs. Europ. J. Pharmacol. **20**, 373—376 (1972a)

Cavero, I., Jandhyala, B. S.: Hemodynamic effects of Δ^9-tetrahydrocannabinol (Δ^9-THC). Fed. Proc. **31**, 505 (1971)

Cavero, I., Lokhandwala, M. F., Buckley, J. P., Jandhyala, B. S.: The effect of (—)-Δ^9-*trans*-tetrahydrocannabinol on myocardial contractibility and venous return in anesthetized dogs. Europ. J. Pharmacol. **29**, 74—82 (1974)

Chesher, G. B., Dahl, C. J., Everingham, M., Jackson, D. M., Marchant-Williams, H., Starmer, G. A.: The effects of cannabinoids on intestinal motality and their antinociceptive effect in mice. Brit. J. Pharmacol. **49**, 588—594 (1973)

Chopra, J. C., Chopra, R. N.: The use of cannabis drugs in India. Bull. Narcot. **9**, 4—29 (1957)

Christensen, H. D., Freudenthal, R. J., Gidley, J. T., Rosenfeld, R., Boegli, G., Testino, L., Brine, D. R., Pitt, C. B., Wall, M. E.: Activity of delta-8 and delta-9-tetrahydrocannabinol and related compounds in the mouse. Science **172**, 165—167 (1971)

Cicero, T. J., Bell, R. D., Wiest, W. G., Allison, J. H., Polakoski, K., Robins, E.: The effects of heroin and methadone use on male sexual function. Personal communication reported to Committee on Problems of Drug Dependence (1975)

Clark, S. C.: Marihuana and the cardiovascular system. Pharmacol. Biochem. Behav. **3**, 299—306 (1975)

Claussen, U., Korte, F.: Über das Verhalten von hanfund von delta-9-6α,10α-trans tetrahydrocannabinol beim Rauchen. Justus Liebigs Ann. Chem. **713**, 162—165 (1968)

Claussen, U., Von Spulak, F., Korte, F.: Zur chemischen Klassifizierung von Pflanzen. XXI. Hashish-X, Cannabichromen, ein neuer Haschisch-Inhalts-Stoff. Tetrahedron **22**, 1477—1479 (1966)

Claussen, U., Von Spulak, F., Korte, F.: Hachisch-XIV, zur Kenntnis der Inhaltsstoffe des Haschisch. Tetrahedron **24**, 1021—1023 (1968)

Cohen, S., Lessin, P. J., Hahn, P. M., Tyrrell, E. D.: A 94-day cannabis study. In: Braude, M., Szara, S. (Eds.): Pharmacology of Marihuana. New York: Raven Press 1976

Cohen, S., Stillman, R. C. (Eds.): The therapeutic aspects of marihuana. A conference at Asilomar, California, Nov. 1975 (in press)

Colburn, R. W., Ng, L. K. Y., Lemberger, L., Kopin, I. J.: Subcellular distribution of Δ^9-tetrahydrocannabinol in rat brain. Biochem. Pharmacol. **23**, 873—877 (1974)

Cousens, K., Di Mascio, A.: Delta-9-THC as an hypnotic: An experimental study of three dose levels. Psychopharmacologia (Berl.) **33**, 355—364 (1973)

Dagirmanjian, R., Boyd, E. S.: Some pharmacological effects of two tetrahydrocannabinols. J. Pharmacol. exp. ther. **135**, 25—33 (1962)

Darley, C. F., Tinklenberg, J. R.: Marihuana and memory. In: Miller, L. L. (Ed.): Marijuana effects of human behavior, pp. 73—102. New York: Academic Press 1974

Davis, W. M., Moreton, J. E., King, W. T., Pace, H. B.: Marihuana on locomotor activity: biphasic effect and tolerance development. Res. Commun. chem. Path. Pharmacol. **3**, 29—35 (1972)

Deakin, S.: Death from taking Indian hemp. Ind. med. Gaz. **71** (1880)

Delini-Stula, A.: Suppression of conditioned avoidance behavior and the development of tolerance to Δ^8-tetrahydrocannabinol and dimethylheptylpran in rats. Pharmacopsychiat. Neuropsychopharmakol. **6**, 189—197 (1973)

Delong, F. L., Levy, B. I.: A model of attention describing the cognitive effects of marijuana. In: Miller, L. L. (Ed.): Marijuana: Effects on human behavior. New York: Academic Press 1974

Deneau, G. A., Kaymakalan, S.: Physiological and psychological dependence to synthetic delta-9-tetrahydrocannabinol (THC) in rhesus monkeys. Pharmacologist **13**, 246 (1971)

Dewey, W. L., Harris, L. S., Howes, J. F., Kennedy, J. S.: Pharmacological effects of some constituents of marijuana. Pharmacologist **11**, 278 (1969)

Dewey, W. L., Harris, L. S., Howes, J. F., Kennedy, S., Anderson, R. N.: A pharmacological investigation of some of the peripheral effects of the constituents of marihuana. Personal communication reported to Committee on Problems of Drug Dependence, 1970a

Dewey, W. L., Harris, L. S., Howes, J. F., Kennedy, J. S., Granchelli, F. E., Pars, H. G., Razdan, R. K.: Pharmacology of some marihuana constituents and two heterocylic analogues. Nature (Lond.) **226**, 1265—1267 (1970b)

Dewey, W. L., Harris, L. S., Kennedy, J. S.: Some pharmacological and toxicological effects of 1-trans-Δ^9 tetrahydrocannabinol in laboratory rodents. Arch. int. Pharmacodyn. **196**, 133—145 (1972a)

Dewey, W. L., Jenkins, J., O'Rourke, T., Harris, L. S.: Trans-Δ^9-tetrahydrocannabinol on behavior and the cardiovascular systems of dogs. Arch. int. Pharmacodyn. **198**, 118—131 (1972b)

Dewey, W. L., Martin, B. R., Beckner, J. S., Harris, L. S.: A comparison of the subcellular distribution of cannabinoids in the brains of tolerant and nontolerant dogs, rats and mice following the injection of radiolabeled Δ^9-THC. In: Nahas, G. G. (Ed.): Marijuana: Chemistry, biochemistry and cellular effects, pp. 349—365. Berlin-Heidelberg-New York: Springer 1976

Dewey, W. L., McMillan, D. E., Harris, L. S., Turk, R. F.: Distribution of radioactivity in brain of tolerant and nontolerant pigeons treated with ^{3}H-delta-9-tetrahydrocannabinol. Biochem. Pharmacol. **22**, 399—405 (1973)

Dewey, W. L., Peng, T. C., Harris, L. S.: The effect of 1-trans-Δ^9-tetrahydrocannabinol on the hypothalamo-hypophyseal-adrenal axis of the rat. Europ. J. Pharmacol. **12**, 383—384 (1970c)

Dewey, W. L., Yonce, L. R., Harris, L. S., Reavis, W. M., Griffin, E. D. Jr., Newley, V. E.: Some cardiovascular effects of *trans*-Δ^9-tetrahydrocannabinol (Δ^9-THC). Pharmacologist **12**, 259 (1970d)

Dezeeuw, R. A., Vree, T. B., Breimer, D. D., Vanginneken, C. A. M.: Cannabivarichromene, a new cannabinoid with a propyl side chain in cannabis. Experientia (Basel) **29**, 260—261 (1973)

Domino, E. F.: Neuropsychopharmacologic studies of marijuana: some synthetic and natural THC derivatives in animals and man. Ann. N.Y. Acad. Sci. **191**, 166—191 (1971)

Dornbush, R. L.: The long term effects of cannabis use. In: Miller, L. L. (Ed.): Marijuana, effects on human behavior. New York: Academic Press 1974

Dorrance, D., Janiger, O., Teplitz, R. L.: In vivo effects of illicit hallucinogens on human lymphocyte chromosomes. J. Amer. med. Ass. **212**, 1488—1491 (1970)

Dren, A. T., Bopp, B. A., Ebert, D. M.: Nitrogen-containing benzopyran analogs; reduction of intraocular pressure in rabbits. Pharmacologist **17**, 267 (1975)

Drew, W. G., Miller, L. L.: Cannabis: Neutral mechanisms and behavior—a theoretical review. Pharmacology **11** 12—32 (1974)

Drug Abuse Council: National survey of marihuana use and attitudes. Washington: Drug Abuse Council 1975

Dubinsky, B., Robichard, R. C., Goldberg, M. F.: Effects of (-)-delta-9-trans-tetrahydrocannabinol and its selectivity in several models of aggressive behavior. Pharmacology **9**, 204—216 (1973)

End, D., Carchman, R. A., Dewey, W. L.: The uptake and distribution of ^{3}H-Δ^9-THC in cultured neural and glial cells. Pharmacologist **18**, 166 (1976)

Ewens, G. F. W.: Insanity following the use of Indian hemp. Ind. med. Gaz. **39**, 401—413 (1904)

Falek, A.: Genetic studies of marihuana: current findings and future directions. In: Tinklenberg, J. R. (Ed.): Marijuana and health hazards. New York: Academic Press 1975

Ferraro, D. P.: Effects of Δ^9-*trans*-tetrahydrocannabinol on simple and complex learned behavior in animals. In: Lewis, M. F. (Ed.): Current research in marihuana. New York: Academic Press 1972

Ferraro, D. P., Grilly, D. M.: Lack of tolerance to delta-9-tetrahydrocannabinol in chimpanzees. Science **179**, 490—492 (1973)

Fetterman, P. S., Keith, E. S., Waller, C. W., Guerrero, O., Doorenbos, N. J., Quimby, M. W.: Mississippi-grown *Cannabis Sativa L.*: preliminary onservations on chemical definition of phenotype and variations in tetrahydrocannabinol content versus, age, sex and plant part. J. pharm. Sci. **60**, 1246—1249 (1971)

Fink, M., Volavka, J., Dornbush, R., Crown, P.: Effects of THC-Δ^9, marijuana, and hashish on EEG, mood and heart rate of volunteers. Psychopharmacologia (Berl.) **26** (Suppl.), 126 (1972)

Fink, M., Volavka, J., Panayiotopoulos, C. P., Stefanis, C.: Quantitative EEG studies of marihuana, Δ^9-THC and hashish in man. In: Braude, M., Szara, S. (Eds.): Pharmacology of marihuana. New York: Raven Press 1976

Fleischman, R. W., Hayden, D. W., Rosenkrantz, H., Braude, M. C.: Teratologic evaluation of Δ^9-tetrahydrocannabinol in mice, including a review of the literature. Teratology **12**, 47—50 (1975)

Foltz, R. L., Fentiman, A. F., Jr., Leighty, E. G., Walter, J. L., Drewes, H. R., Schwartz, W. E., Page, T. F., Jr., Truitt, E. B., Jr.: Metabolite of (1)-trans-Δ^8-tetrahydrocannabinol: identification and synthesis. Science **168**, 844—845 (1970)

Ford, R. D.: Some effects of chronic Δ^9-tetrahydrocannabinal administration on the behavior of pigeons, monkeys, rats and spiders. Ph.D. Thesis, University of North Carolina, p. 150 (1974)

Ford, R. D., Balster, R. L.: The discriminative stimulus properties of Δ^9-tetrahydrocannabinol (Δ^9-THC): Generalization to some metabolites and derivatives. Fed. Proc. **34**, 734 (1975)

Ford, R. D., McMillan, D. E.: Further studies on the behavioral pharmacology of 1-Δ^8 and 1-Δ^9-tetrahydrocannabinol (Δ^8 and Δ^9-THC). Fed. Proc. **31**, 506 (1972)

Frank, I. M., Lessin, P. J., Tyrrell, E. D., Hahn, P. M., Szara, S.: Acute and cumulative effects of marihuana smoking in hospitalized subjects: a 36-day study. In: Braude, M., Szara, S. (Eds.): Pharmacology of marihuana. New York: Raven Press 1976

Frankenheim, J. M., McMillan, D. E., Harris, L. S.: Effects of 1-Δ^9 and 1-Δ^8-trans-tetrahydrocannabinol on schedule-controlled behavior of pigeons and rats. J. Pharmacol. exp. Ther. **178**, 241—252 (1971)

Freudenthal, R. I., Martin, J., Wall, M. E.: Distribution of Δ^9-tetrahydrocannabinol in the mouse. Brit. J. Pharmacol. **44**, 244—249 (1972)

Fried, P. A., McIntyre, D. C.: Electrical and behavioral attenuation of the anticonvulsant properties of Δ^9-THC following chronic administrations. Psychopharmacologia (Berl.) **31**, 215—227 (1973)

Friedrich-Fichtl, J., Spiteller, G.: Neve Cannabinoide-1. Tetrahedron **31**, 479—487 (1975)

Gallager, D. W., Sanders-Bush, E., Sulser, F.: Dissociation between behavioral effects and changes in metabolism of cerebral serotonin following Δ^9-tetrahydrocannabinol. Psychopharmacologia (Berl.) **26**, 337—345 (1972)

Gaoni, Y., Mechoulam, R.: Isolation, structure and partial synthesis of an active constituent of hashish. J. Amer. chem. Soc. **86**, 1646—1647 (1964)

Gaoni, Y., Mechoulam, R.: The isolation and structure of Δ^1-tetrahydrocannabinol and other neutral cannabinoids from hashish. J. Amer. chem. Soc. **93**, 217—224 (1971)

Gayer, H.: Pharmakologische Wertbestimmung von orientalischem Haschisch und Herba cannabis indicae. Naunyn-Schmiedeberg's Arch. exp. Path. Pharmak. **129**, 312—318 (1928)

Geber, W. F., Schramm, L. C.: Effects of marihuana extract on fetal hamsters and rabbits. Toxicol. appl. Pharmacol. **14**, 276—282 (1969)

Gelehrter, I. D.: Lysergic acid diethylamide (LSD) and estrophy of the bladder. J. Pediat. **77**, 1065—1066 (1970)

Gill, E. W., Paton, W. D. M., Pertwee, R. G.: Preliminary experiments on the chemistry and pharmacology of cannabis. Nature (Lond.) **228**, 135—137 (1970)

Gilmour, D. G., Bloom, A. D., Lele, K. P., Robbins, E. S., Maximilliam, C.: Chromosomal abberations in uses of psychoactive drugs. Arch. gen. Psychiat. **24**, 268—272 (1971)

Gonzalez, S. C., Karniol, I. G., Carlini, E. A.: Effect of *cannabis sativa* extract on conditional fear. Behav. Biol. **7**, 83—94 (1972)

Graham, J. D. P., Lewis, M. J., Williams, J.: The effect of Δ^1-tetrahydrocannabinol on the noradrenaline and dopamine content of the brain and heart of the rat. Brit. J. Pharmacol. **52**, 446 (1974)

Graham, J. D. P., Li, D. M. R.: Cardiovascular and respiratory effects of cannabis in cat and rat. Brit. J. Pharmacol. **49**, 1—10 (1973)

Grant, J. D., Gross, S. J., Lomax, P., Wong, R.: Antibody detection of marihuana. Nature (Lond.) **236**, 216—217 (1972)

Green, K., Bowman, K.: Effect of marihuana and derivatives on aqueous humor dynamics in the rabbit. In: Braude, M., Szara, S. (Eds.): Pharmacology of marihuana. New York: Raven Press 1976

Green, K., Pederson, J. E.: Effects of Δ^1-tetrahydrocannabinol on aqueous dynamics and ciliary body permeability in the rabbit. Exp. Eye Res. **15**, 499—509 (1973)
Grilly, D. M., Ferraro, D. P., Braude, M. C.: Observations on the reproductive activity of chimpanzees following long term exposure of marihuana. Pharmacology **11**, 304—307 (1974)
Grinspoon, L.: Marihuana reconsidered. Cambridge: Harvard University Press 1971
Grisham, M. G., Ferraro, D. P.: Biphasic effects of delta-9-tetrahydrocannabinol on variable interval schedule performance in rats. Psychopharmacologia (Berl.) **27**, 163—169 (1972)
Grlic, L. J.: A comparative study on some chemical and biological characteristics of various samples of cannabis resin. Bull. Narcot. **14**, 37 (1962)
Grlic, L.: Recent advances in the chemical research of cannabis. Bull. Narcot. **16**, no. 4, 29 (1964)
Grunfeld, Y., Edery, H.: Psychopharmacological activity of the active constituents of hashish and some related cannabinols. Psychopharmacologia (Berl.) **14**, 200—210 (1969)
Haley, S. L., Wright, P. L., Plank, J. B., Keplinger, M. L., Braude, M. C., Callandra, J. C.: The effect of natural and synthetic Δ^9-tetrahydrocannabinol on fetal development. Toxicol. appl. Pharmacol. **25**, 450 (1973)
Harbison, R. D., Mantilla-Plata, B.: Prenatal toxicity, maternal distribution and placental transfer of tetrahydrocannabinol. J. Pharmacol. exp. Ther. **180**, 446—453 (1972)
Hardman, H. F., Domino, E. F., Seevers, M. H.: The chemistry and pharmacology of certain compounds affecting the central nervous system of animals and man. Clearinghouse for Federal Scientific and Technical Information. AD 707—699, pp. 1—53 (1957)
Hardman, H. F., Domino, E. F., Seevers, M. H.: General pharmacological actions of some synthetic tetrahydrocannabinol derivatives. Pharmacol. Rev. **23**, 295—315 (1971)
Hardman, H. F., Hosko, M. J.: An overview of the cardiovascular-autonomic actions of cannabis. In: Braude, M., Szara, S. (Eds.): Pharmacology of Marihuana. New York: Raven Press 1976
Harmon, J., Alipoulus, M. A.: Gynocomastia in marihuana users. New Engl. J. Med. **287**, 936 (1972)
Harris, L. S.: Marihuana: General and behavioral pharmacology. Pharmacol. Rev. **23**, 285—294 (1971)
Harris, L. S., Munson, A. E., Carchman, R. A.: Antitumor properties of cannabinoids. In: Braude, M., Szara, S. (Eds.): Pharmacology of Marihuana. New York: Raven Press 1976
Harris, L. S., Munson, A. E., Friedman, M. A.: Retardation of tumor growth by Δ^9-tetrahydrocannabinol. Pharmacologist **16**, 259 (1974)
Harris, R. T., Waters, W., McLendon, D.: Behavioral effects in rhesus monkeys of repeated intravenous doses of tetrahydrocannabinol. Psychopharmacologia (Berl.) **26**, 297—306 (1972)
Hecht, F., Beals, R. K., Lees, M. H., Jolly, H., Roberts, P.: Lysergic acid diethylamide and cannabis as possible teratogens in man. Lancet **1968II**, 1087
Henkriksson, B. G., Jarbe, T.: Cannabis induced vocalization in the rat. J. Pharm. (Lond.) **23**, 457—458 (1971)
Hepler, R. S., Frank, I. A.: Marihuana smoking and intraocular pressure. J. Amer. med. Ass. **217**, 1392 (1971)
Hepler, R. S., Frank, I. M., Petrus, R.: Ocular effects of marihuana smoking. In: Braude, M., Szara, S. (Eds.): Pharmacology of Marihuana. New York: Raven Press 1976
Hepler, R. S., Frank, I. M., Ungerleider, J. T.: Pupillary constriction after marihuana smoking. Amer. J. Opthal. **74**, 1185—1190 (1972)
Hepler, R. S., Petrus, R. J.: Experiences with administration of marihuana to glaucoma patients. In: Cohen, S., Stillman, R. C. (Eds.): The therapeutic aspects of marihuana. In press, 1976
Herha, J., Obe, G.: Chromosomal damage in chronical users of cannabis. Pharmacopsychiat. Neuropsychopharmakol. **1**, 328—337 (1974)
Heyndrickx, A., Scheiris, C., Schepens, P.: Toxicological study of a fatal intoxication in man due to cannabis smoking. J. Pharm. Belg. **24**, 371—376 (1969)
Hill, S. Y., Schwin, R., Goodwin, D. W., Powell, B. J.: Marihuana and pain. J. Pharmacol. exp. Ther. **188**, 415—418 (1974)
Hine, B., Friedman, E., Torrelio, M., Gershon, S.: Morphine-dependent rats: Blockade of precipitated abstinence by tetrahydrocannabinol. Science **187**, 443—446 (1975a).
Hine, B., Friedman, E., Torrelio, M., Gershon, S.: Tetrahydrocannabinol-attenuated abstinence and induced rotation in morphine-dependent rats: Possible involvement of dopamine. Neuropharmacology **14**, 607—610 (1975b)

Hirschhorn, I. D., Rosecrans, J. A.: Morphine and delta-9-tetrahydrocannabinol: Tolerance to the stimulus effects. Psychopharmacologia (Berl.) **36**, 243—253 (1974)

Hively, R. L., Mosher, W. A., Hoffman, F. W.: Isolation of trans-Δ^6-tetrahydrocannabinol from marihuana, J. Amer. chem. Soc. **88**, 1832—1833 (1966)

Ho, B. T., Fritchie, G. E., Englert, L. F., McIsaac, W. M., Idanpaan-Heikkila, J. E.: Marihuana: Importance of the route of administration. J. Pharm. (Lond.) **23**, 309—310 (1971 a)

Ho, B. T., Fritchie, G., Kralik, P. M., Englert, L. F., McIsaac, W. M., Idanpaan-Heikkila, J.: Distribution of tritiated-*l*-Δ^9-tetrahydrocannabinol in rat tissues after inhalation. J. Pharm. (Lond.) **22**, 538—539 (1970)

Ho, B. T., Taylor, D., Englert, L. F.: The effect of repeated administration of (-)-Δ^9-tetrahydrocannabinol on the biosynthesis of brain amines. Res. Commun. chem. Path. Pharmacol. **5**, 851—854 (1973)

Ho, B. T., Taylor, D., Englert, L. F.: Effects of Δ^9-tetrahydrocannabinol on the metabolism of ^{3}H-5-hydroxytryptamine and ^{3}H-norepinephrine in the rat brain. Res. Commun. chem. Path. Pharmacol. **1**, 645—650 (1974)

Ho, B. T., Taylor, D., Englert, L. F., McIsaac, W. M.: Neurochemical effects of l-Δ^9-tetrahydrocannabinol in rats following repeated inhalation. Brain Res. **31**, 233—236 (1971 b)

Ho, B. T., Taylor, D., Fritchie, G. E., Englert, L. F., McIsaac, W. M.: Neuropharmacological study of Δ^9 and Δ^8 1-tetrahydrocannabinols in monkeys and mice Brain Res. **38**, 163—170 (1972)

Hollister, L. E.: Marihuana in man; Three years later. Science **172**, 21—29 (1971)

Hollister, L. E.: Structure-activity relationships, in man of cannabis constituents, and homologs and metabolites of Δ^9-tetrahydrocannabinol. Pharmacology **11**, 3—11 (1974)

Hollister, L. E., Gillespie, K. H.: Delta-8- and delta-9-tetrahydrocannabinol. Clin. Pharmacol. Ther. **14**, 353—357 (1973)

Hollister, L. E., Reaven, G. M.: Delta-9-tetrahydrocannabinol and glucose tolerance. Clin. Pharmacol. Ther. **16**, 297—302 (1974)

Hollister, L. E., Richards, R. K., Gillespie, H. K.: Comparison of tetrahydrocannabinol and synhexyl in man. Clin. Pharmacol. Ther. **9**, 783 (1968)

Holtzman, D., Lovell, R. A., Jaffe, J. H., Freedman, D. X.: l-Δ^9-tetrahydrocannabinol: Neurochemical and behavioral effects in the mouse. Science **163**, 1464—1467 (1969)

Hughes, J. E., Steakly, L. P., Bier, M. M.: Marihuana and the diabetic coma. J. Amer. med. Ass. **214**, 1113—1114 (1970)

Indian Hemp Drug Commission Report, 1893—1894, p. 3281. Silver Spring: Thomas Jefferson Publishing Co. (Reprinted 1969)

Isbell, H., Gorodetzky, C. W., Jasinski, D., Claussen, U., Spulak, F., Korte, F.: Effects of (-)-Δ^9-*trans*-tetrahydrocannabinol in man. Psychopharmacologia (Berl.) **11**, 184—188 (1967)

Isbell, H., Jasinski, D.: A comparison of LSD 25 with (-)-Δ^9-trans-tetrahydrocannabinol (THC) and attempted cross tolerance between LSD and THC. Psychopharmacologia (Berl.) **14**, 115—123 (1969)

Johnson, K.: On the mechanism and site of action of (-)-Δ^9-tetrahydrocannabinol in serotonergic neurons. Ph.D. Dissertation, University of Houston, Houston, Texas (1974)

Johnson, K. M., Ho, B. T., Dewey, W. L., Harris, L. S.: In vitro induction of ^{3}H-reserpine binding to subcellular components of the rat forebrain by Δ^9-tetrahydrocannabinol. Neuroscience Abstr. **5**, 241 (1975)

Johnson, S., Domino, E. F.: Some cardiovascular effects of marihuana smoking in normal volunteers. Clin. Pharmacol. Ther. **12**, 762—768 (1971)

Jones, R. T., Benowitz, N.: The 30-day trip: Clinical studies of cannabis tolerance and dependence. In: Braude, M., Szara, S. (Eds.): Pharmacology of Marihuana. New York: Raven Press 1976

Joyce, C. R. B., Curry, S. H. (Eds.): The botany and chemistry of cannabis. London: Churchill 1970

Kabelik, J., Krejci, Z., Santavy, F.: Cannabis as a medicament. Bull. Narcot. **12**, 5—23 (1960)

Karler, R., Cely, W., Turkanis, S. A.: A study of the development of tolerance to an anticonvulsant effect of Δ^9 tetrahydrocannabinol and cannabidiol. Res. Commun. chem. Path. Pharmacol. **9**, 23—39 (1974)

Karniol, I. G., Shirakawa, I., Kasinksi, N., Pfeferman, A., Carlini, E. A.: Cannabidiol interferes with the effects of Δ^9 tetrahydrocannabinol in man. Europ. J. Pharmacol. **28**, 172—177 (1974)

Kaymakcalan, S., Deneau, G. A.: Some pharmacological effects of synthetic Δ^9 tetrahydrocannabinol (THC). Pharmacologist **13**, 247 (1971)

Kaymakcalan, S., Sivil, S.: Lack of tolerance to the bradycardic effect of Δ^9-trans-tetrahydrocannabinol in rats. Pharmacology **12**, 290—295 (1974)

Kaymakcalan, S., Turker, R. K., Turker, M. N.: Analgesic effect of Δ^9-tetrahydrocannabinol in the dog. Psychopharmacologia (Berl.) **35**, 123—128 (1974)

Kennedy, J. S., Waddell, W. J.: Whole-body auto-radiography of the pregnant mouse after administration of 14C-Δ^9-THC. Toxicol. appl. Pharmacol. **22**, 252—258 (1972)

Keplinger, M. L., Wright, P. L., Haley, S. L., Plank, J. B., Braude, M. C., Calandra, J. C.: The effect of natural and synthetic delta-9-tetrahydrocannabinol on reproductive and lactation performance in albino rats. Toxicol. appl. Pharmacol. **25**, 499 (1973)

Kilbey, M. M., Farmer, R., McLendon, D.: Time course of delta-9-tetrahydrocannabinol inhibition of frog-killing behavior in the rat. Presented at the 19th Annual Convention of the Southwestern Psychological Association, Oklahoma City, Oklahoma, 1972

Kilbey, M. M., Harris, R. T., Moore, J. W., Jr.: Increased latency of frog-killing behavior in the rat following administration of delta-9-tetrahydrocannabinol. Personal communication reported to Committee on Problems of Drug Dependence 1831—1840 (1971)

Kilbey, M. M., Moore, J. W., Hall, M.: Delta-9-tetrahydrocannabinol induced inhibition of predatory aggression in the rat. Psychopharmacologia (Berl.) **31**, 157—166 (1973)

Killam, K. F., Killam, E. K.: The action of tetrahydrocannabinol on EEG and photomyoclonic seizures in the baboon. Proc. Fifth int. Congr. Pharmacol. 744 (1972)

Klausner, H. A., Dingell, J. V.: The metabolism and excretion of Δ^9-tetrahydrocannabinol in the rat. Life Sci. **10**, 49—59 (1971)

Kochar, M. S., Hosko, M. J.: Electrocardiographic effects of marihuana. J. Amer. med. Ass. **225**, 25—27 (1973)

Kolodny, R. C.: Research issues in the study of marihuana and male reproductive physiology in humans. In: Tinklenberg, J. C. (Ed.): Marihuana and Health Hazards. New York: Academic Press 1975

Kolodny, R. C., Lessin, P., Toro, G., Masters, W. H., Cohen, S.: Depression of plasma testosterone with acute marihuana administration. In: Braude, M., Szara, S. (Eds.): Pharmacology of Marihuana. New York: Raven Press 1976

Kolodny, R. C., Masters, W. H., Kolodner, R. M., Toro, G.: Depression of plasma testosterone levels after chronic intensive marihuana use. New Engl. J. Med. **290**, 872—874 (1974)

Korte, F., Haag, M., Claussen, U.: Tetrahydrocannabinolcarboxylic acid: a component of hashish. Angew. Chem. Int. Ed. Engl. **4**, 872 (1965)

Korte, F., Sieper, H.: Isolierung von Haschisch-Inhaltsstoffen aus Cannabis sativa non Indica. Justus Liebigs Ann. Chem. **630**, 71—83 (1960)

Korte, F., Sieper, H.: Recent results of hashish analysis. In: Wolstenholme, G. E. W., Knight, J. (Eds.): Hashish: Its Chemistry and Pharmacology. London: Churchill 1965

Kosersky, D. S., Dewey, W. L., Harris, L. S.: Antipyretic, analgesic and anti-inflammatory effects of Δ^9 tetrahydrocannabinol in the rat. Europ. J. Pharmacol. **24**, 1—7 (1973)

Kosersky, D. S., McMillan, D. E., Harris, L. S.: Δ^9 tetrahydrocannabinol and 11-hydroxy-Δ^9 tetrahydrocannabinol: Behavioral effects and tolerance development. J. Pharmacol. exp. Ther. **189**, 61—65 (1974)

Kreuz, D. S., Axelrod, J.: Delta-9-tetrahydrocannabinol: localization in body fat. Science **179**, 391—392 (1973)

Kubena, R. L., Perhack, J. L., Barry, H., III.: Corticosterone elevation mediated centrally by delta-1-tetrahydrocannabinol in rats. Europ. J. Pharmacol. **14**, 89—92 (1971)

Kumar, S., Kunwar, K. B.: Chromasome abnormalities in cannabis addicts. J. Ass. Phycns India **19**, 193—195 (1972)

Lawrence, D. K., Pertwee, R. G.: Brain levels of Δ^1-tetrahydrocannabinol and its metabolites in mice tolerant to the hypothermic effect of Δ^1-tetrahydrocannabinol. Brit. J. Pharmacol. **49**, 373—375 (1973)

Layman, J. M., Milton, A. S.: Distribution of tritium labelled Δ^1-tetrahydrocannibinol in the rat following intraperitoneal administration. Brit. J. Pharmacol. **42**, 308—310 (1972)

LeDain, G.: Interim Report of the Commission of Inquiry into the Non-Medical Use of Drugs. Ottawa: Crown 1970

Lefkowitz, S. S., Yang, C.: Effects of certain highly abused drugs on the immune response in mice. J. Reticuloendoth. Soc. **16**, 25a (1974)

Legator, M. S., Weber, E., Conner, T., Stoeckel, M.: Failure to detect mutagenic effects of Δ^9-THC in dominant lethal test, host mediated assay, blood-urine studies, and cytogenetic evaluation with mice. In: Braude, M., Szara, S. (Eds.): Pharmacology of Marihuana. New York: Raven Press 1976

Lemberger, L., Crabtree, R. E., Rowe, H. M.: 11-Hydroxy-Δ^9 tetrahydrocannabinol. Pharmacology, disposition, and metabolism of a major metabolite of marihuana in man. Science **177**, 62—63 (1972)

Lemberger, L., McMahon, R., Archer, R., Matsumoto, K., Rowe, H.: Pharmacologic effects and physiologic disposition of delta-6_a 10_a-dimethylheptyl tetrahydrocannabinol (DMHP) in man. Clin. Pharmacol. Ther. **15**, 380—386 (1974)

Lemberger, L., Silberstein, S. D., Axelrod, J., Kopin, I. J.: Marihuana: Studies on the disposition and metabolism of delta-9-tetrahydrocannabinol in man. Science **170**, 1320—1322 (1970)

Lemberger, L., Tamarkin, N. R., Axelrod, J., Kopin, I. J.: Delta-9-tetrahydrocannabinol: Metabolism and distribution in long term smokers. Science **173**, 72—73 (1971)

Leonard, B. E.: The effect of Δ^6 tetrahydrocannabinol on biogenic amines and their amino acid precursors in rat brain. Pharmacol. Res. Commun. **3**, 139—145 (1971)

Lessin, P. J., Thomas, S. A.: Assessment of the chronic effects of marihuana on motivation and achievement: A preliminary report. In: Braude, M., Szara, S. (Eds.): Pharmacology of marihuana. New York: Raven Press 1976

Leuchtenberger, C., Leuchtenberger, R.: Morphological and cytochemical effects of marihuana cigarette smoke on epithelioid cells of lung explants from mice. Nature (Lond.) **234**, 227—229 (1971)

Leuchtenberger, C., Leuchtenberger, R.: Cytological and cytochemical studies of the effects of fresh marihuana cigarette smoke on growth and DNA metabolism of animal and human lung cultures. In: Braude, M., Szara, S. (Eds.): Pharmacology of marihuana. New York: Raven Press 1976

Leuchtenberger, C., Leuchtenberger, R., Schwider, A.: Effects of marihuana and tobacco smoke on human lung physiology. (Nature (Lond.) **241**, 137—139 (1973)

Levy, J. A., Munson, A. E., Harris, L. S., Dewey, W. L.: Effect of delta-8 and delta-9-tetrahydrocannabinol on the immune response in mice. Pharmacologist **16**, 259 (1974)

Lewis, M. F. (Ed.): Current research in marihuana. New York-London: Academic Press 1972

Lewis, S. C., Brown, D. J., Forney, R. B.: Absence of tolerance to the hypotensive effects of delta-9-tetrahydrocannabinol in hypertensive rats. Toxicol. appl. Pharmacol. **28**, 7 (1974)

Li, H. C.: An archaeological and historical account of cannabis in China. Econ. Bot. 28, 437—448 (1974)

Littleton, J. M., MacLean, K. I.: The effect of Δ^8 tetrahydrocannabinol (Δ^8THC) on dopamine metabolism in the rat corpus striatum: the influence of environment. Brit. J. Pharmacol. **51**, 117 (1974)

Lomax, P.: Acute tolerance to the hypothermic effect of marihuana in the rat. Res. Commun. chem. Path. Pharmacol. **2**, 159—167 (1971)

Lotter, H. L., Abraham, D. J., Turner, C. E., Knapp, J. E., Schiff, P. L., Skatkin, D. J.: Cannabisativine, a new alkaloid from Cannabis Sativad L. Root. Tetrahedron Letters **1975**, 2815

Luthra, X. K., Rosenkrantz, H.: Cannabinoids: Neurochemical aspects after oral administration to rats. Toxicol. appl. Pharmacol. 27, 158—168 (1974)

Luthra, X. K., Rosenkrantz, H., Heyman, I. A., Braude, M. C.: Differential neurochemistry and temporal pattern in rats treated orally with Δ^9 tetrahydrocannabinol for periods up to six months. Toxicol. appl. Pharmacol. **32**, 418—431 (1975)

Magus, R. D., Harris, L. S.: Carcinogenic potential of marihuana smoke condensate. Fed. Proc. **30**, 279 (1971)

Maitre, L., Staehelin, M., Bein, H. J.: Effect of an extract of Cannabis and of some cannabinols on catecholamine metabolism in rat brain and heart. Agents Actions **1**, 136—143 (1970)

Maitre, L., Waldmeier, C., Baumann, P. A.: Effects of some tetrahydrocannabinols on the biosynthesis and utilization of catecholamines in the rat brain. In: Usdin, E., Snyder, S. (Eds.): Frontiers in Catecholamine Research. London: Pergamon Press 1973

Malit, L. A., Johnstone, R. E., Bourke, D. I., Kulp, R. A., Klein, V., Smith, T. C.: Intravenous Δ^9-tetrahydrocannabinol: Effects on ventilatory control and cardiovascular dynamics. Anesthesiology **42**, 666—673 (1975)
Mann, P. E., Cohen, A. B., Finley, T. N., Ladman, A. J.: Alveolar macrophages. Structural and functional differences between nonsmokers and smokers of marihuana and tobacco. Lab. Invest. **25**, 111—120 (1971)
Manning, F. J.: Acute tolerance to the effects of delta-9-tetrahydrocannabinol on spaced responding by monkeys. Pharmacol. Biochem. Behav. **1**, 665—671 (1973)
Manning, F. J., Elsmore, T. F.: Shock elicited fighting and delta-9-tetrahydrocannabinol. Psychopharmacologia (Berl.) **25**, 218—228 (1972)
Manning, F. J., McDonough, J. H., Jr., Elsmore, T. F., Saller, C., Sodetz, F. J.: Inhibition of normal growth by chronic administration of Δ^9-tetrahydrocannabinol. Science **174**, 424—426 (1971)
Manno, B. R., Manno, J. E., Kilsheimer, G. S., Forney, R. B.: Response of the isolated perfused rat heart to Δ^9-tetrahydrocannabinol. Toxicol. appl. Pharmacol. **17**, 311 (1970)
Manno, J. E., Manno, B. R., Kiplinger, G. F., Forney, R. B.: Motor and mental performance with marihuana: Relationship to administered dose of Δ^9-tetrahydrocannabinol and its interaction with alcohol. In: Miller, L. L. (Ed.): Marihuana: Effects on human behavior. New York: Academic Press 1974
Mantilla-Plata, B., Clewe, C. L., Harbison, R. D.: Teratogenic and mutagenic studies of Δ^9-tetrahydrocannabinol in mice. Fed. Proc. **32**, 746 (1973)
Mantilla-Plata, B., Harbison, R. D.: Influence of alteration of metabolism on induced tetrahydrocannabinol teratogenesis. In: Braude, M., Szara, S. (Eds.): Pharmacology of marihuana. New York: Raven Press 1976
Martin, B. R., Dewey, W. L., Earnhardt, J. T., Adams, M. D., Harris, L. S., Wilson, R. S., May, E. L., Razdan, R.: Pharmacological acitivity of compounds structurally related to tetrahydrocannabinols. (In preparation, 1976a)
Martin, B. R., Dewey, W. L., Harris, L. S., Beckner, J. S.: ^{3}H-delta-9-tetrahydrocannabinol tissue and subcellular distribution in the central nervous system and tissue distribution in peripheral organs of tolerant and nontolerant dogs. J. Pharmacol. exp. Ther. **196**, 128—144 (1976b)
Martin, B. R., Dewey, W. L., Harris, L. S., Beckner, J. S.: ^{3}H-Δ^9-tetrahydrocannabinol distribution in pregnant dogs and their fetuses. (In preparation, 1976c)
Martin, B. R., Dewey, W. L., Harris, L. S., Beckner, J., Wilson, R. S., May, E. L.: Marihuana-like activity of new synthetic tetrahydrocannabinols. Pharmacol. Biochem. Behav. **3**, 849—853 (1975)
Martin, P. A., Thorbunn, M. J., Bryant, S. A.: In vivo and in vitro studies of cytogenetic effects of Cannabis sativa in rats and men. Teratology **9**, 81—86 (1973)
Masur, J., Martz, R. M. W., Carlini, E. A.: Effects of acute and chronic administration of cannabis sativa and (-)-delta-9-tetrahydrocannabinol on the behavior of rats in an open-field arena. Psychopharmacologia (Berl.) **19**, 388—397 (1971)
Masur, J., Martz, R. M. W., Carlini, E. A.: The behavior of workers and nonworker rats under the influence of (-)-Δ^9-trans-tetrahydrocannabinol, chlorpromazine and amylobarbitone. Psychopharmacologia (Berl.) 25, 57—68 (1972)
Matsuyama, S. S., Jarvik L. F., Fu, T., Yen, F. S.: Chromosome studies before and after supervised marijuana smoking. In: Braude, M., Szara, S. (Eds.): Pharmacology of marihuana. New York: Raven Press 1976
Matsuyama, S. S., Yen, F. S., Jarvik, L. F., Fu, T. K.: Marihuana and human chromosomes. Genetics **74**, S175 (1973)
Mazurkiewicz-Kwilecki, I. M., Filczewski, M.: The effects of chronic treatment with Δ^9-tetrahydrocannabinol on catecholamine synthesis in the rat. Psychopharmacologia (Berl.) **33**, 71—79 (1973)
McIsaac, W. M., Fritchie, G. E., Idapaan-Heikkila, J. E., Ho, B. T., Englert, L. F.: Distribution of marihuana in monkey brain and concomitant behavioral effects. Nature (Lond.) **230**, 593—594 (1971)
McMillan, D. E.: Behavioral pharmacology of the tetrahydrocannabinols. In: Thompson, T., Dews, P. (Eds.): Recent advances in behavioral pharmacology. New York: Academic Press 1976 (in press)
McMillan, D. E., Dewey, W. L.: On the mechanism of tolerance to Δ^9-THC. In: Lewis, M. F. (Ed.): Current research in marihuana. New York: Academic Press 1972

McMillan, D. E., Dewey, W. L., Harris, L. S.: Characteristics of tetrahydrocannabinol tolerance. Ann. N.Y. Acad. Sci. **191**, 83—99 (1972)
McMillan, D. E., Dewey, W. L., Turk, R. F., Harris, L. S., McNeil, J.: Blood levels of ^{3}H-delta-9-tetrahydrocannabinol and its metabolites in tolerant and nontolerant pigeons. Biochem. Pharmacol. **22**, 383—397 (1973)
McMillan, D. E., Harris, L. S., Frankenheim, J. M., Kennedy, J. S.: 1-Δ^9-*trans*-tetrahydrocannabinol in pigeons: tolerance to the behavioral effects. Science **169**, 501—503 (1970)
Mechoulam, R.: Marihuana chemistry. Science **168**, 1159—1166 (1970)
Mechoulam, R. (Ed.): Marihuana: chemistry, pharmacology, metabolism and clinical effects. New York-London: Academic Press 1973
Mechoulam, R., Ben-Zvi, Z., Yagnitinsky, B., Shani, A.: A new tetrahydrocannabinolic acid. Tetrahedron Letters 2339—2341 (1969)
Mechoulam, R., Gaoni, Y.: Recent advances in the chemistry of hashish. Fortschr. Chem. Org. Naturst. **25**, 175—213 (1967)
Mendelson, J. H., Kuehnle, J., Ellingboe, J., Babor, T. F.: Plasma testosterone levels before, during and after chronic marihuana smoking. New Engl. J. Med. **291**, 1051—1055 (1974)
Mendelson, J. H., Kuehnle, J., Ellingboe, J., Babor, T. F.: Effects of marijuana on plasma testosterone. In: Tinklenberg, J. C. (Ed.): Marihuana and health hazards. New York: Academic Press 1975
Mendelson, J. H., Kuehnle, J. C., Greenberg, I., Mello, N. K.: The effects of marihuana use on human operant behavior: Individual data. In: Braude, M., Szara, S. (Eds.): Pharmacology of Marihuana. New York: Raven Press 1976
Meyer, R. E.: Psychiatric consequences of marihuana use: the state of the evidence. In: Tinklenberg, J. C. (Ed.): Marihuana and Health Hazards. New York: Academic Press 1975
Miczek, K. A.: Mouse-killing and motor activity effects of chronic Δ^9-tetrahydrocannabinol und pilocarpine. Psychopharmacology **47**, 59—64 (1976)
Mikuriya, T. H.: Historical aspects of cannabis sativa in western medicine. New Phycn **18**, 902—908 (1969)
Miller, L. L. (Ed.): Marihuana: effects on human behavior. New York-San Francisco-London: Academic Press 1974
Miras, C. J.: Some aspects of cannabis action. In: Wolstenholme, G. E. W., Knight, J. (Eds.): Hashish: its chemistry and pharmacology. Boston: Little, Brown and Co. 1965
Moreau, J. J.: Hashish and mental illness. New York: Raven Press 1973
Munson, A. E., Harris, L. S., Friedman, M. A., Dewey, W. L., Carchman, R. A.: Antineoplastic activity of cannabinoids. J. nat. Cancer Inst. **55**, 597—602 (1975)
Munson, A. D., Levy, J. A., Harris, L. S., Dewey, W. L.: Effects of Δ^9-tetrahydrocannabinol on the immune system. In: Braude, M., Szara, S. (Eds.): Pharmacology of marihuana. New York: Raven Press 1976
Nahas, G. G.: Marihuana: Deceptive Weed. New York: Raven Press 1973
Nahas, G. G. (Ed.): Marihuana: Chemistry, Biochemistry and Cellular Effects. Berlin-Heidelberg-New York: Springer 1976
Nahas, G. G., Desoize, B., Armand, J. P., Hsu, J., Morishima, A.: Natural cannabinoids: Apparent depression of nucleic acid and protein synthesis in cultured human lymphocytes. In: Braude, M., Szara, S. (Eds.): Pharmacology of marihuana. New York: Raven Press 1976
Nahas, G. G., Schwartz, I. W., Adamec, J., Manger, W. M.: Tolerance to delta-9-tetrahydrocannabinol in the spontaneously hypertensive rat. Proc. Soc. exp. Biol. (N.Y.) **142**, 58—60 (1973)
Nahas, G. G., Suciu-Foca, N., Armand, J. P., Morishima, A.: Inhibition of cellular mediated immunity in marihuana smokers. Science **183**, 419—420 (1974)
Neumeyer, J. L., Shagoury, R. A.: Chemistry and pharmacology of marihuana. J. pharm. Sci. **60**, 1433—1457 (1971)
Neu, R. L., Powers, R. O., King, S., Gardner, L. I.: Delta-8 and delta-9-tetrahydrocannabinol: Effects on cultured human leukocytes. J. clin. Pharmacol. **10**, 228—230 (1970)
Nichols, W. W., Miller, R. C., Heneen, W., Bradt, C., Hollister, L., Kanter, S.: Cytogenetic studies on human subjects receiving marihuana and delta-9-tetrahydrocannabinol. Mutation Res. **26**, 413—417 (1974)
Noyes, R., Baram, D. A.: Cannabis analgesia. Comprehens. Psychiat. **15**, 531—535 (1974)
Noyes, R., Jr., Brunk, S. F., Baram, D. A., Canter, A.: Analgesic effect of delta-9-tetrahydrocannabinol. J. clin. Pharmacol. **15**, 139—143 (1975)

Ohlsson, A., Abou-Chaar, C. I., Agurell, S., Nilsson, I. M., Olofsson, K., Sandburg, F.: Cannabinoid constituents of male and female *Cannabis Sativa*. Bull. Narcot. **23**, 29—32 (1971)
Ouellet, J., Palaic, D., Albert, J. M., Tetreault, L.: Effect of Δ^9-THC on serotonin, MAO and tryptophan hydroxylase in rat brain. Rev. canad. Biol. **32**, 213—217 (1973)
Pace, H. B., Davis, W. M., Borgen, L. A.: Teratogenesis and marihuana. Ann. N.Y. Acad. Sci. **191**, 123—131 (1971)
Paris, M., Ghirlanda, C., Chigneau, M., Giry, L.: Δ^1-Tetrahydrocannabivarolic acid, new constituent of *Cannabis Sativa*. C.R. Acad. Sci. (Paris) C. **276**, 205—207 (1973)
Pars, H. G.: Symposium on "Approaches to centrally acting drugs derived from the cannabinoid nucleus." Abstr. 68th ACS National Meeting, Atlantic City, N.J. 1974
Paton, W. D. M., Crown, J. (Eds.): Cannabis and its derivatives. London: Oxford University Press 1972
Paton, W. D. M., Pertwee, R. G.: Marihuana. Mechoulam, R. (Ed.). New York-London: Academic Press 1973
Perez-Reyes, M., Timmons, M. C., Lipton, M. A., Davis, K. H., Wall, M. E.: Intravenous injection in man of Δ^9-tetrahydrocannabinol and 11-OH-Δ^9-tetrahydrocannabinol. Science **177**, 633—635 (1972)
Persaud, T. V. N., Ellington, A. C.: Cannabis in early pregnancy. Lancet **1967 II**, 1306
Persaud, T. V. N., Ellington, A. C.: Teratogenic activity of cannabis resin. Lancet **1968 II**, 406—407
Peterson, D. W., Cohen, G. M., Sparber, S. B.: The delay of the behavioral effects of delta-9-tetrahydrocannabinol in rats by 2-diethylaminoethyl-2,2-diphenylvalerate HCl (SKF 525 A). Life Sci. **10**, 1381—1386 (1971)
Petrazilka, T., Haefliger, W., Sikemeier, C.: Synthese von Haschisch: Inhaltsstoffen. Helv. chim. Acta **52**, 1102—1134 (1969)
Phillips, R. N., Brown, D. J., Martz, R. C., Hubbard, J. D., Forney, R. B.: Subacute toxicity of water suspended Δ^9-tetrahydrocannabinol in rats. Toxicol. appl. Pharmacol. **19**, 414 (1971 a)
Phillips, R. N., Turk, R. F., Forney, R. B.: Acute toxicity of Δ^9-tetrahydrocannabinol in rats and mice. Proc. Soc. exp. Biol. (N.Y.) **136**, 260—263 (1971 b)
Pirch, J. H., Cohn, R. A., Barnes, P. R., Barratt, E. S.: Effects of acute and chronic administration of marijuana extract on the rat electrocorticogram. Neuropharmacology **11**, 231—240 (1972 a)
Pirch, J. H., Cohn, R. A., Barnes, P. R., Barratt, E. S.: Tolerance to EEG and behavior effects of marihuana in rats. In: Singh, J. M., Miller, L., Lal, H. (Eds.): Drug addiction: Exp. pharmacol., Vol. 1. Mt. Kisco, N.Y.: Futura Publishing Co., Inc. 1972 b
Pirch, J. H., Osterholm, K. C., Cohn, R. A., Barratt, E. S.: Studies on EEG tolerance to marijuana in the rat. Arch. int. Pharmacodyn. **203**, 213—220 (1973)
Potvin, R. J., Fried, P. A.: Acute and chronic effects on rats of (—) Δ^1-tetrahydrocannabinol on unlearned motor tasks. Psychopharmacologia (Berl.) **26**, 369—378 (1972)
Pradhan, S. N., Bailey, P. T., Ghosh, P.: Some behavioral effects of delta-9-tetrahydrocannabinol in rats. Res. Commun. chem. Pathol. Pharmacol. **31**, 197—204 (1972)
Purnell, W. D., Gregg, J. M.: Delta-9-tetrahydrocannabinol, euphoria and intraocular pressure in man. Ann. Opthal. **7**, 921—923 (1975)
Quimby, M. W., Doorenbos, N. J., Turner, C. E., Masoud, A.: Mississippi-grown marihuana and observed morphological variations. Econ. Bot. **27**, 117—127 (1973)
Rating, D., Broemann, I., Hornecker, H., Kluiol, S., Coper, H.: Effect of subchronic treatment with (—) Δ^8-trans-tetrahydrocannabinol (Δ^8-THC) on food intake, body temperature, hexabarbital sleeping time and hexabarbital elimination in rats. Psychopharmacologia (Berl.) **27**, 349—357 (1972)
Razdan, R. K.: Recent advances in the chemistry of cannabinoids. In: Progress in organic chemistry. London: Butterworths 1973
Razdan, R. K., Dalzell, H. C., Handrick, G. R.: Hashish: A simple one-step synthesis of (—)-Δ^1-tetrahydrocannabinol (THC) from p-Mentha-2,8-dien-1-ol and olivetol. J. Amer. chem. Soc. **96**, 5860—5865 (1974)
Razdan, R. K., Handrick, G. R.: Hashish: A stereospecific synthesis of (—)-Δ^1 and (—)-$\Delta^{1(6)}$-tetrahydrocannabinols. J. Amer. chem. Soc. **92**, 6061—6062 (1970)
Razdan, R. K., Handrick, G. R., Dalzell, H. C.: A one-step synthesis of (—)-Δ^1-tetrahydrocannabinol from chrysanthenol. Experientia (Basel) **31**, 16 (1975)

Razdan,R.K., Puttick,A.J., Zitko,B.A., Handrick,G.R.: Hashish VI: Conversion of (—)-$\Delta^1(^6)$-tetrahydrocannabinol to (—)-$\Delta^1(^7)$-tetrahydrocannabinol. Stability of (—)-Δ^1- and (—)-$\Delta^1(^6)$-tetrahydrocannabinols. Experientia (Basel) **28**, 121—122 (1972)

Razdan,R.K., Zitko,B.A.: Hashish IV: Some acid catalyzed transformations in cannabinoids. Tetrahedron Letters 4947—4950 (1969)

Regelson,W., Butler,J.R., Schulz,J., Kirk,T., Peek,L., Green,M.L., Zakio,O.: Delta-9-THC as an effective antidepressant and appetite stimulating agent in advanced cancer patients. In: Braude,M., Szara,S. (Eds.): Pharmacology of marihuana. New York: Raven Press 1976

Renault,P.F., Schuster,C.R., Heinrich,R., Freeman,D.X.: Marihuana: Standardized smoke administration and dose effect curves on heart rate in humans. Science **174**, 589—591 (1971)

Roft,D.D., Gregg,J.M., Ghia,J., Harris,L.S.: Effects of intravenous Δ^9-THC on experimental and surgical pain responses. J. clin. Pharmacol. and J. New Drugs [In Press] (1976)

Rosenkrantz,H., Braude,M.C.: Comparative chronic toxicities of Δ^9-tetrahydrocannabinol administered orally or by inhalation in rat. In: Braude,M., Szara,S. (Eds.): Pharmacology of marihuana. New York: Raven Press 1976

Rosenkrantz,H., Sprague,R.A., Fleischman,R.W., Braude,M.C.: Oral Δ^9-tetrahydrocannabinol toxicity in rats treated for periods up to six months. Toxicol. appl. Pharmacol. **32**, 399—417 (1975)

Rubin,V. (Ed.): Cannabis and culture. The Hague: Mouton 1975

Rubin,V., Comitas,L.: Ganja in Jamaica: A medical anthropological study of chronic marihuana use. The Hague: Mouton 1975

Sallen,S.E., Zinberg,W.E., Frei,E.: Antiemetic effect of delta-9-tetrahydrocannabinol in patients receiving cancer chemotherapy. New Engl. J. Med. **293**, 795—797 (1975)

Santos,M., Sampaio,M.R.P., Fernandes,N.S., Carlini,E.A.: Effects of *cannabis sativa* (marihuana) on the fighting behavior of mice. Psychopharmacologia (Berl.) **8**, 437—444 (1966)

Schaefer,C.F., Gunn,C.G., Dubowski,K.M.: Normal plasma testosterone concentrations after marihuana smoking. New Engl. J. Med. **292**, 867—868 (1975)

Scheckel,C.L., Baff,E., Dahlen,P., Smart,T.: Behavioral effects in monkeys of racemates of two biological active marihuana constituents. Science **160**, 1467—1469 (1968)

Schildkraut,J.J., Efron,D.H.: The effects of Δ^9-tetrahydrocannabinol on the metabolism of norepinephrine in rat brain. Psychopharmacologia (Berl.) **20**, 191—196 (1971)

Schultes,R.E.: Random thoughts and queries on the botany of cannabis. In: Joyce,C.R.B., Curry,S.H. (Eds.): The botany and chemistry of cannabis. London: Churchill 1970

Schultes,R.E.: Man and marijuana. Natural History **82**, 59—68 (1973)

Secretary,H.E.W.: Marijuana and health reports 1—4. Washington: U.S. Govt. Printing Office 1971—1975

Segelman,A.B., Sofia,R.D., Segelman,F.P., Harakal,J.J., Knoblock,L.C.: *Cannabis sativa* L. (marijuana) V: Pharmacological evaluation of marijuana aqueous extract and volatile oil. J. pharm. Sci. **63**, 962—964 (1974)

Shafer,R.P.: Marihuana: A signal of misunderstanding. First report of the National Commission on Marihuana and Drug Abuse. Washington: U.S. Govt. Printing Office 1972

Shani,A., Mechoulam,R.: A new type of cannabinoid. Synthesis of cannabielsoic acid A by a novel photo-oxidative cyclization. Chem. Commun. 273—274 (1970)

Shapiro,D.: The ocular manifestations of the cannabinols. Ophthalmologica (Basel) **168**, 366—369 (1974)

Shoyama,Y., Oku,R., Yamauchi,T., Nishioka,I.: Cannabis VI: Cannabicyclolic acid. Chem. pharm. Bull. **20**, 1927—1930 (1972)

Sidell,F.R., Pless,J.E., Meitlich,H., Sussman,P., Copelan,H.W., Sim,V.M.: Dimethylheptyl-delta-6_a-10_a-tetrahydrocannabinol: Effects after parenteral administration to man. Proc. Soc. exp. Biol. (N.Y.) **142**, 867—873 (1973)

Silva,M.T.A., Carlini,E.A., Claussen,V., Korte,F.: Lack of cross tolerance in rats among (—)-Δ^9-trans-tetrahydrocannabinol (Δ^9-THC), cannabis extract, mescaline and lysergic acid diethylamide (LSD-25). Psychopharmacologia (Berl.) **13**, 332—340 (1968)

Silverstein,M.J., Lessin,P.J.: Normal skin responses in chronic marihuana users. Science **186**, 740—741 (1974)

Silverstein,M.J., Lessin,P.J.: DCNB skin testing in chronic marihuana users. In: Braude,M., Szara,S. (Eds.): Pharmacology of marihuana. New York: Raven Press 1976

Singer, A.J. (Ed.): Marijuana: Chemistry, pharmacology, and patterns of social use. Ann. N.Y. Acad. Sci. **191**, 1—269 (1971)

Smith, D.E.: Acute and chronic toxicity of marihuana. J. Psychedelic Drugs **2**, 37—41 (1968)

Snyder, S.H.: Uses of marihuana. New York-London-Toronto-Oxford: University Press 1971

Sofia, R.D., Barry, H. III: Acute and chronic effects of Δ^9-tetrahydrocannabinol on food intake by rats. Psychopharmacologia (Berl.) **39**, 213—222 (1974)

Sofia, R.D., Dixit, B.N., Barry, H. III: The effects of Δ^1-tetrahydrocannabinol on serotonin metabolism in the rat brain. Life Sci. **10**, 425—436 (1971)

Sofia, R.D., Knoblock, L.C., Vassar, H.B.: The anti-edema activity of various naturally occurring cannabinoids. Res. Commun. chem. Path. Pharmacol. **6**, 909—918 (1973a)

Sofia, R.D., Nalepa, S.D., Harakal, J.J., Vassar, H.B.: Anti-edema and analgesic properties of Δ^9-tetrahydrocannabinol (THC). J. Pharmacol. exp. Ther. **186**, 646—655 (1973b)

Solomon, D. (Ed.): The marihuana papers. Kansas City-New York-Indianapolis: Bobbs-Merrill 1966

Spaulding, T.C., Ford, R.D., Dewey, W.L., McMillan, D.E., Harris, L.S.: Some pharmacological effects of phenitrone and its interaction with Δ^9-THC. Europ. J. Pharmacol. **19**, 310—317 (1972)

Stearn, W.T.: The cannabis plant: Botanical characteristics. In: Joyce, C.R.B., Curry, S.H. (Eds.): The botany and chemistry of cannabis. London: Churchill 1970

Stefanis, C., Boulougouris, J., Liakos, A.: Clinical and psychophysiological effects of cannabis in long-term users. In: Braude, M., Szara, S. (Eds.): Pharmacology of marihuana. New York: Raven Press 1976

Stenchever, M.A., Allen, M.: The effect of delta-9-tetrahydrocannabinol on the chromosomes of human lymphocytes in vitro. Amer. J. Obstet. Gynec. **114**, 819—821 (1972)

Stenchever, M.A., Kunysz, T.J., Allen, M.: Chromosome breakage in users of marihuana. Amer. J. Obstet. Gynec. **118**, 106—113 (1974)

Tashkin, D.P., Shapiro, B.J., Frank, I.M.: Acute pulmonary physiologic effects of smoked marijuana and oral Δ^9-tetrahydrocannabinol in healthy young men. New Engl. J. Med. **289**, 336—341 (1973)

Tashkin, D.P., Shapiro, B.J., Frandk, I.M.: Acute effects of smoked marijuana and oral delta-9-tetrahydrocannabinol on specific airway conductance in asthmatic subjects. Amer. Rev. resp. Dis. **109**, 420—428 (1974)

Tashkin, D.P., Shapiro, B.J., Ramanna, L., Taplin, G.V., Lee, Y.E., Harper, C.E.: Chronic effects of heavy marihuana smoking on pulmonary function in healthy young males. In: Braude, M., Szara, S. (Eds.): Pharmacology of marihuana. New York: Raven Press 1976

Thompson, G.R., Fleischman, R.W., Rosenkrantz, H., Braude, M.C.: Oral and intravenous toxicity of Δ^9-tetrahydrocannabinol in rhesus monkeys. Toxicol. appl. Pharmacol. **27**, 648—665 (1974a)

Thompson, G.R., Fleischman, R.W., Rosenkrantz, H., Braude, M.C.: Chronic oral toxicity of cannabinoids in monkeys. Abstr. 13th Ann. Meeting Soc. Toxicol. p. 5 (1974b)

Thompson, G.R., Mason, M.M., Rosenkrantz, H., Braude, M.C.: Chronic oral toxicity of cannabinoids in rats. Toxicol. appl. Pharmacol. **25**, 373—390 (1973a)

Thompson, G.R., Rosenkrantz, H., Fleischman, R.W., Braude, M.C.: Effects of Δ^9-tetrahydrocannabinol administered subcutaneously to rabbits for 28 days. Toxicology **4**, 41—51 (1975)

Thompson, G.R., Rosenkrantz, H., Schaeppi, U.H., Braude, M.C.: Comparison of acute oral toxicity of cannabinoids in rats, dogs and monkeys. Toxicol. appl. Pharmacol. **25**, 363—372 (1973b)

Tinklenberg, J.R.: Marijuana and human aggression. In: Miller, L.L. (Ed.): Marijuana: effects on human behavior. New York: Academic Press 1974

Truitt, E.B., Anderson, S.M.: Biogenic amine alterations produced in the brain by tetrahydrocannabinols and their metabolites. Ann. N.Y. Acad. Sci. **191**, 68—73 (1971)

Turk, R.F., Dewey, W.L., Harris, L.S.: Excretion of trans-Δ^9-tetrahydrocannabinol and its metabolites in intact and bile duct-cannulated rats. J. pharm. Sci. **62**, 737—740 (1973)

Turner, C.E., Hadley, K.: Constituents of *Cannabis sativa* L. II: Absence of cannabidiol in an African variant. J. pharm. Sci. **62**, 251—255 (1973)

Turner, C.E., Hadley, K., Getterman, P.S.: Constituents of *Cannabis sativa L.* VI: Propyl homologs in samples of known geographic origin. J. pharm. Sci. **62**, 1739—41 (1973)

Uliss,D.B., Razdan,R.K., Dalzell,H.C., Handrick,G.R.: Hashish: Synthesis of dl-Δ^1-6-*cis*-tetrahydrocannabinol (THC). Tetrahedron Letters 4369—4372 (1975)
Uyeno,E.T.: Δ^9-tetrahydrocannabinol administration during pregnancy of the rat. Proc. W. pharmacol. Soc. **16**, 64—67 (1973)
Vachon,L., Fitzgerald,M.S., Solliday,N.H., Gould,I.A., Gaensler,E.A.: Single-dose effect of marihuana smoke. Bronchial dynamics and respiratory-center sensitivity in normal subjects. New Engl. J. Med. **288**, 985—989 (1973)
Vachon,L., Mikus,P., Morrissey,W., Fitzgerald,M., Gaensler,E.A.: Bronchial effect of marihuana smoke in asthma. In: Braude,M., Szara,S. (Eds.): Pharmacology of marihuana. New York: Raven Press 1976a
Vachon,L., Robins,A., Gaensler,E.A.: Airways response to aersolized delta-9-tetrahydrocannabinol: preliminary report. In: Cohen,S., Stillman,R.C. (Eds.): The therapeutic aspects of marihuana. In press 1976b
Valle,J.R., Lupa,A.J., Barros,G.G.: Pharmacological activity of cannabis according to the sex of the plant. J. Pharm. (Lond.) **20**, 798—799 (1968)
Van Ginneken,C.A.M., Vree,T.B., Breimer,D.D., Thijssen,H.W.H., Van Rossum,J.M.: Cannabinodiol, a new hashish constituent, identified by gas chromatography-mass spectrometry. Proc. int. Symp. gas Chromatog. mass Spectrom. Tamburini, Milan, Italy (1972)
Vollner,L., Bieniek,D., Korte,R.: Haschisch XX Cannabidiviarin, ein neuer Haschisch-Inhaltsstoff. Tetrahedron Letters 145—147 (1969)
Von Spulak,R., Claussen,U., Fehlhaber,H.W., Korte,R.: Haschisch-XIX, Cannabidiolcarbonsaure-Tetrahydrocannabitriol-Ester, ein neuer Haschisch-Inhaltsstoff. Tetrahedron **24**, 5379—5383 (1968)
Vree,T.B., Breimer,D.D., Van Ginneken,C.A.M., Van Rossum,J.M.: Identification of cannabicyclol with a pentyl or propyl side chain by means of combined gas chromatography-mass spectrometry. J. Chromatog. **74**, 124—127 (1972)
Wahlquist,M., Nilsson,I.M., Sandberg,F., Agurell,S., Granstrand,B.: Binding of Δ^1-tetrahydrocannabinol to human plasma proteins. Biochem. Pharmacol. **19**, 2579—2584, (1970)
Wall,M.E.: The in vitro and in vivo metabolism of tetrahydrocannabinol (THC). Ann. N.Y. Acad. Sci. **191**, 23—39 (1971)
Waller,C.W.: Standardized sources of supply for marihuana research. In: Braude,M., Szara,S. (Eds.): The pharmacology of marihuana. New York: Raven Press 1976
Walton,R.P.: Marihuana, America's new drug problem. Philadelphia: J. B. Lippincott Co. 1938
Way,E.L. (Ed.): Marihuana and its surrogates. Pharmacol. Rev. **23**, 263—300 (1971)
Webster,C.D., LeBlanc,A.E., Marshman,J.A., Beaton,J.M.: Acquisition and loss of tolerance to 1-Δ^9-trans-tetrahydrocannabinol in rats on avoidance schedule. Psychopharmacologia (Berl.) **30**, 217—226 (1973)
Weil,A.T., Zinberg,N.E., Nelsen,J.M.: Clinical psychological effects of marihuana in man. Science **162**, 1234—1242 (1968)
Welch,B.L., Welch,A.S., Messiha,F.S., Berger,H.J.: Rapid depletion of adrenal epinephrine and elevation of telencephalic serotonin by (—)-*trans*-Δ^9-tetrahydrocannabinol in mice. Res. Commun. chem. Path. Pharmacol. **2**, 382—391 (1971)
Williams,R.B., Ng,L.K.Y., Lamprecht,F., Roth,K., Kopin I.J.: Δ^9-Tetrahydrocannabinol: A hypotensive effect in rats. Psychopharmacologia (Berl.) **28**, 269—274 (1973)
Willinsky,M.D.: Analytical aspects of cannabis chemistry. In: Mechoulam,R. (Ed.): Marihuana. New York: Academic Press 1973
Wilson,R.S., May,E.L.: Analgesic properties of the tetrahydrocannabinols, their metabolites and analogs. J. med. Chem. **18**, 700—703 (1975)
Wolstenholme,G.E.W., Knight,J. (Eds.): Hashish: Its chemistry and pharmacology. Boston: Little, Brown and Co. 1965
Yagiela,J.A., McCarthy,K.D., Gibb,J.W.: The effect of hypothermic doses of 1-Δ^9-tetrahydrocannabinol on biogenic amine metabolism in selected parts of the rat brain. Life Sci. **14**, 2367—2378 (1974)
Zimmerman,A.M., McClean,D.K.: Action of narcotic and hallucinogenic agents on the cell cycle. In: Zimmerman,A.M., Padilla,G.M., Cameron,I.L. (Eds.): Drugs and the cell cycle. New York: Academic Press 1973
Zitko,B.A., Howes,J.F., Razdan,R.K., Dalzell,B.C., Dalzell,H.C., Sheehan,J.C., Pars,H.G., Dewey,W.L., Harris,L.S.: Water-soluble derivatives of Δ^1-tetrahydrocannabinol. Science **177**, 442—444 (1972)

Author Index

Page numbers in *italics* refer to bibliographie

Abate, F., see Zahn, T.P. 286, *304*
Abbott, W.O., Henry, C.M. 8, *25*
Abdallah, A.H. 131, 134, 138, 139, 141, 143, 173, *181*
Abdallah, A.H., White, H.D. 131, 138, *181*
Abdallah, A.H., White, H.D., Kulkarni, A.S. 87, 134, *181*
Abel, E.L. 410, *415*
Abel, E.L., McMillan, D.E., Harris, L.S. 410, *415*
Abelson, H., Cohen, R., Schrayer, D. 371, *415*
Abelson, R.M., see Harrison, J.M. 98, 99, *206*
Abely, P., Bobin, P., Geier, S. 293, *296*
Abou-Chaar, C.I., see Ohlsson, A. 373, *426*
Abraham, D.J., see Lotter, H.L. 378, *423*
Abraham, G.J.S., see Ahmed, S.S. 179, *182*
Abramson, H.A. 310, *349*
Abramson, H.A., Jarvik, M.E., Gorin, M.H., Hirsch, M.W. 318, *349*
Abramson, H.A., Jarvik, M.E., Hirsch, M.W. 308, *350*
Abramson, H.A., Jarvik, M.E., Hirsch, M.W., Ewald, A.T. 310, *350*
Abramson, H.A., Jarvik, M.E., Kaufman, M.R., Kornetsky, C., Levine, A., Wagner, M. 307, 308, 311, *350*
Abramson, H.A., Rolo, A., Sklarofsky, B., Stache, L. 318, *350*
Abramson, H.A., Sklarofsky, B. 312, *350*
Abramson, H.A., Sklarofsky, B., Baron, M.O., Fremont-Smith, N. 318, *350*
Abramson, H.A., Waxenberg, S.E., Levine, A., Kaufman, M.R., Kornetsky, C. 310, *350*
Abramson, H.A., see Jarvik, M.E. 308, 310, 318, *358–360*
Abreu, E., see Morrison, J.L. 160, *222*
A'Brook, M.F., see Herridge, C.F. 278, *300*
Absava, G.I., Vysotskaya, N.B. 165, *181*
Accornero, F. 70, 96, *181*
Aceto, M.D., see Lynch, V.D. 101, *217*
Acton, J.M., see Yen, T.T. 90, *245*
Adamec, J., see Nahas, G.G. 398, 410, *425*
Adams, M.D., Chait, L.D., Earnhardt, J.T. 399, *415*
Adams, M.D., Earnhardt, J.T., Dewey, W.L., Harris, L.S. 399, *415*
Adams, M.D., see Martin, B.R., 381, 390, *424*
Adams, R.D., Hope, J. 292, *296*
Adey, W.R., Bell, F.R., Dennis, B.J. 315, 316, 341, *350*
Adey, W.R., Porter, R., Walter, D.O., Brown, T.S. 341, *350*
Adler, H.F., Burkhardt, W.L., Ivy, A.C., Atkinson, A.J. 249, *266*
Adler, M.W. 82, 90, *181*
Änggård, E. 68, 78
Änggård, E., Gunne, L.-M. 266, *266*
Änggård, E., Gunne, L.-M., Jönsson, L.E. 295, *296*
Änggård, E., Gunne, L.-M., Jönsson, L.E., Niklasson, F. 14, 15, 18, 21–24, 25, 251, 260, *266*
Änggård, E., Gunne, L.-M., Niklasson, F. 21, 22, *25*
Änggård, E., Jönsson, L.E., Hogmark, A.L., Gunne, L.-M. 8, 9, 14, 15, 18, 19, 21, *25*, 295, *296*
Änggård, E., see Franksson, G. 21, *27*
Änggård, E., see Gunne, L.-M. 24, *28*, 248, 262, 265, 266, *270*
Änggård, E., see Jönsson, L.E. 18, 21, 23, *28*, 265, *271*
Aghajanian, G.K., Bunney, S. 66, *181*
Aghajanian, G.K., Foote, W.E., Sheard, M.H. 330, *350*
Aghajanian, G.K., Haigler, H.J., Bloom, F.E. 330, *350*
Aghajanian, G.K., Weiss, B.L. 330, *350*
Aghajanian, G.K., see Asher, B.M. 79, 89, *183*
Aghajanian, G.K., see Bunney, B.S. 46, 66, *189*
Aghajanian, G.K., see Haigler, H.J. 330, *356*
Aghajanian, G.K., see Foote, W.E. 67, *201*
Aghajanian, G.K., see Freedman, D.X. 320, *355*
Aghajanian, G.K., see Graham, A.W. 65, *204*
Aghajanian, G.K., see Korf, J. 66, *212*
Aghajanian, G.K., see Roth, R.H. 42, 46, 66, *231*
Agid, Y., see Javoy, F. 48, *210*
Agurell, S., Binder, M., Fonseka, K., Gustafsson, B., Lindgren, J.E., Leander, K., Martin, B., Nilsson, I.M., Nordquist, M., Ohlsson, A., Widman, M. 382, *415*
Agurell, S., Dahmen, J., Gustafsson, B., Johansson, I.B., Leander, K., Nilsson, I., Ramsey, C.H., Ryrfeldt, A., Sandberg, F., Widman, M. 382, 384, *415*
Agurell, S., Nilsson, I.M., Ohlsson, A., Sandberg, F. 381, 382, *415*
Agurell, S., see Ohlsson, A. 373, *426*
Agurell, S., see Wahlquist, M. 382, *429*

Ahlskog, J. E., Hoebel, B. G. 135, 136, 139, *181*
Ahmed, S. S., Abraham, G. J. S., Ansari, M. Y. 179, *182*
Aitken, R. C. B., Gedye, J. L. 258, *267*
Albert, A. 277, 278, *296*
Albert, J. M., see Ouellet, J. 395, *426*
Alexander, M., Isaac, W. 90, *181*
Alexander, W. D., see Beckwith, B. E. 129, 130, *185*
Alexandre, J.-M., see Chevillard, C. 41, *192*
Alhava, E. 35, 48, 167, 169, 173, 174, *182*
Alhava, E., Klinge, E. 35, 37, 174, *182*
Alhava, E., Mattila, M. J. 167, *182*
Alipoulus, M. A., see Harmon, J. 400, *420*
Allan, D., Baird, J. R. C., Ellis, K. E. J. 141, 143, 146, *182*
Allbutt, C. 277
Allen, E. V., see Wilbur, D. L. 255, *275*
Allen, M., see Stenchever, M. A. 401, *428*
Allen, R. P., see Safer, D. J. 286, *302*
Alles, G. A., see Piness, G. *302*
Alles, G. A., see Prinzmetal, M. 250, *273*
Alles, G. A. 62, 141, 142, 148, 166, 167, 173, 174, 178, *182*, *267*, 279, 281, 311, *350*
Alles, G. A., Feigen, G. A. 4, 5, *25*
Alles, G. A., Wisegarver, B. B. 13, *25*
Alleva, J. J. 5, 8, *25*
Allison, J. H., see Cicero, T. J. 401, *417*
Allmark, J., Rylander, G. *257*, *267*
Alpern, E. B., Finkelstein, N., Gantt, W. H. *182*
Alphin, R. S., Ward, J. W. 131, 138, *182*
Altman, J. L., Appel, J. B. 314, *350*
Alves, C. N., Carlini, E. A. 387, *415*
Ambrus, C. M., see Harrison, J. W. E. 9, *28*, 157, *206*
Ames, P. L., see Tedeschi, R. E. 87, 154, *239*
Amin, A. H. T., Crawford, T. B. B. Gaddum, J. H. 328, *350*
Ammon, H. P. T., see Estler, C.-J. 87, 161–163, *199*
Anagnoste, B., see Goldstein, M. 9, 15, *28*, 50, *204*
Anania, W., see Miele, E. 87, *221*
Andén, N.-E. 48, 67–69, *182*, 264, 266, *267*
Andén, N.-E., Carlsson, A., Häggendal, J. 34, 42, *182*
Andén, N.-E., Corrodi, H., Fuxe, K. 330, 333, 335, *350*
Andén, N.-E., Corrodi, H., Fuxe, K., Hökfeld, T. 330, 333, 335, *350*
Andén, N.-E., Corrodi, H., Fuxe, K., Meek, J. L. 330, 335, *350*
Andén, N.-E., Dahlström, A., Fuxe, K., Larsson, K. 68, *182*
Andén, N. E., Henning, M., Obianwu, H. 59, *182*
Andén, N.-E., Magnusson, T., Stock, G. 42, *182*
Andén, N.-E., Rubenson, A., Fuxe, K., Hökfelt, T. 68, 69, *182*
Andén, N.-E., Svensson, T. H. 42, *182*
Andén, N.-E., see Butcher, L. L. 118, *189*
Andersen, D. K., see McMahon, E. M. 162, 165, *220*
Anderson, E. W. 247, 248, *267*, 282, *296*
Anderson, E. W., Scott, W. C. 265, *267*
Anderson, K., see D'Encarnacao, P. S. 87, *196*
Anderson, M. M., see Winder, C. V. 62–64, 141, 142, 144, 148, *244*
Anderson, R. N., see Dewey, W. L. *418*
Anderson, S. M., see Truitt, E. B. 391, *428*
Andersson, B., Larsson, S. 136, 140, *182*, 251, *267*
Andres, F., Ohnesorge, F. K., DeVries, R. 161, *182*
Andrew, J. G., Sensenbach, W. 252, *267*
Angel, C., Murphree, O. D., DeLuca, D. C. 109, 117, *182*
Angel, C., see Murphree, O. D. 71, *223*
Angelis, L, de, Soranzio, M. R., Traversa, U., Vertua, R. 61, *182*
Angrist, B. M., Gershon, S. 256, 257, *267*, 282, 292, 293, 295, *296*
Angrist, B. M., Lee, H. K., Gershon, S. 295, *296*
Angrist, B. M., Rotrosen, J., Gershon, S. 319, *350*
Angrist, B. M., Schweitzer, J. W., Gershon, S., Friedhoff, A. J. 284, *296*
Angrist, B. M., Shopsin, B., Gershon, S. 5, 19, *25*, 257, 262, *267*
Angrist, B. M., Shopsin, B., Gershon, S., Wilk, S. 261, *267*
Angrist, B. M., Wilk, S., Gershon, S. 255, *267*
Angrist, B. M., see Schweitzer, J. W. 19, *30*
Angrist, B. M., see Wallach, M. B. 72, *242*
Angrist, B. M., see Willner, J. H. 70, 77, *244*
Anichkov, S. V., Borodkin, Y. S., Zaitsev, Y. V. *182*
Anisman, H., Kokkinidis, L. 100, 101, 107, *183*
Anisman, H., see Remington, G. 106, *229*
Annecker, L., see Castellano, C. 101, *192*
Ansari, M. Y., see Ahmed, S. S. 179, *182*
Antelman, S. M., see Canfield, D. R. 123, *190*
Antonaccio, M. J., Smith, C. B. 142, *183*
Apfelberg, B. 265, *267*, 282, *296*
Appel, J. B. 312, 314, *350*
Appel, J. B., Freedman, D. X. 120, *183*, 315, 320, *351*
Appel, J. B., Freedman, D. X., Filby, Y. M. 314, *351*
Appel, J. B., Lovell, R. A., Freedman, D. X. 315, *351*
Appel, J. B., Whitehead, W. E., Freedman, D. X. 314, *351*
Appel, J. B., see Altman, J. L. 314, *350*
Appel, J. B., see Becker, D. I. 315, *351*
Appel, J. B., see Cameron, O. G. 315, *352*
Appel, J. B., see Dykstra, L. A. 314, *354*
Appel, J. B., see Ferster, C. B. 110, 114, *200*
Appel, J. B., see Freedman, D. X. 314, 319, 320, *355*
Apter, J. T., Pfeiffer, C. C. 337, *351*
Aquilonius, S. M., Sjöström, R. 264, *267*
Aragao, J. A., see Moraes, S. de 63, *222*
Aramaki, H., Tomiyasu, N., Yoshimura, H., Tsukamoto, H. 375, *415*

Arbuthnott, G.W., see Ungerstedt, U. 68, *241*
Archer, R., see Lemberger, L. 385, 387, 399, 400, *423*
Ariens, E.J., see Schoot, J.B. van der 71, 87, 88, *234*
Arlazoroff, A., see Mouriz-Garcia, A. 337, *362*
Armand, J.P., see Nahas, G.G. 402, *425*
Armitage, P., Rushton, R., Steinberg, H. 94, *183*
Arnfred, T., Randrup, A. 77, 78, *183*
Arnold, L.E., Kirilcuk, V., Corson, S.A., Corson, E.O.L. 71, *183*, 250, *267*
Arnold, L.E., Wender, P.H., McCloskey, K., Snyder, S.H. 250, *267*, 286, *296*
Arnold, L.E., see Corson, S.A. 71, 131, 137, *194*
Arnold, L.E., see Huestis, R.D. 286, *300*
Arnold, O.H., Hofmann, G., Leupold-Löwenthal, H. 334, *351*
Aronow, W.S., Cassidy, J. 397, *415*
Aronson, H., Klee, G.D. 308, *351*
Arthur, B., see Zrull, J.P. 286, *304*
Asano, Y., Moroji, T. 35, 165, *183*
Asatoor, A.M., Galman, B.R., Johnson, J.R., Milne, M.D. 7, 15, 21, *25*
Ashcroft, G.W., Eccleston, D., Waddell, J.L. 263, *267*
Asher, B.M., Aghajanian, G.K. 79, 89, *183*
Askew, B.M. 149, 154–156, 167–169, 173, 174, *183*
Ashley, L.G., see Proctor, C.D. 87, *228*
Ashmore, J., see Bewsher, P.D. 162, 165, 175, *186*
Aston, R., Cullumbine, H. 313, *351*
Åström, A. 60, *183*
Atkinson, A.J., see Adler, H.F. 249, *266*
Atrens, D.M., Vietinghoff-Riesch, F. von, Karabetian, A. der, Masliya, E. 121, 122, *183*
Axelrod, J. 4, 5, 7–10, 12, *25*, 34, 42, *183*
Axelrod, J., Brady, R.O., Witkop, B., Evarts, E.V. 333, 334, *351*
Axelrod, J., Hertting, G., Potter, L. 34, 38, 40, *183*
Axelrod, J., Inscoe, J.K. 54, *183*
Axelrod, J., Tomchick, R. 34, 38, 47, *183*
Axelrod, J., Whitby, L.G., Hertting, G. 38, *183*
Axelrod, J., see Cooper, J.R. 17, *27*
Axelrod, J., see Davis, J.M. 14, 15, 21, 22, *27*
Axelrod, J., see Glowinski, J. 4, 10, *28*, 38, 39, 43, 45–48, *204*
Axelrod, J., see Hertting, G. 38, *207*
Axelrod, J., see Kreuz, D.S. 381, *422*
Axelrod, J., see Lemberger, L. 380, *423*
Axelrod, J., see Potter, L.T. 34, 40, *227*
Axelrod, J., see Saavedra, J.M. 332, *363*
Ayhan, J.H., Randrup, A. 71, *183*
Azarnoff, D.L., see Shoeman, D.W. 132, *235*
Azbekyan, S.G., see Lapin, I.P. 87, 154, 173, *214*
Azzaro, A.J., Rutledge, C.O. 44, 55, *183*
Azzaro, A.J., Ziance, R.J., Rutledge, C.O. 44, 45, 50, *183*
Azzaro, A.J., see Rutledge, C.D. 39, 44–47, 50, 55, *232*
Azzaro, A.J., see Ziance, R.J. 39, 44, *246*

Babbini, M., Montanaro, N., Strocchi, P., Gaiardi, M. 77, 87, *183*
Babbini, M., see Montanaro, N. 89, 90, *221*
Babor, T.F., see Mendelson, J.H. 401, *425*
Babulova, A., Cotillo, P., Bonaccorsi, A., Podvalova, I. 59, 60, 143, 144, 147, *183*
Bachelard, H.S., see Manning, D.H. 162, *219*
Bacher, J.A., Jr., see Schulte, J.W. 82, *234*
Bäckström, M., Wetterberg, L. 166, *184*
Bättig, K. 94, 100, 131, 179, *184*
Bättig, K., see Wanner, H.U. 121, 122, *243*
Baez, L.A. 132, 134, *184*
Baff, E., see Scheckel, C.L. 387, *427*
Bagchi, S.P., McGeer, P.L. 47, *184*
Baggot, J.D., Davis, L.E. 20, *25*
Baggot, J.D., Davis, L.E., Neff, C.A. 21, *25*
Bagley, S.K., see Hurst, P.M. 248, *270*
Bahnsen, P., Jacobsen, E., Thesleff, H. 247, 248, 255, 256, *267*
Bahnsen, P., see Jacobsen, E. 247, 248, *270*
Bailey, D.W., see Oliviero, A. 94, *225*
Bailey, P.T., Pradhan, S.N. 315, *351*
Bailey, P.T., see Pradhan, S.N. 388, *426*
Bain, G., see Latz, A. 101, 128, 129, *214*
Bainbridge, J.G. 82, 90, *184*
Baird, J.R.C. 34, 36, *184*
Baird, J.R.C., Lewis, J.J. 34, 35, 36, *184*
Baird, J.R.C., see Allan, D. 141, 143, 146, *182*
Baiz, T.C., see Weiss, S.R. 265, *275*
Baker, J.C., see Fuller, R.W. 11, *27*
Balasse, E., Thys, O. 254, *267*
Balazs, T., see Chappel, C.I. 174, *192*
Baldessarini, R.J. 50, *184*
Baldessarini. R.J., Vogt, M. 49, 50, *184*
Baldessarini, R.J., see Fischer, J.E. 279, *298*
Baldessarini, R.J., see Glowinski, J. 33, 34, *204*
Baldessarini, R.J., see Harris, J.E. 38–40, 47, *206*
Balestrieri, A. 318, *351*
Balestrieri, A., Fontanari, D. 318, *351*
Ballantine, II, P., see Lynch, G.S. 90, *217*
Balster, R.L., Ford, R.D. 388, *415*
Balster, R.L., Schuster, C.R. 124, 125, *184*
Balster, R.L., see Ellinwood, E.H., Jr. 71, 72, 77, *198*
Balster, R.L., see Ford, R.D. 389, *419*
Balster, R.L., see Harris, R.T. 130, *206*
Balzano, E., see Walter, S. 317, 342, 343, *367*
Banarjee, U., Geh, S.L. 87, 96, 173, *184*
Banarjee, U., Linn, G.S. 87, 173, *184*
Bannerjee, B.N., Galbreath, C., Sofia, R.D. 406, *415*

Bapna, J. S., see Menon, M. K. 87, 154, *221*
Baram, D. A., see Noyes, R. Jr., 413, *425*
Baran, L., see Maj, J. 77, 87, *218*
Baran, L. de, see McGaugh, J. L. 115, 117, *220*
Barber, T., see Creaven, P. J. 17, *27*
Barger, G., Dale, H. H. 279, *297*
Barnes, C. D. 177, *184*
Barnes, C. D., Meyers, F. H. 178, *184*
Barnes, P. R., see Pirch, J. H. 410, 411, 413, *426*
Barnett, A., Peschel, E. 173, *184*
Baron, M. O., see Abramson, H. A. 318, *350*
Barratt, E. S., Pray, S. L. 315, 341, *351*
Barratt, E. S., see Pirch, J. H. 87, *227*, 410, 411, 413, *426*
Barrett, M. E., see Dispensa, J. 128, *197*
Barrett, R. J., Leith, N. J., Ray, O. S. 100, 102, 103, 130, *184*
Barrett, R. J., Steranka, L. R. 100, 102, 103, 107, 108, 130, *184*
Barros, G. G., see Valle, J. R. 373, *429*
Barry, H., Buckley, J. P. 98, *184*
Barry, H., Krimmer, E. C. 389, *415*
Barry, H., Miller, N. E. 128, *184*
Barry, H., Wagner, S., Miller, N. E. 115, 116, *184*
Barry, H., see Kubena, R. K. 87, 173, *213*
Barry, H. III, Perhoch, J. L. Kubena, R. K. 400, *415*
Barry, H. III, see Kubena, R. L. 400, *422*
Barry, H. III, see Sofia, R. D. 392, 395, 411, *428*
Barthelemy, C., see Jacob, J. 313, *358*
Bartholini, G., see Pletscher, A. 3, *29*, 53, *227*
Bartholini, G., see Prada, M. Da 54, 55, *228*
Bartholomew, A. A. 250, *267*
Bartolini, A., Pepeu, G. 57, *184*
Bartolini, A., see Deffenu, G. 57, *196*
Bartolini, A., see Domino, E. F. 333, 335, 341, *354*
Bartolini, A., see Nistri, A. 57, *224*
Bartolini, A., see Pepeu, G. C. 57, *226*
Bartter, F. C., see Gill, J. R. 8, 19, *28*
Baselt, R. C., see Gravey, R. H. 264, *268*
Bass, A. D., see Dingell, J. V. 15, *27*
Basso, P. del, Rusca, G., Carpi, A. 41, *184*
Basso, P. del, see Ziegler, H. 77, 87, *246*
Battegay, R., see Ladewig, D. 260, *271*, 285, *300*
Bauer, I., Pickenhain, L. 107, *184*
Bauer, R. H., Duncan, N. C. 102, 103, 129, *185*
Bauer, W. S., Perley, J. E. 167, *185*
Baum, E., North-Diehl, A., Piarroux, M.-C., Thenint, F. Boissier, J.-R. 79, *185*
Baum, T., see Gluckman, M. I. 123, 134, *204*
Baumann, I. R., see Borbély, A. A. 149, 154–156, 158, *188*
Baumann, P., see Ladisch, W. 43, 87, *213*
Baumann, P. A., see Maitre, L. 394, *423*
Baumann, R., see Borbély, A. A. 80, 155, *188*
Bazell, R. 250, *267*
Beaconsfield, F., Ginsburg, J., Rainsbury, R. 397, *415*
Beaulieu, D. M., see Klawans, H. L. 71, 77, *212*
Beaulieu, R. F., see Koella, W. P. 317, *359*
Beals, R. K., see Hecht, F. 408, *420*
Beamish, P., Kiloh, L. G. 282, 291, *297*
Beani, L., Bianchi, C., Santinoceto, L., Marchetti, P. 57, *185*
Beaton, J. M., Bradley, R. J. 128, *185*
Beaton, J. M., LeBlanc, A. E., Webster, C. D. 101, *185*
Beaton, J. M., see Webster, C. D. 410, *429*
Beaubrun, M. H., Knigth, F. 409, *415*
Beauvallet, M. 34, 35, 37, 41, *185*
Beauvallet, M., Fugazza, J., Godefroy, F., Solier, M. 34, 35, 37, 87, 154, 173, *185*
Beauvallet, M., Fugazza, J., Legrand, M. 35, 41, 42, 53, *185*
Beauvallet, M., Fugazza, J., Solier, M. 34, 35, 174, *185*
Beauvallet, M., Halpern, B. N., Fugazza, J., Drudi-Baracco, C. 35, 174, *185*
Beauvallet, M., Legrand, M., Bernard, J., Solier, M. 53, *185*
Beauvallet, M., Legrand, M., Solier, M. 53, *185*
Beauvallet, M., Solier, M. 35, *185*
Beccari, E., see Chistoni, A. 70, *192*
Becker, D. I., Appel, J. B., Freedman, D. X. 315, *351*
Becker, D. P., see Goodman, S. J. 265, *269*
Beckett, A. H., Bélanger, P. M. 11, *25*
Beckett, A. H., Boyes, R. N. Tucker, Q. T. 21, *25*
Beckett, A. H., Brookes, L. G. 3, 12, 14, 21, *25*
Beckett, A. H., Coutts, R. T., Ogunbona, F. A. 11, *25*
Beckett, A. H., Rowland, M. 7, 13, 21, *25*
Beckett, A. H., Rowland, M., Turner, P. 15, 21, *25*
Beckett, A. H., Salmon, J. A., Mitchard, M. 21, 23, *25*
Beckett, A. H., see Wilkinson, G. R. 21, *31*
Beckner, J. S., see Dewey, W. L. 382, 412, *418*
Beckner, J. S., see Martin, B. R. 381–384, 410–412, *424*
Beckwith, B. E., Sandman, C. A., Alexander, W. D. 129, 130, *185*
Beecher, H. K., see Lasagna, L. 248, *271*
Beecher, H. K., see Smith, G. M. 249, *274*
Bein, H. J., see Maitre, L. 393, 394, *423*
Bejerot, N. 282, 291, *297*
Bejrablaya, D., Burn, J. H., Walter, J. M. 142, 143, *185*
Bélanger, P. M., see Beckett, A. H. 11, *25*
Belenky, M. L., Vitolina, M. 149, 154–156, *185*
Belgrade, J., see Schwartzbaum, J. S. 176, *235*
Bell, D. S. 256, 257, *267*, 282, 291–293, 295, 296, *297*
Bell, D. S., Kirkby, R. J., Preson, A. C. 91, 92, *186*

Bell, D. S., Trethowan, W. H. 260, *267*, 282, 291, 294, *297*
Bell, E. C., see Hill, H. E. 116, *208*
Bell, F. R., see Adey, W. R. 315, 316, 341, *350*
Bell, J. A., Martin, W. R. 331, 334, *351*
Bell, J. A., Martin, W. R., Sloan, J. W., Buchwald, W. F. 321, 332, *351*
Bell, J. A., see Martin, W. R. 281, *301*, 348, *361*
Bell, R. D., see Cicero, T. J. 401, *417*
Belleville, R. E. 130, *186*, 308, 311, *351*
Belleville, R. E., see Isbell, H. 308, 310, 311, 318, *358*
Bellward, G., see Drummond, G. I. 58, *197*
Belozertsev, Y. A. 70, 97, *186*
Benakis, A., Thomasset, M. 16, 17, *25*
Bender, L., Cottington, F. 286, *297*
Benesova, O., see Tikal, K. 96, *240*
Benfey, B. G., Varma, D. R. 142, *186*
Benington, F., Morin, R. D. 40, *186*
Benington, F., Morin, R. D., Clark, L. C. 11, *26*
Benington, F., see Smythies, J. R. 15, 18, 19, *30*, 306, 323, 348, *365*
Benn, H. J., see Tripod, J. 70, 87, *241*
Bennett, C. C. 252, *267*
Bennett, I. P., Snyder, S. H. 331, *351*
Benowitz, N., see Jones, R. T. 400, 401, 408, 413, *421*
Benson, W. M., see Uyeno, E. T. 313, *367*
Bente, D., Itil, T. 251, *267*
Ben-Zvi, Z., see Mechoulam, R. 375, *425*
Bercht, C. A. L., Lousberg, R.J.J. Ch., Kuppers, F. J. E. M., Salemink, C. A. 375, *415*
Bercht, C. A. L., Lousberg, R. J. J. Ch., Kuppers, F. J. E. M., Salemink, C. A., Vree, T. B., Rossum, J. M. van 375, *416*
Berde, B., Cerletti, A. 333, *351*
Beregi, L. G., Hugon, P., LeDouarec, J. C., Laubie, M., Duhault, J. 4, *26*
Beretta, C., Locatelli, A. 61, *186*
Bergen, J. R. 315, *351*
Bergen, J. R., Krus, D., Pincus, G. 315, *351*
Bergen, J. R., Perkins, L., Hayano, M. 315, *351*
Bergen, J., see Krus, D. M. 312, *360*
Bergen, J. R., see Koella, W. P. 317, *359*
Berger, B. D. 127, 135, *186*
Berger, B. D., Wise, C. D., Stein, L. 77, 87, 127, 136, 137, 138, *186*
Berger, B. D., see Ebstein, R. P. 44, *198*
Berger, H. J., Brown, C. C., Krantz, J. C. 77, 127, 136, *186*
Berger, H. J., see Welch, B. L. 393, 394, *429*
Bergeron, R. T., see Rumbaugh, C. L. 143, *231*, 252, 265, *273*
Bergner, B. A., see Evans, W. O. 288, *298*
Beringer, K. 306, 310, *351*
Bernard, J., see Beauvallet, M. 53, *185*
Bernardi, D., see Jori, A. 43, 46, 52, 138, 157, *211*
Bernier, A., Sicot, N., LeDouarec, J. C. 136, *186*
Bernstein, B. M., Latimer, C. N. 99, 106, *186*
Bernstein, T. M. 290, *297*
Berry, M. J., Poyser, R. H., Robertson, M. I. 146, *186*
Berseus, S. 254, *267*
Bert, J., see Gastaut, H. 251, *269*
Bertino, J. R., Klee, G. D. Collier, D., Weintraub, W. 312, *351*
Bertino, J. R., Klee, G. D., Weintraub, W. 312, *351*
Bertino, J., see Klee, G. D. 308, 312, *359*
Bertino, J. R., see Cervoni, P. 335, *352*
Besser, G. M., Butler, P. W. P., Landon, J., Rees, L. 255, *267*
Bessirard, D., see Drudi-Baracco, C. 173, *197*
Bessirard, D., see Halpern, B. N. 36, 173, *206*
Besson, M. J., Cheramy, A., Feltz, P., Glowinski, J. 44, 47, *186*
Besson, M. J., Cheramy, A., Gauchy, C., Musacchio, J. 47, *186*
Besson, M. J., Cheramy, A., Glowinski, J. 44, 47, *186*
Besson, M. J., see Gauchy, C. 44, *202*
Besson, M. J., see Philippu, A. 44, *226*
Bett, W. R. 280, 283, *297*
Beuthin, F. C., Miya, T. S., Blake, D. E., Bousquet, W. F. 88, 154, *186*
Bevan, P., Bradshaw, C. M., Roberts, M. H. T., Szabadi, E. 338, *351*
Bewsher, P. D., Hillmann, C. C. Ashmore, J. 162, 165, 175, *186*
Beyer, J., see Philippu, A. 39, *226*
Beyer, K. H. 254, *267*
Bhagat, B. 62, 63, 144, *186*
Bhagat, B., Shideman, F. E. 146, *186*
Bhagat, B., Wheeler, N. 179, *186*
Bialer, I., see Winsberg, B. G. 286, *303*
Bianchi, C., Marazzi-Uberti, E. 102, *186*
Bianchi, C., see Beani, L. 57, *185*
Bicher, H. I., Mechoulam, R. 389, *416*
Bickford, R. G., see Schwartz, B. E. 312, 341, 342, *364*
Bidder, T. G., see Carlson, N. J. 128, 139, *191*
Biel, J. H. 3, *26*
Bienek, D., see Vollner, L. 375, *429*
Bier, M. M., see Hughes, J. E. 403, *421*
Bierwagen, M. E., see Buyniski, J. P. 62, 70, 141, 142, *190*
Bignami, G. 98, *187*, 314, *352*
Bignami, G., Gatti, G. L. 112, *187*
Binder, M., see Agurell, S. 382, *415*
Bindra, D., see Krebs, H. 65, 137, *213*
Bioulac, B., see Gauchy, C. 44, *202*
Birmingham, M. K. 398, *416*
Birren, J. E., see Kay, H. 179, *211*
Biscardi, A. M., Carpi, A., Orsingher, O. A. 41, *187*
Bisgard, E., see Will, A. 141–143, *244*
Bishop, M. P., see Heath, R. G. 292, *299*
Bishop, P. O., Burke, W., Hayhow, W. R. 336, *352*

Bishop, P. O., Field, G., Hennessy, B. L., Smith, J. R. 336, *352*
Bivens, L. W., see Ray, O. S. 106, 110–112, 114, 118, *229*
Bivens, L. W., Ray, O. S. 110–112, 114, *187*
Bizzi, A., Bonaccorsi, A., Jespersen, S., Jori, A., Garattini, S. 43, 46, 131, 141, 143, 147, 149, 154, 155, 156, 163, 164, *187*, 251, *268*
Bizzi, A., Garattini, S., Veneroni, E. 163, *187*
Bizzi, A., see Marsh, J. B. 164, *219*
Black, M. D., Woods, J. H., Domino, E. F. 411, *416*
Blackham, A., Spencer, P. S. J. 87, 154, 173, *187*
Blackman, H. J., see Clark, W. C. 168, 169, 173, 174, 175, *193*
Blake, D. E., see Beuthin, F. C. 88, 154, *186*
Blanton, F. L., see Weiss, B. 167, 169, 173, *243*
Blaschko, H. 45, *187*
Blaschko, H., Richter, D., Schlossman, H. 10, *26*, 45, *187*
Blaschko, H., Strömblad, B. C. R. 45, 46, *187*
Blaschko, H., see Pratesi, P. 10, *29*, 45, 46, *228*
Blessing, J. A., see Stone, C. A. 146, *238*
Bliss, E. L., see Clark, L. D. 318, *353*
Bloom, A. S., Dewey, W. L., Harris, L. S., Brosius, K. 390, *416*
Bloom, A. S., Johnson, K. M., Bowman, F. J., Dewey, W. L. 392–394, *416*
Bloom, A. S., Johnson, K. M., Dewey, W. L., Harris, L. S. 393, 394, *416*
Bloom, A. D., see Gilmour, D. G. 402, *419*
Bloom, F. E., Costa, E., Salmoiraghi, G. C. 329, *352*
Bloom, F. E., see Aghajanian, G. K. 330, *350*
Bloomberg, W. 258, *268*, 282, *297*
Bloomberg, W., see Prinzmetal, M. 250, 258, 273, 280, *302*
Blum, J. S., see Blum, R. A. 90, 166, *187*
Blum, R. A., Blum, J. S., Chow, K. L. 90, 166, *187*
Blundell, J. E. 140, *187*
Blundell, J. E., Leshem, M. B. 136, 137, *187*
Boakes, R. J., Bradley, P. B., Briggs, I., Dary, A. 329, *352*
Boakes, R. J., Bradley, P. B., Candy, J. M. 65, *187*
Boardman, W. K., Goldstone, S., Lhamon, W. T. 308, *352*
Bobin, P., see Abely, P. 293, *296*
Bocknik, S. E., Kulkarni, A. S. 142, *187*
Bocknik, S. E., see Reitz, R. H. 39, 106, *229*
Boegli, G., see Christensen, H. D. 383, *417*
Bösser, T. H., Joyce, D., Summerfield, A. 129, *187*
Böszörményi, Z. 311, *352*
Böszörményi, Z., Dér, P., Nagy, T. 311, 343, *352*
Boff, E., see Heise, G. A. 99, *207*
Boff, E., see Scheckel, C. L. 103, 106, *233*
Bogaert, B., see Plane, P. 128, *227*
Bogdanski, D. F., Pletscher, A., Brodie, B. B., Udenfriend, S. 328, *352*
Boggan, W. O., see Neill, D. B. 79, 94, *223*
Bohdanecky, S., Jarvik, M. E. 98, 102, 130, *187*
Bohner, B., see McLean, J. R. 87, *220*
Boissier, J.-R., Etevenon, P., Piarroux, M.-C., Simon, P. 79, *187*
Boissier, J.-R., Hirtz, J., Dumont, C., Gérardin, A. 13, *26*
Boissier, J.-R., Simon, P., Guernet, M., Tillement, J.-P. 56, *187*
Boissier, J.-R., Simon, P., Lwoff, J. M. 173, *187*
Boissier, J.-R., see Baum, E. 79, *185*
Boissier, J. R., see Etevenon, P. R. 340, *354*
Boissier, J.-R., see Simon, P. 34, 53, 77, 87, *236*
Bolt, A. G., Mulligan, B. M., Graham, G. 164, *187*
Bonaccorsi, A. 146, *187*
Bonaccorsi, A., see Babulova, A. 59, 60, 143, 144, 147, *183*
Bonaccorsi, A., see Bizzi, A. 43, 46, 131, 141, 143, 147, 149, 154–156, 163, 164, *187*, 251, *268*
Bonaccorsi, A., see Jespersen, S. 143, 147, 154, 173, *210*
Bonhoff, G., Lewrenz, H. 33, *187*, 247, 251, 254, 259, 260, 263, *268*, 281, 288, *297*
Bonnie, R. J., Whitebread, C. H. II 371, *416*
Bonvallet, M., see Hiebel, G. 176, *207*
Booth, D. A. 137, 139, *187*
Booth, D. A., see Panksepp, J. 136, 138, 139, *226*
Bopp, B. A., see Dren, A. T. 415, *418*
Borbély, A. A., Baumann, I. R., Waser, P. G. 149, 154–156, 158, *188*
Borbély, A. A., Huston, J. P., Baumann, R. 80, 155, *188*
Borbély, A. A., see Huston, J. P. 80, 90, 155, *209*
Borberg, S. 80, *188*
Borella, L. E. 16, *26*
Borella, L. E., Herr, F. 87, 88, *188*
Borella, L. E., Herr, F., Wojdan, A. 16, *26*, 87, 88, 154, *188*
Borella, L. E., Paquette, R., Herr, F. 87, 134, *188*
Borgen, L. A., Davis, W. M., Pace, H. B. 406, *416*
Borgen, L. A., see Pace, H. B. 406, 407, *426*
Borgstedt, H. H., see Richards, K. C. 311, *363*
Borodkin, Y. S., see Anichkov, S. V. *182*
Boulougouris, J., see Stefanis, C. 409, *428*
Boulton, A. A., see Philips, S. R. 332, *362*
Boulo, R., Rapin, J. R., Lebas, M., Jacquot, C. 37, 68, *188*
Bourke, D. I., see Malit, L. A. 397, *424*
Bourne, G. H., see Nandy, K. 333, *362*
Bouquet, J. 373, 403, *416*
Bousquet, W. F., see Beuthin, F. C. 88, 154, *186*
Bouvet, D., see Javoy, F. 48, *210*
Bovet, D., Gatti, G. L. 102, *188*
Bovet, D., Oliverio, A. 101, *188*
Bovet, D., see Keleman, K. 100, 102, *211*
Bovet, D., see Sansone, M. 101, *232*
Bowden, D. M., see German, D. C. 121, 123, *203*
Bowen, M., see Bradley, C. 286, *297*

Bowers, M.B.Jr. 312, *352*
Bowling, C., see Pradhan, S.N. 121, *228*
Bowling, E., see Fish, C.H. 288, *299*
Bowmann, F.J., see Bloom, A.S. 392–394, *416*
Bowman, K., see Green, K. 414, 415, *419*
Boxhall, E., see Southgate, P.J. 77, 98, *236*
Boyajy, L.D., see White, R.P. 177, *244*
Boyd, E.S., see Dagirmanjian, R. 399, 400, *417*
Boyd, E.S., see Slater, I.H. 334, *365*
Boyes, R.N., see Beckett, A.H. 21, *25*
Braceland, F.J., see Rome, H.P. 292, *302*
Bradbard, G., see Rapoport, J.L. 286, *302*
Bradham, G.B. 285, *297*
Bradley, C. 280, 286, *297*
Bradley, C., Bowen, M. 286, *297*
Bradley, C., Key, B.J. 252, *268*
Bradley, D.W.H., Joyce, D., Murphy, E.H., Nash, B.M., Porsolt, R.D., Summerfield, A., Twyman, W.A. 94, *188*
Bradley, P.B. 65, *188*
Bradley, P.B., Elkes, J. 176, *188*, 341, 347, *352*
Bradley, P.B., Flugel, F., Hoch, P.H. 306, *352*
Bradley, P.B., Hance, A.J. 341, *352*
Bradley, P.B., Key, B.J. 176, *188*, 341, *352*
Bradley, P.B., Wolstencroft, J.H. 65, *188*
Bradley, P.B., see Boakes, R.J. 65, *187*, 329, *352*
Bradley, P.B., see Elkes, J. 176, *198*, 343, *354*
Bradley, P.B., see Phillips, M.I. 77, 87, 93, 94, *227*
Bradley, P.B., see Vane, J.R. 313, 331, 332, 335, 342, *367*
Bradley, R.J., see Beaton, J.M. 128, *185*
Bradley, R.J., see Smythies, J.R. 15, 18, 19, *30*, 101, 128, *236*, 323, 328, *365*
Bradshaw, C.M., Roberts, M.H.T., Szabadi, E. 338, *352*
Bradshaw, C.M., see Bevan, P. 338, *351*
Bradt, C., see Nichols, W.W. 402, *425*
Brady, J.V. 115, *188*
Brady, J.V., see Gollub, L.R. 120, *204*
Brady, J.V., see Hodos, W. 100, 114, *208*
Brady, R.O., see Axelrod, J. 333, 334, *351*
Branch, M.N. 115, *188*
Branch, M.N., Gollup, L.R. 112, 113, 116, *188*
Brand, U., see Menge, H.G. 77, *220*
Brandon, S., Smith, D. 289, *297*
Brandon, S., see Kiloh, L.G. 282, 289, 291, *300*
Braude, M.C., Szara, S. 372, 384, *416*
Braude, M.C., see Fleischman, R.W. 407, 408, *419*
Braude, M.C., see Grilly, D.M. 407, *420*
Braude, M.C., see Haley, S.L. 406, 407, *420*
Braude, M.C., see Keplinger, M.L. 406, *422*
Braude, M.C., see Luthra, X.K. 405, *423*
Braude, M.C., see Rosenkrantz, H. 401, 404, 405, *427*
Braude, M.C., see Thompson, G.R. 403, 404, 405, 410, *428*
Braun, M., see Feinberg, I. 252, 260, *269*, 286, *298*
Braunwald, E., see Chidsey, C.A. 8, *26*, 41, 143, *192*
Braunwald, E., see Harrison, D.C. 146, *206*
Brawley, P., Duffield, I.C. 306, *352*
Brazier, M.A.B. 251, *268*
Brecher, E.M. 372, *416*
Breckenridge, B.M., Norman, J.H. 58, *188*
Breda, J.B., Carlini, E.A., Sader, N.F.A. 128, *188*
Breese, G.R., Cooper, B.R., Smith, R.D. 79, 87, 88, *188*
Breese, G.R., Kopin, I.J., Weise, V.K. 35, 37, 52, *188*
Breese, G.R., see Chase, T.N. 330, *352*
Breese, G.R., see Cooper, B.R. 122, 123, *194*
Breese, G.R., see Hollister, A.S. 79, 87, 88, 89, *208*
Breimer, D.D., see Dezeeuw, R.A. 375, *418*
Breimer, D.D., see Ginneken, C.A.M. Van 375, *429*
Breimer, D.D., see Vree, T.B. *429*
Breitner, C. 291, *297*
Breteau, M., see Simon, P. 34, 53, 87, *236*
Brewster, J.M., see Siegel, R.K. 317, *365*
Brezinova, V., see Oswald, I. 257, *272*
Bridger, W.H., see Stoff, D.M. 314, *365*
Bridges, S.R., see Martin, W.R. 316, 332, *361*
Bridges, S.R., see Sloan, J.W. 332, *365*
Bridges, J.W., Gorrod, J.W., Parke, D.V. 13, *26*
Briggs, I., see Boakes, R.J. 329, *352*
Briggs, M., see Oswald, I. 257, *272*
Briggs, M., see Steel, J.M. 257, *274*
Bright, L.D., see Martin, W.R. 281, *301*
Brill, N.O., Crumpton, E., Frank, I.M., Hochman, J.S., Lontax, P., McGlothin, W.H., West, L.J. 403, *416*
Brimblecombe, R.W. 313, 338, *352*
Brine, D.R., see Christensen, H.D. 383, *417*
Brittain, R.T. 158, *188*
Brittain, R.T., Jack, D., Spencer, P.S.J. 154, 173, *189*
Brittain, R.T., see Kruk, Z.L. 149, 158, *213*
Broadie, L., see Wenzel, D. 90, *244*
Brobeck, J.R., Larsson, S., Reyes, E. 137, *189*
Brodie, B.B., Cho, A.K., Gessa, G.L. 9, 15, 19, *26*, 34, 35, 50, 51, 52, 154, 163, 164, *189*
Brodie, B.B., Cho, A.K., Stefano, F.J.E., Gessa, G.L. 34, 35, 51, 63, 139, 149, 150, 154, 156, 157, 163, 164, *189*
Brodie, B.B., Costa, E., Groppetti, A., Matsumoto, C. 34, 50, *189*
Brodie, B.B., Gillette, J.R., LaDu, B.N. 9–11, *26*
Brodie, B.B., Shore, P.A. 70, *189*
Brodie, B.B., see Bogdanski, D.F. 328, *352*
Brodie, B.B., see Chang, C.C. 147, 148, *192*
Brodie, B.B., see Cooper, J.R. 17, *27*
Brodie, B.B., see Gessa, G.L. 50, 51, 149, 154, 156, 157, 163, 164, *203*
Broekkamp, C.L.E., Honig, W.M.M., Pauli, A.I., Rossum, J.M.van 137, *189*

Broemann, I., see Rating, D. 410, *426*
Broermann, I., see Ziem, M. 139, 155, 157, *246*
Broermann, J., see Rating, D. 17, *30*
Bronzino, J. D., see Stern, W. C. 342, *365*
Brookes, L. G., see Beckett, A. H. 3, 12, 14, 21, *25*
Brooks, E., see Rapoport, J. L. 286, *302*
Brosius, K., see Bloom, A. S. 390, *416*
Brown, A. M., Julian, T. 155, *189*
Brown, B. B. 82, 87, *189*
Brown, C. C., see Berger, H. J. 77, 127, 136, *186*
Brown, C. T. 293, *297*
Brown, C. W., see Searle, L. V. 70, *235*
Brown, D. J., Miller, F. N., Longnecker, D. E., Greenwald, E. K., Harris, P. D., Forney, R. B. 399, *416*
Brown, D. J., see Lewis, S. C. 398, *423*
Brown, D. J., see Phillips, R. N. 404, *426*
Brown, H. 110, 111, 119, *189*, 397, *416*
Brown, H., Richards, R. K. 115, *189*
Brown, T. S., see Adey, W. R. 341, *350*
Brownlee, G., Williams, G. W. 173, *189*
Bruce, R. B., Maynard, W. R., Jr. 21, *26*
Bruderer, H., see Pletscher, A. 3, *29*, 53, 56, *227*
Brücke, F., Gogolák, G., Stumpf, Ch. 340, *352*
Brunecker, G., see Sai-Halász, A. 311, *363*
Brunfelter, B., see Ross, S. D. 49, 50, *231*
Bruni, E., see Russek, M. 138, 139, *232*
Brunk, S. F., see Noyes, R., Jr. 413, *425*
Brunner, H., see Maitre, L. von 51, 147, *218*
Brus, I., see Lewander, T. 140, 150, 157, 158, *216*
Bryant, S. A., see Martin, P. A. 402, *424*
Buchsbaum, M., Wender, P. 286, *297*
Buchwald, W. F., see Bell, J. A. 321, 332, *351*
Buchwald, W. F., see Martin, W. R. 316, 332, *361*
Buchwald, W. F., see Sloan, J. W. 332, *365*
Buck, C. J., see Holland, G. F. 3, *28*
Buckley, J. P., see Barry, H. 98, *184*
Buckley, J. P., see Cavero, I. 397, 398, 399, *417*
Buening, M. K., Gibb, J. W. 48, *189*
Bueno, O. F. A., Carlini, E. A. 409, *416*
Buffoni, F., Della Corte, L. 45, *189*
Bunce, M. E., see Wolf, H. H. 154, 155, 173, 174, *244*
Bunnell, S., see Shulgin, A. T. 311, 328, *364*
Bunney, B. S., Aghajanian, G. K. 66, *189*
Bunney, B. S., Aghajanian, G. K., Roth, R. H. 66, *189*
Bunney, B. S., Walters, J. R., Roth, R. H., Aghajanian, G. K. 46, 66, *189*
Bunney, B. S., see Korf, J. 66, *212*
Bunney, S., see Aghajanian, G. K. 66, *181*
Burch, N. R., see Greiner, T. 310, *356*
Burgen, A. S. V., Iversen, L. L. 38, *189*
Burger, A. 306, *352*
Burke, J. C., see Horovitz, Z. P. 97, *209*
Burke, J. C., see Rubin, B. 338, *363*
Burke, W., see Bishop, P. O. 336, *352*
Burkhardt, W. L., see Adler, H. F. 249, *266*
Burkhard, W. P., see Pletscher, A. 3, *29*, 53, 56, *227*
Burks, T. F., see Palmer, G. C. 333, *362*
Burks, T. F., see Terwelp, D. R. 149, 150, 154, 158, 159, *240*
Burn, J. H., Hobbs, R. 173, *189*
Burn, J. P., Rand, M. J. 33, 59, 143, 144, 146, *189*
Burn, J. H., see Bejrablaya, D. 142, 143, *185*
Burn, J. H., see Fleckenstein, A. 58, *201*
Burnett, L. S., see Stolk, J. M. 162, 175, *238*
Burril, D. Y., see Goetzl, F. R. 251, *269*
Burrill, D. Y., see Ivy, A. C. 288, *300*
Burton, R. M., Sodd, M. A., Goldin, A. 179, *189*
Buser, P., see Viala, D. 69, *242*
Bussato, P. A., see Moraes, S. de 60, *222*
Butcher, L. L., Andén, N.-E. 118, *189*
Butcher, L. L., Butcher, S. G. Larsson, K. 98, *189*
Butcher, L. L., see Schlechter, J. M. 87, *234*
Butcher, S. G., see Butcher, L. L. 98. *189*
Butler, J. R., see Regelson, W. 413, 414, *427*
Butler, P. W. P., see Besser, G. M. 255, *267*
Butters, N. 314, *352*
Buus Lassen, J. 87, *190*
Buxbaum, D. M. 390, *416*
Buyniski, J. P., Smith, M. L., Bierwagen, M. E. 62, 70, 141, 142, *190*
Byers, L. W., see Heath, R. G. 292, *299*
Byrd, L. D. 110, 112, 116, 140, *190*
Byrne-Quinn, E., Grover, R. F. 143, *190*

Caccia, S., Ceccetti, G., Garattini, S., Jori, A. 36, 43, 156, 157, *190*
Caccia, S., see Jori, A. 8, 9, 19, *28*
Cadilhac, J., see Passouant, P. 341, *362*
Cahn, J., Herold, M. 93, 154, 177, *190*
Caldwell, D. F., see Domino, E. F. 102, *197*
Caldwell, D. R., see Levis, D. J. 314, *360*
Caldwell, J., Dring, L. G., Williams, R. T. 9, 12, *26*
Caldwell, J., Sever, P. S. 3, *26*, 154, 157, *190*
Caldwell, J., Sever, P. S., Trelinski, M. 154, *190*
Caldwell, J., see Sever, P. S. 8, 9, 18, *30*, 138, 154, 157, *235*
Caldwell, J., see Williams, R. T. 5, 14, *31*
Calhoun, D. W., see Stone, G. C. 109, *238*
Calil, H. M., see Silva, M. T. A. 313, 314, *365*
Callandra, J. C., see Haley, S. L. 406, 407, *420*
Calandra, J. C., see Keplinger, M. L. 406, *422*
Callaway, E., III., see Klee, G. D. 308, 312, *359*
Callaway, E., see Spilker, B. 252, *274*
Calverley, J. R., see Mattson, R. 263, *272*

Cameron, J. S., Sprecht, P. G., Wendt, G. R. 247, *268*
Cameron, O. G., Appel, J. B. 315, *352*
Cameron, W. M., see Pedden, J. R. 148, *226*
Campbell, B. A., Fibiger, H. C. 91, *190*
Campbell, B. A., see Fibiger, H. C. 91, *200*
Campbell, B. A., see Mabry, P. D. 87, *217*
Campbell, B. A., see Lynch, G. S. 90, *217*
Campbell, B. A., see Lytle, L. D. 138, *217*
Campbell, D. B. 21, *26*
Campbell, D., see Patil, P. N. 60, *226*
Campbell, E. H., see Hanson, H. M. 111, 115, 116, *206*
Campbell, J. C., Seiden, L. S. 114, 120, *190*
Campbell, J. F., see Krebs, H. 64, 137, *213*
Campbell, J. T., see Owens, J. E. 110, 117, *225*
Campbell, M., see Fish, B. 310, *355*
Campo, R. A. 411, *416*
Canafax, D. M., see Craigmill, A. L. 87, 154, 173, *195*
Candy, J. M., see Boakes, R. J. 65, *187*
Canfield, D. R., Antelman, S. M., Lippa, A. S., Fischer, A. E. 123, *190*
Cannon, C., see Cowan, F. F. 60, 62, *194*
Cannon, E., see Gillin, J. C. 320, *356*
Canter, A., see Noyes, R., Jr. 413, *425*
Cantwell, D. P., see Satterfield, J. H. 286, *303*
Cappell, H., Ginsberg, R., Webster, C. D. 115, 116, *190*
Cappell, H., LeBlanc, A. E. 127, 135, *190*
Cappell, H., Pliner, P. 385, *416*
Cappell, H., see LeBlanc, A. E. 127, *215*
Carakushansky, G., Neu, R. L., Gardner, L. I. 408, *416*
Carchman, R. A., Harris, L. S., Munson, A. E. 402, *416*
Carchman, R. A., see End, D. 382, *418*
Carchman, R. A., see Harris, L. S. 402, *420*
Carchman, R. A., see Munson, A. E. 402, *425*
Carder, B., Olsen, J. 387, *416*
Cardo, B. 100, 101, *190*
Carenzi, A., see Costa, E. 58, *194*
Carey, J. T., Mandel, J. 282, 288, *297*
Carey, R. J. 120, 127, 130, *190*
Carey, R. J., Goodall, E. B. 112, 127, *190*
Carey, R. J., Goodall, E. B., Procopio, G. F. 110, 117, 122, *190*
Carey, R. J., Kritkausky, R. P. 114, 115, *190*
Carey, R. J., Salim, A. P. 90, 136, *190*
Carey, R. J., see Friedman, H. 314, *355*
Carl, G. P., see Turner, W. D. 247, *274*
Carlini, E. A. 388, 389, 409, *416*
Carlini, E. A., Gonzalez, C. 96, *190*
Carlini, E. A., Hamaoui, A., Martz, R. M. 387, *416*
Carlini, E. A., Masur, J. 387, *417*
Carlini, E. A., see Alves, C. N. 387, *415*
Carlini, E. A., see Breda, J. B. 128, *188*
Carlini, E. A., see Bueno, O. F. A. 409, *416*
Carlini, E. A., see Gonzalez, S. C. 389, *419*
Carlini, E. A., see Karniol, I. G. 317, *421*
Carlini, E. A., see Lima, I. M. 165, *216*
Carlini, E. A., see Masur, J. 388, 410, *424*
Carlini, E. A., see Santos, M. 387, *427*
Carlini, E. A., see Silva, M. T. A. 320, *365*, 409, *427*
Carlisle, H. J. 136, *190*
Carlisle, H. J., Reynolds, R. W. 136, *190*
Carlisle, H. J., see Thode, W. F. 137, *240*
Carlson, N. J., Doyle, G. A., Bidder, T. G. 128, 139, *191*
Carlsson, A. 34, 42, *191*
Carlsson, A., Corrodi, H., Fuxe, K., Hökfelt, T. 38, 54, *191*
Carlsson, A., Fuxe, K., Hamberger, B., Lindqvist, M. 38, 39, 42, *191*
Carlsson, A., Fuxe, K., Hökfelt, T. *191*
Carlsson, A., Lindqvist, M., Dahlström, A., Fuxe, K., Masuoka, D. 34, 36, 38, 39, 42, 53, *191*
Carlsson, A., Lindvist, M., Fuxe, K., Hamberger, B. 38, 39, 42, *191*
Carlsson, A., Waldeck, B. 38, 40, *191*
Carlsson, A., see Andén, N.-E. 34, 42, *182*
Carlton, P. L. 15, *26*, 99, 106, 114, *191*
Carlton, P. L., Didamo, P. 99, 106, *191*
Carlton, P. L., Wolgin, D. L. 120, 138, *191*
Carlton, P. I., see Horovitz, Z. P. 122, *208*
Carmo, M. M., see Moraes, S. de 60, *222*
Carpenter, R. S., see Lindsley, D. F. 316, 342, *360*
Carpi, A., see Basso, P. del 41, *184*
Carpi, A., see Biscardi, A. M. 41, *187*
Carr, C. B., see Griffith, J. D. 285, *299*
Carr, L. A., Moore, K. E. 43, *191*
Carr, L. A., see Moore, K. E. 87, *222*
Carter, M. F., see Shulgin, A. T. 323, *364*
Carvalho, V. F., see Moraes, S. de 60, 63, *222*
Carves, F. B., see Hitzmann, R. J. 35, 36, 37, 49, 80, 91, *208*
Case, Q., McAndrew, J. B. 263, *268*
Cassens, G., see Pick 121, *227*
Cassidy, J., see Aronow, W. S. 397, *415*
Castellano, C. 128, 129, *191*
Castellano, C., Sansone, E. M., Renzi, P., Annecker, L. 101, *192*
Castellano, C., see Sansone, M. 102, *232*
Castells, C., see Gastaut, H. 343, *355*
Catania, A. C., see Cook, L. 100, 110, 112, 116, *194*
Cattabeni, F., Racagni, G., Groppetti, A. 51, 52, *192*
Cattabeni, F., Revuelta, A., Costa, E. *56*, *192*
Cattabeni, F., see Morgan, C. D. 51–53, 56, *222*
Cattel, J. P., see Hoch, P. H. 336, *357*
Cavanaugh, J. H., Griffith, J. D., Oates, J. A. 9, 19, *26*
Cavanaugh, J., see Griffith, J. D. 256, 261, *269*, *270*, 282, 293, 295, *299*

Cavanaugh, J. H., see Rangno, R.E. 8, *30*
Cavero, I., Buckley, J. P., Jandhyala, B. S. 397, 398, 399, *417*
Cavero, I., Ertel, R., Buckley, J. P., Jandhyala, B. S. 398, *417*
Cavero, I., Jandhyala, B. S. 398, *417*
Cavero, I., Lokhandwala, M. F., Buckley, J. P., Jandhyala, B. S. 399, *417*
Cavness, C., see Feinberg, I. 252, 260, *269*, 286, *298*
Ceccetti, G., see Caccia, S. 36, 43, 156, 157, *190*
Cely, W., see Karler, R. 410, *421*
Cerletti, A. 377, *352*
Cerletti, A., see Berde, B. 333, *351*
Cerletti, A., see Neuhold, K. 313, 315, 337, *362*
Cerletti, A., see Rothlin, E. 307, 313, *363*
Cerletti, A., see Taeschler, M. 313, 314, *366*
Cerletti, A., see Weidmann, H. 334, 335, *367*
Cervoni, P., Bertino, J. R., Geiger, L. E. 335, *352*
Cervoni, P., see Geiger, L. E. 334, *355*
Cession-Fossion, A. 41, 62, 146, 148, *192*
Chait, L. D., see Adams, M. D. 399, *415*
Chambers, W. F., see Hanna, C. 176–178, *206*
Champlain, J. de, see Feltz, P. 66, *200*
Chance, M. R. A. 96, 166, 167, 168, *191*
Chance, M. R. A., Mansour, T. E. 61, *192*
Chang, C. C. *192*
Chang, C. C., Costa, E., Brodie, B. B. 147, 148, *192*
Chapel, J. L. 294, *297*
Chaplin, E., see Colville, K. I. 179, *193*
Chapman, K. W., see Vogel, V. H. 255, *274*
Chapman, L. F., Walter, R. D. 343, *352*
Chapman, L. F., see Richards, N. 310, *363*
Chappel, C. I., Rona, G., Balazs, T., Gaudry, R. 174, *192*
Chase, G. D., see Neff, N. 334, *362*
Chase, T. N., Breese, G. R., Kopin, I. J. 330, *352*
Chase, T. N., see Ng, K. Y. 44, *224*
Chaurand, J. P., see Vergnes, M. 97, *242*
Checchinato, D., see Will, B. 109, *244*
Cheimat, R., see Simon, P. 77, 87, *236*
Chen, K. K., Schmidt, C. F. 279, *297*
Cheney, C., see Schwartz, A. S. 337, *364*
Cheng, H. C., Long, J. P. 61, *192*
Chenoweth, M. B., see Schneider, C. W. 128, *234*
Cheramy, A., see Besson, M. J. 44, 47, *186*
Cheramy, A., see Gauchy, C. 44, *202*
Chéramy, A., see Jones, B. E. 57, *210*
Chernov, H., Furness, P., Partyka, D., Plummer, A. J. 167, *192*
Chesher, G. B., Dahl, C. J., Everingham, M., Jackson, D. M., Marchant-Williams, H., Starmer, G. A. 390, *417*
Chesher, G. B., see Hawke, D. M. 61, *207*
Chessick, R. D., see Lal, H. 35, 36, 54, 174, *214*
Chessick, R. D., see Wikler, A. 308, *367*
Chevillard, C., Alexandre, J.-M. 41, *192*
Chevillard, C., Duchène, N., Alexandre, J.-M. 41, *192*
Cheymol, M. J., Levassort, C. 154, *192*
Chidsey, C. A., Harrison, D. C., Braunwald, E. 8, *26*, 41, 143, *192*
Chidsey, C. A., see Harrison, D. C. 146, *206*
Chiel, H., Yehuda, S., Wurtman, R. J. 158, *192*
Chigneau, M., see Paris, M. 375, *426*
Chistoni, A., Beccari, E. 70, *192*
Chiueh, C. C., Moore, K. E. 43, 44, *192*
Chiueh, C. C., see Gudetsky, G. A. 44, *205*
Chmouliovsky, M., see Frommel, E. 173, *201*
Cho, A. K., Lindeke, B., Jenden, D. J. 11, *26*
Cho, A. K., Lindeke, B., Sum, C. Y. 11, *26*
Cho, A. K., see Brodie, B. B. 9, 15, 19, *26*, 34, 35, 50–52, 63, 139, 149, 150, 154, 156, 163, 164, *189*
Cho, A. K., see Clay, G. A. 50, 51, *193*
Cho, A. K., see Lindeke, B. 11, *29*
Cho, J. B., see Proctor, C. D. 173, *228*
Cholden, L. 306, *353*
Cholden, L. S., Kurland, A., Savage, C. 318, *353*
Chopra, J. C., Chopra, R. N. 413, *417*
Chopra, R. N., see Chopra, J. C. 413, *417*
Chorover, S., see Jarvik, M. E. 317, *359*
Chow, K. L., see Blum, R. A. 90, 166, *187*
Chow, M. I., see Horovitz, Z. P. 122, *208*
Christensen, H. D., Freudenthal, R. J., Gidley, J. T., Rosenfeld, R., Boegli, G., Testino, L., Brine, D. R., Pitt, C. B., Wall, M. E. 383, *417*
Christian, S. T., see Martin, W.R. 332, *361*
Christie, J. E., Crow, T. J. 68, 69, *193*
Chrusciel, M., see Chrusciel, T. L. 3, *26*
Chrusciel, T. L., Chrusciel, M. 3, *26*
Chrusciel, T. L., see Herman, Z. S. 35, 37, 53, 87, 91, 92, *207*
Chu, H., Opitz, K., Intemann, E. 161, *193*
Chu, H., see Opitz, K. 161, *225*
Cicala, G. A., Kremer, E. 102, 103, *193*
Cicero, T. J., Bell, R. D., Wiest, W. G., Allison, J. H., Polakoski, K., Robins, E. 401, *417*
Cimbura, G. 312, *353*
Cinnamon, A. D., see Mogenson, G. J. *221*
Ciofalo, V. B., see Goldberg, M. E. 106, *204*
Citron, B. P., Halpern, M., McCarron, M., Lundberg, G. D., McCormick, R., Pincus, I. J., Tatter, D., Haverback, B. J. 261, *268*
Claghorn, J. L., see Lu, T.-C. 134, 138, 139, *217*
Clark, F. C., Steele, B. J. 110, 112, 113, *193*
Clark, L. D., Bliss, E. L. 318, *353*

Clark, L. C., see Benington, F. 11, *26*
Clark, L. C., see Smythies, J. R. 15, 18, 19, *30*
Clark, L. C., Jr., see Smythies, J.R. 323, *365*
Clark, L. D., see Swinyard, E.A. 175, 177, *239*
Clark, S. C. 398, *417*
Clark, W.C., Blackman, H. J., Preston, J. E. 168, 169, 173, 174, 175, *193*
Clark, W.G., see Menon, M. K. 43, 46, 87, 173, *221*
Clausen, J., see Landis, C. 308, 310, *360*
Claussen, U., Korte, F. 373, *417*
Claussen, U., Spulak, F. von, Korte, F. 375, *417*
Claussen, U., see Isbell, H. 380, 385, *421*
Claussen, U., see Korte, F. 375, 377, *422*
Claussen, U., see Silva, M.T.A. 320, *365*
Claussen, U., see Spulak, R. von 375, *429*
Claussen, V., see Silva, M.T.A. 409, *427*
Clay, G.A., Cho, A. K., Roberfroid, M. 50, 51, *193*
Clay, G.A., see Gessa, G. L. 50, 51, 149, 154, 156, 157, 163, 164, *203*
Clayton, R. N., Grossman, L., Mayeda, T. K. 285, *297*
Cleek, D. M., see Himmelsbach, C. K. 283, 295, *300*
Clements, T. H., see Martin, W.R. 332, *361*
Clements, T. H., see Sloan, J.W. 332, *365*
Clewe, C. L., see Mantilla-Plata, B. 407, *424*
Clineschmidt, B.V., Lotti, V.J. 339, *353*
Clineschmidt, B. V., McGuffin, J. C., Werner, A.B. 132, 134, *193*
Clymer, N. V., Seifter, J. 70, *193*
Cockrill, B. L., see Platt, D.S. 17, *29*
Cohen, A. B., see Mann, P. E. 401, *424*
Cohen, G. M., see Peterson, D.W. 388, *426*
Cohen, L., see Kahn, E. 286, *300*
Cohen, M., Lal, H. 173, *193*
Cohen, N.J., Douglas, V. I., Morgenstern, G. 250, *268*
Cohen, R., see Abelson, H. 371, *415*
Cohen, R., see Loon, G. R. van 165, *217*
Cohen, S. 251, *268*, 306, *353*
Cohen, S., Lessin, P. J., Hahn, P. M., Tyrrell, E. D. 408, *417*
Cohen, S., Stillman, R. C. 413, *417*
Cohen, S., see Edwards, A. E. 308, *354*
Cohen, S., see Kolodny, R. C. 400, 408, *422*
Cohen, Y., see ElGuedri, H. 52, *198*
Cohen, Y., see Jacquot, C. 48, 51, *209*, *210*
Cohen, Y., see ElGuedri, H. 9, 19, *27*, 41, 49, 51, *229*
Cohn, M. M., see Quinn, G. P. 21, *30*
Cohn, R. A., see Pirch, J. H. 87, *227*, 410, 411, 413, *426*
Colburn, R. W., Ng, L. K. Y., Lemberger, L., Kopin, I. J. 382, *417*
Cole, J., Glees, P. 90, *193*
Cole, J. M., Pieper, W.A. 317, *353*
Cole, J. O. 289, *297*
Cole, J. O., Freedman, A. M., Friedhoff, A. J. 306, *353*
Cole, J. O., see Finnerty, R. J. 250, *269*
Cole, J. O., see Tecce, J. J. 248, *274*
Cole, S. O. 109, 111, 129, 136, 138, 139, 140, *193*, 250, *268*
Cole, S. O., Gay, P. E. 90, 97, 139, 140, *193*
Collier, D., see Bertino, J. R. 312, *351*
Collier, H. O. J. 344, 345, *353*
Collier, H. O. J., see Vane, J. R. 313, 331, 332, 335, 342, *367*
Colville, K. I., Chaplin, E. 179, *193*
Comitas, L., see Rubin, V. 409, *427*
Commarato, M. A., Lum, B. K. B. 142–144, 146, *193*
Connell, P. H. 18, *26*, 251, 262, 263, *268*, 282, 288, 291, 292, 294, *298*
Conner, T., see Legator, M. S. 402, *423*
Conners, C. K. 250, *268*
Conners, C. K., Eisenberg, L. 286, *298*
Conney, A. H. 17, *26*
Connor, J. D., see Pardridge, W. M. 20, *29*
Connors, C. K., see Epstein, L.C. 286, *298*
Consolo, S., Dolfini, F., Garattini, S., Valzelli, L. 15, 16, *26*
Consolo, S., Garattini, S., Ghielmetti, R., Valzelli, L. 17, *27*, 36, *193*
Consolo, S., Garattini, S., Valzelli, L. 97, 168, *193*
Consolo, S., Ladinsky, H., Garattini, S. 57, *193*
Consolo, S., Ladinsky, H., Peri, G., Garattini, S. 57, *194*
Consolo, S., see Valzelli, L. 16, *31*, 150, 154, *241*
Contrera, J. F., see Goldstein, M. 9, *28*
Cook, D. A., see Thiessen, P. N. 312, *366*
Cook, J. D., Schanberg, S. M. 35, 38, 43, 46, 71, 80, *194*
Cook, J. D., see Schanberg, S. M. 35, 38, 43, 80, 81, *233*
Cook, L., Catania, A. C. 100, 110, 112, 116, *194*
Cook, L., Weidley, E. 314, *353*
Cook, L., see Kelleher, R. T. 110–114, 116, *211*
Cook, L., see Tedeschi, R. E. 87, 154, *239*
Cools, A. R. 77, 78, *194*
Cools, A. R., Rossum, J. M. van 77, 78, *194*
Cooper, B. R., Cott, J. M., Breese, G. R. 122, 123, *194*
Cooper, B. R., see Breese, G. R. 79, 87, 88, *188*
Cooper, B. R., see Hollister, A. S. 79, 87–89, *208*
Cooper, J. R., Axelrod, J., Brodie, B. B. 17, *27*
Cooper, P. D., Walters, G. C. 348, *353*
Cooper, S. J., Joyce, D., Summerfield, A. 123, 130, *194*
Copelan, H. W., see Sidell, F. R. 385, 387, *427*
Coper, H., Lison, H., Rommelspacher, H., Schulze, G., Strauss, S. 149, 154, 159, *194*
Coper, H., see Magour, S. 91, 92, 138, 139, 164, 173, *218*
Coper, H., see Rating, D. 410, *426*
Coper, H., see Ziem, M. 139, 155, 157, *246*
Corne, S. J., Pickering, R.W. 312, *353*
Corne, S. J., see Vane, J. R. 313, 331, 332, 335, 342, *367*
Correa, D. L., see Todorov, J. C. 111, *240*

Corrodi, H., Fuxe, H., Hökfeldt, T. 34–36, 48, 66, *194*
Corrodi, H., Fuxe, K., Ljungdahl, A., Ögren, S.-O. 77, 87, *194*, 264, *268*
Corrodi, H., see Andén, N.-E. 330, 333, 335, *350*
Corrodi, H., see Carlsson, A. 38, 54, *191*
Corson, E. O., see Arnold, L. E. 71, *183*
Corson, E. O. L., see Arnold, L. E. 250, *267*
Corson, E. O. L., see Corson, S. A. 71, 131, 137, *194*
Corson, S. A., Corson, E. O. L., Kirilcuk, V., Arnold, L. E. 71, 131, 137, *194*
Corson, S. A., see Arnold, L. E. 71, *183*, 250, *267*
Costa, E. 329, *353*
Costa, E., Carenzi, A., Guidotti, A., Revuelta, A. 58, *194*
Costa, E., Groppetti, A. 9, 15, 18, 19, *27*, 34, 35, 48, 50, 51, 78, *194*
Costa, E., Groppetti, A., Naimzada, M. K. 34, 35, 37, 48, 52, 53, 90, 132, 157, *194*
Costa, E., Groppetti, A., Revuelta, A. 56, *194*
Costa, E., see Bloom, F. E. 329, *352*
Costa, E., see Brodie, B. B. 34, 50, *189*
Costa, E., see Cattabeni, F. 56, *192*
Costa, E., see Chang, C. C. 147, 148, *192*
Costa, E., see Groppetti, A. 16, 17, *28*, 34–36, 48, 50, 51, 63, 90, 157, *205*
Costa, E., see Lin, R. C. *360*
Costa, E., see Morgan, C. D. 39, 45, 51–53, 56, *222*
Costall, B., Naylor, R. J. 72, 77–79, 89, *194*
Costall, B., Naylor, R. J., Olley, J. E. 77–79, *194*
Costall, B., Naylor, R. J., Wright, T. 77, 78, *194*
Costall, B., see Naylor, R. J. 71, 77, *223*
Cotillo, P., see Babulova, A. 59, 60, 143, 144, 147, *183*
Cott, J. M., see Cooper, B. R. 122, 123, *194*
Cottington, F., see Bender, L. 286, *297*
Cousens, K., DiMascio, A. 414, *417*
Coutts, R. T., see Beckett, A. H. 11, *25*
Cowan, A., see Wray, R. 92, 94, *245*
Cowan, F. F., Cannon, C., Koppanyi, T., Maengwyn-Davis, G. D. 60, 62, *194*
Cowan, F. F., see Maengwyn-Davies, G. D. 62, 143, *217*
Cox, B., Tha, S. J. 77, *195*
Cox, R. H., Maickel, R. P. 101, *195*
Cox, R. H., Jr., see Maickel, R. P. 20, *29*, 53, 56, 82, 99, 106, 128, 131, 134, 140, 179, *218*
Cox, R. H., see Miller, F. P. 87, *221*
Cox, R. H., Jr., see Segal, D. S. 99, 106, *235*
Coyle, J. T., Snyder, S. H. 39, 40, *195*
Coyle, J. T., see Horn, A. S. 39, *208*
Coyle, J. T., see Snyder, S. H. 72, 82, *236*
Crabai, F., see Mulas, A. 94, *223*
Crabtree, J. M., Moyer, K. E. 96, *195*
Crabtree, R. E., see Lemberger, L. 383, *423*
Craig, A. L., Kupferberg, H. J. 149, 155, 156, 167, *195*
Craigmill, A. L., Canafax, D. M., Curties, F. R. 87, 154, 173, *195*
Craigmill, A. L., Kupferberg, H. J., Miller, J. W. 155, 156, *195*
Craver, B. N., see Rubin, B. 338, *363*
Cravey, R. H., Baselt, R. C. 264, *268*
Cravey, R. H., Reed, D. 264, *268*
Crawford, M. A., see Milne, M. D. 7, *29*
Crawford, T. B. B., see Amin, A. H. T. 328, *350*
Creaven, P. J., Barber, T. 17, *27*
Creaven, P. J., Barber, T., Roach, M. K. 17, *27*
Creese, I., Iversen, S. D. 79, 87, 89, *195*
Creveling, C. R., Daly, J. W., Witkop, B., Udenfriend, S. 9, 18, *27*
Creveling, C. R., see Daly, J. W. 4, *27*
Crighel, E., Stoica, E. 342, *353*
Crighel, E., see Kreindler, A. 342, *360*
Crighel, E., see Mison-Crighel, N. 342, *362*
Crismon, C. A., Tainter, M. L. 62, *195*
Croughan, J. L., see Kay, D. C. 315, 335, 341, 342, *359*
Crow, T. J. 121–123, *195*
Crow, T. J., Gillbe, C. 69, *195*
Crow, T. J., see Christie, J. E. 68, 69, *193*
Crowder, W. F., see Pickens, R. W. 127, *227*
Crowley, T. J. 96, 97, *195*
Crowley, T. P., Sisemore, D. D. 130, *195*
Crown, J., see Paton, W. D. M. 372, *426*
Crumpton, E., see Brill, N. O. 403, *416*
Csáki, L., see Pfeiffer, A. K. 49, *226*
Cucinell, S. A., see Faraj, B. A. 3, *27*
Cuello, A. C., see Horn, A. S. 40, *208*
Cuenca, E., see Laporte, J. 60, *214*
Cuenca, E., see Valdecasas, F. G. 69, 80, 146, 147, *241*
Cuevas, P., Fine, J. 279, *298*
Cugurra, F., see Vaccari, A. 61, *241*
Cullumbine, H., see Aston, R. 313, *351*
Curry, G. A. 265, *268*
Curry, S. H., see Joyce, C. R. B. 372, *421*
Curtis, D. R., Davis, R. 336, 337, *353*
Curties, F. R., see Craigmill, A. L. 87, 154, 173, *195*
Czlonkowski, A., see Kostowski, W. 80, 97, *213*

Dagirmanjian, R., Boyd, E. S. 399, 400, *417*
Dahl, C. J., see Chesher, G. B. 390, *417*
Dahlen, P., see Scheckel, C. L. 387, *427*
Dahlström, A., see Andén, N.-E. 68, *182*
Dahlström, A., see Carlsson, A. 34, 36, 38, 39, 42, 53, *191*
Dahmen, J., see Agurell, S. 382, 384, *415*
Daigneault, E. A., see White, R. P. 176, *244*
Dale, H. H., see Barger, G. 279, *297*
Dallemagne, G. H. 99, *195*
Dalrymple, S. D., Stretch, R. 100, 106, *195*
Daly, C., see Dill, D. B. 254, *268*
Daly, D. D., see Yoss, R. E. 285, *304*

Daly, J. W., Creveling, C. R., Witkop, B. 4, *27*
Daly, J. W., see Creveling, C. R. 9, 18, *27*
Dalzell, H. C., see Razdan, R. K. 379, 380, *426*
Dalzell, H. C., see Uliss, D. B. 379, *429*
Dalzell, B. C., see Zitko, B. A. 380, *429*
Dandiya, P. C., see Gupta, B. D. 93, *205*
Dandiya, P. C., see Kulkarni, S. K. 93, *213*
Dandiya, P. C., see Menon, M. K. 87, 154, 173, *221*
Daniel, F., see Zetler, G. 77, *245*
Dannenburg, W. N., Kardian, B. C. 164, *195*
D'Arcy, P. F., Spurling, N. W. 169, *195*
Darley, C. F., Tinklenberg, J. R. 385, *417*
Dary, A., see Boakes, R. J. 329, *352*
Dasgupta, S. R., see Haley, T. J. 315, 316, *356*
Davenport, G. R., see Palmer, G. C. 130, *226*
Davey, M. J., Farmer, J. B. 62, *195*
Davidoff, E. 280, *298*
Davidoff, E., Reifenstein, E. C. 248, 250, 254, 255, *268*
Davidoff, E., see Reifenstein, E. C. 247, *273*, 282, *302*
Davidson, H. A. 290, *298*
Davies, I. J., 265, *268*
Davies, J. A., Jackson, B., Redfern, P. H. 102, *195*
Davis, G. D. 87, *196*
Davis, J. M., Janowsky, D. S. 18, *27*
Davis, J. M., Kopin, I. K., Lemberger, L., Axelrod, J. 14, 15, 21, 22, *27*
Davis, J. M., see Janowsky, D. S. 287, *300*
Davis, J. M., see Lemberger, L. 16, *29*, 77, *215*
Davis, J. P., see Toman, J. E. P. 176, *240*
Davis, K. H., see Perez-Reyes, M. 380, *383*, *426*
Davis, K. H., see Slater, I. H. 334, *365*
Davis, L. E., see Baggot, J. D. 20, 21, *25*
Davis, M., Sheard, M. H. 313, *353*
Davis, R. A., Horlington, M. 179, *196*
Davis, R., see Curtis, D. R. 336, 337, *353*
Davis, W. M., Logston, D. G., Hickenbottom, J. P. 130, 173, 175, *196*
Davis, W. M., Moreton, J. E., King, W. T., Pace, H. B. 410, 413, *417*
Davis, W. M., Smith, S. G. 124, 125, *196*
Davis, W. M., see Borgen, L. A. 406, *416*
Davis, W. M., see Pace, H. B. 406, 407, *426*
Day, M. D. 62, 63, 144, 147, *196*
Day, M. D., Rand, M. J. 59, 60, 62, 63, 141, 142, 144, 146, 147, *196*
Day, W. A., see Fuller, R. W. 11, *27*
Dayton, P. G., see Faraj, B. A. 3, *27*
Deakin, S. 403, *417*
Dean, H. G., Hughes, I. E. 147, *196*
Dear, E. M. A., Malcolm, J. L. 339, *353*
DeBaran, L., Gogolak, G., Longo, V. G., Stumpf, Ch. 340, *353*
DeCarolis, A. S., see Florio, V. 339, 340, 342, *355*
Deegan, J., see Kelleher, R. T. 110–113, 116, *211*
Defeo, J. J., see Lal, H. 97, *214*
Deffenu, G., Bartolini, A., Pepeu, G. 57, *196*
Deffenu, G., see Nistri, A. 57, *224*
Deffenu, G., see Pepeu, G. C. 57, *226*
DelAngel, R. A., see Dolfini, E. 36, *197*
DeLaTorre, J. C. 333, *353*
Delay, J., Lhermitte, F., Verdeaux, G., Verdeaux, J. 339, *353*
Delay, J., Pichot, P., Lemperière, T., Sadoun, R. 293, *298*
Delini-Stula, A. 411, *417*
Delini-Stula, A., Morpurgo, C. 79, *196*
Dell, P., see Hiebel, G. 176, *207*
DellaCorte, L., see Buffoni, F. 45, *189*
Delong, F. L., Levy, B. I. 385, *417*
Delphaut, J. 173, *196*
DeLuca, D. C., see Angel, C. 109, 117, *182*
DeLuca, D. C., see Murphree, O. D. 71, *223*
DelRio, J., Fuentes, J. A. 77, 94, *196*
DelRio, J., see Fuentes, J. A. 43, 46, *201*
DeMaar, E. W. J., Williams, H. L., Miller, A. I., Pfeiffer, C. C. 312, *353*
Dement, W., Rechtschaffen, A., Gulevich, G. 285, *298*
Dement, W., see Ferguson, J. 96, 97, *200*
Dement, W. C., see Rechtschaffen, A. 285, *302*
Denber, H. C. B., see Merlis, S. 343, *362*
D'Encarnacao, P. S., Anderson, K. 87, *196*
D'Encarnacao, P. S., D'Encarnacao, P. D., Tapp, J. T. 87, *196*, 264, *268*
D'Encarnacao, P. D., see D'Encarnacao, P. S. 87, *196*, 264, *268*
Deneau, G. A., Kaymakalan, S. 413, *417*
Deneau, G. A., Yanagita, T., Seevers, M. H. 124, 126, *196*
Deneau, G. A., see Kaymakcalan, S. 410, *422*
Denefield, B. A., see Proctor, C. D. 87, *228*
Dengler, H. J., Spiegel, H. E., Titus, E. O. 38, *196*
Denhoff, E., see Laufer, M. W. 286, *300*
Dennis, B. J., see Adey, W. R. 315, 316, 341, *350*
Depoilly, F., see Schmitt, H. 146, *234*
Depoortere, H., Loew, D. M. 339, *353*
Depoortere, H., see Loew, D. M. 339, *360*
Dér, P., see Böszörményi, Z. 311, 343, *352*
Derks, P., see Teitelbaum, P. 99, 115, 116, *240*
Derouaux, G. 253, *268*
DeShon, H. J., see Rinkel, M. 308, 343, *363*
Desoize, B., see Nahas, G. G. 402, *425*
Detrick, L. E., Millikan, R., Modern, F. S., Thienes, C. H. 60, 62, 141, 142, 146, 148, *196*
DeVries, R., see Andres, F. 161, *182*
Dewar, A. J., Winterburn, A. K. 164, 165, *196*
Dewey, W. L., Harris, L. S., Howes, J. F., Kennedy, J. S. 381, 409, *418*

Dewey, W.L., Harris, L.S., Howes, J.F., Kennedy, S., Anderson, R.N. *418*
Dewey, W.L., Harris, L.S., Howes, J.F., Kennedy, J.S., Granchelli, F.E., Pars, H.G., Razdan, R.K. 399, *418*
Dewey, W.L., Harris, L.S., Kennedy, J.S. 389–391, 397, 403, *418*
Dewey, W.L., Jenkins, J., O'Rourke, T., Harris, L.S. 386, 390, 399, 411, 412, *418*
Dewey, W.L., Martin, B.R., Beckner, J.S., Harris, L.S. 382, 412, *418*
Dewey, W.L., McMillan, D.E., Harris, L.S., Turk, R.F. 412, *418*
Dewey, W.L., Peng, T.C., Harris, L.S. 400, *418*
Dewey, W.L., Yonce, L.R., Harris, L.S., Reavis, W.M., Griffin, E.D., Jr., Newley, V.E. 398, *418*
Dewey, W.L., see Adams, M.D. 399, *415*
Dewey, W.L., see Bloom, A.S. 390, 392–394, *416*
Dewey, W.L., see End, D. 382, *418*
Dewey, W.L., see Johnson, K.M. 392, *421*
Dewey, W.L., see Kosersky, D.S. 391, *422*
Dewey, W.L., see Levy, J.A. 402, *423*
Dewey, W.L., see Martin, B.R. 381–384, 390, 410–412, *424*
Dewey, W.L., see McMillan, D.E. 411, 412, *424*, *425*
Dewey, W.L., see Munson, A.D. 402, *425*
Dewey, W.L., see Spaulding, T.C. 386, *428*
Dewey, W.L., see Turk, R.F. 380, *428*
Dewey, W.L., see Zitko, B.A. 380, *429*
Dewhurst, W.G., Marley, E. 70, 95, 96, 176, 177, *196*, 331, *353*
Dews, P.B. 82, 110–113, *197*
Dews, P.B., Morse, W.H. 98, *197*
Dezeeuw, R.A., Vree, T.B., Breimer, D.D., Van Ginneken, C.A.M. 375, *418*
Dhawan, B.N. 338, *353*
Dhawan, B.N., Gupta, G.P. 313, 336, *354*
Dhawan, B.N., see Srimal, R.C. 77, *237*
Diab, I.M., Freedman, D.X., Roth, L.J. 334, *354*
Diaz, J.-L., Huttunen, M.O. 53, 55, 56, *197*
DiCarlo, R., Edel, S., Randrianarisoa, H., Mandel, P. 165, *197*
Didamo, P., see Carlton, P.L. 99, 106, *191*
Dienske, H., see Hart, M. 97, *206*
Dill, D.B., Johnson, R.E., Daly, C. 254, *268*
Dill, R.E., see Little, M.D. 313, *360*
Dille, J.M., see Elder, J.T., Jr. 316, *354*
Dille, J.M., see Gogerty, J.H. 320, 328, *356*
Dille, J.M., see Horita, A. 337, *357*
Di Mascio, A., Greenblatt, M., Hyde, R.W. 310, *354*
Di Mascio, A., see Cousens, K. 414, *417*
Di Mascio, A., see Meyer, R.E. 253, *272*
Dingell, J.V., Bass, A.D. 15, *27*
Dingell, J.V., Owens, M.L., Norwich, M.R., Sulser, F. 82, 87, *197*
Dingell, J.V., see Klausner, H.A. 380, *422*
Dingell, J.V., see Miller, K.W. 87, 106, *221*
Dingell, J.V., see Sulser, F. 15, 16, *30*, 87, 88, *239*
Dinnen, A. 265, *268*
Dispensa, J., Barrett, M.E. 128, *197*
DiStefano, V., Leary, D.E. 334, *354*
DiStefano, V., see Little, K.D. 334, *360*
Ditman, K.S., Whittlesey, J.R.B. 308, *354*
Ditzion, B.R., see Paul, M.I. 57, *226*
Divac, I. 77, 79, *197*
Dixit, B.N., see Sofia, R.D. 392, 395, *428*
Dixon, A.K. 307, 313, 315, *354*
Dixon, W.E. 277
Dockens, W.S., see Schuster, C.R. 99, 103, 107, 112, 114, 120, *235*
Doda, M., György, L., Nador, K. 59, *197*
Dolfini, E., Del Angel, R.A., Garattini, S., Valzelli, L. 36, *197*
Dolfini, E., Garattini, S., Valzelli, L. 35, 36, 149, 154 155, 167, 168, *197*
Dolfini, E., Kobayashi, M. 156, *197*
Dolfini, E., Tansella, M., Valzelli, L., Garattini, S. 155, *197*
Dolfini, E., see Jori, A. 43, 46, 52, *211*
Dolfini, E., see Valzelli, L. 16, *31*, 154, *242*
Dolfini, F., see Consolo, S. 15 16, *26*
Dolfini, L., see Rosso, R. 17, *30*
Domer, F.R., Longo, V.G. 340, *354*
Dominic, J.A., Moore, K.E. 87, 88, *197*
Dominic, J.A., see Moore, K.E. 87, 88, *222*
Domino, E.F. 388, 392, 397, *418*
Domino, E.F., Bartolini, A. 333, 335, 341, *354*
Domino, E.F., Caldwell, D.F., Henke, R. 102, *197*
Domino, E.F., Olds, M.E. 57, 122, *197*
Domino, E.F., Wilson, A. 57, *197*
Domino, E.F., see Black, M.D. 411, *416*
Domino, E.F., see Hardman, H.F. 398–400, 410, *420*
Domino, E.F., see Hitzemann, R.J. 35–37, 49, 56, 80, 91, *208*
Domino, E.F., see Johnson, S. 397, *421*
Domino, E.F., see Vasko, M.R. *242*
Donabedian, A.S., see Pinkston, J.O. 62, 141, 142, *227*
Done, A.K., see Espelin, D.E. 266, *269*, 288, 294, *298*
Dornbush, R.L. 409, *418*
Dornbush, R., see Fink, M. 410, *418*
Doorenbos, N.J., see Fetterman, P.S. 373, 374, *418*
Doorenbos, N.J., see Quimby, M.W. 374, *426*
Dorff, J.E., see Kornetsky, C. 249, *271*
Dorrance, D., Janiger, O., Teplitz, R.L. 402, *418*
Dorris, R.L., Shore, P.A. 45, *197*
Dorris, R.L., see Enna, S.J. 44, 45, *199*
Dorris, R.L., see Shore, P.A. *235*
Dosi, I., see Klissiunis, N. 154, *212*
Doty, B.A., Doty, L.A. 102, *197*

Doty, L. A., see Doty, B. A. 102, *197*
Dougharty, J. A., see Pickens, R. 125, 126, *227*
Douglas, V., see Weiss, G. 286, *303*
Douglas, V. I., see Cohen, N. J. 250, *268*
Douglas, V. I., see Sykes, D. H. 250, *274*
Downing, O. A. 64, *197*
Doyle, G. A., see Carlson, N. J. 128, 139, *191*
Draskoczy, P. R., see Schildkraut, J. J. 261, *273*, 294, *303*
Drawbaugh, R. B., see Heffner, T. G. 110–112, 116, *207*
Dreisbach, R. H., Pfeiffer, C. 277, *298*
Dreiser, T. 277, *298*
Drell, H. J., see Monroe, R. R. 265, *272*, 283, *301*
Dren, A. T., Bopp, B. A., Ebert, D. M. 415, *418*
Dresse, A. D., see Janssen, P. A. J. 77, *210*
Drew, W. G., Miller, L. L. 397, *418*
Drewes, H. R., see Foltz, R. L. 383, *419*
Dring, L. G., Smith, R. L., Williams, R. T. 5, 7, 8, 10, 14, *27*
Dring, L. G., see Caldwell, J. 9, 12, *26*
Dring, L. G., see Sever, P. S. 8, 9, 18, *30*
Dring, L. G., see Smith, R. L. 8, *30*
Dring, L. G., see Williams, R. T. 5, 14, *31*
Drudi-Baracco, C., Halpern, B. N., Bessirard, D. 173, *197*
Drudi-Baracco, C., see Beauvallet, M. 35, 174, *185*
Drudi-Baracco, C., see Halpern, B. N. 36, 173, 174, *206*
Drummond, G. I., Bellward, G. 58, *197*
Drybanski, A., see Herman, Z. S. 35, 37, 53, 87, 91, 92, *207*
Duarte-Escalante, O., Ellinwood, E. H., Jr. 174, *198*
Duarte-Escalante, O., see Ellinwood, E. H., Jr. 81, *198*
Dubinsky, B., Robichard, R. C. Goldberg, M. F. 387, *418*
Dubnick, B., Ruchi, E. W., Salama, A. I. 54, *198*
Dubnick, B., see Möller-Nielsen, I. 45, 46, *221*
Dubnick, B., see Robichaud, R. C. 93, *230*
Dubowski, K. M., see Schaefer, C. F. 401, *427*
Duchène, N., see Chevillard, C. 41, *192*
Dudderidge, H. J., Gray, J. A. 128, *198*
Duffield, I. C., see Brawley, P. 306, *352*
Duhault, J., Verdavaienne, C. 3, *27*, 56, *198*
Duhault, J., see Beregi, L. G. 4, *26*
Dumont, C., see Boissier, J. R. 13, *26*
Duncan, L. J. P., see Munro, J. F. 285, *301*
Duncan, L. J. P., see Seaton, D. A. 259, *273*, 285, *303*
Duncan, N. C., see Bauer, R. H. 102, 103, 129, *185*
Duncan, S., see Powell, D. A. 97, *228*
Dunleavy, D. L. F., MacLean, A. W., Oswald, I. 252, *268*
Dunleavy, D. L. F., Oswald, I., Strong, J. A. 255, *268*
Dunleavy, D. L. F., see Lewis, S. A. 257, *271*
Dunleavy, D. L. F., see Oswald, I. 257, *272*
Durden, D. A., see Phillips, S. R. 332, *362*
Durek, B., see Maj, J. 81, *218*
Dureman, I., Scholander, T. 248, *268*
Dutta, S. N., see Pradhan, S. N. 114, *228*
Duvernoy, W. F. 250, *268*
Dvornik, D., see Zsoter, T. 163, *246*
Dykstra, L. A., Appel, J. B. 314, *354*

Eades, C. G., see Martin, W. R. 316, 317, 321–323, 328, 331, 335, 336, 345, *361*
Earnhardt, J. T., see Adams, M. D. 399, *415*
Earnhardt, J. T., see Martin, B. R. 381, 390, *424*
Easterbrook, J. A., see Eysenck, H. J. 252, *269*
Ebaugh, F. G. 286, *298*
Ebert, D. M., see Dren, A. T. 415, *418*
Eble, J. N., Rudzik, A. D. 62, 143, 148, *198*
Eble, J. N., see Rudzik, A. D. 146, *231*
Ebstein, B. S., see Ebstein, R. P. 44, *198*
Ebstein, R. P., Ebstein, B. S., Samuel, D., Berger, B. D. 44, *198*
Eccleston, D., see Ashcroft, G. W. 263, *267*
Echinard-Garin, P., see Jacob, J. 313, *358*
Edel, S., see DiCarlo, R. 165, *197*
Edeleano, L. 279, *298*
Edelberg, R., see Greiner, T. 310, *356*
Edery, H., see Grunfeld, Y. 387, *420*
Eddy, N. B., Halbach, H., Isbell, H., Seevers, M. H. 257, *269*
Edison, G. 250, 257, *269*
Edwards, A. E., Cohen, S. 308, *354*
Edwards, G., see Hawks, D. 282, 291, *299*
Efron, D. H. 306, *354*
Efron, D. H., Holmstedt, B., Kline, N. S. 306, *354*
Efron, D. H., see Schildkraut, J. J. 391, 393, 395, *427*
Ehrich, W. E., Krumbhaar, E. B. 70, 96, 131, 139, 167, 173, *198*
Ehrich, W. E., Lewy, F. H., Krumbhaar, E. B. 70, 166, 175, *198*, 282, *298*
Ehrlich, P. 282
Eisenberg, L. 250, *269*
Eisenberg, L., see Conners, C. K. 286, *298*
Eleftheriou, B. E., see Oliviero, A. 94, *225*
El Guedri, H., Jacquot, C., Rapin, J., Cohen, Y. 9, 19, *27*, 52, *198*
El Guedri, H., see Jacquot, C. 51, *210*
Eliasson, M., Michanek, A., Meyersson, B. J. 69, 98, *198*
Elkes, C., see Elkes, J. 176, *198*, 343, *354*
Elkes, J., Elkes, C., Bradley, P. B. 176, *198*
Elkes, J., see Bradley, P. B. 176, *188*, 341, 347, *352*
Ellen, P., see Liberson, W. T. 129, *216*
Ellingboe, J., see Mendelson, J. H. 401, *425*
Ellington, A. C., see Persaud, T. V. 406, 407, *426*
Ellinwood, E. H., Jr. 70, 81, 127, *198*, 262, *269*, 291–293, *298*
Ellinwood, E. H., Jr., Balster, R. L. 71, 72, 77, *198*
Ellinwood, E. H., Jr., Duarte-Escalante, O. 81, *198*
Ellinwood, E. H., Jr., Escalante, O. 36, 54, 57, 70, 81, 97, 173, 178, *198*

Ellinwood, E. H., Jr., Sudilovsky, A., Grabowy, R. 178, *198*
Ellinwood, E. H., Jr., Sudilovsky, A., Nelson, L. M. 77, 81, 178, *198*
Ellinwood, E. H., Jr., see Duarte-Escalante, O. 174, *198*
Ellinwood, E. H., Jr., see Escalante, O. D. 174, *199*
Ellinwood, E. H., Jr., see Martin, J. C. 127, *219*
Ellis, K. E. J., see Allan, D. 141, 143, 146, *182*
Ellison, T., Gutzait, L., Loon, E. J. van 5, 8, 14, *27*
Ellison, T., Okun, R., Silverman, A., Siegel, M. 15, 18, 20, 21, *27*
Ellison, T., Siegel, M., Silverman, A. G., Okun, R. 81, 173, *199*
Elder, J. T., Jr., Dille, J. M. 316, *354*
Elder, J. T., Shellenberger, M. K. 338, *354*
Elkes, J. 347, *354*
Elkes, J., Elkes, C., Bradley, P. B. 343, *354*
Elsmore, T. F., see Manning, F. J. 388, 389, 404, *424*
El-Toumi, A., see Sletten, I. W. 250, *274*
El-Yousef, M. K., see Janowsky, D. S. 287, *300*
Emde, H. 280, *298*
End, D., Carchman, R. A., Dewey, W. L. 382, *418*
Englert, L. F., see Ho, B. T. 391, 393–396, *421*
Englert, L. F., see McIsaac, W, M. 381, *424*
Enna, S. J., Dorris, R. L., Shore, P. A. 44, 45, *199*
Enna, S. J., Shore, P. A. 40, 41, *199*
Ensberg, D. L., see Handley, C, A. 288, *299*
Ensor, C. R., see McLean, J. R. 87, *220*
Epstein, A. N. 136, 140, *199*
Epstein, L. C., Lasagna, L., Connors, C. K., Rodriguez, A. 286, *298*
Epstein, L. C., see Lasagna, L. 250, *271*
Ernst, A. M. 77, 87, *199*
Ertel, R., see Cavero, I. 398, *417*
Escalante, O. D., Ellinwood, E. H., Jr. 174, *199*
Escalante, O., see Ellinwood, E. H. Jr. 36, 54, 57, 70, 81, 97, 173, 178, *198*
Esecover, H., see Malitz, S. 310, 343, *361*
Espelin, D. E., Done, A. K. 266, *269*, 288, 294, *298*
Estler, C.-J., Ammon, H. P. T. 87, 161, 162, *199*
Estler, G.-J., Ammon, H. P. T. Fickl, W., Fröhlich, H.-N. 87, 162, 163, *199*
Estler, C.-J., Mitznegg, P. 162, *199*
Estler, K. J., see Heim, F. 161, *207*
Etevenon, P. 176, *199*
Etevenon, P., see Boissier, J.-R. 79, *187*
Etevenon, P. R., Boissier, J. R. 340, *354*
Euler, U. S. von, Lishajko, F. 38, 41 *199*
Evangelista, A. M., Gattoni, R. C., Izquierdo, I. 102, 129, *199*
Evans, A. O. 288, *298*
Evans, H. L. 99, 106, 111, 116, 118, *199*
Evans, H. L., Ghiselli, W. B., Patton, R. A. 112, 113, 118, *199*
Evans, W. O. 288, *298*
Evans, W. O., Bergner, B. A. 288, *298*
Evans, W. O., Smith, R. P. 247, *269*, 288, *298*
Evarts, E. V. 317, 336, 341, 342, *354*
Evarts, E. V., Landau, W., Freygang, W., Jr., Marshall, W. H. 336, *354*
Evarts, E. V., see Axelrod, J. 333, 334, *351*
Evarts, E. V., vee Kies, M. W. 310, *359*
Everett, G. M., Thoman, J. E. P., Smith, A. H. 70, 77, *199*
Everett, G. M., Yellin, T. O. 35, 37, 45, 54, 56, *199*
Everingham, M., see Chesher, G. B. 390, *417*
Evetts, K. D., Uretsky, N. J., Iversen, L. L., Iversen, S. D. 87, 89, 96, *199*
Ewald, A. T., see Abramson, H. A. 310, *350*
Ewald, A. T., see Jarvik, M. E. 308, *359*
Ewens, G. F. W. *418*
Ewing, P. L., Moore, B. M., Moore, W. T. 128, 129, *199*
Eybl, V., see Mayer, O. 77, 93, *219*
Eysenck, H. J., Easterbrook, J. A. 252, *269*
Eysenck, H. J., see Trouton, D. 249, *274*
Ezoe, T., see Utena, H. 46, 70, 81, 161, 173, *241*

Faber, J., see Roth, B. 250, *273*, 286, *302*
Fabing, H. D. 310, *354*
Fabing, H. D., Hawkins, J. R. 310, *355*
Fähndrich, C. H., see Magour, S. 91, 92, 138, 139, 164, 173, *218*
Faidherbe, J., Richelle, M., Schlaug, J. 117, *200*
Faillace, L. A., Snyder, S. H., Weingartner, H. 311, *355*
Faillace, L., see Snyder, S. H. 4, *30*, 311, 312, 328, *365*
Falb, A., Schmidt, G., Schwabe, U. 163, *200*
Falek, A. 408, *418*
Falkenberg, G. 348, *355*
Fanelli, O. 91, 92, 138, *200*
Fang, H. C. H., see Rumbaugh, C. L. 143, *231*, 252, 265, *273*
Faraj, B. A., Israili, Z. H., Perel, J. M., Jenkins, M. L., Holtzman, S. G., Cucinell, S. A., Dayton, P. G. 3, *27*
Farmer, J. B., see Davey, M. J. 62, *195*
Farmer, R., see Kilbey, M. M. 387, *422*
Farnebo, L.-O. 38, 39, 40, 44, *200*
Farrow, J. T., Vunakis, H. van 334, *355*
Fassina, G. 163, 164, *200*
Fassina, G., see Santi, R. 163, *232*
Fawas, G., Simaan, J. 63, *200*
Federico, R., see Raiteri, M. 45, *228*
Fehlhaber, H. W., see Spulak, R. von 375, *429*
Feigen, G. A., see Alles, G. A. 4, 5, *25*
Feinberg, I., Hibi, S., Braun, M., Cavness, C., Westerman, G., Small, A. 252, 260, *269*, 286, *298*
Feinstein, A. R. 285, *298*
Feldberg, W., Lang, W. J. 154, 158, *200*
Feldman, J. M., see McMahon, E. M. 162, 165, *220*
Feldman, R. S., see Liberson, W. T. 129, *216*
Feller, D. R., Malspeis, L. 18, *27*
Feller, D. R., see Finger, K. F. 164, *200*

Fellows, E.J., see Leonard, D.A. 138, *215*
Fellows, E.J., see Tedeschi, R.E. 87, 154, *239*, 313, 332, 339, *366*
Felsinger, J.M. von, see Lasagna, L. 248, *271*
Feltz, P. 66, *200*
Feltz, P., Champlain, J. de 66, *200*
Feltz, P., see Besson, M.J. 44, 47, *186*
Fenard, S., see Schmitt, H. 247, *273*
Fennel, W., see Fischman, M.W. 278, *298*
Fentiman, A.F., Jr., see Foltz, R. L. 383, *419*
Ferguson, J., Dement, W. 96, 97, *200*
Fernandes, N.S., see Santos, M. 387, *427*
Ferraro, D.P. 410, *418*
Ferraro, D.P., Grilly, D.M. 412, *418*
Ferraro, D.P., see Grilly, D.M. 407, *420*
Ferraro, D.P., see Grisham, M. G. 389, *420*
Ferrendelli, J.A., Kinscherf, D. A., Kipnis, D.M. 58, *200*
Ferrer, S., see Gastaut, H. 343, *355*
Ferris, R.M., Maxwell, R.A. 39, 44, *200*
Ferris, R.M., Tang, F.L.M., Maxwell, R.A. 38, 39, 40, 44, *200*
Ferster, C.B., Appel, J.B., Hiss, R.A. 110, 114, *200*
Fessock, L., see MacDonnell, M.F. 97, *217*
Fetterman, P.S., Keith, E.S., Waller, C.W., Guerrero, O., Doorenbos, N.J., Quimby, M.W. 373, 374, *418*
Feudis, F.V. de 162, *200*
Feudis, F.V. de, Marks, J.H. 168, *200*
Fibiger, H.C. 79, 87, 91, 134, 135, *200*
Fibiger, H.C., Fibiger, H.P., Zis, A.P. 79, 87, *200*
Fibiger, H.C., McGeer, E.G. 48, *200*
Fibiger, H.C., Trimbach, C., Campbell, B.A. 91, *200*
Fibiger, H.C., see Campbell, B, A. 91, *190*
Fibiger, H.C., see McGeer, E.G. 72, 77, *220*
Fibiger, H.C., see Phillips, A.G. 121, 122, *226*
Fibiger, H.P., see Fibiger, H.C. 79, 87, *200*
Fickl, W., see Estler, C.-J. 87, 162, 163, *199*
Field, G., see Bishop, P.O. 336, *352*
Filby, Y.M., see Appel, J.B. 314, *351*
Filczewski, M., see Mazurkiewicz-Kwilecki, I.M. 393, 394, *424*
Fine, J., see Cuevas, P. 279, *298*
Finger, K.F., Page, J.G., Feller, D.R. 164, *200*
Fink, M., Shapiro, D.M., Itil, T. M. 251, *269*
Fink, M., Volavka, J., Dornbush, R., Crown, P. 410, *418*
Fink, M., Volavka, J., Panayiotopoulos, C.P., Stefanis, C. 409, *419*
Finkelstein, N., see Alpern, E.B. *182*
Finley, T.N., see Mann, P.E. 401, *424*
Finnerty, R.J., Soltys, J.J. Cole, J.O. 250, *269*
Finney, L.A., see Langfitt, T.W. 339, 341, 342, *360*
Fischer, A.E., see Canfield, D.R. 123, *190*
Fischer, E., Heller, B. 70, 77, *200*
Fischer, J.E., Baldessarini, R.J. 279, *298*
Fischer, J.E., James, J.H. 279, *298*
Fischer, J.E., see Kopin, I.J. 63, *212*
Fischer, K.C., Wilson, W.P. 250, *269*
Fischer, R. 347, *355*
Fischer, R., see Goldman, H. 335, *356*
Fischman, M.W., Schuster, C. R., Resnekov, L., Shick, F.E., Krasnegor, N.A., Fennel, W., Freedman, D.X. 278, *298*
Fischman, V.S., see Kramer, J. C., 257, 258, 260, 263, *271*
Fish, B. 250, *269*, 287, *299*
Fish, B., Campbell, M., Shapiro, T., Floyd, A., Jr. 310, *355*
Fish, C.H., Bowling, E. 288, *299*
Fisher, C., see Rechtschaffen, A. 285, *302*
Fishman, V.S., see Kramer, J.C. 282, 290, 291, 293, *300*
Fitz-Gerald, F.L. 70, 96, *200*
Fitzgerald, M.S., see Vachon, L. 414, *429*
Flaming, D., see Spooner, C.E. 70, 95, 96, 131, 176, *237*
Flataker, L., see Winter, C.A. 314, *368*
Fleckenstein, A., Burn, J.H. 58, *201*
Fleckenstein, A., Stöckle, D. 58, *201*
Flegin, O.T., Morgan, D.H., Oates, J.A., Shand, D.G. 147, *201*
Fleischmann, R.W., Hayden, D.W., Rosenkrantz, H., Braude, M.C. 407, 408, *419*
Fleischman, R.W., see Rosenkrantz, H. 404, 427,
Fleischman, R.W., see Thompson, G.R. 403, 405, *428*
Flemenbaum, A. 266, *269*
Fleming, R.M., see Menon, M.K. 43, 46, 87, 173, *221*
Fleming, W.W., see Trendelenburg, U, 59, 146, *241*
Florio, V., Lipparini, F., De Carolis, A.S., Longo, V.G. 339, 340, 342, *355*
Floyd, A., Jr., see Fish, B. 310, *355*
Flugel, F., see Bradley, P.B. 306, *352*
Fodor, M., see Pfeiffer, A.K. 49, 87, 161, 177, *226*
Fog, R. 70, 71, 77, 78, *201*
Fog, R., Pakkenberg, H. 78, *201*
Fog, R., Randrup, A., Pakkenberg, H. 77–79, *201*
Follenfant, M.J., Robson, R.D. 59, 147, *201*
Folstein, M.F., see Robinson, R.G. 285, *302*
Foltz, R.L., Fentiman, A.F., Jr., Leighty, E.G., Walter, J.L., Drewes, H.R., Schwartz, W. E., Page, T.F., Jr., Truitt, E.B., Jr. 383, *419*
Fonseka, K., see Agurell, S. 382, *415*
Fontanari, D., see Balestrieri, A. 318, *351*
Foote, W.E., Sheard, M.H., Aghajanian, G.K. 67, *201*
Foote, W.E., see Aghajanian, G. K. 330, *350*
Ford, R.D. 410, 411, *419*
Ford, R.D., Balster, R.L. 389, *419*
Ford, R.D., McMillan, D.E. 411, *419*
Ford, R.D., see Balster, R.L. 388, *415*
Ford, R.D., see Spaulding, T.C. 386, *428*

Foree, D. P., Moretz, F. H., McMillan, D. E. 115, 116, *201*
Foreman, R. L., Siegel, F. P., Mrtek, R. G. 11, *27*
Forney, R. B., see Brown, D. J. 399, *416*
Forney, R. B., see Garriott, J. C. 87, *202*
Forney, R. B., see Khan, A. U. 163, *211*
Forney, R. B., see Lewis, S. C. 398, *423*
Forney, R. B., see Manno, B. R. 385, 399, *424*
Forney, R. B., see Phillips, R. N. 403, 404, 407, *426*
Forrer, G. R., Goldner, R. D. 310, *355*
Fosset, M., see Simon, P. 77, 87, *236*
Fowble, J., see Miller, D. D. *221*
Fowler, P. F., see Tedeschi, D. H. 96, 97, *239*
Fox, F. W. 285, *299*
Fox, S. S. 109, *201*
Foxwell, M. H., Funderburk, W. H., Ward, J. W. 137, *201*
Franchi, G., see Rosso, R. 17, *30*
Frank, G. 283, 290, *299*
Frank, I. A., see Hepler, R. S. 414, *420*
Frank, I. M., Lessin, P. J., Tyrrell, E. D., Hahn, P. M., Szara, S. 409, *419*
Frank, I. M., see Brill, N. O. 403, *416*
Frank, I. M., see Hepler, R. S. 408, 409, 414, *420*
Frank, I. M., see Tashkin, D. P. 414, *428*
Frank, R. M., see Wong, D. T. 49, *245*
Franke, H. 247, *269*
Frankenhäuser, M., Post, B. 255, *269*
Frankenheim, J. M., McMillan, D. E., Harris, L. S. 388, 389, *419*
Frankenheim, J. M., see McMillan, D. E. 388, 411, *425*
Franklin, M. R. 8, *27*
Frankova, S. 96, *201*
Franksson, G., Änggård, E. 21, *27*
Fraser, H. F., see Isbell, H. 308, 310, 311, 318, *358*
Freedman, A. M., see Cole, J. O. 306, *353*
Freedman, D. X. 287, *299*, 330, *355*
Freedman, D. X., Aghajanian, G. K., Ornitz, E. M., Rosner, B. S. 320, *355*
Freedman, D. X., Appel, J. B., Hartman, F. R., Molliver, M. E. 314, 319, 320, *355*
Freedman, D. X., Giarman, N. J. 330, *355*
Freedman, D. X., see Appel, J. B. 120, *183*, 314, 315, 320, *351*
Freedman, D. X., see Becker, D. I. 315, *351*
Freedman, D. X., see Diab, I. M. 334, *354*
Freedman, D. X., see Fischman, M. W. 278, *298*
Freedman, D. X., see Holtzman, D. 391, 393, 394, *421*
Freedman, D. X., see Renault, P. F. 397, *427*
Freedman, D. X., see Rosecrans, J. A. 330, *363*
Freedman, L., see Shapiro, S. L. 138, 139, *235*
Freeman, H. 310, *355*
Freeman, H., see Krus, D. M. 312, *360*
Freeman, J. J., Miller, K. W., Sulser, F. 87, *201*
Freeman, J. J., Sulser, F. 50, 51, 87, *201*
Freeman, J. J., see Miller, K. W. 87, 106, *221*
Frei, E., see Sallen, S. E. 414, *427*
Fremont-Smith, N., see Abramson, H. A. 318, *350*
Freudenthal, R. I., Martin, J., Wall, M. E. 381, *419*
Freudenthal, R. J., see Christensen, H. D. 383, *417*
Frey, H.-H. 132, 178, *201*
Frey, H.-H., Kampmann, E. 178, *201*
Frey, H.-H., Magnussen, M. P. 87, *201*
Frey, H.-H., Schulz, R. 132, 134, *201*
Frey, H.-H., see Görlitz, B.-D. 53, 55, 179, *204*
Frey, H.-H., see Nielsen, C. K. 45, 61, 146, *224*
Freygang, W., Jr., see Evarts, E. V. 336, *354*
Fried, P. A., McIntyre, D. C. 410, *419*
Fried, P. A., see Potvin, R. J. 410, *426*
Friedenberg, S. 282, 289, *299*
Friedhoff, A. J., see Angrist, B. M. 284, *296*
Friedhoff, A. J., see Cole, J. O. 306, *353*
Friedhoff, A. J., see Hollister, L. E. 311, 323, *357*
Friedhoff, A. J., see Schweitzer, J. W. 19, *30*
Friedman, E., see Hine, B. 413, *420*
Friedman, E., see Wallach, M. B. 316, 320, 342, *367*
Friedman, H., Carey, R. J. 314, *355*
Friedman, M. A., see Harris, L. S. 402, *420*
Friedman, M. A., see Munson, A. E. 402, *425*
Friedman, S. A., Hirsch, S. E. 310, *355*
Friedrich-Fichtl, J., Spiteller, G. 375, *419*
Fritchie, G. E., see Ho, B. T. 393–396, *421*
Fritchie, G. E., see McIsaac, W. M. 381, *424*
Fröhlich, H.-N., see Estler, C.-J. 87, 162, 163, *199*
Frommel, E., Chmouliovsky, M. 173, *201*
Fry, W., see Kelleher, R. T. 110–113, 116, *211*
Fryö, B., see Schuberth, J. 54–56, *234*
Fu, T. K., see Matsuyama, S. S. 402, *424*
Fudema, J. J., see Proctor, C. D. 173, *228*
Fuentes, J. A., Del Rio, J. 43, 46, *201*
Fuentes, J. A., see Del Rio, J. 77, 94, *196*
Fugazza, J., see Beauvallet, M. 34, 35, 37, 41, 42, 53, 87, 154, 173, 174, *185*
Fujimori, M., Himwich, H. E. 176, *201*, 340, *355*
Fulginiti, S., Orsingher, O. A. 49, 106, *202*
Fulginiti, S., see Orsingher, O. A. 102, 106, 107, *225*
Fuller, G. C., see Lal, H. 179, *214*
Fuller, J. L., see Wimer, R. E. 93, 94, *244*
Fuller, R. W. 45, *202*
Fuller, R. W., Hines, C. W. 168, 173, *202*
Fuller, R. W., Hines, C. W., Mills, J. 3, *27*, 56, *202*
Fuller, R. W., Molloy, B. B., Rousch, B. W., Hauser, K. M. 5, *27*
Fuller, R. W., Perry, K. W., Baker, J. C., Parli, J., Lee, N., Day, W. A., Molloy, B. B. 11, *27*

Fuller, R. W., Schaffer, R. J., Roush, B. W., Molloy, B. B. 49, 53, 56, *202*
Fuller, R. W., Shaw, W. N., Molloy, B. B. 156, *202*
Fuller, R. W., Snoddy, H. D., Molloy, B. B. 5, *28*
Fuller, R. W., Walters, C. P. 45, 46, *202*
Fuller, R. W., see Shaw, W. N. 163, 164, *235*
Fuller, R. W., see Wong, D. T. 49, 54, 56, *245*
Funatogawa, S. 54, 77, 96, *202*
Funderburk, W. H., Hazelwood, J. C., Ruckart, R. T., Ward, J. W. 132, 134, *202*
Funderburk, W. H., see Foxwell, M. H. 137, *201*
Furlani, A., see Vaccari, A. *241*
Furness, P., see Chernov, H. 167, *192*
Fuster, J. M. 317, *355*
Fuxe, K., Hamberger, B., Malmfors, T. 39, 54, *202*
Fuxe, K., Hökfelt, T., Ungerstedt, U. 68, 79, *202*
Fuxe, K., Ungerstedt, U. 37–39, 43, 54, 67–69, 77–79, 87, *202*
Fuxe, K., see Andén, N.-E. 68, 69, *182*, 330, 333, 335, *350*
Fuxe, K., see Carlsson, A. 34, 36, 38, 39, 42, 53, 54, *191*
Fuxe, H., see Corrodi, H. 34–36, 48, 66, 77, 87, *194*, 264, *268*

Gabba, A. K., see Gupta, B. D. 93, *205*
Gabrio, B. W., see Salomon, K. 310, *364*
Gaddum, J. H. 332, *355*
Gaddum, J. H., Hameed, K. A. 328, 329, *355*
Gaddum, J. H., Picarelli, Z. P. 329, *355*
Gaddum, J. H., Vogt, M. 315, 316, 329, *355*
Gaddum, J. H., see Amin, A. H. T. 328, *350*
Gaensler, E. A., see Vachon, L. 414, *429*
Gaiardi, M., see Babbini, M. 77, 87, *183*
Gajda, L., see Maj, J. 149, 167, *218*
Galambos, E., Pfeifer, A. K., György, L., Monlár, J. 87, *202*
Galambos, E., see Pfeiffer, A. K. 87, 173, *226*
Galbreath, C., see Bannerjee, B. N. 406, *415*
Galdston, I. 282, *299*
Gallagher, D. W., Sanders-Bush, E., Sulser, F. 389, 395, *419*
Gallagher, N. I., Knight, W. A. 259, *269*
Galman, B. R., see Asatoor, A. M. 7, 15, 21, *25*
Gangloff, H., see Killam, E. K. 176, 177, *212*
Ganguly, D. K. 53, *202*
Ganong, W. F., see Loon, G. R. van 165, *217*
Gantt, W. H., see Alpern, E. B. *182*
Gaoni, Y., Mechoulam, R. 375, 378, 379, 384, *419*
Gaoni, Y., see Mechoulam, R. 373–375, *425*
Garattini, S., Ghetti, V. 306, *355*
Garattini, S., see Bizzi, A. 43, 46, 131, 141, 143, 147, 149, 154–156, 163, 164, *187*, 251, *268*
Garattini, S., see Caccia, S. 36, 43, 156, 157, *190*
Garattini, S., see Consolo, S. 15–17, *26*, *27*, 36, 57, 97, 168, *193*, *194*
Garattini, S., see Dolfini, E. 35, 36, 149, 154, 155, 167, 168, *197*
Garrattini, S., see Hajos, G. J. T. 132, 149, 150, 163, *206*
Garattini, S., see Jori, A. 36, 43, 46, 80, 90, 137, 154, 155, 156, 163, *211*
Garattini, S., see Samanin, R. 134–136, *232*
Garattini, S., see Valzelli, L. 16, *31*, 97, 154, 167, *242*
Gardner, E. L. 314, *355*
Gardner, L. I., see Carakus-hansky, G. 408, *416*
Gardner, L. I., see Neu, R. L. 401, *425*
Gardocki, J. F., Schuler, M. E., Goldstein, L. 87, 167, 173, *202*
Gardocki, J. F., see Goldstein, L. 178, *204*
Garfinkel, B. D., Webster, C. D., Sloman, L. 286, *299*
Garriott, J. C., King, L. J., Forney, R. B., Hughes, H. W. 87, *202*
Gastaut, H., Ferrer, S., Castells, C. 343, *355*
Gastaut, H., Plas, R., Bert, J. 251, *269*
Gatti, G. L., see Bignami, G. 112, *187*
Gatti, G. L., see Bovet, D. 102, *188*
Gattoni, R. C., see Evangelista, A. M. 102, 129, *199*
Gauchy, C., Bioulac, B., Cheramy, A., Besson, M. J., Glowinski, J., Vincent, J. D. 44, *202*
Gauchy, C., see Besson, M. J. 47, *186*
Gauchy, C., see Jones, B. E. 57, *210*
Gaudry, R., see Chappel, C. I. 174, *192*
Gay, P. E., see Cole, S. O. 90, 97, 139, 140, *193*
Gayer, H. 386, *419*
Geber, W. F., Schramm, L. C. 407, *419*
Gedye, J. L., see Aitken, R. C. B. 258, *267*
Gegoretti, L., see Giberti, F. 312, *356*
Geh, S. L., see Banarjee, U. 87, 96, 173, *184*
Geier, S., see Abely, P. 293, *296*
Geiger, L. E., Cervoni, P. 334, *355*
Geiger, L. E., see Cervoni, P. 335, *352*
Gelder, M. G., Vane, J. R. 331, *355*
Gelehrter, I. D. 408, *419*
Geller, I., Seifter, J. 115, *203*
George, D. J., Wolf, H. H. 167, 169, *203*
Gerald, M. C., Riffee, W. H. 177, 178, *203*
Gérardin, A., see Boissier, J. R. 13, *26*
Gerber, J., see Stretch, S. 124, 126, *238*
Gericke, O. L. 265, *269*
Gerkens, J. F., McCulloch, M. W., Wilson, J. 147, *203*
German, D. C., Bowden, D. M. 121, 123, *203*
Gershon, S., see Angrist, B. M. 5, 19, *25*, 255–257, 261, 262, *267*, 282, 284, 292, 293, 295, *296*, 319, *350*
Gershon, S., see Ho, A. K. S. 57, *208*
Gershon, S., see Hine, B. 413, *420*
Gershon, S., see Klingenstein, R. J. 77, *212*
Gershon, S., see Sanghvi, I. 77, *232*
Gershon, S., see Schweitzer, J. W. 19, *30*

Gershon, S., see Singer, G. 77, 134, *236*
Gershon, S., see Wallach, M. B. 70, 72, 77, 78, 80, 176, *242*, 316, 320, 342, *367*
Gershon, S., see Willner, J. H. 70, 77, *244*
Gessa, G. L., Clay, G. A., Brodie, B. B. 50, 51, 149, 154, 156, 157, 163, 164, *203*
Gessa, G. L., see Brodie, B. B. 9, 15, 19, *26*, 34, 35, 50–52, 63, 139, 149, 150, 154, 156, 163, 164, *189*
Gessa, G., see Tagliamonte, A. 49, 53, 55, 56, 138, *239*
Gessner, P. K., Page, I. H. 314, *355*
Getterman, P. S., see Turner, C. E. 374, *428*
Gey, K. F., see Pletscher, A. 3, *29*, 53, 56, *227*
Gey, K. F., see Thoenen, H. 9, *31*, 50, 52, 63, *240*
Geyer, M. A., Segal, D. S. 88, *203*
Geyer, M. A., see Knapp, S. 56, *212*
Geyer, M. A., see Segal, D. S. 72, 78, 87, *235*
Ghetti, V., see Garattini, S. 306, *355*
Ghezzi, D., see Samanin, R. 134–136, *232*
Ghia, J., see Roft, D. D. 413, *427*
Ghielmetti, R., see Consolo, S. 17, *27*, 36, *193*
Ghirlanda, C., see Paris, M. 375, *426*
Ghiselli, W. B., see Evans, H. L. 112, 113, 118, *199*
Ghosh, M. N., Parvathy, S. 138, *203*
Ghosh, P., see Pradhan, S. N. 388, *426*
Giacalone, E., see Valzelli, L. 97, 167, *242*
Giarman, N. J., see Freedman, D. X. 330, *355*
Gibb, J. W., see Buening, M. K. 48, *189*
Gibb, J. W., see Koda, L. Y. 37, 48, *212*
Gibb, J. W., see Yagiela, J. A. 393–395, *429*
Gibbs, F. A., Maltby, G. L. 251, *269*
Giberti, F., Gegoretti, L. 312, *356*
Gibson, W. C., see Peretz, D. I. 311, 316, 328, *362*
Gidley, J. T., see Christensen, H. D. 383, *417*
Gill, E. W., Paton, W. D. M., Pertwee, R. G. 389, 410, *419*
Gill, J. R., Mason, D. T., Bartter, F. C. 8, 19, *28*
Gillbe, C., see Crow, T. J. 69, *195*
Gillin, J. C., Cannon, E., Magyar, R., Schwartz, M., Wyatt, R. J. 320, *356*
Gillespie, H. K., see Hollister, L. E. 247, *270*, 311, 319, *357*, 380, 385, *421*
Gillette, J. R., see Brodie, B. H. 9–11, *26*
Gilmour, D. G., Bloom, A. D., Lele, K. P., Robbins, E. S., Maximilliam, C. 402, *419*
Ginn, S. A., Hofman, L. B. 288, *299*
Ginneken, C. A. M. Van, Vree, T. B., Breimer, D. D., Thijssen, H. W. H., Rossum, J. M. Van 375, *429*
Ginneken, C. A. Van, see Henderson, P. T. 11, 14, *28*
Ginneken, C. A. M. Van, see Vree, T. B. *429*
Ginocchio, S., see Lal, H. 173, *214*
Ginsberg, M. D., Hertzmann, M., Schmidt-Nowara, W. W. 264, *269*, 288, *299*
Ginsberg, R., see Cappell, H. 115, 116, *190*
Ginsburg, J., see Beaconsfield, F. 397, *415*
Ginzel, K. H. 335, 336, *356*
Ginzel, K. H., Mayer-Gross, W. 318, *356*
Giry, L., see Paris, M. 375, *426*
Giurea, C., see Mourevieff-Lesuisse, F. 130, *223*
Glees, P., see Cole, J. 90, *193*
Glick, S. D. 90, 117, 127, 129, *203*
Glick, S. D., Greenstein, S. 68, *203*
Glick, S. D., Jarvik, M. E. 115, 118, *203*
Glick, S. D., Jerussi, T. P. 117, *203*
Glick, S. D., Jerussi, T. P., Waters, D. H., Green, J. P. 34, 37, 57, 68, *203*
Glick, S. D., Marsanico, R. G. 89, 90, 119, *203*
Glick, S. D., Müller, R. U. 110, 140, *203*
Glick, S. D., Nakamura, R. K., Jarvik, M. E. 90, 102, 107, *203*
Glick, S. D., Waters, D. H., Milloy, S. 34, *203*
Glick, S. D., see Jerussi, T. P. 68, *210*
Glisson, S. N., see Votava, Z. 315, 333, *367*
Glow, P. H. 314, *356*
Glowinski, J. 43, 44, 48, 55, *203*
Glowinski, J., Axelrod, J. 10, *28*, 38, 43, 48, *204*
Glowinski, J., Axelrod, J., Iversen, L. 4, 10, *28*, 38, 39, 43, 45, 46, 47, *204*
Glowinski, J., Baldessarini, R. J. 33, 34, *204*
Glowinski, J., Iversen, L., Axelrod, J. 38, 43, 46, *204*
Glowinski, J., see Besson, M. J. 44, 47, *186*
Glowinski, J., see Gauchy, C. 44, *202*
Glowinski, J., see Javoy, F. 35, 48, 49, *210*
Glowinski, J., see Jones, B. E. 57, *210*
Glowinski, J., see Philippu, A. 44, *226*
Gluckman, M. I., Baum, T. 123, 134, *204*
Godefroy, F., see Beauvallet, M. 34, 35, 37, 87, 154, 173, *185*
Görlitz, B.-D., Frey, H.-H. 53, 55, 179, *204*
Götestam, K. G., Gunne, L. M. 249, *269*
Goetz, C., Klawans, H. L. 77, 78, *204*
Goetz, L., see Weiner, W. J. 77, 78, *243*
Goetzl, F. R., Burril, D. Y., Ivy, A. C. 251, *269*
Goetzl, F. R., see Ivy, A. C. 33, *209*, 247, 251, 253, 254, *270*, 288, *300*
Gogerty, J. H., Dille, J. M. 320, 328, *356*
Gogerty, J. H., see Horita, A. 331, 338, *357*
Gogolák, G., Liebeswar, G., Sumpf, Ch. *204*
Gogolák, G., see Brücke, F. 340, *352*
Gogolák, G., see DeBaran, L. 340, *353*
Gogolák, G., see Stumpf, Ch. 340, *366*
Gokhale, S. D., Gulati, O. D., Udwadia, B. P. 146, 147, *204*
Gold, D., Reinert, H. 64, 142, *204*
Goldberg, L. 252, 257, *269*, 291, *299*
Goldberg, M. E., Ciofalo, V. B. 106, *204*

Goldberg, M. E., Salama, A. I. 167, 173, *204*
Goldberg, M. E., see Malick, J. B. 123, *218*
Goldberg, M. E., see Robichaud, R. C. 93, *230*
Goldberg, M. E., see Salama, A. I. 34, 35, 46, 53, 90, 97, *232*
Goldberg, M. F., see Dubinsky, B. 387, *418*
Goldberg, S. R. 126, *204*
Goldberg, S. R., see Hoffmeister, F. 124, *208*
Goldberg, S. R., see Schlichting, U. U. 124–126, *234*
Goldin, A., see Burton, R. M. 179, *189*
Goldman, H., Fischer, R., Nicolov, N., Murphy, S. 335, *356*
Goldman, M., see Latz, A. 101, 128, 129, *214*
Goldner, R. D., see Forrer, G. R. 310, *355*
Goldstein, L., Gardocki, J. F., Mundschenk, D. C., O'Brien, G. 178, *204*
Goldstein, L., Murphree, H. B., Pfeiffer, C. C. 251, *269*
Goldstein, L., Murphree, H. B., Sugerman, A. A., Pfeiffer, C. C., Jenney, E. H. 343, *356*
Goldstein, L., see Gardocki, J. F. 87, 167, 173, *202*
Goldstein, L., see Munoz, C. 177, *223*
Goldstein, L., see Pfeiffer, C. 251, *272*
Goldstein, M., Anagnoste, B. 9, 15, *28*, 50, *204*
Goldstein, M., Contrera, J. F. 9, *28*
Goldstein, M., McKereghan, M. R., Lauber, E. *9*, *28*, 48, 50, *204*
Goldstein, N. P., see Kies, M. W. 310, *359*
Goldstone, S., see Boardman, W. K. 308, *352*
Gollub, L. R., Brady, J. V. 120, *204*
Gollub, L. R., Mann, W. G., Jr. 98, 110, 117, *204*
Gollub, L. R., see Branch, M. N. 112, 113, 116, *188*
Gollub, L. R., see MacPhail, R. C. 138, *217*
Gomer, F. E., Jakubczak, L. F. 117, 129, *204*
Gomez, A. B., see Trendelenburg, U. 59, 146, *241*
Gomez, B. A. S., see Trendelenburg, U. 142, *241*
Gonzalez, C., see Carlini, E. A. 96, *190*
Gonzalez, S. C., Karniol, I. G., Carlini, E. A. 389, *419*
Gonzalez, R. I., see Tainter, M. L. 177, *239*
Gonzalez-Barón, S., Jiménez-Vargas, J., Lopez, G. G. 98, *204*
Gonzalez-Vegas, J. A. 329, *356*
Goodall, E. B., see Carey, R. J. 110, 112, 117, 122, 127, *190*
Goodell, H., see Richards, N. 310, *363*
Goodman, S. J., Becker, D. P. 265, *269*
Goodwin, F. K., see Post, R. M. 287, *302*
Goodwin, D. W., see Hill, S. Y. 413, *420*
Gorayeb, S. R. P., see Todorov, J. C. 111, *240*
Gordon, E. K., see Kopin, I. J. 50, *212*
Gordon, R. J., see McKenzie, G. M. 68, 78, *220*
Gorelick, D. A., see Stoff, D. M. 314, *365*
Gorgels, J. P. M. C., see Vree, T. B. 11, 14, *31*
Gorin, M. H., see Abramson, H. A. 318, *349*
Gorodetzky, C. W., Isbell, H. 310, *356*
Gorodetzky, C. W., see Isbell, H. 310, *358*, 380, 385, *421*
Gorrod, J. W. 5, *28*
Gorrod, J. W., see Bridges, J. W. 13, *26*
Gottsegen, G. 282, *299*
Gould, I. A., see Vachon, L. 414, *429*
Grabowska, M., see Maj, J. 149, 167, *218*
Grabowy, R., see Ellinwood, E. H., Jr. 178, *198*
Graeff, F. G., see Oliveira, L. de 112, 118, *225*
Graeff, F. G., see Todorov, J. C. 111, *240*
Graham, A. W., Aghajanian, G. K. 65, *204*
Graham, G., see Bolt, A. G. 164, *187*
Graham, J. D. P., Lewis, M. J., Williams, J. 393, 394, *419*
Graham, J. D. P., Li, D. M. R. 398, 411, *419*
Graham, J. M., Schreiber, R. A., Zemp, J. W. 177, *204*
Grahn, H. V. 284, *299*
Gram, T. E. 12, *28*
Grana, E., Lilla, L. 45–47, *204*
Grana, E., Sossi, D. 87, *204*
Granchelli, F. E., see Dewey, W. L. 399, *418*
Grandjean, J. L., see L'Huillier, J. 167, 173, *216*
Grandjean, J. L., see Nakajima, H. 154, *223*
Grandjean, J. L., see Thuillier, J. 150, 154, *240*
Granstrand, B., see Wahlquist, M. 382, *429*
Grant, J. D., Gross, S. J., Lomax, P., Wong, R. 378, *419*
Grant, L. D., see Neill, D. H. 90, 136, *223*
Gray, J. A., see Dudderidge, H. J. 128, *198*
Graybiel, A., see Wood, C. D. 288, *304*
Greden, J. F. 277, *299*
Green, A. L. 45, 47, *204*
Green, J. P., see Glick, S. D. 34, 37, 57, 68, *203*
Green, J. P., see Kang, S. 348, *359*
Green, K., Bowman, K. 414, 415, *419*
Green, K., Pederson, J. E. 414, *420*
Green, M. L., see Regelson, W. 413, 414, *427*
Green, T., Harvey, J. A. 111, 119, *205*
Green, W. J. 343, *356*
Greenberg, I., see Mendelson, J. H. 385, 409, *425*
Greenberg, I., see Pihl, R. O. 90, *227*
Greenberg, I., see Wayner, M. J. 112, 140, *243*
Greenberg, M. J. 329, *356*
Greenblatt, E. N., Osterberg, A. C. 155, 166, 169, 174, *205*
Greenblatt, M., see DiMascio, A. 310, *354*
Greenfield, E. J., see Proctor, C. D. 166, 173, *228*
Greengard, P., Kebabian, J. W. 65, *205*
Greenhard, P., see Quinn, G. P. 21, *30*
Greenstein, S., see Glick, S. D. 68, *203*
Greenwald, E. K., see Brown, D. J. 399, *416*
Gregg, J. M., see Purnell, W. D. 414, *426*
Gregg, J. M., see Roft, D. D. 413, *427*
Greiner, T., Burch, N. R., Edelberg, R. 310, *356*
Greving, H. 263, *269*

Griffin, E. D., Jr., see Dewey, W. L. 398, *418*
Griffin, W., see Matsumoto, C. 149, 154, *219*
Griffith, J. D. 282, 284, 291, *299*
Griffith, J. D., Carr, C. B., Jasinski, D. R. 285, *299*
Griffith, J. D., Cavanaugh, J., Held, J., Oates, J. A. 256, 261, *269*, *270*, 282, 293, 295, *299*
Griffith, J. D., Cavanaugh, J., Oates, J. A. 282, 293, 295, *299*
Griffith, J. D., Nutt, J. G., Jasinski, D. R. 249, 253, *270*
Griffith, J. D., see Cavanaugh, J. H. 9, 19, *26*
Griffith, J. D., see Jasinski, D. R. 285, *300*
Griffiths, D., Wahlsten, D. 100, *205*
Grilly, D. M., Ferraro, D. P., Braude, M. C. 407, *420*
Grilly, D. M., see Ferraro, D. P. 412, *418*
Grinspoon, L. 372, *420*
Grisham, M. G., Ferraro, D. P. 389, *420*
Grlic, L. J. 373, *420*
Groppetti, A., Costa, E. 16, 17, *28*, 34–36, 50, 51, 63, *205*
Groppetti, A., Misher, A., Naimzada, M., Revuelta, A., Costa, E. 48, 90, 157, *205*
Groppetti, A., Zambotti, F., Mantegazza, P. 36, 37, *205*
Groppetti, A., see Brodie, B. B. 34, 50, *189*
Groppetti, A., see Cattabeni, F. 51, 52, *192*
Groppetti, A., see Costa, E. 9, 15, 18, 19, 27, 34, 35, 37, 48, 50–53, 56, 78, 90, 132, 157, *194*
Gross, S. J., see Grant, J. D. 378, *419*
Grossman, L., see Clayton, R. N. 285, *297*
Grossman, S. P., Sclafani, A. 33, 98, *205*
Grossman, S. P., see Neill, D. B. 79, 90, 94, 134, 136, *223*
Grover, R. F., see Byrne-Quinn, E. 143, *190*
Groves, P. M., Rebec, G. V., Segal, D. S. 65, 67, *205*
Grown, P., see Fink, M. 410, *418*
Gruber, C. M., Weese, M. de 251, *270*
Grüter, H., see Opitz, K. 161, 166, *225*
Grunfeld, Y., Edery, H. 387, *420*
Guareschi, A. 251, *270*
Gudetsky, G. A., McCall, R. B., Chiueh, C. C., Moore, K. E. 44, *205*
Guernet, M., see Boissier, J.-R. 56, *187*
Guernet, M., see Simon, P. 34, 53, 87, *236*
Guerrero, O., see Fetterman, P. S. 373, 374, *418*
Guidotti, A., see Costa, E. 58, *194*
Guilleux, H., Peterfalvi, M. 68, *205*
Gulati, O. D., see Gokhale, S. D. 146, 147, *204*
Guldberg, H. C., see Marsden, C. A. 68, 69, *219*
Gulevich, G., see Dement, W. 285, *298*
Gumulka, W., see Samanin, R. 331, *364*
Gunn, C. G., see Schaefer, C. F. 401, *427*
Gunn, C. G., Jr., see Vogt, M. 335, 341, *367*
Gunn, J. A., Gurd, M. R. 62, 166, *205*
Gunn, J. A., Gurd, M. R., Sachs, I. 70, 141, 142, 144, 174, *205*
Gunne, L.-M. 14, *28*, 253, 263, *270*
Gunne, L.-M., Änggård, E. 24, *28*, 248, 262, 266, *270*
Gunne, L.-M., Änggård, E., Jönnsson, L. E. 265, *270*
Gunne, L.-M., Holmgren, P., Lindquist, O., Saldeen, T. 254, 264, *270*
Gunne, L.-M., Lewander, T. 34–37, 41, 42, *205*
Gunne, L. M., see Änggård, E. 8, 9, 14, 15, 18, 19, 21–24, *25*, 251, 260, 266, *266*, 295, *296*
Gunne, L. M., see Götestam, K. G. 249, *269*
Gunne, L.-M., see Jönsson, L. E. 18, 21, 23, *28*, 261, 265, *271*, 295, *300*
Gunne, L.-M., see Jonsson, J. 77, 154, *210*
Gupta, B. D., Dandiya, P. C., Gupta, M. L., Gabba, A. K. 93, *205*
Gupta, B. D., Holland, H. C. 100–102, *205*
Gupta, G. P., see Dhawan, B. N. 313, 336, *354*
Gupta, M. L., see Gupta, B. D. 93, *205*
Gura, S. J., see Raevskii, K. S. 173, 174, *228*
Gurd, M. R., see Gunn, J. A. 62, 70, 141, 142, 144, 166, 174, *205*
Gustafsson, B., see Agurell, S. 382, 384, *415*
Gustavson, J. W., see Schwartzbaum, J. S. 176, *235*
Guttman, E. 280, 282, *299*
Guttmann, E., Maclay, W. S. 307, *356*
Guttmann, E., Sargant, W. 247, 255, 258, *270*
Guttmann, E., see Peoples, S. A. 253, *272*, 280, *302*
Gutzait, L., see Ellison, T. 5, 8, 14, *27*
Guynet, P., see Jones, B. E. 57, *210*
Guzman, M., see Villarreal, J. E. 87, *242*
Gwynn, H. B., Yater, W. M. 247, *270*
Gyermek, L. 329, *356*
György, L., see Doda, M. 59, *197*
György, L., see Galambos, E. 87, *202*
György, L., see Pfeiffer, A. K. 49, 87, 161, 173, 177, *226*

Haag, M., see Korte, F. 375, 377, *422*
Haar, T. A. Von der 287, *303*
Haas, H. T. A. 149, 150, 154, *205*
Hada, H., see Utena, H. 46, 70, 81, 161, 161, 173, *241*
Hadick, D. G., see Overstreet, D. H. 135, *225*
Hadley, K., see Turner, C. E. 374, 380, *428*
Haefely, W., Jalfre, M., Monachon, M.-A. 178, *205*
Haefely, W., see Pieri, L. 333, *362*
Haefely, W., see Thoenen, H. 9, *31*, 49, 50, 52, 63, *240*
Haefliger, W., see Petrazilka, T. 379, *426*
Häggendal, J., Hamberger, B. 38, 39, *206*
Häggendal, J., see Andén, N.-E. 34, 42, *182*
Haertzen, C. A. 308, *356*
Haertzen, C. A., see Hill, H. E. 308, 309, *357*
Haertzen, C. A., see Wikler, A. 308, *367*
Haffner, F. 149, 178, *205*
Hahn, P. M., see Cohen, S. 408, *417*
Hahn, P. M., see Frank, I. M. 409, *419*
Haigler, H. J., Aghajanian, G. K. 330, *356*

Haigler, H.J., see Aghajanian, G.K. 330, *350*
Hajos, G.J.T., Garrattini, S. 132, 149, 150, 163, *206*
Halbach, A., see Matussek, N. 312, *361*
Halbach, H., see Eddy, N.B. 257, *269*
Hale, A.R., Reed, A.F. 310, *356*
Haley, S.L., Wright, P.L., Plank, J.B., Keplinger, M.L., Braude, M.C., Callandra, J.C. 406, 407, *420*
Haley, S.L., see Keplinger, M.L. 406, *422*
Haley, T.J. 253, *270*, 313, 316, *356*
Haley, T.J., Dasgupta, S.R. 315, 316, *356*
Haley, T.J., Rutschmann, J. 333, *356*
Hall, M., see Kilbey, M.M. 387, *422*
Halliwell, G., Quinton, R.H., Williams, F.E. 77, *206*
Halliwell, G., see Quinton, R.M. 71, 77, *228*
Halpern, B.N., Drudi-Baracco, C., Bessirard, D. 36, 173, *206*
Halpern, B.N., Morard, J.C., Drudi-Baracco, C. 174, *206*
Halpern, B.N., see Beauvallet, M. 35, 174, *185*
Halpern, B.N., see Drudi-Baracco, C. 173, *197*
Halpern, M., see Citron, B.P. 261, *268*
Hamaoui, A., see Carlini, E.A. 387, *416*
Hamar, C., see Rapin, J. 41, 51, *229*
Hamberger, B., Malmfors, T. 38, *206*
Hamberger, B., see Carlsson, A. 38, 39, 42, *191*
Hamberger, B., see Carlsson, M. 38, 39, 42, *191*
Hamberger, B., see Fuxe, K. 39, 54, *202*
Hamberger, B., see Häggendal, J. 38, 39, *206*
Hameed, K.A., see Gaddum, J.H. 328, 329, *355*
Hamilton, A.E., see Horita, A. 333, 338, *357*
Hamilton, C.L. 314, *356*
Hamilton, C.L., Wilpizeski, C. 314, *357*
Hamon, M., see Javoy, F. 49, *210*
Hampton, W.H. 251, *270*, 291, *299*
Hance, A.J., see Bradley, P.B. 341, *352*
Handley, C.A., Ensberg, D.L. 288, *299*
Handley, C.A., Sweeney, H.M. 148, *206*
Handley, S.L., see Sayers, A.C. 77, *233*
Handlon, J.H., see Szara, S. 311, *366*
Handrick, G.R., see Razdan, R.K. 379, 380, *426*, *427*
Handrick, G.R., see Uliss, D.B. 379, *429*
Hanington, E. 279, *299*
Hanington, E., see Sandler, M. 279, *303*
Hanna, C. 62, 141, 142, *206*
Hanna, C., Upton, P.D., Chambers, W.F. 176–178, *206*
Hansen, P.H. 255, *270*
Hanson, H.M., Witoslawski, J.J. Campbell, E.H. 111, 115, 116, *206*
Hanson, L.C.F. 101, 106, *206*
Harakal, J.J., see Sofia, R.D. 391, *428*
Harakal, J.J., see Segelman, A.B. 390, *427*
Harbison, R.D., Mantilla-Plata, B. 407, *420*
Harbison, R.D., see Mantilla-Plata, B. 407, 408, *424*
Hardinge, M.G., see Peterson, D.I. 175, *226*
Hardman, H.F., Domino, E.F., Seevers, M.H. 398–400, 410, *420*
Hardman, H.F., Hosko, M.J. 398, *420*
Harik, S.I., see North, R.B. 72, 82, *224*
Harmon, J., Alipoulus, M.A. 400, *420*
Harper, C.E., see Tashkin, D.P. 409, *428*
Harrie, J.R. 277, *299*
Harris, J.E., Baldessarini, R.J. 38–40, 47, *206*
Harris, L.S. 403, *420*
Harris, L.S., Munson, A.E., Carchman, R.A. 402, *420*
Harris, L.S., Munson, A.E., Friedman, M.A. 402, *420*
Harris, L.S., see Abel, E.L. 410, *415*
Harris, L.S., see Adams, M.D. 399, *415*
Harris, L.S., see Bloom, A.S. 390, 393, 394, *416*
Harris, L.S., see Carchman, R.A. 402, *416*
Harris, L.S., see Dewey, W.L. 381, 382, 386, 389, 390, 391, 397–400, 403, 409, 411, 412, *418*
Harris, L.S., see Frankenheim, J.M. 388, 389, *419*
Harris, L.S., see Johnson, K.M. 392, *421*
Harris, L.S., see Kosersky, D.S. 388, 391, 410, *422*
Harris, L.S., see Levy, J.A. 402, *423*
Harris, L.S., see Magus, R.D. 401, *423*
Harris, L.S., see Martin, B.R. 381–384, 390, 410–412, *424*
Harris, L.S., see McMillan, D.E. 388, 411, 412, *425*
Harris, L.S., see Munson, A.E. 402, *425*
Harris, L.S., see Roft, D.D. 413, *427*
Harris, L.S., see Spaulding, T.C. 386, *428*
Harris, L.S., see Turk, R.F. 380, *428*
Harris, L.S., see Zitko, B.A. 380, *429*
Harris, P.D., see Brown, D.J. 399, *416*
Harris, R.T., Balster, R.L. 130, *206*
Harris, R.T., Waters, W., McLendon, D. 410, *420*
Harris, R.T., see Jones, C.N. 130, 131, *210*
Harris, R.T., see Kilbey, M.M. 387, *422*
Harris, R.T., see Water, W.H. 130, *243*
Harris, S.C., Ivy, A., Searle, L.M. 131, 138, *206*, 250, *270*, 280, *299*
Harris, W.C., see Pickens, R. 124, 125, *227*
Harrison, D.C., Chidsey, C.A., Braunwald, E. 146, *206*
Harrison, D.C., see Chidsey, C.A. 8, *26*, 41, 143, *192*
Harrison, J.M., Abelson, R.M. 98, 99, *206*
Harrison, J.W.E., Ambrus, C.M., Ambrus, J.L. 9, *28*, 157, *206*
Hart, E.R., see Marrazzi, A.S. 339, *361*
Hart, M., Dienske, H., Udo de Haes, H.A., Noach, E.L. 97, *206*
Hartmann, A.M., Hollister, L.E. 310, *357*
Hartman, A.M., see Hollister, L.E. 308, 310, *357*

Hartmann,E., see Schildkraut, J.J. 261, *273*, 294, *303*
Hartmann,E., see Watson,R. 294, *303*
Hartman,F.R., see Freedman, D.X. 314, 319, 320, *355*
Hartung,W.H., Munch,J.C. 141, *206*
Harvey,J.A., see Green,T. 111, 119, *205*
Harvey,J.U., Todd,C.W., Howard,J.W. 265, *270*
Harvey,S.C., Sulkowski,T.S., Weenig,D.J. 41, 62, 63, 141–144, 146, *206*
Hasegawa,A.T., see Wang,R.I. H. 167, *243*
Hasselager,E., Rolinski,Z., Randrup,A. 77, 87, 88, 93, *207*
Hasselager,E., see Scheel-Krüger,J. 55, 149, 155, 157, *233*
Hatch,R.C. 177, 179, *207*
Hauschild,F. 70, *207*, 263, *270*, 280, *299*
Hauser,K.M., see Fuller,R.W. 5, *27*
Hauty,G.T., Payne,R.B. 249, *270*
Hauty,G.T., see Payne,R.B. 249, *272*
Haverback,B.J., see Citron,B.P. 261, *268*
Hawke,D.M., Chesher,G.B. 61, *207*
Hawkins,J.R., see Fabing,H.D. 310, *355*
Hawks,D., Mitcheson,M., Ogborne,A., Edwards,G. 282, 291, *299*
Hayano,M., see Bergen,J.R. 315, *351*
Hayden,D.W., see Fleischman, R.W. 407, 408, *419*
Hayhow,W.R., see Bishop,P.O. 336, *352*
Hazelwood,J.C., see Funderburk,W.H. 132, 134, *202*
Hearst,E. 111, 117, *207*
Hearst,E., Vane,J.R. 111, 114, *207*
Hearst,E., Whalen,R.E. 101, 103, *207*
Heath,R.G., Nessehoff,W., Bishop,M.P., Byers,L.W. 292, *299*
Heath,R.G., see Monroe,R.R. 343, *362*
Hecht,F., Beals,R.K., Lees,M. H., Jolly,H., Roberts,P. 408, *420*
Heffner,T.G., Drawbaugh,R.B., Zigmond,M.J. 110–112, 116, *207*
Heil,G.C., see Leonhard,D.A. 138, *215*
Heim,F., Leuschner,F., Estler, K.J. 161, *207*
Heim,R., Wasson,R.G. 305, *357*
Heimburger,M., see Valette,G. 62, 63, 141, *241*
Heimstra,N.W. 70, 96, 97, *207*
Heimstra,N.W., McDonald,A. 71, 90, 96, *207*
Heinrich,R., see Renault,P.F. 397, *427*
Heise,G.A., Boff,E. 99, *207*
Heise,G.A., Lilie,N.L. 115, *207*
Held,J., see Griffith,J.D. 256, 261, *269*, *270*, 282, 293, 295, *299*
Heller,B., see Fischer,E. 70, 77, *200*
Hemsworth,B.A., Neal,M.J. 57, *207*
Henderson,J.K., see Stark,P. 121, 122, *237*
Henderson,P.T., Vree,T.B., Ginneken,C.A., van, Rossum,J.M.van, 11, 14, *28*
Hendley,E.D., Snyder,S.H. 38, 40, *207*
Hendry,D.P., Toller,C.van 115, *207*
Heneen,W., see Nichols,W.W. 402, *425*
Henke,R., see Domino,E.F. 102, *197*
Hennessy,B.L., see Bishop,P.O. 336, *352*
Henning,M., see Andén,N.-E. 59, *182*
Henriksson,B.G., Jarbe,T. 388, *420*
Henry,C.M., see Abbott,W.O. 8, *25*
Hensel,D., see Smetzer,D.L. 142, *236*
Hepler,R.S., Frank,I.A. 414, *420*
Hepler,R.S., Frank,I.M., Petrus,R. 408, 409, 414, *420*
Hepler,R.S., Frank,I.M., Ungerleider,J.T. *420*
Hepler,R.S., Petrus,R.J. *420*
Herha,J., Obe,G. 401, *420*
Herman,Z.S. 77, *207*
Herman,Z.S., Kmieciak-Kolada,K., Drybanski,A., Sokola,A., Trzeciak,H., Chrusciel,T.L. 35, 37, 87, 91, *207*
Herman,Z.S., Trzeciak,H., Chrusciel,T.L., Kmieciak-Kolada,K., Drybanski,A., Sokola,A. 35, 53, 92, *207*
Hernodon,J.G., Neill,D.B. 98, *207*
Herold,M., see Cahn,J. 93, 154, 177, *190*
Heron,W.T., see Skinner,B.F. 112, 117, *236*
Herr,F., see Borella,L.E. 16, *26*, 87, 88, 134, 154, *188*
Herridge,C.F., A'Brook,M.F. 278, *300*
Herrnstein,R.J., see Morse,W. H. 110, 116, *223*
Hertting,G., Axelrod,J., Whitby,G.L. 38, *207*
Hertting,G., see Axelrod,J. 34, 38, 40, *183*
Hertzmann,M., see Ginsberg, M.D. 264, *269*, 288, *299*
Herz,A. 98, *207*
Hewson,R., see Yelnosky,J. 147, *245*
Heyman,I.A., see Luthra,X.K. 405, *423*
Heyndrickx,A., Scheiris,C., Schepens,P. 403, *420*
Hibi,S., see Feinberg,I. 252, 260, *269*, 286, *298*
Hibbitts,J.T., see Hundley,J.M. 247, 248, *270*
Hickenbottom,I.P., see Davis, W.M. 130, 143, 175, *196*
Hiebel,G., Bonvallet,M., Huvé,P., Dell,P. 176, *207*
High,J.P., see Horovitz,Z.P. 97, *209*
Higuchi,H., Matsuo,T., Shimamoto,K. 36, *208*
Hill,D., see Wada,J.A. 101, *242*
Hill,D.F., see Hungen,K.von 333, *367*
Hill,H.E., Bell,E.C., Wikler,A. 116, *208*
Hill,H.E., Haertzen,C.A., Wolbach,A.B.,Jr., Miner, E.J. 308, 309, *357*
Hill,H.E., see Wikler,A. 308, *367*
Hill,H.F., Horita,A. 149, 154, 157, *208*
Hill,H.F., see Jones,C.N. 130, 131, *210*
Hill,H.F., see Horita,A. 149, *208*, 337, 338, *357*
Hill,J. 250, *270*

Hill, R., see Hulme, E. C. 47, 56, 134, *209*
Hill, S. Y., Schwin, R., Goodwin, D. W., Powell, B. J. 413, *420*
Hillman, C. C., see Bewsher, P. D. 162, 165, 175, *186*
Himmelsbach, C. K. 344, *357*
Himmelsbach, C. K., Cleek, D. M. Lloyd, B. J. Jr. 283, 295, *300*
Himwich, H. E. 306, *357*
Himwich, H. E., see Fujimori, M. 176, *201*, 340, *355*
Himwich, H. E., see Rinaldi, F. 176, *230*, 339, 340, *363*
Himwich, H. E., see Schweigerdt, A. K. 340, *364*
Himwich, H. E., see Shimizu, A. 178, *235*
Himwich, H. E., see Takeo, Y. 176, *239*, 340, *366*
Himwich, H. E., see Votava, Z. 315, 333, *367*
Hine, B., Friedman, E., Torrelio, M., Gershon, S. 413, *420*
Hine, B., see Wallach, M. B. 320, *367*
Hines, C. W., see Fuller, R. W. 3, *27*, 56, 168, 173, *202*
Hirsch, M. W., see Abramson, H. A. 308, 310, 318, *350*
Hirsch, M. W., see Jarvik, M. E. 308, 310, 318, *358*
Hirsch, S. E., see Friedman, S. A. 310, *355*
Hirschhorn, I. D., Winter, J. C. 314, 315, *357*
Hirschhorn, I. D., Rosecrans, J. A. 413, *421*
Hirtz, J., see Boissier, J. R. 13, *26*
Hishikawa, Y., Ida, H., Nakai, K., Kaneko, Z. 286, *300*
Hiss, R. A., see Ferster, C. B. 110, 114, *200*
Hitomi, M., see Wilson, M. C. 124, 125, *244*
Hitzemann, R. J., Loh, H. H. 52, 164, *208*
Hitzemann, R. J., Loh, H. H., Carves, F. B., Domino, E. F. 35–37, 49, 80, 91, *208*
Hitzemann, R. J., Loh, H. H., Domino, E. F. 56, *208*
Hitzemann, R., see Loh, H. 164, *217*
Hitzeman, R. J., see Tseng, L. F. 44, 46, *241*
Hively, R. L., Mosher, W. A., Hoffman, F. W. 375, *421*
Ho, A. K. S., Gershon, S. 57, *208*
Ho, B. T., Fritchie, G. E., Englert, L. F., McIsaac, W. M., Idanpaan-Heikkila, J. E. *421*
Ho, B. T., Fritchie, G. E., Kralik, P. M., Englert, L. F., Mc Isaac, W. M., Idapan-Heikkila, J. *421*
Ho, B. T., Taylor, D., Englert, L. F. 391, 393, 395, 396, *421*
Ho, B. T., Taylor, D., Englert, L. F., McIsaac, W. M. 393–396, *421*
Ho, B. T., Taylor, D., Fritchie, G. E., Englert, L. F., McIssac, W. M. 393–396, *421*
Ho, B. T., see Johnson, K. M. 392, *421*
Ho, B. T., see Huang, J.-T. 87, 146, *209*
Ho, B. T., see Lu, T.-C. 92, *217*
Ho, B. T., see McIsaac, W. M. 381, *424*
Ho, J. E., see Scott, J. P. 155, *235*
Hobbs, R., see Burn, J. H. 173, *189*
Hobson, J. A. 342, *357*
Hoch, P. H. 312, *357*
Hoch, P. H., Cattell, J. P., Pennes, H. H. 336, *357*
Hoch, P. H., see Bradley, P. B. 306, *352*
Hoch, P. H., see Malitz, S. 310, 343, *361*
Hochman, J. S., see Brill, N. O. 403, *416*
Hodos, W., Ross, G. S., Brady, J. V. 100, 114, *208*
Hoebel, B. G., see Ahlskog, J. E. 135, 136, 139, *181*
Höhn, R., Lasagna, L. 167–169, 174, *208*
Hökfelt, T., see Andén, N.-E. 68, 69, *182*, 333, 335, *350*
Hökfelt, T., see Carlsson, A. 38, 54, *191*
Hökfeldt, T., see Corrodi, H. 34–36, 48, 66, *194*
Hökfelt, T., see Fuxe, K. 68, 79, *202*
Hoffer, A., Osmond, H. 306, *357*
Hoffman, F. W., see Hively, R. L. 375, *421*
Hoffmann, K. P., see Straschill, M. 65, *238*
Hoffmeister, F., Goldberg, S. R. 124, *208*
Hoffmeister, F., see Schlichting, U. U. 124–126, *234*
Hoffström, I., Orrenius, S. 12, *28*
Hofmann, G., see Arnold, O. H. 334, *351*
Hofman, L. B., see Ginn, S. A. 288, *299*
Hogmark, A. L., see Änggård, E. 8, 9, 14, 15, 18, 19, 21, *25*, 295, *296*
Holland, G. F., Buck, C. J., Weissman, A. 3, *28*
Holland, H. C., see Gupta, B. D. 100–102, *205*
Holley, J. R., see Powell, D. A. 97, *228*
Hollister, A. S., Breese, G. R., Cooper, B. R. 79, 87–89, *208*
Hollister, L. E. 250, *270*, 277, *300*, 306, 310, *357*, 382, 385, 414, *421*
Hollister, L. E., Friedhoff, A. J. 311, 323, *357*
Hollister, L. E., Gillespie, H. K. 247, *270*, 380, *421*
Hollister, L. E., Hartman, A. M. 308, 310, *357*
Hollister, L. E., Macnicol, M. F., Gillespie, H. K. 311, 319, *357*
Hollister, L. E., Reaven, G. M. 380, 400, *421*
Hollister, L. E., Richards, R. K., Gillespie, H. K. 385, *421*
Hollister, L. E., Sjoberg, B. M. 307, *357*
Hollister, L. E., see Hartman, A. M. 310, *357*
Hollister, L., see Nichols, W. W. 402, *425*
Hollister, L., see Snyder, S. H. 4, *30*, 311, 328, *365*
Hollunger, G., Persson, S. A. 333, *357*
Holm, T., Huus, I., Kopf, R., Nielsen, I. M., Petersen, D. V. 138, *208*
Holmgren, P., see Gunne, L.-M. 254, 264, *270*
Holmstedt, B., see Efron, D. H. 306, *354*
Holtzman, D., Lovell, R. A., Jaffe, J. H., Freedman, D. X. 391, 393, 394, *421*
Holtzman, S. G. 87, 106, 134, 154, *208*
Holtzman, S. G., Jewet, R. E. 134, *208*
Holtzman, S. G., see Faraj, B. A. 3, *27*
Honecker, H., see Rating, D. 17, *30*
Honecker, H., see Rommelspacher, H. 8, *30*
Honig, W. M. M., see Broekkamp, C. L. E. 137, *189*

Hope, J., see Adams, R. D. 292, *296*
Hopkins, J. T., see Schmidt, M. J. 57, 58, *234*
Horita, A., Dille, J. M. 337, *357*
Horita, A., Gogerty, J. H. 331, 338, *357*
Horita, A., Hamilton, A. E. 333, 338, *357*
Horita, A., Hill, H. F., 149, *208*, 337, 338, *357*
Horita, A., see Hill, H. F. 149, 154, 157, *208*
Horita, A., see Matsumoto, C. 147, 148, *219*
Horita, A., see Roszell, D. K. 338, *363*
Horlington, M., see Davis, R. A. 179, *196*
Horn, A. S., Coyle, J. T., Snyder, S. H. 39, *208*
Horn, A. S., Cuello, A. C., Miller, R. J. 40, *208*
Horn, A. S., Snyder, S. H. 39, 40, *208*
Horn, A. S., see Snodgrass, S. R. 332, *365*
Horn, G., McKay, J. M. 337, *358*
Horn, G., see McKay, J. M. 337, *361*
Hornecker, H., see Rating, D. 410, *426*
Horng, J. S., see Wong, D. T. 49, 54, 56, *245*
Horovitz, Z. P., Chow, M. I., Carlton, P. I. 122, *208*
Horovitz, Z. P., Piala, J. J., High, J. P., Burke, J. C., Leaf, R. C. 97, *209*
Horovitz, Z. P., see Leaf, R. C. 97, *215*
Horst, D., see Kies, M. W. 310, *359*
Horst, W. D., see Kopin, I. J. 63, *212*
Hosko, M. J., see Hardman, H. F. 398, *420*
Hosko, M. J., see Kochar, M. S. 397, *422*
Houser, V. P. 100, 107, *209*
Howard, J. W., see Harvey, J. K. 265, *270*
Howes, J. F., see Dewey, W. L. 381, 399, 409, *418*
Howes, J. F., see Zitko, B. A. 380, *429*
Hoyer, I., Zwieten, P. A. van 142, *209*, 247, *270*
Hrbek, J., see Navratil, J. 128, *223*
Hsu, J., see Nahas, G. G. 402, *425*
Huang, J.-T., Ho, B. T. 87, 146, *209*
Hubbard, J. D., see Phillipps, R, N. 404, *426*
Hubin, P., Servais, J. 248, *270*
Hucker, H. B. Michniewicz, B. M., Rhodes, R. E. 9, 11, *28*
Huerlimann, A., see Thoenen, H. 9, *31*, 49, 50, 52, 63, *240*
Huestis, R. D., Arnold, L. E., Smeltzer, D. J. 286, *300*
Hughes, F. W., see Garriott, J. C. 87, *202*
Hughes, F. W., see Khan, A. U. 163, *211*
Hughes, J. E., Steakly, L. P., Bier, M. M. 403, *421*
Hughes, I. E., see Deab, H. G. 147, *196*
Hugon, P., see Beregi, L. G. 4, *26*
Hulme, E. C., Hill, R., North, M., Kibby, M. R. 47, 56, 134, *209*
Humoller, F. L., see Mahler, D. J. 320, *361*
Hundley, J. M., Krautz, J. C., Hibbitz, J. T. 247, 248, *270*
Hungen, K. von, Roberts, S., Hill, D. F. 333, *367*
Hurkmans, J. A., see Rossum, J. M. van 70, 71, 87, 88, *231*
Hurkmans, J. A. Th., see Schoot, J. B. van der 71, 87, 88, *234*
Hurst, P. M., Bagley, S. K. 248, *270*
Huston, J. P., Borbély, A. A. 80, 90, 155, *209*
Huston, J. R., see Borbély, A. A. 80, 155, *188*
Hutchins, D. A., Rogers, K. J. 87, 162, *209*
Hutchins, D. A., see Rogers, K. J. 162, *230*
Huttunen, M. O., see Diaz, J.-L. 53, 55, 56, *197*
Huus, I., see Holm, T. 138, *208*
Huxley, A. 305, *358*
Huvé, P., see Hiebel, G. 176, *207*
Hyde, R. W., see DiMascio, A. 310, *354*
Hyde, R. W., see Rinkel, M. 308, 343, *363*
Hyde, R. W., see Salvatore, S. 308, *364*

Ida, H., see Hishikawa, Y. 286, *300*
Idanpaan-Heikkila, J. E., see Ho, B. T. *421*
Idanpaan-Heikkila, J. E., see McIsaac, W. M. 381, *424*
Ide-Johanson, L., see Schwartzbaum, J. S. 176, *235*
Igisu, T., see Takasaki, K. 62, *239*
Ihli, M., see Schuster, J. 143, *235*
Inghe, G. 257, *270*
Ingvar, D. H., Soderberg, U. 341, *358*
Innes, I. R. 61, *209*
Innes, I. R., Kohli, J. D. 61, *209*
Innes, I. R., Nickerson, M. 249, 251, 253, 254, 259, *270*
Innis, N. K., see Wuttke, W. 112, *245*
Inscoe, J. K., see Axelrod, J. 54, *183*
Intemann, E., see Chu, H. 161, *193*
Irwin, S., Slabok, K., Thomas, G. 87, 90, *209*
Isaac, W. 90, *209*
Isaac, W., Troelstrup, R. 90, *209*
Isaac, W., see Alexander, M. 90, *182*
Isaac, W., see Seegal, R. F. 90, *235*
Isachenko, V. B., Nikulcheva, N. G., Nargorney, V. A. 164, *209*
Isbell, H. 310, *358*
Isbell, H., Belleville, R. E., Fraser, H. F., Wikler, A., Logan, C. R. 308, 310, 311, 318, *358*
Isbell, H., Fraser, H. F., Wikler, A., Belleville, R. E. 318, *358*
Isbell, H., Gorodetzky, C. W. 310, *358*
Isbell, H., Gorodetzky, C. W., Jasinski, D., Claussen, U., Spulak, F., Korte, F. 380, 385, *421*
Isbell, H., Jasinski, D. R. 319, *358*, 385, *421*
Isbell, H., Logan, C. R. 312, *358*
Isbell, H., Logan, C. R., Miner, E. J. 310, 312, 318, 330, *358*
Isbell, H., Rosenberg, D. E., Miner, E. J., Logan, C. R. 319, *358*
Isbell, H., Wolbach, A. B., Wikler, A., Miner, E. J. 318, *358*
Isbell, H., see Eddy, N. B. 257, 269
Isbell, H., see Gorodetzki, C. W. 310, *356*
Isbell, H., see Rosenberg, D. E. 253, 254, 256, 261, *273*, 310, 311, 318, 319, *363*

Isbell, H., see Vogel, V. H. 255, *274*
Isbell, H., see Wolbach, A. B., Jr. 310, 318, *368*
Island, D., see Rangno, R. E. 8, *30*
Israili, Z. H., see Faraj, B. A. 3, *27*
Itil, T., see Bente, D. 251, *267*
Itil, T. M., see Fink, M. 251, *269*
Ito, K., see Onodera, I. 165, *225*
Iversen, L. L. 38, 42, *209*
Iversen, L. L., see Burgen, A. S. V. 38, *189*
Iversen, L. L., see Evetts, K. D. 87, 89, 96, *199*
Iversen, L. L., see Glowinski, J. 4, 10, *28*, 38, 39, 43, 45–47, *204*
Iversen, L. L., see Lew, C. 87, *215*
Iversen, S. D. 90, *209*
Iversen, S. D., Wilkinson, S., Simpson, B. 79, 90, *209*
Iversen, S. D., see Creese, I. 79, 87, 89, *195*
Iversen, S. D., see Evetts, K. D. 87, 89, 96, *199*
Iversen, S. D., see Lew, C. 87, *215*
Iversen, S. D., see Robbins, T. 94, 114, *230*
Iversen, S. D., see Simpson, B. A. 79, 89, *236*
Ivy, A. C., Goetzl, F. R. 33, *209*, 247, 251, 253 254, *270*
Ivy, A. C., Goetzl, F. R., Burril, D. Y. 288, *300*
Ivy, A. C., Krasno, L. R. 247, 251, 253, 254, 255, 259, *270*
Ivy, A. C., see Adler, H. F. 249, *266*
Ivy, A. C., see Goetzl, F. R. 251, *269*
Ivy, A. C., see Harris, S. C. 131, 138, *206*, 250, *270*, 280, *299*
Iwata, H., Watanabe, K., Nikishawa, T., Ohashi, K. 80, 142, *209*
Izquierdo, I., Vasquez, B. J., Nasello, A. G. 176, *209*
Izquierdo, I., see Evangelista, A. M. 102, 129, *199*
Izquierdo, I., see Nasello, A. G. 165, *223*
Izquierdo, I., see Vasquez, A. J. 106, *242*

Jaanus, S. D., see Rubin, R. P. 41, *231*
Jack, D., see Brittain, R. T. 154, 173, *189*
Jackson, B., see Davies, J. A. 102, *195*
Jackson, C. O. 283, *300*
Jackson, D. M., see Chesher, G. B. 390, *417*
Jacob, J. 338, *358*
Jacob, J., Lafille, C. 315, 338, *358*
Jacob, J., Lafille, C., Loiseau, G., Echinard-Garin, P., Barthelemy, C. 313, *358*
Jacob, J., Michaud, G. 179, *209*
Jacobsen, E. 306, *358*
Jacobsen, E., Bahnsen, P., Wollstein, A. 247, 248, *270*
Jacobsen, E., see Bahnsen, P. 247, 248, 255, 256, *267*
Jaquet, C., see Rapin, J. R. 49, *229*
Jacquot, C., L'Hermite, P., Cohen, Y., Valette, G. 51, *209*
Jacquot, C., L'Hermite, P., Rapin, J. R., Cohen, Y. 51, *210*
Jacquot, C., Rapin, J., El Guedri, H., Cohen, Y. 51, *210*
Jacquot, C., Rapin, J., Renault, H., Cohen, Y. 48, *210*
Jacquot, C., see Boulu, R. 37, 68, *188*
Jacquot, C., see El Guedri, H. 9, 19, *27*, 52, *198*
Jacquot, C., see Rapin, J. 41, 51, *229*
Jacquot, C., see Valette, G. 62, 63, 141, *241*
Jaffe, J. H., Sharpless, S. K. 344, *358*
Jaffe, J. H., see Holtzman, D. 391, 393, 394, *421*
Jakubczak, L. F., see Gomer, F. E. 117, 129, *204*
Jalfre, M., see Haefely, W. 178, *205*
James, G. W. B. 255, *270*
James, J. H., see Fischer, J. E. 279, *298*
Jandhyala, B. S., see Cavero, I. 397–399, *417*
Jané, J., see Laporte, J. 60, *214*
Janiger, O., see Dorrance, D. 402, *418*
Janowsky, D. S., El-Yousef, M. K., Davis, J. M., Sekerke, J. H. 287, *300*
Janowsky, D. S., see Davis, J. M. 18, *27*
Janssen, P. A. J. 15, *28*
Janssen, P. A. J., Niemegeers, J. E., Schellekens, K. H. L. 77, 81, *210*
Janssen, P. A. J., Niemegeers, J. E., Schellekens, K. H. L., Lenaerts, F. M. 77, *210*
Janssen, P. A. J., Niemegeers, J. E., Schellekens, K. H. L., Dresse, A. D., Lenaerts, F. M., Pinchard, A., Schaper, W. K. A., Nueten, J. M. van, Verbruggen, F. J. 77, *210*
Janssen, P. A. J., see Niemegeers, C. J. E. 98, 100, 101, 106, *224*
Jarbe, T., see Henriksson, B. G. 388, *420*
Jaros, O., see Sercl, M. 251, *273*
Jarrard, L. E. 314, *358*
Jarvik, L. F., see Matsuyama, S. S. 402, *424*
Jarvik, M. E., Abramson, H. A., Hirsch, M. W. 308, 310, 318, *358*
Jarvik, M. E., Abramson, H. A., Hirsch, M. W., Ewald, A. T. 308, *359*
Jarvik, M. E., Chorover, S. 317, *359*
Jarvik, M. E., see Abramson, H. A. 307, 308, 310, 311, 318, *350*
Jarvik, M. E., see Bodanecky, S. 98, 102, 130, *187*
Jarvik, M. E., see Glick, S. D. 90, 102, 107, 115, 118, *203*
Jarvik, M. E., see Siegel, R. K. 317, *365*
Jasinski, D. R., Nutt, J. G. 285, 288, *300*
Jasinski, D. R., Nutt, J. G., Griffith, J. D. 285, *300*
Jasinski, D. R., see Griffith, J. D. 249, 253, *270*, 285, *299*
Jasinski, D. R., see Isbell, H. 319, *358*, 380, 385, *421*
Jasinski, D. R., see Martin, W. R. 253, 254, *272*, 278, 281, 285, *301*
Javoy, F., Agid, Y., Bouvet, D., Glowinski, J. 48, *210*
Javoy, F., Hamon, M., Glowinski, J. 49, *210*
Javoy, F., Thierry, A. M., Kety, S. S., Glowinski, J. 35, *210*
Jellinek, P. 149, *210*
Jenden, D. J., see Cho, A. K. 11, *26*
Jenden, D. J., see Schuberth, J. 261, *273*
Jenkins, J., see Dewey, W. L. 386, 390, 399, 411, 412, *418*

Jenkins, M. L., see Faraj, B. A. 3, *27*
Jenney, E. H., see Goldstein, L. 343, *356*
Jennings, W. A., Jr. 93, *210*
Jerussi, T. P., Glick, S. D. 68, *210*
Jerussi, T. P., see Glick, S. D. 34, 37, 57, 68, 117, *203*
Jespersen, J., Bonaccorsi, A. *210*
Jespersen, S., Bonaccorsi, A. 143, 147, 154, 173, *210*
Jespersen, S., Scheel-Krüger, J. 132, 134, *210*
Jespersen, S., see Bizzi, A. 43, 46, 141, 143, 147, 149, 154–156, 163, 164, *187*, 251, *268*
Jewett, R. E., Norton, S. 178, *210*
Jewett, R. E., see Holtzman, S. G. 134, *208*
Jiménez-Vargas, J., see Gonzalez-Baron, S. 98, *204*
Job, W. M., see Kulkarni, A. S. 102, 129, *213*
Jönsson, L.-E. 254, 261, 265, *271*
Jönsson, L.-E., Änggård, E., Gunne, L.-M. 18, 21, *28*, 265, *271*
Jönsson, L.-E., Gunne, L.-M. 295, *300*
Jönsson, L.-E., Gunne, L.-M., Änggård, E. 23, *28*, 265, *271*
Jönsson, L.-E., Lewander, T., Gunne, L.-M. 261, *271*
Jönsson, L.-E., Schuberth, J., Sundwall, A. 261, *271*
Jönsson, L. E., see Änggård, E. 8, 9, 14, 15, 18, 19, 21–24, *25*, 251, 260, *266*, 295, *296*
Jönsson, L. E., see Gunne, L.-M. 265, *270*
Johansson, I. B., see Agurell, S. 382, 384, *415*
John, E. R., Wenzel, B. M., Tschirgi, R. D. 106, *210*
Johnson, F. H., Waite, K. 129, *210*
Johnson, G. A., see Rudzik, A. D. 177, 178, *231*
Johnson, G. E., see Wood, D. R. 287, *304*
Johnson, J., Milner, G. 291, *300*
Johnson, J. E., see Kriskó, I. 264, *271*, 288, *300*
Johnson, J. R., see Asatoor, A, M, 7, 15, 21, *25*
Johnson, K. 394, 395, *421*
Johnson, K. M., Ho, B. T., Dewey, W. L., Harris, L. S. 392, *421*
Johnson, K. M., see Bloom, A. S. 392, 393, 394, *416*
Johnson, R. E., see Dill, D. B. 254, *268*
Johnson, R. H., see Sulaiman, W. R. 255, *274*
Johnson, S. 277
Johnson, S., Domino, E. F. 397, *421*
Johnson, S. A., see Maickel, R. P. 93, *218*
Johnson, V. S., see Smythies, J. R. 101, 128, *236*
Johnsson, V. S., see Smythies, J. R. 15, 18, 19, *30*
Johnstone, R. E., see Malit, L. A. 397, *424*
Johnston, V. S., see Smythies, J. R. 323, 328, *365*
Jolly, H., see Hecht, F. 408, *420*
Jonas, W., Scheel-Krüger, J. 42, 80, *210*
Jonas, W., see Randrup, A. 42, 77, 80, *228*
Jonas, W., see Scheel-Krüger, J. 77, *233*
Jones, B. E., Guynet, P., Chéramy, A., Gauchy, C., Glowinski, J. 57, *210*
Jones, C. N., Hill, H. F., Harris, R. T. 130, 131, *210*
Jones, H. S., see Oswald, I. 260, *272*
Jones, L. W., Wallis, E. S. 279, *300*
Jones, R. T., Benowitz, N. 400, 401, 408, 413, *421*
Jonsson, J. 8, *28*
Jonsson, J., Gunne, L.-M. 77, 154, *210*
Jonsson, J., Lewander, T. 17, 20, *28*, 132, *210*
Jonsson, J., see Lewander, T. 76, 78, 86, 132, *216*
Jori, A., Bernardi, D. 43, 46, 52, 138, 157, *211*
Jori, A., Caccia, S. 8, 9, 19, *28*
Jori, A., Dolfini, E. 43, 46, 52, *211*
Jori, A., Dolfini, E., Tognoni, G., Garattini, S. 43, 46, *211*
Jori, A., Garattini, S. 36, 43, 80, 90, 137, 154–156, 163, *211*
Jori, A., see Bizzi, A, 43, 46, 131, 141, 143, 147, 149, 154, 155, 156, 163, 164, *187*, 251, *268*
Jori, A., see Caccia, S. 36, 43, 156, 157, *190*
Joyce, C. R. B., Curry, S. H. 372, *421*
Joyce, D., see Bösser, T. H. 129, *187*
Joyce, D., see Bradley, D. W. H. 94, *188*
Joyce, D., see Cooper, S. J. 123, 130, *194*
Joyce, D., see Porsolt, R. D. 94, *227*
Julian, T., see Brown, A. M. 155, *189*

Kabelik, J., Krejci, Z., Santavy, F. 373, *421*
Kahn, E., Cohen, L. 286, *300*
Kaiser, C., see Zirkle, C. L. 4, *31*
Kalant, H., Kalant, O. J. 281, 288, *300*
Kalant, O. J. 253, 257, 262, 264, 265, *271*, 277, 282, 287–289, 291, 292, *300*
Kalant, O. J., see Kalant, H. 281, 288, *300*
Kaldor, A., see Skolnik, J. 57, *236*
Kamano, D. K., see Powell, B. J. 102, *227*
Kammen, D. P. van, Murphy, D. L. 287, *303*
Kampmann, E., see Frey, H.-H. 178, *201*
Kampmann, E., see Nielsen, C. K. 45, *224*
Kane, F. J., Keeler, M. H. Reifler, C. B. 265, *271*
Kaneko, M., see Takasaki, K. 62, *239*
Kaneko, Z., see Hishikawa, Y. 286, *300*
Kang, S., Green, J. P. 348, *359*
Kanter, S., see Nichols, W. W. 402, *425*
Kaplan, E. R., Sapeika, N., Zwarenstein, H. 163, *211*
Kaplan, E. R., see Sapeika, N. 163, *233*
Kaplan, G., see Zalis, E. G. 166, 175, *245*
Kapturkiewicz, Z., see Maj, J. 77, 82, 87, *218*
Karabetian, A., der, see Atrens, D. M. 121, 122, *183*
Karczmar, A. G., see Richardsen, D. 70, 80, *230*
Kardian, B. C., see Dannenburg, W. N. 164, *195*
Karler, R., Cely, W., Turkanis, S. A. 410, *421*
Karli, P. 97, *211*
Karniol, I. G., Shirakawa, I., Kasinski, N., Pfeferman, A., Carlini, E. A. 397, *421*
Karniol, I. G., see Gonzalez, S. C. 389, *419*

Kasinski, N., see Karniol, I. G. 397, *421*
Kasirsky, G. I., Zaidi, I. H., Tansy, M. F. 166, 173, 174, *211*
Kato, N., see Utena, H. 46, 70, 81, 161, 173, *241*
Kaufman, M. R., see Abramson, H. A. 307, 308, 310, 311, *350*
Kaufman, M. R., see Levine, A. 308, *360*
Kaufman, J. S., see Rangno, R. E. 8, *30*
Kay, D. C., Martin, W. R., Croughan, J. L. 315, 335, 341, 342, *359*
Kay, H., Birren, J. E. 179, *211*
Kaymakcalan, S., Deneau, G. A. 410, *422*
Kaymakcalan, S., Sivil, S. 398, *422*
Kaymackalan, S., Turker, R. K., Turker, M. N. 390, *422*
Kaymakalan, S., see Deneau, G. A. 413, *417*
Kebabian, J. W., see Greengard, P. 65, *205*
Keeler, M. H. 310, *359*
Keeler, M. H., see Kane, F. J. 265, *271*
Keesey, R. E., Powley, T. L. 285, *300*
Keith, E. S., see Fetterman, P. S. 373, 374, *418*
Keleman, K., Bovet, D. 100, 102, *211*
Kelfer, D. A., Rosen, A. J. 129, *211*
Kelleher, R. T., Cook, L. 110, 114, *211*
Kelleher, R. T., Fry, W., Deegan, J., Cook, L. 110–113, 116, *211*
Kelleher, R. T., Morse, W. H. 98, 110, 112, 116, 117, *211*
Kelleher, R. T., see Rutledge, C. O. 110, 112, 118, *232*
Keller, D. L., Umbreit, W. W. 312, *359*
Kelley, N. M., see Roszkowski, A. P. 131, 132, 139, *231*
Kennedy, J. S., Waddell, W. J. 381, *422*
Kennedy, J. S., see Dewey, W. L. 381, 389–391, 397, 399, 403, 409, *418*
Kennedy, J. S., see McMillan, D. E. 388, 411, *425*
Kennedy, M. S., see Terwelp, D. R. 149, 150, 154, 158, 159, *240*
Keplinger, M. L., Wright, P. L., Haley, S. L., Plank, J. B., Braude, M. C., Calandra, J. C. 406, *422*
Keplinger, M. L., see Haley, S. L. 406, 407, *420*
Kerley, T. L., see Weaver, L. C. 167, *243*
Kessler, E. K., see Kornetsky, C. 249, *271*
Ketteler, H. J., see McGrath, W. R. 59, 87, *220*
Kety, S. S., see Javoy, F. 35, *210*
Keup, W. 306, *359*
Kewitz, H., Reinert, H. 59, 64, *211*
Key, B. J. 316, 342, *359*
Key, B. J., Marley, E. 176, *211*
Key, B. J., see Bradley, C. 252, *268*
Key, B. J., see Bradley, P. B. 176, *188*, 341, *352*
Khan, A. U., Forney, R. B., Hughes, F. W. 163, *211*
Khavari, K. A. 115, *211*
Khazan, N., McCash, D. 339, 340, *359*
Khazan, N., Sawyer, C. H. 178, *211*, 340, *359*
Kibby, M. R., see Hulme, E. C. 47, 56, 134, *209*
Kido, R., see Umemoto, M. 122, *241*
Kiernan, C. C. 114, *211*
Kies, M. W., Horst, D., Evarts, E. V., Goldstein, N. P. 310, *359*
Kiessig, H. J. 149, *211*
Kiessig, H. J., Orzechowski, G. 179, *211*
Kikuchi, M., see Machiyama, Y. 70, 81, *217*
Kilbey, M. M., Farmer, R., McLendon, D. 387, *422*
Kilbey, M. M., Harris, R. T., Moore, J. W., Jr. 387, *422*
Kilbey, M. M., Moore, J. W., Hall, M. 387, *422*
Killam, E. K. 178, *211*, 306, *359*
Killam, E. K., Gangloff, H., Konigsmark, B., Killam, K. F. 176, 177, *212*
Killam, E. K., see Killam, K. F. 341, 342, *359*, 410, *422*
Killam, E. K., see Lindsley, D. F. 316, 342, *360*
Killam, K. F., Killam, E. K. 341, 342, *359*, 410, *422*
Killam, K. F., see Killam, E. K. 176, 177, *212*
Killam, K. F., see Lindsley, D. F. 316, 342, *360*
Kiloh, L. G., Brandon, S. 282, 289, 291, *300*
Kiloh, L. G., see Beamish, P. 282, 291, *297*
Kilsheimer, G. S., see Manno, B. R. 399, *424*
King, A. R., Martin, I. L., Melville, K. A. 314, *359*
King, L. J., see Garriott, J. C. 87, *202*
King, S., see Neu, R. L. 401, *425*
King, W. T., see Davis, W. M. 410, 413, *417*
Kinscherf, D. A., see Ferrendelli, J. A. 58, *200*
Kiplinger, G. F., see Manno, J. E. 385, *424*
Kipnis, D. M., see Ferrendelli, J. A. 58, *200*
Kirby, M. J., Turner, P. 162, *212*
Kirilcuk, V., see Arnold, L. E. 71, *183*, 250, *267*
Kirilcuk, V., see Corson, S. A. 71, 131, 137, *194*
Kirk, R., see Storz, H. 254, *274*
Kirk, T., see Regelson, W. 413, 414, *427*
Kirkby, R. J., see Bell, D. S. 91, 92, *186*
Kirtland, H. B., see Tainter, M. L. 177, *239*
Kits, F. P., see Praag, H. M. Van 3, *31*
Kjellberg, B., Randrup, A. 70, 94, 96, 98, *212*
Klages, W., see Panse, F. 278, *302*
Klaus, M., see Weiss, G. 286, *303*
Klausner, H. A., Dingell, J. V. 380, *422*
Klawans, H. L., Moses, H., Beaulieu, D. M. 71, 77, *212*
Klawans, H. D., Rubovits, R. 264, *271*
Klawans, H. L., Rubovits, R., Patel, B. C., Weiner, W. J. 70, 77, *212*
Klawans, H. L., see Goetz, C. 77, 78, *204*
Klawans, H. L., see Weiner, W. J. 77, 78, *243*
Klee, G. D., Bertino, J., Callaway, E. III, Weintraub, W. 312, *359*
Klee, G. D., Bertino, J., Weintraub, W., Callaway, E. III 308, *359*
Klee, G. D., see Aronson, H. 308, *351*
Klee, G. D., see Bertino, J. R. 312, *351*
Klee, G. D., see Silverstein, A. B. 308, *365*

Klein, V., see Malit, L. A. 397, *424*
Kleinrok, Z., see Przegalinski, E. 87, 154, 173, *228*
Kline, N. S., see Efron, D. H. 306, *354*
Klinge, E., see Alhava, E. 35, 37, 174, *182*
Klingenstein, R. J., Wallach, M. B., Gershon, S. 77, *212*
Klinger, W. 285, *300*
Klissiunis, N., Dosi, I. 154, *212*
Klopfenstein, M. H., see Stone, G. C. 109, *238*
Kluiol, S., see Rating, D. 410, *426*
Kluver, H. 307, *359*
Kmieciak-Kolada, K., see Herman, Z. S. 35, 37, 53, 87, 91, 92, *207*
Knapp, J. E., see Lotter, H. L. 378, *423*
Knapp, P. H. 255, *271*, 281, 287, 289, 294, *300*
Knapp, S., Mandell, A. J., Geyer, M. A. 56, *212*
Knapp, S., see Mandell, A. J. 47, 48, 57, *218*
Knee, S. T., Razani, J. 292, *300*
Knight, F., see Beaubrun, M. H. 409, *415*
Knight, J., see Wolstenholme, G. E. W. 372, *429*
Knight, W. A., see Gallagher, N. I. 259, *269*
Knoblock, L. C., see Segelman, A. B. 390, *427*
Knoblock, L. C., see Sofia, R. D. 391, *428*
Knoll, B., see Knoll, J. 81, *212*
Knoll, J., Vajnovszky, B. 82, *212*
Knoll, J., Vizi, E. S., Knoll, B. 81, *212*
Knoll, J., see Török, T. 161, *240*
Knoll, J., see Vizi, E. S. 146, 161, *242*
Knowler, W. C., Ukena, T. E. 109, 140, *212*
Knutsson, R. A., see Zalis, E. G. 166, 174, 175, *245*
Kobayashi, M., see Dolfini, E. 156, *197*
Kobayshi, T., see Tsuyussaki, T. 92, *241*
Kochar, M. S., Hosko, M. J. 397, *422*
Koda, L. Y., Gibb, J. W. 37, 48, *212*
Koe, B. K., see Weissman, A. 77, 106, 134, *243*, 264, *275*
Koella, W. P., Beaulieu, R. F., Bergen, J. R. 317, *359*
Koella, W. P., Wells, C.-H. 340, 342, *359*
Koella, W. P., see Smythies, J. R. 340, *365*
Koelle, G. B. 306, *359*
Koerker, R. L., see Wagner, L. A. 41, 49, *242*
Kohler 278
Kohli, J. D., see Innes, I. R. 61, *209*
Kokkinidis, L., see Anisman, H. 100, 101, 107, *183*
Kolodner, R. M., see Kolodny, R. C. 400, *422*
Kolodny, R. C. 401, *422*
Kolodny, R. C., Lessin, P., Toro, G., Masters, W. H., Cohen, S. 400, 408, *422*
Kolodny, R. C., Masters, W. H., Kolodner, R. M., Toro, G. 400, *422*
Komenda, S., see Navratil, J. 128, *223*
Konigsmark, B., see Killam, E. K. 176, 177, *212*
Konzett, H., see Rothlin, E. 307, *363*
Kopf, R., see Holm, T. 138, *208*
Kopin, I. J. 42, *212*
Kopin, I. J., Fischer, J. E., Musacchio, J. M., Horst, W. D., Weise, W. K. 63, *212*
Kopin, I. J., Gordon, E. K. 50, *212*
Kopin, I. J., see Breese, G. R. 35, 37, 52, *188*
Kopin, I. J., see Chase, T. N. 330, *352*
Kopin, I. J., see Colburn, R. W. 382, *417*
Kopin, I. K., see Davis, J. M. 14, 15, 21, 22, *27*
Kopin, I. J., see Lemberger, L. 16, *29*, 77, *215*, 380, *423*
Kopin, I. J., see Ng, K. Y. 44, *224*
Kopin, I. J., see Williams, R. B. 398, *429*
Koppanyi, T., Maling, H. M. 149, 154, 168, *212*
Koppanyi, T., see Cowan, F. F. 60, 62, *194*
Koppanyi, T., see Maling, H. M. 149, 154–156, *218*
Koppanyi, T., see Maengwyn-Davies, G. D. 62, 143, *217*
Korein, J., Musacchio, J. M. 343, *360*
Korenyi, C., Perry, G. F., Whittier, J. R. 251, *271*
Korf, J., Bunney, B. S., Aghajanian, G. K., 66, *212*
Korf, J., see Praag, H. M. Van 3, *31*
Kornetsky, C., Mirsky, A. F., Kessler, E. K., Dorff, J. E. 249, *271*
Kornetsky, C., see Abramson, H. A. 307, 308, 310, 311, *350*
Kornetsky, C., see Latz, A. 101, 128, 129, *214*
Korte, F., Haag, M., Claussen, U. 375, 377, *422*
Korte, F., Sieper, H. 373, 375, *422*
Korte, F., see Claussen, U. 373, 375, *417*
Korte, F., see Isbell, H. 380, 385, *421*
Korte, F., see Silva, M. T. A. 320, *365*, 409, *427*
Korte, R., see Spulak, R. von 375, *429*
Korte, R., see Vollner, L. 375, *429*
Kosersky, D. S., McMillan, D. E., Harris, L. S. 388, 410, *422*
Kosersky, D. S., Dewey, W. L., Harris, L. S. 391, *422*
Kosman, M. E. 129, *212*
Kosman, M. E., Unna, K. R. 91, *213*, 258, *271*
Kostowski, W. 341, *360*
Kostowski, W., Czlonkowski, A. 80, 97, *213*
Kotin, J. K., see Post, R. M. 287, *302*
Kralik, P. M., see Ho, B. T. *421*
Kramer, J. C. 254, 258, 259, *271*
Kramer, J. C., Fischman, V. S., Littlefield, D. C. 257, 258, 260, 263, *271*, 282, 290, 291, 293, *300*
Kraml, M., see Zsoter, T. 163, *246*
Krantz, J. C., see Berger, H. J. 77, 127, 136, *186*
Krasnegor, N. A., see Fischman, M. W. 278, *298*
Krasno, L. R., see Ivy, A. C. 247, 251, 253–255, 259, *270*
Kraupp, O., see Stühlinger, W. von 143, *238*
Krautz, J. C., see Hundley, J. M. 247, 248, *270*
Krebs, H., Bindra, D. 65, 137, *213*
Krebs, H., Bindra, D., Campbell, J. F. 65, 137, *213*
Kreindler, A., Crighel, E., Sotirescu, N. 342, *360*
Kreinik, C. J., see Schwartzbaum, J. S. 176, *235*
Krejci, Z., see Kabelik, J. 373, *421*

Kremer, E., see Cicala, G. A. 102, 103, *193*
Kreuz, D. S., Axelrod, J. 381, *422*
Krieckhaus, E. E. 101, 107, *213*
Krieckhaus, E. E., Miller, N. E., Zimmerman, P. 102, 103, *213*
Krimmer, E. C., see Barry, H. 389, *415*
Krip, G., see Vasquez, A. J. 177, *242*
Krise, G. M., see McDowell, A. A. 91, *220*
Kriskó, I., Lewis, E., Johnson, J. E. 264, *271*, 288, *300*
Kritkausky, R. P., see Carey, R. J. 114, 115, *190*
Krivanek, J. A., McGaugh, J. L. 129, *213*
Krivitskaya, G. N., see Popova, E. N. 174, *227*
Krivoy, W. A. 333, 334, *360*
Krnjević, K., Phillis, J. W. 338, *360*
Kruk, Z. L. 132, 134, 149, 154, 158, *213*
Kruk, Z. L., Brittain, R. T. 149, 158, *213*
Krumbhaar, E. B., see Ehrich, W. E. 70, 96, 131, 139, 166, 167, 173, 175, *198*, 282, *298*
Krus, D. M., Wappner, S., Bergen, J., Freeman, H. 312, *360*
Krus, D., see Bergen, J. R. 315, *351*
Ksir, C. J., see Maickel, R. P. 53, 56, 82, 99, 106, 131, 134, 140, *218*
Kubena, R. K., Barry, H. 87, 173, *213*
Kubena, R. K., see Barry, H. III 400, *415*
Kubena, R. L., Perhack, J. L., Barry, H., III. 400, *422*
Kuczenski, R. T., see Mandell, A. J. 47, *218*
Kuehnle, J. C., see Mendelson, J. H. 385, 401, 409, *425*
Kuhlo, W. 250, *271*
Kulkarni, A. S. 102, *213*
Kulkarni, A. S., Job, W. M. 102, 129, *213*
Kulkarni, A. S., see Abdallah, A. H. 87, 134, *181*
Kulkarni, A. S., see Bocknik, S. E. 142, *187*
Kulkarni, A. S., see Reitz, R. H. 39, 106, *229*
Kulkarni, S. K., Dandiya, P. C. 93, *213*
Kulp, R. A., see Malit, L. A. 397, *424*
Kumar, R. 94, 98, *213*
Kumar, S., Kunwar, K. B. 402, *422*
Kunwar, K. B., see Kumar, S. 402, *422*
Kunysz, T. J., see Stenchever, M. A. 401, *428*
Kupferberg, H. J., see Craig, A. L. 149, 155, 156, 167, *195*
Kupferberg, H. J., see Craigmill, A. L. 155, 156, *195*
Kupietz, S., see Winsberg, B. G. 286, *303*
Kuppers, F. J. E. M., see Bercht, C. A. L. 375, *415*, *416*
Kurihara, K., see Thuillier, J. 150, 154, *240*
Kurland, A., see Cholden, L. S. 318, *353*

Labhardt, F., see Ladewig, D. 260, *271*
Lackey, R. W., see Tainter, M. L. 177, *239*
Ladefoged, O. 154, *213*, 338, *360*
Lader, M. 247, *271*
Ladewig, D., Battegay, R. 285, *300*
Ladewig, D., Battegay, R., Labhardt, F. 260, *271*
Ladinsky, H., see Consolo, S. 57, *193*, *194*
Ladisch, W., Baumann, P. 43, 87, *213*
Ladisch, W., Volbehr, H., Matussek, N. 43, 71, *213*
Ladman, A. J., see Mann, P. E. 401, *424*
LaDu, B. N., Mandel, H. G., Way, E. L. 12, *28*
LaDu, B. N., see Brodie, B. B. 9–11, *26*
Laduena, F. P., see Tainter, M. L. 177, *239*
Lafille, E., see Jacob, J. 313, 315, 338, *358*
La Flore, J. E., see Rosen, A. J. 100, *230*
Lagerspetz, K. Y. H., Lagerspetz, K. M. J. 168, *213*
Lagerspetz, K. M. J., see Lagerspetz, K. Y. H. 168, *213*
Laidlaw, P. P. 316, *360*
Lal, H. 130, *213*
Lal, H., Chessick, R. D. 35, 36, 54, 174, *214*
Lal, H., Defeo, J. J., Thut, P. 97, *214*
Lal, H., Ginocchio, S., Shefner, A. 173, *214*
Lal, H., Nesson, B., Smith, N. 96, 179, *214*
Lal, H., O'Brien, J., Puri, S. K. 77, 96, *214*
Lal, H., Puri, S. K., Fuller, G. C. 179, *214*
Lal, H., Roffman, M. 131, *214*
Lal, H., see Cohen, M. 173, *193*
Lal, H., see Roffman, M. 102, 106, 130, 131, *230*
Lal, S., Missala, K., Sourkes, T. L. 77, *214*
Lal, S., Sourkes, T. L. 72, 77, *214*
Lal, S., Sourkes, T. L., Missala, K. 77, *214*
Lal, S., see Papeschi, R. 80, 91, *226*
Lamprecht, F., see Rapoport, J. L. 286, *302*
Lamprecht, F., see Williams, R. B. 398, *429*
Lanciault, G., Wolf, H. H. 177, 179, *214*
Landau, W., see Evarts, E. V. 336, *354*
Landis, C., Clausen, J. 308, 310, *360*
Landon, J., see Besser, G. M. 255, *267*
Lang, W. J., see Feldberg, W. 154, 158, *200*
Langfitt, T. W., Finney, L. A. 339, 341, 342, *360*
Langs, R. J., see Linton, H. B. 308, *360*
La Pidus, J. B., see Patil, P. N. 60, 63, *226*
Lapin, I. P., Oxenkrug, G. G., Azbekyan, S. G. 87, 154, 173, *214*
Lapin, I. P., Schelkunov, E. L. 77, 87, 106, 173, *214*
Laporte, J., Jané, J., Cuenca, E., Rodriguez, L., Valdecasas, F. G. 60, *214*
Lariviere, E. W., see Moore, K. E. 34–36, *222*
Larousse, C., see Simon, P. 34, 53, 87, *236*
Larsson, K., see Andén, N.-E. 68, *182*
Larsson, K., see Butcher, L. L. 98, *189*
Larsson, S., see Andersson, B. 136, 140, *182*, 251, *267*
Larsson, S., see Brobeck, J. R. 137, *189*
Lasagna, L. 285, *300*
Lasagna, L., Epstein, L. C. 250, *271*
Lasagna, L., Felsinger, J. M. von, Beecher, H. K. 248, *271*
Lasagna, L., McCann, W. P. 173, *214*

Lasagna, L., see Epstein, L. C. 286, *298*
Lasagna, L., see Höhn, R. 167–169, 174, *208*
Lasagna, L., see Okun, R. 175, *224*
Lasagna, L., see Tormey, J. 92, 138, 139, *240*
Lat, J. 72, 82, 94, *214*
Laties, V. G. 109, 115, *214*
Laties, V. G., Weiss, B. 112, 113, 115, *214*
Laties, V. G., see Weiss, B. 109, 114, 118, 167, 169, 173, *243*, 249, *275*
Latimer, C. N., see Bernstein, B. M. 99, 106, *186*
Latz, A., Bain, G., Goldman, M., Kornetski, C. 101, 128, 129, *214*
Latz, A., Kornetsky, C., Bain, G. 101, *214*
Lauber, E., see Goldstein, M. 9, *28*, 48, 50, *204*
Laubie, M., see Beregi, L. G. 4, *26*
Laubie, M., see LeDouarec, J. C. 3, *29*
Lauener, H. 115, *214*
Laufer, M. W., Denhoff, E. 286, *300*
Laufer, M. W., Denhoff, E., Solomans, G. 286, *300*
Laurent, J. M. 277, *301*
Laverty, R., Sharman, D. F. 36, 37, 43, 46, 54, 55, 70, *214*
Laverty, R., see Taylor, K. M. 94, 106, *239*
Lawlor, R. B., Trivedi, M. C., Yelnosky, J. 5, *28*, 132, 138, *215*
Lawlor, R. B., see Yelnosky, J. 132, 134, 140, *245*
Lawrence, D. K., Pertwee, R. G. 410, *422*
Lawrence, W. S., see Schulte, J. W. 82, *234*
Lawrenz, H., see Bonhoff, G. 281, 288, *297*
Layman, J. M., Milton, A. S. 381, *422*
Leaf, R. C., Learner, L., Horovitz, Z. P. 97, *215*
Leaf, R. C., see Horovitz, Z. P. 97, *209*
Leake, C. D. 33, *215*, 247, 251, 258, *271*, 279, 280, *301*
Leander, J. D., McMillan, D. E. 110, 112, 113, 115, 116, *215*
Leander, K., see Agurell, S. 382, 384, *415*
Learner, L., see Leaf, R. C. 97, *215*
Leary, D. E., see DiStefano, V. 334, *354*
Leary, D. E., see Little, K. D. 334, *360*
Leary, D. E., see Slater, I. H. 334, *365*
Lebas, M., see Boulu, R. 37, 68, *188*
LeBlanc, A. E., Cappell, H. 127, *215*
LeBlanc, A. E., see Beaton, J. M. 101, *185*
LeBlanc, A. E., see Cappell, H. 127, 135, *190*
LeBlanc, A. E., see Webster, C. D. 410, *429*
LeDain, G. 372, *422*
LeDouarec, J. C., Neveu, C. 4, *29*, 249, *271*
LeDouarec, J. C., Schmitt, H., Laubie, M. 3, *29*
LeDourarec, J. C., see Beregi, L. G. 4, *26*
LeDouarec, J. C., see Bernier, A. 136, *186*
Lee, C. T., see Scott, J. P. 155, *235*
Lee, H. K., see Angrist, B. M. 295, *296*
Lee, N., see Fuller, R. W. 11, *27*
Lee, Y. E., see Tashkin, D. P. 409, *428*
Lees, M. H., see Hecht, F. 408, *420*
Lefkowitz, S. S., Yang, C. 402, *423*
Legator, M. S., Weber, E., Conner, T., Stoeckel, M. 402, *423*
Legrand, M., see Beauvallet, M. 35, 41, 42, 53, *185*
Leibowitz, S. F. 137, 139, 140, *215*
Leighty, E. G., see Foltz, R. L. 383, *419*
Leith, N. J., see Barrett, R. J. 100, 102, 103, 130, *184*
Leitz, F. H. 38, *215*
Leitz, F. H., Stefano, F. J. E. 40, 45, 47, *215*
Lele, K. P., see Gilmour, D. G. 402, *419*
Lemberger, L., Crabtree, R. E., Rowe, H. M. 383, *423*
Lemberger, L., McMahon, R., Archer, R., Matsumoto, K., Rowe, H. 385, 387, 399, 400, *423*
Lemberger, L., Silberstein, S. D., Axelrod, J., Kopin, I. J. 380, *423*
Lemberger, L., Tamarkin, N. R., Axelrod, J., Kopin, I. J. 380, *423*
Lemberger, L., Witt, E. D., Davis, J. M., Kopin, I. J. 16, *29*, 77, *215*
Lemberger, L., see Colburn, R. W. 382, *417*
Lemberger, L., see Davis, J. M. 14, 15, 21, 22, *27*
Lemmer, B. 54, 55, *215*
Lemperière, T., see Delay, J. 293, *298*
Lenaerts, F. M., see Janssen, P. A. J. 77, *210*
Leonard, B. E. 34, 35, 37, 48, 53–56, *215*, 396, *423*
Leonard, B. E., Neuhoff, V. 164, *215*
Leonard, B. E., Shallice, S. A. 34, 35, 37, 38, 49, 53, 56, 57, *215*
Leonard, B. E., Tonge, S. R. 333, *360*
Leonard, B. E., Wiseman, B. D. 94, *215*
Leonard, B. E., see Osborne, N. N. 54, *225*
Leonard, D. A., Heil, G. C., Fellows, E. J. 138, *215*
Leshem, M. B., see Blundell, J. E. 136, 137, *187*
Leslie, G. B., Maxwell, D. R. 77, 173, *215*
Lesses, M. F., Myerson, A. 250, *271*, 280, *301*
Lessin, P. J., Thomas, S. A., 409, *423*
Lessin, P., see Kolodny, R. C. 400, 408, *422*
Lessin, P. J., see Cohen, S. 408, *417*
Lessin, P. J., see Frank, I. M. 409, *419*
Lessin, P. J., see Silverstein, M. J. 402, *427*
Leuchtenberger, C., Leuchtenberger, R. 401, *423*
Leuchtenberger, C., Leuchtenberger, R., Schwider, A. 401, *423*
Leuchtenberger, R., see Leuchtenberger, C. 401, *423*
Leupold-Löwenthal, H., see Arnold, O. H. 334, *351*
Leuschner, F., see Heim, F. 161, *207*
Levassort, C., see Cheymol, M. J. 154, *192*
Levi, G., see Raiteri, M. 45, *228*
Levin, A. 258, *271*

Levine, A., Abramson, H. A., Kaufman, M. R., Markham, S. 308, *360*
Levine, A., see Abramson, H. A. 307, 308, 310, 311, *350*
Levine, J. M., see Miller, R. E. 40, 96, *221*
Levis, D. J., Caldwell, D. F. 314, *360*
Levy, C. K., see Smythies, J. R. 340, *365*
Levy, B. I., see Delong, F. L. 385, *417*
Levy, J. A., Munson, A. E., Harris, L. S., Dewey, W. L. 402, *423*
Levy, J. A., see Munson, A. D. 402, *425*
Lew, C., Iversen, S. D., Iversen, L. L. 87, *215*
Lewander, T. 9, 15–17, 19, 20, *29*, 34–37, 41–43, 46, 49, 50, 51, 52, 55, 63 70, 72, 77, 81, 82, 87, 131, 132, 134, 138, 139, 154, 157, 173, *215*
Lewander, T., Jonsson, J. 76, 78, 86, 132, *215*
Lewander, T., Moliis, G., Brus, I. 140, 150, 157, 158, *216*
Lewander, T., see Gunne, L.-M. 34–37, 41, 42, *205*
Lewander, T., see Jönsson, L.-E. 261, *271*
Lewander, T., see Jonsson, J. 17, 20, *28*, 132, *210*
Lewin, L. 305, 306, 310, *360*
Lewis, A. 277, 285, *301*
Lewis, E., Jr. 292, *301*
Lewis, E., see Kriskó, I. 264, *271*, 288, *300*
Lewis, J. J., Petten, G. R. van 161, *216*
Lewis, J. J., Pollack, D. 161, *216*
Lewis, J. J., see Baird, J. R. C. 34–36, *184*
Lewis, M. F. 372, *423*
Lewis, M. J., see Graham, J. D. P. 393, 394, *419*
Lewis, S. A. 252, *271*
Lewis, S. A., Oswald, I., Dunleavy, D. L. F. 257, *271*
Lewis, S. A., see Oswald, I. 257, *272*
Lewis, S. C., Brown, D. J., Forney, R. B. 398, *423*
Lewman, L. V. 254, *272*
Lewrenz, H., see Bonhoff, G. 33, *187*, 247, 251, 254, 259, 260, 263, *268*
Lewy, F. H., see Ehrich, W. E. 70, *166*, 175, *198*, 282, *298*
Lhamon, W. T., see Boardman, W. K. 308, *352*
L'Hermite, P., see Jacquot, C. 51, *209*, *210*
Lhermitte, F., see Delay, J. 339, *353*
L'Huillier, J., Grandjean, J. L., Thullier, J. 167, 173, *216*
L'Huillier, J., see Nakajima, H. 154, *223*
Li, D. M. R., see Graham, J. D. P. 398, 411, *419*
Li, H. C. *423*
Liakos, A., see Stefanis, C. 409, *428*
Liberson, W. T., Ellen, P., Feldman, R. S. 129, *216*
Lichtenheld, E. R., see Schwartz, B. E. 341, 342, *364*
Liddon, S. C., Satran, R. 292, *301*
Liddon, S. C., see Okun, R. 179, *224*
Liebert, R. S., Werner, H., Wapner, S. 310, *360*
Liebeswar, G., see Gololák, G. *204*
Liebman, J. 142, *216*
Liesiene, V. A. 176, *216*
Lilie, N. L., see Heise, G. A. 115, *207*
Lilla, L., see Grana, E. 45, 46, 47, *204*
Lima, I. M., Luiz, R., Carlini, E. A. 165, *216*
Lin, G. S., see Banarjee, U. 87, 173, *184*
Lin, R. C., Ngai, S. H., Costa, E. *360*
Lindeke, B., Cho, A. K., Thomas, T. L., Michelsen, L. 11, *29*
Lindeke, B., see Cho, A. K. 11, *26*
Lindgren, J. E., see Agurell, S. 382, *415*
Lindsley, D. F., Carpenter, R. S., Killam, E. K., Killam, K. F. 316, 342, *360*
Lindquist, M., see Carlsson, A. 34, 36, 38, 39, 42, 53, *191*
Lindquist, O., see Gunne, L.-M. 254, 264, *270*
Linsmayer, M., see Matussek, N. 71, 77, *219*
Linton, H. B., Langs, R. J. 308, *360*
Lippa, A. S., see Canfield, D. R. 123, *190*
Lipparini, F., see Florio, V. 339, 340, 342, *355*
Lipton, M. A., see Perez-Reyes, M. 380, 383, *426*
Lishajko, F., see Euler, U. S. von 38, 41, *199*
Lison, H., see Coper, H. 149, 154, 159, *194*
Little, B. C., see Zahn, T. P. 286, *304*
Little, H. J., Rees, J. M. 179, *216*
Little, K. D., DiStefano, V., Leary, D. E. 334, *360*
Little, M. D., Dill, R. E. 313, *360*
Littlefield, D. C., see Kramer, J. C. 257, 258, 260, 263, *271*, 282, 290, 291, 293, *300*
Littleton, J. M. 35, 36, 37, 48, 167, 169, 174, *216*
Littleton, J. M., MacLean, K. I. 396, *423*
Ljungdahl, A., see Corrodi, H. 77, 87, *194*, 264, *268*
Llewellyn, R. C., see Monroe, R. R. 343, *362*
Lloyd, B. J. Jr., see Himmelsbach, C. K. 283, 295, *300*
Lloyd, G. 289, *301*
Lloyd, J. T., Walker, D. R. H. 265, *272*
Locatelli, A., see Beretta, C. 61, *186*
Lockett, M. F. 58, 141, 142, *217*
Löffelholz, K., Muscholl, E. 40, *217*
Löfstrandh, S., see Morgan, C. D. 39, 45, 54, 56, *222*
Loeser, A., see Opitz, K. 161, 163, 164, 166, *225*
Loew, D. M., Depoortere, H., Vigouret, J. M. 339, *360*
Loew, D. M., see Depoortere, H. 339, *353*
Logan, C. R., see Isbell, H. 308, 310–312, 318, 319, 330, *358*
Logan, C. R., see Rosenberg, D. E. 318, *363*
Logston, D. G., see Davis, W. M. 130, 173, 175, *196*
Loh, H., Hitzemann, R., Stolman, S. 164, *217*
Loh, H. H., see Hitzemann, R. J. 35–37, 49, 52, 56, 80, 91, 164, *208*
Loh, H. H., see Tseng, L. F. 44, 46, 87, *241*
Loiseau, G., see Jacob, J. 313, *358*
Lokhandwala, M. F., see Cavero, I. 399, *417*
Loman, J., Rinkel, M., Myerson, A. 247, *272*
Lomax, P. 410, *423*
Lomax, P., see Grant, J. D. 378, *419*

Long, J. P., see Cheng, H. C. 61, *192*
Longnecker, D. E., see Brown, D. J. 399, *416*
Longo, V. G. *360*
Longo, V. G., Silvestrini, B. 176, 179, *217*
Longo, V. G., see DeBaran, L. 340, *353*
Longo, V. G., see Domer, F. R. 340, *354*
Longo, V. G., see Florio, V. 339, 340, 342, *355*
Longo, V. G., see McGaugh, J. L. 115, 117, *220*
Longo, V. G., see Sadowski, B. 115, 117, *232*
Longo, V. G., see Ziegler, H. 77, 87, *246*
Lontax, P., see Brill, N. O. 403, *416*
Loon, E. J. van, see Ellison, T. 5, 8, 14, *27*
Loon, G. R. van, Scapagnini, U., Cohen, R., Ganong, W. F. 165, *217*
Lopez, G. G., see Gonzalez-Barón, S. 98, *204*
Lotter, H. L., Abraham, D. J., Turner, C. E., Knapp, J. E., Schiff, P. L., Skatkin, D. J. 378, *423*
Lotti, V. J., see Clineschmidt, B. V. 339, *353*
Lousberg, R. J. J. Ch., see Bercht, C. A. L. 375, *415*, *416*
Lovell, R. A., see Appel, J. B. 315, *351*
Lovell, R. A., see Holtzman, D. 391, 393, 394, *421*
Lovell, R. A., see Rosecrans, J. A. 330, *363*
Lowe, G., Williams, D. I. 314, *360*
Lu, T.-C., Claghorn, J. L. Schoolar, J. C. 134, 138, 139, *217*
Lu, T.-C., Ho, B. T., McIsaac, W. M. 92, *217*
Lu, K.-H., Meites, J. 166, *217*
Luby, E., see Rodin, E. 343, *363*
Luca, N., see Mison-Crighel, N. 342, *362*
Ludden, C. T., see Stone, C. A. 146, *238*
Luiz, R., see Lima, I. M. 165, *216*
Lum, B. K. B., see Commarato, M. A. 142–144, 146, *193*
Lundberg, G. D., see Citron, B. P. 261, *268*
Lundberg, G. D., see Zalis, E. G. 166, 174, 175, *245*
Lundborg, P. 40, *217*
Lundborg, P., Waldeck, B. 40, 50, *217*
Lundborg, P., see Obianwu, H. O. 49, 50, 148, *224*
Lundy, R. O., see Proctor, C. D. 166, 173, *228*
Lupa, A. J., see Valle, J. R. 373, *429*
Luthra, X. K., Rosenkrantz, H. 405, *423*
Luthra, X. K., Rosenkrantz, H., Heyman, I. A., Braude, M. C. 405, *423*
Lwoff, J. M., see Boissier, J.-R. 173, *187*
Lynch, G. S., Ballantine, II, P., Campbell, B. A. 90, *217*
Lynch, V. D., Aceto, M. D., Thomas, R. K. 101, *217*
Lyon, M. 99, 121, *217*
Lyon, M., Randrup, A. 72, 99, 121, *217*
Lyon, M., see Nielsen, E. B. *224*
Lytle, L. D., Moorcroft, W. H., Campbell, B. A. 138, *217*

Mabry, P. D., Campbell, B. A. 87, *217*
MacCuish, A. C., see Munro, J. F. 285, *301*
MacDonnell, M. F., Fessock, L. 97, *217*
Machiyama, Y., Utena, H., Kikuchi, M. 70, 81, *217*
Macko, E., see Tedeschi, D. H. 96, 97, *239*
MacLay, W. S., see Guttmann, E. 307, *356*
MacLean, A. R., see Wilbur, D. L. 255, *275*
MacLean, A. W., see Dunleavy, D. L. F. 252, *268*
MacLean, K. I., see Littleton, J. M. 396, *423*
Macnicol, M. F., see Hollister, L. E. 311, 319, *357*
MacPhail, R. C., Gollub, L. R. 138, *217*
Maehly, A. C., see Orrenius, S. 264, *272*
Maengwyn-Davies, G. D., Cowan, F. F., Koppanyi, T, 62, 143, *217*
Maengwyn-Davies, G. D., Koppanyi, T. 62, 143, *217*
Maengwyn-Davies, G. D., see Cowan, F. F. 60, 62, *194*
Magnussen, M. P., see Frey, H.-H. 87, *201*
Magnussen, M. P., see Nielsen, C. K. 45, *224*
Magnusson, T., see Andén, N.-E. 42, *182*
Magos, L. 80, *217*
Magour, S., Coper, H., Fähndrich, C. H. 91, 92, 138, 139, 164, 173, *218*
Magus, R. D., Harris, L. S. 401, *423*
Magyar, R., see Gillin, J. C. 320, *356*
Mahler, D. J., Humoller, F. L. 320, *361*
Mahler, K., see Zetler, G. 77, *246*
Ma Hung, *278*
Maickel, R. P., Cox, R. H., Jr., Ksir, C. J., Snodgrass, W. R., Miller, F. P. 53, 56, 82, 99, 106, 131, 134, 140, *218*
Maickel, R. P., Cox, R. H., Jr., Miller, F. P., Segal, D. S., Russel, R. W. 20, *29*, 99, 128, 179, *218*
Maickel, R. P., Johnson, S. A. 93, *218*
Maickel, R. P., Webb, R. W. *218*
Maickel, R. P., see Cox, R. H. 101, *195*
Maickel, R. P., see Miller, F. P. 87, *221*
Maickel, R. P., see Segal, D. S. 99, 106, *235*
Maickel, R. P., see Sokol, G. H. 173, *236*
Maitre, L. von, Brunner, H. 51, 147, *218*
Maitre, L., Staehelin, M., Bein, H. J. 393, 394, *423*
Maitre, L., Waldmeier, C., Baumann, P. A. 394, *423*
Maj, J., Durek, B., Palider, W. 81, *218*
Maj, J., Grabowska, M., Gajda, L., Michaluk, J. 149, 167, *218*
Maj, J., Przegalinski, E. 77, 87, *218*
Maj, J., Sowinska, H., Baran, L. 87, *218*
Maj, J., Sowinska, H., Baran, L., Kapturkiewicz, Z. 77, 87, *218*
Maj, J., Sowinska, H., Kapturkiewicz, Z., Sarnek, J. 82, 87, *218*
Major, C. T., Pleuvry, B. J. 179, *218*
Malcolm, J. L., see Dear, E. M. A. 339, *353*
Malick, J. B., Goldberg, M. E. 123, *218*
Maling, H. M., Koppanyi, T. 149, 154–156, *218*

Maling, H. M., Williams, M. A., Koppanyi, T. 149, 154, 156, *218*
Maling, H. M., see Koppanyi, T. 149, 154, 168, *212*
Malit, L. A., Johnstone, R. E., Bourke, D. I., Kulp, R. A., Klein, V., Smith, T. C. 397, *424*
Malitz, S., Esecover, H., Wilkens, B., Hoch, P. H. 310, 343, *361*
Malmfors, T., see Fuxe, K. 39, 54, *202*
Malmfors, T., see Hamberger, B. 38, *206*
Malmnäs, C. O. 313, *361*
Malorny, G., Orzechowski, G. 62, 141, 142, *218*
Malspeis, L., see Fuller, D. R. 18, *27*
Maltby, G. L., see Gibbs, F. A. 251, *269*
Mandel, H. G., see LaDu, B. N. 12, *28*
Mandel, I. J., see Stoff, D. M. 314, *365*
Mandel, J., see Carey, J. T. 282, 288, *297*
Mandel, P., see DiCarlo, R. 165, *197*
Mandell, A. J., Knapp, S. 48, 57, *218*
Mandell, A. J., Knapp, S., Kuczenski, R. T., Segal, D. S. 47, *218*
Mandell, A. J., Morgan, M. 48, 57, *219*
Mandell, A. J., see Knapp, S. 56, *212*
Mandell, A. J., see Segal, D. S. 91, 92, *235*
Manger, W. M., see Nahas, G. G. 398, 410, *425*
Mankiewics, J. 283
Mann, G. V. 285, *301*
Mann, P. E., Cohen, A. B., Finley, T. N., Ladman, A. J. 401, *424*
Mann, P. J. G., Quastel, J. H. 10, *29*, 45, 47, *219*
Mann, W. G., Jr., see Gollub, L. R. 98, 110, 117, *204*
Mannerheim, J. E., see Oswald, I. 260, *272*
Manning, D. H., Strang, R. H., Bachelard, H. S. 162, *219*
Manning, F. J. 410, *424*
Manning, F. J., Elsmore, T. F. 388, 389, *424*
Manning, F. J., McDonough, J. H., Jr., Elsmore, T. F., Saller, C., Sodetz, F. J. 404, *424*
Manno, B. R., Manno, J. E., Kilsheimer, G. S., Forney, R. B. 399, *424*
Manno, B. R., see Manno, J. E. 385, *424*
Manno, J. E., Manno, B. R., Kiplinger, G. F., Forney, R. B. 385, *424*
Manno, J. E., see Manno, B. R. 399, *424*
Mansour, T. E. 61, *219*
Mansour, T. E., see Chance, M. R. A. 61, *192*
Mantegazza, P., Müller, E. E., Naimzada, M. K., Riva, M. 87, 134, 154, 173, *219*
Mantegazza, P., Naimzada, M. K., Riva, M. 87, 91, 134, 138, 154, 156, 163, 169, 173, *219*
Mantegazza, P., see Gropetti, A. 36, 37, *205*
Mantegazza, P., see Riva, M. 139, 149, *230*
Mantilla-Plata, B., Clewe, C. L., Harbison, R. D. 407, *424*
Mantilla-Plata, B., Harbison, R. D. 408, *424*
Mantilla-Plata, B., see Harbison, R. D. 407, *420*
Marazzi-Uberti, E., see Bianchi, C. 102, *186*
Marchant-Williams, H., see Chesher, G. B. 390, *417*
Marchbanks, R. M. 334, *361*
Marchetti, P., see Beani, L. 57, *185*
Marcus, R. J., see Winters, W. D. 178, *244*, 341, 342, 348, *368*
Marczynski, T. J. 342, *361*
Margules, D. L. 122, *219*
Markham, S., see Levine, A. 308, *360*
Marks, H. H. 285, *301*
Marley, E. 60, *219*
Marley, E., Stephenson, J. D. 176, 177, *219*
Marley, E., see Dewhurst, W. G. 70, 95, 96, 176, 177, *196*, 331, *353*
Marley, E., see Key, B. J. 176, *211*
Marley, E., see Vane, J. R. 313, 331, 332, 335, 342, *367*
Marks, J. H., see Feudis, F. V. de 188, *200*
Maron, L., see Rechtschaffen, A. 252, *273*
Marrazzi, A. S. 339, *361*
Marrazzi, A. S., Hart, E. R. 339, *361*
Marriott, A. S. 94, *219*
Marsanico, R. G., see Glick, S. D. 89, 90, 119, *203*
Marsden, C. A., Guldberg, H. C. 68, 69, *219*
Marsh, J. B., Bizzi, A. 164, *219*
Marshall, W. H., see Evarts, E. V. 336, *354*
Marshman, J. A., see Webster, C. D. 410, *429*
Martin, B., see Agurell, S. 382, *415*
Martin, B. R., Dewey, W. L., Earnhardt, J. T., Adams, M. D., Harris, L. S., Wilson, R. S., May, E. L., Razdan, R. 381, 390, *424*
Martin, B. R., Dewey, W. L. Harris, L. S., Beckner, J. S. 381–383, 412, *424*
Martin, B. R., Dewey, W. L., Harris, L. S., Beckner, J., Wilson, R. S., May, E. L., 384, 410, 411, *424*
Martin, B. R., see Dewey, W. L. 382, 412, *418*
Martin, I. L., see King, A. R. 314, *359*
Martin, J. C., Ellinwood, E. H., Jr. 127, *219*
Martin, J., see Freudenthal, R, I. 381, *419*
Martin, L. K., see Powell, B. J. 102, *225*
Martin, P. A., Thorbunn, M. J., Bryant, S. A. 402, *424*
Martin, W. R. 344, 348, *361*
Martin, W. R., Eades, C. G. 316, 317, 321–323, 328, 331, 335, 336, 345, *361*
Martin, W. R., Sloan, J. W. 310, 331, 336, 348, *361*
Martin, W. R., Sloan, J. W., Buchwald, W. F., Bridges, S. R. 316, 332, *361*
Martin, W. R., Sloan, J. W., Buchwald, W. F., Clements, T. H. 332, *361*
Martin, W. R., Sloan, J. W., Christian, S. T., Clements, T. H. 332, *361*
Martin, W. R., Sloan, J. W., Sapira, J. D., Jasinski, D. R. 253, 254, *272*, 278, 281, 285, *301*
Martin, W. R., Sloan, J. W., Vaupel, D. B., Bell, J. A., Nozaki, M. 348, *361*
Martin, W. R., Vaupel, D. B., Sloan, J. W., Bell, J. A., Nozaki, M., Bright, L. D. 281, *301*
Martin, W. R., see Bell, J. A. 321, 331, 332, 334, *351*

Martin, W.R., see Kay, D.C. 315, 335, 341, 342, *359*
Martin, W.R., see Nozaki, M. 316, 317, 321, *362*
Martin, W.R., see Sloan, J.W. 332, *365*
Martin, W.R., see Vaupel, D.B. 321, 322, 331, *367*
Martz, R.C., see Phillips, R.N. 404, *426*
Martz, R.M., see Carlini, E.A. 387, *416*
Martz, R.M.W., see Masur, J. 388, 410, *424*
Marty, R., see Verdeaux, G. 339, 340, *367*
Masaki, T. 282, 291, *301*
Masliya, E., see Atrens, D.M. 121, 122, *183*
Mason, D.T., see Gill, J.R. 8, 19, *28*
Mason, M.M., see Thompson, G.R. 404, 410, *428*
Masoud, A., see Quimby, C.E. 374, *426*
Masuoka, D., see Carlsson, A. 34, 36, 38, 39, 42, 53, *191*
Masur, J., Martz, R.M.W., Carlini, E.A. 388, 410, *424*
Masur, J., see Carlini, E.A. 387, *417*
Masur, J., see Terada, C.W. 96, *240*
Masters, W.H., see Kolodny, R.C. 400, 408, *422*
Mastrangelo, A.S., see Yelnosky, J. 146, 147, *245*
Matsumoto, C., Griffin, W. 149, 154, *219*
Matsumoto, C., Horita, A. 147, 148, *219*
Matsumoto, C., Shaw, W.N. 154, 163, *219*
Matsumoto, C., see Brodie, B.B. 34, 50, *189*
Matsumoto, C., see Shaw, W.N. 163, 164, *235*
Matsumoto, K., see Lemberger, L. 385, 387, 399, 400, *423*
Matsuo, T., see Higuchi, H. 36, *208*
Matsuyama, S.S., Jarvik, L.F., Fu, T., Yen, F.S. 402, *424*
Matsuyama, S.S., Yen, F.S., Jarvik, L.F., Fu, T.K. 402, *424*
Matthes, K. 253, *272*
Matthies, H., see Ott, T. 53, *225*
Matthies, H., see Pohle, W. 63, *227*
Mattila, M.J., see Alhava, E. 167, *182*
Mattis, P.A., see Tedeschi, R.E. 87, 154, *239*
Mattson, R., Calverley, J.R. 263, *272*
Matussek, N., Halbach, A. 312, *361*
Matussek, N., Linsmayer, M. 71, 77, *219*
Matussek, N., see Ladisch, W. 43, 71, *213*
Maurer, J.I., see Taylor, R.L. 312, *366*
Maximilliam, C., see Gilmour, D.G. 402, *419*
Maxwell, D.R. 173, *219*
Maxwell, D.R., see Leslie, G.B. 77, 173, *215*
Maxwell, R.A., Plummer, A.J., Povalski, H., Schneider, F. 146, *219*
Maxwell, R.A., Povalski, H., Plummer, A.J. 146, *219*
Maxwell, R.A., see Ferris, R.M. 38–40, 44, *200*
May, E.L., see Martin, B.R. 381, 384, 390, 410, 411, *424*
May, E.L., see Wilson, R.S. 384, 390, *429*
Mayeda, T.K., see Clayton, R.N. 285, *297*
Mayer, O., Eybl, V. 77, 93, *219*
Mayer, S.R., see Southgate, P.J. 77, 98, *236*
Mayer-Gross, W., see Ginzel, K.H. 318, *356*
Maynard, W.R., Jr., see Bruce, R.B. 21, *26*
Mazurkiewicz-Kwilecki, I.M., Filczewski, M. 393, 394, *424*
McAllister, C., see Segal, D.S. 72, 78, 87, *235*
McAndrew, J.B., see Case, Q. 263, *268*
McCall, R.B., see Gudetsky, G.A. 44, *205*
McCann, W.P., see Lasagna, L. 173, *214*
McCarron, M., see Citron, B.P. 261, *268*
McCartney, M, see McLean, J.R. 34, 35, 53, *220*
McCarthy, D.A., see McLean, J.R. 87, *220*
McCarthy, K.D., see Yagiela, J.A. 393–395, *429*
McCash, D., see Khazan, N. 339, 340, *359*
McClean, D.K., see Zimmerman, A.M. 402, *429*
McCloskey, K., see Arnold. L.E. 250, *267*, 286, *296*
McConnell, W.B. 291, *301*
McConnell, W.B., McIlwaine, R.J. 282, 291, *301*
McCormick, R., see Citron, B.P. 261, *268*
McCormick, R., see Rumbaugh, C.L. 143, *231*, 252, 265, *273*
McCubin, J.W., see Salmoiraghi, G. 335, *364*
McCulloch, M.W., see Gerkens, J.F. 147, *203*
McCullough, D.O., Milberg, J.N., Robison, S.M. 51, 52, 149, 150, *220*
McDonald, A., see Heimstra, N.W. 71, 90, 96, *207*
McDonough, J.H., Jr., see Manning, F.J. 404, *424*
McDowell, A.A., Ziller, H.H., Krise, G.M. 91, *220*
McGaugh, J.L., Baran, L.de, Longo, V.G. 115, 117, *220*
McGaugh, J.L., see Krivanek, J.A. 129, *213*
McGeer, E.G., Fibiger, H.C., Wickson, V. 72, 77, *220*
McGeer, E.G., McGeer, P.L. 47, *220*
McGeer, E.G., see Fibiger, H.C. 48, *200*
McGeer, E.G., see Wada, J.A. 101, *242*
McGeer, P.L., see Bagchi, S.P. 47, *184*
McGeer, P.L., see McGeer, E.G. 47, *220*
McGeer, P.L., see Wada, J.A. 101, *242*
McGill, J.S., see Yelnosky, J. 146, 147, *245*
McGlothin, W.H., see Brill, N.O. 403, *416*
McGrath, W.R., Ketteler, H.J. 59, 87, *220*
McGuffin, J.C., see Clineschmidt, B.V. 132, 134, *193*
McGuire, L.E., see Pickens, R. 124–126, *227*
McHugh, P.R., see Robinson, R.G. 285, *302*
McIlwaine, R.J., see McConnel, W.B. 282, 291, *301*
McIntyre, D.C., see Fried, P.A. 410, *419*
McIsaac, W.M., Fritchie, G.E., Idapaan-Heikkila, J.E., Ho, B.T., Englert, L.F. 381, *424*
McIsaac, W.M., see Ho, B.T. 393–396, *421*
McIsaac, W.M., see Lu, T.-C. 92, *217*
McKay, J.M., Horn, G. 337, *361*

McKay, J. M., see Horn, G. 337, *358*
McKay, R. H. G. 259, *272*
McKearney, J. W. 106, 110–112, 116, 118, *220*
McKenzie, G. M., Gordon, R. J., Viik, K. 68, 78, *220*
McKenzie, G. M., Soroko, F. E. 82, 177, *220*
McKenzie, G. M., Szerb, J. C. 44, *220*
McKereghan, M. R., see Goldstein, M. 9, *28*, 48, 50, *204*
McLaren-Hume, M., see Stunkard, A. 285, *303*
McLean, J. R., Ensor, C. R., McCarthy, D. A., Bohner, B., Potoczak, D. 87, *220*
McLean, J. R., McCartney, M. 34, 35, 53, *220*
McLendon, D., see Harris, R. T. 410, *420*
McLendon, D., see Kilbey, M. M. 387, *422*
McMahon, E. M., Andersen, D. K., Feldman, J. M., Schanberg, S. M. 162, 165, *220*
McMahon, R. E. 12, *29*
McMahon, R., see Lemberger, L. 385, 387, 399, 400, *423*
McMahon, R. E., see Parli, C. J. 9, 11, 16, *29*
McManamy, M. C., Schube, P. G. 277, *301*
McManamy, M. C., see Schube, P. G. 282, *303*
McMillan, D. E. 110–113, 115, 116, 118, *220*, 388, *424*
McMillan, D. E., Dewey, W. L. 411, *424*
McMillan, D. E., Dewey, W. L., Harris, L. S. 411, *425*
McMillan, D. E., Dewey, W. L., Turk, R. F., Harris, L. S., McNeil, J. 412, *425*
McMillan, D. E., Harris, L. S., Frankenheim, J. M., Kennedy, J. S. 388, 411, *425*
McMillan, D. E., see Abel, E. L. 410, *415*
McMillan, D. E., see Dewey, W. L. 412, *418*
McMillan, D. E., see Ford, R. D. 411, *419*
McMillan, D. E., see Foree, D. P. 115, 116, *201*
McMillan, D. E., see Frankenheim, J. M. 388, 389, *419*
McMillan, D. E., see Kosersky, D. S. 388, 410, *422*
McMillan, D. E., see Leander, J. D. 110, 112, 113, 115, 116, *215*
McMillan, D. E., see Spaulding, T. C. 386, *428*
McNeill, J. H., Muschek, L. D. 58, *220*
McNeill, J., see McMillan, D. E. 412, *425*
Mead, P. G. 129, *220*
Mechoulam, R. 372, 374, 375, 378–380, *425*
Mechoulam, R., Ben-Zvi, Z., Yagnitinsky, B., Shani, A. 375, *425*
Mechoulam, R., Gaoni, Y. 373–375, *425*
Mechoulam, R., see Bicher, H. I. 389, *416*
Mechoulam, R., see Gaoni, Y. 375, 378, 379, 384, *419*
Mechoulam, R., see Shani, A. 375, *427*
Medek, A., see Navratil, J. 128, *223*
Meek, J. L., see Andén, N.-E. 330, 335, *350*
Meginnis, R. F. 110, 117, *220*
Meisch, R., see Pickens, R. 124–126, *227*
Meier, R., see Tripod, J. 70, 87, *241*
Meites, J., see Lu, K.-H. 166, *217*
Meitlich, H., see Sidell, F. R. 385, 387, *427*
Meldrum, B. S., Naquet, R. 317, 343, *361*
Mello, N. K., see Mendelson, J. H. 385, 409, *425*
Melville, K. A., see King, A. R. 314, *359*
Mendels, J. 310, *361*
Mendelson, J. H., Kuehnle, J., Ellingboe, J., Babor, T. F. 401, *425*
Mendelson, J. H., Kuehnle, J. C., Greenberg, I., Mello, N. K. 385, 409, *425*
Menge, H. G., Brand, U. 77, *220*
Mennear, J. H. 173, *221*
Mennear, J. H., Rudzik, A. D. 167, 173, *221*
Mennear, J., see Rudzik, A. D. 173, *231*
Menendez, S., see Sletten, I. W. 250, *274*
Menon, M. K., Clark, W. G., Fleming, R. M. 43, 46, 87, 173, *221*
Menon, M. K., Dandiya, P. C. 154, 173, *221*
Menon, M. K., Dandiya, P. C., Bapna, J. S. 87, 154, *221*
Menon, M. K., Fleming, R. M., Clark, W. G. 43, *221*
Menon, M. K., Subramanian, K., Menon, N. K., Sharma, S. N. 132, 134, *221*
Menon, N. K., see Menon, M. K. 132, 134, *221*
Merlis, S., Denber, H. C. B. 343, *362*
Merlis, S., see Turner, W. J. 310, *366*
Messiha, F. S., see Welch, B. L. 393, 394, *429*
Meyer, N. J., see Yoss, R. E. 285, *304*
Meyer, R. E. 385, *425*
Meyer, R. E., Di Mascio, A., Stifler, L. 253, *272*
Meyerhoff, J. L., see Snyder, S. H. 72, 82, *236*
Meyers, F. H., see Barnes, C. D. 178, *184*
Meyerson, B. J. 69, 98, *221*
Meyersson, B. J., see Eliasson, M. 69, 98, *198*
Michaluk, J., see Maj, J. 149, 167, *218*
Michanek, A., see Elliasson, M. 69, 98, *198*
Michaud, G., see Jacob, J. 179, *209*
Michelsen, L., see Lindeke, B. 11, *29*
Michniewicz, B. M., see Hucker, H. B. 9, 11, *28*
Mickle, W. A., see Monroe, R. R. 343, *362*
Miczek, K. A. 96, *221*, 388, *425*
Miele, E., Satta, M., Anania, W. 87, *221*
Mikuriya, T. H. 413, *425*
Mikus, P., see Vachon, L. 414, *429*
Milberg, J. N., see McCullough, 51, 52, 149, 150, *220*
Miller, A. I., see De Maar, E. W. J. 312, *353*
Miller, A. T., see Stowe, F. R. 136, *238*
Miller, D. D., Fowble, J., Patil, P. N. *221*
Miller, D. S., Parsonage, S. 285, *301*
Miller, E., Nieburg, H. A. 247, *272*
Miller, E. K., see Miller, R. C. 307, 314, *362*
Miller, F. N., see Brown, D. J. 399, *416*
Miller, F. P., Cox, R. H., Maickel, R. P. 87, *221*
Miller, F. P., see Maickel, R. P. 20, *29*, 53, 56, 82, 99, 106, 128, 131, 134, 140, 179, *218*
Miller, H., see Piness, G. *302*

Miller, J. W., see Craigmill, A. L. 155, 156, *195*
Miller, K. W., Freeman, J. J., Dingell, J. V., Sulser, F. 87, 106, *221*
Miller, K. W., see Freeman, J. J. 87, *201*
Miller, L. L. 372, 384, *425*
Miller, L. L., see Drew, W. G. 397, *418*
Miller, N. E. 111, 135, 139, *221*
Miller, N. E., see Barry, H. 115, 116, 128, *184*
Miller, N. E., see Krieckhaus, E. E. 102, 103, *213*
Miller, R. B., see Tedeschi, D. H. 96, 97, *239*
Miller, R. C., Miller, E. K. 307, 314, *362*
Miller, R. C., see Nichols, W. W. 402, *425*
Miller, R. E., Levine, J. M., Mirsky, I. A. 40, 96, *221*
Miller, R. J., see Horn, A. S. 40, *208*
Millikan, R., see Detrick, L. E. 60, 62, 141, 142, 146, 148, *196*
Milloy, S., see Glick, S. D. 34, *203*
Mills, A. W., see Pick, C. G. 121, *227*
Mills, J., see Fuller, R. W. 3, *27*, 56, *202*
Mills, S. L., see Proctor, C. D. 173, *228*
Milne, M. D., Scribner, B. H., Crawford, M. A. 7, *29*
Milne, M. D., see Asatoor, A. M. 7, 15, 21, *25*
Milner, G., see Johnson, J. 291, *300*
Milner, P., see Olds, J. 121, *225*
Milner, P., see Zucker, I. 140, *246*
Milton, A. S., see Layman, J. M. 381, *422*
Milyukas, V. I., see Skaringa, A. P. 176, *236*
Miner, E. J., see Hill, H. E. 308, 309, *357*
Miner, E. J., see Isbell, H. 310, 312, 318, 319, 330, *358*
Miner, E. J., see Rosenberg, D. E. 253, 254, 256, 261, *273*, 310, 311, 318, 319, *363*
Miner, E. J., see Wolbach, A. B., Jr. 310, 318, *368*
Minkowsky, W. L. 129, *221*
Miras, C. J. 406, 409, *425*
Mirsky, A. F., see Kornetsky, C. 249, *271*
Mirsky, I. A., see Miller, R. E. 40, 96, *221*
Misher, A., see Groppetti, A. 48, 90, 157, *205*
Mison-Crighel, N., Luca, N., Crighel, E. 342, *362*
Missala, K., see Lal, S. 77, *214*
Missere, G., see Tonini, G. 100, 107, *240*
Mitchard, M., see Beckett, A. H. 21, 23, *25*
Mitchell, J., see Yelnosky, J. 147, *245*
Mitchell, K. J., see Proctor, C. D. 173, *228*
Mitchell, S. A., see Rechtschaffen, A. 285, *302*
Mitcheson, M., see Hawks, D. 282, 291, *299*
Mitoma, C., see Uyeno, E. T. 317, *367*
Mitskis, A. M., see Skaringa, A. P. 176, *236*
Mitznegg, P., see Estler, C.-J. 162, *199*
Miya, T. S., see Beuthin, F. C. 88, 154, *186*
Miyahara, J. T., see Swinyard, E. A. 175, 177, *239*
Modell, W. 250, 258, 259, *272*, 285, *301*
Modell, W., Reader, G. G. 279, *301*
Modern, F. S., see Detrick, L. E. 60, 62, 141, 142, 146, 148, *196*
Möller-Nielsen, I., Dubnick, B. 45, 46, *221*
Mogenson, G. J. *221*
Mogenson, G. J., Cinnamon, A. D., Stevenson, J. A. *221*
Mogenson, G. J., see Russek, M. 138, *232*
Moi, K., see Winters, W. D. 178, *244*
Moisset, B., Welch, B. L. 90, *221*
Moliis, G., see Lewander, T. 140, 150, 157, 158, *216*
Molinengo, L., Ricci-Gamalero, S. 101, 106, *221*
Molliver, M. E., see Freedman, D. X. 314, 319, 320, *355*
Molloy, B. B., see Fuller, R. W. 5, 11, *27*, *28*, 49, 53, 56, 156, *202*
Monachon, M.-A., see Haefely, W. 178, *205*
Monlár, J., see Galambos, E. 87, *202*
Monroe, R. R., Drell, H. J. 265, *272*, 283, *301*
Monroe, R. R., Heath, R. G. 343, *362*
Monroe, R. R., Heath, R. G., Mickle, W. A., Llewellyn, R. C. 343, *361*
Montagu, J. D. 251, *272*
Montanaro, N., Babbini, M. 89, 90, *221*
Montanaro, N., see Babbini, M. 77, 87, *183*
Moog, E., see Zetler, G. 77, 154, *246*
Moorcroft, W. H., see Lytle, L. D. 138, *217*
Moore, B. M., see Ewing, P. L. 128, 129, *199*
Moore, J. W., Jr., see Kilbey, M. M. 387, *422*
Moore, K. E. 34–36, 162, 167, 169, 173, 174, *221*
Moore, K. E., Carr, L. A., Dominic, J. A. 87, *222*
Moore, K. E., Dominic, J. A. 88, *222*
Moore, K. E., Lariviere, E. W. 34–36, *222*
Moore, K. E., Rech, R. H. 106, *222*
Moore, K. E., Sawdy, L. C., Shaul, S. R. 162, 173, 175, *222*
Moore, K. E., Thornburg, J. E. 87, *222*
Moore, K. E., see Carr, L. A. 43, *191*
Moore, K. E., see Chiueh, C. C. 43, 44, *192*
Moore, K. E., see Dominic, J. A. 87, 88, *197*
Moore, K. E., see Gudetsky, G. A. 44, *205*
Moore, K. E., see Rech, R. H. 106, *229*
Moore, K. E., see Thornburg, J. E. 39, 82, 87, *240*
Moore, K. E., see Voigtländer, P. F. von 44, 68, *242*
Moormann, J., Opitz, K. 179, *222*
Moraes, S. de, Bussato, P. A., Carvalho, F. V. 60, *222*
Moraes, S. de, Carmo, M. M., Carvalho, V. F. 60, *222*
Moraes, S. de, Carvalho, V. F. 60, 63, *222*
Moraes, S. de, Carvalho, V. F. de, Aragao, J. A. 63, *222*
Moraes, S. de, Carvalho, V. F. de, Wehrle, R. D. 60, *222*
Morard, J. C., see Halpern, B. N. 174, *206*
Moreau, J. J. 384, *425*
Moreton, J. E., see Davis, W. M. 410, 413, *417*
Moretz, F. H., see Foree, D. P. 115, 116, *201*
Morgan, C. D., Cattabeni, F., Costa, E. 51–53, 56, *222*

Morgan, C. D., Löfstrandh, S., Costa, E. 39, 45, 54, 56, *222*
Morgan, D. H., see Flegin, O. T. 147, *201*
Morgan, M., see Mandell, A. J. 48, 57, *219*
Morgane, P. J., see Stern, W. C. 342, *365*
Morgenstern, G., see Cohen, N. J. 250, *268*
Morgenstern, G., see Sykes, D. H. 250, *274*
Morgenstern, N. L., see Weiss, S. R. 265, *275*
Mori, K., see Winters, W. D. 341, 342, 348, *368*
Morin, R. D., see Benington, F. 11, *26*, 40, *186*
Morin, R. D., see Smythies, J. R. 15, 18, 19, *30*, 306, 323, 348, *365*
Morishima, A., see Nahas, G. G. 402, *425*
Moroji, T., see Asano, Y. 35, 165, *183*
Moroji, T., see Onodera, I. 165, *225*
Morpurgo, C., Theobald, W. 77, 87, 96, 97, 106, 154, 173, *222*
Morpurgo, C., see Delini-Stula, A. 79, *196*
Morpurgo, C., see Valzelli, L. 16, *31*, 150, 154, *241*
Morrison, C. F. 114, *222*
Morrison, J. L., Abreu, E. 160, *222*
Morrissey, W., see Vachon, L. 414, *429*
Morse, R. M., see Swanson, D. W. 285, *303*
Morse, W. H., Herrnstein, R. J. 110, 116, *223*
Morse, W. H., see Dews, P. B. 98, *197*
Morse, W. H., see Kelleher, R. T. 98, 110, 112, 116, 117, *211*
Moses, H., see Klawans, H. L. 71, 77, *212*
Mosher, W. A., see Hively, R. L. 375, *421*
Mourevieff-Lesuisse, F., Giurea, C. 130, *223*
Mouriz-Garcia, A., Schmidt, R., Arlazoroff, A. 337, *362*
Moyer, K. E., see Crabtree, J. M. 96, *195*
Mrtek, R. G., see Foreman, R. L. 11, *27*
Mudge, G. H., see Weiner, J. M. 7, *31*
Müller, E. E., see Mantegazza, P. 87, 134, 154, 173, *219*
Müller, H., Schoetensack, W. 77, 87, 173, *223*
Mueller, M. R., see Rumehart, D. E. 96, *231*
Müller, P. J., Vernikos-Danellis, J. 168, *223*
Müller, R. U., see Glick, S. D. 110, 140, *203*
Mulas, A., Crabai, F., Pepeu, G. 94, *223*
Mulligan, B. M., see Bolt, A. G. 164, *187*
Munch, J. C., see Hartung, W. H. 141, *206*
Mundschenk, D. C., see Goldstein, L. 178, *204*
Mundy, J., see Yelnosky, J. 147, *245*
Munkvad, I., Pakkenberg, H., Randrup, A. 71, 78, *223*
Munkvad, I., Randrup, A. 77, *223*
Munkvad, I., see Randrup, A. 70–72, 77, 82, 87, 88, 94, 95, 96, 99, *228*, *229*, 263, 264, *273*
Munoz, C., Goldstein, L. 177, *223*
Munro, J. F., MacCuish, A. C., Wilson, E. M., Duncan, L. J. P. 285, *301*
Munson, A. E., Harris, L. S., Friedman, M. A., Dewey, W. L., Carchman, R. A. 402, *425*
Munson, A. E., see Carchman, R. A. 402, *416*
Munson, A. E., see Harris, L. S. 402, *420*
Munson, A. E., see Levy, J. A. 402, *423*
Munson, A. D., Levy, J. A., Harris, L. S., Dewey, W. L. 402, *425*
Murphree, H. B., see Goldstein, L. 251, *269*, 343, *356*
Murphree, H. B., see Pfeiffer, C. 251, *272*
Murphree, O. D., DeLuca, D. C., Angel, C. 71, *223*
Murphree, O. D., see Angel, C. 109, 117, *182*
Murphy, D. L., see Kammen, D. P. van 287, *303*
Murphy, E. H., see Bradley, D. W. H. 94, *188*
Murphy, S., see Goldman, H. 335, *356*
Musacchio, J., see Besson, M. J. 47, *186*
Musacchio, J. M., see Kopin, I. J. 63, *212*
Musacchio, J. M., see Korein, J. 343, *360*
Muschek, L. D., see McNeill, J. H. 58, *220*
Muscholl, E. 58, *223*
Muscholl, E., see Löffelholz, K. 40, *217*
Muskens, M. J., see Vree, T. B. 11, 14, *31*
Muskus, A., see Trendelenburg, U. 59, 142, 146, *241*
Musto, D. F. 278, *301*
Myerson, A. 248, *272*, 280, 284, 289, *301*
Myerson, A., Ritvo, M. 280, *301*
Myerson, A., see Lesses, M. F. 250, *271*, 280, *301*
Myerson, A., see Loman, J. 247, *272*
Myerson, A., see Schube, P. G. 282, *303*

Nador, K., see Doda, M. 59, *197*
Naess, K. 254, *272*
Nagasaka, see Sano, I. 291, *303*
Nagy, T., see Böszörményi, Z. 311, 343, *352*
Nahas, G. G. 372, 384, *425*
Nahas, G. G., Desoize, B., Armand, J. P., Hsu, J., Morishima, A. 402, *425*
Nahas, G. G., Schwartz, I. W., Adamec, C., Manger, W. M. 398, 410, *425*
Nahas, G. G., Suciu-Foca, N., Armand, J. P., Morishima, A. 402, *425*
Nahorski, S. R., Rogers, K. J. 161, 162, *223*
Naik, P. Y., see Sethy, V. H. 87, *235*
Naimzada, M. K., see Costa, E. 34, 35, 37, 48, 52, 53, 90, 132, 157, *194*
Naimzada, M. K., see Groppetti, A. 48, 90, 157, *205*
Naimzada, M. K., see Mantegazza, P. 87, 91, 134, 138, 154, 156, 163, 169, 173, *219*
Naimzada, M. K., see Riva, M. 139, 149, *230*
Nakai, K., see Hishikawa, Y. 286, *300*
Nakajima, H., Grandjean, J. L., L'Huillier, J., Thullier, J. 154, *223*
Nakajima, H., see Thuillier, J. 150, 154, *240*
Nakamura, K., Shimamoto, K. 59, *223*
Nakamura, R. K., see Glick, S. D. 90, 102, 107, *203*

Nalepa, S. D., see Sofia, R. D. 391, *428*
Nandy, K., Bourne, G. H. 333, *362*
Naquet, R., see Meldrum, B. S. 317, 343, *361*
Naquet, R., see Walter, S. 317, 342, 343, *367*
Naranjo, C., Shulgin, A. T., Sargent, T. 311, 328, *362*
Naranjo, C., see Shulgin, A. T. 311, 312, 328, *364*
Nargorney, V. A., see Isachenko, V. B. 164, *209*
Nasello, A. G., Izquierdo, I. 165, *223*
Nasello, A. G., see Izquierdo, I. 176, *209*
Nash, B. M., see Bradley, D. W. H. 94, *188*
Nash, C. B., see White, R. P. 70, 77, 176, *244*
Nasmyth, P. A. 165, *223*
Nathanson, M. H. 247, 248, 250, *272*, 280, *301*
Naumenko, E. V. 165, *223*
Navratil, J., Medek, A., Hrbek, J., Komenda, S. 128, *223*
Naylor, R. J., Costall, B. 71, 77, *223*
Naylor, R. J., Olley, J. E. 68, 79, 89, *223*
Naylor, R. J., see Costall, B. 72, 77, 78, 79, 89, *194*
Neal, M. J., see Hemsworth, B. A. 57, *207*
Neff, C. A., see Baggot, J. D. 21, *25*
Neff, N., Rossi, G. V., Chase, G. D., Rabinowitz, J. L. 334, *362*
Neill, D. B., Boggan, W. O., Grossman, S. P. 79, 94, *223*
Neill, D. B., Grant, L. D., Grossman, S. P. 90, 136, *223*
Neill, D. B., Grossman, S. P. 134, *223*
Neill, D. B., see Hernodon, J. G. 98, *207*
Nelsen, J. M., see Weil, A. T. 385, *429*
Nelson, L. M., see Ellinwood, E. H., Jr. 77, 81, 178, *198*
Nessehoff, W., see Heath, R. G. 292, *299*
Nesson, B., see Lal, H. 96, 179, *214*
Neu, R. L., Powers, R. O., King, S., Gardner, L. I. 401, *425*
Neu, R. L., see Carakushansky, G. 408, *416*
Neuburg, J., Thut, P. D. 87, *223*
Neuhoff, V., see Leonhard, B. E. 164, *215*
Neuhoff, V., see Osborne, N. N. 54, *225*
Neuhold, K., Taeschler, M., Cerletti, A. 313, 315, 337, *362*
Neumeyer, J. L., Shagoury, R. A. 374, *425*
Neuru, E. N., see Tainter, M. L. 177, *239*
Neveu, C., see Le Douarec, J. C. 4, *29*, 249, *271*
Nevisimalova, S., see Roth, B. 250, *273*, 286, *302*
Newley, V. E., see Dewey, W. L. 398, *418*
Newman, L. M. 121–123, *223*
Ney, P. G. 287, *301*
Ng, K. Y., Chase, T. N., Kopin, I. J. 44, *224*
Ng, L. K. Y., see Colburn, R. W. 382, *417*
Ng, L. K. Y., see Williams, R. B. 398, *429*
Ngai, S. H., see Lin, R. C. *360*
Nichols, R. E., Patterson, C. S., Walaszek, E. J. 161, 162, *224*
Nichols, W. W., Miller, R. C., Heneen, W., Bradt, C., Hollister, L., Kanter, S. 402, *425*
Nickerson, M., see Innes, I. R. 249, 251, 253, 254, 259, *270*
Nicol, E., see Richards, R. 169, *230*
Nicolov, N., see Goldman, H. 335, *356*
Nieburg, H. A., see Miller, E. 247, *272*
Nieford, K. A. 4, *29*, 262, *272*
Nielsen, C. K., Frey, H.-H. 61, 146, *224*
Nielsen, C. K., Magnussen, M. P., Kampmann, E., Frey, H.-H. 45, *224*
Nielsen, E. B., Lyon, M. *224*
Nielsen, E. G. 282
Nielsen, I. M., see Holm, T. 138, *208*
Nielson, H. C., see Sharp, J. C. 135, 136, *235*
Niemegeers, C. J. E., Verbruggen, F. J., Janssen, P. A. J. 101, 106, *224*
Niemegeers, C. J. E., Verbruggen, F. J., Wauqier, A., Janssen, P. A. J. 98, 100, 101, *224*
Niemegeers, C. J. E., see Wauquier, A. 121, 122, *243*
Niemegeers, C. J. E., see Janssen, P. A. J. 77, 81, *210*
Nieschulz, O. 278, *301*
Niijima, A. 138, *224*
Nikishawa, T., see Iwata, H. 80, 142, *209*
Niklasson, F., see Änggård, E. 14, 15, 18, 21–24, *25*, 251, 260, *266*
Nikulcheva, N. G., see Isachenko, V. B. 164, *209*
Nilsson, I. M., see Agurell, S. 381, 382, 384, *415*
Nilsson, I. M., see Ohlsson, A. 373, *426*
Nilsson, I. M., see Wahlquist, M. 382, *429*
Nilsson, J. L. G., see Agurell, S. 382, 384, *415*
Ninteman, F. W., see Poschel, B. P. H. 106, 122, 123, *227*
Nishioka, I., see Shoyama, Y. 375, *427*
Nishizaea, T., see Takasaki, K. 62, *239*
Nistri, A., Bartolini, A., Deffenu, G., Pepeu, G. 57, *224*
Noach, E. L., see Hart, M. 97, *206*
Nordquist, M., see Agurell, S. 382, 384, *415*
Norman, J. H., see Breckenridge, B. M. 58, *188*
North, M., see Hulme, E. C. 47, 56, 134, *209*
North, R. B., Harik, S. I., Snyder, S. H. 72, 82, *224*
North-Diehl, A., see Baum, E. 79, *185*
Norton, S. 70, 93, *224*
Norton, S., see Jewett, R. E. 178, *210*
Norwich, M. R., see Dingell, J. V. 82, 87, *197*
Norwich, M. R., see Sulser, F. 87, 88, *239*
Novelli, A. N., Tainter, M. L. 82, *224*
Novick, L., Pihl, R. 90, *224*
Noyes, R., Baram, D. A. 413, *425*
Noyes, R., Jr., Brunk, S. F., Baram, D. A., Canter, A. 413, *425*
Nozaki, M., Vaupel, D. B., Martin, W. R. 316, 317, 321, *362*
Nozaki, M., see Martin, W. R. 281, *301*, 348, *361*
Nozaki, M., see Vaupel, D. B., 316, 317, 321, 322, 335, *367*
Nueten, J. M. van, see Janssen, P. A. J. 77, *210*
Nutt, J. G., see Griffith, J. D. 249, 253, *270*
Nutt, J. G., see Jasinski, D. R. 285, 288, *300*
Nybäck, H., see Schuberth, J. 54, 55, 56, *234*

Oates, J., see Rangno, R. E. 8, *30*
Oates, J. A., see Cavanaugh, J. H. 9, 19, *26*
Oates, J. A., see Flegin, O. T. 147, *201*
Oates, J. A., see Griffith, J. D. 256, 261, *269*, *270*, 282, 293, 295, *299*
Obe, G., see Herha, J. 401, *420*
Obianwu, H. O. 59, 63, 147, 148, *224*
Obianwu, H. O., Stitzel, R., Lundborg, P. 49, 50, 148, *224*
Obianwu, H., see Andén, N.-E. 59, *182*
O'Brien, G., see Goldstein, L. 178, *204*
O'Brien, J., see Lal, H. 77, 96, *214*
O'Dea, K., Rand, M. J. 173, *224*
Ögren, S.-O., see Corrodi, H. 77, 87, *194*, 264, *268*
Ökrös, I., see Pfeiffer, A. K. 49, *226*
Offermeier, J., Potgieter, B. 82, 87, *224*
Ogata, A. 280, *301*
Ogborne, A., see Hawks, D. 282, 291, *299*
Ogiu, K., see Takagi, H. 341, *366*
Ogle, K. N., see Yoss, R. E. 285, *304*
Ognjanov, V., see Sletten, I. W. 250, *274*
Ogunbona, F. A., see Beckett, A. H. 11, *25*
Ohashi, K., see Iwata, H. 80, 142, *209*
Ohler, E. A., Wakerlin, G. E. 141, *224*
Ohlsson, A., Abou-Chaar, C. I., Agurell, S., Nilsson, I. M., Olofsson, K., Sandburg, F. 373, *426*
Ohlsson, A., see Agurell, S. 381, 382, *415*
Ohnesorge, F. K., see Andres, F. 161, *182*
Oi, S., see Yasunobu, K. T. 45, *245*
Oku, R., see Shoyama, Y. 375, *427*
Okun, R., Liddon, S. C., Lasagna, L. 179, *224*
Okun, R., see Ellison, T. 15, 18, 20, 21, *27*, 81, 173, *199*
Olds, J. 109, 117, 121, 122, 123, *224*
Olds, J., Milner, P. 121, *225*
Olds, M. E. 121, *225*
Olds, M. E., see Domino, E. F. 57, 122, *197*
Oliveira, L. de, Graeff, F. G. 112, 118, *225*
Oliviero, A. 101, 106, *225*
Oliviero, A., Eleftheriou, B. E., Bailey, D. W. 94, *225*
Oliverio, A., see Bovet, D. 101, *188*
Olley, J. E., see Costall, B. 77, 78, 79, *194*
Olley, J. E., see Naylor, R. J. 68, 79, 89, *223*
Olofsson, K., see Ohlsson, A. 373, *426*
Olsen, J., see Carder, B. 387, *416*
Ong, B. H. 288, *302*
Onodera, I., Ito, K., Moroji, T., Yamashita, I. 165, *225*
Opitz, K. 4, *29*, 56, 163, *225*
Opitz, K., Chu, H., Grüter, H. 161, *225*
Opitz, K., Grüter, H., Loeser, A. 161, 166, *225*
Opitz, K., Loeser, A. 163, 164, *225*
Opitz, K., see Chu, H. 161, *193*
Opitz, K., see Moormann, J. 179, *222*
Ornitz, E. M., see Freedman, D. X. 320, *355*
O'Rourke, T., see Dewey, W. L. 386, 390, 399, 411, 412, *418*
Orrenius, S., Maehly, A. C. 264, *272*
Orrenius, S., see Hoffström, I. 12, *28*
Orsingher, O. A., Fulginiti, S. 102, 106, 107, *225*
Orsingher, O. A., see Biscardi, A. M. 41, *187*
Orsingher, O. A., see Fulginiti, A. 49, 106, *202*
Orzechowski, G. 45, 62, 63, 141, 142, *225*
Orzechowski, G., see Kiessig, H. J. 179, *211*
Orzechowski, G., see Malorny, G. 62, 141, 142, *218*
Osborne, N. N., Neuhoff, V., Leonard, B. E. 54, *225*
Osmond, H., see Hoffer, A. 306, *357*
Osterberg, A. C., see Greenblatt, E. N. 155, 166, 169, 174, *205*
Osterholm, K. C., see Pirch, H. 87, *227*, 410, *426*
Ostwald, P., see Rockwell, D. 292, *302*
Oswald, I. 252, *272*
Oswald, I., Jones, H. S., Mannerheim, J. E. 260, *272*
Oswald, I., Lewis, S. A., Dunleavy, D. L. F., Brezinowa, V., Briggs, M. 257, *272*
Oswald, I., Thacore, V. R. 260, *272*
Oswald, I., see Dunleavy, D. L. F. 252, 255, *268*
Oswald, I., see Lewis, S. A. 257, *271*
Otis, L. S., see Sharpe, L. G. 317, *364*
Otis, L. S., see Uyeno, E. T. 317, *367*
Ott, T., Schmitt, M., Pohle, W., Matthies, H. 53, *225*
Ouellet, J., Palaic, D., Albert, J. M., Tetreault, L. 395, *426*
Overstreet, D. H., Hadick, D. G., Russell, R. W. 135, *225*
Owens, J. E. 110, 111, 117, *225*
Owens, J. E., Campbell, J. T. 110, 117, *225*
Owens, L., see Sulser, F. 87, *239*
Owens, M., see Sulser, F. 87, 88, *239*
Owens, M. C., see Sulser, F. 15, 16, *30*
Owens, M. L., see Dingell, J. V. 82, 87, *197*
Oxenkrug, G. G., see Lapin, I. P. 87, 154, 173, *214*

Paasonen, M. K., Vogt, M. 54, 57, *225*
Pace, H. B., Davis, W. M., Borgen, L. A. 406, 407, *426*
Pace, H. B., see Borgen, L. A. 406, *416*
Pace, H. B., see Davis, W. M. 410, 413, *417*
Page, I. H., see Gessner, P. K. 314, *355*
Page, I. H., see Salmoiraghi, G. C. 335, *364*
Page, I. H., see Twarog, B. M. 328, *366*
Page, J. G., see Finger, K. F. 164, *200*
Page, T. F., Jr., see Foltz, R. L. 383, *419*
Pakkenberg, H., see Fog, R. 77, 78, 79, *201*
Pakkenberg, H., see Munkvad, I. 71, 78, *223*
Palaic, D., see Ouellet, J. 395, *426*
Palatucci, D. M. 263, *272*
Palider, W., see Maj, J. 81, *218*
Palladin, A. V., Rybina, A. A. 161, *225*
Palmer, G. C. 58, *225*

Palmer, G. C., Burks, T. F. 333, *362*
Palmer, G. C., Davenport, G. R., Ward, J. W. 130, *226*
Panayiotopoulos, C. P., see Fink, M. 409, *419*
Panda, J. N., Ray, A. K. 165, *226*
Panksepp, J., Booth, D. A. 136, 138, 139, *226*
Panse, F., Klages, W. 278, *302*
Papeschi, R., Randrup, A. 77, *226*
Papeschi, R., Randrup, A., Lal, S. 80, 91, *226*
Paquette, R., see Borella, L. E. 87, 134, *188*
Pardridge, W. M., Connor, J. D. 20, *29*
Paris, M., Ghirlanda, C., Chigneau, M., Giry, L. 375, *426*
Parke, D. V., see Bridges, J. W. 13, *26*
Parke, H. C., see Winder, C. V. 62–64, 141, 142, 144, 148, *244*
Parli, C. J., McMahon, R. E. 11, *29*
Parli, C. J., Wang, N., McMahon, R. E. 9, 11, 16, *29*
Parli, J., see Fuller, R. W. 11, *27*
Parmar, S. S. 45, 46, *226*
Parmley, L. F., see Zalis, E. 264, *275*, 288, *304*
Pars, H. G. *426*
Pars, H. G., see Dewey, W. L. 399, *418*
Pars, H. G., see Zitko, B. A. 380, *429*
Parsonage, S., see Miller, D. S. 285, *301*
Partyka, D., see Chernov, H. 167, *192*
Parvathy, S., see Ghosh, M. N. 138, *203*
Passouant, P., Passouant-Fontaine, Th., Cadilhac, J. 341, *362*
Passouant-Fontaine, Th., see Passouant, P. 341, *362*
Patel, B. C., see Klawans, H. L. 70, 77, *212*
Patil, P. N., LaPidus, J. B., Campbell, D., Tye, A. 60, *226*
Patil, P. N., LaPidus, J. B., Tye, A. 60, 63, *226*
Patil, P. N., see Miller, D. D. *221*
Paton, D. M. 345, *362*
Paton, W. D. M. *362*
Paton, W. D. M., Crown, J. 372, *426*
Paton, W. D. M., Pertwee, R. G. 284, *426*
Paton, W. D. M., see Gill, E. W. 389, 410, *419*
Pattee, C. J., see Pinter, E. J. 163, *227*, 254, *272*
Patterson, C. S., see Nichols, R. E. 161, 162, *224*
Patton, R. A., see Evans, H. L. 112, 113, 118, *199*
Pauk, G. L., see Paul, M. I. 57, *226*
Paul, M. I., Pauk, G. L., Ditzion, B. R. 57, *226*
Pauli, A. I., see Broekkamp, C. L. W. 137, *189*
Pauly, J. E., see Scheving, L. E. 167, *233*
Pawan, G. L. S. 4, *29*
Pawlowski, A. A. 314, *362*
Payne, R. B., Hauty, G. T. 249, *272*
Payne, R. B., see Hauty, G. T. 249, *270*
Pazzagli, A., Pepeu, G. 128, *226*
Peczenik, O., Zei, M. 173, *226*
Pedden, J. R., Tainter, M. L., Cameron, W. M. 148, *226*
Pederson, J. E., see Green, K. 414, *420*
Peek, L., see Regelson, W. 413, 414, *427*
Peng, T. C., see Dewey, W. L. 400, *418*
Pennes, H. H., see Hoch, P. H. 336, *357*
Peoples, S. A., Guttmann, E. 253, *272*, 280, *302*
Pepeu, G., Bartolini, A. 57, *226*
Pepeu, G. C., Bartolini, A., Deffenu, G. 57, *226*
Pepeu, G., see Bartolini, A. 57, *184*
Pepeu, G., see Deffenu, G. 57, *196*
Pepeu, G., see Mulas, A. 94, *223*
Pepeu, G., see Nistri, A. 57, *224*
Pepeu, G., see Pazzagli, A. 128, *226*
Perel, J. M., see Faraj, B. A. 3, *27*
Peretz, D. I., Smythies, J. R., Gibson, W. C. 311, 316, 328, *362*
Perez-Cruet, J., see Tagliamonte, A. 49, 53, 55, 56, *239*
Perez-Reyes, M., Timmons, M. C., Lipton, M. A., Davis, K. H., Wall, M. E. 380, 383, *426*
Perhack, J. L., see Kubena, R. L. 400, *422*
Perhoch, J. L., see Barry, H. III 400, *415*
Peri, G., see Consolo, S. 57, *194*
Perkins, L., see Bergen, J. R. 315, *351*
Perley, J. E., see Bauer, W. S. 167, *185*
Perry, G. F., see Korenyi, C. 251, *271*
Perry, K. W., see Fuller, R. W. 11, *27*
Persaud, T. V. N., Ellington, A. C. 406, 407, *426*
Persson, S. A., see Hollunger, G. 333, *357*
Persson, T. 48, *226*
Pertwee, R. G., see Gill, E. W. 389, 410, *419*
Pertwee, R. G., see Lawrence D. K. 410, *422*
Pertwee, R. G., see Paton, W. D. M. 384, *426*
Peschel, E., see Barnett, A. 173, *184*
Pescor, F. T., see Wikler, A. 308, *367*
Peterfalvi, M., see Guilleux, H. 68, *205*
Peters, D. A. V. 330, 333, *362*
Peters, N. J., see Wang, R. I. H. 167, *243*
Petersen, D. V., see Holm, T. 138, *208*
Petersen, M. C., see Schwartz, B. E. *364*
Peterson, B. H., Somerville, D. M. 293, *302*
Peterson, D. I., Hardinge, M. G., Tilton, B. E. 175, *226*
Peterson, D. W., Cohen, G. M., Sparber, S. B. 388, *426*
Peterson, D. W., see Sparber, S. B. 110, 111, 113, 119, *237*
Petrazilka, T., Haeflinger, W., Sikemeier, C. 379, *426*
Petrus, R., see Hepler, R. S. 408, 409, 414, *420*
Petryshyn, W. R., see Satinder, K. P. 100, *233*
Petsche, H., see Stumpf, Ch. 340, *366*
Petten, G. R. van, see Lewis, J. J. 161, *216*
Pfeferman, A., see Karniol, I. G. 397, *421*
Pfeiffer, A. K. 161, *226*
Pfeiffer, A. K., Csáki, L., Fodor, M., György, L., Ökrös, I. 49, *226*
Pfeiffer, A. K., Galambos, E., György, L. 87, 173, *226*
Pfeiffer, A. K., György, L., Fodor, M. 87, 161, 177, *226*
Pfeifer, A. K., see Galambos, E. 87, *202*
Pfeiffer, A. K., see Skolnik, J. 57, *236*

Pfeiffer, C., Goldstein, L., Murphree, H. B., Sugerman, A. 251, *272*
Pfeiffer, C., see Dreisbach, R. H. 277, *298*
Pfeiffer, C. C., see Apter, J. T. 337, *351*
Pfeiffer, C. C., see DeMaar, E. W. J. 312, *353*
Pfeiffer, C. C., see Goldstein, L. 251, *269*, 343, *356*
Pflanz, M., Schrader, A. 251, *272*
Philippu, A., Beyer, J. 39, *226*
Philippu, A., Glowinski, J., Besson, M. J. 44, *226*
Philippü, A., see Schumann, H. J. 41, *235*
Philips, S. R., Durden, D. A., Boulton, A. A. 332, *362*
Phillips, A. G., Fibiger, H. C. 121, 122, *226*
Phillips, M. I., Bradley, P. B. 77, 87, 93, 94, *227*
Phillips, R. N., Brown, D. J., Martz, R. C., Hubbard, J. D., Forney, R. B. 404, *426*
Phillips, R. N., Turk, R. F., Forney, R. B. 403, 407, *426*
Phillis, J. W., see Krnjević, K. 338, *360*
Piala, J. J., see Horovitz, Z. P. 97, *209*
Piala, J. J., see Rubin, B. 338, *363*
Piarroux, M.-C., see Baum, E. 79, *185*
Piarroux, M.-C., see Boissier, J.-R. 79, *187*
Picarelli, Z. P., see Gaddum, J. H. 329, *355*
Pichot, P., see Delay, J. 293, *298*
Pick, Cassens, G., Mills, A. W. 121, *227*
Pickenhain, L., see Bauer, I. 107, *184*
Pickering, R. W., see Corne, S. J. 312, *353*
Pickens, R. 124, 125, *227*
Pickens, R., Harris, W. C. 124, 125, *227*
Pickens, R., Meisch, R. A., Dougharty, J. A. 125, 126, *227*
Pickens, R., Meisch, R., McGuire, L. E. 124, 125, 126, *227*
Pickens, R., Thompson, T., Yokel, R. A. 124, *227*
Pickens, R., see Thompson, T. 124, 126, *240*
Pickens, R., see Yokel, R. A. 5, *31*, 124, 125, 138, *245*
Pickens, R. W., Crowder, W. F. 127, *227*
Pickworth, W. B., Sharpe, L. G. 339, *362*
Pieper, W. A., see Cole, J. M. 317, *353*
Pieri, L., Pieri, M., Haefely, W. 333, *362*
Pieri, M., see Pieri, L. 333, *362*
Pihl, R. O., Greenberg, I. 90, *227*
Pihl, R., see Novick, L. 90, *224*
Pinchard, A., see Janssen, P. A. J. 77, *210*
Pincus, G., see Bergen, J. R. 315, *351*
Pincus, I. J., see Citron, B. P. 261, *268*
Piness, G., Miller, H., Alles, G. A. *302*
Pinkston, J. O., Pinkston, J. L., Donabedian, A. S. 62, 141, 142, *227*
Pinkston, J. L., see Pinkston, J. O. 62, 141, 142, *227*
Pinter, E. J., Pattee, C. J. 163, *227*, 254, *272*
Pirch, J. H., Cohn, R. A., Barnes, P. R., Barratt, E. S. 87, *227*, 410, 411, 413, *426*
Pirch, J. H., Osterholm, K. C., Cohn, R. A., Barratt, E. S. 410, *426*
Pirola, C., see Riva, M. 139, 149, *230*
Pitt, C. B., see Christensen, H. D. 383, *417*
Plane, P., Bogaert, B. 128, *227*
Plank, J. B., see Haley, S. L. 406, 407, *420*
Plank, J. B., see Keplinger, M. L. 406, *422*
Plas, R., see Gastaud, H. 251, *269*
Platt, D. S., Cockrill, B. L. 17, *29*
Pless, J. E., see Sidell, F. R. 385, 387, *427*
Pletscher, A., Bartholini, G., Bruderer, H., Burkard, W. P., Gey, K. F. 3, *29*, 53, *227*
Pletscher, A., Burkhard, W. P., Bruderer, H., Gey, K. F. 56, *227*
Pletscher, A., Prada, M., Da, Burkhard, W. P. 56, *227*
Pletscher, A., see Bogdanski, D. F. 328, *352*
Pletscher, A., see Prada, M. Da 54, 55, *228*
Pleuvry, B. J. 179, *227*
Pleuvry, B. J., see Major, C. T. 179, *218*
Pliner, P., see Cappell, H. 385, *416*
Plotnikoff, N. 102, *227*
Plummer, A. J., see Chernov, H. 167, *192*
Plummer, A. J., see Maxwell, R. A. 146, *219*
Podvalova, I., see Babulova, A. 59, 60, 143, 144, 147, *183*
Pohle, W., Matthies, H. 63, *227*
Pohle, W., see Ott, T. 53, *225*
Polakoski, K., see Cicero, T. J. 401, *417*
Pollack, D., see Lewis, J. J. 161, *216*
Pontrelli, E. 265, *272*
Popova, E. N., Vavilov, A. M., Krivitskaya, G. N., Tumanov, V. P. 174, *227*
Porsolt, R. D., Joyce, D., Summerfield, A. 94, *227*
Porsolt, R. D., see Bradley, D. W. H. 94, *188*
Porter, C. C., see Stone, C. A. 146, *238*
Porter, R., see Adey, W. R. 341, *350*
Porter, P. B., see Sharp, J. C. 135, 136, *235*
Poschel, B. P. H. 111, 117, *227*
Poschel, B. P. H., Ninteman, F. W. 106, 122, 123, *227*
Possanza, G. J., see White, R. P. 70, 77, 176, *244*
Post, B., see Frankenhäuser, M. 255, *269*
Post, R. M. 282, 287, *302*
Post, R. M., Kotin, J. K., Goodwin, F. K. 287, *302*
Poteliakhoff, A., Roughton, B. C. 265, *273*
Potgieter, B., see Offermeier, J. 82, 87, *224*
Potoczak, D., see McLean, J. R. 87, *220*
Potter, L. T., Axelrod, J. 34, 40, *227*
Potter, L., see Axelrod, J. 34, 38, 40, *183*
Potts, J. L., see Proctor, C. D. 87, 166, 173, *228*
Potvin, R. J., Fried. P. A. 410, *426*
Povalski, H., see Maxwell, R. A. 146, *219*
Powell, B. J., Martin, L. K. Kamano, D. K. 102, *227*
Powell, B. J., see Hill, S. Y. 413, *420*
Powell, D. A., Walters, K., Duncan, S., Holley, J. R. 97, *228*
Powers, R. O., see Neu, R. L. 401, *425*

Powley, T. L., see Keesey, R. E. 285, *300*
Poyser, R. H., see Berry, M. J. 146, *186*
Praag, H. M. van 285, *302*
Praag, H. M. Van, Korf, J., Wandenberg, F. Van, Kits, F. P. 3, *31*
Prabhu, V. G. 173, *228*
Prabhu, V. G., see Proctor, C. D. 173, *228*
Prada, M. Da, Bartholini, G., Pletscher, A. 54, 55, *228*
Prada, M. Da, see Pletscher, A. 56, *227*
Pradhan, S. N., Bailey, P. T., Ghosh, P. 388, *426*
Pradhan, S. N., Bowling, C. 121, *228*
Pradhan, S. N., Dutta, S. N. 144, *228*
Pradhan, S. N., see Bailey, P. T. 315, *351*
Pratesi, P., Blaschko, H. 10, *29*, 45, 46, *228*
Pray, S. L., see Barratt, E. S. 315, 341, *351*
Preson, A. C., see Bell, D. S. 91, 92, *186*
Preston, J. E., see Clark, W. C. 168, 169, 173, 174, 185, *193*
Prinzmetal, M., Alles, G. A. 250, *273*
Prinzmetal, M., Bloomberg, W. 250, 258, *273*, 280, *302*
Procopio, G. F., see Carey, R. J. 110, 117, 122, *190*
Proctor, C. D., Cho, J. B., Lundy, R. O., Mills, S. L., Mitchell, K. J. 173, *228*
Proctor, C. D., Potts, J. L., Ashley, L. G., Denefield, B. A. 87, *228*
Proctor, C. D., Potts, J. L., Lundy, R. O., Greenfield, E. J. 166, 173, *228*
Proctor, C. D., Ridlon, S. A., Fudema, J. J., Prabhu, V. G. 173, *228*
Prokop, H. 278, *302*
Przegalinski, E., Kleinrock, Z. 87, 154, 173, *228*
Przegalinski, E., see Maj, J. 77, 87, *218*
Püllen, C. 247, *273*
Pugh, C. E. M., Quastel, J. H. 45, *228*
Puri, S. K., see Lal, H. 77, 96, 179, *214*
Purnell, W. D., Gregg, J. M. 414, *426*
Purpura, D. P. 339, 342, 346, 347, *362*
Puttick, A. J., see Razdan, R. K. 379, *427*
Pytlyk, P. J., see Weiss, S. R. 265, *275*

Quastel, J. H., see Mann, P. J. G. 10, *29*, 45, 47, *219*
Quastel, J. H., see Pugh, C. E. M. 45, *228*
Quimby, M. W., Doorenbos, N. J., Turner, C. E., Masoud, A. 374, *426*
Quimby, M. W., see Fetterman, P. S. 373, 374, *418*
Quinn, G. P., Cohn, M. M., Reid, M. B., Greenhard, P., Weiner, M. 21, *30*
Quinn, P. O., Rapoport, J. L. 286, *302*
Quinn, P. O., see Rapoport, J. L. 286, *302*
Quinton, R. M., Halliwell, G. 71, 77, *228*
Quinton, R. H., see Halliwell, G. 77, *206*

Rabinowitz, J. L., see Neff, N. 334, *362*
Racagni, G., see Cattabeni, F. 51, 52, *192*
Raevskii, K. S., Gura, S. J. 173, 174, *228*
Rahman, H. 129, 140, *228*
Rahman, H., see Rensch, B. 129, *230*
Rainsbury, R., see Beaconsfield, F. 397, *415*
Raiteri, M., Levi, G., Federico, R. 45, *228*
Ramanna, L., see Tashkin, D. P. 409, *428*
Ramsey, C. H., see Agurell, S. 382, 384, *415*
Rand, M. J., Trinker, F. R. 146, *228*
Rand, M. J., see Burn, J. P. 33, 59, 143, 144, 146, *189*
Rand, M. J., see Day, M. D. 59, 60, 62, 63, 141, 142, 144, 146, 147, *196*
Rand, M. J., see O'Dea, K. 173, *224*
Randrianarisoa, H., see DiCarlo, R. 165, *197*
Randrup, A., Jonas, W. 42, 77, 80, *228*
Randrup, A., Munkvad, I. 70–72, 77, 82, 87, 88, 94, 95, 96, 99, *228*, 263, 264, *273*
Randrup, A., Munkvad, I., Udsen, P. 77, *229*
Randrup, A., Scheel-Krüger, J. 77, *229*
Randrup, A., see Arnfred, T. 77, 78, *183*
Randrup, A., see Ayhan, J. H. 71, *183*
Randrup, A., see Fog, R. 77, 78, 79, *201*
Randrup, A., see Hasselager, E. 77, 87, 88, 93, *207*
Randrup, A., see Kjellberg, B. 70, 94, 96, 98, *212*
Randrup, A., see Lyon, M. 72, 99, 121, *217*
Randrup, A., see Munkvad, I. 71, 77, 78, *223*
Randrup, A., see Papeschi, R. 77, 80, 91, *226*
Randrup, A., see Schiørring, E. 96, *234*
Rangno, R. E., Kaufman, J. S., Cavanaugh, J. H., Island, D., Watson, J. T., Oates, J. 8, *30*
Rapin, J., Hamar, C., Cohen, Y. 41, *229*
Rapin, J., Hamar, C., Jacquot, Ch., Cohen, Y., Valette, G. *229*
Rapin, J. R., Jacquet, C., Cohen, Y., Valette, G. 49, *229*
Rapin, J., Jacquot, C., Hamar, C., Cohen, Y. 41, 51, *229*
Rapin, J. R., see Boulu, R. 37, 68, *188*
Rapin, J., see ElGuedri, H. 9, 19, *27*, 52, *198*
Rapin, J. R., see Jacquot, C. 48, 51, *210*
Rapoport, J. L., Quinn, P. O., Bradbard, G., Riddle, K. D., Brooks, E. 286, *302*
Rapoport, J. L., Quinn, P. O., Lamprecht, F. 286, *302*
Rapoport, J. L., see Quinn, P. O. 286, *302*
Raskind, R., see Weiss, S. R. 265, *275*
Rating, D., Broemann, I., Hornecker, H., Kluiol, S., Coper, H. 410, *426*
Rating, D., Honecker, H., Broermann, J., Strauss, S. 17, *30*
Rawlin, J. W. 283, *302*
Ray, A. K., see Panda, J. N. 165, *226*
Ray, O. S. *229*
Ray, O. S., Bivens, L. W. 106, 110–112, 114, 118, *229*
Ray, O. S., see Barrett, R. J. 100, 102, 103, 130, *184*
Ray, O. S., see Bivens, L. W. 110–112, 114, *187*
Ray, O. S., see Stein, L. 121, *237*

Razani, J., see Knee, S.T. 292, *300*
Razdan, R.K. 374, 375, 380, *426*
Razdan, R.K., Dalzell, H.C., Handrick, G.R. 379, *426*
Razdan, R.K., Handrick, G.R. 379, *426*
Razdan, R.K., Handrick, G.R., Dalzell, H.C. 379, 380, *426*
Razdan, R.K., Puttick, A.J., Zitko, B.A., Handrick, G.R. 379, *427*
Razdan, R.K., Zitko, B.A. 375, *427*
Razdan, R.K., see Dewey, W.L. 399, *418*
Razdan, R., see Martin, B.R. 381, 390, *424*
Razdan, R.K., see Uliss, D.B. 379, *429*
Razdan, R.K., see Zitko, B.A. 380, *429*
Reader, G.G., see Modell, W. 279, *301*
Reaven, G.M., see Hollister, L. E. 380, 400, *421*
Reavis, W.M., see Dewey, W.L. 398, *418*
Rebec, G.V., see Groves, P.M. 65, 67, *205*
Rech, R.H. 101–103, 106, 107, *229*
Rech, R.H., Moore, K.E. 106, *229*
Rech, R.H., Stolk, J.M. 82, 87, 88, 106, *229*
Rech, R.H., Thut, P.D. 92, 107, *229*
Rech, R.H., see Moore, K.E. 106, *222*
Rech, R.H., see Stolk, J.M. 87, 162, 166, 167, 173, 175, *238*
Rech, R.H., see Tilson, H.A. 92, 107, 110, 119, *240*
Rechtschaffen, A., Maron, L. 252, *273*
Rechtschaffen, A., Wolpert, E.A., Dement, W.C., Mitchell, S.A., Fisher, C. 285, *302*
Rechtschaffen, A., see Dement, W. 285, *298*
Redfern, P.H., see Davies, J.A. 102, *195*
Redman, C.E., see Stark, P. 121, 122, *237*
Reed, A.F., see Hale, A.R. 310, *356*
Reed, D., see Cravey, R.H. 264, *268*
Rees, J.M., see Little, H.J. 179, *216*
Rees, L., see Besser, G.M. 255, *267*
Regelson, W., Butler, J.R., Schulz, J., Kirk, T., Peek, L., Green, M.L., Zakio, O. 413, 414, *427*
Reid, M.B., see Quinn, G.P. 21, *30*
Reid, W.D. 53, 55, 154, 157, *229*
Reif, E.C., see Schulte, J.W. 82, *234*
Reifenstein, E.C., Davidoff, E. 247, *273*, 282, *302*
Reifenstein, E.C., see Davidoff, E. 248, 250, 254, 255, *268*
Reifler, C.B., see Kane, F.J. 265, *271*
Reigle, R.G., see Wolf, H.H. 177, *245*
Reimann, H.A. 277, *302*
Reimherr, F.W., see Wood, D. R. 287, *304*
Reinert, H. 62, 64, 141, 142, 144, 146, 148, *229*
Reinert, H., see Gold, D. 64, 142, *204*
Reinert, H., see Kewitz, H. 59, 64, *211*
Reiter, P. 248, *273*
Reitz, R.H., Bocknik, S.E., Kulkarni, A.S. 39, 106, *229*
Reivich, M., see Snyder, S.H. 334, *365*
Remington, G., Anisman, H. 106, *229*
Renault, H., see Jacquot, C. 48, *210*
Renault, P.F., Schuster, C.R., Heinrich, R., Freeman, D.X. 397, *427*
Rensch, B., Rahman, H. 129, *230*
Renyi, A.L., see Ross, S.B. 39, 49, 50, 54, 77, 154, *230*, *231*
Renzi, P. 101, *230*
Renzi, P., see Castellano, C. 101, *192*
Renzi, P., see Sansone, M. 101, 102, 103, *232*
Resnekov, L., see Fischman, M. W. 278, *298*
Revuelta, A., see Cattabeni, F. 56, *192*
Revuelte, A., see Costa, E. 56, 58, *194*
Revuelta, A., see Groppetti, A. 48, 90, 157, *205*
Reyes, E., see Brobeck, J.R. 137, *189*
Reynolds, R.W. 136, *230*
Reynolds, R.W., see Carlisle, H. J. 136, *190*
Rhodes, R.E., see Hucker, H.B. 9, 11, *28*
Rice, D.L., see Zrull, J.P. 286, *304*
Ricci-Gamalero, S., see Molinengo, L. 101, 106, *221*
Riccioni, M.L., see Tonini, G. 100, 107, *240*
Richards, D.W., see Water, W. H. 130, *243*
Richards, K.C., Borgstedt, H. H. 311, *363*
Richards, N., Chapman, L.F., Goodell, H., Wolff, H.G. 310, *363*
Richards, R., Nicol, E., Young, P. 169, *230*
Richards, R.K., see Brown, H. 115, *189*
Richards, R.K., see Hollister, L.E. 385, *421*
Richardson, D., Karczmar, A.G., Scudder, C.L. 70, 80, *230*
Richelle, M. 112, 117, 118, *230*
Richelle, M., see Faidherbe, J. 117, *200*
Richelson, E., see Snyder, S.H. 348, *365*
Richter, D., see Blaschko, H. 10, *26*, 45, *187*
Riddell, D., Szerb, J.C. 44, *230*
Riddle, K.D., see Rapoport, J.L. 286, *302*
Ridlon, S.A., see Proctor, C.D. 173, *228*
Riffee, W.H., see Gerald, M.C. 177, 178, *203*
Rimm, A., see Wang, R.I.H. 167, *243*
Rinaldi, F., Himwich, H.E. 176, *230*, 339, 340, *363*
Rinkel, M., DeShon, H.J., Hyde, R.W., Solomon, H.C. 308, 343, *363*
Rinkel, M., see Loman, J. 247, *272*
Ritter, S., Stein, L. 122, *230*
Ritvo, M., see Myerson, A. 280, *301*
Riva, M., Naimzada, M.K., Pirola, C., Mantegazza, P. 139, 149, *230*
Riva, M., see Mantegazza, P. 87, 91, 134, 138, 154, 156, 163, 169, 173, *219*
Rizzotti, G. 277, *302*
Roach, M.K., see Creaven, P.J. 17, *27*
Robbins, E.S., see Gilmour, D. G. 402, *419*
Roberfroid, M., see Clay, G.A. 50, 51, *193*

Roberts, M. H. T., Straughan, D. W. 329, 338, *363*
Roberts, M. H. T., see Bevan, P. 338, *351*
Roberts, M. H. T., see Bradshaw, C. M. 338, *352*
Roberts, P., see Hecht, F. 408, *420*
Roberts, S., see Hungen, K. von 333, *367*
Robertson, M. I., see Berry, M. J. 146, *186*
Robichard, R. C., see Dubinsky, B. 387, *418*
Robichaud, R. C., Goldberg, M. E., Dubnick, B. 93, *230*
Robins, A., see Vachon, L. 414, *429*
Robins, E., see Cicero, T. J. 401, *417*
Robbins, T., Iversen, S. D. 94, 114, *230*
Robinson, G. A., see Schmidt, M. J. 57, 58, *234*
Robinson, J. 289, 290, *302*
Robinson, R. G., Mc Hugh, P. R., Folstein, M. F. 285, *302*
Robison, S. M., see Mc Cullough, D. O. 51, 52, 149, 150, *220*
Robson, R. D., see Follenfant, M. J. 59, 147, *201*
Rockland, L. H., see Szara, S. 311, *366*
Rockwell, D., Ostwald, P. 292, *302*
Rodin, E., Luby, E. 343, *363*
Rodriques-Zendejas, A. M., see Russek, M. 136, *232*
Rodriguez, A., see Epstein, L. C. 286, *298*
Rodriguez, L., see Laporte, J. 60, *214*
Roffler-Tarlov, S., Sharman, D. F., Tegerdine, P. 43, 46, *230*
Roffman, M., Lal, H. 102, 106, 130, 131, *230*
Roffman, M., see Lal, H. 131, *214*
Roft, D. D., Gregg, J. M., Ghia, J., Harris, L. S. 413, *427*
Rogers, K. J., Hutchins, D. A. 162, *230*
Rogers, K. J., see Hutchins, D. A. 87, 162, *209*
Rogers, K. J., see Nahorski, S. R. 161, 162, *223*
Rolinski, Z. 96, 97, *230*
Rolinski, Z., Scheel-Krüger, J. 87, 93, *230*
Rolinski, Z., see Hasselager, E. 77, 87, 88, 93, *207*
Rollins, D. E., see Wolf, H. H. 177, *245*
Rolo, A., see Abramson, H. A. 318, *350*
Rome, H. B., Braceland, F. J. 292, *302*
Rome, H. P., see Schwarz, B. E. 312, *364*
Rommelspacher, H., Honecker, H., Schulze, G., Strauss, S. M. 8, *30*
Rommelspacher, H., see Coper, H. 149, 154, 159, *194*
Rona, G., see Chappel, C. I. 174, *192*
Rose, K., see Seaton, D. A. 259, *273*
Rosecrans, J. A., Lovell, R. A., Freedman, D. X. 330, *363*
Rosecrans, J. A., see Hirschhorn, I. D. 413, *421*
Rosecrans, J. A., see Schechter, M. D. 130, *233*, 315, *364*
Rosell, S. 259, *273*
Rosen, A. J., La Flore, J. E. 100, *230*
Rosen, A. J., see Kelfer, D. A. 129, *211*
Rosenberg, D. E., Isbell, H., Miner, E. J. 310, 311, *363*
Rosenberg, D. E., Isbell, H., Miner, E. J., Logan, C. R., 318, *363*
Rosenberg, D. E., Wolbach, A. B., Jr., Miner, E. J., Isbell, H. 253, 254, 256, 261, *273*, 319, *363*
Rosenberg, D. E., see Isbell, H. 319, *358*
Rosenberg, P. 250, *273*, 280, *302*
Rosenfeld, R., see Christensen, H. D. 383, *417*
Rosenkrantz, H., Braude, M. C. 401, 405, *427*
Rosenkrantz, H., Sprague, R. A., Fleischman, R. W., Braude, M. C. 404, *427*
Rosenkrantz, H., see Fleischman, R. W. 407, 408, *419*
Rosenkrantz, H., see Luthra, X. K. 405, *423*
Rosenkrantz, H., see Thompson, G. R. 403–405, 410, *428*
Rosenthal, D., see Szara, S. 311, *366*
Rosner, B. S., see Freedman, D. X. 320, *355*
Ross, C. A., see Stone, C. A. 146, *238*
Ross, G. S., see Hodos, W. 100, 114, *208*
Ross, S. B., Renyi, A. L. 39, 49, 50, 54, 77, 154, *230*
Ross, S. B., Renyi, A. L., Brunfelter, B. 49, 50, *231*
Rossi, G. V., see Neff, N. 334, *362*
Rosso, R., Dolfini, L., Franchi, G. 17, *30*
Rossum, J. M. van 33, 34, 71, 88, 132, *231*
Rossum, J. M. van, Hurkmans, J. A. Th. 87, *231*
Rossum, J. M. van, Schott, J. B. van der, Hurkmans, J. A. 70, 71, 87, 88, *231*
Rossum, J. M. van, Simons, F. 3, 4, 20, *30*, 82, 139, *231*
Rossum, J. M. van, see Bercht, C. A. L. 375, *416*
Rossum, J. M. van, see Broekkamp, C. L. W. 137, *189*
Rossum, J. M. van, see Cools, A. R. 77, 78, *194*
Rossum, J. M. Van, see Ginneken, C. A. M. Van 375, *429*
Rossum, J. M. Van, see Henderson, P. T. 11, 14, *28*
Rossum, J. M. van, see Schoot, J. B. van der 71, 87, 88, *234*
Rossum, J. M. Van, see Vree, T. B. 7, 11, 12, 14, 18, 20, 21, *31*, *429*
Roszell, D. K., Horita, A. 338, *363*
Roszkowski, A. P., Kelley, N. M. 131, 132, 139, *231*
Roth, B., Faber, J., Nevisimalova, S., Tosovsky, J. 250, *273*, 286, *302*
Roth, K., see Williams, R. B. 398, *429*
Roth, L. J., see Diab, I. M. 334, *354*
Roth, R. H., Walters, J. R., Aghajanian, G. K. 42, 46, 66, *231*
Roth, R. H., see Bunney, B. S. 46, 66, *189*
Roth, W. T. *363*
Rothlin, E. 307, 315, 316, 335, *363*
Rothlin, E., Cerletti, A. 307, 313, *363*
Rothlin, E., Cerletti, A., Konzett, H., Schalch, W. R., Taeschler, M. 307, *363*
Rothlin, E., see Stoll, A. 333, *366*
Rotrosen, J., see Angrist, B. 319, *350*
Roughton, B. C., see Poteliakhoff, A. 265, *273*
Roush, B. W., see Fuller, R. W. 5, 27, 49, 53, 56, *202*

Rovetta, P. 342, *363*
Rowe, H., see Lemberger, L. 383, 385, 387, 399, 400, *423*
Rowland, C. R., see Wolf, H. H. 177, *245*
Rowland, M. 21, 23, *30*
Rowland, M., see Beckett, A. H. 7, 13, 15, 21, *25*
Royce, J. R., see Satinder, K. P. 70, *233*
Rubenson, A., see Andén, N.-E. 68, 69, *182*
Rubin, B., Piala, J. J., Burke, J. C., Craver, B. N. 338, *363*
Rubin, R. P. 41, 45, *231*
Rubin, R. P., Jaanus, S. D. 41, *231*
Rubin, V. 371, 413, *427*
Rubin, V., Comitas, L. 409, *427*
Rubovits, R., see Klawans, H. L. 70, 77, *212*, 264, *271*
Ruchi, E. W., see Dubnick, B. 54, *198*
Ruckart, R. T., see Funderburk, W. H. 132, 134, *202*
Rudzik, A. D., Eble, J. N. 146, *231*
Rudzik, A. D., Johnson, G. A. 177, 178, *231*
Rudzik, A. D., Mennear, J. 173, *231*
Rudzik, A. D., see Eble, J. N. 62, 143, 148, *198*
Rudzik, A. D., see Mennear, J. H. 167, 173, *221*
Rumbaugh, C. L., Bergeron, R. T., Fang, H. C. H., McCormick, R. 252, 265, *273*
Rumbaugh, C. L., Bergeron, R. T., Scanlan, R. L., Teal, J. S., Segall, H. D., Fang, H. C. H., McCormick, R. 143, *231*
Rumbaugh, C. L., Bergeron, R. T., Scanlan, R. L., Teal, J. S., Segall, H. D., Fang, H. D., McCormick, R. 252, 265, *273*
Rumelhart, D. E., Mueller, M. R. 96, *231*
Rusca, G., see Basso, P. del 41, *184*
Rushton, R., Steinberg, H. 94, *231*
Rushton, R., Steinberg, H., Tinson, Ch. 94, *231*
Rushton, R., see Armitage, P. 94, *183*
Rushton, R., see Steinberg, H. 94, *238*
Russek, M. 138, 139, *231*
Russek, M., Bruni, E. 138, 139, *232*
Russek, M., Rodriques-Zendejas, A. M., Teitelbaum, P. 136, *232*
Russek, M., Stevenson, J. A. F. 138, *232*
Russek, M., Stevenson, J. A. F., Mogenson, G. J. 138, *232*
Russel, R. W., see Maickel, R. P. 20, *29*, 99, 128, 179, *218*
Russell, R. W., see Overstreet, D. H. 135, *225*
Rutledge, C. D. 46, 47, *232*
Rutledge, C. D., Azzaro, A. J., Ziance, R. J. 39, 44–47, 50, 55, *232*
Rutledge, C. O. 10, *30*
Rutledge, C. O., Kelleher, R. T. 110, 112, 118, *232*
Rutledge, C. O., see Azzaro, A. J. 44, 45, 50, 55, *183*
Rutledge, C. O., see Wenger, G. R. 8, *31*, 45, 46, *244*
Rutledge, C. O., see Ziance, R. J. 39, 44, 46, *246*
Rutschmann, J., see Haley, T. J. 333, *356*
Rutschmann, J., see Stoll, A. 333, *366*
Ryall, R. W. 58, 59, *232*
Rybina, A. A., see Palladin, A. V. 161, *225*
Rylander, G. 263, *273*, 290, *302*
Rylander, G., see Allmark, J. 257, *267*
Ryrfeldt, A., see Agurell, S. 382, 384, *415*

Saavedra, J. M., Axelrod, J. 332, *363*
Sachs, I., see Gunn, J. A. 70, 141, 142, 144, 174, *205*
Sader, N. F. A., see Breda, J. B. 128, *188*
Sadoun, R., see Delay, J. 293, *298*
Sadowski, B., Longo, V. G. 115, 117, *232*
Safer, D. J. 312, *363*
Safer, D. J., Allen, R. P. 286, *302*
Sai-Halász, A., Brunecker, G., Szára, S. 311, *363*
Salama, A. I., Goldberg, M. E. 34, 35, 46, 53, 90, 97, *232*
Salama, A. I., see Dubnick, B. 54, *198*
Salama, A. I., see Goldberg, M. E. 167, 173, *204*
Saldeen, T., see Gunne, L.-M. 254, 264, *270*
Salemink, C. A., see Bercht, C. A. L. 375, *415*, *416*
Salim, A. P., see Carey, R. J. 90, 136, *190*
Sallen, S. E., Zinberg, W. E., Frei, E. 414, *427*
Saller, C., see Manning, F. J. 404, *424*
Salmoiraghi, G. C., McCubin, J. W., Page, I. H. 335, *364*
Salmoiraghi, G. C., see Bloom, F. E. 329, *352*
Salmon, J. A., see Beckett, A. H. 21, 23, *25*
Salomon, K., Gabrio, B. W., Thale, T. 310, *364*
Salva, J. A., see Valdecasas, F. G. 69, 80, 146, 147, *241*
Salvatore, S., Hyde, R. W. 308, *364*
Samach, M., see Willner, J. H. 70, 77, *244*
Samanin, R., Ghezzi, D., Valzelli, L., Garattini, S. 134–136, *232*
Samanin, R., Valzelli, L., Gumulka, W. 331, *364*
Sampaio, M. R. P., see Santos, M. 387, *427*
Samuel, D., see Epstein, R. P. 44, *198*
Sanan, S., Vogt, M. 34, 36, *232*
Sandberg, F., see Agurell, S. 381, 382, 384, *415*
Sandberg, F., see Wahlquist, M. 382, *429*
Sandburg, F., see Ohlsson, A. 373, *426*
Sanders-Bush, E., Sulser, F. 56, *232*
Sanders-Bush, E., see Gallagher, D. W. 389, 395, *419*
Sanders-Bush, E., see Strada, S. J. 38, 43, 46, *238*
Sanders-Bush, E., see Sulser, F. 3, *30*, 33, 34, *239*
Sandler, M., Youdin, M. B. H., Hanington, E. 279, *303*
Sandman, C. A., see Beckwith, B. E. 129, 130, *185*
Sanghvi, I., Gershon, S. 77, *232*
Sanghvi, I., see Singer, G. 77, 134, *236*
Sano, I., Nagasaka 291, *303*
Sansone, E. M., see Castellano, C. 101, *192*
Sansone, M., Bovet, D. 101, *232*
Sansone, M., Renzi, P. 101, 103, *232*
Sansone, M., Renzi, P., Castellano, C. 102, *232*
Santavy, F., see Kabelik, J. 373, *421*
Santi, R., Fassina, G. 163, *232*
Santinoceto, L., see Beani, L. 57, *185*

Santos, M., Sampaio, M. R. P., Fernandes, N. S., Carlini, E. A. 387, *427*
Sapeika, N., Zwarenstein, H., Kaplan, E. R. 163, *233*
Sapeika, N., see Kaplan, E. R. 163, *211*
Sapeika, N., see Zwarenstein, H. 163, *246*
Sapira, J. D., see Martin, W. R. 253, 254, *272*, 278, 281, 285, *301*
Sargant, W., see Guttmann, E. 247, 255, 258, *270*
Sargent, F., see Shulgin, A. T. 4, *30*, 311, 312, 328, *364*
Sargent, T., see Naranjo, C. 311, 328, *362*
Saruek, J., see Maj, J. 82, 87, *218*
Satinder, K. P. 100, 101, *233*
Satinder, K. P., Petryshyn, W. R. 100, *233*
Satinder, K. P., Royce, J. R., Yeudall, L. T. 70, *233*
Satran, R., see Liddon, S. C. 292, *301*
Satta, M., see Miele, E. 87, *221*
Satterfield, J. H., Cantwell, D. P., Saul, R. E., Yusin, A. 286, *303*
Saul, R. E., see Satterfield, J. H. 286, *303*
Savage, C., see Cholden, L. S. 318, *353*
Sawdy, L. C., see Moore, K. E. 162, 173, 175, *222*
Sawyer, C. H., see Khazan, N. 178, *211*, 340, *359*
Sawyer, C. H., see Vogt, M. 335, 341, *367*
Sayers, A. C. 77, 78, *233*
Sayers, A. C., Handley, S. L. 77, *233*
Sayers, A. C., Spencer, P. S. J. 77, *233*
Scanlan, R. L., see Rumbaugh, C. L. 143, *231*, 252, 265, *273*
Scapagnini, U., see Loon, G. R. van 165, *217*
Schaefer, C. F., Gunn, C. G., Dubowski, K. M. 401, *427*
Schaeppi, U. H., see Thompson, G. R. 403, 404, *428*
Schaffer, R. J., see Fuller, R. W. 49, 53, 56, *202*
Schalch, W. R., see Rothlin, E. 307, *363*
Schalch, W. R., see Stoll, A. 333, *366*
Schallek, W., Walz, D. 176, *233*
Schanberg, S. M., Cook, J. D. 35, 38, 43, 80, 81, *233*
Schanberg, S. M., see Cook, J. D. 35, 38, 43, 46, 71, 80, *194*
Schanberg, S. M., see McMahon, E. M. 162, 165, *220*
Schanker, L. S. 20, *30*
Schaper, W. K. A., see Janssen, P. A. J. 77, *210*
Schechter, M. D., Rosecrans, J. A. 130, *233*, 315, *364*
Schechter, M. D., Winter, J. C. 314, *364*
Scheckel, C. L., Baff, E., Dahlen, P., Smart, T. 387, *427*
Scheckel, C. L., Boff, E. 103, 106, *233*
Scheel-Krüger, J. 42, 71, 72, 77, 80, 87, 88, *233*
Scheel-Krüger, J., Hasselager, E. 55, 149, 155, 157, *233*
Scheel-Krüger, J., Jonas, W. 77, *233*
Scheel-Krüger, J., see Jespersen, S. 132, 134, *210*
Scheel-Krüger, J., see Jonas, W. 42, 80, *210*
Scheel-Krüger, J., see Randrup, A. 77, *229*
Scheel-Krüger, J., see Roliński, Z. 87, 93, *230*
Scheiris, C., see Heyndrickx, A. 403, *420*
Schelkunov, E. L. 77, *233*
Schelkunov, E. L., see Lapin, I. P. 77, 87, 106, 173, *214*
Schellekens, K. H. L., see Janssen, P. A. J. 77, 81, *210*
Schepens, P., see Heyndrickx, A. 403, *420*
Scheving, L. E., Vedral, D. F., Pauly, J. E. 167, *233*
Schiff, P. L., see Lotter, H. L. 378, *423*
Schilder, P. 286, *303*
Schildkraut, J. J., Efron, D. H. 391, 393, 395, *427*
Schildkraut, J. J., Watson, R., Draskoczy, P. R., Hartmann, E. 261, *273*, 294, *303*
Schildkraut, J. J., see Watson, R. 294, *303*
Schiørring, E. 70, 96, *233*
Schiørring, E., Randrup, A. 96, *234*
Schlaug, J., see Faidherbe, J. 117, *200*
Schlechter, J. M., Butcher, L. L. 87, *234*
Schlichting, U. U., Goldberg, S. R., Wuttke, W., Hoffmeister, F. 124–126, *234*
Schlossman, H., see Blaschko, H. 10, *26*, 45, *187*
Schmidt, C. F., see Chen, K. K. 279, *297*
Schmidt, D. E., see Schmidt, M. J. 57, 58, *234*
Schmidt, G., see Falb, A. 163, *200*
Schmidt, M. J., Hopkins, J. T., Schmidt, D. E., Robinson, G. A. 57, 58, *234*
Schmidt, R., see Mouriz-Garcia, A. 337, *362*
Schmidt-Nowara, W. W., see Ginsberg, M. D. 264, *269*, 288, *299*
Schmitt, H. 132, 134, *234*
Schmitt, H., Schmitt, H. 59, 63, 141, 144, 146, 147, *234*
Schmitt, H., Schmitt, H., Depoilly, F. 146, *234*
Schmitt, H., Schmitt, H., Fenard, S. 247, *273*
Schmitt, H., see Le Douarec, J. C. 3, *29*
Schmitt, H., see Schmitt, H. 59, 63, 141, 144, 146, 147, *234*, 247, *273*
Schmitt, M., see Ott, T. 53, *225*
Schnackenberg, R. C. 286, *303*
Schneider, C. 313, *364*
Schneider, C. W., Chenoweth, M. B. 128, *234*
Schneider, F. H. 41, *234*
Schneider, F., see Maxwell, R. A. 146, *219*
Schneider, F. H., see Wagner, L. A. 41, 49, *242*
Schoenfield, R. I., Zigmond, M. J. 119, *243*
Schoetensack, W., see Müller, H. 77, 87, 173, *223*
Scholander, T., see Dureman, I. 248, *268*
Schoolar, J. C., see Lu, T.-C. 134, 138, 139, *217*
Schoot, J. B., van der, Ariens, E. J., Rossum, J. M. van, Hurkmans, J. A. Th. 71, 87, 88, *234*
Schoot, J. B. van der, see Rossum, J. M. van 70, 71, 87, 88, *231*
Schrader, A., see Pflanz, M. 251, *272*
Schramm, L. C., see Geber, W. F. 407, *419*
Schrayer, D., see Abelson, H. 371, *415*
Schreiber, R. A., see Graham, J. M. 177, *204*

Schroeder, L. A., see Tang, A. H. 313, 335, 338, 339, 340, *366*
Schrold, J. 95, *234*
Schrold, J., Squires, R. F. 54, 55, 70, 95, 96, *234*
Schube, P. G., McManamy, M. C., Trapp, C. E., Myerson, A. 282, *303*
Schube, P. G., see McManamy, M. C. 277, *301*
Schuberth, J., Fryö, B., Nybäck, H., Sedvall, C. 54–56, *234*
Schuberth, J., Jenden, D. J. 261, *273*
Schuberth, J., Sedvall, G. 56, *234*
Schuberth, J., see Jönsson, L.-E. 261, *271*
Schuler, M. E., see Gardocki, J. F. 87, 167, 173, *202*
Schulte, J. W., Reif, E. C., Bacher, J. A., Jr., Lawrence, W. S., Tainter, M. L. 82, *234*
Schultes, R. E. 305, *364*, 373, 413, *427*
Schulz, J., see Regelson, W. 413, 414, *427*
Schulz, R., see Frey, H.-H. 132, 134, *201*
Schulze, G., see Coper, H. 149, 154, 159, *194*
Schulze, G., see Rommelspacher. H. 8, *30*
Schumann, H. J., Philippü, A. 41, *235*
Schuster, C. R., Dockens, W. S., Woods, J. H. 99, 103, 107, 112, 114, 120, *235*
Schuster, C. R., Zimmermann, J. 92, 114, 120, *235*
Schuster, C. R., see Balster, R. L. 124, 125, *184*
Schuster, C. R., see Fischman, M. W. 278, *298*
Schuster, C. R., see Renault, P. F. 397, *427*
Schuster, C. R., see Wilson, M. C. 124–126, *244*
Schuster, C. R., see Zimmerman, J. 120, *246*
Schuster, J., Ihli, M. 143, *235*
Schusterman, R. J., see Sharpe, L. G. 317, *364*
Schwabe, U., see Falb, A. 163, *200*
Schwartz, A. S., Cheney, C. 337, *364*
Schwarz, B. E., Bickford, R. G., Rome, H. P. 312, *364*
Schwartz, B. E., Sem-Jacobsen, C. W., Petersen, M. C. *364*
Schwartz, B. E., Wakim, K. G., Bickford, R. G., Lichtenheld, E. R. 341, 342, *364*
Schwartz, I. W., see Nahas, G. G. 398, 410, *425*
Schwartz, M., see Gillin, J. C. 320, *356*
Schwartz, W. E., see Foltz, R. L. 383, *419*
Schwartzbaum, J. S., Ide-Johanson, L., Belgrade, J. 176, *235*
Schwartzbaum, J. S., Kreinik, C. J., Gustavson, J. W. 176, *235*
Schweigerdt, A. K., Himwich, M. E. 340, *364*
Schweigerdt, A. K., Stewart, A. H., Himwich, H. E. 340, *364*
Schweitzer, J. W., Friedhoff, A. J. *30*
Schweitzer, J. W., Friedhoff, A. J., Angrist, B. M., Gershon, S. 19, *30*
Schweitzer, J. W., see Angrist, B. M. 284, *296*
Schwider, A., see Leuchtenberger, C. 401, *423*
Schwin, R., see Hill, S. Y. 413, *420*
Sclafani, A., see Grossman, S. P. 33, 98, *205*
Scott, A. M., see Seaton, D. A. 259, *273*
Scott, J. P., Lee, C. T., Ho, J. E. 155, *235*
Scott, P. D., Willcox, D. R. C. 289, *303*
Scott, W. C., see Anderson, E. W. 265, *267*
Scoville, W. B., see Young, D. 282, *304*
Scribner, B. H., see Milne, M. D. 7, *29*
Scudder, C. L., see Richardson, D. 70, 80, *230*
Searle, L. M., see Harris, S. C. 131, 138, *206*, 250, *270*, 280, *299*
Searle, L. V., Brown, C. W. 70, *235*
Seaton, D. A., Duncan, L. J. P. 285, *303*
Seaton, D. A., Duncan, L. J. P., Rose, K., Scott, A. M. 259, *273*
Secretary, H. E. W. 371, 372, 384, 388, 400, *427*
Sedvall, C., see Schubert, J. 54–56, *234*
Seegal, R. F., Isaac, W. 90, *235*
Seevers, M. H. *273*
Seevers, M. H., see Deneau, G. 124, *196*
Seevers, M. H., see Eddy, N. B. 257, *269*
Seevers, M. H., see Hardman, H. F. 398–400, 410, *420*
Segal, D. S., Cox, R. H., Jr., Stern, W. G., Maickel, R. P. 99, 106, *235*
Segal, D. S., McAllister, C., Geyer, M. A. 72, 78, 87, *235*
Segal, D. S., Mandell, A. J. 91, 92, *235*
Segal, D. S., see Geyer, M. A. 88, *203*
Segal, D. S., see Groves, P. M. 65, 67, *205*
Segal, D. S., see Maickel, R. P. 20, *29*, 99, 128, 179, *218*
Segal, D. S., see Mandell, A. J. 47, *218*
Segal, E. 111, 114, 116, *235*
Segall, H. D., see Rumbaugh, C. L. 143, *231*, 252, 265, *273*
Segelman, A. B., Sofia, R. D., Segelman, F. P., Harakal, J. J., Knoblock, L. C. 390, *427*
Segelman, F. P., see Segelman, A. B. 390, *427*
Seiden, L. S., see Campbell, J. C. 114, 120, *190*
Seifter, J., see Clymer, N. V. 70, *193*
Seifter, J., see Geller, I. 115, *203*
Seifter, J., see Stein, L. 121–123, *237*
Sekerke, J. H., see Janowsky, D. S. 287, *300*
Seliger, D. L. 167, *235*
Sem-Jacobsen, C. W., see Schwartz, B. E. *364*
Sensenbach, W., see Andrew, J. G. 252, *267*
Senta, T., see Smetzer, D. L. 142, *236*
Sercl, M., Zdrahal, L., Jaros, O. 251, *273*
Sethy, V. H., Naik, P. Y., Sheth, U. K. 87, *235*
Servais, J., see Hubin, P. 248, *270*
Sever, P. S., Caldwell, J., 138, 157, *235*
Sever, P. S., Caldwell, J., Dring, L. G., Williams, R. T. 8, 9, 18, *30*
Sever, P. S., Caldwell, J., Williams, R. T. 154, 157, *235*
Sever, P. S., see Caldwell, J. 3, *26*, 154, 157, *190*

Shafer, R. P. 371, 372, 403, 408, *427*
Shagass, C. 343, *364*
Shagoury, R. A., see Neumeyer, J. L. 374, *425*
Shallice, S. A., see Leonhard, B. E. 34, 35, 37, 38, 49, 53, 56, 57, *215*
Shand, D. G., see Flegin, O. T. 147, *201*
Shani, A., Mechoulam, R. 375, *427*
Shani, A., see Mechoulam, R. 375, *425*
Shapiro, B., see Tashkin, D. P. 409, 414, *428*
Shapiro, D. 414
Shapiro, D. M., see Fink, M. 251, *269*
Shapiro, S. L., Freedman, L. 138, 139, *235*
Shapiro, T., see Fish, B. 310, *355*
Sharma, S. N., see Menon, M. K. 132, 134, *221*
Sharman, D. F., see Laverty. R. 36, 37, 43, 46, 55, 70, *214*
Sharman, D. F., see Roffler-Tarlov, S. 43, 46, *230*
Sharp, J. C., Nielson, H. C., Porter, P. B. 135, 136, *235*
Sharpe, L. G., Otis, L. S., Schusterman, R. J. 317, *364*
Sharpe, L. G., see Pickworth, W. B. 339, *362*
Sharpless, S. K., see Jaffe, J. H. 344, *358*
Shaul, S. R., see Moore, K. E. 162, 173, 175, *222*
Shaw, E., see Woolley, D. W. 312, 328, 329, 332, *368*
Shaw, W. N., Fuller, R. W., Matsumoto, C. 163, 164, *235*
Shaw, W. N., see Fuller, R. W. 156, *202*
Shaw, W. N., see Matsumoto, C. 154, 163, *219*
Sheard, M. H. 96, 97, *235*
Sheard, M. H., see Aghajanian, G. K. 330, *350*
Sheard, M. H., see Davis, M. 313, *353*
Sheard, M. H., see Foote, W. E. 67, *201*
Sheehan, J. C., see Zitko, B. A. 380, *429*
Shefner, A., see Lal, H. 173, *214*
Shellenberger, M. K., see Elder, J. T. 338, *354*
Sherwood, L. W., see Tainter, M. L. 177, *239*
Shet, U. K., see Sethy, V. H. 87, *235*
Shick, F. E., see Fishman, M. W. 278, *298*
Shideman, F. E., see Bhagat, B. 146, *186*
Shillito, E. E. 91, *235*
Shimamoto, K., see Higuchi, H. 36, *208*
Shimamoto, K., see Nakamura, K. 59, *223*
Shimizu, A., Himwich, H. E. 178, *235*
Shirakawa, I., see Karniol, I. G. 397, *421*
Shoeman, D. W., Sirtori, C. R., Azarnoff, D. L. 132, *235*
Shopsin, B., see Angrist, B. M. 5, 19, *25*, 257, 261, 262, *267*
Shore, P. A., Dorris, R. L. *235*
Shore, P. A., see Brodie, B. B. 70, *189*
Shore, P. A., see Dorris, R. L. 45, *197*
Shore, P. A., see Enna, S. J. 40, 41, 44, 45, *199*
Shorvon, H. J. 284, *303*
Shoyama, Y., Oku, R., Yamauchi, T., Nishioka, I. 375, *427*
Shulgin, A. T. 311, 328, *364*
Shulgin, A. T., Bunnell, S., Sargent, T. III 311, 328, *364*
Shulgin, A. T., Carter, M. F. 323, *364*
Shulgin, A. T., Sargent, T. 4, *30*, 311, 328, *364*
Shulgin, A. T., Sargent, T., Naranjo, C. 311, 312, 328, *364*
Shulgin, A. T., see Naranjo, C. 311, 328, *362*
Sicot, N., see Bernier, A. 136, *186*
Sidell, F. R., Pless, J. E., Meitlich, H., Sussman, P., Copelan, H. W., Sim, V. M. 385, 387, *427*
Sidman, M. 113, *236*
Siegel, F. P., see Foreman, R. L. 11, *27*
Siegel, M., see Ellison, T. 15, 18, 20, 21, *27*, 81, 173, *199*
Siegel, P., Sterling. T. P. 138, *236*
Siegel, R. K., Brewster, J. M., Jarvik, M. E. 317, *365*
Sieper, H., see Korte, F. 373, 375, *422*
Sikemeier, C., see Petrazilka, T. 379, *426*
Silberstein, S. D., see Lemberger, L. 380, *423*
Silva, M. T. A., Calil, H. M. 313, 314, *365*
Silva, M. T. A., Carlini, E. A., Claussen, U., Korte, F. 320, *365*, 409, *427*
Silverman, A. P. 70, 93, 97, *236*
Silverman, A. G., see Ellison, T. 15, 18, 20, 21, *27*, 81, 173, *199*
Silverstein, A. B., Klee, G. D. 308, *365*
Silverstein, M. J., Lessin, P. J. 402, *427*
Silverstone, J. T., Stunkard, A. J. 250, *274*, 285, *303*
Silvestrini, B., see Longo, V. G. 176, 179, *217*
Sim, V. M., see Sidell, F. R. 385, 387, *427*
Simaan, J., see Fawas, G. 63, *200*
Simon, P., Cheimat, R., Fosset, M., Boissier, J.-R. 77, 87, *236*
Simon, P., Larousse, C., Boissier, J.-R. 87, *236*
Simon, P., Tillement, J.-P., Larousse, C., Breteau, M., Guernet, M., Boissier, J.-R. 34, 53, 87, *236*
Simon, P., see Boissier, J.-R. 56, 79, 173, *187*
Simons, F., see Rossum, J. M. van 3, 4, 20, *30*, 82, 139, *231*
Simonyi, J., Szentgyörgyi, D. 149, *236*
Simpson, B. A., Iversen, S. D. 79, 89, *236*
Simpson, B., see Iversen, S. D. 79, 90, *209*
Singer, A. J. 372, *428*
Singer, G., Sanghvi, I., Gershon, S. 77, 134, *236*
Sirtori, C. R., see Shoeman, D. W. 132, *235*
Sisemore, D. D., see Crowley, T. P. 130, *195*
Sivadjian, J. 317, *365*
Sivil, S., see Kaymakcalan, S. 398, *422*
Sjoberg, B. M., see Hollister, L. E. 307, *357*
Sjoerdsma, A., Studnitz, W. von 9. 19, *30*
Sjöström, R., see Aquilonius, S. M. 264, *267*
Skaringa, A. P., Milyukas, V. I., Mitskis, A. M. 176, *236*
Skatkin, D. J., see Lotter, H. L. 378, *423*
Skinner, B. F., Heron, W. T. 112, 117, *236*
Sklarofsky, B., see Abramson, H. A. 312, 318, *350*
Skolnik, J., Pfeiffer, A. K., Kaldor, A. 57, *236*
Slabok, K., see Irwin, S. 87, 90, *209*
Slater, I. H., Davis, K. H., Leary, D. E., Boyd, E. S. 334, *365*

Sletten, I. W., Ognjanov, V., Menendez, S., Sundland, D., El-Toumi, A. 250, *274*
Sloan, J. W., Martin, W. R., Clements, T. H., Buchwald, W. F., Bridges, S. R. 332, *365*
Sloan, J. W., see Bell, J. A. 321, 332, *351*
Sloan, J. W., see Martin, W. R. 253, 254, *272*, 278, 281, 285, *301*, 310, 316, 331, 332, 336, 348, *361*
Sloman, L., see Garfinkel, B. D. 286, *299*
Small, A., see Feinberg, I. 252, 260, *269*, 286, *298*
Smart, J. V., Sneddon, J. M., Turner, P. 253, *274*
Smart, J. V., Turner, P. 253, *274*
Smart, T., see Scheckel, C. L. 387, *427*
Smeltzer, D. J., see Huestis, R. D. 286, *300*
Smetzer, D. L., Senta, T., Hensel, D. 142, *236*
Smith, A. H., see Everett, G. M. 70, 77, *199*
Smith, C. B. 35, 54, 71, 87, 88, 110, 112, 118, 142, *236*
Smith, C. B., see Antonaccio, M. J. 142, *183*
Smith, C. B., see Villarreal, J. E. 87, *242*
Smith, C. M., Walaszek, E. J. 333, *365*
Smith, D. E. 257, *274*, 282, 288, *303*, 403, *428*
Smith, D., see Brandon, S. 289, *297*
Smith, G. M., Beecher, H. K. 249, *274*
Smith, J. R., see Bishop, P. O. 336, *352*
Smith, L. C. 265, *274*, 282, *303*
Smith, M. L., see Buyniski, J. P. 62, 70, 141, 142, *190*
Smith, N., see Lal, H. 96, 179, *214*
Smith, R. C. 282, 288, *303*
Smith, R. D., see Breese, G. R. 79, 87, 88, *188*
Smith, R. L., Dring, L. G. 8, *30*
Smith, R. L., see Dring, L. G. 5, 7, 8, 10, 14, *27*
Smith, R. P., see Evans, W. O. 247, *269*, 288, *298*
Smith, S. G., see Davis, W. M. 124, 125, *196*
Smith, T. C., see Malit, L. A. 397, *424*
Smythies, J. R. 306, 313, 348, *365*
Smythies, J. R., Benington, F., Morin, R. D. 306, 348, *365*
Smythies, J. R., Bradley, R. J., Johnston, V. S., Benington, F., Morin, R. D., Clark, L. C., Jr. 323, *365*
Smythies, J. R., Johnsson, V. S., Bradley, R. J. 101, 128, *236*, 328, *365*
Smythies, J. R., Johnsson, V. S., Bradley, R. J., Benington, F., Morin, R. D., Clark, L. C. 15, 18, 19, *30*
Smythies, J. R., Koella, W. P., Levy, C. K. 340, *365*
Smythies, J. R., see Peretz, D. I. 311, 316, 328, *362*
Sneddon, J. M., see Smart, J. V. 253, *274*
Snoddy, H. D., see Fuller, R. W. 5, *28*
Snodgrass, S. R., Horn, A. S. 332, *365*
Snodgrass, W. R., see Maickel, R. P. 53, 65, 82, 99, 106, 131, 134, 140, *218*
Snyder, S. H. 5, *30*, 262, *274*, 372, 413, *428*
Snyder, S. H., Faillace, L., Hollister, L. 4, *30*, 311, 328, *365*
Snyder, S. H., Reevich, M. 334, *365*
Snyder, S. H., Richelson, E. 348, *365*
Snyder, S. H., Taylor, K. M., Coyle, J. T., Meyerhoff, J. L. 72, 82, *236*
Snyder, S. H., Weingartner, H., Faillace, L. A. 312, *365*
Snyder, S. H., see Arnold, L. E. 250, *267*, 286, *296*
Snyder, S. H., see Bennett, I. P. 331, *351*
Snyder, S. H., see Coyle, J. T. 39, 40, *195*
Snyder, S. H., see Faillace, L. A. 311, *355*
Snyder, S. H., see Hendley, E. D. 38, 40, *207*
Snyder, S. H., see Horn, A. S. 39, 40, *208*
Snyder, S. H., see North, R. B. 72, 82, *224*
Snyder, S. H., see Taylor, K. M. 5, *30*, 34, 35, 37–39, 72, 82, *239*, 257, 266, *274*
Sodd, M. A., see Burton, R. M. 179, *189*
Soderberg, U., see Ingvar, D. H. 341, *358*
Sodetz, F. J., see Manning, F. J. 404, *424*
Sofia, R. D. 90, *236*
Sofia, R. D., Barry, H. III 411, *428*
Sofia, R. D., Dixit, B. N., Barry, H. III 392, 395, *428*
Sofia, R. D., Knoblock, L. C., Vassar, H. B. 391, *428*
Sofia, R. D., Nalepa, S. D., Harakal, J. J., Vassar, H. B. 391, *428*
Sofia, R. D., see Bannerjee, B. N. 406, *415*
Sofia, R. D., see Segelman, A. B. 390, *427*
Sokol, G. H., Maickel, R. P. 173, *236*
Sokola, A., see Herman, Z. S. 35, 37, 53, 87, 91, 92, *207*
Solliday, N. H., see Vachon, L. 414, *429*
Solier, M., see Beauvallet, M. 34, 35, 37, 53, 87, 154, 173, 174, *185*
Solomans, G., see Laufer, M. W. 286, *300*
Solomon, D. 372, *428*
Solomon, H. C., see Rinkel, M. 308, 343, *363*
Soltys, J. J., see Finnerty, R. J. 250, *269*
Somerville, D. M., see Peterson, B. H. 293, *302*
Soranzio, M. R., see Angelis, L. de 61, *182*
Soroko, F. E., see McKenzie, G. M. 82, 177, *220*
Sossi, D., see Grana, E. 87, *204*
Sotirescu, N., see Kreindler, A. 342, *360*
Soulairac, A., see Soulairac, M.-L. 98, *236*
Soulairac, M.-L. 98, *236*
Soulairac, M.-L., Soulairac, A. 98, *236*
Sourkes, T. L., see Lal, S. 72, 77, *214*
Southgate, P. J., Mayer, S. R., Boxhall, E., Wilson, A. 77, 98, *236*
Sowinska, H., see Maj, J. 77, 82, 87, *218*
Sparber, S. B., Peterson, D. W. 110, 111, 113, 119, *237*
Sparber, S. B., Tilson, H. A. 43, 55, 110, 111, 119, *237*
Sparber, S. B., Tilson, H. A., Peterson, D. W. 111, *237*
Sparber, S. B., see Peterson, D. W. 388, *426*
Sparber, S. B., see Tilson, H. A. 43, 110–113, 119. 120, *240*, 314, 320, 330, *366*
Spaulding, T. C., Ford, R. D., Dewey, W. L., McMillan, D. E., Harris, L. S. 386, *428*

Speck, L. B. 335, 339, *365*
Speiser, Z., see Weinstock, M. 87, *243*
Spencer, P. S. J. 154, *237*
Spencer, P. S. J., Turner, T. A. R. 177, 178, *237*
Spencer, P. S. J., see Blackham, A. 87, 154, 173, *187*
Spencer, P. S. J., see Brittain, R. T. 154, 173, *189*
Spencer, P. S. J., see Sayers, A. C. 77, *233*
Spengler, J. 87, 134, *237*
Spengler, J., Waser, P. 131, *237*
Spiegel, H. E., see Dengler, H. J. 38, *196*
Spilker, B., Callaway, E. 252, *274*
Spiteller, G., see Friedrich-Fichtl, J. 375, *419*
Spooner, C. E., Flaming, D. 70, 95, 96, 131, 176, *237*
Spooner, C. E., see Tice, P. 173, *240*
Spooner, C. E., see Winters, W. D. 178, *244*, 341, 342, 348, *368*
Sprague, R. A., see Rosenkrantz, H. 404, *427*
Sprecht, P. G., see Cameron, J. S. 247, *268*
Spriggs, T. L. B. 147, *237*
Spulak, R. von, Claussen, U., Fehlhaber, H. W., Korte, R. 375, *429*
Spulak, F. von, see Claussen, U. 375, *417*
Spulak, F., see Isbell, H. 380, 385, *421*
Spurling, N. W., see D'Arcy, P. F. 169, *195*
Squires, R. F., see Schrold, J. 54, 55, 70, 95, 96, *234*
Srimal, R. C., Dhawan, B. N. 77, *237*
Sroufe, L. A., Stewart, M. A. 287, *303*
Stache, J., see Abramson, H. A. 318, *350*
Staehelin, J. E. 282, *303*
Staehelin, M., see Maitre, L. 393, 394, *423*
Stark, P., Totty, C. W. 137, *237*
Stark, P., Turk, J. A., Redman, C. E., Henderson, J. K. 121, 122, *237*
Starke, K. 147, *237*
Starmer, G. A., see Chesher, G. B. 390, *417*
Stearn, W. T. 373, *429*
Steakly, L. P., see Hughes, J. E. 403, *421*
Steel, J. M., Briggs, M. 257, *274*
Steele, B. J., see Clark, F. C. 110, 112, 113, *193*
Stefanis, C., Boulougouris, J., Liakos, A. 409, *428*
Stefanis, C., see Fink, M. 409, *419*
Stefano, F. J. E., see Brodie, B. B. 34, 35, 51, 63, 139, 149, 150, 154, 156, 163, 164, *189*
Stefano, F. J. E., see Leitz, F. H. 40, 45, 47, *215*
Stein, L. 10, 15, *30*, 121, 122, 123, *237*
Stein, L., Ray, O. S. 121, *237*
Stein, L., Seifter, J. 121–123, *237*
Stein, L., Wise, C. D. 44, 121, 122, 123, *237*
Stein, L., see Berger, B. D. 77, 87, 127, 136–138, *186*
Stein, L., see Ritter, S. 122, *230*
Stein, L., see Wise, C. D. 121, 123, *244*
Steinberg, H., Rushton, R., Tinson, C. 94, *238*
Steinberg, H., see Armitage, P. 94, *183*
Steinberg, H., see Rushton, R. 94, *231*
Steiner, J. E., Sulman, F. G. 340, *365*
Steiner, S. S., Stokely, S. N. 121, *238*
Stenchever, M. A., Allen, M. 401, *428*
Stenchever, M. A., Kunysz, T. J., Allen, M. 401, *428*
Stepanek, J. 142, 148, 165, *238*
Stephenson, J. D., see Marley, E. 176, 177, *219*
Steranka, L. R., see Barrett, R. J. 100, 102, 103, 107, 108, 130, *184*
Sterling, T. P., see Siegel, P. 138, *236*
Stern, S., see Tagliamonte, A. 49, 53, 55, 56, *239*
Stern, W. C., Morgane, P. J., Bronzino, J. D. 342, *365*
Stern, W. G., see Segal, D. S. 99, 106, *235*
Stevenson, J. A., see Mogenson, G. J. *221*
Stevenson, J. A. F., see Russek, M. 138, *232*
Stewart, A. H., see Schweigerdt, A. K. 340, *364*
Stewart, G. A. 180, *238*
Stewart, M. A., see Sroufe, L. A. 287, *303*
Stifler, L., see Meyer, R. E. 253, *272*
Stille, G. 178, *238*
Stillman, R. C., see Cohen, S. 413, *417*
Stitzel, R., see Obianwu, H. O. 49, 50, 148, *224*
Stock, G., see Andén, N.-E. 42, *182*
Stoeckel, M., see Legator, M. S. 402, *423*
Stöckle, D., see Fleckenstein, A. 58, *201*
Stoff, D. M., Mandel, I. J., Gorelick, D. A., Bridger, W. H. 314, *365*
Stoica, E., see Crighel, E. 342, *353*
Stokely, S. N., see Steiner, S. S. 121, *238*
Stolk, J. M., Burnett, L. S., Rech, R. H. 162, 175, *238*
Stolk, J. M., Rech, R. H. 87, 166, 167, 173, *238*
Stolk, J. M., see Rech, R. H. 82, 87, 88, 106, *229*
Stoll, A., Rothlin, E., Rutschmann, J., Schalch, W. R. 333, *366*
Stoll, W. A. 305, 310, *366*
Stolman, S., see Loh, H. 164, *217*
Stone, G. C. 99, *238*
Stone, G. C., Calhoun, D. W., Klopfenstein, M. H. 109, *238*
Stone, C. A., Ross, C. A., Wenger, H. C., Ludden, C. T., Blessing, J. A., Totaro, J. A., Porter, C. C. 146, *238*
Storz, H., Kirk, R. 254, *274*
Stowe, F. R., Miller, A. T. 136, *238*
Strada, S. J., Sanders-Bush, E., Sulser, F. 38, 43, 46, *238*
Straschill, M., Hoffmann, K. P. 65, *238*
Straub, W. 345, *365*
Straughan, D. W., see Roberts, M. H. T. 329, 338, *363*
Strang, R. H., see Manning, D. H. 162, *219*
Strauss, S., see Coper, H. 149, 154, 159, *194*
Strauss, S., see Rating, D. 17, *30*
Strauss, S., see Ziem, M. 139, 155, 157, *246*
Strauss, S. M., see Rommelspacher, H. 8, *30*
Stretch, R., Gerber, G. J. 126, *238*
Stretch, R., see Dalrymple, S. D. 100, 106, *195*
Stretch, S., Gerber, J., Wood, M. 124, 126, *238*
Strocchi, P., see Babbini, M. 77, 87, *183*

Strömblad, B. C. R., see Blaschko, H. 45, 46, *187*
Strong, J. A., see Dunleavy, D. L. F. 255, *268*
Strubelt, O., Vogelberg, W., Zetler, G. 87, *238*
Stubbs, D. A., Thomas, J. R. 114, *238*
Studnitz, W. von, see Sjoerdsma, A. 9, 19, *30*
Stühlinger, W. von, Turnheim, K., Kraupp, O. 143, *238*
Stümpf, C. 176, *238*
Stumpf, Ch. 340, *366*
Stumpf, Ch., Petsche, H., Gogolak, G. 340, *366*
Stumpf, Ch., see Brücke, F. 340, *352*
Stumpf, Ch., see DeBaran, L. 340, *353*
Stumpf, Ch., see Gogolák, G. *204*
Stunkard, A., McLaren-Hume, M. 285, *303*
Stunkard, A. J., see Silverstone, J. T. 250, *274*, 285, *303*
Subramanian, K., see Menon, M. K. 132, 134, *221*
Suciu-Foca, N., see Nahas, G. G. 402, *425*
Sudilovsky, A., see Ellinwood, E. H., Jr., 77, 81, 178, *198*
Sugerman, A. A., see Goldstein, L. 343, *356*
Sugerman, A., see Pfeiffer, C. 251, *272*
Sugrue, M. F. 315, 333, *366*
Sulaiman, W. R., Johnson, R. H. 255, *274*
Sulkowski, T. S., see Harvey, S. C. 41, 62, 63, 141–144, 146, *206*
Sulman, F. G., see Steiner, J. E. 340, *365*
Sulser, F., Dingell, J. V. 16, *30*, 87, 88, *239*
Sulser, F., Owens, M. C., Dingell, J. V. 15, 16, *30*, 87, *239*
Sulser, F., Owens, M., Norvich, M. R., Dingell, J. V. 87, 88, *239*
Sulser, F., Sanders-Bush, E. 3, *30*, 33, 34, *239*
Sulser, F., see Dingell, J. V. 82, 87, *197*
Sulser, F., see Freeman, J. J. 50, 51, 87, *201*
Sulser, F., see Gallagher, D. W. 389, 395, *419*
Sulser, F., see Miller, K. W. 87, 106, *221*
Sulser, F., see Sanders-Bush, E. 56, *232*
Sulser, F., see Strada, S. J. 38, 43, 46, *238*
Sulser, F., see Taylor, W. A. 8, 9, 19, *31*, 35, 43, 52, *239*
Sum, C. Y., see Cho, A. K. 11, *26*
Summerfield, A., see Bösser, T. H. 129, *187*
Summerfield, A., see Bradley, D. W. H. 94, *188*
Summerfield, A., see Cooper, S. J. 123, 130, *194*
Summerfield, A., see Porsolt, R. D. 94, *227*
Sundland, D., see Sletten, I. W. 250, *274*
Sundwall, A., see Jönsson, L.-E. 261, *271*
Sussman, P., see Sidell, F. R. 385, 387, *427*
Svensson, T. H. 5, *30*, 36, 39, 42, 82, 87, 88, *239*
Svensson, T. H., Thieme, G. 81, *239*
Svensson, T. H., see Andén, N.-E. 42, *182*
Swanson, D. W., Weddige, R. L., Morse, R. M. 285, *303*
Sweeney, H. M., see Handley, C. A. 148, *206*
Swinyard, E. A., Clark, L. D., Miyahara, J. T., Wolf, H. H. 175, 177, *239*
Sykes, D. H., Douglas, V. I., Morgenstern, G. 250, *274*
Sykes, D., see Weiss, G. 286, *303*
Syme, G. J., see Syme, L. A. 96, *239*
Syme, L. A., Syme, G. J. 96, *239*
Szabadi, E., see Bevan, P. 338, *351*
Szabadi, E., see Bradshaw, C. M. 338, *352*
Szara, S. 311, *366*
Szara, S., Rockland, L. H., Rosenthal, D., Handlon, J. H. 311, *366*
Szara, S., see Braude, M. C. 372, 384, *416*
Szara, S., see Frank, I. M. 409, *419*
Szára, S., see Sai-Halász, A. 311, *363*
Szentgyörgyi, D., see Simonyi, J. 149, *236*
Szerb, J. C., see McKenzie, G. M. 44, *220*
Szerb, J. C., see Riddell, D. 44, *230*

Tabin, T., see Vycklicky, L. 69, *242*
Taeschler, M., Cerletti, A. 313, *366*
Taeschler, M., Weidmann, H., Cerletti, A. 314, *366*
Taeschler, M., see Neuhold, K. 313, 315, 337, *362*
Taeschler, M., see Rothlin, E. 307, *363*
Tagliamonte, A., Tagliamonte, P., Gessa, G. L. 138, 239
Tagliamonte, A., Tagliamonte, P., Perez-Cruet, J., Gessa, G. 49, 53, 55, 56, *239*
Tagliamonte, A., Tagliamonte, P., Perez-Cruet, J., Stern, S., Gessa, G. 49, 53, 55, 56, *239*
Tagliamonte, P., see Tagliamonte, A. 49, 53, 55, 56, 138, *239*
Tainter, E. G., see Tainter, M. L. 177, *239*
Tainter, M. L. 82, 91, 92, 131, *239*
Tainter, M. L., Tainter, E. G., Sherwood, L. W., Neuru, E. N., Lackey, R. W., Laduena, F. P., Kirtland, H. B., Gonzalez, R. I. 177, *239*
Tainter, M. L., see Crismon, C. A. 62, *195*
Tainter, M. L., see Novelli, A. N. 82, *224*
Tainter, M. L., see Pedden, J. R. 148, *226*
Tainter, M. L., see Schulte, J. W. 82, *234*
Takacs, L. 291, *303*
Takagi, H., Yamamoto, S., Takaori, S., Ogiu, K. 341, *366*
Takaori, S., see Takagi, H. 341, *366*
Takasaki, K., Nishizaca, T., Igisu, T., Kaneko, M. 62, *239*
Takeo, Y., Himwich, H. E. 176, *239*, 340, *366*
Takeo, Y., see Yui, T. 339, 340, *368*
Tamarkin, N. R., see Lemberger, L. 380, *423*
Tang, A. H., Uhlinger, B. J., Schroeder, L. A. 313, 335, 338–340, *366*
Tang, F. L. M., see Ferris, R. M. 38–40, 44, *200*
Tansella, M., see Dolfini, E. 155, *197*
Tansella, M., see Valzelli, L. 16, *31*, 154, *242*
Tansy, M. F., see Kasirsky, G. I. 166, 173, 174, *211*
Taplin, G. V., see Tashkin, D. P. 409, *428*

Tapp, J.T., see D'Encarnacao, P.S. 87, *196*, 264, *268*
Tashkin, D.P., Shapiro, B., Frank, I.M. 414, *428*
Tashkin, D.P., Shapiro, B., Ramanna, L., Taplin, G.V., Lee, Y.E., Harper, C.E. 409, *428*
Tatetsu, S. 262, 263, *274*, 291, *303*
Tatter, D., see Citron, B.P. 261, *268*
Taylor, B. 50, *239*
Taylor, D., see Ho, B.T. 391, 393–396, *421*
Taylor, K.B. 9, *30*
Taylor, K.M., Laverty, R. 94, 106, *239*
Taylor, K.M., Snyder, S.H. 5, *30*, 34, 35, 37–39, 72, 82, *239*, 257, 266, *274*
Taylor, K.M., see Snyder, S.H. 72, 82, *236*
Taylor, R.L., Maurer, J.I., Tinklenberg, J.R. 312, *366*
Taylor, W.A., Sulser, F. 8, 9, 19, *31*, 35, 43, 54, *239*
Teal, J.S., see Rumbaugh, C.L. 143, *231*, 252, 265, *273*
Tecce, J.J., Cole, J.O. 248, *274*
Tedeschi, D.H., Fowler, P.F., Miller, R.B., Macko, E. 96, 97, *239*
Tedeschi, D.H., Tedeschi, R.E., Fellows, E.J. 313, 332, 339, *366*
Tedeschi, D.H., see Tedeschi, R.E. 87, 154, *239*
Tedeschi, R.E., Tedeschi, D.H., Ames, P.L., Cook, L., Mattis, P.A., Fellows, E.J. 87, 154, *239*
Tedeschi, R.E., see Tedeschi, D.H. 313, 332, 339, *366*
Tegerdine, P., see Roffler-Tarlov, S. 43, 46, *230*
Teitelbaum, P., Derks, P. 99, 115, 116, *240*
Teitelbaum, P., see Russek, M. 136, *232*
Tenen, S., see Weissman, A, 77, 106, 134, *243*, 264, *275*
Teplitz, R.L., see Dorrance, D. 402, *418*
Terada, C.W., Masur, J. 96, *240*
Terwelp, D.R., Kennedy, M.S., Burks, T.F. 149, 150, 154, 158, 159, *240*
Testino, L., see Christensen, H.D. 383, *417*
Tetreault, L., see Ouellet, J. 395, *426*
Tha, S.J., see Cox, B. 77, *195*
Thacore, V.R., see Oswald, I. 260, *272*
Thale, T., see Salomon, K. 310, *364*
Thenint, F., see Baum, E. 79, *185*
Theobald, W., see Morpurgo, C. 77, 87, 96, 97, 106, 154, 173, *222*
Thesleff, H., see Bahnsen, P. 247, 248, 255, 256, *267*
Thieme, G., see Svensson, T.H. 81, *239*
Thienes, C.H., see Detrick, L.E. 60, 62, 141, 142, 146, 148, *196*
Thierry, A.M., see Javoy, F. 35, *210*
Thiessen, P.N., Cook, D.A. 312, *366*
Thijssen, H.W.H., see Ginneken, C.A.M. van 375, *429*
Thode, W.F., Carlisle, H.J. 137, *240*
Thoenen, H., Hürlimann, A., Gey, K.F., Haefely, W. 9, *31*, 50, 52, 63, *240*
Thoenen, H., Hürlimann, A., Haefely, W. 49, 50, *240*
Thörner, R., see Zetler, G. 77, *246*
Thoman, J.E.P., see Everett, G.M. 70, 77, *199*
Thomas, G., see Irwin, S. 87, 90, *209*
Thomas, J.R., see Stubbs, D.A. 114, *238*
Thomas, R.K., see Lynch, V.D. 101, *217*
Thomas, S.A., see Lessin, P.J. 409, *423*
Thomas, T.L., see Lindeke, B. 11, *29*
Thomasset, M., see Benakis, A. 16, 17, *25*
Thompson, G.R., Fleischman, R.W., Rosenkrantz, H., Braude, M.C. 403, 405, *428*
Thompson, G.R., Mason, M.M., Rosenkrantz, H., Braude, M.C. 404, 410, *428*
Thompson, G.R., Rosenkrantz, H., Fleischman, R.W., Braude, M.C. 405, *428*
Thompson, G.R., Rosenkrantz, H., Schaeppi, U.H., Braude, M.C. 403, 404, *428*
Thompson, T., Pickens, R. 124, 126, *240*
Thompson, T., see Pickens, R. 124, *227*
Thor, D.H. 96, *240*
Thorbunn, M.J., see Martin, P.A. 402, *424*
Thornburg, J.E., Moore, K.E. 39, 82, 87, *240*
Thornburg, J.E., see Moore, K.E. 87, *222*
Thuillier, J., Nakajima, H., Grandjean, J.L., Kurihara, K. 150, 154, *240*
Thullier, J., see L'Huillier, J. 167, 173, *216*
Thullier, J., see Nakajima, H. 154, *223*
Thut, P., see Lal, H. 97, *214*
Thut, P.D., see Neuburg, J. 87, *223*
Thut, P.D., see Rech, R.H. 92, 107, *229*
Thys, O., see Balasse, E. 254, *267*
Tice, P., Spooner, C.E. 173, *240*
Tikal, K., Benesova, O. 96, *240*
Tillement, J.-P., see Boissier, J.-R. 56, *187*
Tillement, J.-P., see Simon, P. 34, 53, 87, *236*
Tilson, H.A., Rech, R.H. 92, 107, 110, 119, *240*
Tilson, H.A., Sparber, S.B. 43, 110–113, 119, 120, *240*, 314, 320, 330, *366*
Tilson, H.A., see Sparber, S.B. 43, 55 110, 111, 119, *237*
Tilton, B.E., see Peterson, D.I. 175, *226*
Timmons, M.C., see Perez-Reyes, M. 380, 383, *426*
Tinklenberg, J.R. 385, 388, *428*
Tinklenberg, J.R., see Darley, C.F. 385, *417*
Tinklenberg, J.R., see Taylor, R.L. 312, *366*
Tinson, Ch., see Rushton, R. 94, *231*
Tinson, C., see Steinberg, H. 94, *238*
Titus, E.O., see Dengler, H.J. *38*, *196*
Tobias, J., see Winsberg, B.G. 286, *303*
Todd, C.W., see Harvey, J.K. 265, *270*
Todorov, J.C., Gorayeb, S.R.P., Correa, D.L., Graeff, F.G. 111, *240*
Török, T. 87, *240*
Török, T., Vizi, E.S., Knoll, J. 161, *240*
Tognoni, G., see Jori, A. 43, 46, *211*
Toller, C. van, see Hendry, D.P. 115, *207*

Tom, H., see Zsoter, T. 163, *246*
Toman, J. E. P., Davis, J. P. 176, *240*
Tomchick, R., see Axelrod, J. 34, 38, 47, *183*
Tomiyasu, N., see Aramaki, H. 375, *415*
Tonge, S. R., see Leonard, B. E. 333, *360*
Tonini, G., Riccioni, M. L., Missere, G. 100, 107, *240*
Tormey, J., Lasagna, L. 92, 138, 139, *240*
Torrelio, M., see Hine, B. 413, *420*
Toro, G., see Kolodny, R. C. 400, 408, *422*
Tosovsky, J., see Roth, B. 250, *273*, 286, *302*
Totaro, J. A., see Stone, C. A. 146, *238*
Totty, C. W., see Stark, P. 137, *237*
Trapp, C. E., see Schube, P. G. 282, *303*
Trapp, C. E., see Ulrich, H. 280, *303*
Traversa, U., see Angelis, L. de 61, *182*
Trelinski, M., see Caldwell, J. 154, *190*
Trellee, M. 277, *303*
Trendelenburg, U. 58, *241*
Trendelenburg, U., Gomez, B. A. S., Nuskus, A. 142, *241*
Trendelenburg, U., Muskus, A., Fleming, W. W., Gomez, A. B. 59, 146, *241*
Trethowan, W. H., see Bell, D. S. 260, *267*, 282, 291, 294, *297*
Trimbach, C., see Fibiger, H. C. 91, *200*
Trinker, F. R., see Rand, M. J. 146, *228*
Tripod, J. 87, *241*
Tripod, J., Benn, H. J., Meier, R. 70, 87, *241*
Trivedi, M. C., see Lawlor, R. B. 5, *28*, 132, 138, *215*
Troelstrup, R., see Isaac, W. 90, *209*
Trouton, D., Eysenck, H. J. 249, *274*
Trowbridge, J., see Wayner, M. J. 112, 140, *243*
Truitt, E. B., Anderson, S. M. 391, *428*
Truitt, E. B.,, Jr., see Foltz, R. L. 383, *419*
Trzeciak, H., see Herman, Z. S. 35, 37, 53, 87, 91, 92, *207*
Tschirgi, R. D., see John, E. R. 106, *210*
Tseng, L. F., Hitzeman, R. J., Loh, H. H. 44, 46, *241*
Tseng, L. F., Loh, H. H. 87, *241*
Tseng, L. F., Walaszek, E. J. 77, *241*
Tsukamoto, H., see Aramaki, H. 375, *415*
Tsuyussaki, T., Kobayshi, T, 92, *241*
Tucker, Q. T., see Beckett, A. H. 21, *25*
Tumanov, V. P., see Popova, E. N. 174, *227*
Tumanov, V. P., see Vtyurin, B. V. 174, *242*
Turk, J. A., see Stark, P. 121, 122, *237*
Turk, R. F., Dewey, W. L., Harris, L. S. 380, *428*
Turk, R. F., see Dewey, W. L. 412, *418*
Turk, R. F., see McMillan, D. E. 412, *425*
Turk, R. F., see Phillips, R. N. 403, 407, *426*
Turkanis, S. A., see Karler, R. 410, *421*
Turker, M. N., see Kaymackalan, S. 390, *422*
Turner, C. E., Hadley, K. 374, 380, *428*
Turner, C. E., Hadley, K., Getterman, P. S. 374, *428*
Turner, C. E., see Lotter, H. L. 378, *423*
Turner, C. E., see Quimby, M. W. 374, *426*
Turner, P. 253, *274*
Turner, P., see Beckett, A. H. 15, 21, *25*
Turner, P., see Kirby, M. J. 162, *212*
Turner, P., see Smart, J. V. 253, *274*
Turner, T. A. R., see Spencer, P. S. J. 177, 178. *237*
Turner, W. D., Carl, G. P. 247, *274*
Turner, W. J., Merlis, S. 310, *366*
Turnheim, K., see Stühlinger, W. von 143, *238*
Twarog, B. M., Page, I. H. 328, *366*
Twyman, W. A., see Bradley, D. W. H. 94, *188*
Tye, A., see Patil, P. N. 60, 63, *226*
Tyrrell, E. D., see Frank, I. M. 409. *419*
Tyrrell, E. D., see Cohen, S. 408, *417*
Udenfriend, S., see Bogdanski, D. F. 328, *352*
Udenfriend, S., see Creveling, C. R. 9, 18, *27*
Udo de Haes, H. A., see Hart, M. 97, *206*
Udsen, P., see Randrup, A. 77, *229*
Udwadia, B. P., see Gokhale, S. D. 146, 147, *204*
Uehling, B. S. 91, 92, 109, 119, *241*
Ueki, S., see Yamamoto, T. 313, *368*
Uhlinger, B. J., see Tang, A. H. 313, 335, 338–340, *366*
Ukena, T. E., see Knowler, W. C. 109, 140, *212*
Uliss, D. B., Razdan, R. K., Dalzell, H. C., Handrick, G. R. 379, *429*
Ulrich, H., Trapp, C. E., Vidgoff, B. 280, *303*
Umbreit, W. W., see Keller, D. L. 312, *359*
Umemoto, M., Kido, R. 122, *241*
Uncapher, B., see Wasson, H. W. 252, *275*
Ungerleider, J. T., see Hepler, R. S. *420*
Ungerstedt, U. 65, 66, 68, *241*
Ungerstedt, U., Arbuthnott, G. W. 68, *241*
Ungerstedt, U., see Fuxe, K. 37–39, 43, 54, 67–69, 77–79, 87, *202*
Unna, K. R., see Kosman, M. E. 91, *213*, 258, *271*
Upton, P. D., see Hanna, C. 176–178, *206*
Uretsky, N. J., see Evetts, K. D. 87, 89, 96, *199*
Ursin, H. 315, 341, *366*
Utena, H. 54, 57, 70, 81, 91, 173, *241*
Utena, H., Ezoe, T., Kato, N., Hada, H. 46, 70, 81, 161, 173, *241*
Utena, H., see Machiyama, Y. 70, 81, *217*
Uyeno, E. T. 313, *367*, 406, *429*
Uyeno, E. T., Benson, W. M. 313, *367*
Uyeno, E. T., Otis, L. S., Mitoma, C. 317, *367*

Vaccari, A. 61, *241*
Vaccari, A., Cugurra, F. 61, *241*
Vaccari, A., Vertua, R. 61, *241*
Vaccari, A., Vertua, R., Furlani, A. *241*

Vachon, L., Fitzgerald, M. S., Solliday, N. H., Gould, I. A., Gaensler, E. A. 414, *429*
Vachon, L., Mikus, P., Morrissey, W., Fitzgerald, M., Gaensler, E. A. 414, *429*
Vachon, L., Robins, A., Gaensler, E. A. 414, *429*
Vajnovszky, B., see Knoll, J. 82, *212*
Valdecasas, F. G., Salva, J. A., Cuenca, E. 69, 80, 146, 147, *241*
Valdecasas, F. G., see Laporte, J. 60, *214*
Valette, G., Jacquot, C., Heimburger, M. 62, 63, 141, *241*
Valette, G., see Jacquot, C. 51, *209*
Valette, G., see Rapin, J. R. 49, *229*
Valle, J. R., Lupa, A. J., Barros, G. G. 373, *429*
Valzelli, L., Consolo, S., Morpurgo, C. 16, *31*, 150, 154, *241*
Valzelli, L., Dolfini, E., Tansella, M., Garattini, S. 16, *31*, 154, *242*
Valzelli, L., Giacalone, E., Garattini, S. 97, 167, *242*
Valzelli, L., see Consolo, S. 15, 16, *26*, *27*, 36, 97, 168, *193*
Valzelli, L., see Dolfini, E. 35, 36, 149, 154, 155, 167, 168, *197*
Valzelli, L., see Samanin, R. 134–136, *232*, 331, *364*
Van Ginneken, C. A. M., see Dezeeuw, R. A. 375, *418*
Vane, J. R. 61, *242*, 331, *367*
Vane, J. R., Collier, H. O. J., Corne, S. J., Marley, E., Bradley, P. B. 313, 331, 332, 335, 342, *367*
Vane, J. R., see Gelder, M. G. 331, *355*
Vane, J. R., see Hearst, E. 111, 114, *207*
Varma, D. R., see Benfey, B. G. 142, *186*
Vasko, M. R., Domino, L. E., Domino, E. F. *242*
Vasquez, A. J., Izquierdo, I, 106, *242*
Vasquez, A. J., Krip, G. 177, *242*
Vasquez, B. J., see Izquierdo, I. 176, *209*
Vassar, H. B., see Sofia, R. D. 391, *428*
Vaupel, D. B., Martin, W. R. 321, 331, *367*
Vaupel, D. B., Nozaki, M. 316, 317, 321, 335, *367*
Vaupel, D. B., Nozaki, M., Martin, W. R. 321, 322, *367*
Vaupel, D. B., see Martin, W. R. 281, *301*, 348, *361*
Vaupel, D. B., see Nozaki, M. 316, 317, 321, *362*
Vavilov, A. M., see Popova, E. N. 174, *227*
Vedernikov, Y. P. 77, 173, 179, *242*
Vedral, D. F., see Scheving, L. E. 167, *233*
Veneroni, E., see Bizzi, A. 163, *187*
Verbruggen, F. J., see Janssen, P. A. J. 77, *210*
Verbruggen, F. J., see Niemegeers, C. J. E. 98, 100, 101, 106, *224*
Verdavienne, C., see Duhault, J. 3, *27*, 56, *198*
Verdeaux, G., Marty, R. 339, 340, *367*
Verdeaux, G., see Delay, J. 339, *353*
Vergnes, M., Chaurand, J. P. 97, *242*
Verhave, T. 99, 101, 102, *242*
Vernikos-Danellis, J., see Müller, P. J. 168, *223*
Vertua, R., see Angelis, L. de 61, *162*
Vertua, R., see Vaccari, A. 61, *241*
Viala, D., Buser, P. 69, *242*
Vidgoff, B., see Ulrich, H. 280, *303*
Vietinghoff-Riesch, F. von, see Atrens, D. M. 121, 122, *183*
Vigouret, J. M., see Loew, D. M. 339, *360*
Viik, K., see McKenzie, G. M. 68, 78, *220*
Villarreal, J. E., Guzman, M., Smith, C. B. 87, *242*
Vincent, J. D., see Gauchy, C. 44, *202*
Vitolina, M., see Belenky, M. L. 149, 154–156, *185*
Vizi, E. S., Knoll, B., Knoll, J. 146, 161, *242*
Vizi, E. S., see Knoll, J. 81, *212*
Vizi, E. S., see Török, T. 161, *240*
Vogel, V. H., Isbell, H., Chapman, K. W. 255, *274*
Vogt, M., Gunn, C. G., Jr., Sawyer, C. H. 335, 341, *367*
Vogt, M., see Baldessarini, R. J. 49, 50, *184*
Vogt, M., see Gaddum, J. H. 315, 316, 329, *355*
Vogt, M., see Paasonen, M. K. 54, 57, *225*
Vogt, M., see Sanan, S. 34, 36, *232*
Vogelberg, W., see Strubelt, O. 87, *238*
Voigtländer, P. F., von, 66, *242*
Voigtländer, P. F., von, Moore, K. E. 44, 68, *242*
Volavka, J., see Fink, M. 409, 410, *418*, *419*
Volbehr, H., see Ladisch, W. 43, 71, *213*
Vollner, L., Bieniek, D., Korte, R. 375, *429*
Votava, Z., Glisson, S. N., Himwich, H. E. 315, 333, *367*
Vree, T. B. 3, 20, 21, *31*
Vree, T. B., Breimer, D. D., Ginneken, C. A. M. van, Rossum, J. M. van *429*
Vree, T. B., Gorgels, J. P. M. C., Muskens, M., Rossum, J. M. Van 11, 14, *31*
Vree, T. B., Muskens, M. J., Rossum, J. M. Van 11, 14, *31*
Vree, T. B., Rossum, J. M. Van 7, 12, 18, 20, 21, *31*
Vree, T. B., see Bercht, C. A. L. 375
Vree, T. B., see Dezeeuw, R. A. 375, *418*
Vree, T. B., see Ginneken, C. A. M. van 375, *429*
Vree, T. B., see Henderson, P. T. 11, 14, *28*
Vtyurin, B. B., Tumanov, V. P. 174, *242*
Vuillon-Caciuttolo, G., see Walter, S. 317, 342, 343, *367*
Vunakis, H. van, see Farrow, J. T. 334, *355*
Vycklicky, L., Tabin, T. 69, *242*
Vysotskaya, N. B., see Absava, G. I. 165, *181*

Wada, J. A., Wrinch, J., Hill, D., McGeer, P. L., McGeer, E. G. 101, *242*
Waddel, J. L., see Ashcroft, G. W. 263, *267*
Waddel, W. J., see Kennedy, J. S. 381, *422*

Wagner, L. A., Koerker, R. L., Schneider, F. H. 41, 49, *242*
Wagner, M., see Abramson, H. A. 307, 308, 311, *350*
Wagner, S., see Barry, H. 115, 116, *184*
Wahlsten, D., see Griffiths, D. 100, *205*
Wahlström, G., Widerlöv, E. 178, *242*
Wahlström, G., see Widerlöv, E. 178, *244*
Wahlquist, M., Nilsson, I. M., Sandberg, F., Agurell, S., Granstrand, B. 382, *429*
Waite, K., see Johnson, F. H. 129, *210*
Wakerlin, G. E., see Ohler, E. A. 141, *224*
Wakim, K. G., see Schwartz, B. E. 341, 342, *364*
Walaszek, E. J., see Nichols, R. E. 161, 162, *224*
Walaszek, E. J., see Smith, C. W. 333, *365*
Walaszek, E. J., see Tseng, L. F. 77, *241*
Waldeck, B., see Carlsson, A. 38, 40, *191*
Waldeck, B., see Lundborg, P. 40, 50, *217*
Waldmeier, C., see Maitre, L. 394, *423*
Walker, D. R. H., see Lloyd, J. T. 265, *272*
Wall, M. E. 384, *429*
Wall, M. E., see Christensen, H. D. 383, *417*
Wall, M. E., see Freudenthal, R. I. 381, *419*
Wall, M. E., see Perez-Reyes, M. 380, 383, *426*
Wallach, M. B., Angrist, B. M., Gershon, S. 72, *242*
Wallach, M. B., Friedman, E., Gershon, S. 316, 320, 342, *367*
Wallach, M. B., Gershon, S. 70, 77, 78, 80, 176, *242*
Wallach, M. B., Hine, B., Gershon, S. 320, *367*
Wallach, M. B., see Klingenstein, R. J. 77, *212*
Wallach, M. B., see Willner, J. H. 70, 77, *244*
Wallach, M. B., see Winters, W. D. 178, *244*, 341, 342, 348, *368*
Waller, C. W. 373, *429*
Waller, C. W., see Fetterman, P. S. 373, 374, *418*
Wallis, E. S., see Jones, L. W. 279, *300*
Wallis, G. G. 265, *274*
Walls, E. K., see Wishart, T. B. 136, 140, *244*
Walter, D. O., see Adey, W. R. 341, *350*
Walter, J. L., see Foltz, R. L. 383, *419*
Walter, J. M., see Bejrablaya, D. 142, 143, *185*
Walter, R. D., see Chapman, L. F. 343, *352*
Walter, S., Balzano, E., Vuillon-Caciuttolo, G., Naquet, R. 317, 342, 343, *367*
Walters, C. P., see Fuller, R. W. 45, 46, *202*
Walters, G. C., see Cooper, P. D. 348, *353*
Walters, J. R., see Bunney, B. S. 46, 66, *189*
Walters, J. R., see Roth, R. H. 42, 46, 66, *231*
Walters, K., see Powell, D. A. 97, *228*
Walton, R. P. 386, 403, *429*
Walz, D., see Schallek, W. 176, *233*
Wandenberg, F. Van, see Praag, H. M. Van 3, *31*
Wang, N., see Parli, C. J. 9, 11, 16, *29*
Wang, R. I. H., Hasegawa, A. T. Peters, N. J., Rimm, A. 167, *243*
Wanner, H. U., Bättig, K. 121, 122, *243*
Wappner, S., see Krus, D. M. 312, *360*
Wapner, S., see Liebert, R. S. 310, *360*
Ward, J. W., see Alphin, R. S. 131, 138, *182*
Ward, J. W., see Foxwell, M. H. 137, *201*
Ward, J. W., see Funderburk, W. H. 132, 134, *202*
Ward, J. W., see Palmer, G. C. 130, *226*
Warstadt, D. 248, *274*
Waser, P. G., see Borbély, A. A. 149, 154–156, 158, *188*
Waser, P., see Spengler, J. 131, *237*
Wasson, H. W., Uncapher, B. 252, *275*
Wasson, R. G. 305, *367*
Wasson, R. G., see Heim, R. 305, *357*
Watanabe, K., see Iwata, H. 80, 142, *209*
Water, W. H., Richards, D. W., Harris, R. T. 130, *243*
Waterman, F. A. 161, *243*
Waters, D. H., see Glick, S. D. 34, 37, 57, 68, *203*
Waters, W., see Harris, R. T. 410, *420*
Watson, J. T., see Rangno, R. E. 8, *30*
Watson, R., Hartmann, E., Schildkraut, J. J. 294, *303*
Watson, R., see Schildkraut, J. J. 261, *273*, 294, *303*
Waud, S. P. 282, *303*
Wauquier, A., Niemegeers, C. J. 121, 122, *243*
Wauquier, A., see Niemegeers, C. J. E. 98, 100, 101, *224*
Waxenberg, S. E., see Abramson, H. A. 310, *350*
Way, E. L. 372, *429*
Way, E. L., see LaDu, B. N. 12, *28*
Wayner, M. J., Greenberg, I., Trowbridge, J. 112, 140, *243*
Weaver, L. C., Kerley, T. L. 167, *243*
Webb, R. W., see Maickel, R. P. *218*
Weber, E., see Legator, M. S. 402, *423*
Webster, C. D., LeBlanc, A. E., Marshman, J. A., Beaton, J. M. 410, *429*
Webster, C. D., see Beaton, J. M. 101, *185*
Webster, C. D., see Cappell, H. 115, 116, *190*
Webster, C. D., see Garfinkel, B. D. 286, *299*
Weddige, R. L., see Swanson, D. W. 285, *303*
Weenig, D. J., see Harvey, S. C. 41, 62, 63, 141–144, 146, *206*
Weese, M. De, see Gruber, C. M. 251, *270*
Wehrle, R. D., see Moraes, S. de 60, *222*
Weidley, E., see Cook, L. 314, *353*
Weidmann, H., Cerletti, A. 334, 335, *367*
Weidmann, H., see Taeschler, M. 314, *366*
Weil, A. T., Zinberg, N. E., Nelsen, J. M. 385, *429*
Weiner, J. M., Mudge, G. H. 7, *31*
Weiner, M., see Quinn, G. P. 21, *30*
Weiner, N. 33, 34, *243*
Weiner, W. J., Goetz, L., Westheimer, R., Klawans, H. L. 77, 78, *243*
Weiner, W. J., see Klawans, H. L. 70, 77, *212*

Weingartner, H., see Faillace, L.A. 311, *355*
Weingartner, H., see Snyder, S. H. 312, *365*
Weinstock, M., Speiser, Z. 87, *243*
Weintraub, W., see Bertino, J.R. 312, *351*
Weintraub, W., see Klee, G.D. 308, 312, *359*
Weis, J. 142, 154–156, 160, 161, *243*
Weischer, M.-L. 96, *243*
Weise, V.K., see Breese, G.R. 35, 37, 52, *188*
Weise, W.K., see Kopin, I.J. 63, *212*
Weiss, B. 112, 114, *243*, 249, *275*
Weiss, B., Laties, V.G. 109, 114, 118, *243*, 249, *275*
Weiss, B., Laties, V.G., Blanton, F.L. 167, 169, 173, *243*
Weiss, B.L., see Aghajanian, G.K. 330, *350*
Weiss, B., see Laties, V.G. 112, 113, 115, *214*
Weiss, G., Werry, J., Klaus, M., Douglas, V., Sykes, D. 286, *303*
Weiss, S.R., Raskind, R., Morgenstern, N.L., Pytlyk, P.J., Baiz, T.C. 265, *275*
Weissman, A. 87, 99, 106, 109, 117, *243*
Weissman, A., Koe, B.K. 77, 106, *243*
Weissman, A., Koe, B.K., Tenen, S. 77, 106, 134, *243*, 264, *275*
Weissman, A., see Holland, G.F. 3, *28*
Wejdan, A., see Borella, L. 16, *26*
Welch, A.-M., see Welch, B. 35, 36, 37, 54, 96, 97, 168, *243*
Welch, A.S., see Welch, B.L. *243*, 393, 394, *429*
Welch, B., Welch, A.-M. 35, 36, 37, 54, 96, 97, 168, *243*
Welch, B.L., Welch, A.S. *243*
Welch, B.L., Welch, A.S., Messiha, F.S., Berger, H.J. 393, 394, *429*
Welch, B.L., see Moisset, B. 90, *221*
Wells, B. 306, *367*
Wells, C.-H., see Koella, W.P. 340, 342, *359*
Welsh, A.L. 250, 259, *275*
Wender, P.H., see Arnold, L.E. 250, *267*, 286, *296*
Wender, P.H., see Buchsbaum, M. 286, *297*
Wender, P.H., see Zahn, T.P. 286, *304*
Wendt, G.R., see Cameron, J.S. 247, *268*
Wenger, G.R., Rutledge, C.O. 8, *31*, 45, 46, *244*
Wenger, H.C., see Stone, C.A. 146, *238*
Wentink, E.A. 112, 118, *244*
Wenzel, B.M., see John, E.R. 106, *210*
Wenzel, D., Broadie, L. 90, *224*
Werner, A.B., see Clineschmidt, B.V. 132, 134, *193*
Werner, H., see Liebert, R.S. 310, *360*
Werry, J., see Weiss, G. 286, *303*
West, L.J., see Brill, N.O. 403, *416*
Westerbeke, E.J., see White, R.P. 70, 77, 176, *244*
Westerman, G., see Feinberg, I. 252, 260, *269*, 286, *298*
Westheimer, R., see Weiner, W.J. 77, 78, *243*
Westman, J.C., see Zrull, J.P. 286, *304*
Wettenberg, L., see Bäckström, M. 166, *184*
Whalen, R.E., see Hearst, E. 101, 103, *207*
Wheeler, N., see Bhagat, B. 179, *186*
Whitby, L.G., see Axelrod, J. 38, *183*
Whitby, G.L., see Hertting, G. 38, *207*
White, H.D., see Abdallah, A.H. 87, 131, 134, 138, *181*
White, R.P., Boyajy, L.D. 177, *244*
White, R.P., Daigneault, E.A. 176, *244*
White, R.P., Nash, C.B., Westerbeke, E.J., Possanza, G.J. 70, 77, 176, *244*
Whitebread, C.H. II, see Bonnie, R.J. 371, *416*
Whitehead, W.E., see Appel, J.B. 314, *351*
Whittier, J.R., see Korenyi, C. 251, *271*
Whittlesey, J.R.B., see Ditman, K.S. 308, *354*
Wickson, V., see McGeer, E.G. 72, 77, *220*
Widerlöv, E., Wahlström, G. 178, *244*
Widerlöv, E., see Wahlström, G. 178, *242*
Widman, M., see Agurell, S. 382, 384, *415*
Wiest, W.G., see Cicero, T.J. 401, *417*
Wikler, A. 306, 343, *367*
Wikler, A., Haertzen, C.A., Chessick, R.D., Hill, H.E., Pescor, F.T. 308, *367*
Wikler, A., see Hill, H.E. 116, *208*
Wikler, A., see Isbell, H. 308, 310, 311, 318, *358*
Wilbur, D.L., MacLean, A.R., Allen, E.V. 255, *275*
Wilk, S., see Angrist, B.M. 255, 261, *267*
Wilkens, B., see Malitz, S. 310, 343, *361*
Wilkinson, G.R., Beckett, A.H. 21, *31*
Wilkinson, S., see Iversen, S.D. 79, 90, *209*
Will, A., Bisgard, E. 141–143, *244*
Will, B., Checchinato, D. 109, *244*
Willcox, D.R.C., see Scott, P.D. 289, *303*
Williams, D.I., see Lowe, G. 314, *360*
Williams, F.E., see Halliwell, G. 77, *206*
Williams, G.W., see Brownlee, G. 173, *189*
Williams, H.L., see DeMaar, E.W.J. 312, *353*
Williams, J., see Graham, J.D.P. 393, 394, *419*
Williams, M.A., see Maling, H.M. 149, 154, 156, *218*
Williams, R.B., Ng, L.K.Y., Lamprecht, F., Roth, K., Kopin, I.J. 398, *429*
Williams, R.T., Caldwell, J., Dring, L.G. 5, 14, *31*
Williams, R.T., see Caldwell, J. 9, 12, *26*
Williams, R.T., see Dring, L.G. 5, 7, 8, 10, 14, *27*
Williams, R.T., see Sever, P.S. 8, 9, 18, *30*, 154, 157, *235*
Willinsky, M.D. 374, 375, *429*
Willner, J.H., Samach, M., Angrist, B.M., Wallach, M.B., Gershon, S. 70, 77, *244*
Wilpizeski, C., see Hamilton, C.L. 314, *357*
Wilson, A., see Domino, E.F. 57, *197*
Wilson, A., see Southgate, P.J. 77, 98, *236*
Wilson, E. 283,
Wilson, E.M., see Munro, J.F. 285, *301*
Wilson, J., see Gerkens, J.F. 147, *203*

Wilson, M. C., Hitomi, M., Schuster, C. R. 124, 125, *244*
Wilson, M. C., Schuster, C. R. 124, 126, *244*
Wilson, R. S., May, E. L. 384, 390, *429*
Wilson, R. S., see Martin, B. R. 381, 384, 390, 410, 411, *424*
Wilson, W. P., see Fischer, K. C. 250, *269*
Wilson, W. S. 161, *244*
Wimer, R. E., Fuller, J. L. 93, 94, *244*
Winder, C. V., Anderson, M. M., Parke, H. C. 62–64, 141, 142, 144, 148, *244*
Winder, P. H., see Wood, D. R. 287, *304*
Winsberg, B. G., Bialer, I., Kupietz, S., Tobias, J. 286, *303*
Winter, C. A., Flataker, L. 314, *368*
Winter, J. C. 314, 315, 320, *368*
Winter, J. C., see Hirschhorn, I. D. 314, 315, *357*
Winter, J. C., see Schechter, M. D. 314, *364*
Winterburn, A. K., see Dewar, A. J. 164, 165, *196*
Winters, W. D. 348, *368*
Winters, W. D., Mori, K., Wallach, M. B., Marcus, R. J., Spooner, C. E. 178, *244*, 341, 342, 348, *368*
Wise, C. D., Stein, L. 121, 123, *244*
Wise, C. D., see Berger, B. D. 77, 87, 127, 136, 137, 138, *186*
Wise, C. D., see Stein, L. 44, 121–123, *237*
Wisegarver, B. B., see Alles, G. A. 13, *25*
Wishart, T. B., Walls, E. K. 136, 140, *244*
Wiseman, B. D., see Leonhard, B. E. 94, *215*
Witkop, B., see Axelrod, J. 333, 334, *351*
Witkop, B., see Creveling, C. R. 9, 18, *27*
Witkop, B., see Daly, J. W. 4, *27*
Witoslawski, J. J., see Hanson, H. M. 111, 115, 116, *206*
Witt, E. D., see Lemberger, L. 16, *29*, 77, *215*
Wojdan, A., see Borella, L. E. 87, 88, 154, *188*
Wolbach, A. B., Jr., Isbell, H., Miner, E. J. 318, *368*
Wolbach, A. B., Jr., Miner, E. J., Isbell, H. 310, *368*
Wolbach, A. B., Jr., see Hill, H. E. 308, 309, *357*
Wolbach, A. B., see Isbell, H. 318, *358*
Wolbach, A. B., see Rosenberg, D. E. 253, 254, 256, 261, *273*, 319, *363*
Wolf, H. H., Bunce, M. E. 154, 155, 173, 174, *244*
Wolf, H. H., Rollins, D. E., Rowland, C. R., Reigle, R. G. 177, *245*
Wolf, H. H., see George, D. J. 167, 169, *203*
Wolf, H. H., see Lanciault, G. 177, 179, *214*
Wolf, H. H., see Swinyard, E. A. 175, 177, *239*
Wolff, H. G., see Richards, N. 310, *363*
Wolgin, D. L., see Carlton, P. L. 120, 138, *191*
Wollstein, A., see Jacobsen, E. 247, 248, *270*
Wolpert, E. A., see Rechtschaffen, A. 285, *302*
Wolstencroft, J. H., see Bradley, P. B. 65, *188*
Wolstenholme, G. E. W., Knight, J. 372, *429*
Wolthuis, O. L. 102, 129, 130, *245*
Wong, D. T., Frank, R. M. Van, Horng, J. S., Fuller, R. W. 49, *245*
Wong, D. T., Horng, J. S., Fuller, R. W. 54, 56, *245*
Wong, R., see Grant, J. D. 378, *419*
Wood, C. D., Graybiel, A. 288, *304*
Wood, D. R., Reimherr, F. W., Winder, P. H., Johnson, G. E. 287, *304*
Wood, M., see Stretch, S. 124, 126, *238*
Woods, J. H., see Black, M. D. 411, *416*
Woods, J. H., see Schuster, C. R. 99, 103, 107, 112, 114, 120, *235*
Woolley, D. W. 306, *368*
Woolley, D. W., Shaw, E. 312, 328, 329, 332, *368*
Wray, R., Cowan, A. 92, 94, *245*
Wright, P. L., see Haley, S. L. 406, 407, *420*
Wright, P. L., see Keplinger, M. L. 406, *422*
Wright, T., see Costall, B. 77, 78, *194*
Wrinch, J., see Wada, J. A. 101, *242*
Wurtman, R. J., see Chiel, H. 158, *192*
Wurtman, R. J., see Yehuda, S. 149, 154, 155, 158, 159, *245*
Wuttke, W. 112, 116, *245*
Wuttke, W., Innis, N. K. 112, *245*
Wuttke, W., see Schlichting, U. U. 124, 125, 126, *234*
Wyatt, R. J., see Gillin, J. C. 320, *356*

Yagi, B. 91, *245*
Yagiela, J. A., McCarthy, K. D., Gibb, J. W. 393–395, *429*
Yagnitinsky, B., see Mechoulam, R. 375, *425*
Yamamoto, S., see Takagi, H. 341, *366*
Yamamoto, T., Ueki, S. 313, *368*
Yamashita, I., see Onodera, I. 165, *225*
Yamauchi, T., see Shoyama, Y. 375, *427*
Yanagita, T., see Deneau, G. 124, *196*
Yang, C., see Lefkowitz, S. S. 402, *423*
Yasunobu, K. T., Oi, S. 45, *245*
Yater, W. M., see Gwynn, H. B. 247, *270*
Yehuda, S., Wurtman, R. J. 149, 154, 155, 158, 159, *245*
Yehuda, S., see Chiel, H. 158, *192*
Yellin, T. O., see Everett, G. M. 35, 37, 45, 54, 56, *199*
Yelnosky, J., Hewson, R., Mundy, J., Mitchell, J. 147, *245*
Yelnosky, J., Lawlor, R. B. 132, 134, 140, *245*
Yelnosky, J., McGill, J. S., Mastrangelo, A. S. 146, 147, *245*
Yelnosky, J., see Lawlor, R. B. 5, *28*, 132, 138, *215*
Yen, F. S., see Matsuyama, S. S. 402, *424*
Yen, T. T., Acton, J. M. 90, *245*
Yeudall, L. T., see Satinder, K. P. 70, *233*
Yokel, R. A., Pickens, R. 5, *31*, 124, 125, 138, *245*
Yokel, R. A., see Pickens, R. 124, *227*
Yonce, L. R., see Dewey, W. L. 398, *418*
Yoshimura, H., see Aramaki, H. 375, *415*
Yoss, R. E. 285, *304*
Yoss, R. E., Daly, D. D. 285, *304*
Yoss, R. E., Meyer, N. J., Ogle, K. N. 285, *304*

Youdin, M. B. H., see Sandler, M. 279, *303*
Young, D., Scoville, W. B. 282, *304*
Young, J. H. 278, *304*
Young, P., see Richards, R. 169, *230*
Yui, T., Takeo, Y. 339, 340, *368*
Yusin, A., see Satterfield, J. H. 286, *303*

Zahn, T. P., Abate, F., Little, B. C., Wender, P. H. 286, *304*
Zaidi, I. H., see Kasirsky, G. I. 166, 173, 174, *211*
Zaitsev, Y. V., see Anichkov, S. V. *182*
Zakio, O., see Regelson, W. 413, 414, *427*
Zalis, E. G., Kaplan, G. 166, *245*
Zalis, E. G., Kaplan, G., Lundberg, G. D., Knutsson, R. A. 175, *245*
Zalis, E. G., Lundberg, G. D., Knutsson, R. A. 166, 174, 175, *245*
Zalis, E. G., Parmley, L. F., Jr. 264, *275*, 288, *304*
Zambotti, F., see Groppetti, A. 36, 37, *205*
Zarcone, V. 285, 286, *304*
Zdrahal, L., see Sercl, M. 251, *273*
Zei, M., see Peczenik, O. 173, *226*
Zemp, J. W., see Graham, J. M. 177, *204*
Zetler, G. 77, 154, *245*
Zetler, G., Mahler, K., Daniel, F. 77, *246*
Zetler, G., Moog, E. 77, 154, *246*
Zetler, G., Thörner, R. 77, *246*
Zetler, G., see Strubelt, O. 87, *238*
Ziance, R. J., Azzaro, A. J., Rutledge, C. O. 39, 44, *246*
Ziance, R. J., Rutledge, C. O. 39, 44, 46, *246*
Ziance, R. J., see Azzaro, A. J. 44, 45, 50, *183*
Ziance, R. J., see Rutledge, C. D. 39, 44–47, 50, 55, *232*
Ziegler, H., Basso, P. del, Longo, V. G. 77, 87, *246*
Ziem, M. 71, *246*
Ziem, M., Coper, H., Broermann, I., Strauss, S. 139, 155, 157, *246*
Zieve, L. 92, *246*
Zigmond, M. J., see Heffner, T. G. 110–112, 116, *207*
Zigmond, M. J., see Schoenfield, R. I. 119, *234*
Ziller, H. H., see McDowell, A. A. 91, *220*
Zimmerman, A. M., McClean, D. K. 402, *429*
Zimmerman, J., Schuster, C. R. 120, *246*
Zimmerman, J., see Schuster, C. R. 92, 114, 120, *235*
Zimmerman, P., see Krieckhaus, E. E. 102, 103, *213*
Zinberg, N. E., see Weil, A. T. 385, *429*
Zinberg, W. E., see Sallen, S. E. 414, *427*
Zirkle, C. L., Kaiser, C. 4, *31*
Zis, A. P., see Fibiger, H. C. 79, 87, *200*
Zitko, B. A., Howes, J. F., Razdan, R. K., Dalzell, B. C., Dalzell, H. C., Sheehan, J. C., Pars, H. G., Dewey, W. L., Harris, L. S. 380, *429*
Zitko, B. A., see Razdan, R. K. 375, 379, *427*
Zrull, J. P., Westman, J. C., Arthur, B., Rice, D. L. 286, *304*
Zsoter, T., Tom, H., Kraml, M., Dvornik, D. 163, *246*
Zucker, I., Milner, P. 140, *246*
Zwarenstein, H., Sapeika, N. 163, *246*
Zwarenstein, H., see Kaplan, E. R. 163, *211*
Zwarenstein, H., see Sapeika, N. 163, *233*
Zwieten, P. A. van 142, *246*
Zwieten, P. A. van, see Hoyer, I. 142, *209*, 247, *270*

Subject Index

Absinthe 278
Abuse potential, amphetamine 257
Aceperone 73, 83, 97
Acetozolamide 178
Acetylcholine (ACh) 56, 123, 261, 329, 333
Acetylcholine esterase (AChE) 57, 405
d-1-Acetyl-lysergic acid diethylamide 330
l-Acetyl-lysergic acid diethylamide 336
Acetylsalicylic acid 153
Addiction Research Center Inventory (ARCI)
  LSD Group Variability Scale 308
  LSD Significant Scale 308
Adenosine triphosphatase (ATPase) 161
Adenyl cyclase 333
Adipsia 117
  amphetamine 140
Adrenalectomy 144
Adrenal gland 41
  Demedullation 316
Adrenaline (see catecholamines) 34
α-Adrenergic blocking agents 93, 103
Adrenergic transmission 33
Adrenochrome 339
Adrenocorticotropic hormone (ACTH) 255, 400
Aggregation 36
Aggression
  amphetamine 96
  Δ^9-THC 386, 387
Agroclavine 340
Alcohol 292
Alprenolol 83
Amantadine 40, 43, 44, 76, 86, 89, 172
γ-Aminobutyric acid (GABA) 57, 329, 342
Aminorex 7
Amitriptyline 12, 16, 75, 104, 152
Amphetamine 6, 7, 9, 33—180, 277—296, 313, 315, 319, 331, 332, 338, 339, 342
  absorption 21
  abstinence 91
  abuse 281—284
  abuse potential 257
  acquisition of conditioned avoidance response 101
  ACTH 165
  acute intoxication 288
  adrenalectomy 144, 163
  adrenal glands 34
  aggression 96
  aliphatic hydroxylation 5, 9
  analgesia 179, 251, 288
  anorexia 131—140, 250
    brain stimulation 137
    intracerebral injection 137
  antagonists 265
  aromatic hydroxylation 5, 8
  asthenia 280
  ATP 161
  auditory threshold 252
  aversive effects 126, 127
  avoidance behavior 98—108
  basal ganglia 79
  basal glanglion lesions 68
  behavioral effects 70, 95
    birds 95
  behavioral incompatibility 139
  blood levels 23
  blood pressure 141, 253
  brain CA release 43, 44
  brain DA 36
    release 42
    uptake 39
  brain NA 34
    uptake 38
  brain NE release 42
  brain stem reticular formation 65
  CA
    release 40
    turnover 48
    uptake 38
  carbohydrate metabolism 162
  catatonia 280
  caudate nucleus 66
  cerebral blood flow 252
  cerebral circulation 143
  cerebrovascular accidents 265
  cholinergic transmission 56
  choreic syndrome 263

Amphetamine
chronic administration 255—261
anorexia 259
cardiovascular effects 261
EEG 260
excitation 264
hormones 261
hyperkinesis 258
metabolic effects 261
mood 255—257
motor activity 92
motor effects 91
neurotransmitters 261
REM sleep 260
shock 264
sleep 258
stereotyped behavior 263
temperature 261
toxicity 173
wakefulness 258
circadian rhythm 167
COMT 47
conditioned avoidance response 98—100
conditioned emotional response 115—117
conditioning 128
continuous reinforcement 109
corticosteroids 165
critical flicker frequency 253
cross tolerance 120
N-dealkylation 5, 12
deamination 55
delirium 293
dependence 277—296
amnesia 293
delirium 294
depression 294
hyperpyrexia 294
hypertension 294
mood 255—257, 289—294
paranoid delusions 293
treatment 294
depression 248, 257, 280
depressor response 141
differential reinforcement 113
discriminant behavior 114, 117
dysphoria 248, 256, 257
ECG 254
EEG 176—179, 251
effect in man 23
effect of brain lesions on
anorexia 135
motor activity 89
operant behavior 118
effect of chronic administration on
avoidance behavior 107
operant behavior 119
emotional depression 287
enuresis 255
euphoria 248, 256
evoked responses 252
excretion 21, 22
exploratory behavior 94
extensor reflex 69
extravascular distribution 21
fixed interval reinforcement 112
fixed ratio reinforcement 109—111
flexor reflex 69
food deprivation 138
genetics 94
growth hormone 166
guinea pig
bladder 61
ileum 60
heart
CA 34
-lung preparation 142
NE 34
rate 142
hoarding behavior 140
hormones 255
5-HT
inactivation 55
release 54
storage 53
synthesis 55
turnover 55
uptake 54
hunger center 135, 137
p-hydroxynorephedrine storage 50
hyperkinesis 249, 280
hyperkinetic children 280
hyperkinetic syndromes of children 286
hyperthermia
age 155
ambient temperature 155
brain lesions 155
endocrines 156
free fatty acids 156
strain 155
tolerance 157
hypothalamus 65
hypothermia 158
insulin 165
intracranial self-stimulation 121—124
iris 60
laws 283
libido 251
lipid metabolism 163, 164
locus ceruleus 65
man 247—266, 277—296
MAO inhibitor 45—47
maze
learning 129
performance 127
medical indications 284—288

metabolism 3—24, 254
  age 17
  aggregation 17
  dependence 18
  neoplasms 17
  psychosis 18
  sex 17
  species 14
  stereoselectivity 13
  urinary pH 15
mood 248
motor activity 81, 88, 90
  isomers 82
narcolepsy 282, 285
nictitating membrane 58
nigrostriatal lesions 67
nucleic acid metabolism 165
obesity 280, 284, 285
open field behavior 93
operant behavior 71, 98—128
N-oxidation 5, 13
oxidative deamination 5, 9—12
oxygen consumption 160, 161
paranoia 257, 258
paranoid psychosis 262, 282
pathology 173
performance 179, 249
peripheral resistance 143
pharmacokinetics 3—24
  species 20
pineal 166
pKa 6, 7, 20
plasma
  CA 41
  levels 20
polycythemia 255
prolactin 166
protein metabolism 164
psychosis 287, 291—296
  experimental 295
  metabolic hypothesis 18
pulmonary circulation 143
pupils 253
raphé lesions 68
raphé nuclei 67
rat
  fundus 61
  inferior eyelid 58
regulations 283
reinforced operant behavior 108—121
release 49
REM sleep 250, 252
respiration 148, 254
salivation 149
satiety center 135—137
sedative-hypnotic interactions 89
seizures 177
self-administration 124—126
self-stimulation 123
  tolerance 123
septum 107
sexual behavior 97
sleep 178
social behavior 95—98
species differences 91
state dependent learning 130
stereoisomerism 5
stereotypic behavior 71—81, 124
  brain sites 78
  catecholamines 71—81
  thalamus 79
  tolerance 80
strain differences 90
structure-activity relationships 3—5
substantia nigra 66
  lesions 68
superior cervical ganglion 34
sympathetic ganglion 64
sympathetic nerves 38
  CA release 40
tachyphylaxis 62, 63, 148
temperature 149—160, 254
thyroid 161, 166
thyroidectomy 163
tolerance 119, 255—261
  adipsia 140
  anorexia 139
toxicity 262
  aggregation 166—169
  ambient temperature 168
  anxiety and fear 175
  hyperthermia 174
  neuromuscular blockade 175
  social stress 167
  strains 167
tyrosine hydroxylase 47
uptake 49
urinary CA 41
variable interval reinforcement 111
wakefulness 250
water intake 140
p-Cl-amphetamine 153
Amphetamine-barbiturate combinations 126
Amygdala 44, 79, 97, 155
Amylobarbitone 128, 129, 253
Analeptics 288
Analgesia, amphetamine 179, 288
Angiotensin 146
Anorexia 117, 281
  amphetamine 131—140, 250
Anorexigenics 280
Antabuse 292
Antidepressants 281, 286
Antipyrine 338
Apomorphine 313, 333, 336, 338
Appetite, amphetamine 250

Area postrema 136
Arecoline 75, 78
Ascorbic acid 153
Atropine 17, 60, 61, 75, 85, 96, 105, 123, 146, 152, 171, 176, 288, 312, 313, 316, 335, 340, 397
Avoidance behavior, amphetamine 98—108
Azapetin 150

Barbiturate-amphetamine combinations 126
Barbiturates 76, 85, 94, 260, 292, 313
Basal ganglia 89
Benactyzine 75, 85
Benanserine 74, 312
Benzedrine 280
Benzhexol 75
Benzoic acid 18
Benzphetamine 6, 162
Benztropine 75, 95
1-Benzyl-2-methyl-5-hydroxytryptamine 332
Bethanidine 147
Blood-brain barrier 20
Bradykinin 333
Brain proteins 405
Brain stem 39, 54, 65, 67, 68, 176
Bretylium 147, 170
D-2-Bromlysergic acid diethylamide (BOL) 61, 74, 83, 93, 151, 170, 310, 313—318, 320, 329, 330, 332—340
Bufotenine 310, 312, 313, 317, 330, 334—336, 339, 340
Bulbocapnine 76, 86, 153
Butoxamine 163

Caffeine 76, 277, 286, 332
Cannabichromene 375, 376, 378
Cannabichromenic acid 376
Cannabicitran 375, 376
Cannabicyclol 375, 376, 378
Cannabicyclolic acid 376
Cannabidiol 375, 376, 378, 385, 402
Cannabidiolic acid 376
Cannabidiorocal 377
Cannabidivarol 377
Cannabidivatrol 375
Cannabielsoic acid 376
Cannabigerol 375, 376, 378
Cannabigerolic acid 376
Cannabinodiol 376
Cannabinoids (see cannabis, marihuana)
  biosynthesis 378—380
  numbering systems 374
Cannabinol 375, 376, 378, 385
Cannabinolic acid 376
Cannabiorcol 377
Cannabis (see marihuana)
  aggression 386
  central nervous system effects 384—386
  chemistry 374
  development 406
  extract, tolerance 411
  preparations 372, 373
  toxicity 403
Cannabis sativa (see marihuana and tetrahydrocannabinol) 371—415
Cannabisativine 378
Cannabitriol 375
Cannabitriol ester 377
Cannabivarol 377
Caramiphen 75
Carbachol 171
Caryophyllene 378
Cataplexy 285
Catecholamines 107
  amphetamine
    anorexia 131—135
    behavior 103—107, 117—119
    brain 38—40, 42—45
    cardiovascular 141, 142, 144—146, 148
    cerebral circulation 143
    EEG 177
    inactivation 45—47
    metabolism 161—164
    motor activity 82—91
    release 40—45
    self-stimulation 121—126
    stereotypy 72—78
    storage 34, 38
    toxicity 169—174
    turnover 48—53
  hyperthermia 150—154, 156, 157
  hypothermia 159
  toxicity 169—173
  turnover 48
Catechol-O-methyl-transferase (COMT) 42, 47
Caudate 44, 48, 55, 57, 66, 68, 78, 79
Caudate nucleus 89
Cerebellum 39, 165
Cerebral blood flow, amphetamine 252
Cerebral cortex 39, 53, 55, 57, 78, 165
Cerebrovascular accidents, amphetamine 265
Cerveau isolé 176
Chanoclavine 310
Chloraciazine 73, 83
Chloral hydrate 76
Chlordane 172
Chlordiazepoxide 17, 76, 85, 94, 97, 109, 134, 135, 171
Chlorimipramine 75, 132, 134, 152
Chlorisondamine 152, 163

p-Chloroamphetamine 3, 11, 159
p-Chlorophenylalanine (pCPA) 69, 84, 88, 93, 95, 98, 104, 131, 133, 150, 151, 161, 169, 171, 177—179, 315, 316, 332, 338
Chlorpheniramine 75, 95
Chlorphentermine 7, 281
Chlorpromazine 16, 48, 57, 68, 73, 78, 83, 92, 94, 97, 104, 117, 118, 122, 123, 126, 131, 133, 145, 150, 159, 162, 163, 170, 177, 266, 312, 315—317, 321, 329, 332, 334—336, 338, 340, 342, 343
Chlorprothixene 73, 150
Chlothiapine 73
Choline 261, 378
Choline acetylase 57
Cholinesterase 333
Choreic syndrome, amphetamine 263
Chronic administration, amphetamine
  avoidance behavior 107
  libido 260
  performance 258
  wakefulness 258
Chrysanthinol 379
Cinanserin 61, 83, 133, 145, 338, 339
Cinobufotenin 172
Clozapine 73, 83
Cocaine 17, 20, 50, 51, 61, 105, 123, 143, 146, 152, 278, 288, 291, 292
Coffee 277
Cola 277
Conditioned avoidance response 98—100
Conditioned behavior, Δ^9-THC 388
Conditioned emotional response, amphetamine 115—117
Conditioning, amphetamine 128
Continuous reinforcement, amphetamine 109
Convulsions (see seizures)
Corticosteroids 292
Corticosterone 400
Cortisol 315
Cross tolerance
  amphetamine 120
  LSD 318—321
Curare 153, 338
Cyanide, potassium 335
Cyclazocine 313
Cyclic adenosine-3′,5′-monophosphate (cAMP) 57
Cyclic guanosine-3′,5′-monophosphate (cGMP) 58
Cyproheptadine 74, 83, 95, 131, 133, 145, 177, 313, 316, 317, 321, 331—333, 335, 336, 338, 339

Decamethonium 153
Denervation supersensitivity 344
Deoxyglucose 169
2-Deoxyglucose 172
Deoxyribonucleic acid (DNA) 401
Desipramine 15, 16, 51
Desmethylimipramine 40, 42, 50, 51, 68, 71, 75, 84, 85, 88, 104, 134, 150, 152, 163, 171, 286, 315
Dexamethasone 164
Diazepam 17, 76, 85, 118, 133, 146, 171
Dibenamine 83, 104, 133, 170, 177, 315, 338
Dibenzyline 150
Dibozane 83, 145
Dichlorisoproterenol 73, 163, 170
Diethyldithiocarbamate 74, 84, 88, 123, 132, 151, 161, 171, 178, 179
Diethylpropion 7, 12, 253, 257—260, 281
Diethyltryptamine (DET) 311, 313, 314, 326, 335, 338, 340
Differential reinforcement, amphetamine 113
β-*β*-Difluoroamphetamine 7
Digitaline 172
Dihydroagroclavine 340
Dihydroelymoclavine 340
Dihydroergotamine 73, 83, 150, 159, 336, 338
Dihydro-lysergic acid diethylamine (dihydro-LSD) 327
2,3-Dihydro-lysergic acid diethylamide (2,3-dihydro-LSD) 310
Dihydroxyphenylserine 88
8,11-Dihydroxy-Δ^9-tetrahydrocannabinol 382
Diisopropyl fluorophosphate (DFP) 135
Dimethoxyamphetamine (DMA) 316, 321, 322, 324, 335
2,5-Dimethoxyamphetamine (2,5-DMA) 317, 333
2,5-Dimethoxy-4-ethylamphetamine (DOET) 312, 313, 325
2,5-Dimethoxy-4-methylamphetamine (DOM) 4, 311, 313, 316, 320, 325, 330, 335, 339, 340, 342
3,4-Dimethoxy-*β*-phenethylamine (DMPEA) 311, 314, 315, 324, 330, 339
2′-(3-Dimethylaminopropylthio)cinnamanilide 329
Dimethylheptylpyran (DMHP) 385, 399
  tolerance 411
Dimethyltryptamine (DMT) 311—314, 317, 318, 320, 321, 323, 326, 330, 332—336, 338—340, 342—343
4,*α*-Dimethyl-*m*-tyramine 38
2,4-Dinitrochlorobenzene 402
Diphenhydramine 75, 95, 313
Diphenylhydantoin 17, 178
Diplacine 153
Dipropyltryptamine (DPT) 311, 326
Discriminate avoidance behavior 100
Disulfiram 17, 74, 88, 132, 133, 143, 145, 178, 292

Ditran 159
Dopa 279, 333, 338
l-Dopa 74, 75, 77, 96, 104, 126, 131, 151, 179
Dopamine (DA) (see catecholamines) 261, 333, 337, 346, 391
  motor activity 88
Dopamine-β-hydroxylase 9, 51, 72, 338
  inhibitors 118

Electroencephalograph (EEG), amphetamine 176—179, 251
Elymoclavine 310, 340
Endocrines, Δ^9-tetrahydrocannabinol 400
Endotoxic shock 279
Enuresis, amphetamine 255
Ephedrine 6, 12, 21, 63, 250, 254, 278, 279, 292, 332
Epinephrine (E) (see catecholamines) 105, 339, 391, 397
Ergometrine 310, 337
Ergonovine 334
Ergotamine 142, 337
Ethanol 17, 86, 94, 118, 169, 172
Ethylamphetamine 12
N-Ethyl-3-piperidyl benzylate 319
σ-Ethyltryptamine 332, 335
Ethynyl estradiol 76
Excitation syndrome, amphetamine 264
Exploratory behavior, amphetamine 94
Extensor reflex 69

β-Farnesene 378
Fenfluramine 7, 12, 46, 76, 86, 89, 132, 134, 136, 146, 153, 156, 159, 164, 169, 172, 249, 251, 253, 257, 258, 260, 262, 281
Fenproporex 146
Fixed interval reinforcement, amphetamine 112
Fixed ratio reinforcement, amphetamine 109—111
FLA-63 74, 84, 88, 97, 123
Flexor reflex 69
6-Fluorodiethyl tryptamine (6-FDET) 311
Fluphenazine 73, 83, 104, 170
Free fatty acids 156, 163, 164
Frontal cortex 90
  lesions 98, 119, 136

Genetic differences, exploratory behavior 94
Globus pallidus 68, 78, 79, 89
Glucose 164
Glutamate, Na 329
Glutamic acid 342
Glutamine 342
Glycine 329
Guanethidine 40, 86, 144—145, 147—148, 161, 170—171, 338
Guanozane 163
Guinea pig
  bladder 61
  ileum 60
  vas deferens 60

Hallucination 384
Haloperidol 16, 44, 57, 68, 73, 78, 83, 103, 104, 133, 142, 150, 159, 170, 312, 316, 333, 338
Halothane 153
Harmaline 338
Harmine 328
Hexamethonium 61, 143, 146, 152, 163, 338
Hippocampus 37, 53, 78, 155, 165
D,L-Homocysteic acid 329
Homovanillic acid (HVA) 43, 46, 52, 80, 157, 255, 261
Hunger center, amphetamine 135
Hydergine 336, 338
Hydralazine 146, 316
Hydroxyamphetamine 6
N-Hydroxyamphetamine 11
p-Hydroxyamphetamine 4, 8, 45, 46, 51, 52, 60, 62, 63, 76, 78, 96, 100, 111, 113, 139, 141, 150, 151, 156, 162, 164
  uptake 51
γ-Hydroxybutyric acid 42, 43
6-Hydroxy-dimethyltryptamine 313
6-Hydroxydopamine (6-OHDA) 40, 51, 79, 88, 89, 94, 105, 107, 118, 135, 153, 159, 164, 169, 172
p-Hydroxyephedrine 51, 141, 164
5-Hydroxyindoleacetic acid (5-HIAA) 261, 330, 392, 395
p-Hydroxynorephedrine 9, 19, 45, 50—53, 62, 63, 139, 141, 145, 151, 157
8-Hydroxy-Δ^9-tetrahydrocannabinol 382
8-α-Hydroxy-Δ^9-tetrahydrocannabinol 382
8-β-Hydroxy-Δ^9-tetrahydrocannabinol 382
11-Hydroxy-Δ^9-tetrahydrocannabinol 383, 389
11-Hydroxy-Δ^9-tetrahydrocannabinol 382, 383, 385, 388
  tolerance 410
4-Hydroxytryptamine (4-HT) 337
5-Hydroxytryptamine (5-HT) 61, 88, 98, 131, 169, 172, 177, 179, 261, 314, 328, 329, 332, 333, 335—339, 391, 392, 394
  Δ^9-THC 394
6-Hydroxytryptamine (6-HT) 337
7-Hydroxytryptamine (7-HT) 337
5-Hydroxytryptophan (5-HTP) 75, 84, 131, 312, 314, 331, 332, 337, 338
Hypnogogic hallucinations 285
Hyperkinesis, amphetamine 249
Hyperkinetic children 249

Hypothalamus 39, 44, 54, 65, 67, 69, 122—123, 135—137, 165, 400
amphetamine, unit activity 137
Hyoscyamine 75, 152

Imipramine 16, 75, 85, 93, 98, 104, 123, 134, 152, 171, 250, 286, 338
INPEA (see nifenalol)
Intracranial self-stimulation, amphetamine 121—124
Iprindole 51, 75, 85, 123, 134
Iproniazid 75, 85, 104, 133, 145, 152, 171, 312, 332
Isocarboxazid 104
Isolpropoxamine 163
d-Iso-lysergic acid amide 310
d-Isolysergic diethylamide 340

Khat 277, 292

Lateral geniculate body 337
Laws, amphetamine 283
Learning, amphetamine 129
Levallorphan 313
Levoamphetamine 286
Libido, amphetamine 260
Limbic system 176
Limonene 378
Lithium 76, 85, 161, 266
Locus ceruleus 65, 122
Lynestrol 86, 153, 172
d-Lysergic acid 340
d-Lysergic acid amide 310
d-Lysergic acid diethylamide (LSD) 61, 67, 85, 93, 120, 145, 150, 172, 249, 256, 258, 305, 312—314, 326, 329, 330, 332, 339, 340
abuse potential 348
ACh 333
amphetamine 328
amygdala 341, 343
antagonists 312
autonomic effects 307
basal ganglia 341
behavioral effects 307
rat 313
bradykinin 333
CA 333
cardiovascular changes 335
cat 315, 320
caudate nucleus 334
cerebellum 334
cerebral cortex 334, 338—343
C-fiber reflex 332
characteristics 321, 343
chemoreceptive trigger zone 336
choroid plexus 334
clinical grades 308
colliculi 334
cortical evoked potentials 339—343
criteria 321
criterion drugs 322
cross tolerance 318—321
DA 333
diencephalon-thermal regulation 337
discriminative stimulus 314
distribution 333
dog 316, 321
EEG 331, 339—343, 347
activation 341
cat 341—342
man 343
monkey 342—343
rabbit 339
rat 339
encephale isolé 340
entorhinal cortex 341
goat 317
guinea pig 317
hippocampus 340—343
hypothalamus 334
indoleamines 328
lateral geniculate 336
limbic system 334
LSD-like hallucinogens 305—348
chemical structure 323—327
LSD Significant Scale 309
lysergic acid amines 328
man 318
MAO inhibitor 339
mechanisms of action 346
mesencephalon 336, 341
metabolism 315
midbrain raphé 334
mode of action 328—334, 347
monkey 317
mouse 312
NE 333
olivary body 334
periaqueductal gray 334
phenethylamines 323
physiologic changes 307, 310
pons medulla 335
pupils 337
rabbit 315, 320
raphé nuclei 341
rat 313—315, 319, 320
recruitment 341
reticular formation 334, 341
seizures 339
septum 340, 343
serotonin antagonist theory 328
sites of action 334—343
sleep 339—343
sleep loss 312

d-Lysergic acid diethylamide (LSD)
somatomotor changes 307
spinal cord 331, 334
subjective effects 306—309
clinical grade 308
Substance A 333
Substance P 333
synaptosomes 334
tegmentum 334
temperature 337, 338
thalamus 341
time contraction 347
tolerance 318—321, 336, 344
tryptamine 321
tryptamine agonist theory 331—332
d-Lysergic acid dimethylamide 340
Lysergic acid monoethylamide 336
d-Lysergic acid monoethylamide (LAE) 313, 340
d-Lysergic acid morpholide 340
Lysergic acid pyroledide 336

Mamillothalamic tract 101
Marihuana (see tetrahydrocannabinol) 371—415
abuse 371
incidence 371
chemistry 374
chronic toxicity 409
cortisol 400
cytotoxicity 401
extracts
development 407
tolerance 410
gonadotropins 400
pharmacology 385
prolactin 400
thyroxine 400
toxicity 404
Maze performance, amphetamine 127
Mazindol 281
Mecamylamine 41, 86, 123, 146, 163, 171
Medial forebrain bundle 119, 122
Medulla 39, 44, 55
β-Melanocyte stimulating hormone (β-MSH) 333
Melatonin 337
Memory, amphetamine 129
Memory consolidation, amphetamine 129
Mephentermine 63, 280
inhalers 284
Meprobamate 17, 76, 85, 134, 171, 340
Mescaline 120, 305, 306, 312, 314—316, 318, 320, 324, 328—330, 334—340, 342, 343
subjective effects 306
tolerance 318
visual changes 306
Mesencephalon 53, 155, 176
Mesolimbic pathway 121
Mestranol 86, 153, 172
Metabolism, amphetamine 3—24
Metaraminol 40, 145, 170
Methacholine 85
Methadone 313
Methamphetamine 6, 7, 12, 21, 35, 37, 39, 43, 46, 51—54, 69, 91, 100, 111, 112, 116, 118, 122, 128—130, 132, 138, 140, 141, 150, 154, 164, 165, 176, 178, 179, 251, 254, 257, 258, 262, 279, 280, 283, 284, 286, 292
inhalers 284
Methergoline 61, 95, 133, 339
Methioplegium 171
Methoxamine 331, 345
p-Methoxyamphetamine 19
4-Methoxy-dimethyltryptamine 313
5-Methoxy-dimethyltryptamine 314, 340
5-Methoxy-*N*,*N*′-dimethyltryptamine 338
3-Methoxy-4-hydroxyphenylethylglycol (MHPG) 255, 261
5-Methoxy-3,4-methylene-dioxyamphetamine (MMDA) 311, 316, 317, 325
5-Methoxy-α-methyltryptamine 332
5-Methoxytryptamine 340
3-Methoxytyramine (MTA) 43
Methylatropine 75, 152
4-Methyl-5-(β-chloroethyl)-thiazol 172
α-Methyldopa 74, 144, 145, 162, 170, 332
α-Methyl-dopamine 42
3,4-Methylene-dioxyamphetamine (MDA) 311, 316, 317, 321, 325, 333
Methylergometrine 337
1-Methyl-*d*-lysergic acid butanolamide 330
d-1-Methyl-lysergic acid diethylamide 340
1-Methyl-lysergic acid diethylamide 318, 333
Methylphenidate 3, 6, 105, 146, 250, 254, 258, 281, 285—288, 292, 332
α-Methyl-noradrenaline 40, 42, 177
1-(3-Methylphenoxy)-3-isopropylamine-2-propanol 151
11-Methyl-Δ^8-tetrahydrocannabinol, tolerance 411
α-Methyltryptamine 332
α-Methyl-tyrosine 44, 57, 69, 72, 74, 77, 78, 84, 97, 98, 104, 106, 118, 123, 126, 132, 133, 140, 143, 145, 150, 151, 154, 161, 163, 170, 177, 179, 265, 315, 316, 333, 338
α-Methyl-*m*-tyrosine 75, 78, 85, 133, 145, 171, 333
α-Methyl-*p*-tyrosine 45
Methysergide 61, 74, 78, 83, 93, 95, 123, 133, 145, 159, 170, 177, 310, 312, 315, 316, 318, 327, 329—333, 335, 337—340, 343
Metoclopramide 73
Midbrain tegmentum 122

Minor tranquilizers 292
Mode of action, LSD 328—334
Monoamine oxidase 10, 45—47
inhibitors 42, 62, 106, 142, 144, 154, 169
Mood, amphetamine, chronic 255—257
Morphine 61, 76, 96, 172, 313, 336, 338, 397
Motion sickness 288
Motor activity, amphetamine 81—93

Nalorphine 312
Naloxone 86, 105, 134, 153
4-(*l*-Naphthylvinyl)pyridine 105
Narcolepsy 279
amphetamine 282, 285
Narcotic antagonist 390
Nethalide 170
Neuraminidase 61
Neuroleptics 93
Nialamide 17, 39, 42, 44, 75, 85, 145, 171
Nicotine 152
Nicotinic acid 312
Nigrostriatal bundle (pathway) 37, 39, 44, 66, 79, 121
Nifenalol 151, 159, 162
Nikethamide 332
Nitrazepam 76, 85
n-Nonecasane 378
Noradrenaline (NA) (see catecholamines) 88
Norephedrine 9, 18, 151, 156
Norepinephrine (NE) 9, 257, 261, 329, 333, 337, 338, 346, 391, 394
Δ^9-THC 393
9-*Nor*-9β-hydroxy-hexahydrocannabinol 390
Normetanephrine 43, 396
Norpseudoephedrine 6, 277
Nortriptyline 16, 75, 152
Nucleus accumbens 79

Obesity, amphetamine 284, 285
Octopamine 145
Oil of wormwood 278
Olfactory bulb 53, 79
Open field behavior, amphetamine 93
Operant behavior
adipsia 117
amphetamine 98—128
anorexia 117
conditioned emotional response 115—117
Optic tract 337
Orphenadrine 75
Ouabain 172
Oxazepam 76, 85
Oxotremorine 75
Oxypertine 73

Papaverine 397
Paramethoxyamphetamine (PMA) 312, 324, 333
Paranoid psychosis, amphetamine 262, 282
Pargyline 75, 85, 104, 133, 145, 171, 330
Pemoline 286
Penfluridol 83
Pentamethazene 171
Pentobarbital 105, 118, 134, 171, 337, 338, 340
Pentolinium 146, 152
Pentylenetetrazol 177, 178, 332
Performance, amphetamine, chronic 258
Perphenazine 73, 83, 104, 150, 170
Petit mal epilepsy 288
Peyote 305, 306
Pharmacologic redundancy 344
β-Phellandrene 378
Phenazetine 153
Phencyclidine 312, 313
Phenelzine 17, 145, 152, 171, 265
β-Phenethylamine 141, 323, 345
Phenglutarimide 75
Pheniprazine 85, 151, 312
Phenmetrazine 6, 20, 21, 253, 254, 258, 260, 281, 292
Phenobarbital 16, 125, 171, 178
Phenothiazines 288
Phenoxybenzamine 57, 61, 73, 88, 97, 103, 104, 145, 150, 159, 161, 162, 170, 177, 266, 315—317, 329, 331, 335, 336, 338
Phentermine 4, 6, 11, 258
Phentolamine 57, 73, 83, 133, 145, 150, 159, 161, 170, 177
Phenylacetone oxime 12
β-Phenylisopropyl-hydrazine 332
Phenylpropanolamine 18
Phosphatidylcholine 164
Phosphotidylethanolamine 164
Physical dependence, Δ^9-THC 412
Physostigmine 75, 105, 176, 178, 340
Pilocarpine 75, 85, 152
Pimozide 73, 83, 95—97, 103, 104, 133, 140, 150, 158, 159, 266, 338
Pimpamperone 104
Piperidine 378
Piperoxane 133, 142
Pipradol 332
Pons 53
Pontine geniculate occipital (PGO) spike 178
Posterior hypothalamus 79
Practolol 73, 83
Prednisone 312
Prefrontal cortex 136
Prenylamine 163
Prindolol 73, 83, 133
Procaine 172
Prochlorperazine 16, 332, 339

Progesterone 43, 86, 312, 315
Promazine 73, 105
Promethazine 16, 75, 105, 152
Pronethanol 73, 83, 162, 163, 170, 315
Propericiazine 150
Propranolol 57, 73, 83, 93, 132, 133, 145, 150, 159, 161—163, 170, 266, 315, 316
Propoxyphene 12
α-Propyldopacetamide 98, 330
Propylhexidrine 284
Prostaglandin 153, 154, 163
Protriptyline 16, 40, 50, 68
Psilocin 310, 312, 316, 317, 326, 334—338, 342
Psilocybin 310, 313, 315, 318, 320, 326, 330, 334, 335, 337, 338, 340, 343
Psychedelic amphetamines 281
Punding 287
Pure Food and Drug Act 278
Putamen 68, 78, 79, 89
Pyamidone 153

Rabbit ileum 60
Raphé nuclei 67, 136, 330
Rapid eye movement (REM) sleep 96, 178
amphetamine 250, 252
Red oil (hashish extract) 86
Reinforces operant behavior, amphetamine 108—121
Reserpine 17, 39—41, 44, 50, 59—61, 66, 69, 74, 78, 84, 96, 103, 104, 118, 123, 131—133, 142—145, 147, 151, 154, 159, 161—163, 170, 179, 312—316, 330, 333, 338, 340, 342
Respiration, amphetamine 254
Ribonucleic acid (RNA) 165, 402, 405

Salicylates 169
Ca 153, 172
Satiety center, amphetamine 135
Scopolamine 75, 85, 94, 95, 105, 152, 171, 312, 316, 319
Sebacinyl-bis-choline 335
Secobarbital 17
Sedative-hypnotics 135
Seizures, amphetamine 177
Self-administration, amphetamine 124
α-Selinene 378
Septum 57, 90, 100, 107, 122, 136, 155, 333
Serotonin (5-HT) (see 5-hydroxytryptamine)
Sexual behavior, amphetamine 97
Shock, amphetamine 264
SKF-525A 17, 51, 76, 86, 153, 172
Sleep
amphetamine 178
paralysis 285
slow wave 178
Social behavior, amphetamine 95—98
Sodium succinate 313

Sotalol 338
Sparteine 172
Speed binge 258
Spinal cord, LSD 334
Spiramide 73, 83, 94, 95, 97
Spiriperidol 68
State dependent learning 120
amphetamine 130
Stereoisomerism, amphetamine 5
Stereotyped behavior 71—81
amphetamine 263
STP (also DOM) (see 2,5-dimethoxy-4-methylamphetamine)
Striatum 53, 54, 78, 90, 155
Subjective effects
amphetamine 248, 249, 265
cannabidiol 385
cannabinol 385
DMPH 385
LSD 306—309
Δ^{6a}-THC 385
Δ^{8}-THC 385
Δ^{9}-THC 385
Substance A 333
Substance P 57, 333, 334
Substantia nigra 66, 68, 78, 79, 89, 122
Sympathectomy 316

Tachyphylaxis 62
Tea 277
Telencephalon 39, 54
Testosterone 315, 408
Tetrabenazine 42, 74, 78, 103—104, 118, 123, 133, 143—145, 147, 151, 170, 315
Tetrahydrocannabinol, isomers 379
Δ^{8}-Tetrahydrocannabinol 375, 376, 378, 382, 389, 392, 398, 401
analgesia 389
cardiovascular effects 398
chronic toxicity 404
DA 396
5-HIAA 396
5-HT 396
NE 396
normetanephrine 396
tolerance 410
toxicity 403
Δ^{9}-Tetrahydrocannabinol 86, 153, 172, 313, 319, 320, 374—378
absorption 380
abstinence 408
ACh 392
aggression 386, 387
amotivational syndrome 409
analgesia 389, 390, 413
angina 397
anti-emesis 414
antineoplastic activity 402

biogenic amines 391—397
bronchodilation 414
cardiovascular effects 397, 398
central nervous system effects 384
chromosomal abnormalities 401
chronic toxicity 404, 405
cytotoxicity 401
DA 391
levels 394
synthesis 394
turnover 394
dependence 408
development 406
distribution 380—382
DNA 401
dog 386
E 391
EKG 397
endocrines 400
glaucoma 414
growth hormone 400
half-life 380
5-HIAA 395
5-HT 391
levels 394
synthesis 395
turnover 395
immune reactions 402
insulin 400
liver function 404
metabolism 382, 383
monkey 387
muricide 387
NE 391
levels 393
synthesis 393
turnover 393
neoplasms 402
normetanephrine 396
numbering system 374
operant behavior 388
pharmacology 384, 386
rabbit 386—387
physical dependence 412
protein synthesis 402
reticuloendothelial system 402
RNA 402
stereoisomerism 378
subjective effects 384, 385
teratogenesis 408
testosterone 400
tolerance 408—410
toxicity 403
Tetrahydrocannabivarol 17, 377
Tetrahydro-β-napthylamine 279
Tetramethylammonium 146, 152
Thalamus 78, 90, 165
Thioproperazine 73
Thiordazine 73, 83, 338
Thujune 278
Tolerance
amphetamine 319
cannabis 409—412
DMT 318
JB-318 319
LSD 318—321, 336
mescaline 318
methysergide 318
scopolamine 319
Δ^9-THC 319, 409, 410
Toxicity
amphetamine 166—175
cannabis 403
LSD 348
Toxifirin 153
Tranylcypromine 75, 145, 151, 171, 315
Tremorine 85, 105, 152, 169, 171
Tricyclic antidepressants 106
Trifluoperazine 73, 78, 83, 97, 150, 170, 332, 337, 339
Trigonellin 378
Trihexyphenidyl 152
Trimethoxyamphetamine (TMA) 313, 316, 317, 325
3,4,5-Trimethoxyamphetamine 4, 311
2,3,4-Trimethoxyphenethylamine (2,3,4-TMPEA) 315
Trimipramine 152
Tripelennamine 316
Tryptamine 61, 310, 313, 316, 321, 326, 330—332, 334—336, 339, 340, 345, 346
brain levels 332
spinal cord 332
Tryptophan 56, 330
Tryptophan hydroxylase 56, 330
Tyramine 59, 63, 105, 145, 148, 159, 279
Tyrosine 333
mBr-Tyrosine 74
mI-Tyrosine 74
Tyrosine hydroxylase 47, 77, 333

U-22, 394A 313, 335, 338, 340
Urethane 86, 105

Variable interval reinforcement, amphetamine 111
Ventral tegmentum 330

Wakefulness, amphetamine 258

Xanthines 277
Xylocholine 145, 147
2,6-Xylyl-choline ether bromide (TM-10) 145

Yohimbine 73, 133, 142, 328

Handbuch der experimentellen Pharmakologie / Handbook of Experimental Pharmacology

Heffter-Heubner, New Series

Volume 23:
Neurohypophysial Hormones and Similar Polypeptides

Volume 24:
Diuretica

Volume 25:
Bradykinin, Kallidin and Kallikrein

Volume 26:
Vergleichende Pharmakologie von Überträgersubstanzen in tiersystematischer Darstellung

Volume 27:
Anticoagulantien

Volume 28: Part 1
Concepts in Biochemical Pharmacology I
Part 2
Concepts in Biochemical Pharmacology II
Part 3
Concepts in Biochemical Pharmacology III

Volume 29:
Oral wirksame Antidiabetika

Volume 30:
Modern Inhalation Anesthetics

Volume 31:
Antianginal Drugs
Out of print

Volume 32: Part 1
Insulin I
Part 2
Insulin II

Volume 33:
Catecholamines

Volume 34:
Secretin, Cholecystokinin-Pancreozymin and Gastrin

Volume 35: Part 1
Androgene I
Part 2
Androgens II and Antiandrogens/Androgene II und Antiandrogene

Volume 36:
Uranium-Plutonium-Transplutonic Elements

Volume 37:
Angiotensin

Volume 38: Part 1
Antineoplastic and Immunosuppressive Agents I
Part 2
Antineoplastic and Immunosuppressive Agents II

Volume 39:
Antihypertensive Agents

Volume 40:
Organic Nitrates

Volume 41:
Hypolipidemic Agents

Volume 42:
Neuromuscular Junction

Volume 43:
Anabolic-Androgenic Steroids

Volume 44:
Heme and Hemoproteins

Volume 45:
Drug Addiction I and II

Volume 46:
Fibrinolytics and Antifibrinolytics (in preparation)

Volume 47:
Drug-Reception Inactivation (in preparation)

Volume 48:
Arthropod Venoms

Volume 49:
Ganglionic Transmission (in preparation)

Volume 50:
Inflammation and Anti-inflammatory Drugs (in preparation)

Springer-Verlag
Berlin
Heidelberg
New York

MIX
Papier aus verantwortungsvollen Quellen
Paper from responsible sources
FSC® C105338

If you have any concerns about our products,
you can contact us on
ProductSafety@springernature.com

In case Publisher is established outside the EU,
the EU authorized representative is:
Springer Nature Customer Service Center GmbH
Europaplatz 3, 69115 Heidelberg, Germany

Printed by Libri Plureos GmbH
in Hamburg, Germany